FOURTH EDITION

LIVESTOCK FEEDS
AND FEEDING

Richard O. Kellems, Ph.D.
D. C. Church, Ph.D.

PRENTICE HALL, Upper Saddle River, New Jersey 07458

Library of Congress Cataloging-in-Publication Data

Kellems, Richard O.
    Livestock feeds and feeding / Richard O. Kellems, D.C. Church. — 4th ed.
        p.   cm.
    Rev. ed. of: Livestock feeds and feeding / D.C. Church, 3rd ed.
    c1991.
    Includes index.
    ISBN 0–13–241795–2
    1. Feeds.   2. Animal feeding.   3. Animal nutrition.   I. Church,
D. C.   II. Church, D. C.   Livestock feeds and feeding.   III. Title.
SF95.K32   1998
636.08′4—dc21   96-46382
                                                                      CIP

Tables from *Nutrient Requirements of Dairy Cattle: Update 1989, Sixth Revised Edition,*
© 1989 by the National Academy of Sciences, Washington DC; *Nutrient Requirements
of Goats: Angora, Dairy, and Meat Goats in Temperate and Tropical Countries,* © 1981 by
the National Academy of Sciences; *Nutrient Requirements of Swine, Ninth Revised
Edition, 1988,* © 1988 by the National Academy of Sciences; *Nutrient Requirements of
Poultry, Eighth Revised Edition, 1994,* © 1994 by the National Academy of Sciences; *Nutrient Requirements of Horses, Fifth Revised Edition, 1989,* © 1989 by the National Academy of Sciences; *Nutrient Requirements of Beef Cattle, Sixth Revised Edition, 1996,*
© 1996 by the National Academy of Sciences; *Nutrient Requirements of Sheep, Sixth Revised Edition, 1985,* © 1985 by the National Academy of Sciences. Acknowledgments to
CAB International, Wallingford, Oxon, UK for permission to reproduce tables and figures from *The Nutrient Requirements of Ruminant Livestock.*

Acquisitions editor: Charles Stewart
Director of production and manufacturing: Bruce Johnson
Managing editor: Mary Carnis
Editorial / production supervision and interior design: Tally Morgan,
    WordCrafters Editorial Services, Inc.
Cover design: Miguel Ortiz
Cover photo: Courtesy of Richard O. Kellems
Manufacturing buyer: Marc Bove / Ilene Sanford
Marketing manager: Debbie Yarnell

©1998, 1991, 1984, 1977 by Prentice-Hall, Inc.
Simon & Schuster/A Viacom Company
Upper Saddle River, New Jersey 07458

Printed in the United States of America
10 9 8 7 6 5 4 3 2 1

ISBN 0-13-241795-2

PRENTICE-HALL INTERNATIONAL (UK) LIMITED, *London*
PRENTICE-HALL OF AUSTRALIA PTY. LIMITED, *Sydney*
PRENTICE-HALL CANADA INC., *Toronto*
PRENTICE-HALL HISPANOAMERICANA, S.A., *Mexico*
PRENTICE-HALL OF INDIA PRIVATE LIMITED, *New Delhi*
PRENTICE-HALL OF JAPAN, INC., *Tokyo*
SIMON & SCHUSTER ASIA PTE. LTD., *Singapore*
EDITORA PRENTICE-HALL DO BRASIL, LTDA., *Rio de Janeiro*

CONTENTS

PREFACE

This book will provide the reader with a basic understanding of the principles of nutrition, feedstuffs, and practical livestock feeding. Many changes have occurred since its last publication with respect to international relations, scientific knowledge, politics, and in other areas. The countries of the world have dropped many of their trade restrictions and more of a true world market has developed. Now the challenge is to efficiently produce livestock products that can compete on this world market. Since the cost of feeding is normally the largest single cost associated with producing livestock, a basic understanding of nutrition and feeding is essential for any individual who will be involved in the production of livestock today.

This book provides a brief historical perspective relating to the development of modern livestock production. The ability to prove a safe, nutritious food supply for the world's people has been the driving force behind the development of modern agriculture, with the wellbeing of the animals and profitability being major considerations. Many advances in genetics and nutrient utilization have taken place in recent years and more will be made. It is important to have an understanding of the effect that various nutrients have on animal health and performance. Economical production requires understanding of feed processing, feeding practices, and animal management by those involved in various types of livestock operations.

This book will provide the reader with a basic understanding of the importance of nutrition as it relates to livestock production and how various feedstuffs can be utilized to provide the required nutrients. Appendix tables have been included that give nutritional information for the commonly used feedstuffs, as well as the nutrient requirement guidelines for the domestically important livestock species. The chapters relating to specific species provide information relating to feeding and management and explain how they are used with these different species. It is our sincere hope that this book will enable to reader to gain a knowledge of the principles and practices used in feeding livestock.

Richard O. Kellems
D. C. Church

CONTRIBUTORS

Mark S. Aseltine
Dairy Nutritionist and Management
   Consultant
Modesto, California
B.S. California Polytechnic State—
   San Luis Obispo
M.S. and Ph.D. Oregon State University

Jan Bowman, Ph.D.
Beef Cattle Nutritionist
Animal and Range Sciences
Montana State University
B.S., M.S., and Ph.D. University of Missouri

D. C. Church
Professor, Emeritus
Oregon State University
B.S. Kansas State University
M.S. University of Idaho
Ph.D. Oklahoma State University

Gary L. Cromwell, Ph.D.
Professor of Animal Science Department
University of Kentucky

Glenn Duff, Ph.D.
Ruminant Nutritionist
Clayton Livestock Research and Extension
   Center
New Mexico State University

Michael Galyean, Ph.D.
Professor of Animal Science
West Texas A&M University
Canyon, Texas

Danny M. Hooge
Consulting Poultry Nutritionist
Eagle Mountain, Utah
B.S. University of Tennessee
M.S. Brigham Young University
Ph.D. Texas A&M University

Diane A. Hirakawa
Senior Vice President Research and
   Development IAMS Company
B.S., M.S. and Ph.D. University of Illinois
   at Urbana-Champaign

Hugo Varela-Alvarez
Animal Nutritionist
Ph.D. Penn State University

Ed Huston, Ph.D.
Professor of Animal and Range Sciences
Texas Agricultural Experiment Station
   at San Angelo

Richard O. Kellems
Animal/Dairy Nutritionist
Brigham Young University
B.S. Brigham Young University
M.S. and Ph.D. Oregon State University

Rodney Kott
Extension Sheep Specialist
Montana State University
B.S. and M.S. Texas A&M University
Ph.D. New Mexico State University

Laurie Lawrence
Equine Nutritionist
University of Kentucky

David Schingoethe
Professor of Dairy Science
South Dakota State University
B.S. and M.S. University of Illinois
Ph.D. Michigan State University

Bok Sowell
Beef Cattle Nutritionist
Range Science Department
Montana State University

LIVESTOCK FEEDS AND FEEDING

1

LIVESTOCK FEEDING

Richard O. Kellems and D. C. Church

Dramatic changes in the feeding and production of livestock have occurred over the centuries. In the United States and other areas of the world, the inhabitants have shifted from a primarily nomadic existence to an agrarian and then to a primarily urban type of existence. The first inhabitants were normally nomadic hunters and gatherers, initially moving from area to area on foot; then they started to domesticate animals, such as horses, or build boats that allowed them more mobility so that they could collect food from a wider area. Then a shift occurred over time to a more agrarian type of living style, in which dependence on hunting lessened, and they started to produce crops that could be used as food and domestic animals to serve as a source of meat. As a result of these changes, people became less dependent on what they could hunt or gather for their food supply and more dependent on what they could produce. The primitive methods used required that they devote a major portion of their time to producing enough food to survive. As the population increased, the size of the communities increased, and people became less involved in food production and started to have other types of employment, so the proportion of the people involved in food production started to decline; today less that 2% of the population in the United States is involved in producing enough food for everyone else. Tremendous changes have occurred in how livestock are produced and fed as

production systems have changed from subsistence to very specialized large-scale operations, such as found in the United States and other developed countries. Populations can still be found in these various stages of agricultural transition in different areas of the world today.

As these changes occurred, knowledge relating to how livestock should be fed has increased. Initially, individual producers most likely shared their experiences with one another, such as when a certain plant was grazed their animal grew well, or when another plant was consumed their animal became sick or died. These type of exchanges most likely caused people to feed their animals differently and improved animal performance. Most of the technical information relating to the feeding of livestock has been developed from countless experiments carried out worldwide, primarily during the past 150 years. Governments established public institutions (land grant universities in the United States and research centers in other parts of the world) to assist the agrarian sector to develop a knowledge base that would support the producers so that they could produce a safe and adequate food supply. Initially, very little research was conducted by privately sponsored institutions or individuals, but today more and more research is being conducted by private companies that are trying to develop products or systems that will increase production efficiency, which can then be

1

marketed by these companies. In the United States and other developed countries, livestock production systems have been developed to a very high level. Even though the science of animal nutrition and feeding has been advanced significantly over time, our knowledge is still not complete and advances continue.

As the population and standard of living continue to increase throughout the world, the demand for animal products will continue to increase. The challenge to livestock producers and nutritionists is to increase the efficiency of conversion of feedstuffs into animal products, which includes the use of more alternative feedstuffs, such as straws and agricultural by-products, so that an adequate supply of animal products can be provided. During the past decades we have seen dramatic improvement in crop production, even though the amount of land being farmed has changed very little (Table 1-1). Similar increases in production of animal products has also occurred during that same period of time. Food and feed production vary each year depending on the following: government regulatory practices; price being received for the products and projected future price; environmental conditions, particularly weather; and the cost of various inputs, such as fertilizer, equipment, labor, and availability of money. The challenge is to understand how we can utilize available natural resources more efficiently to ensure that an adequate food supply can be provided for the world's population.

This book will provide the reader with a basic understanding of the feeding and nutritional practices associated with modern livestock production, specifically by providing the reader with a basic understanding of how nutrients are digested and utilized, the function of various nutrients, an understanding of the nutritional value of various feedstuffs, how pro-

cessing effects nutrient utilization, specific nutrient requirements for various livestock species, practical feeding practices used for various species of livestock, and a basic understanding of how rations are formulated.

BASIS FOR SUCCESSFUL FEEDING OF LIVESTOCK

The fundamental principles relating to the economical feeding of livestock were outlined by Professor W. O. Atwater (2) in 1878:

> The right feeding of stock, then, is not merely a matter of so much hay and grain and roots, but rather of so much water, starch, gluten, etc., of which they are composed. To use fodder economically, we must so mix and deal it out that the ration shall contain just the amounts of the various nutrients needed for maintenance and for the particular form of production that is required. . . . [W]e must consider: What is the chemical composition of our fodder materials? How many pounds of . . . [nutrient] . . . are contained in a hundred pounds of hay, clover, potatoes, meal, etc.? Of these various ingredients of food, what proportion of each [is] digestible and consequently nutritious? What part does each of these food ingredients play in the animal economy? . . . How much of each do different animals, as oxen and cows, need for maintenance of life and production of meat, milk, etc.? And finally, how may different kinds of fodder be mixed and fed so that the digestible material shall be most fully digested and utilized, and the least quantity wasted?

This statement by Atwater, which was published more than 100 years ago, is in most respects still applicable to the concepts used in feeding livestock today. In addition to the factors outlined by Atwater, more recent advances in our nutritional knowledge have occurred. Understanding of the interactions between nutrient utilization and various other factors, such as disease, parasites, and toxic substances in feed, has been elucidated. For example, we now know that it is cost effective to eliminate internal parasites in order to increase performance and maximize the efficiency of nutrient utilization. Management practices such as increasing frequency of feeding have been shown to be profitable in some situations and can be used to maximize nutrient intake and thus increase performance. Many feedstuffs currently fed to livestock contain antiquality factors, such as the trypsin inhibitor found in soybeans, which, if fed to poultry without being inacti-

TABLE 1-1

World coarse grain (in million metric tons)

	Coarse Grains[a]	
	Tons per Hectare	Total
1966/67	1.61	508.9
1976/77	2.02	692.0
1986/87	2.45	822.4
1994/95	2.76	858.4

Source: *Grain World Market and Trade Report*, USDA. Oct 1994 (1).

[a]Coarse grains consist of corn, sorghum, barley, and oats.

vated, will impair performance; but certain amounts can be fed to cattle without having an adverse effect. Our understanding has increased as to how to inactive these antiquality factors and to what species of livestock they can be fed without observing detrimental effects. The nutrient requirements of our various livestock species continue to change as genetic advances are made. The nutrient requirements for a dairy cow producing 45 kilograms (kg) of milk per day are different than those for a cow producing 10 kg of milk per day. The same is true for a broiler that takes only 7 weeks to reach market size versus one that takes 12 weeks to reach the same size. With the rapid advances in biotechonolocial currently being made, the use of new products, such as bovine somatotropin (Bst), will change how nutrients are utilized by the animal and thus their nutrient requirements. So the feeding of livestock is a very dynamic and ever changing discipline and has become more of a science and less of an art as our knowledge relating to nutrition continues to increase.

As a result of genetic advances in livestock, the conversion of feed nutrients into animal products continues to improve. Table 1-2 shows the improvements that have been made in efficiency in feed conversion in poultry; these are a result of a better understanding of nutrition and improvements in growth rates resulting from genetic advances. Fewer cows are now required to produce more milk, which increases efficiency of production (Table 1-2). High-producing animals require less feed per unit of gain, primarily because less of the nutrients that are being consumed are being used for maintenance and because they require less time to reach their desired market size. Since the primary dietary constituent used for maintenance is energy and the primary constituent found in animal products is protein, dietary protein levels will need to be increased as the genetic productive potential of livestock increases, as will the dietary concentration of other nutrients. Thus nutrient requirements are continually changing as genetic progress is being made.

FEED CONSUMPTION EFFECT ON DIETARY NUTRIENT DENSITY

Feed consumption has a dramatic effect on dietary nutrient specifications. Feed consumption can be affected by a number of different factors, such as environmental temperature, humidity, and stress. Adjustments need to be made to feeding programs as a result of changes in consumption. If for some reason an animal consumes 10% less feed than projected, then, in order to achieve the project level of performance, it will be necessary to increase the nutrient density of the diet to compensate for the reduction in consumption. For instance, if the diet was formulated to contain 10% crude protein, then if feed were reduced by 10% the diet would have to contain 11.1% [10%/0.9 (decimal equivalency of 90%) = 11.1% CP] in order to have the same intake of crude protein.

Another change that has occurred is that the nutrient content of feedstuffs is changing as production practices and genetics changes are made in these crops. As the yield of a crop increases, often the trace mineral content decreases; so to satisfy the requirement of the animal, these minerals need to be fortified at a higher level. The nutrient content of alfalfa grown in an area where only three crops are produced may differ from alfalfa grown in an area where seven or eight crops are harvested. Climatic conditions can affect the nutrient characteristics of a crop, especially those produced by dryland operations. Watering and fertilization practices can alter the nutrient characteristics of a crop. Farming practices often affect the nutrient content of feedstuffs, such as the excessive application of manure or

TABLE 1-2

Changes in pounds of feed required per pound of gain in poultry and milk production per cow and total milk production in the United States

	Feed Required per Pound of Gain For Poultry	Milk Production (lb/cow/yr)	Total Milk Production per Year (million lb)
1965	3.3	8,522	127,000
1976	3.2	10,354	115,458
1986	3.0	13,031	143,667
1994	2.8	15,704	150,704

Source: USDA. *Agricultural Statistics,* 1994 (3).

fertilizer to land, which often increases the nitrate content of the crop being harvested, which could be toxic to the livestock being fed.

Recent genetic improvements of many of the crops used in feeding livestock have increased the digestibility and reduced the amount of antiquality factors associated with these crops. For example, the genetic improvements made in alfalfa has dramatically increased its crude protein content, decreased acid detergent fiber (cellulose and lignin), and, as a result of these changes, increased the digestibility of the energy-yielding components of alfalfa. Plant breeders have reduced the antiquality factors in several crops that are extensively used as feeds for livestock. Rapeseed is one example of a crop that has had the antiquality factor (goitrogenic compounds) reduced by plant breeders; these changes have been so dramatic that the name of the resultant crop has been changed to canola. Canola is primarily produced as an oil source, but the residue remaining after the oil is extracted is marketed as canola meal. The above-mentioned changes can be monitored using laboratory procedures, but continued updating of these changes in the nutritional characteristics of various feeds are required of the livestock producer in order to effectively optimize the use of these feedstuffs in various feeding programs.

Atwater pointed out that an animal has a requirement for a specified amount of specific nutrients. Common practice is to list the nutrient requirements for an animal expressed as a concentration in the diet, rather than as an amount of a particular nutrient. For example, the crude protein requirement for a beef steer that is being finished in a feedlot could be indicated to be 11%, but a more precise method of describing the requirement is to specify how many kilograms or grams of crude protein are required each day to achieve a certain performance level, thus allowing for differences in feed consumption to be taken into account. If a steer consumes 10 kg of a ration containing 11% crude protein, it would consume 1.1 kg of crude protein per day. If you wanted the same level of intake of crude protein by another group of steers that were only consuming 9 kg of ration, the crude protein content of the ration would need to be increased to 12.2% ($10 \times 0.11 = 1.1$ kg CP/9 kg $\times$ 100 = 12.2% CP). Since the intake can vary among groups of animals, it is becoming more common to specify nutrient requirements as an amount instead of as a proportion (%) of the ration.

Many processing (pelleting, grinding, chopping, etc.) and management practices (feeding frequency, lighting regimes, etc.) are being implemented that alter feed consumption and change the concentration of specific nutrients in the diet. An individual involved in feeding livestock in a modern operation must have an understanding of the effect feed processing and management practices have on livestock performance. For example, if a roughage crop is chopped or ground and pelleted and fed to livestock, nutrient consumption is increased in some situations dramatically, which means that more forage could be fed instead of grains, and performance could be maintained. Utilization of the nutrient components found in cereal grains is increased dramatically by rolling and grinding. Many antiquality factors can be inactivated by using various processing methods so that other nutrient components of these products can be more efficiently utilized. Processing techniques have been developed that have changed the physical characteristics of products so that the desired consumption of various nutrients can be achieved, such as salt blocks, molasses blocks, protein blocks, and liquid supplements, or where the physical form restricts consumption and livestock have free access (range, grazing situations, etc.) to these products.

Those involved in formulating rations need to have an understanding of the harmful substances found in various feedstuffs. For example, cottonseed meal derived from certain varieties of cotton can have a relatively high amount of gossypol associated with it. Gossypol is relatively toxic to monogastric animals (calves, swine, poultry); thus, normally, cottonseed meal is restricted or not used in the rations for these animals. If the person formulating rations for feedlot cattle had the choice between rapeseed or canola meal (same family as rapeseed) that had the same nutrient contents, then he or she would need to know that the rapeseed meal contains higher levels of glucosinolates, which have a negative impact on performance; so the canola would be the preferred supplemental protein to use.

When formulating or evaluating feed tags, it is useful to have general knowledge of the nutrient contents and feeding characteristics of various feedstuffs. For example, if corn that is low in the amino acid lysine is the primary ingredient in a pig starter ration, then a protein supplement like soybean meal, which is high in lysine, or a synthetic lysine source would need to be added if the lysine requirement of the animal is to be met. Often the dietary calcium level is increased in a ration when most of the calcium is being derived from alfalfa, since the calcium in alfalfa has a lower biological availability.

Understanding how agricultural by-products, derived from the production of other food products for human consumption, can be effectively utilized in feeding livestock is becoming more critical, because the lands used for agricultural production are not increasing and in some areas of the world are declining. So it is becoming more and more important to utilize all the resources generated by photosynthesis more effectively. In countries like the United States, where cereal grains are reasonably priced, they will continue to be the foundation of our livestock feeding operations; but in other countries where cereal grains cannot be economically used to feed livestock, by-products will need to be used.

It is certainly necessary to have some knowledge of supplemental feedstuffs—those that are used in relatively small amounts for specific reasons or to make up for inadequacies associated with the other ingredients that are being used. Synthetic sources of amino acids and vitamins are examples of these types of products. It is also necessary to have a knowledge of the different nonnutritive additives that are in common usage—medicants, antibiotics, hormones, and others.

Last, but certainly not least, the cost of different sources of nutrients must be given due consideration. Feed is the major cost associated with the production of all livestock and often accounts for between 50% and 75% of total costs; thus it becomes easy to see how the cost of the individual feedstuffs and rations may determine whether an operation will stay in business or go bankrupt.

The previous sections have highlighted some of the factors that affect the feeding of livestock and for which a beginning student needs to gain some appreciation. These factors and concepts will be discussed in other chapters in more detail so that gradual understanding can be developed.

REFERENCES

1. USDA. 1994. *Grain world market and trade report* (Oct).
2. Atwater, W. O. 1877–88. *Report of work of the Agricultural Experiment Station.* Middletown, CN: Agricultural Experiment Station.
3. USDA. 1994. *Agricultural statistics,* U.S. Government Printing Office, Washington, DC, 20402.

2

THE GASTROINTESTINAL TRACT AND NUTRIENT UTILIZATION

Richard O. Kellems and D. C. Church

This chapter provides a brief overview of the anatomy and physiology of the various types of digestive tracts that are commonly found in the domesticated animal species. An understanding of the **gastrointestinal** (GI) **tract** and the process by which it converts the nutrients contained in the various feeds consumed into forms that can be used by an animal is important to an individual involved in the feeding and management of livestock. Feeding management can affect nutrient digestibility and efficiency of feed utilization, subsequent profitability, and the well-being of an animal, so an understanding of these processes is important. Additional information relating to specific species is given in the chapters in Part III of this book and other books (1, 2).

The GI tract is vitally important to an animal because it allows it to convert complex nutrient sources, such as proteins, carbohydrates, lipids, vitamins, and minerals, into forms that can be absorbed and used by the animal. Digestion and absorption are terms associated with this conversion process. **Digestion** is defined as the process of converting the complex nutrients found associated with a feed into forms that can be absorbed by the animal. Digestion

can entail all or some of the following processes. Modification of the physical form of the food occurs by such processes as chewing (or **mastication**) or the muscular contractions of the GI tract. Action of digestive secretions, such as the chemical action of hydrochloric acid (HCl) in the stomach, or the action of bile (secreted by liver) into the small intestine, or digestive action of the enzymes secreted into the GI tract are all essential to the digestive process. Microbial organisms present in the rumen and large intestine play an important role in the digestive process for some species. The ultimate objective of these various digestive processes is to convert the complex dietary nutrients into compounds that can be absorbed by the animal.

Absorption is the process by which the digested nutrients cross the cellular lining (membranes) of the GI tract. Specific physiological mechanisms are involved in the absorption of various nutrients, such as passive absorption (going from a higher concentration to a lower concentration) or active absorption (pumping into the cell often against a concentration gradient). Once the nutrients have been absorbed, they are transported from the digestive tract to various tissues of the animal's body,

6

where they are metabolized and used by the animal for various bodily functions.

CLASSIFICATION OF VARIOUS DIGESTIVE SYSTEMS

The dramatic variations among the GI tracts of the common domesticated livestock species are related to the types of diets that they consume and utilize. Generally, animals are classified into groups based on their type of diet, which provides some indication as to the type of GI tract that they have. **Herbivores** are animals that consume primarily plant materials, **carnivores** eat other animals, and **omnivores** eat a combination of plants and animal matter. There are many subgroups to which animals can be classified; for example, with birds there are those that are fish eaters, insect eaters, fruit eaters, and so on.

When animals are classified based on their digestive physiology, it is common to describe them as having either a simple stomach, often referred to as a **monogastric** (or **nonruminant**) or **ruminant** animal. Swine is an example of a monogastric animal that is omnivorous, which means it consumes both plant and animal matter (Fig. 2-1). Poultry (broilers, layers, turkeys) are omnivorous; they have a complex foregut (three sections that replace the normal stomach) and a relatively simple intestinal tract (Fig. 2-2). Dogs and cats are monogastric animals that are carnivores. Horses and mules are monogastric animals, but they are herbivorous animals; they have a rather large and complex large intestine, which allows them to digest plant materials quite efficiently. Rabbits are another example of a monogastric animal that is a herbivore, with a complex large intestine (Fig. 2-3). Ruminant animals are able to consume and digest plant materials and are classified as herbivores; they include cattle, sheep, goats, deer, elk, and many other wild species.

The relative shape and size of the various components of the GI tract vary considerably among species. In swine, for example, the stomach is relatively large, with a capacity in the adult of about 6 to 8 liters, which is approximately 30% of the total capacity of the GI tract.

CHARACTERISTICS AND FUNCTION OF DIGESTIVE TRACT

The anatomy and organs associated with various species of animals and livestock share some

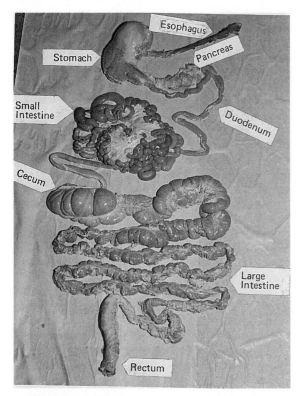

FIGURE 2-1 Digestive tract of swine. Photo by D. C. Church.

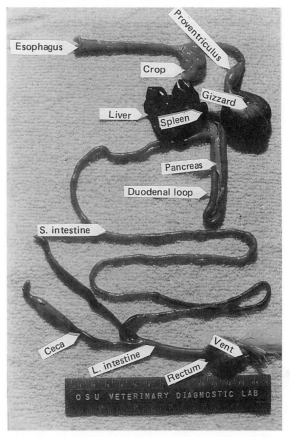

FIGURE 2-2 Digestive tract of chicken. Photo by Don Helfer, Oregon State University Diagnostic Laboratory.

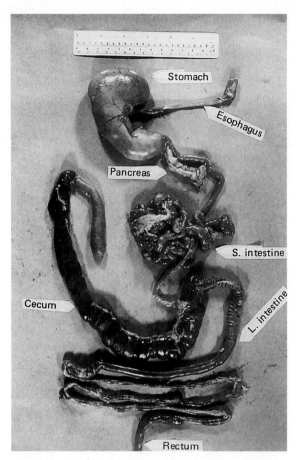

FIGURE 2-3 Digestive tract of rabbit. Photo by D. C. Church.

characteristics and common features and have the same basic digestive functions, even though the sizes of the various anatomical components of the digestive tract vary among species giving each its own unique characteristics. This section provides a brief overview of the processes associated with the various components associated with most GI tracts. Unique differences between monogastric and ruminant GI tracts are discussed in a later section of this chapter. A more in depth coverage of the specific functions of the various components of the GI tract can be found in other sources (3).

MOUTH

The initial component of the GI tract of mammalian and avian species is the **mouth** and its associated structures (beak, teeth, lips, tongue), which are used for grasping and masticating food. The primary function of the mouth is the prehension and preparation of food to enter the GI tract, where the nutrients will be subjected to the digestive process and converted into forms that can be absorbed and utilized by the

animal for various functions. There is wide variation among species with respect to how they consume their food. Cattle are herbivores that use their tongues to wrap around food and pull it into their mouths, while other herbivores, such as sheep, use their lips to selectively consume their food. Thus sheep can be more selective in what they consume and, as a result, the food consumed is higher in nutrient content than for cattle when they are allowed access to similar feeds, especially when they are grazing. Both cattle and sheep will only masticate (chew) their food to a limited extent before ingesting it, but will subsequently regurgitate the coarser constituents consumed and remasticate them (cud chewing). This process is called **rumination.**

Herbivorous species, such as the horse, have incisor teeth adapted to nipping off plant materials and molars with flat surfaces used to grind plant fibers. Using both vertical and lateral movements of the jaws, horses effectively shred fibrous plant materials. Ruminants, on the other hand, have no upper incisors and depend on an upper dental pad and lower incisors for biting off plant materials. Omnivorous species, such as swine, use their incisor teeth primarily to bite off pieces of the food and the molar teeth to masticate the food before swallowing; the tongue is used relatively little. Avian species have no teeth; thus the beak and/or claws serve to reduce food to a size that can be swallowed. In carnivorous species, such as the canine, the teeth are adapted to the tearing of muscle and bone, while the pointed molars are adapted for the crushing bones and to a limited extent the mastication of food.

During the process of **mastication** (chewing), **saliva** is added, primarily from three bilateral pairs of salivary glands present in the mouth. Before ingesting, the food is mixed with saliva and formed into a bolus in the mouth; the bolus is coated with saliva (lubricated) and then swallowed. Saliva has other functions, such as keeping the mouth moist, aiding in the taste mechanisms, providing digestive enzymes and acting as a buffer, that all play vital roles in the initial digestive processes.

ESOPHAGUS

The esophagus is a tube that allows the bolus formed in the mouth to be transported (swallowed) to the initial compartment of the GI tract. The central nervous system controls the contractions of the muscular lining of the esophagus, which transports the bolus to the stomach in monogastric or to the reticulorumen in ruminant animals. These controlled muscular contractions of the esophagus in monogastric animals move the bolus to the stomach, whereas in the ruminants the bolus can be moved in both directions. The process of moving the bolus from the reticulorumen to the mouth is called **rumination** and allows the animal to chew its cud, which aids in the digestion of fibrous feed components that have been consumed. There is considerable variation in the length of the esophagus among different species.

STOMACH (GLANDULAR STOMACH)

The **stomach** is often referred to as the glandular stomach because it is lined with specialized secretory tissues that secrete a variety of substances involved in the digestive process; the relative proportion of these tissues varies considerably among regions of the stomach and among species, which accounts in part for the differences observed in their ability to digest different dietary components, especially the protein components. The differences that exist with respect to the relative proportion of these different regions that are found in different species are shown in Fig. 2-4. The muscular lining of the stomach varies among species, but generally it is well developed. Contractions of the muscles lining the stomach cause the bolus to be mixed with the gastric secretions being produced by the tissues lining the various regions of the stomach, and the digestive processes are initiated.

The regions of the stomach are as follows: The first region found in the stomach is the **nonglandular region** in which no digestive secretions or absorption occurs. The second region is the **cardiac region,** which is lined by epithelium cells that secrete primarily mucin. The third region is the **fundic region,** which has three types of cells: **parietal cells,** which secrete **hydrochloric acid** (HCl), **neck chief cells,** which secrete **mucin,** and **body chief cells,** which secrete **pepsinogen, rennin,** and **lipase.** The presence of food in the stomach

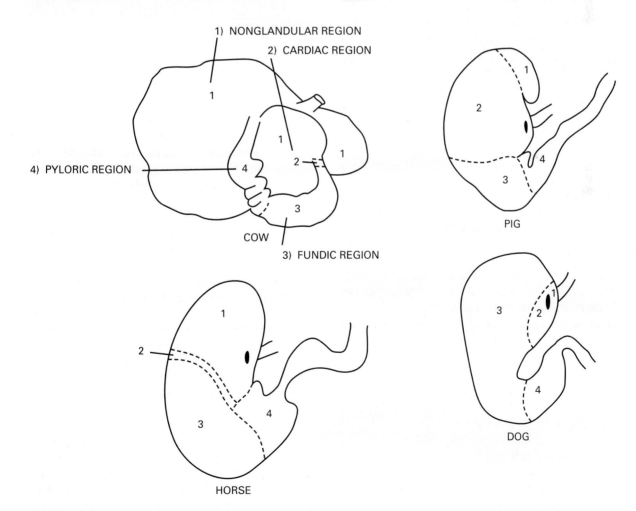

FIGURE 2-4 Tissue types in the glandular stomachs of various species.

causes HCl and enzymes to be secreted, and when the muscular lining of the stomach contracts, this causes the acid and enzymes to be mixed with the ingested food, which becomes acidified (pH 2 to 2.5) and the digestive process starts. **Mucin** coats the lining of the stomach and prevents the tissues from being digested by the digestive secretions produced by the other tissues found in the stomach. Region 4 has only the neck chief and body chief cells. When feed has been consumed, a sequence of events causes the various tissues to secrete and convert pepsinogen into pepsin (active form), so the animal only digests the ingested food and not the lining of its stomach. When food is not present, HCl is not produced and the pepsinogen is not converted into pepsin. This prevents the stomach from digesting itself. This partially digested mixture, which has a jellylike consistency and which is leaving the stomach and entering the small intestine, is referred to as **chyme.**

SMALL INTESTINE

The acidic chyme enters the **small intestine,** which is the primary site for enzymatic digestion in the GI tract, which results from enzymes produced by the animal. The small intestine is composed of three sections; the first section is the **duodenum,** followed by the **jejunum** and **ileum.** In the duodenum the pH of the acidic chyme is neutralized to a pH of 6.8 to 7.0 by **bile,** which is produced by the liver. In most species, bile is stored in the **gall bladder** (the horse has no gall bladder) and is released when food is present in the duodenum via the bile duct. Besides its neutralizing action, bile is also involved in the **emulsification** of fats, which is essential for the efficient digestion and absorption of dietary fats. The digestive enzymes produced by the pancreas, which are involved in protein, carbohydrate, and fat digestion, are secreted into the duodenum. Additional enzymes are also secreted by tissues lining the duodenum, which are primarily involved in the final digestive steps of converting proteins and carbohydrates into amino acids and monosaccharides (simple sugars), which can then be absorbed by the animal. The small intestine is lined with fingerlike projections called **villi** (Fig. 2-5), which increase both the surface area and absorption efficiency. When ulcers are present in this region, the GI tract has a reduced ability to digest and absorb nutrients.

Muscular contractions of the muscular lining of the small intestine continue to mix the

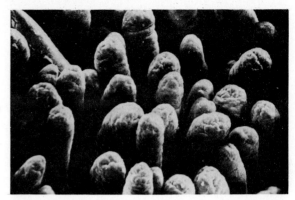

FIGURE 2-5 Scanning electron micrograph showing the intestinal villi of the baby pig. Courtesy of H. Moon, USDA, ARS, National Animal Disease Center, Ames, IA.

ingested food and move it down the GI tract. Enzymatic digestion continues as the ingesta pass into the jejunum and ileum. The relative lengths of these sections of the small intestine vary among different species. The jejunum and ileum sections are also the primary sites for absorption of nutrients in the GI tract of monogastric animals.

LARGE INTESTINE

Normally, most of the nutrients contained in the feed consumed by the animal have been digested and absorbed prior to passing into the **large intestine,** which consist of the cecum, colon, and rectum. Microbial digestion is the primary mode of digestion that occurs in this region. The microbial population starts to proliferate in the ileum and continues in the cecum and colon regions of the large intestine. The specific type and number of microorganisms depends on the amount of undigested food constituents that are passing out of the small intestine and varies depending on the diet and species being fed. Absorption of some organic acids and other organic compounds such as ammonia occurs in the cecum and large intestine. The relative length, diameter, and extent of sacculation vary considerably among species. The large intestine tends to be much larger in herbivorous species, such as horses and rabbits (Fig. 2-3). Both the horse and the rabbit have a voluminous cecum, which serves as a site in which microbial fermentation can occur. Large amounts of water are absorbed and resorbed by the large intestine, which appears to be a major function of the cecum and large (or long) sacculated colons (3). There is considerable variation among species with respect to the amount of water removed from the feces before being eliminated (e.g., between cattle and sheep). The remaining undigestible components are then eliminated via the rectum.

OTHER FUNCTIONS OF THE GI TRACT

In addition to the GI tract's major activities relating to digestion and absorption, it is also a major route for the excretion of many compounds that need to be eliminated by the animal. The liver is a very active site of detoxification of many toxic compounds found in plants, microbes, or drugs that may be administered to the animal. The liver also excretes many mineral elements in addition to the detoxified compounds that need to be eliminated. These compounds are normally excreted with the bile. In the large intestine, some net excretion of mineral elements, especially calcium, magnesium, and phosphorus, may occur depending on the status of the animal and the level in the diet.

Another important activity that occurs in the GI tract is the synthesis of specific nutrients by the microbial organisms. These organisms synthesize the water-soluble vitamins and a number of other essential organic compounds, such as amino acids, proteins, different carbohydrates, and some lipids. The amount of absorption of nutrients, such as vitamins, that occurs past the ileum is assumed to be limited. In some animals, particularly rodents (rabbits and rats), this potential lack of absorption is circumvented by the practice of coprophagy. **Coprophagy** is the term used to describe the practice of an animal consuming its own feces, which allows the essential nutrients resulting from microbial fermentation to be passed through the GI tract and subjected to the digestive and absorptive processes again. Rabbits and rats produce night feces, which are believed to originate mainly from the cecum and have a high vitamin and perhaps essential amino acid content. This practice enables an animal to survive on diets that would otherwise not have sufficient vitamins and essential amino acids to support life. Depending on whether the microbial synthesis is in the rumen or large intestine (pre- or post-absorptive site) or whether the animal practices coprophagy, dietary requirements for some specific nutrients may be completely eliminated in some species, such as the requirement for water-soluble vitamins in ruminant species.

AVIAN SPECIES

In avian species (Fig. 2-2), the crop, proventriculus, and gizzard replace the simple stomach found in other monogastric species. Even here there are variations among different types of birds; for example, most insect-eating or fish-eating species have no crop. When a crop is present, the ingested food goes directly to it and the crop acts as a temporary storage site; in many species this structure has a great capacity for expansion. The proventriculus of birds is similar to the stomach, where gastric secretions (mucin, HCl, and pepsinogen) are produced.

The gizzard is the next section and is characterized by having a very tough muscular lining. The muscular lining of the gizzard contracts, and when adequate amounts of grit (small stones) are present, grinds up the ingested food, similar to the chewing action that occurs in the mouth of other monogastric animals. Birds have a relatively long small intestine, followed by two rather large ceca and a very short section of large intestine. Birds also differ from mammals in that, instead of having separate excretory ducts for urine and feces, their urinary secretions are combined with the feces before being eliminated.

RUMINANT SPECIES

In ruminant species, there are major modifications of the GI tract relating to the stomach region. The stomach is divided into four compartments: the reticulum, rumen, omasum, and abomasum (Fig. 2-6). In a few species the stomach has only three compartments, such as the camel and related species, which are classified as pseudoruminants. Ingested food is subjected to very extensive pregastric microbial fermentation. Most of the ingesta are fermented by the microbes (60% to 75%) before being exposed to typical gastric and intestinal digestive processes; thus, this is a very different system from that of a typical monogastric animal.

The four compartments composing the stomach of a ruminant animal are relatively large as compared to the other sections of the GI tract. The stomach and its contents of a pig amount to about 4% of its body weight, whereas the stomach of sheep and cattle are about 25% to 28% of their body weight. When expressed as a percentage of the total GI tract, the pig stomach is approximately 14% of the total, while the stomach of sheep and cattle is 37% and 45%, respectively.

The first three compartments of the ruminant stomach are lined with a cell type (stratified squamous epithelium) not normally associated with an absorptive surface. This is the

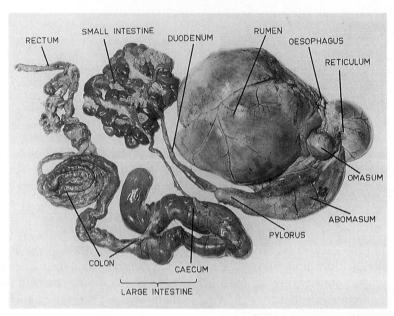

FIGURE 2-6 Digestive tract of a sheep. Courtesy of CSIRO, Canberra, Australia.

same type of tissue that covers the external surface of most mammals. Since the lining is only one or two cells thick, absorption can occur in this region.

The first compartment of a ruminant stomach is the **reticulum,** which has a unique appearance similar to a honeycomb. In this region of the stomach, sharp objects such as nails or wire may penetrate the lining of the stomach and cause a condition known as **hardware disease,** which allows the GI tract contents to leak into the parenteral cavity. The second compartment is the **rumen** or paunch, the largest of the four compartments, which is lined with tonguelike projections called **papillae,** which are underlain by an extensive capillary system that allows for efficient absorption to occur. The density and size of the papillae vary among species and by location in the rumen. The rumen has several strong, muscular pillars (groups) that divide it into several sacs. These muscles contract in a rhythmic manner, causing the ingesta in the rumen to be mixed and partitioned, with the coarse, fibrous feeds subsequently being regurgitated and rechewed by the animal. The third compartment is the **omasum,** which consists of a spherical-shaped organ containing various-sized leaves that extend into the lumen and prevent undigested materials from leaving the rumen and entering the abomasum. The **abomasum** is the fourth compartment and has the same function as the glandular stomach in monogastric species. Like other herbivorous species, the small intestine is similar to that found in monogastric species, and the large intestine is considerably larger than found in omnivorous species, but smaller than in horses or rabbits.

The process of regurgitating ingested food that needs further mastication is referred to as **rumination.** This is the process by which the animal causes the undigested coarse foods to be collected in the reticulum and then formed into a bolus and moved back up the esophagus to the mouth for further chewing. Ruminants may spend 8 hours or more per day ruminating, the amount of time depending on the nature of their diet. Consumption of coarse, fibrous diets results in more time being spent ruminating as compared to diets that are higher in concentrates or are finely ground.

Eructation (belching of gas) is another process that is unique and important to a ruminant animal. When microbial fermentation occurs in the rumen, large amounts of gases (primarily carbon dioxide and methane) are produced, which need to be eliminated or the animal will bloat. This is accomplished by contractions of the muscles lining the rumen to force the gases into the reticulum and then into the esophagus, which dilates and allows the gases to be mixed with inspired air in the lungs and exhaled through the nostrils. Bloat normally occurs when a froth is formed in the rumen, particularly after legume (alfalfa, clover, etc.) species or a high-concentrate diet has been consumed. If present in the area where the esophagus enters the rumen, froth inhibits the eructation process, which would prevent inhalation of froth into the lungs and the occurrence of pneumonia. Under some feeding conditions, bloat may be a very severe problem in cattle, resulting in reduced performance and many deaths, for example when alfalfa is grazed.

A mutually beneficial **symbiotic** relationship (a relationship that is beneficial to both the host animal and the microbial population) exists between ruminant species and the microbial organisms that are found in their reticulorumen. Ruminant animals maintain a very favorable environment for microbial activity and survival in the reticulorumen, which is composed of the reticulum and rumen. The environment is moist, temperature is maintained within very narrow limits, the feed being consumed provides the substrate for the microbes to grow on, and the end products of digestion, are continually removed, making the reticulorumen an ideal place for microorganisms to grow.

Many bacterial types may be found in the reticulorumen and typical bacterial counts range from 25 to 80 billion/milliliter (ml). Besides bacteria, some 35 or more species of ciliated protozoa have been isolated from the contents of the reticulorumen of animals being fed a wide variety of feedstuffs. Protozoa numbers vary widely, but typical counts are between 20,000 and 500,000/ml. These organisms are much larger than bacteria; although their numbers are less, they account for about the same amount of microbial protoplasm as the bacteria found in the reticulorumen.

The type of organisms present depend on the type of feed consumed. If high-forage diets are consumed, which are high in cellulose and hemicellulose, then the microorganisms that digest these substrates will proliferate. When diets high in cereal grains are consumed, the organisms that digest starch proliferate and are in highest concentrations. Normally, the maximum number of microorganisms will be present 2 to 3 hours (h) after a high-concentrate (grain–starch) diet has been consumed and 4 to 5 h after a high-roughage (cellulose–hemicellulose) diet has been consumed. The level of intake will influence the number of organisms present. When feed consumption is high, the number of organisms will be high, and when feed consumption is low, the numbers will decline. As microbial fermentation occurs, volatile fatty acids, ammonia, and other prod-

ucts are produced, which are end products of digestion, and these end products are continually being absorbed across the rumen wall so that the ideal environment can be maintained for the microorganisms.

When the microbial population thrives in the reticulorumen, it greatly affects the nutrient requirements of the host animal. Fibrous feeds, such as hays and silages, are digested more efficiently by the microbes in the rumen than in the large intestine or cecum. Cellulose and hemicellulose can only be digested by microbial enzymes, because animals do not produce these enzymes. In addition, bacteria can use simple forms of nitrogen, such as ammonia or urea, to synthesize their cellular proteins. This reduces the dependence of the animal on high-quality dietary protein sources and allows them to use compounds such as urea as a dietary protein replacement. A further benefit is that the microbes synthesize adequate amounts of the vitamins, except vitamins A, D, and E; as a result, the animal is not dependent on a dietary source of these vitamins, with rare exceptions.

The moisture and pH of the rumen are maintained in part by vast amounts of saliva being secreted, as much as 150 or more liters per day in a mature cow. Saliva contains large amounts of sodium bicarbonate, which acts as a buffer and neutralizes the acids produced in the rumen, which aids in maintaining an optimal rumen pH. Saliva is also important in maintaining an optimal moisture content for the microbes in the reticulorumen.

A major disadvantage of microbial fermentation is that most high-quality dietary protein sources, which have a well-balanced amino acid profile, are partially degraded to produce ammonia, which is resynthesized into microbial protein (intermediate quality protein) when dietary energy is adequate. This downgrading of dietary protein is wasteful. The readily degradable carbohydrates, such as sugars and starches, are rapidly and completely degraded into **volatile fatty acids** (acetic, propionic, and butyric), which are absorbed and used as the primary energy source by ruminant animals. While these acids are readily used by the animal's tissues, the efficiency of their use is less than if the original carbohydrates were digested and absorbed in the small intestine. Furthermore, in the microbial fermentation process as much as 8% to 10% of the energy consumed is converted to methane, which is eliminated by the eructation process and not utilized by the animal as an energy source. The overall effect of rumen fermentation is that these animals can survive and do well on less complex and lower-quality diets than monogastric species, but they utilize good-quality dietary ingredients less efficiently than do monogastric animals. These differences in the type of feeds being consumed and how they are digested by a ruminant animal illustrate in a simplified manner why **feed conversion** of ruminants is low as compared to that of monogastric species. Feed conversion (weight of feed consumed per unit of product produced) for a ruminant is often twice or more than that of monogastric species.

When a ruminant animal is born it has a nonfunctional rumen and depends on the digestion occurring in the abomasum and small intestine for its nutrients supply. An anatomical peculiarity of ruminant species is that they have a structure called the reticular (or esophageal) groove. When a young ruminant animal suckles or drinks milk, the reticulorumen groove will close, forming a tube that connects the esophagus with the omasum. This allows the milk to bypass the reticulorumen and go directly to the omasum and then into the abomasum, where it will be digested. Thus the milk being consumed bypasses the rumen and escapes bacterial fermentation. When the animal starts to consume solid food (grain or hay), the reticulum and the rumen will start to develop and will become fully functional by about 8 weeks in lambs and kids and 6 to 9 months in domestic bovines.

RELATIONSHIP BETWEEN TYPE OF GI TRACT AND TYPE OF DIET REQUIRED

The preceding discussion of the anatomy and function of the GI tract indicates why differences exist with respect to the types of diets that can be effectively utilized by different species. For example, the GI tract of avian species is relatively short and simple and does not provide for extensive fermentation in the large intestine; some fermentation does occur in the ceca, but relatively much less than in most other species. Consequently, avian species do not have the ability to effectively utilize large amounts of fibrous plant materials in their diets. For most birds to consume their needed nutrients, they must be fed diets relatively low in fiber and moderate to high in readily digestible carbohydrates, such as starches. This rules out the use of feed ingredients high in fiber, such as alfalfa hay, if feed consumption and efficiency are to be maximized in poultry.

In omnivorous species such as swine, the GI tract can utilize relatively more fiber than avian species, but much less than herbivorous species. Some ground legumes can be used in dry rations of adult swine or they can be grazed on pastures for a portion of their sustenance. Wild pigs depend on other food items, such as nuts and roots, in addition to **herbage,** but their productivity on this type of diet is much lower than economically feasible for a commercial swine operation.

As an example of a nonruminant herbivorous species, the horse can survive and do well on plant materials of much lower quality than that required by swine. Hard-working horses clearly require higher levels of dietary energy consumption. This may be accomplished by feeding more of a high-quality hay, but often requires that grain be fed.

Ruminant animals are well adapted to diets that consist primarily of fibrous plant materials. Although cattle can be fed diets that are high in concentrates, if this practice is carried on for any great length of time (several months), digestive disturbances are likely to occur and the level of feeding management is increased dramatically. Except for recent times when there have been grain surpluses, cattle and other domestic ruminants have traditionally been fed high-fiber diets, pastures or coarse fodder that has little food value for poultry, swine, or humans. Current trends indicate that a return to these types of feeding systems may occur. Although greater dependence on roughages will reduce productivity, it will probably result in fewer problems with the animals and, perhaps, a more efficient overall utilization of our total agricultural resources.

SUMMARY

An understanding of the differences that exist among the GI tracts of the various domesticated livestock species leads to an understanding of why differences in effective diet types exist among different species. For example, animals that can use large amounts of forages have either a stomach or large intestine that allows for either pregastric microbial fermentation (ruminants) or postgastric microbial fermentation (horses or rabbits). Omnivorous species such as swine have less complex digestive tracts; thus they only have a limited ability to digest fibrous plant carbohydrates. Animals with simple stomachs and intestines such as poultry must be fed diets that are highly digestible and contain high-quality nutrient sources.

The GI tract is a unique system that provides the means by which an animal accomplishes the following: digests the food that has been consumed; absorbs essential nutrients that are subsequently used for various bodily functions; conserves water and synthesizes essential vitamins and other nutrients required by the animal.

REFERENCES

1. Church, D. C. 1988. *The ruminant animal.* Englewood Cliffs, NJ: Prentice Hall.

2. Pond, W. G., D. C. Church, and K. R. Pond. 1995. *Basic animal nutrition and feeding.* 4th ed. New York: Wiley.

3. Swenson, M. J., and W. O. Reece, eds. 1993. *Dukes' physiology of domestic animals.* 11th ed. Ithaca, NY: Cornell University Press.

3

NUTRIENTS: THEIR METABOLISM AND FEEDING STANDARDS

Richard O. Kellems and D. C. Church

Nutrition is one of the major biological sciences. The definition of nutrition is the series of processes by which an animal takes in and assimilates feed components for promoting growth, milk, or fiber production and replacing worn or injured tissues. The volume of information relating to nutrition is extensive. This chapter will provide the reader with an overview of the functions of the various nutrients, feeding standards, and analytical procedures for assessing feedstuffs and a familiarity with the associated terminology. A more in depth coverage of these specific topics can be found in references listed at the end of the chapter (1, 2, 3, 4).

WATER

Water is often overlooked and not considered as a nutrient when formulating rations for livestock, but water is extremely important to animals and composes 71% to 73% of the fat-free animal's body weight. Water plays an essential role in a number of functions vital to an animal, such as digestion, nutrient transport, waste excretion, and temperature regulation. Even though water requirements for livestock

are not listed, it is assumed that animals have free access to a good-quality water supply; if they do not, their performance and well-being will be impaired.

FUNCTIONS

Water is involved in numerous vital functions. It acts as a solvent for many different biological systems. Food that is consumed is mixed with water, and this allows the digestive secretions that are water soluble to transform the food into end products that can be absorbed and utilized by the animal. Water is used as the media to transport materials in the body via the blood and other body fluids to the sites where they will be metabolized. Water is also involved in the transport of waste products eliminated by the animal's body via the urine or digestive tract. It absorbs a large amount of heat when it evaporates (changes from a liquid to a gas), which allows an animal to use it to cool its body.

SOURCES

An animal can derive water in a number of different ways. Drinking is the most common source by which an animal intakes water. Common practice is to have free-choice drinking water available to animals at all times, and facilities must be designed to accomplish this. Varying amounts of **free water** are also contained in the feedstuffs that an animal consumes and can account for a considerable amount of water intake. Free water is not chemically bound to the feed, but is just the moisture associated with the feed. The free-water content of feedstuffs is quite variable. For example, most dry feedstuffs contain between 9% and 13% free water (grains, hays, etc.). Corn silage is an example of a feedstuff with a high amount of free water; it contains between 65% and 75%. Thus an animal consuming 20 lb of corn silage per day will consume 13 to 15 lb of water or 1.5 to 1.8 gal of water per day. Another source of water is **metabolic water;** this is water that is chemically bound and is released when nutrients or body tissues are broken down by metabolic processes (primarily oxidative) occurring in the cells of the animal's body. Some species, such as certain desert rodents, can satisfy their water requirements with metabolic water and therefore have to consume very little drinking water. Metabolic water production among different nutrient sources varies (grams of water per gram of nutrient); carbohydrates yield 0.6, proteins yield 0.4, and fats yield 1.0. Generally, animals found in an arid environment get a higher percentage of their total water requirements from sources other than drinking water as compared to animals found in more humid regions.

LOSSES

Body water losses can occur in a number of different ways: through urine, feces, lungs, skin surface, and in milk. The kidney eliminates a considerable amount of water in the urine. Some animals conserve their body water by having kidneys that produce concentrated urine; other species produce very dilute urine, which increases their water losses. Normally, species found in regions that have limited water will generally be more likely to concentrate their urine than species that are accustomed to plentiful amounts of water. Water losses increase when diets are fed that are higher in protein (increased urea production increases urinary excretions) or high in various mineral salts, in high-fiber diets, and when food consumption is increased. If the water supply is limited, food consumption should be cut back before water becomes a restrictive factor. When water con-

sumption is reduced, feed consumption and performance will be reduced, especially on high-producing animals.

The amount of water lost in feces varies among species; when they are consuming similar diets, cattle eliminate a lot of water in their feces, while sheep conserve their body water by resorbing water in the large intestine and producing a dryer feces. Often the water content in the feces reflects the water content in the food being consumed; for example, cattle consuming pastures high in water content (spring) produce feces that are higher in water content than when consuming feed off dryer pastures (summer and fall).

Evaporative water losses (skin and lungs) are increased greatly in warm or hot environments because the animal must increase evaporative losses to maintain body temperature within a region that allows the animal to survive. Increased air movement (wind) increases water losses and requirements, while high humidity decreases water losses and reduces an animal's ability to cool itself, because of the reduction of evaporative losses.

Lactating animals have a higher requirement for water because of the amount eliminated in the milk. For example, a dairy cow producing 100 lb of milk per day eliminates approximately 87 lb of water in the milk. Milk production will be dramatically depressed when water of adequate quality and amount is not available.

WATER QUALITY

Water quality will often affect feed consumption and animal health, because it will normally reduce water consumption and feed consumption. A general definition for good-quality water is that it contain less than 2,500 mg/l (0.25%) of dissolved solids. Depending on the type of dissolved solids present, as much as 15,000 mg/l (1.5%) may be tolerated, but normally the palatability of the water is reduced and production will likely be depressed at this level. Levels of salts, such as nitrates, fluorine, and other heavy metals, may become toxic prior to affecting palatability. Levels of 100 to 200 parts per million (ppm) nitrates are potentially toxic, and 1 g of sulfate per liter may result in diarrhea. Other materials that may affect palatability or be toxic include pathogenic microorganisms, algae and protozoa, hydrocarbons, pesticides, and many industrial chemicals, which are sometimes found in water supplies.

The most common minerals present in water include chloride (Cl), sodium (Na), calcium (Ca), magnesium (Mg), sulfates (SO_4) and bicarbonate (HCO_3). The minerals present depend on the soil type and the water source. The tolerance of animals to alkaline (salts dissolved) water depends on factors such as water requirements, species, age, physiological condition, season of the year, and salt content of total diet. Generally, water containing more than 1% NaCl (common salt) is generally not considered good quality, because that is about the maximum that cattle and sheep can tolerate without decreasing production. Some animals, such as chickens and swine, will not tolerate this level of salt. Thus, if there is a question with respect to water quality, the water should be tested, especially for dairy, broiler, layer, and swine operations.

WATER AS A SOURCE FOR MINERALS

Water is normally overlooked as a source of nutrients, specifically the minerals that may be dissolved in it. Mineral content varies tremendously between water sources. If average values are used for mineral content, computations indicate that cattle might receive from 20% to 40% of their NaCl requirements, 7% to 28% of Ca needs, 6% to 9% of Mg needs, and 20% to 45% of their sulfur requirements (5). Therefore, water supplies should be considered when formulating mineral supplies, especially in areas where the amounts of dissolved solids are high.

REQUIREMENTS

Water requirements vary considerably among species and are related to the type of diet they consume and the environmental conditions under which they are being maintained. It is important to have an adequate supply of acceptable-quality water available for animals. More detailed information relating to water requirements is given in chapters 13 through 23.

CARBOHYDRATES

The primary component found in livestock feeds is carbohydrates. Carbohydrates are considered to be a renewable resource, because when they are consumed by animals they are converted into carbon dioxide, which is then used through photosynthesis to form carbohydrates again by plants. The plants synthesize many different carbohydrates; the primary building block found in most plant materials is glucose, which is the primary subunit found in most carbohydrates. In plant material, carbohydrates comprise up to 70% of the dry matter of forages and as much as 80% in cereal grains. Carbohydrates serve as a source of energy or bulk in the diet (forages), but no specific requirements are listed for carbohydrates.

TYPES OF CARBOHYDRATES

Chemically, carbohydrates are composed of carbon (C), hydrogen (H), and oxygen (O). The simplest form of a carbohydrate is a monosaccharide, which contains either five or six carbons and is often referred to as a sugar. When monosaccharide contains five carbons, it is referred to as a pentose, and if it contains six carbons, it is a hexose. When two monosaccharides are combined, we have a disaccharide, and when three or more monosaccharides are linked, a polysaccharide. Examples of the various types of carbohydrates associated with feeds are listed in Table 3-1.

TABLE 3-1
Components of plant carbohydrates

Monosaccharides (composed of 1 sugar molecule)
Pentoses (five-carbon sugars)
Arabinose
Ribose
Xylose
Hexoses (six-carbon sugars)
Fructose
Galactose
Glucose
Mannose

Disaccharides (composed of two sugar molecules)	
Cellobiose	glucose–glucose[a]
Lactose	glucose–galactose[a]
Maltose	glucose–glucose[a]
Sucrose	glucose–fructose[a]

Polysaccharides (contain multiple sugars molecules)	
Pentosans (contain pentose sugars)	
Araban	arabinose[a]
Xylan	xylose[a]
Hexosans (contain hexose sugars)	
Cellulose	glucose[a]
Glycogen	glucose[a]
Inulin	fructose[a]
Starch	glucose[a]
Mixed polysaccharides	
Gums	pentoses and hexoses
Hemicellulose	pentoses and hexoses
Pectins	pentoses and hexoses

[a]Sugars contained as subunits.

Glucose (also referred to as dextrose) and fructose are the most common simple sugars (monosaccharides) found in feedstuffs. They occur as the simple sugars in both plant and animal tissues, but only in low concentrations. Fructose is converted to glucose in the animal body and is therefore metabolized as glucose.

The disaccharides (sugars containing two units of simple sugars) and polysaccharides (containing numerous units of simple sugars) are present in plant materials in much higher concentrations than the simple sugars. Sucrose (common table sugar) is a combination of glucose and fructose and is found in plants such as sugar cane and sugar beets. Lactose (milk sugar) contains glucose and galactose and is found only in milk. Maltose is an intermediate breakdown product produced when starches are being digested and contains two glucose subunits.

Starch and cellulose are the two polysaccharides found in plants in the highest concentrations. Starch is found in grains, tubers, and other roots. Cellulose is found in high concentrations in forages. Both starches and cellulose have glucose subunits that are linked by different chemical bonds. Starches can be digested by the enzymes produced by an animal, but cellulose can only be digested by the enzymes produced by the microbial organisms present in the GI tract.

Other carbohydrates such as gums, pectins, and hemicellulose are found in varying amounts in some plant material. Like cellulose, these compounds are digested by the microbial organisms present in the GI tract, and instead of being broken down into simple sugars, they are converted into organic acids, which are used as the primary energy in herbivorous and ruminant species.

ABSORPTION AND METABOLISM

Dietary carbohydrates must be converted into simple sugars (monosaccharides) before they can be absorbed by a monogastric animal. A very small amount of amylase, enzymes that convert starch into maltose, is present in the saliva of some species. Amylase is the primary enzyme involved in carbohydrate digestion in monogastric animals; it is produced by the pancreas and secreted into the duodenum of the small intestine. Other enzymes capable of hydrolyzing the disaccharides are produced by the mucosal lining of the duodenum. There is considerable variation with respect to the amounts of carbohydrate-digesting enzymes produced by

different species. Therefore, some species can be fed diets high in carbohydrates, while others need carbohydrates at a lower level. When the carbohydrate level in the diet exceeds the ability that the animal has to digest it, diarrhea may occur, because when undigested carbohydrates reach the large intestine and microbial organisms start to digest them, by-products are produced that cause water to be absorbed into the GI tract, resulting in diarrhea.

Once simple sugars are formed, they are absorbed rapidly by the small intestine. Some studies indicate that glucose and galactose are absorbed more rapidly than some of the pentoses, such as xylose and arabinose. After absorption, a high proportion of simple sugars are converted to glucose prior to being transported to the animal's tissues to be metabolized as an energy source. The animal's body stores very little energy as carbohydrate, but some glucose is converted into glycogen, which is used as a rapidly released form of energy during times of excessive muscular activity. Glycogen is also used to maintain blood glucose levels within a relatively narrow range.

Mammals never developed the ability to produce enzymes capable of digesting cellulose, hemicellulose, and other carbohydrates found in the fibrous portions of feedstuffs. The digestion of these completed carbohydrates occurs in the rumen and large intestine (cecum and colon) as a result of microbial action. A wide variety of microbes found in the GI tract are capable of digesting fibrous carbohydrates. The horse is the monogastric animal that can digest high-fiber diets the best. Swine and rabbits are intermediate, and poultry, dogs, and cats digest very little of the fibrous components found in feedstuffs.

When young ruminant animals are born, they are functionally a monogastric animal. When they begin to eat solid food, they gradually develop a bacterial and protozoal population in the rumen. The digestive by-products (volatile fatty acids) produced by the microbial organism cause the rumen to mature and develop the ability to digest complex carbohydrates, such as cellulose and hemicellulose. Once the rumen has developed, most carbohydrate digestion will occur in this region of the GI tract. Anaerobic microorganisms digest the starches, sugars (readily available carbohydrates), cellulose, and other polysaccharides to produce carbon dioxide, water, heat, and volatile fatty acids (primarily acetic, propionic, and butyric). These volatile fatty acids are absorbed through the rumen wall or the GI tract wall and provide a source of energy for the animal. High-cellulose diets produce more acetic acid and high-cereal-grain diets produce more propionic acid. When high-grain (starch) diets are fed, substantial amounts of starch may pass out of the rumen without being digested and subsequently be digested in the small intestine. In ruminant animals the amount of amylase as well as other enzymes secreted into the small intestine is quite low, which limits the animals ability to digest starch in the small intestine.

Ruminant animals are the most efficient of the herbivorous species at digesting fibrous carbohydrates (cellulose and hemicellulose). These carbohydrates are usually retained in the rumen for some period of time, as long as 6 to 10 days by cattle fed straw, although higher-quality forages (alfalfa hay) are retained for a much shorter period of time. At any rate, forages are retained long enough for the microbes to digest them. A combination of microbial digestion and chewing during the rumination process causes the particle sizes to be reduced so that they will be able to pass out of the rumen.

Lignin is a compound that is associated with fibrous feedstuffs and dramatically affects the animal's ability to digest cellulose and hemicellulose. Lignin is not a carbohydrate; it is an undigestible polyphenolic polymer that combines with the cellular wall components of plants. The lignin content of plants increase as they mature, and this is the primary reason that the digestibility decreases as the plant ages. Lignin is found in higher concentrations in the stems than the leaves.

PROTEIN

Protein is found in the highest concentration of any nutrient, except water, in all living organisms and animals. All cells synthesize proteins, and without protein synthesis life could not exist. Thousands of different proteins are found in various tissues. Proteins range from the very insoluble, such as feather, hairs, wool, and hooves, to highly soluble, such as plasma globulins. Proteins are large molecules ranging in weight from 35,000 to several hundred thousand grams. Proteins have a variety of unique functions within an animal's body, ranging from protecting the body (hair, skin), digesting food (enzymes), metabolizing nutrients in the animal cells (enzymes), stimulating growth (hor-

mones), and defending the animal against invading organisms (immunoglobulins).

Proteins are long chains of **amino acids** that have been linked together. An amino acid is a compound that contains both an amino ($-NH_2$) and a (COOH) group attached to a carbon skeleton. The physical and chemical characteristics of proteins are derived from their amino acid sequence and the subsequent linkages formed between the different amino acids and other compounds. For example, some proteins may contain minerals (hemoglobin, which contains iron) and casein, which is one of the proteins found in milk and contains large amounts of P (phosphorus). Other proteins may contain carbohydrates (glycoproteins) or lipids (lipoproteins).

The production of proteins is regulated by the genetic materials contained in the nucleus of the animal's cells. This genetic material is commonly referred to as DNA (deoxyribonucleic acid), which is transferred from one generation to another, which gives them their unique characteristics. The DNA controls all protein synthesis occurring in an animal's body. The DNA specifies in what order amino acids will be linked together when forming a specific protein. If for some reason adequate amounts of amino acids are not being provided to the animal, protein synthesis cannot produce the protein. The amino acids required for cellular protein synthesis are supplied in the diet or result from digestive processes occurring in the GI tract.

Proteins are composed of amino acids. More than 200 naturally occurring compounds have been classified as amino acids that are of plant and animal origin. Most proteins found in plants and animals are only composed of about 20 amino acids. To synthesize these proteins, a plant or animal needs a source of amino acids. The primary difference between plants and animals with respect to their amino acid requirements is that plants are capable of synthesizing all the amino acids that they require from inorganic nitrogen sources, such as ammonia (NH_2), nitrate, or other nitrogen-containing compounds. Many microorganisms also have this capability, but higher animals are not capable of synthesizing all the amino acids required by their various tissues. Therefore, a dietary source of amino acids must be provided for most animals.

AMINO ACIDS SOURCES

The dietary amino acids are subdivided into two categories: (1) **essential amino acids** (or nondispensable) and (2) **nonessential amino acids** (or dispensable). Essential amino acids are those that the animal cannot produce in adequate amounts to satisfy its requirements. Nonessential amino acids are those that the tissues of the animal can synthesize in adequate amounts. Both groups of amino acids are listed in Table 3-2.

Numerous trials with rats, mice, dogs, pigs, and chicks have been conducted to define the amino acid requirements for these different species. Such studies have been done using purified diets that do not contain amino acids, such as starches, sugar(s), fats (lard or corn oil), purified vitamins, and mineral sources; then various levels of individual amino acids are added. These results have been used to establish amino acid feeding guidelines for these species. This process is continuing as new genetic lines having different amino acid requirements for different species continue to be developed. For example, these studies indicate that arginine is required in the diet of some species for maximum growth, but not for maintenance; neither is it required by young calves. Because there are situations in which one or more amino acids may be essential or give an added response (in growth or production) when included in the diet, some authors prefer to list the amino acids as essential, semiessential, or as nonessential. Semiessential amino acids are those that, under certain conditions, are not synthesized in adequate amounts to satisfy the requirements of the animal. Table 3-3 lists these amino acids.

There is considerable variation among feedstuffs with respect to their amino acid content; this variation is related to the different

TABLE 3-3

Classification of amino acids

Essential	Semiessential	Nonessential
Isoleucine	Arginine	Alanine
Leucine	Cystine	Aspartic acid
Lysine	Glycine	Citrulline
Methionine	Histidine	Glutamic acid
Phenylalanine	Proline	Hydroxyproline
Threonine	Tyrosine	Serine
Tryptophan		
Valine		
Taurine[a]		

[a]Amino acid essential for cats.

types of proteins found in these feedstuffs. In practical diets the amino acids most likely to be deficient are lysine, methionine, and tryptophan. The reason for this is because cereal grains such as corn and milo are very low in these amino acids, and diets using these grains as the primary energy source usually need to be supplemented with protein sources that contain higher levels of these amino acids, such as soybean meal. If a specific amino acid required by an animal to synthesize a protein is not available, the protein cannot be synthesized, and this amino acid is referred to as a **limiting amino acid.** A limiting amino is one that is not being provided in adequate amounts to satisfy the protein synthesis processes of an animal. For example, if corn makes up the major portion of a diet, then, most likely, lysine will be the limiting amino acid for the diet. Thus, when a protein molecule is being elongated and a lysine is needed, no lysine is available to continue the elongation process and the protein synthesis process will stop. If the level of lysine provided is increased, then the protein elongation process continues until another amino acid becomes limiting; then the protein synthesis stops and the amino acid limiting protein synthesis becomes the limiting amino acid. This continues until the ideal balance of amino acids is provided and none of the amino acids is limiting protein synthesis.

Ruminant animals and some herbivores do not have the same dietary requirements for amino acids as do monogastric species. The reason for this is that the microbial population in the GI tract, primarily in the rumen and large intestine, synthesizes microbial protein, which then can be digested, and thus provides amino acids that can subsequently be absorbed and utilized by the animal. These microbial organisms can synthesize both essential and

TABLE 3-2

Essential and nonessential amino acids

Essential	Nonessential
Arginine	Alanine
Histidine	Aspartic acid
Isoleucine	Citrulline
Leucine	Cystine
Lysine	Glutamic acid[a]
Methionine	Glycine[a]
Phenylalanine	Hydroxyproline
Threonine	Proline[a]
Tryptophan	Serine
Valine	Tyrosine
Taurine[b]	

[a]Amino acids required in addition to the essential amino acids by chick for optimal growth.

[b]Additional amino acid required by cats.

nonessential amino acids from simple compounds, such as urea or ammonia and organic acids that are derived from microbial digestion of carbohydrates. Ruminant animals can survive on diets containing no protein, but that contain a nitrogen source that can be converted into microbial protein. Evidence gathered with high-producing dairy cattle indicates that adequate levels of lysine and methionine may not be provided by the microbial protein being produced in the rumen, and other dietary sources need to be provided (2).

BIOLOGICAL VALUE OF PROTEIN SOURCES

The **biological value (BV)** of a protein source depends on two factors: (1) how well the animal's digestive process can convert it into amino acids and (2) how the resultant amino acid balance compares with the animal's requirement for amino acids. The relative biological value can be simply defined as the portion of amino acids that when consumed are retained by the animal. As the amount retained increases, the biological value of the protein source increases.

An animal must be able to convert the dietary protein source into a source of amino acids that can be subsequently absorbed. Generally, most protein sources used in feeding animals will be between 75% and 80% digestible. In simple-stomached animals, dietary proteins are converted into amino acids by the digestive secretions in the stomach (HCl, pepsin) and small intestine (chymotrypsin, trypsin, etc.). The amino acids can then be absorbed in the anterior portion of the small intestine, transported to the cells via the blood, and used to synthesize various proteins. The ability of an animal to digest various protein sources varies considerably. Protein associated with feathers, hooves, and the like, are slowly digested, unless they have been properly processed; therefore, their value as a protein source is relatively low. On the other hand, the protein associated with soybean meal is almost completely converted into amino acid and is highly digestible. Many plant protein sources contain various inhibitors that affect protein utilization (Ch. 8).

Amino acids are readily absorbed in the anterior portion of the small intestine. The balance of amino acids being absorbed has an effect on the biological value of the protein source being fed. The animal requires certain amino acids to synthesize the proteins that it needs to support its body functions at the desired level. If the balance of amino acids that is being absorbed does not provide what is needed by the animal, protein synthesis will be reduced, animal performance will be depressed, and the biological value of the protein source will be low. If the amino acids being absorbed are more similar to what the animal needs, the biological value for the protein source will be higher. The proportions of available amino acids present in the protein source affect its biological value, especially the essential amino acids. Whole egg protein has a BV of about 100; meat proteins, 72 to 79; cereal proteins, 50 to 65; and gelatin, 12 to 16.

Other measures of protein adequacy are the protein efficiency ration (PER) and net protein value (NPV). PER is by definition the number of grams of body weight gain of an animal per unit of protein consumed. NPV measures efficiency of growth by comparing body N resulting from feeding a test protein with that resulting from feeding a comparable group of animals a protein-free diet for the same period of time. NPV can also be computed by multiplying the digestibility of a protein by the BV. More information relating to the assessment of protein utilization can be found in other sources (1).

The blending of feedstuffs to improve the balance of amino acid in the final diet takes advantage of the differences in amino acid content that exist in feedstuffs, which is referred to as an **associative effect** between feeds. Often feedstuffs are blended to take advantage of this associative effect so that a more ideal balance of amino acids is obtained in the final diet at the most economical price; so it is important to have some understanding of the levels of essential amino acids that are found in commonly used feedstuffs.

When excess amino acids are absorbed by an animal, they will be metabolized and used as a source of energy. When this occurs, the amino (NH_2) group will be removed from the amino acid (**deamination**) and either transferred to another carbon skeleton to form a nonessential amino acid or converted to urea and excreted in the urine. The remaining carbon structure is then used to generate a source of energy that can be used by the body tissues.

In general, protein quality is less important to ruminant animals than to simple-stomached species. In the rumen a high proportion of dietary proteins is hydrolyzed by rumen microbes to amino acids, many of which are further degraded to organic acids, ammonia, and carbon dioxide. The free ammonia in the rumen is utilized by bacteria to synthesize new amino acids essential for their function. The bacteria, in turn, may be ingested by protozoa, which go through the same cycle of degradation and resynthesis of proteins. Eventually, bacterial, protozoal, and undegraded dietary proteins pass into the lower intestinal tract, where they are converted into amino acids and absorbed. Bacterial and protozoal proteins are generally lower in BV than are the high-quality proteins found in egg and milk, but they are of higher quality than many plant sources. Thus the tendency is to degrade the value of very high quality proteins and upgrade the low-quality dietary proteins. This process allows ruminant animals to convert some nonprotein nitrogen compounds such as urea into microbial protein, which can be converted into a source of amino acids. Further information on this topic is presented in Ch. 8.

DIETARY PROTEIN REQUIREMENTS

The dietary requirement for monogastric and avian species is for essential amino acids. There is substantial volumes of information on the amino acid content of all major feedstuffs, but chemical analyses do not tell us how much of a given amino acid will be digested and absorbed. Thus the requirements (given in various appendix tables) are usually expressed in amounts of total protein, with additional information on some of the limiting amino acids.

Requirements are always highest (in terms of concentration in the diet) for young, rapidly growing animals. The needs decrease as the growth rate declines. Requirements are lowest for adult animals in a maintenance situation. They are increased during pregnancy and increased markedly during periods of peak lactation or egg production. Further details will be given in chapters 13 through 23.

PROTEIN DEFICIENCY

A deficiency of protein can result from one or more reasons: (1) one or more amino acids are limiting and (2) dietary protein level is inadequate. Signs of protein deficiency include poor growth rate and reduced N retention by the body, poor utilization and lower consumption of feed, lowered birth weights, often accompanied by high infant mortality, reduced milk or egg production, and infertility in both males and females. The severity of the symptoms is normally related to the severity of the deficiency. From a practical point of view, insufficient protein is most noticeable in young, rapidly growing animals, lactating females, or laying hens. Normally, protein deficiency is accompanied by deficiencies of one or more other nutrients and less energy, in particular for herbivore species.

Subclinical deficiencies (those that cannot be diagnosed by examination of the animal) are probably relatively common in many countries. One reason is that proteins are expensive feed ingredients to purchase, and the tendency of livestock feeders is to reduce the level or quality fed if possible. Sometimes, milk deficiencies can be detected by lowered blood proteins, reduced growth rates, and so on, but it is usually difficult to be quite sure that the animal is deficient. Blood urea nitrogen can be helpful in accessing the protein status of an animal. Only in the case of lysine is there a specific sign of a deficiency. In black-feathered turkeys, a lysine deficiency produces a white barring of the primary flight feathers.

EXCESS DIETARY PROTEIN

Herbivorous animals in a natural habitat normally encounter excess protein only during periods of lush growth of vegetation in the spring months (or early in the rainy season in the tropics). With confined domestic animals, it is not a common problem because of the costs associated with protein supplements. Most studies do not suggest any marked adverse effects from consuming excess protein, particularly if the protein is of adequate quality and consumption is not continued for long periods of time. There is some indication that fertility in dairy cattle may decline when dietary protein is excessive (6).

Toxicity in monogastric animals does not seem to be a problem when excess dietary protein is consumed when animals have access to an adequate supply of water. Problems can occur when nonprotein nitrogen sources are mistakenly incorporated into their rations.

When urea or other nonprotein nitrogen sources are used in feeding ruminant animals, toxicity and death can occur if fed without an adequate supply of carbohydrates such as starch or sugars. Often this situation arises when an animal is consuming low-quality forages, which results in a relatively high rumen pH. In this condition, urea is rapidly hydrolyzed to ammonia, the ammonia is absorbed quickly and exceeds the amount that can be detoxified by the liver, and the level may become toxic and death may occur. Toxicity may also occur when animals are not adequately adapted to urea or if feeds have not been properly mixed, allowing the consumption of excess amounts of urea.

LIPIDS

Lipids (fats) are organic compounds that are characterized by the fact that they are insoluble in water, but soluble in an organic solvent (benzene, ether, etc.). A variety of different types of compounds is found in both plant and animals, all of which are involved in important biochemical or physiological functions. Chemically, lipids range from fats and oils to complex sterols. Glycolipids are combinations of carbohydrates and lipids and essential for normal cellular processes to occur. Lipoproteins are important constituents of cells. Phospholipids are compounds that contain phosphorus and fatty acids and are constituents of cellular membranes. Sterols range from compounds such as vitamin D to cholesterol and likewise are involved in maintaining essential functions. Fats serve as a concentrated form of stored energy; 1 g of fat yields about 2.25 times (9.45 kcal) as much energy as carbohydrates (4.1 kcal) when it is completely combusted.

Nutritionally, lipids and fats are used to provide energy and a source of essential fatty acids in the diet. Fats are composed of fatty acids of varying lengths combined with a glycerol molecule. Fatty acids consist of chains of carbon atoms ranging in length from 2 to 24. When all the bonds on the carbon atom are taken up by a hydrogen, the fatty acid is referred to as saturated. If one or more double bonds are present, then the fatty acid is unsaturated. Fat sources can be classified as either fats or oils, the two being differentiated by their physical consistency at room temperature. Oils are liquid and fats are solids at room temperature. Normally, oils are composed of short chains of fatty acids or unsaturated fatty acids, and fats contain saturated or longer-chained fatty acids. Some of the most common fatty acids found in fat sources are listed in Table 3-4.

Most fatty acids found in animal tissues are straight chained and contain an even number of carbons. The reason for this is that when fatty acids are being synthesized in an animal's body the metabolic process of elongation adds two carbons at a time; therefore, the carbon length is always an even number. Fatty acids having odd carbon lengths or branched chains are often produced by microorganisms. The microbial action in the GI tract of ruminant animals often causes their body fats to contain

TABLE 3-4

Various fatty acids

Fatty Acid	Abbreviation[a]
Saturated acids	
Acetic	C2:0
Propionic	C3:0
Butyric	C4:0
Caproic	C6:0
Caprylic	C8:0
Capric	C10:0
Lauric	C12:0
Myristic	C14:0
Palmitic	C16:0
Stearic	C18:0
Arachidic	C20:0
Lignoceric	C24:0
Unsaturated acids	
Palmitoleic	C16:1
Oleic	C18:1
Linoleic	C18:2
Linolenic	C18:3
Arachidonic	C20:4

[a]For acetic acid, the C2 means that it contains two carbon atoms and :0 means that it has no double bonds.

these types of acids. Fats found in plants and animals are generally in the triglyceride form, but some are found as diglycerides (two fatty acids attached to a glycerol), with a molecule of galactose (monosaccharide) attached to a glycerol.

DIGESTION AND METABOLISM

In monogastric (simple-stomached animals) the primary site of fat digestion is the small intestine. The combined action of bile and pancreatic lipase digests the dietary fats. Bile is produced by the liver and is secreted into the duodenum; its primary function is to emulsify the fat, thus increasing the surface area, which increases the fat exposure to the enzyme lipase. The lipase produced by the pancreas then hydrolyzes the triglyceride, yielding fatty acids and glycerol. The fatty acids and glycerol are then complexed with bile and absorbed. Once in the tissues lining the GI tract, they are transformed back into triglycerides and transported to the cells of the animal to be metabolized. Fats containing long chains are transported via the lymphatic system, which combines with the blood system just prior to the heart. Short-chained fatty acids are transported via the circulating blood system. The absorption of fatty acids generated by the digestive process is quite high. For example, one study with chicks evaluating different fats found the following: soybean oil, 96%; corn oil, 94%; lard, 92%; beef tallow, 70%; and fish oil, 88% absorption rate. Generally, oils (liquid at room temperature) are absorbed more completely than fats (solid at room temperature). Adequate amounts of fats are also essential for absorption of fat-soluble vitamins (A, D, E, K).

Fats are used by the animal to synthesize various compounds required by the animal's body or they are stored in fat deposits in the animal's body. Animals store fats as an energy reserve so that during times of inadequate energy consumption the fat stores can be mobilized and used as an energy source. The end product of energy metabolism is carbon dioxide, water, heat, and ATP (adenosine triphosphate). When animals are metabolizing large amounts of fats to provide the energy that they need, abnormal conditions may arise that cause the energy-metabolizing process to shut down, and ketones are produced. If moderate amounts of ketones are produced, they can be further metabolized by the tissues and used as an energy source; but excessive amounts are detrimental to the animal and can be detected in the urine, milk, and lungs.

The microbes found in ruminant animals are capable of altering dietary fatty acids. Fats are ingested in amounts typical for common feedstuffs (2% to 8%), of which a high portion are unsaturated. A large proportion of the unsaturated fats will be saturated by the microbes present in the rumen. High levels of fat in the diet or fats that have been treated in a manner that protects them from rumen action will prevent the saturation process from occurring. Large amounts of fats may be synthesized by an animal's body, even when the dietary intake is quite low. Normally, this occurs when dietary energy intake is in excess of what the animal requires to carry out its body functions; then the excess energy is converted to fats, which is the form in which excess energy is stored. The mechanism of fat synthesis is relatively complex, but in simple terms it can be said that two carbon units that have been derived from excess carbohydrates or proteins are combined to form a fatty acid, which is then combined with glycerol to form a triglyceride, which is then stored in the adipose tissue of the animal.

ESSENTIAL FATTY ACIDS

Essential fatty acids are fatty acids that the animal requires, but which it cannot synthesize in adequate amounts to meet its needs. Monogastric animals require that essential fatty acids be provided at a rate of approximately 1% of the diet. The two fatty acids that are considered to be essential for mammalian species are linoleic (C18:2) and linolenic (C18:3). Both of these fatty acids contain 18 carbons and have two or three double bonds, with the first double bond inserted six carbons from the fatty acid's terminal end, so they are ω-6 (omega 6) fatty acids. Fish and some other aquatic species have essential fatty acid requirements for fatty acids that have the first double bond inserted three carbons from the terminal end of the fatty acid, which are ω-3 (omega 3) fatty acids. For example, linoleic acid can be either C18:2ω6 or C18:2ω3, which can be used as an essential fatty acid source for mammals and fish, respectively, but are different fatty acids. Essential fatty acids are important because they are an integral part of the lipid–protein structure of the cell membranes, and they appear to be important in the structure of prostaglandins, hormone-like compounds that are required for the biochemical processes to function normally within the cell. Deficiency symptoms resulting from insufficient consumption of essential fatty acids have been demonstrated in pigs, chickens, calves, dogs, mice, and guinea pigs. Signs of a deficiency are scaly skin and necrosis of the tail, reduction in growth and reproductive performance, edema, subcutaneous hemorrhages, and poor feathering in chickens. It is rather puzzling that no deficiencies have been observed in adult ruminants fed purified diets containing no fat, which seems to be because the rumen microorganisms are capable of producing adequate amounts of essential fatty acids.

From a feeding perspective, deficiencies of essential fatty acids do not seem to be a common problem in domestic animals except for poultry. Fortunately, the essential fatty acids are distributed widely among most of the common feedstuffs used in their diets. Both corn and soybean oils are excellent sources of linoleic and linolenic acids.

COMPOSITION OF BODY FAT

Body fat deposits in monogastric animals species often reflect the type and amount of dietary fat that is being consumed. It has been known for many years that the fatty acid composition of the fat deposits reflects what the animal is consuming. If an unsaturated fat is fed, the fat deposits become more unsaturated, and if its level in the diet increases, this further increases the amount found in the fat deposits. It has been known that if an oil is fed to poultry or swine then the carcass fat becomes less saturated, softer, and has a lower melting point, which will dramatically change its processing characteristics. If fish oil is fed, flavor associated with the oil will be present in the meat also.

Minor changes in the diet do not have a marked influence on body fat deposits of ruminant animals, primarily because of the effect of the rumen microorganisms on dietary fats. However, it is possible to feed protected fat sources and alter the fatty acid composition of adipose tissue and milk fat.

MINERALS

Minerals are inorganic components of the diet; they are solid, crystalline elements that cannot be decomposed or synthesized by chemical reactions. Dietary mineral sources can be classified based on the concentrations found in an animal's body. Macrominerals or major minerals are found in concentrations that exceed 100 ppm and include calcium (Ca), phosphorus (P), chlorine (Cl), magnesium (Mg), potassium (K), sodium (Na), and sulfur (S). Microminerals or trace minerals are found in concentrations of less than 100 ppm and include chromium (Cr), cobalt (Co), copper (Cu), fluorine (F), iron (Fe), iodine (I), manganese (Mn), molybdenum (Mo), nickel (Ni), selenium (Se), silicon (Si), and zinc (Zn).

The ash content (total mineral content) of an animal's carcass is about 3.5% of the carcass for cattle. The largest amounts of minerals are found associated with the skeleton system and provide its structural rigidity. Ca, P, and Mg are the major minerals found in bone. Other minerals, such as Na, Zn, Mo, and Mn, are also found in the skeletal system. Calcium represents about 46% and P about 29% of the total minerals in an animal's body. K, S, Na, Cl, and Mg together account for about 24%, while the essential trace elements constitute less than 0.3% of the total. Calcium is also the major component of egg shells.

Most of the mineral elements are also involved in complex biochemical reactions. Those involved in enzyme activity include Ca, Mg, Fe, Co, Mn, Mo, and Zn. Iron is an essential constituent of hemoglobin in the blood and myoglobin in muscle tissues. Cobalt is a structural component of vitamin B12. Iodine is a component of thyroid hormone. Other minerals such as Ca, K, Mg, and Na are involved in activity of the nervous system. Sodium, K, and Cl are involved in the regulation of osmotic pressure and pH in an animal's body fluids. The preceding only highlights some of the functions that minerals are involved in, a more complete description of their functions is provided elsewhere (7).

ABSORPTION AND METABOLISM

Dietary mineral elements are converted into their ionic (free form) and absorbed from the GI tract by using either active or passive methods. **Active absorption** means that the mineral element is pumped by the intestinal wall from the lumen into the intestinal cells. Mineral elements that are actively absorbed include Ca, P, and Na. Often active absorption occurs against a concentration gradient, which means that the mineral is being pumped from a lower concentration to a higher concentration and energy needs to be expended to accomplish this. However, most minerals are absorbed in a passive manner: the element diffuses across the lining of the GI tract, going from a higher to a lower concentration. Thus with passive absorption the concentration of an element in the feed and in the body greatly affects the amount absorbed.

Mineral elements are absorbed primarily in the ionic form. Some digesta components may bind (chelate) minerals and make them unavailable for absorption. Phytates, oxalates, and fats are compounds that may bind certain mineral elements and thereby reduce their availability to the animal. Minerals sometimes interfere with the utilization of other essential elements. For example, excess Ca is particularly a problem, because it interferes with P and Zn absorption.

Other factors may influence the absorptive efficiency of minerals. Young animals are more efficient than older animals in absorbing minerals. Thus, even though the cellular requirements for minerals are lower for older animals, dietary minerals may have to be increased because the animals absorb less. The form of the element (organic versus inorganic) and the pH of the intestinal tract can also affect absorption.

REQUIREMENTS AND DEFICIENCIES

Mineral requirements for the various species are given in the appendix tables. Some information is also given in Chs. 13 through 23 for the various domesticated species of animals. More in depth information is contained in other reference books (1, 2, 8).

The essential macrominerals most commonly deficient or imbalanced in livestock rations are Ca, Mg, Na, and P. Clinical signs and symptoms vary somewhat from species to species, so only a general discussion of these signs will be given here. A deficiency of Ca or P (or vitamin D) or an imbalance (ratio outside 1 to 7:1 of Ca:P) can result in rickets in young animals, which is manifested by inadequate mineralization of the bones, crooked legs, and enlarged joints. In older animals the minerals are withdrawn from the bones, resulting in osteoporosis, the condition in which the mineral content of the bone is reduced and the bones become more porous, weaker and more subject to fractures. Mild to severe P deficiencies are not uncommon, particularly in grazing species of animals. It often results in the animal developing a depraved appetite (pica), which is manifested by an animal chewing bones, rocks, boards, and other abnormal objects. Phosphorus deficiency has a marked effect on reproduction and growth. Salt (NaCl) is the most widely fed mineral. When salt is not fed and the soil or water supplies do not contain much Na, deficiencies may occur. Clinical signs are a craving for salt, emaciation, listlessness, and poor performance. Potassium is found in high concentrations in forages, so a deficiency will only occur when high levels of grains are being fed. Deficiency symptoms for K are similar to those for Na. Magnesium deficiency is not common in most livestock species. Various factors prevent normal utilization of Mg during cool, cloudy weather in the early spring or late winter or during similar light and temperature conditions, which results in a Mg deficiency. The plants contain adequate amounts of Mg, but it is bound in a form unavailable to the animals, and a Mg deficiency will occur, which is referred to as grass tetany or Mg tetany. The symptoms of a Mg deficiency is irritability, convulsions, and coma and death in more advanced stages. Mature animals are more likely to be affected, particularly lactating females.

With regard to the microminerals, Fe is always deficient for very young pigs. A deficiency occurs because body reserves of newborn pigs are low, young pigs grow very rapidly, and milk contains an inadequate amount of Fe. The results will be anemia (insufficient hemoglobin in the blood), which prevents the animal from being able to transport the required amount of oxygen to the tissues of the body (see Ch. 19). With chicks and other poultry species, Mn may be deficient; the sign is perosis, a condition in which the hock joint is enlarged and deformed such that the bird has difficulty moving. Iodine may be deficient for most species, depending on the source of their feed. An iodine deficiency results in goiter, an enlargement of the thyroid gland. Cobalt is deficient in some soil types, and plant grown on these soils are also deficient in Co. A Co deficiency causes a deficiency of vitamin B12 since Co is a structural component of B12, and when Co is not present, the vitamin cannot be produced. Animals will have retarded growth and a listless appearance. In some situations, Cu may be deficient, or a deficiency may result from excessive amounts of Mo or sulfate being present, which depress Cu utilization. Typical signs of a Cu deficiency are depigmentation of the hair (lighter than normal hair color), partial paralysis of the rear quarters, and other manifestations depending on the age of the affected animal. Selenium deficiency occurs in many regions where the soils are low in Se and is an important problem for domestic livestock, especially those that are being grazed. The clinical sign of Se deficiency is nutritional muscular dystrophy (white muscle disease), which occurs primarily in young animals. Some of the muscles of these animals appear whitish because of increased Ca–P salt deposits. It becomes very painful or difficult for the animal to contract their muscles, and because of reduced usage, the muscles start to degenerate. High death rates are common in newly born animals with a severe deficiency. Zinc deficiencies are also relatively common and are increased when high dietary levels of Ca are being consumed. Signs of a Zn deficiency are parakeratosis, which is a dermatitis manifested by itching, scaly skin lesions.

All deficiencies will, sooner or later, affect animal performance (growth, lactation, egg production, etc.), even though the appearance of the animal may not be affected. In some cases, multiple deficiencies of nutrients may occur, such as protein–energy, energy–mineral, or energy–vitamin. Mild deficiencies may be difficult to detect because the only effect on the animal may be a reduction in performance below what would be normally expected. Such situations require careful evaluation by individuals trained in such matters to determine what the problem may be.

TOXICITY

As a general rule, mineral toxicity is much less of a problem than mineral deficiencies. Water containing high levels of some alkali salts may be toxic, but such water would not normally be consumed if a better-quality water is available. Of the macromineral, NaCl is often a problem for poultry and swine, but is not often a problem for other species.

Of the microminerals, F can occur in toxic amounts if sufficient amounts of rock phosphates containing F are being fed over a period of time (months or years). Consumption of vegetation grown downwind from a mill processing ores high in F may result in toxicity. Fluorine toxicity results in enlarged, soft bones, teeth that wear off much more rapidly than normal, difficulty in walking, and generally poor performance. Copper toxicity can occur in young animals, especially when mineral supplements are being fed that are designed for other species. For example, if a mineral supplement that was formulated for cattle was mistakenly fed to sheep for a period of time, the Cu may prove to be toxic to the sheep. Copper toxicity affects the liver and can cause the red blood cells to be destroyed, resulting in relatively high death rates. Plants may be grown on soils that are high in Se, and some plants have been found to accumulate Se; in both cases the Se levels cause a toxicity to occur manifested by elongated hooves, loss of tail and mane hair in horses, and difficulty in walking.

VITAMINS

Vitamins are organic substances that are required by animal tissues in very small amounts. All vitamins are essential for animal tissues, but some species of animals are able to synthesize certain vitamins in their tissues, or they are able to utilize vitamins synthesized by the microorganisms in their GI tract. Consequently, vitamins needed in the diet vary from species to species. For example, humans, guinea pigs, monkeys, and some other species require vitamin C (ascorbic acid), but most animal species do not, because they can synthesize adequate amounts to satisfy their needs.

Vitamins are classified as either water soluble or fat soluble. The water-soluble vitamins are C (ascorbic acid) and the B-complex vitamins: thiamin (B1), riboflavin (B2), niacin, pyridoxine (B6), pantothenic acid, folic acid, cyanocobalamin (B12), biotin, choline, inositol, and para-aminobenzoic acid (PABA). The fat-soluble vitamins are vitamin A (retinal or retinoic acid) or its precursor, carotene; vitamin D, of which there are several forms; vitamin E, or alpha-tocopherol; and vitamin K, which also has several active forms.

FUNCTIONS

The primary functions of many water-soluble vitamins are as cofactors (a substance that is required for an enzyme to be active) of enzymes.

The fat-soluble vitamins do not act as cofactors for enzymes, but are involved in other processes that are essential to bodily functions. Vitamin A, for example, is involved in vision and with the maintenance of the epithelial cells (these cells line the body surface and cavities). Vitamin D is involved in Ca absorption and bone deposition. Vitamin E functions as a metabolic antioxidant. Vitamin K is required for normal blood clotting to occur.

TISSUE DISTRIBUTION

The major storage site for most vitamins is the liver, with lesser amounts being stored in the kidney, spleen, and other tissues and organs. Most are stored bound to specific proteins. Vitamins stored in these tissues are released at the rates necessary to maintain a relatively constant level in the blood.

The presence of vitamins in milk is important because milk often provides the sole food source for newborn animals. Colostrum is especially high in vitamin content, thus ensuring an adequate intake for the young mammal early in life.

ABSORPTION AND METABOLISM

Vitamins are absorbed primarily from the small intestine. The B-complex vitamins and vitamin K are synthesized by the microorganisms in the large intestine in monogastric species and in the rumen and large intestine of ruminant animals. Ruminants, with one or two exceptions, have no dietary requirements for these vitamins because sufficient quantities are produced in the rumen and absorbed from the intestinal tract.

The ability of monogastric animals to absorb vitamins synthesized in their intestinal tract varies with the vitamin and the species of animal. Vitamin K is synthesized and absorbed so efficiently that it is almost impossible to produce a deficiency in anything other than poultry. Pantothenic acid and B12 on the other hand are synthesized in the intestine, but little of the synthesized vitamin is absorbed; consequently, monogastric species require a dietary source. Swine absorb enough of the folacin from the gut to satisfy most of its needs, while poultry absorb none of it and are entirely dependent on dietary sources. In the upper intestine the absorption of fat-soluble vitamins is less efficient than for water-soluble vitamins. Fat-soluble vitamins are more efficiently absorbed when there is a dietary fat and adequate bile is present in the GI tract.

A number of vitamins are consumed in forms that have no vitamin activity, but are subsequently transformed to an active form of the vitamin. Vitamin D is a unique vitamin that is found in plants as ergosterol, which is a precursor and must be converted to the active form of vitamin D2 (calciferol) in the skin of an animal's body when it is exposed to ultraviolet light. Most mammals can use D2, but D3 (7-dehydrocholesterol) has a much higher activity in avian species and is normally the vitamin D source used in their diets. However, in the animal body both D2 and D3 are converted by the liver to a more active form, which in turn is metabolized to another compound in the kidney; the latter compound (1,25-dihydroxycholecalciferol) is involved in Ca metabolism and absorption.

Plants produce many different carotenoid pigments, but only a few can be converted to vitamin A. Various different types of carotene produced by plants serve as the primary precursor of vitamin A. The carotene is converted in the wall of the small intestine or the liver to the active form of the vitamin.

VITAMIN DEFICIENCIES

Many of the signs of the various vitamin deficiencies are similar. These include anorexia (poor appetite), reduced growth, dermatitis, weakness, and muscular incoordination. Some vitamins deficiencies cause additional specific symptoms (1, 2). For example, vitamin A deficiency can cause various kinds of blindness, including night, color, and total blindness. Vitamin D deficiency causes rickets and related bone disorders, and vitamin K deficiency causes hemorrhaging in the tissues. If niacin is absent, lesions develop on the tongue, lips, and mouth.

Vitamins that may be deficient under practical conditions often vary among different classes of livestock and with the age of the animal. With ruminants the main concern is with vitamin A, possibly with vitamin E, and to a limited extent with vitamin D in specific situations. Swine producers must be concerned about dietary requirements for riboflavin, niacin, pantothenic acid, B12, and choline, as well as vitamin A, D, and E. Poultry raisers must monitor the intake of all the vitamins except ascorbic acid, inositol, and PABA.

DIETARY ENERGY

From a quantitative standpoint, energy is required in the highest amounts in an animal's diet. Feeding standards used for formulating rations for all species are based on some measure of energy, with additional requirements for protein or amino acids, essential fatty acids, vitamins, and minerals. Animals derive their energy from the dietary organic components being consumed. These organic components are digested and absorbed and subsequently oxidized in the cells of the animal. Carbohydrates are normally used to provide the bulk of the energy that an animal requires, because of the relative low cost per unit of energy. The different organic compounds (carbohydrates, proteins, fats) consumed by animals yield different amounts of energy; thus it is necessary to discuss the terminology that is used to describe energy.

TERMINOLOGY

Energy is defined as the capacity to do work. In nutritional applications it is the amount of heat that is produced when completely oxidized in the body, or loss of energy from the body. In different parts of the world, energy is measured using different units: calories, British thermal units (BTUs), or joules. In the United States the calorie (cal), kilocalorie (kcal), and megacalorie (Mcal) are commonly used in animal nutrition. European countries use the joule. A calorie is the amount of heat required to raise the temperature of 1 gram (g) of water 1°C (= 4.1855 joules). A kilocalorie equals 1,000 cal, and a megacalorie (or therm) equals 1,000 kcal or 1,000,000 cal.

Various terms are used to describe the energy value associated with a feedstuff. The diagram in Fig. 3-1 shows the various terms that are used to describe the energy content of feedstuffs. Definitions for each of these feedstuffs are given next.

GROSS ENERGY (GE) Gross energy is the amount of heat produced when a feed is completely oxidized (burnt). The instrument used to perform this measurement is called a bomb calorimeter. A known amount of sample is placed in the bomb calorimeter, and then oxygen is used to fill the chamber so that the sample will be completely oxidized. The GE content of a feed determines its total energy content, but has little correlation with the portion of the energy that is available to an animal. For example, the energy contents of corn, grain, and wood are very similar, but the ways that they are used as an energy source by an animal are quite different.

DIGESTIBLE ENERGY (DE) Digestible energy is a measure of the amount of energy

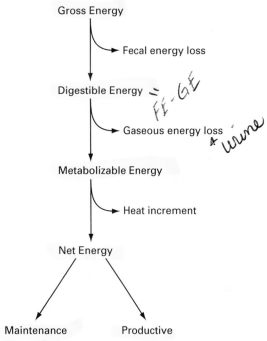

Gross Energy

→ Fecal energy loss

Digestible Energy ≐ *FE - GE*

+ urine

→ Gaseous energy loss

Metabolizable Energy

→ Heat increment

Net Energy

Maintenance   Productive

FIGURE 3-1 Flow diagram of energy terms.

apparently absorbed from a feed after it has been consumed. This value is obtained by subtracting fecal energy from GE consumed. It is not strictly a measure of absorbed energy, because some of the energy of fecal excretion is derived from tissues lining the digestive tract being sloughed off, rather than from undigested food. Digestible energy can be determined relatively easily by performing a digestion trial in which a known amount of the dietary component being evaluated is fed, and feces are collected and the GE content of the feed and feces is determined. DE = GE − fecal energy. Digestible energy values have been developed for quite a number of species and are widely used in the United States and other countries as well.

METABOLIZABLE ENERGY (ME) Metabolizable energy is determined by subtracting energy losses in urine and combustible gases from DE consumed. In addition to collecting feces, urine and gaseous losses must be collected and their energy contents determined. It is slightly more accurate than DE with respect to estimating the amount of energy that is available for use by the animal, but it is more expensive to determine. It is the common energy term used for formulating diets for avian species, and it is commonly used for other domestic species in European countries.

NET ENERGY (NE) Net energy is determined by subtracting energy losses resulting from rumen fermentation and tissue metabolism from ME. This energy term most accurately predicts the amount of energy that is going to be available for use by the animal for maintenance and productive functions. Net energy values have only been determined on a few feedstuffs, and the bulk of the NE values presented in the appendix tables have been computed using equations. If an animal is maintained in an environment outside its normal comfort zone [a temperature range in which nutrients are not required to be metabolized (oxidized) to increase or decrease the body temperature], NE values will be different than if determined within the comfort zone. Nevertheless, NE values are widely used in formulating diets for various ruminant species. Net energy values are also available for maintenance (NEm), gain (NEg) and milk production (NEl).

TOTAL DIGESTIBLE NUTRIENTS (TDN) Total digestible nutrients is a method used for many years for estimating the energy content of a feed. This method sums all the fractions that are digestible. A food is fed to an animal, and the amount of each different component that makes up this food is determined; then the amount that subsequently ends up in the feces is determined. Based on this, the amount digested is determined. The TDN is computed by summing the following: TDN = digestible crude protein + digestible crude fiber + digestible nitrogen-free extract (starch and sugars) + 2.25 digestible ether extract (fat). The ether extract is multiplied by 2.25 in an attempt to adjust its energy value to reflect its higher caloric density (fat = 9.1 cal/g and carbohydrates = 4.1 cal/g). The chief criticism of TDN is that it tends to overvalue roughages as compared to ME or NE methods. Nevertheless, it is still widely used, and some nutritionists feel that it is as accurate as the ME or NE method, especially when being used in formulating diets for animals that are out of the comfort zone. The TDN is very similar to DE, but DE and NE are more commonly used currently in the United States.

FACTORS AFFECTING ENERGY METABOLISM

Energy utilization is a very complex topic and is beyond the scope of this book. But it is necessary to clarify a few points to familiarize the reader with the topic of energy utilization. In the process of digesting and metabolizing energy, the greatest loss is that associated with fecal losses. This represents the components associated with the diet that the animal is unable to digest. The components in the diet greatly influence the amount of feces produced. For example, monogastric species generally consume more digestible diets than herbivorous species; thus the digestibility of the energy will usually be higher. Young animals, particularly mammals, also are fed diets that are more digestible than are adults, so digestibility will be higher. High-quality diets fed to poultry or swine may be digested to the extent of 85% or more; and on the other extreme, poor-quality diets such as straw fed to ruminants may be less than 35% digestible. The level of feed consumption also influences the digestibility of the dietary components in some species. In ruminants, digestibility decreases as the level of feed intake increases (increased passage rate), but there is little effect in other species. Factors that disturb the normal intestinal or stomach functions, such as diarrhea, presence of toxins, and parasite infections, often result in a reduction in digestibility. Digestibility is often enhanced in various species by using proper feeding and feed-processing procedures.

Other losses that occur are associated with metabolism losses. Energy lost through urine and microbial methane production in the GI tract will amount to about 10% of the GE in ruminants, but less in most monogastric species. Losses after absorption can vary greatly, depending on level of intake, quality of the diet, and other factors. Heat is produced as a result of microbial fermentation in the GI tract (**heat of fermentation**). While such heat can be used to maintain body temperature, if excess is produced the animal is not capable of storing the heat by chemical means such as synthesis of fat, so it must be dissipated and is lost. Heat is also produced when nutrients are oxidized. This is referred to as **heat increment**. Proteins (amino acids) have a large heat increment associated with their metabolism, followed by carbohydrates and then fats. As with the heat of fermentation, such heat can be used to warm the body, but it cannot be stored. Thus both the heat of fermentation and the heat increment are detrimental when the animal is in a heat-stressed situation.

Energy requirements are affected by many different factors: age, species, activity level, production level, environmental conditions, nutrient deficiencies, and a number of other factors. However, for a healthy animal, energy requirements are directly related to body surface area; this is so because heat is lost or gained in proportion to the area exposed. If body weight is multiplied by a factorial power (0.75 is commonly used), this provides a reasonably good estimate relating to weight and surface area and is referred to as **metabolic body weight** (see 1, 3). This method can be used in accurately computing the energy requirements of species that have wide differences in weight (elephant versus hummingbird).

Various management practices can influence the surface area of an animal and change the rate of heat loss. Fore example, shearing sheep increases their susceptibility to cold stress, but reduces heat stress. Winter hair coats provide more insulation, as does a thick layer of body fat under the skin. In dry areas, sprinkling animals with water during hot periods increases evaporation and heat removal and reduces the amount of heat stress experienced by the animal.

All homotherms (species that maintain a fairly constant body temperature) operate within a comfort zone. Within this temperature range, the animal does not need to increase oxidation of nutrients to either cool or warm itself. The comfort zone is affected greatly by humidity, because evaporative losses from the body surface and lungs are reduced dramatically when humidity is high, and the animal will not be able to dissipate excess heat. Wind, on the other hand, increases heat loss in a cold climate and causes the animal to produce more heat in order to maintain body temperature. In a hot, dry environment, wind helps to increase the evaporative losses. Normally, activity or an increased level of intake increases body heat production and lowers the temperature at which an animal is comfortable. Immersion in cold water greatly increases the rate of body heat loss. If an animal is immersed in cold water for any length of time, body temperature can decrease to the point at which the animal cannot survive.

Certain feeds have high heat increments associated with them, such as straw. So when straw is fed to cattle during cool weather it will generate more heat increment, which helps the animal to maintain its body temperature. If straw is fed to cattle during periods of heat stress, this causes excessive heat buildup, which the animal then has to dissipate somehow or its body temperature will rise.

ENERGY DEFICIENCY

Many wild species go through alternating periods of energy surplus, adequacy, and deficiency as the seasons change and the availability of food changes. This also may happen in free-ranging domestic animals, but usually the extremes of deficiency are less severe because of supplementary feed supplied by the livestock owner. Such extremes should not occur with animals in confinement.

When an energy deficiency occurs, the animal must mobilize body reserves (fat) of energy to sustain itself; this results in a loss of body fat, weight loss, and emaciation. Periods of energy deficiency can have a negative effect on an animal. A pregnancy may be interrupted by reabsorption or abortion, milk or egg production may be decreased drastically, and fertility may be reduced. Energy deficiency is usually accompanied or caused by a deficiency of other nutrients, but the overriding importance of body energy need may prevent the symptoms of these other deficiencies from being observed.

FEEDING STANDARDS

Feeding standards or guidelines have been developed for each of the domestically important livestock species and define the amounts of nutrients required by animals for various functions. The use of these standards dates to the early 1800s. These standards have been gradually developed over the years to the point where nutrient requirements for farm animals may be specified with a reasonable degree of accuracy, particularly for growing chicks and pigs. Although there are still many situations for which nutrient needs of animals cannot be specified with great accuracy, nutrient needs of some domestic animals have been defined more completely than those of humans. This is primarily due to the simple fact that people do not lend themselves to the types of experimentation required to collect good quantitative data.

In the United States the most widely used standards are those published by the various committees of the National Research Council (NRC) (4) under the auspices of the National Academy of Science. These standards (see the appendix tables) are revised and reissued at intervals of a few years. In England, the standards in use are put out by the Agricultural Research Council (ARC) (3). Other countries have similar bodies that update information and make recommendations on animals' nutrient requirements.

TERMINOLOGY USED IN FEEDING STANDARDS

Feeding standards are usually expressed in quantities of nutrients required per day or as a percentage in a diet, the former being used for animals given exact quantities of a diet and the latter more commonly used when rations are being fed **ad libitum** or free choice. With respect to the various nutrients, most are expressed in weight units, percentage, or ppm (parts per million). Some vitamins (A, D, E) are often expressed in international units. Protein requirements are commonly expressed as **crude protein** (CP) (estimated based on the nitrogen content of the feed) and are sometimes expressed as **digestible protein** (DP), when adequate information is available on amino acids quantities in monogastric animals. Dietary energy requirements are expressed in a variety of different ways. The NRC uses ME for poultry; DE, ME, or TDN for swine; DE, ME, and TDN for sheep; and NEm and NEg for growing and fattening cattle, as well as ME and TDN. For dairy cattle, values are given as DE, ME, TDN, NEm, and NEg, with NEl being used for lactating cows. The ARC uses ME almost exclusively, with energy expressed in terms of MJ (megajoules) rather than Mcal (megacalories). Other European standards are based on starch equivalents, Scandinavian feed units, and so forth. Regardless of the units used, feeding standards are based on some estimate of animal needs and have been derived from data obtained from a great many experimental studies done under a wide variety of conditions with a diverse list of feed ingredients.

DEFICIENCIES ASSOCIATED WITH FEEDING STANDARDS

A discussion of the means and methods of developing feeding standards is beyond the scope of this book; more in depth information can be found in references 1, 3, and 4. Feeding standards provide a useful base for the nutritionist when formulating rations or estimating the requirements for a specific class of animal. Feeding standards should be used as a guide, but not considered as the precise requirement. The NRC recommendations are specified in terms believed by the committees to be minimum requirements for an average population of animals of a given species, age, weight, and productive status. Some earlier versions were called allowances and included an additional amount on top of what was believed to be required. It is well known that animal requirements vary considerable, even within a relatively uniform herd. For example, a dietary protein intake may be satisfactory for most animals of a given species in a given situation, but will not be sufficient for a few of the more rapidly growing animals, while it will be excessive for animals growing at a slower rate. Given our present production methods, NRC recommendations usually would be the most feasible basis of feeding. In our current system the poor producers are culled and the high producers are provided with additional allowances, which enable them to express their genetic potential.

It is obvious from published literature that management and feeding methods may alter an animal's requirements or efficiency of nutrient utilization apart from known breed differences in nutrient metabolism. In addition, most current recommendations provide no basis for increasing intake in severe weather or reduction in milk climates. The effect of climate may be very great. For example, recent data show that pregnant cows with no shelter have metabolic rates that are 18% to 36% higher than cows provided with shelter.

Comment is required relating to the values listed in NRC publications for NEm, NEg for beef cattle, and NEl lactating dairy cattle. If one looks at the tables relating to feed composition in the respective NRC publications, NE values are given for almost every feedstuff listed. The reader should be aware that most of these NE values are estimates that have been computed from other energy values, using such factors as TDN, DE, or ME. Only a few feedstuffs have actually had NE values determined for them. The accuracy of these values is not increased over that of the original data and is most likely decreased somewhat.

No allowances for various stresses, such as disease, parasitism, injury, temperature, or humidity, are given in the recommendations. Beneficial effects of additives, hormones, or feed preparatory methods are not always considered when developing feeding standards. Thus many variables may alter nutrient needs and nutri-ent utilization, and these variables are not normally built into the feeding standards.

VARIOUS FUNCTIONS OF NUTRIENTS

The remainder of this chapter will be devoted to a general discussion that relates the effect of various productive functions on nutrient requirements. This should give the reader a better understanding of nutrient requirements as affected by growth, fattening, reproduction, lactation, and work.

MAINTENANCE Maintenance may be defined as the condition in which an animal is neither gaining nor losing body energy (or other nutrients). In the case of productive animals there are only a few times when a true maintenance situation is approached. It is closely or approximately attained in adult male breeding animals other than during the breeding season and, perhaps, for a few days or weeks in adult females following the cessation of lactation and before pregnancy increases requirements substantially. Maintenance is commonly used as a reference point for assessing nutritional needs, which are subsequently used as a basis for formulating rations.

Nutrient needs are minimal during maintenance. Under range conditions during the dry periods of the year or during winter months, an animal may find it necessary to expend through movement a considerable amount of energy just to obtain enough plant material to meet its needs, as opposed to the amount expended when forage growth is more lush, which causes the animals maintenance requirements to be different during these two times.

GROWTH AND DEVELOPMENT

Growth, as measured by increase in body weight, is usually maximized during early life. When growth is expressed as a percentage increase in body weight, it is normally very high until the animal reaches puberty; then it declines at an increasing rate until the animal reaches its mature weight. As an animal grows, different tissues and organs develop at different rates, and these differences in development have an effect on the animal's nutrient requirements. Growth rate probably decreases because hormonal stimulus is decreased, because young animals cannot physiologically continue to consume as much food per unit of body weight to support this level of growth.

Nutrient requirements per unit of body weight or metabolic size (body weight raised to the 0.75 power) are greatest for very young animals. These needs gradually decline as growth rate declines and an animal approaches its mature weight. In young mammals, nutrient requirements are so high that they must be fed diets that are highly digestible to achieve maximum growth rates. As the young mammal grows, the quality of the diet generally decreases as more of the food consumed is from nonmilk sources, which results in lower digestibility of the food components being consumed and less efficient production.

Nutrient deficiencies show up rapidly in young animals, particularly when the young are dependent in the early stages of life on tissue reserves obtained while in utero. With few exceptions, tissue reserves in newborn animals are inadequate to support growth for an extended period of time. Milk or other foods consumed by a young animal may be inadequate in several required nutrients, so a deficiency may develop until a diet adequate in these nutrients is fed. The young pig is an example. Iron reserves are very low and body reserves are rapidly depleted; because milk is a poor source of iron, young pigs often become anemic unless supplemented or injected with iron. Newborn animals need to receive colostrum as soon as possible after birth. The colostrum not only provides essential nutrients, but it also provides immunoglobulins that transfer immunity for different diseases from the dam to its offspring.

From a production standpoint, nutrient requirements per unit of gain are least and gross efficiency (total production divided by total food consumption) is greatest when animals are growing at their maximum rates. However, net efficiency (total production divided by nutrient needs above maintenance) may not be altered greatly. As performance rate increases, efficiency also increases, because a small portion of the nutrients consumed are used for maintenance and a larger portion for the growth functions. In some situations it may not be desirable or economical to attempt to achieve maximal gain. For example, if we want to market a milk-fed veal calf at an early age, maximal growth is desired. On the other hand, if the calf is being grown as a replacement heifer, less than maximal gain is satisfactory and substantially cheaper.

When an animal's growth rate is suppressed, this may result in permanently stunting the animal. It is possible to maintain young animals for a period of time during which they do not increase body energy reserves, yet if other nutrients are adequate, they will continue to increase in stature. Following a period of subnormal growth due to energy restriction, most young animals gain weight at a faster than normal rate when placed on an adequate ration. This phenomenon is referred to as **compensatory growth.** For example, when young calves are wintered on diets that are adequate in nutrients other than energy and then placed on pastures adequate in nutrients in the spring or in the feedlot, weight gains occur at a very rapid rate initially. Efficiency for the total period and especially for a given amount of gain is greater. However, if the cost of the ration is excessive, it might be more profitable to use cheaper feedstuffs, reduce the performance, and defer marketing the animals until a later date.

LACTATION

High levels of milk production require the highest levels of nutrient consumption, except for a sustained period of extremely high exercise. High-producing cows and goats typically produce milk with a dry matter content equivalent to four- to fivefold that of their body weight each year, with some animals being as high as sevenfold. High-producing cows give so much milk that it is impossible for them to consume enough feed to prevent weight loss during the peak period of lactation.

Milk of most domestic species runs 80% to 88% water; thus water is a critical nutrient needed to sustain the lactation function. The requirements for all nutrients are increased during lactation and are directly related to the level of production. Milk components can either be supplied in the animal's diet or provided by mobilization of body tissues. All recognized nutrients are secreted to some extent in milk, although the major components of milk are fat, protein, lactose (milk sugar), and a substantial amount of ash, primarily Ca and P.

Milk production varies widely among and within species. In cows, peak yield usually occurs between 60 and 90 days after parturition and then gradually declines at a rate of between 8% and 10% per month. Nutrient requirements peak at the same time that milk production peaks, but normally feed consumption peaks later. Thus high-producing animals normally have to rely on body reserves of energy and protein during these periods of inadequate dietary intake. Milk composition, especially in ruminant species, has been shown to be affected by the composition of the ration fed. This is particularly true for butterfat and, to a lesser extent, protein and lactose. If the fiber level is raised in a dairy cow's ration, normally the butterfat percentage will increase. In monogastric species, changes in diet seem to have less of an effect on milk composition.

Limiting water or energy intake of the lactating cow (and probably any other species) results in a marked drop in milk production, whereas protein restriction has a less noticeable effect, especially if the shortage is only for a short period of time. Although deficiencies of minerals do not affect milk composition markedly, they result in rapid depletion of the lactating animal's body reserves. Extended Ca depletion causes the skeletal system to be less dense and weakened. The need for elements such as Cu, Fe, and Se increases during lactation, even above the levels that are associated with the milk that is being produced. This is because these minerals act as cofactors to the processes that are involved in producing the milk components. Often the effect of nutrient deficiencies occurring during lactation carry over into pregnancy and the next lactation.

REPRODUCTION

The nutrient needs of animals for reproduction are generally considerably less critical than during rapid growth or heavy lactation, but they are certainly more critical than for maintenance. If nutrient deficiencies occur prior to breeding, the result may be sterility, low fertility, silent estrous, or failure to establish or maintain pregnancy.

It has been demonstrated many times that underfeeding (of energy or protein) during growth results in delayed sexual maturity and that both underfeeding and overfeeding (of energy) will usually result in reduced fertility as compared to animals fed on a medium intake. Of the two, overfeeding is usually more detrimental to fertility.

Energy needs for most species during pregnancy are more critical during the last one-third of the pregnancy. Deposition of nutrients in fetal tissues indicates that only a relatively small percentage of the total nutrient requirements are used for fetal tissue growth, even late in gestation. These differences may be associated with increases in other bodily functions associated with pregnancy. The metabolic rate of pregnant animals is higher; thus requirements to support metabolic processes are higher. The basal metabolic rate in a cow is about 1.5 times that of a nonpregnant identical twin. Fortunately, pregnant animals have greater appetites, which allows them to consume similar diets as nonpregnant animals, but as a result of consuming more food they also consume more nutrients.

Inadequate nutrient consumption by the dam during pregnancy may have several different effects depending on the species of animal, the extent of malnutrition, the specific nutrient involved, and the stage of pregnancy. Protein is more critical for development of the fetus in the late stages of development than in early stages, as is true for Ca, P, and other minerals and vitamins. With a moderate deficiency, fetal tissues tend to have priority over the dam's tissues; thus body reserves of the dam may be withdrawn to nourish the fetus. However, if the deficiency is severe, usually a partial depletion will occur and then such detrimental effects as resorption of the fetus, abortion, malformed young, or birth of dead, weak, or undersized young occurs. Often this has a long-term effect on the mother. When the dam's tissues are depleted of critical nutrients, tissue storage in the young animal is almost always low, nutrient concentrations in the colostrum are reduced, milk production declines, and survival of the young is much less likely than when the nutrition of the dam is at an adequate level.

WORK

Studies with animals indicate that work (physical exercise) results in an increased energy demand and is proportional to the level of work being done. Dietary carbohydrates are utilized as energy sources in mammals and are more efficiently used than fats or proteins. Dietary requirements for protein have been shown to only increase slightly, if any, when horses are exercised. Nitrogen excretion has been shown to increase, resulting from muscular activity.

When appreciable sweating occurs, work increases the need for Na and Cl, which is lost in this process. Phosphorus intake should be increased during work, as it is a vital nutrient in many energy-yielding reactions. Likewise, the B vitamins involved in energy metabolism, particularly thiamin, niacin, and riboflavin, probably should be increased as the level of exercise is increased, although more research is needed in this area.

METHODS FOR ASSESSING NUTRIENT CONTENT OF FEEDS

The reader needs to have some idea of the common methods and terms used when analyzing and describing the nutrient content of feeds. Only a brief overview of this topic will be given here. For a more detailed coverage of this subject, the reader is referred to reference 9. Thousands of different tests could be performed on feed samples. These can be lumped broadly into qualitative and quantitative tests, which provide information relating to various substances present in a feed sample.

In recent years there have been marked improvements in instrumentation, much of which is automated, for doing analyses on nutrients. Consequently, tests can be performed more rapidly and less expensively. The discussion that follows relates to the more common terms used to describe the nutrient fractions found associated with feeds.

Proximate analysis is the most common analysis performed on feed samples. Proximate analysis consists of a series of analyses performed in an effort to estimate the nutrient characteristics of a feed and consists of the following: dry matter, crude protein, ether extract, crude fiber, ash, and nitrogen-free extract. These various analyses are described next.

DRY MATTER (DM)

Water is a useful substance to an animal; its presence in feed acts as a dilutent. Therefore, most feeds are analyzed and the data presented either on a dry matter (DM) basis or an as-is basis, with information shown on its moisture content. Dry matter is determined by drying the sample in an oven of one type or another. Microwave ovens can be used to speed up the drying process. Older procedures use ovens with temperatures ranging from 60° to 105°C; these required 24 hours or more, depending on the nature of the product being dried.

CRUDE PROTEIN (CP)

Protein level is estimated by a procedure called the Kjeldahl method, which measures the N content of the feed, irrespective of its source. The feed sample is digested in hot, concentrated sulfuric acid, which converts all the carbon-containing compounds to carbon dioxide, and N is trapped and subsequently measured and expressed as percent of N. The rationale behind this procedure is that all proteins contain nitrogen, but all N-containing compounds are not proteins. The average protein content of most feedstuffs is 16%, but this varies somewhat among different feeds depending on their amino acid composition. Therefore, by multi-plying the percent of N by 6.25 (100/16 = 6.25) the protein content of a feed can be estimated. This is why some feedstuffs composed of nitrogen-containing compounds other than protein can have CP contents that are greater than 100%. For example, urea, which contains no protein, but does contain approximately 46% N, would have a CP content of 287.5% (46% N × 6.25 = 287.5% CP).

The N form estimated by Kjeldahl analysis could be from urea, an insoluble protein such as uncooked feathers (nearly completely undigestible), or a high-quality protein in milk. The Kjeldahl procedure only provides an estimate of the quantity of N present and does not provide any indication of quality. The advantages of the procedure are that it is relatively rapid and repeatable; it has been used for over 100 years and most people are familiar with it.

If more precise information is needed, many different types of analyses can be performed. Automated methods are available for analyzing for amino acids, but they require quite a bit of time and are expensive. Analytical procedures are also available for determining types of proteins, solubility of proteins, and proteins that are bound to components that make them unavailable to an animal.

ETHER EXTRACT (EE) (CRUDE FAT)

Crude fat is the product resulting from extracting a feed with ethyl ether or some other organic solvent or combination of solvents, such as chloroform and ethyl alcohol. A small sample of feed is placed in a specially designed container, and the solvent is dripped through the sample, which removes the fats and other soluble substances. The ether extract fraction can contain many other substances other than true fats. For example, most plants leaves are coated with a waxy material that, although soluble in ether, is not a fat and is essentially undigestible and of no nutritional value to animals. As with the crude protein analysis, the results obtained using this procedure are quantitative rather than qualitative in nature. No information is provided relating to the fatty acid composition. Methods are available that will determine fatty acids and the other components making up the ether extract (EE) (9).

CRUDE FIBER (CF)

The analysis for crude fiber was developed many years ago. It involves boiling (refluxing) a known amount of a ground feed sample (usually fat-extracted first) in a weak acid solution, filtrating and boiling in a weak solution of alkali, and filtering and drying; the residue remaining is the CF. This is an attempt to simulate the digestive processes occurring in the stomach and intestine of an animal. This procedure estimates the less digestible fractions of a feed sample, which include cellulose, hemicellulose, xylans, lignin, and other components associated with fibrous carbohydrates. This procedure has a number of disadvantages. It is slow, tedious, and not very repeatable, and the information is less applicable to some feeds than to others. The reason for the latter statement is that some hemicelluloses are dissolved by the chemical treatment and, if protein is bound to the lignin or other chemicals in an insoluble form, it will also show up in the crude fiber fraction. Unfortunately, most state regulatory agencies still require crude fiber analyses to be reported on commercial feed tags (Ch. 5).

ASH

Inorganic minerals are estimated by the ash value, which is determined by burning a feed at a temperature of 350° to 600°C until nothing is left but metallic oxides or contaminants such as rocks and soil. No qualitative information is provided, so the specific mineral components are not determined. Additional procedures, such as atomic absorption and spectrophometry, are used to determine the individual minerals. Results for some of the trace elements are not too precise because of their very low concentrations, which make analytical results less accurate.

NITROGEN-FREE EXTRACT (NFE)

This fraction is calculated and not actually determined using a laboratory procedure. The NFE is an estimate of the readily available carbohydrates (sugars, dexrins, starches), which are highly utilizable. The following formula is used to determine the NFE content of a feed sample.

$$NFE = 100 - (\text{crude protein} + \text{crude fat} + \text{crude fiber} + \text{ash})$$

This estimate was made because there were no quick, simple analytical methods (at the time) for starch, which must be hydrolyzed to sugar, and then an analysis must be done for sugars.

The values of NFE for grains and other components high in sugar and starch are a reliable estimate of the readily available carbo-

hydrates, but they do not work well with feeds that are high in hemicellulose, when the crude fiber determinations should be used.

NEUTRAL-DETERGENT FIBER (NDF) AND ACID-DETERGENT FIBER (ADF)

These methods are becoming more commonly used and are replacing the CF procedure, because they more accurately define the carbohydrate components associated with plant materials. These determinations are not part of the proximate analysis. Methods using detergents have been developed to overcome some of the problems with crude fiber analysis. If a feed is extracted with the appropriate neutral-detergent solution, the solution will solubilize (remove) materials that are essentially the same as the contents of the cell, thus dividing the feed into fractions of cell contents and cell walls. Neutral-detergent fiber and cell wall content are synonymous terms. Most of the soluble materials in the cell contents (proteins, lipids, sugars, starches, pectins, and others) are highly utilizable by animals of all types, whereas the cell walls are composed of the fibrous carbohydrates, lignin, heat-damaged protein, and silica. Then the residue is extracted with an acid detergent solution, which removes the hemicellulose and leaves all the other components. Hemicellulose is intermediate in digestibility between starches, sugars, and cellulose, so the NDF and ADF procedures provide a more accurate estimate of how well a feed will be digested by an animal.

ENERGY

The energy content of a feed is obtained by using an instrument called an oxygen bomb calorimeter. A small sample of known weight is placed in the oxygen bomb calorimeter and filled with oxygen under pressure. The bomb is then placed in a container of water of known volume, the sample is ignited, and the change in water temperature is monitored. In this way the energy content of the sample can be determined. To accurately estimate how efficiently the energy will be utilized, digestion trials must be conducted.

SUMMARY

Animals require an adequate source of good-quality water for maximum levels of production. At times water may contain too many minerals, salts, or other contaminants that, if high enough, will impair animal performance. Drinking water can also be an important source of some minerals, reducing the amount required to be provided in diets.

Carbohydrates makes up the majority of most diets, except for the carnivores, but no specific carbohydrate requirement is listed. Sugars and starches are very efficiently utilized by monogastric and ruminant animals. Fibrous carbohydrates such as hemicellulose and cellulose are utilized only because of digestion in the GI tract by microbial organisms; animals do not produce the enzyme required to digest these compounds.

Animals need a dietary source of either amino acids (protein) or nitrogen. Ruminant animals can use largely nonprotein N sources, but the monogastric animals need adequate amounts of essential amino acids (amino acids that they cannot synthesize in adequate amounts). Other amino acids needed for protein synthesis can usually be synthesized in adequate amounts by the animal's tissues.

Lipids are normally only found in small amounts in an animal's diet. There are two essential fatty acids, but only in very rare situations are these likely to be inadequate when normal feedstuffs are used to prepare a ration.

Many different mineral elements are required in animal diets. Those macrominerals that may not be adequate are Ca, P, and Mg. Of the microminerals, Cu, Fe, I, Mn, and Zn are usually most likely to be inadequate in the diet. Vitamins may also be a problem in some diets. Vitamin A is usually thought to be limiting for many species. Vitamin D may be deficient at times. With the water-soluble vitamins (B complex), the most likely problems occur with poultry, which may need to be supplemented with a number of these vital nutrients.

Energy is often limiting, especially with grazing animals, but should not be a problem normally with animals being fed in confinement. Energy utilization can be affected markedly by a deficiency of other nutrients and by feed processing. Feeding standards are intended to define the nutrient needs of domestic animals of different species and at different ages or in various production situations. They are redefined from time to time as new research is conducted in an effort to improve them. There are a number of inadequacies in most standards, but they serve as a most useful base from which to project animal needs when formulating rations. Common nutritional analyses of feedstuffs were discussed, and some of the problems associated with their estimating accuracy were pointed out.

REFERENCES

1. Pond, W. G., D. C. Church, and K. R. Pond. 1995. *Basic animal nutrition and feeding.* 4th ed. New York: Wiley.

2. Church, D. C., ed. 1988. *The ruminant animal.* Englewood Cliffs, NJ: Prentice Hall.

3. ARC. 1980. *The nutrient requirements of ruminant livestock.* London, UK: Agr. Res. Council, Commonwealth Agr. Bureaux.

4. NRC. *Nutrient requirements of domestic animals.* Washington, DC: National Academy Press. Publications are available on all domestic animals, including cats, dogs, laboratory animals, warm and coldwater fishes, nonhuman primates, mink, and foxes.

5. NRC. 1980. *Mineral tolerance of domestic animals.* Washington, DC: National Academy Press.

6. Church, D. C., ed. 1979. *Digestive physiology and nutrition of ruminants.* Vol. 2: Nutrition. 2d ed. Corvallis, OR: O & B Books.

7. Swenson, M. J., and W. O. Reece, eds. 1993. *Dukes' physiology of domestic animals.* 11th ed. Ithaca, NY: Cornell University Press.

8. Underwood, E. J. 1981. *The mineral nutrition of livestock.* 2d ed. London, UK: Commonwealth Agr., Bureaux.

9. AOAC. 1995. *Official methods of analysis.* 16th ed. Washington, DC: Assoc. of Official Analytical Chemists.

4

FEEDSTUFFS

Richard O. Kellems and D. C. Church

Throughout the world a wide variety of feed-stuffs is available for feeding animals, and these depend on what can be grown in a particular region. Many feedstuffs used for feeding animals are by-products of food production for humans. Well over 2,000 different feedstuffs have been characterized to some extent as animal feeds, not counting varietal differences in various forages and grains.

Since the number of feedstuffs is so great it is feasible only to provide general information relating to their classification. More detailed information is provided on specific feedstuffs in Chs. 6, 7, and 8 of this book. Information is given in Chs. 13 through 25 regarding the utilization and suitability of various feedstuffs for the specific animals discussed in these chapters.

More information on nutrient composition of feedstuffs can be found in the literature, although many times the information may not be as complete as one would wish. At least four books in English deal almost entirely with compositional data. They include the NRC publication (1), which gives data on feedstuffs utilized in North America and other temperate climates. A second book on this topic is published in Utah (2). The third is published by the University of Florida (3) and provides information on Latin American feedstuffs. The fourth source is published by the FAO (4) and provides information on tropical feeds. In addition to these sources, numerous texts dealing with domestic animals present information relating to various feedstuffs. There is also an extensive listing of commonly used feedstuffs contained in the NRC and ARC publications, as well as publications of other countries relating to feeding guidelines.

CLASSIFICATION OF FEEDSTUFFS

A **feedstuff** can be defined as any component of a diet (ration) that serves some useful function. Most feedstuffs provide a source of one or more nutrients, such as energy, protein, minerals, or vitamins, that are required by an animal. Some feed ingredients may be included not only to provide a nutrient, but to modify the diet's characteristics, for example, to emulsify fats, provide bulk (reduce physical density), reduce oxidation of readily oxidizable nutrients, or provide flavor, color, desirable odor, or other factors that enhance its acceptability. Medicinal compounds are usually excluded from lists of feedstuffs. Most descriptive information relating to feedstuffs gives an International Feed Number, which indicates how a feedstuff has been categorized. The International Feed Identification System classifies feedstuffs into eight general categories. The first digit of the International Feed Identification Number (IFN) indicates the major category to which a feedstuff has been classified. The eight major categories are 1, dry roughage; 2, pasture and range grasses; 3, ensiled roughages; 4, high energy concentrate; 5, protein sources; 6, minerals; 7, vitamins; and 8, additives. So one can tell something about the feedstuff by looking at its International Feed Identification Number.

The reason for classifying a feedstuff into a particular category is apparent, but additional clarification is needed for categories 4 and 5. To be categorized as 4, the feedstuff must have a TDN (total digestible nutrient) content that exceeds 70% and contain less than 18% crude fiber, but sometimes a feedstuff will not receive a 4 classification when its TDN exceeds 70%. For example, well-eared corn silage will be higher in TDN than 70%, but will still be categorized as a silage (IFN = 3). The designation of 5 is reserved for protein sources. To be classified as a protein source, the feedstuff must contain a minimum of 20% crude protein (CP) when expressed on a dry matter basis (all the moisture removed). Normally, any feedstuff that contains above 20% CP is given the 5 designation, but again there are some exceptions. For example, alfalfa hay containing 23% CP is listed as a dry roughage (IFN = 1). Generally, the International Feed Number provides the individual with a good idea of the characteristics of a feed sample. There are a number of subcategories listed for each major classification. The following table is the NRC listing of the different categories of feedstuffs (1).

International Feed Number

1 Category Roughages
   Dry forages and roughages
      Hay
         Legume (alfalfa)
         Grass–legume
         Nonlegume
      Straw and chaff
      Fodder, stover
      Other feedstuffs containing greater than
      18% crude fiber category
         Corn cobs
         Cottonseed hulls, gin trash
         Milling by-products
         Shells and hulls
         Sugar cane bagasse
         Paper, wood by-products
2 Pasture, range plants, and plants fed green
      Grazed plants
         Growing plants
         Dormant plants
      Soilage or greenchop
      Cannery and food crop residues
3 Silages and haylages
      Corn, sorghum
      Grass, grass–legume, legume
      Miscellaneous ensiled materials
4 Energy feeds (less than 18% crude fiber and greater than 70% total digestible Nutrients, but less than 20% crude protein)
      Cereal grains
      Milling by-products of cereal grains

      Beet and citrus pulp
      Molasses of various types
      Seed and mill screenings
      Animal, marine, and vegetable fats
      Roots and tubers, fresh or ensiled
      Miscellaneous
5 Protein supplements (containing greater than 20% crude protein)
      Animal, avian, and marine sources
      Milk and milk by-products
      Legume seeds
      Dehydrated legume plants
      Milling by-products of grains
      Brewery and distillery by-products
      Single-cell sources (bacteria, yeast, algae)
      Nonprotein nitrogen (urea, ammonia, biuret, etc.)
6 Minerals supplements
7 Vitamins supplements
8 Nonnutritive additives
      Antimicrobials, antifungals, antibiotics
      Antioxidants
      Probiotics
      Buffers
      Colors and flavors
      Emulsifying agents
      Enzymes
      Hormones
      Medicines
      Miscellaneous

The commercial feed trade uses collective terms on feed labels (see Ch. 5) instead of listing a long string of feed ingredients. An individual must be aware that there is some difference in the terminology used as compared to what is listed by the NRC (1). The following lists the collective terms approved by the Association of American Feed Control Officials (AFCO) (5) for use on labels or feed tags.

ANIMAL PROTEIN PRODUCTS may include one or more of the following:

Animal blood dried
Animal by-product meal
Buttermilk, condensed
Buttermilk dried
Casein

Casein, dried hydrolyzed
Cheese rind
Crab meal
Fish by-product
Fish liver and glandular meal
Fish meal
Fish protein concentrate
Fish residue meal
Fish solubles, condensed
Fish solubles, dried
Fleshing hydrolysate
Hydrolyzed hair
Hydrolyzed leather meal
Hydrolyzed poultry by-product aggregate
Hydrolyzed poultry feathers
Meat and bone meal
Meat and bone meal tankage
Meat meal
Meat meal tankage
Meal solubles, dried
Milk albumin, dried
Milk, dried whole
Milk protein, dried
Poultry by-products
Poultry by-product meal
Poultry hatchery by-product
Shrimp meal
Skimmed milk, condensed
Skimmed milk, condensed cultured
Skimmed milk, dried
Skimmed milk, dried cultured
Whey, condensed
Whey, condensed cultured
Whey, condensed hydrolyzed
Whey, dried
Whey, dried hydrolyzed
Whey product, condensed
Whey product, dried
Whey solubles, condensed
Whey solubles, dried

FORAGE PRODUCTS may include one or more of the following:

Alfalfa leaf meal
Alfalfa meal, dehydrated
Alfalfa hay, ground
Alfalfa meal, suncured
Coastal bermuda grass hay

Corn plant, dehydrated
Dehydrated silage (ensilage pellets)
Flax plant product
Ground grass
Lespedeza meal
Lespedeza stem meal
Soybean hay ground

GRAIN PRODUCTS may include one or more of the following in any of the normal forms, such as whole, ground, cracked, screen cracked, flaked, kibbled, toasted, or heat processed.

Barley
Corn
Corn feed meal
Grain sorghums
Mixed feed oats
Oats
Rice (ground brown, ground paddy, ground rough, broken or chipped)
Rice, brown
Rye
Triticale
Wheat

PLANT PROTEIN PRODUCTS may include one or more of the following:

Algae meal
Beans
Canola meal
Coconut meal
Cottonseed cake
Cottonseed flakes
Cottonseed meal
Cottonseed meal, low gossypol
Cottonseed, whole pressed
Guar meal
Linseed meal
Peanut meal
Peas
Rapeseed meal
Safflower meal
Soy protein concentrate
Soybean feed
Soybean, ground
Soybean, heat processed
Soybean meal
Soybean meal kibbled

Soy flour
Soy grits
Sunflower meal
Sunflower meal, dehulled
Yeast, active dry
Yeast, brewers
Yeast, culture
Yeast, dried
Yeast, grain distillers dried
Yeast, molasses distillers dried
Yeast, primary dried
Yeast, torula dried

PROCESSED GRAIN BY-PRODUCTS
may include one or more of the following:

Aspirated grain fractions
Brewers dried grains
Buckwheat middlings
Condensed distillers, solubles
Condensed fermented corn extractives with
    germ meal bran
Corn bran
Corn flour
Corn germ meal (wet and dry milled)
Corn gluten feed
Corn gluten meal
Corn grits
Distillers dried grains
Distillers dried grains, solubles
Distillers dried solubles
Flour
Grain sorghum germ cake
Grain sorghum germ meal
Grain sorghum grits
Grain sorghum mill feed
Hominy feed
Malt sprouts
Oat groats
Oat meal, feeding
Pearl barley by-products
Peanut skins
Rice bran
Rice polishings
Rye middlings
Sorghum grain flour, gelatinized
Sorghum grain flour, partially aspirated,
    gelatinized
Wheat bran

Wheat feed flour
Wheat germ meal
Wheat germ meal, defatted
Wheat middlings
Wheat mill run
Wheat red dog
Wheat shorts

ROUGHAGE PRODUCTS may include
one or more of the following:

Almond hulls, ground
Apple pectin pulp, dried
Apple pomace, dried
Bagasses
Barley hulls
Barley mill by-product
Beet pulp, dried
Buckwheat hulls
Citrus meal, dried
Citrus pulp, dried
Citrus seed meal
Corn cob fractions
Corn plant pulp
Cottonseed hulls
Flax straw by-product
Husks
Malt hulls
Oat hulls
Oat mill by-product
Oat mill by-product, clipped
Peanut hulls
Rice hulls
Rice mill by-product
Rye mill run
Soybean hulls
Soybean mill feed
Soybean mill run
Sunflower hull
Straw, ground
Tomato pomace, dried

MOLASSES PRODUCTS may include
one or more of the following:

Beet molasses
Beet molasses, dried product
Beet pulp, dried, molasses
Cane molasses
Citrus molasses

Condensed molasses, fermentation solubles
Molasses, distillers condensed solubles
Molasses, distillers dried solubles
Molasses yeast, condensed solubles
Starch molasses

The IFN categorizes feeds into groups based on the major nutrient provided, while the AFCO list primarily attempts to classify a feed based on its source (5). Both of these methods of classifying feedstuffs serve useful purposes. The IFN is most often used when formulating diets (rations) for animals. The AFCO method is used commercially to satisfy the regulations requiring that a list of feed ingredients must be given on the label or feed tag. An understanding of both methods is essential for an individual involved in feeding and management of livestock.

ESTIMATING NUTRITIONAL VALUE OF A FEED

Various methods are available for evaluating feeds for livestock. The ultimate objective of all these methods is to estimate how well the nutrients contained in these various feedstuffs will be utilized for various functions (maintenance, growth, etc.) and if any antinutritional factors are present that will influence an animal's performance. Some of these methods will be discussed in this section.

CHEMICAL ANALYSIS

Chemical analysis procedures attempt to subdivide the components associated with a feed into different general groups [water, carbohydrates, proteins, lipids (fats), minerals, vitamins] and to estimate the relative amount that is present. Chapter 3 described in detail the various chemical methods that are used.

Many different chemical methods can be used in assessing specific nutrients (6), such as one of the vitamins or a given mineral. In the evaluation of a feedstuff, the **Proximate Analysis** or some modification of it is most commonly used. While this procedure has received much criticism from nutritionists, it is still the most widely used. Often other procedures are used in conjunction with proximate analysis, especially to enhance the assessment of the carbohydrate fraction of a given feed.

The primary problem with chemical procedures is how well they actually estimate utilization by an animal. The problems are that utilization varies for the components measured by these different procedures for different feedstuffs, and it may be different for different animal species or age groups. Although various equations have been developed to predict utilization (Ch. 6), the fact is that often they cannot predict accurately. For this reason, it is often necessary to use other methods that are more costly and time consuming to obtain an accurate estimate of the feeding value of a feedstuff.

DIGESTION AND BALANCE TRIALS

In digestion trials the technique requires that both feed consumption and fecal excretion be measured over a period of time. This allows a measurement of digestibility to be made for the various fractions contained in a feed. The formula used for calculating digestibility is

$$\text{Digestibility (\%)} = \frac{\text{nutrient consumption} - \text{fecal excretion}}{\text{nutrient consumption}} \times 100$$

The number derived from this equation is the coefficient of apparent digestibility for a given nutrient or component of the feed. "Apparent" is used because this information does not give a measure of true digestibility, because an animal will be sloughing off tissues and excreting various things into the GI tract even when no food is being consumed.

All the commonly used feedstuffs have been evaluated using digestion trials. With respect to feed utilization and metabolism, the greatest loss occurs as a result of incomplete digestion; the magnitude of this loss depends on the animal species, level of feeding, and how the feed has been processed. When performed correctly, digestion trials provide a more accurate estimate of the nutritive value of a feed than do chemical analyses. Digestion trials often give results that are slightly lower for digestibility than what really occurs (true digestibility) for any component that is excreted by an animal into the GI tract. On the other hand, if it is realized that some diet organic matter may be converted to gases, particularly to methane in ruminants, and that these gas losses are not accounted for in a digestion trial, we obtain an overestimate of nutritive value.

Balance trials differ from digestion trials in that they account for losses associated with urine and respiration and sometimes those associated with the skin, so a more precise measure of intake and excretion can be obtained. This provides a more accurate measure of nutrient retention. For example, if urea is fed to a beef steer, the digestibility of the CP (nitrogen-containing fraction) will approach 100%, because it will be completely digested and absorbed by the animal. Unfortunately, if there is not an adequate source of dietary energy, the nitrogen released from the urea will not be converted into microbial protein, but rather into ammonia. When excessive amounts of ammonia are produced in the rumen, the ammonia is absorbed by the animal, converted by the liver to urea, and excreted, resulting in a net retention of nitrogen of near zero. On the other hand, if enough dietary energy is present to produce microbial protein and the microbial protein is converted into amino acids, which are absorbed and used to synthesize tissues, then the net retention of the nitrogen could be as high as 50% to 70%. These results can be greatly influenced by the state that the animal is in: negative (losing weight), positive (gaining weight), and at equilibrium (maintenance). More information relating to digestion and balance trials is given in references 7 and 8.

FEEDING TRIALS

Feeding trials are used extensively to evaluate feedstuffs, dietary nutrient levels, additives, feed processing methods, and the like. The trial may be designed to evaluate the growth of young animals, egg production, wool growth, or milk production. Often diets are formulated so that maximum performance can be achieved, so a feedstuff's characteristics should be assessed in conditions approaching those of a practical feeding situation. Trials can also be conducted to assess a feed in less demanding situations related to feeding mature animals for maintenance or pregnancy.

A typical trial for evaluating a feedstuff, such as a new protein supplement, would be to include graded amounts in a diet to replace a supplemental protein source of known quality. For example, varying amounts (25%, 50%, 75%, or 100%) of soybean meal could be replaced by another supplemental protein source in a broiler diet. The diet is then fed to the broilers and data collected on growth and feed conversion. Other data might also be collected, such as carcass quality or general health. Trials of this nature quickly show how the new product compares. We can readily find out if animals will eat the diet containing the untested ingredient, if there is any marked effect on production (body weight gain), or if there is evidence of undesirable effects. Often the response observed with small amounts of the unknown ingredient may be satisfactory, but replacement of all the standard ingredients is apt to result in less desirable performance. This type of response is not always the case, but is very common.

Other trials might be conducted to evaluate the effect of an essential nutrient. Poor performance may result when none is added to the diet, but a dramatic increase is observed when it is added. The animal's response with further addition will normally taper off and may eventually decline if too much is added. By conducting a trial in which graded levels are added, it is possible to determine the optimum level that should be added to a diet.

There are many different types of experiments with different objectives and responses. It might seem logical to conduct a digestion or balance trial before doing a feeding trial, but feeding trials are usually conducted first, sometimes even before chemical analysis is performed. Feeding trials are often subject to criticism, mainly because animal responses may be affected by other factors, especially when measuring performance parameters such as growth or milk production. Feeding trials by themselves do not provide the type of data needed to explain the differences observed in performance. Further experimentation will usually sort out these differences.

PHYSICAL APPEARANCE

The appearance of many feedstuffs often provides clues as to its relative feeding value. For example, for roughages such as alfalfa, visual appraisal for color will indicate if it has been cured (dried) properly, which has a dramatic effect on its feeding value. It may be possible to see or smell mold or to see differences in the relative number of leaves or stem size; all these factors will influence its feeding value. In addition, undesirable weeds or other plants may be present. Smell can often provide a reasonable estimate of the quality of a hay. Unfortunately, none of these are good qualitative estimates, but they are useful nonetheless.

With grains or other concentrated feeds, it may be possible to observe things that may have some effect on feeding value, for example, shriveled seeds, presence of other grains, weed seeds, molds, or foreign material such as hair, dirt, dust, rodent droppings, and so on. With processed feeds, we can sometimes see that rolled grains have been poorly rolled, or in pelleted concentrates that there is a high percentage of fines (finely ground material), or in meals that the texture is too fine and dusty. Visual appraisal is not quantitative, but it is an effective way of quickly assessing the processing and general characteristics of a feed or ration.

CULTURAL PRACTICES

A general knowledge of the soil fertilizing practices used for a practical crop often provides insight into potential nutritional problems. For example, if we know that a given crop is grown in an area that is deficient in a particular mineral and this deficiency has not been adjusted for in the fertilizer (P or I), it can be expected that the plant materials are low in these nutrients. Or if the crop comes from an area where soil levels of an element are very high, potential toxicity might be anticipated to some animals from elements such as selenium or molybdenum. If high N fertilization has been used and there has been a shortage of water for the plants, an undesirable and possibly toxic accumulation of nitrates might be anticipated.

Knowledge as to the stage of maturity at which a crop was harvested often gives a clue as to its relative feeding value. Most grasses, for example, decrease rapidly in feeding value as they mature. Thus, if hay is harvested at a mature stage, the feeding value will be lower than if it was harvested at a less mature stage.

Other factors may affect the feeding value of a given feed. Has the feed been treated with an insecticide or is there insect damage? How long has the feed been stored? Has the feed been heated during storage? Information relating to these and other factors may be used to assess the feeding value of a feed.

COST OF FEEDSTUFFS

In any commercial operation, consideration must be given to the cost of the feedstuffs used when formulating rations. This does not mean that ration composition should be dramatically altered every time feed prices change. Often an individual that is responsible for formulating rations will have a list of feedstuffs that can be interchanged depending on their cost. For example, either corn or milo can be used as the primary energy source in poultry rations depending on their market price. Although discussed in more detail in other chapters, a few brief comments are in order here.

Feed costs normally are not the same throughout the year. For crops that are harvested annually, the price is usually lowest at harvest and tends to increase thereafter until the next harvest. This increase is not normally caused by inflation, but rather by increased storage and additional handling costs. Another reason is that the moisture content of the grain or hay will normally decrease somewhat when stored, which requires that the producer must increase its cost in order to have the same revenue. Costs also increase with distance from where the feed is grown or processed. It is a common practice in the United States to price soybean meal at a location in Indiana. Thus, if the user lives close by, transportation costs will be low; however, if the user lives in California, transportation costs will account for a substantial portion of the delivered price.

ESTIMATING ENERGY VALUES OF FEEDSTUFFS

The actual energy content of a feedstuff is not normally determined, but rather estimated. Energy estimates for feedstuffs are commonly computed using results obtained from proximate analyses. Several different equations are available for computing various energy values (DE, ME, NE, TDN). An example of the equation used to estimate the TDN (total digestible nutrients) content of a hay sample is given in Fig. 6-10. The laboratory performing your analyses can provide you with the formula that they use. Additional information on this topic is given in the various NRC publications for the individual species and in reference 9.

NUTRIENT CONTENT OF MAJOR FEEDSTUFF GROUPS

A brief introduction will be given on relative nutrient contents of major feed ingredients. These are shown in Table 4-1. Except for fats and some mineral supplements, all feedstuffs provide several nutrients for animals, but often they are a much more important source for one type of nutrient than for others. For example, oil seed meals (soybean meal, cottonseed meal, etc.), plant proteins, and animal proteins (meat meal, blood meal, etc.) are used primarily as sources of protein and amino acids. However, these products also supply energy, minerals, and vitamins in varying amounts.

VARIABILITY IN FEEDSTUFF COMPOSITION

It is important to realize that variations exist in feedstuffs for animals. Situations have occurred in which the formulation of the ration was not changed, but animal performance changed dramatically when a new batch of one of the

TABLE 4-1

Nutrient content of major feedstuff groups

Feedstuff Group	Relative Value[a]						
			Minerals		Vitamins		
	Protein	Energy	Macro	Trace	Fat Soluble	Water Soluble	Bulk
High-quality hay	+++	++	++	++	+++	+	+++
Low-quality hay	+	+	+	+	−	−	++++
Cereal grains	++	+++	+	+	+	+	+
Grain millfeeds	++	++	++	++	+	++	++
Feeding fats	−	++++	−	−	−	−	−
Molasses	+	+++	++	++	−	+	−
Fermentation products	+++	++	+	++	−	++++	+/−
Oil seed proteins	+++	+++	++	++	+	++	+
Animal proteins	++++	+++	+++	+++	++	+++	+

[a]Relative values are indicated by the number of + (provided) or − (not provided). The feeding value (nutrient content and availability) of any product depends on many different factors, which are discussed in Chs. 6 through 10.

ingredients in the ration was fed. There are numerous examples in the literature to illustrate the variation that occurs, but unfortunately the problem is often not recognized by the individuals concerned with feeding animals.

A good example of this variation occurring in milo and corn is shown in Table 4-2. These data show the percentage of each grade received by a large feedlot in Texas. Grade 1 is the highest quality, and the Sample has low test weight and many broken kernels. Unfortunately, information relating grade to nutritional quality is not well documented. It may be that there is relatively little difference, but, on the other hand,

there could be a big difference. The presence of foreign materials may be expected to reduce nutritive value. Other factors, such as broken grains or small seeds, make it more difficult to produce a uniform product when processing methods such as steam flaking are used.

Another example of the variability to be expected in milo is shown in Fig. 4-1. These data were accumulated by a beef feedlot in Arizona. The graph shows the tremendous range in protein and starch content of milo and the correlations with bushel weight. Note that the protein content (moisture-free basis) ranges from a low of 7.4% to a high of almost 16%. Starch

TABLE 4-2

Grades of grain received at a feedlot in Texas (expressed as a percent of loads in each grade)

Grade	Month					
	April	May	June	July	August	Sept
Milo						
1	5	4	2	1	1	1
2	40	44	41	55	41	41
3	37	40	54	38	51	40
4	6	5	3	4	4	9
5	0	0	0	0	0	0
Sample	12	7	0	2	3	9
	100	100	100	100	100	100
Corn						
1	6	1	4	4	0	0
2	31	32	40	21	9	2
3	21	20	13	21	25	5
4	13	11	15	15	4	10
5	12	12	15	18	10	14
Sample	17	24	13	21	52	69
	100	100	100	100	100	100

Source: Fleming (8).

Grade 1 has the highest test weight and Sample has many damaged kernels.

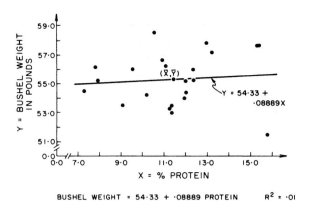

BUSHEL WEIGHT vs. MOISTURE-FREE PROTEIN IN ARIZONA MILO

BUSHEL WEIGHT = 54.33 + .08889 PROTEIN $R^2 = .01$

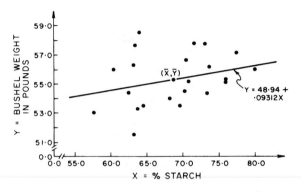

BUSHEL WEIGHT vs. MOISTURE-FREE STARCH IN ARIZONA MILO

BUSHEL WEIGHT = 48.94 + .09312 STARCH $R^2 = .095$

FIGURE 4-1 Relations between bushel weight of milo and the protein and starch content of the grain. From Cardon (10).

ranged from a low of 55.6% to a high of 80%. The tremendous range in these two important nutrients must have an effect on the nutritive value of the milo.

A final example of variation in legume–grass forage is shown in Table 4-3. Note the tremendous range observed in most of the listed nutrients. The author points out that nutritionists should probably have less concern with energy than some of the other nutrients, because the energy content (TDN) varies much less. For example, there was a 7.5-fold difference in crude protein, about a 10-fold difference for most of the major mineral elements, and about a 45-fold difference for most of the trace minerals.

These illustrations show clearly that average book values for nutrient composition may be very misleading when used in the formulation of animal diets. When appropriate analytical data are not available, little can be done about it except to try to find more accurate analytical values from the region where the feedstuff originated.

OTHER FACTORS EFFECTING FEED QUALITY

Many different environmental situations can cause damage and reduction in the feeding value of a feedstuff prior to harvest, during harvest, or during storage. Consequently, the reader should be familiar with some of the problems.

High humidity and rainfall will delay curing (drying) of forage cut hay. If drying is delayed, particularly at a relatively high temperature, it allows enzymes in the plant tissues to continue to be active after cutting (see the section on hay and silage making in Ch. 6). This action results in some metabolism of sugars and conversion of amino acids to nonprotein compounds. Thus the hay will contain less protein and readily available carbohydrates and have a lower feeding value. Heating usually occurs, especially if the forage is tightly packed, which

TABLE 4-3

Range in nutritive content (dry basis) of mixed legume–grass forage

Nutrients	Mean	Range	Fold Difference
TDN, %	59.4	51–71.7	1.4
Crude protein, %	16.4	5.5–40.3	7.5
Potassium, %	2.26	0.42–9.63	22.9
Calcium, %	1.02	0.01–2.61	261.0
Phosphorus, %	0.29	0.07–0.74	10.6
Magnesium, %	0.22	0.07–0.75	10.7
Sulfur, %	0.23	0.04–0.38	9.5
Manganese, ppm	48.1	6.0–265	44.2
Iron, ppm	222	10.0–2,599	259.9
Copper, ppm	13.1	2.0–9.2	46.0
Zinc, ppm	27.2	8.0–300	37.5

Source: Adams (11).

will cause a portion of the carbohydrates and proteins to be complexed and converted into a form that will not be utilized by the animal. Excess exposure to sunlight results in a rapid loss of carotenes (provitamin for vitamin A) from green plant tissues. All these result in some type of loss in nutritive value.

Cereal grains and other crops may be damaged by insects, rodents, birds, and other species before harvest or during storage. Contamination of stored grains by rodent feces, urine, hairs, and other foreign matter makes the feed less palatable and increases the chance of spreading disease. If beetles or larvae are present in large enough numbers, the damage may be very severe. This can normally be prevented by cleaning and treating storage bins with various chemicals to inhibit insect infestations prior to storing.

MOLD AND OTHER FUNGI

Generally, all feedstuffs provide a suitable medium for various molds and fungi if temperature and moisture conditions allow growth to occur. Some molds produce toxins that can be detrimental to animals. Although mold growth can be detected, it is not necessarily an indication that toxins are present, but it is an indication that the feed has been stored improperly. Sometimes toxins levels can be high enough to cause problems when no noticeable mold is present.

More than 100 different molds have been shown to grow on feeds if the proper conditions are provided, and many of these produce toxins (mycotoxins). About 20 of these toxins have been shown to affect animal performance. Molds require four things in order to grow: moisture, acceptable temperature, acceptable substrate (something to grow on), and an environment that contains adequate levels of oxygen. Only in isolated cases have toxicities actually been diagnosed (12). The most important impact on animals is most likely associated with subclinical toxicity (no acute symptoms are observed), resulting in lowered productivity (reduced weight gains, lower milk production, etc.), impaired reproductive performance, and reduction in immunity to diseases. Some states and countries require that some of the more suspectible feeds be tested for mycotoxins and that a certificate indicating the levels found accompany the shipment.

Molds may grow on a crop while it is still growing in the field. For example, scab (fusarium fungi) has been found growing on barley, wheat, and other cereal grains and some grasses. This fungi causes the kernels to appear bleached, shriveled, and often covered with a pinkish mold. This mold produces two potent mycotoxins, zearalenone and vomitoxin. Zearalenone interferes with the estrous cycle. Vomitoxin causes vomiting. Ruminant animals are generally less affected than swine or poultry, because the microorganisms in the rumen will partially detoxicify these mycotoxins. Cereal grains, especially rye, may also be infested with ergot, a fungus that produces a variety of clinical effects (13).

Many of the toxins from molds and other fungi are heat stable and can survive at pelleting temperatures and in other feed-processing methods. The only sure way to control mold is to have the moisture level of the feed low enough to prevent mold growth. Several mold inhibitors can be satisfactorily used in high-moisture feeds (propionic acid, for example). Generally, very little can be done to make a moldy feed satisfactory to feed. Treatment of corn and cottonseed with anhydrous ammonia has been shown to be an effective way of destroying the mycotoxins and is commonly done in some areas where mold growth is a problem.

Molds may be a serious economic problem if feed is stored in a manner so that the moisture level is relatively high. The moisture of most feeds must be below 14% to be safe to store. In some cases, feeds will mold when the moisture level is 10% or less. There are exceptions, of course; molasses, a high-moisture feed, does not mold because of its high sugar content, which causes the osmotic pressure of the molasses to be high enough that molds cannot survive, although they might be observed growing on the surface of molasses. Failure to cool pelleted or flaked feeds and to remove moisture added while processing will invariably result in mold if the feeds are stored for several days because of condensation of moisture in the bin.

Most molds do not grow at freezing temperatures. The optimum temperature is between 20° and 30°C. Temperatures above 50° to 55°C will inactivate most molds. It has been shown that *Aspergillus candidus* and *A. flavus* predominate when grain is heated to temperatures up to 50° to 55°C. Above this temperature, thermophilic (heat-loving) fungi predominate, and then thermophilic bacteria, which may grow at temperatures up to 70° to 75°C. The combination of mold infestation, oxidation of feed components, and long-term heating reduces the nutritional value of a feed dramatically (see Table 6-12) and also creates a potential fire hazard.

Molds require oxygen in order to grow; this is the reason that feeds will normally have more mold growth on their surfaces and around the edges. During the ensiling process, the rapid fermentation by bacteria of the readily fermentable carbohydrates depletes the oxygen supply, thus reducing the potential for mold growth. The reduction in pH (increased acidity) will also have a preservative effect.

Toxins produced by molds may cause clinical symptoms ranging from photosensitivity of the skin to death. New test kits are making it easier to determine in the field if a particular batch of feed contains toxins. Some mycotoxins fluoresce when exposed to ultraviolet light, but a number of other things also fluoresce that are not toxic. The question that the feeder still has to ask is whether to feed moldy feed to animals. There are marked differences among animal species with respect to susceptibility to a given mold toxin. Generally, younger animals are more sensitive than older animals.

The *Feed Additive Compendium* (14) has suggested the following criteria for determining whether to feed spoiled feed: (1) Type of spoilage (yeast, blight, mold). (2) Extent of spoilage: less than 10%, chancy; 10% to 50%, definitely risky; over 50%, discard the feed. (3) Age and species of livestock to be fed: young, fast-growing animals are most adversely affected within a species; breeding animals may not be severely affected, but the developing embryos are endangered; in decreasing order, trout, ducks, geese, turkeys, chickens, fish, swine, other simple-stomached animals, cattle, and sheep are affected. (4) Willingness to accept the toxic effects that might occur. (5) Estimated value of the questionable feedstuff minus the possible detrimental effects compared to cost of good feedstuff. (6) What past experience has taught you; send a sample to a laboratory for analysis.

FEED MANUFACTURING TERMINOLOGY

The following terms are commonly used in feed manufacturing. Many of these terms will be used in succeeding chapters when discussing the characteristics of the various feedstuffs. These terms are defined by the Association of American Feed Control Officials (AFCO) (5).

Additive An ingredient or combination of ingredients added to the basic feed mix or parts thereof to fulfill a specific need. Usually used in micro quantities and requires careful handling and mixing.

Antibiotics A class of drug. They are usually synthesized by a living microorganism and in proper concentration inhibit the growth of other microorganisms.

Artificially dried (Process) Moisture having been removed by other than natural means.

Aspirated, aspirating Having removed chaff, dust, or other light materials by use of air.

Balanced A term that may be applied to a diet, ration, or feed having all known required nutrients in proper amount and proportion based upon recommendations of recognized authorities in the field of animal nutrition, such as the Nation Research Council, for a given set of physiological animal requirements. The species for which it is intended and the functions such as maintenance or maintenance plus production (growth, fetus, fat, milk, eggs, wool, feathers, or work) shall be specified.

Blending (Process) To mingle or combine two or more ingredients of feed. It does not imply a uniformity of dispersion.

Blocked, blocking (Process) Having agglomerated individual ingredients or mixtures into a large mass.

Blocks (Physical form) Agglomerated feed compressed into a solid mass cohesive enough to hold its form and weighing over 2 pounds, and generally weighing 30 to 50 pounds.

Brand name Any word, name, symbol, or device or any combination thereof identifying the commercial feed of a distributor and distinguishing it from that of others.

By-product (Part) Secondary products produced in addition to the principal product.

Cake (Physical form) The mass resulting from the pressing of seeds, meat, or fish in order to remove oils, fats, or other liquids.

Carriers An edible material to which ingredients are added to facilitate uniform incorporation of the latter into feeds. The active particles are absorbed, impregnated, or coated into or onto the edible material in such a way as to physically carry the active ingredient.

Chaff (Part) Glumes, husks, or other seed covering together with other plant parts separated from seed in threshing or processing.

Cleaned, cleaning (Process) Removal of material by such methods as scalping, aspirating, magnetic separation, or by any other method.

Cleaning (Part) Chaff, weed seeds, dust, and other foreign matter removed from cereal grains.

Clipped, clipping (Process) Removal of the ends of whole grain.

Commercial feed The term commercial feed (as defined in the Uniform Feed Bill) means all materials distributed for use as feed or for mixing in feed, for animals other than humans except as follows: (A) Option A—Unmixed seed, whole or processed, made directly from the entire seed. Option B—Unmixed or unprocessed whole seeds. Hay, straw, stover, silage, cobs, husks, and hulls (a) when unground and (b) when unmixed with other materials. (B) Individual chemical compounds when not mixed with other materials.

Complete feed A nutritionally adequate feed for animals other than humans, which by specific formula is compounded to be fed as the sole ration and is capable of maintaining life and/or promoting production without any additional substance being consumed except water.

Concentrate A feed used with another to improve the nutritive balance of the total and intended to be further diluted and mixed to produce a supplement or a complete feed.

Condensed, condensing (Process) Reduced to denser form by removal of moisture.

Conditioned, conditioning (Process) Having achieved predetermined moisture characteristics and/or temperature of ingredients or a mixture of ingredients prior to further processing.

Cooked, cooking (Process) Heated in the presence of moisture to alter chemical and/or physical characteristics or to sterilize.

Cooled, cooling (Process) Temperature reduced by air movement, usually accompanied by a simultaneous drying action.

Cracked, cracking (Process) Particle size reduced by a combined breaking and crushing action.

Crumbled, crumbling (Process) Pellets reduced to granular form.

Crumbles (Physical form) Pelleted feed reduced to granular form.

Cubes (Physical form) See *Pellets*.

Cubes, range (Physical form) See *Pellets* and *Range cubes*.

Dehulled, dehulling (Process) Having removed the outer covering from grains or other seeds.

Dehydrated, dehydrating (Process) Having been freed of moisture by thermal means.

Diet Feed ingredient or mixture of ingredients, including water, that is consumed by animals.

Dilute (Physical form) An edible substance used to mix with and reduce the concentration of nutrients and/or additives to make them more acceptable to animals, safer to use, and more capable of being mixed uniformly in a feed. (It may also be a carrier.)

Dressed, dressing (Process) Made uniform in texture by breaking or screening of lumps from feed and/or the application of liquid(s).

Dried, drying (Process) Materials from which water or other liquid has been removed.

Drug (As defined by FDA as applied to feed) A substance (a) intended for use in the diagnosis, cure, mitigation, treatment, or prevention of disease in humans or other animals, or (b) a substance other than food intended to affect the structure or any function of the body of humans or other animals.

Dust (Part) Fine, dry pulverized particles of matter usually resulting from the cleaning or grinding of grain.

Emulsifier A material capable of causing fat or oils to remain in liquid suspension.

Evaporated, evaporating (Process) Reduced to denser form; concentrated as by evaporation or distillation.

Expanded, expanding (Process) Subjected to moisture, pressure, and temperature to gelatinize the starch portion. When extruded, its volume is increased, due to abrupt reduction in pressure.

Extracted, mechanical (Process) Having removed fat or oil from materials by heat and mechanical pressure. Similar terms are expeller extracted, hydraulic extracted, and "old process."

Extracted, solvent (Process) Having removed fat or oil from materials by organic solvents. Similar term: "new process."

Extruded (Process) A process by which feed has been pressed, pushed, or protruded through orifices under pressure.

Feed(s) Edible material(s) consumed by animals that contributes energy and/or nutrients to the animal's diet (usually refers to animals rather than humans).

Feed additive concentrate (Part) (As defined by FDA) An article intended to be further diluted to produce a complete feed or a feed additive supplement and not suitable for offering as a supplement or for offering free choice without dilution. It contains, among other things, one or more additives in amounts in a suitable feed base such that from 10 to 100 pounds of concentrate must be diluted to produce 1 ton of a complete feed. A feed additive concentrate is unsafe if fed free choice or as a supplement because of danger to the health of the animal or because of the production of residues in the edible products from food-producing animals in excess of the safe levels established.

Feed additive premix   (As defined by FDA) An article that must be diluted for safe use in a feed additive concentrate, feed additive supplement, or complete feed. It contains, among other things, one or more additives in high concentration in a suitable feed base such that up to 100 pounds must be diluted to produce 1 ton of complete feed. A feed additive premix contains additives at levels for which safety to the animal has not been demonstrated and/or that may result when fed undiluted in residues in the edible products from food-producing animals in excess of the safe levels established.

Feed additive supplement   (As defined by FDA) An article for the diet of an animal that contains one or more food additives and is intended to be (a) further diluted and mixed to produce a complete feed, or (b) fed undiluted as a supplement to other feeds, or (c) offered free choice with other parts of the rations separately available. *Note:* A feed additive supplement is safe for the animal and will not produce unsafe residues in the edible products from food-producing animals if fed according to directions.

Feed grade   Suitable for animal consumption.

Feed mixture   See *Formula feed.*

Feedstuff   See *Feed(s).*

Fines   (Physical form) Any material that will pass through a screen whose openings are immediately smaller than the specified minimum crumble size of pellet diameter.

Flaked, flaking   (Process) See *Rolled.*

Flakes   (Physical form) An ingredient rolled or cut into flat pieces with or without prior steam conditioning.

Flour   (Part) Soft, finely ground and bolted meal obtained from the milling of cereal grains, other seeds, or products. It consists essentially of the starch and gluten of the endosperm.

Food(s)   When used in reference to animals, this is synonymous with feed(s). See *Feed(s).*

Formula feed   Two or more ingredients proportioned, mixed, and processed according to specifications.

Free choice   A feeding system by which animals are given unlimited access to the separate components or groups of components constituting the diet.

Gelatinized, gelatinizing   (Process) Having had the starch granules completely ruptured by a combination of moisture, heat, and pressure, and, in some instances, mechanical shear.

Grain   (Part) Seed from cereal plants.

GRAS   Abbreviation for the phrase "Generally Recognized as Safe." A substance that is generally recognized as safe by experts qualified to evaluate the safety of the substance for its intended use.

Grits   (Part) Coarsely ground grain from which the bran and germ have been removed, usually screened to uniform particle size.

Groats   Grain from which the hulls have been removed.

Ground, grinding   (Process) Reduced in particle size by impact, shearing, or attrition.

Heat processed, heat processing   (Process) Subjected to a method or preparation involving the use of elevated temperatures with or without pressure.

Hulls   (Part) Outer covering of grain or other seed.

Ingredient, feed ingredient   A component part or constituent of any combination or mixture making up a commercial feed.

Kibbled, kibbling   (Process) Cracked or crushed baked dough or extruded feed that has been cooked prior to or during the extrusion process.

Mash   (Physical form) A mixture of ingredients in meal form. Similar term: mash feed.

Meal   (Physical form) An ingredient that has been ground or otherwise reduced in particle size.

Medicated feed   Any feed that contains drug ingredients intended or represented for the cure, mitigation, treatment, or prevention of diseases of animals other than humans or that contains drug ingredients intended to affect the structure or function of the body of animals other than humans.

Microingredients   Vitamins, minerals, antibiotics, drugs, and other materials normally required in small amounts and measured in milligrams, micrograms, or parts per million (ppm).

Mill by-product   (Part) A secondary product obtained in addition to the principal product in milling practice.

Mill dust   (Part) Fine feed particles of undetermined origin resulting from handling and processing feed and feed ingredients.

Mill run (Part) The state in which a material comes from the mill, ungraded and usually uninspected.

Mineralize, mineralized (Process) To supply, impregnate, or add inorganic mineral compounds to a feed ingredient or mixture.

Mixing (Process) To combine by agitation two or more materials to a specific degree of dispersion.

Pearled, pearling (Process) Dehulled grains reduced by machine brushing into smaller smooth particles.

Pellet, soft (Physical form) Similar terms: high molasses pellets. Pellets containing sufficient liquid to require immediate dusting and cooling.

Pelleted, pelleting (Process) Having agglomerated feed by compacting it and forcing it through die openings.

Pellets (Physical form) Agglomerated feed formed by compacting and forcing through die openings by a mechanical process. Similar terms: pelleted feed, hard pellet.

Premix A uniform mixture of one or more microingredients with diluent and/or carrier. Premixes are used to facilitate uniform dispersion of the microingredients in a larger mix.

Premixing (Process) The preliminary mixing of ingredients with diluents and/or carriers.

Product (Part) A substance produced from one or more substances as a result of chemical or physical change.

Protein (Part) Any of a large class of naturally occurring complex combinations of amino acids.

Pulverized, pulverizing (Process) See *Ground, grinding*.

Range cake (Physical form) See *Cake*.

Range cubes (Physical form) Large pellets designed to be fed on the ground. Similar term: range wafer.

Ration The amount of the total feed provided to one animal over a 24-hour period.

Rolled, rolling (Process) Having changed the shape and/or size of particles by compressing between rollers. It may entail tempering or conditioning.

Scalped, scalping (Process) Having removed larger material by screening.

Scratch (Physical form) Whole, cracked, or coarsely cut grain. Similar terms: scratch grain, scratch feed.

Screened, screening (Process) Having separated various-sized particles by passing them over and/or through screens.

Self-fed A feeding system where animals have continuous free access to some or all component(s) of a ration, either individually or as mixtures.

Separating (Process) Classification of particle size, shape, and/or density.

Separating, magnetic (Process) Removing ferrous materials by magnetic attraction.

Sizing (Process) See *Screened, screening*.

Steamed, steaming (Process) Having treated ingredients with steam to alter physical and/or chemical properties. Similar terms: steam cooked, steam rendered, tanked.

Supplement A feed used with another to improve the nutritive balance or performance of the total and intended to be (a) fed undiluted as a supplement to other feeds, or (b) offered free choice with other parts of the ration separately available, or (c) further diluted and mixed to produce a complete feed.

Tempered, tempering (Process) See *Conditioned, conditioning*.

Toasted (Process) Browned, dried, or parched by exposure to a fire or to gas or electric heat.

Trace minerals Mineral nutrients required by animals in micro amounts only (measured in milligrams per pound of feed or smaller amounts).

Vitamins Organic compounds that function as parts of enzyme systems essential for the transmission of energy and the regulation of metabolism of the body.

Wafer (Physical form) A form of agglomerated feed based on fibrous ingredients in which the finished form usually has a diameter or cross section measurement greater than its length.

Wafered, wafering (Process) Having agglomerated a feed of a fibrous nature by compressing into a form usually having a diameter or cross section measurement greater than its length.

SUMMARY

Classification of feedstuffs can be made according to the primary nutrient that they provide in the diet, such as energy, protein, or roughage, or according to their origin; the latter method is commonly used by feed manufacturers. Feedstuffs vary considerably in composition, especially those that have not been standardized in a manufacturing process. Various methods are used to estimate the feeding value of a feedstuff. Laboratory analysis involving chemical fractionation of the feed is often performed, and these results are used in equations to predict the relative feeding value. Animal trials are used when more accurate data are desired, which include digestion and feeding trials. Differences in soil mineral content as well as fertilization practices will affect the nutrient content of a feedstuff. Nutrient content of feedstuffs can also be affected by changes occurring during storage and spoilage. Molds and bacterial decomposition are especially important when moisture levels are elevated in a feed.

The terminology used by the feed manufacturing industry as defined by the Association of American Feed Control Officials has been given. These terms will be used in subsequent chapters.

REFERENCES

1. NRC. 1982. *United States-Canadian tables of feed composition* 3rd ed. Washington, DC: National Academy Press.

2. Fonnesbeck, P. V., et al. 1984. *IFI tables of feed composition.* Logan, UT; International Feedstuffs Institute, Utah Agr. Expt. Std.

3. McDowell, I. R., et al. 1974. *Latin American tables of feed composition.* Gainesville, FL: Dept. of Animal Sci., Univ. of Florida.

4. Gohl, B. 1975. *Tropical feeds.* Rome: Food and Agr. Organ. of the United Nations.

5. AFCO. 1995. *Official publication 1995.* Washington, DC: Amer. Feed Control Officials.

6. AOAC. 1995. *Official methods of analysis.* 16th ed. Washington, DC: Assoc. of Official Analytical Chemists.

7. Pond, W. G., D. C. Church, and K. R. Pond. 1995. *Basic animal nutrition and feeding.* 4th. ed. New York: Wiley.

8. Fleming, B. 1975. *Beef* 11(6):60.

9. Fonnesbeck, P. V., M. F. Wardeh, and L. E. Harris. 1984. *Bulletin.* 508. Logan, UT: Utah Agr. Expt. Sta.

10. Cardon, B. P. 1975. Personal communication.

11. Adams, R. S. 1975. *Feedstuffs* 47(22):22.

12. Wilcox, R. A. 1988. In: *Feed additive compendium.* Minnetonka, MN: Miller.

13. Church, D. C. 1979. *Digestive physiology and nutrition of ruminants.* Vol. 2: Nutrition. Corvallis, OR: O & B Books.

14. In: *Feed additive compendium.* 1996. Minnetonka, MN: Miller.

5

FEED LAWS AND LABELING

Richard O. Kellems and D. C. Church

Regulations and laws relating to feed manufacturing and distribution have been established to safeguard the health of man and animals. These regulations and laws were also established to ensure orderly commerce, provide protection to the consumer, and protect the feed industry from unfair competition and deceptive practices. In the United States the first regulations were implemented in 1909. Normally, these regulations are administrated in each state by its Department of Agriculture. In other countries, other agencies may administer these laws. The manufacture and distribution of commercial feeds, mineral supplements, liquid supplements, premixes, or other feed supplements are controlled by these agencies. A company that is commercially producing a feed is regulated by these laws. Feed regulations need to be universally applied so that each regulated party is required to follow the same rules. Any person who violates the rules must be brought into compliance or, as a last resort, eliminated from the marketplace. Feed regulations should also include a means by which those in violation of the regulations are made known to the public, including other industry members.

In the United States most feed regulation agencies use the Model Feed Bill as the basis for their regulations. The Model Feed Bill was developed by the Association of American Feed Control Officials (AFCO) and endorsed by the American Feed Industry Association and the Pet Food Institute. The AFCO is an organization composed of individuals representing the regulatory agencies involved in regulating commercial feeds in the various states and countries of North America. The basic goal of the AFCO is to provide a mechanism for developing and implementing uniform and equitable laws, regulations standards, and definitions and enforcement policies for regulating the manufacture, labeling, distribution, and sale of animal feeds, resulting in safe, effective, and useful feeds. The Model Feed Bill covers a variety of different subjects, such as appropriate definitions, registration of brand feed names, labeling, and other appropriate regulations.

For a company to market a feed product in a given state, it must register the product with the appropriate state agency. This normally involves submitting a label and paying an annual fee. The state agency is then responsible for monitoring the feed product to ensure that it complies with the information listed on the label. The use of uniform laws and regulations greatly facilitates interstate trade of manufactured feeds.

LABELING REGULATIONS

Consumers are concerned about what they are purchasing, and label disclosure can serve to inform them about product content. Then, armed with information about competing products, the consumer can make an informed purchase decision. Labeling regulations must include the required, optional, and prohibited label information. Unsubstantiated performance claims can be just as misleading to an unwary consumer as unsubstantiated nutrient claims; therefore, a need exists for including prohibited label information in the regulations. Labeling must also direct the consumer in the proper use of the product. Directions may be as simple as naming the species for which the feed is intended or providing a lengthy explanation of the feed rate. Feed regulations should assure that the minimum information is provided in order for the product to be correctly used.

The following information should be present on a label: identify the product, inform the purchaser and/or user of the nature of the product and its intended purpose, provide instruction on how to use the product, and convey any particular cautions pertaining to the product's use. Labels are required to be attached to individual bags of feed or to be available for inspection when feed is sold in bulk lots, except for custom formula feeds. Minor differences exist from state to state, but labels generally contain the following information: product name and brand name, if any; purpose statement; guaranteed analysis; list of ingredients; directions for use; warnings, if any; name and address of manufacturer; and net weight. Custom feeds (those prepared for a specific customer) are to be accompanied by a label, delivery slip, or other shipping document containing the following information:

1. Name and address of the manufacturer
2. Name and address of the purchaser
3. Date of delivery
4. Product name and brand name, if any
5. Net weight of each ingredient used in the mixture
6. Adequate directions for use of feeds containing drugs or other ingredients to assure their safe use
7. Such precautionary statements as are necessary for safe and effective use of the feed

Labels on human food items require that ingredients be listed starting with the item found in highest concentration in the mixture and continuing with those found in smaller quantities. This practice is more or less followed by feed manufacturers, although the Model Feed Bill does not specify this and most states

probably do not require it. Collective feed names are used by most feed manufacturers. One reason for this is that it avoids the necessity of reregistration and preparation of new labels if the use of one or more ingredients is discontinued. It also allows the manufacturer to avoid listing every specific ingredient. The commonly used groups are animal protein products, forage products, grain products, plant protein products, processed grain by-products, roughage products, and molasses products.

The guaranteed analysis found on the feed label normally requires that specific information be provided, and this differs somewhat among species. A partial listing of the information required by the Model Feed Bill is shown in Table 5-1. In addition to these data, some states require that the percentage of roughage ingredients be shown if more than 5% is in the feed. With liquid supplements, some states require, in addition to the other information given, a maximum moisture and minimum total sugar content. When NPN is used, the maximum percent included must be listed. NPN compounds are not added to any feed except that intended for ruminant animals and, when fed in excess, can be toxic. Generally, if the formulation contains more than 8.75% crude protein as NPN or exceeds one-third of the total crude protein, the label must give adequate directions to assure safe usage. Normally, this requires that a precautionary statement be included: "CAUTION: Use as directed." Several states require that any feed formulation containing urea have a statement reading "For Ruminant Use Only" or "Feed Only to Ruminants."

Most species are subdivided into different classes, and these various classifications need to be listed on the label. The following lists the different classes specified for swine; similar classifications exist for the other species.

1. PreStarter: 2 to 11 lb
2. Starter: 11 to 44 lb
3. Grower: 44 to 110 lb
4. Finisher: 110 to 242 lb (market)
5. Gilts, sows, and adult boars
6. Lactating gilts and sows

Details on differences from state to state may be found in the *Feed Additive Compendium* (2). Differences exist between nonmedicated and medicated feed labels. The following table gives the label requirements for nonmedicated and medicated feed.

Nonmedicated Feed	Medicated Feed
Brand name, if any, and product name	Brand name, if any, and product name
Purpose statement	The word "Medicated"
Guaranteed analysis	Purpose statement
List of ingredients	Drug purpose statement
Directions for use	Guaranteed analysis
Warning (if any)	List of ingredients
Name and address of manufacturer	Directions for use
Net weight	Warning (if any)
	Name and address of manufacturer
	Net weight

A hypothetical example of a typical label for nonmedicated feed is shown in Figure 5-1. Figure 5-2 shows a typical label for medicated feed.

MEDICATED FEEDS

A medicated feed is defined as any animal feed that contains one or more drugs at any level and includes (1) medicated complete feed that is intended to be the sole ration for an animal, (2) medicated supplements that are safe for direct consumption by the intended animal and can be offered in a free-choice feeding plan, and (3) medicated concentrates that are to be mixed with other feed materials to make either a supplement or a complete feed before being offered to the intended animal. A high percentage of commercial feeds contain some type of additive that would be classed as a drug.

In addition to the usual labeling requirements, the addition of an additive classed as a drug requires (per FDA regulations) that the word "medicated" appear directly following and below the product name. The purpose of medication must be stated, the names and amounts of the active ingredients must be listed, a warning statement must be included (when required by the FDA) listing the minimum withdrawal period required before slaughter for human food, and warnings against misuse and appropriate directions must be given for use of the feed. An example of a typical medicated label is shown in Fig. 5-2. The FDA requires that all medicated feed be labeled. If bagged, each bag must be labeled appropriately. Custom-mixed feeds are subject to the same basic labeling requirements as registered feeds.

In the case of premixes (mixes that will be diluted with other feedstuffs prior to being fed), the label must state the intended mixing rations, the resultant drug levels, and the purpose of the final mixed medication. Any instructions and warning statements must also appear. In the case of drugs for which the FDA requires a

TABLE 5-1

Guarantees required by feed type under the 1995 AFCO model feed bill

Nutrient	Dairy	Beef	Swine	Poultry	Sheep	Equine	Fish	Rabbits
Crude protein, min %	Yes	Yes	Yes	Yes	Yes	Yes	Yes	Yes
NPN, max %	a	a	No	No	a	No	No	No
Lysine, min %	No	No	Yes	Yes	No	No	No	No
Methionine, min %	No	No	No	Yes	No	No	No	No
Crude fat, min %	Yes	Yes	Yes	Yes	Yes	Yes	Yes	Yes
Crude fiber, max %	Yes	Yes	Yes	Yes	Yes	Yes	Yes	d
ADF, max %	Yes	No	No	No	No	No	No	No
Calcium, min and max %	Yes	Yes	Yes	Yes	Yes	Yes	No	Yes
Phosphorus, min %	Yes	Yes	Yes	Yes	Yes	Yes	Yes	Yes
Salt, min and max %	No	Yes	Yes	Yes	Yes	No	No	b
Sodium, min and max %	No	b	b	b	b	No	No	No
Magnesium, min %	No	No	No	No	No	No	No	No
Potassium, min %	No	Yes	No	No	No	No	No	No
Copper, min ppm	No	No	No	No	c	Yes	No	No
Selenium, min ppm	Yes	No	Yes	No	Yes	Yes	No	No
Zinc, min ppm	No	No	Yes	No	No	Yes	No	No
Vitamin A, IU/lb	a	a	No	No	a	a	No	a

Source: Modified from AFCO, 1995. Official Publication (1).

aGuarantee for equivalent crude protein from nonprotein nitrogen and guarantee for vitamin A, required only when added.

bSodium guarantee required only when total sodium exceeds that furnished by the maximum salt guarantee.

cCopper guarantee for sheep required when added or level exceeds 20 ppm.

dRabbit feeds require minimum and maximum crude fiber guarantee.

GREEN BIRD BEEF FEED
Pasture Extender for Beef Cattle
Guaranteed Analysis

Crude protein (min) 12.0%
    (This includes not more than 2.9% equivalent crude protein
    from nonprotein nitrogen)
Crude fat (min) 2.0%
Crude fiber (max) 10.0%
Calcium (min) 0.5%
Calcium (max) 1.0%
Phosphorus (min) 0.5%
Salt (min) 11.0%
Salt (max) 13.2%
Potassium (min) 0.4%
Vitamin A (min) . . . . 10,000 IU/lb

Ingredients Statement
Grain Products, Plant Protein Products, Molasses Products, Processed Grain By-Products, Urea, Vitamin A Supplement, Vitamin D-3 Supplement, Vitamin E Supplement, Calcium Carbonate, Monocalcium Phosphate, Salt, Manganous Oxide, Ferrous Sulfate, Copper Oxide, Magnesium Oxide, Zinc Oxide, Cobalt Carbonate, Ethylenediamine Dihydiodide, Potassium Chloride.

Feeding Directions: Self-feed to beef cattle on pasture. Feed 4 to 6 pounds per head per day as a pasture extender. Provide plenty of fresh, clean water at all times.

Caution: Use as directed. Observe cattle daily and monitor intake. Do not feed additional salt.

Manufactured by:
Green Bird Feed Mill
City, State, Zip Code
NET WT 50 LB (22.67 kg)

FIGURE 5-1 Example of nonmedicated feed label.

form 1900 (see Ch. 10), the feed manufacturer must file these forms and receive FDA approval before selling such medicated feed.

Feed manufacturers producing medicated feeds are expected to become familiar with the Good Manufacturing Practice (GMP) regulations. These regulations have been developed and published in the *Federal Register* for different classes of feeds. The manufacturer is responsible for being familiar with these practices and for taking periodic samples of feed and submitting them for analysis to determine if the drugs are present at the specified levels.

A problem that occasionally occurs is the use of medicated feeds at concentrations not approved or in combinations not approved or feeding them to animals other than those on the FDA approved list. These situations may occur when a veterinarian and/or nutritionist requests a "special mix" of a medicated feed for a client. If any of the parties knowingly produces and feeds an illegal mix, they are legally liable.

SUMMARY

Most states in the United States subscribe to use of the Model Feed Bill, which was developed to promote uniformity in laws controlling feed manufacturing and the sales and labeling of manufactured feed. Labels are required to be attached to individual bags of feed or to be available for inspection when feed is sold in bulk lots. The label must provide the information required for the particular type of feed in the state in which it is sold, usually including the minimum crude protein content, the maximum crude fiber and ash content (and/or Ca), and, if present, the content of nonprotein nitrogen compounds. In addition, if medicated feed is involved, the label must state the names and amounts of the active ingredients, it must include a warning statement (if required by the FDA) listing the withdrawal period required before slaughter for human food, and it must give warnings against misuse and appropriate directions for use of the feed.

GREEN BIRD SUPER PIG FEED
MEDICATED
For Starter Pigs Weighing 11 to 44 pounds

Administer to swine in a complete feed for reduction of the incidence of cervical abscesses; treatment of bacterial enteritis (salmonellosis or necrotic enteritis caused by *Salmonella chloreaesuis* and vibrionic dysentery); maintenance of weight gains in the presence of atrophic rhinitis; increased rate of weight gain and improved feed efficiency up to 6 weeks postweaning.

Active Drug Ingredients

Chlortetracycline . . . . 100 grams/ton
Sufathiazole 0.011%
Penicillin (from procaine penicillin) . . . . 50 grams/ton

Guaranteed Analysis

Crude protein (min) 20%
Lysine (min) 1.2%
Crude fat (min) 4%
Crude fiber (max) 4%
Calcium (min) 0.8%
Calcium (max) 1.3%
Phosphorus (min) 0.65%
Salt (min) 0.35%
Salt (max) 0.5%
Selenium (min) . . . . 0.3 ppm
Zinc (min) . . . . 150 ppm

Ingredients Statement

Grain Products, Plant Protein Products, Processed Grain By-product, Dried Whey, Calcium Lignin Sulfonate, Animal Fat, Vitamin A Supplement, D-Activated Animal Sterol (source of Vitamin D), L-Lysine, Riboflavin Supplement, Choline Chloride, Biotin, Thiamine Mononitrate, Pyridoxine Hydrochloride, Vitamin Supplement, Menadione, Sodium Bisulfite Complex (source of Vitamin K Activity), Folic Acid, Ethoxyquin (a preservative), Ground Limestone, Dicalcium Phosphate, Salt, Copper Sulfate, Manganous Oxide, Zinc Oxide, Iron Sulfate, Cobalt Carbonate, Calcium Iodate, Sodium Selenite.

Feeding Directions

Feed as the complete ration to starter pigs weighing 11 to 44 pounds.
Warning: Withdraw 7 days prior to slaughter.

Manufactured By:
Green Bird Feed Mill
City, State, Zip Code
NET WT 50 LB (22.67 kg)

FIGURE 5-2 Example of medicated feed label.

REFERENCES

1. Association of Feed Control Officials, 1995. Washington, D.C.

2. *Feed additive compendium.* 1996. Minnetonka, MN: Miller.

6

ROUGHAGES

Richard O. Kellems and D. C. Church

Plant crops of various types have been utilized extensively as feed resources for animals for centuries. Plants are considered to be a renewable resource which is produced by transforming solar energy using the photosynthesis process into nutrient sources that can be utilized by animals. These plant materials primarily provide a dietary carbohydrate source for herbivorous animals and are commonly referred to as **forages** or **roughages.** Forage is defined as the total plant material available to be consumed by an animal. Roughage is a term most often used by animal feeders and nutritionists to describe those dietary components that are characterized by being high in fiber (cellulose). Another term often used by individuals involved in management of wildlife is **herbage,** which is plant material that does not include the seeds or roots and can be utilized as a food by herbivorous animals. The terms forages and roughages are often used interchangeably to describe plant materials that are relatively high in structural carbohydrates (making up the rigid cell walls of plants), which contain high amounts of cellulose and hemicellulose (Ch. 3). These feeds require the action of microbial digestion in the GI tract in order to be digested efficiently. Ruminant animals (cattle, sheep, deer, elk, etc.) or animals having extensive microbial digestion occurring in the large intestine (horse, rabbit, etc.) perform best on diets high in roughage, because the microorganisms are able to digest the cellulose and hemicellulose. Livestock producers use roughages extensively as a food supply for domesticated ruminant species (beef, dairy, sheep, goats, etc.), as well as some monogastric animals (horse, limited extent swine, etc.). Roughages are generally low in readily available carbohydrates (starches, sugars) as compared to cereal grains and many other feedstuffs.

It is important to realize the importance of the grazing animal species with respect to the world food supply. Approximately one-third of Earth's surface is land (34 billion acres or 13.7 billion ha). Of this land, only 3% to 4% is utilized for urban and industrial purposes, while about 10% is being farmed. Forest lands, some of which may be utilized by animals, account for another 28% to 30% of the land. Approximately 15% of the total land is nonproductive and ranges from deserts to land covered by ice in the Arctic and Antarctic regions. The remaining 40% of the total land area is comprised of rangeland, which is more suitable for grazing than cultivation. Rangeland is composed of grasslands, savannas, scrublands, most deserts, tundra, alpine communities, coastal marshes, and wet meadows. Thus it is obvious that production of materials useful to humans (food, fiber, clothing, etc.) can only be achieved from a large portion of the world's land by grazing animals, both domestic and wild.

Duel use of rangelands in the western part of the United States and in other parts of

TABLE 6-1

Harvested forage production in the United States (in million of tons)

	Year		
	1980	1985	1994
All hays	143.1	149.0	157.0
Alfalfa and alfalfa mixture	83.7	85.3	85.9
All other hays	59.4	63.7	71.1

USDA (1).

the world, when properly managed, has increased the carrying capacity for both domesticated livestock and wild species, thus providing more animal products for human consumption. Continued research is needed in order to optimize the use of these resources for the benefit of humans.

In the United States, approximately 36% of the feed consumed by livestock is provided by pastures. An additional 18% is harvested and preserved as hay, so roughages account for approximately 54% of all feed fed to livestock in the United States. Table 6-1 provides a summary of the amount of forage crops that have been harvested in the United States in recent years. The proportion of roughages fed to animals in the nondeveloped countries of the world is much higher. Pastures and rangelands can be a viable, sustainable means of utilizing the land resources that are available and at the same time producing a high-quality food source for the human population.

CHARACTERISTICS OF ROUGHAGES

Two basic types of photosynthetic processes have been found to occur in plants. The basic difference between these two processes is the intermediate compound that is produced. The C3 process produces an intermediate compound that contains three carbons and the C4 process produces a four-carbon compound. The anatomical features of the C3 and C4 plants differ, and as a result they are adapted to different climatic conditions. The C4 plants are more heat tolerant and grow better at hot temperatures than the C3 plants. The C3 plants can start growing at around 0°C; for most, maximum growth occurs between 20° and 25°C. C4 plants grow very little at 10°C and have their maximum growth at 35° to 40°C. Some exceptions do occur with C3 plants; for example, alfalfa has a wider growing range (5° to 30°C). Another general

characteristic is that C4 plants have root systems that extend deeper into the ground than C3 plants; therefore, they are better adapted to conditions of water stress. Plants adapted to temperate regions (C3) normally store more carbohydrates (such as starches) than do C4 plants. The C3 plants can then draw on their carbohydrate reserves when they start to grow again, as alfalfa does.

NATURE OF ROUGHAGES

To most livestock feeders, roughage is a bulky feed that has a low weight per unit of volume. For example, the same weight of corn grain will occupy less space (less bulky) than the same weight of alfalfa hay (more bulky). This is probably the best means of classifying a feedstuff as a roughage, but any means of classification has its limitations. Feedstuffs classified as roughages have a high crude fiber (CF) content, and the digestibility of the other components, such as protein and energy, is generally lower. The NRC (2) uses the following criteria to classify a feedstuff as a roughage: when it contains greater than 18% crude fiber and less than 70% total digestible nutrients (TDN).

FACTORS THAT AFFECT ROUGHAGE UTILIZATION

Roughages contain high amounts of plant cell walls, which can vary considerably among sources. The **neutral detergent fiber** (NDF) fraction estimates the cell wall content of a feedstuff. The major components found in the cell wall are cellulose, hemicellulose, pectin, silica, and lignin; other components occur in lesser amounts. The amounts of these different components change as a plant matures, which changes the digestibility of the roughage. **Lignin** is a component normally found in grass and legumes (alfalfa, clover, etc.); it is associated with the structural carbohydrate fraction. Lignin is considered to be completely indigestible. Complexes are formed (lignin–cellulose/hemicellulose) that block enzymes from digesting the cellulose and hemicellulose. Lignin is also bound to some of the plant proteins. As a grass or legume matures, the lignin content increases, thus reducing the digestibility of the roughage. This is the primary reason why grasses and legumes decline in their feeding value as they mature. Grazing strategies have been developed that allow the manager to keep pastures and rangelands in an immature state so that they are more digestible and have higher nutritive value. Generally, grasses adapted to warm or tropical climates have higher lignin and cellulose contents than do cool-season grasses. Legumes tend to have higher concentrations of lignin than do grasses, and the highest concentration is found in the stem portion. Mature, dry cereal straws normally have the highest levels of lignin found in animal feeds, with the content usually being 10% to 15%.

The protein, mineral, and vitamin content can vary dramatically among roughages. Legumes can contain as high as 25% crude protein, but approximately one-third of this is **nonprotein nitrogen** (NPN), while other roughages such as straw only contain 2% to 4% crude protein. Mineral content is quite variable and ranges from being inadequate to excellent in different roughages. Potassium (K) is the mineral found in the highest concentrations and exceeds an animal's requirements in legumes (2.6% to 1.5%) and grasses (3.6% to 1.0%) and can be increased by applying fertilizer. Legumes are good sources of calcium (Ca) (2.3% to 1.1%), whereas grasses are moderate sources (approximately 0.65%). Magnesium (Mg) is another mineral that is supplied in good to adequate levels in both legumes (0.51% to 0.24%) and grasses (0.22% to 0.11%). Phosphorus (P) is a mineral found in moderate to low amounts in legumes (0.38% to 0.18%) and grasses (0.54% to 0.2%). The micromineral content varies dramatically depending on plant species, soil type, and the fertilization practices being used.

The nutrition quality of roughages ranges from very good in lush young grass, legumes, and high-quality silages to very poor in straws, hulls, and some browses. The primary objective of a good forage management system is to maximize the output of available nutrients being produced. Most of the time this is not possible, since forage production changes dramatically with the season and adequate numbers of animals are not available when availability of forages is at its highest. Since most areas cannot grow forage crops year around, a portion of the forage crop needs to be diverted so that it can be fed during the winter season when inadequate forage is available. Often forage nutrients are diverted by putting it up as either hay or silage, which can then be fed during the periods when inadequate forage is being produced. Forage crops are sometimes left standing so that they can be consumed by the animals during the periods of time when forages are not being produced. This can reduce costs.

As a forage matures its feeding value declines. The nutritional value of poor-quality roughages can often be improved by properly supplementing them. For example, if straw is being fed, then normally CP is the nutrient that is limiting its utilization. This is because the microorganisms that are digesting the carbohydrates (cellulose and hemicellulose) associated with the straw require a source of nitrogen or CP in order to grow. Straw does not provide adequate amounts of nitrogen or CP, so a CP-containing supplement will improve the utilization of the straw.

PASTURES AND RANGELANDS

Grazing animals utilize many different types of vegetation. There is a wide variation with respect to type of forage or herbage consumed by domesticated and wildlife, ranging from improved irrigated pastures to sagebrush on desert rangelands. Plant breeders have continued to make improvements in varieties of forage crops (grasses and legumes) to increase their productivity and versatility. Through good grazing management and reseeding with improved varieties, the carrying capacity of rangeland can be dramatically increased for both domesticated animals and wildlife. Herbage can be subdivided into the following categories:

Grasses members of the family Gramineae, which has over 6,000 different species. *Cool-season grasses* (timothy, sweetclover, etc.): grasses that make their best growth in the spring and fall. *Warm-season grasses* (bermudagrass, switchgrass, etc.): grasses that grow slowly in the early spring and grow most actively in early summer, setting seed in summer or fall.

Legumes members of the family Leguminosae, which has over 14,000 species. These plants have the ability to convert nitrogen present in the air into crude protein forms present in the plant.

Forbes primarily broadleaf, nonwoody plants (many different species).

Browse woody plants consumed to some extent by ruminants and horses, particularly the selective grazers, such as sheep, deer, and goats.

GRASSES

Grasses include not only all the wild and cultivated species used by grazing animals, but also all the cereal grains, which account for the majority of the agriculturally important crops produced in the world. There are many different grasses, adapted to grow on a wide range of soils types and climatic conditions. It is beyond the scope of this text to cover in depth the information relating to grasses. For more information, consult references (3 through 7). Many types of grasses are suitable for growing in different regions, extending from the tundra of the Arctic to the tropical regions of Africa. The following is a listing of grasses that grow well in the temperate regions of the world: smooth bromegrass, fescue, bluegrass, timothy, reed canarygrass, ryegrass, wheatgrass, bluestem, switchgrass, redtop, orchard grass, Bahia grass, carpetgrass, Bermuda grass, Johnson grass, and Dallis grass. Some of these grasses are adapted to drier or wetter and hotter or cooler conditions than are others. These grasses are normally the **C3 type** of plants.

In tropical and subtropical regions, the following grass species grow well: buffelgrass, rhodegrass, kikuyugrass, pangola grass, paragrass, napiergrass, sudan grass, elephant grass, greenleaf desmodium, green panicgrass, guinea grass, kikuyugrass, koa haole, molassesgrass and tropical kudzu. These grasses are primarily of the **C4 type.** Generally, their protein content is lower than in C3 grasses, which often accounts for inadequate dietary protein intake for animals consuming these grasses. Generally, a larger portion of their carbohydrates are of the structural type (cellulose and hemicellulose) and therefore they are less digestible and have lower feeding values.

Land that is suitable for the production of forages for animals varies widely with respect to temperature (subarctic, temperate, subtropical, tropical), water availability (wet, humid, semiarid, arid), and seasonality of rainfall (summer, winter, aseasonal). Therefore, forages will grow at different times depending on the conditions that they are exposed to. For example, a temperate grass starts to grow at temperatures as low as 0°C, if water and other nutrients are adequate, and reaches its maximal growth rate at 20°C, but growth will be depressed as the temperature raises higher. On the other hand, a tropical grass does not start to grow until the temperature is 10° to 15°C and reaches its maximum growth rate between 30° and 35°C.

In some areas, both domesticated and wild animals move as the climatic conditions change, which causes the forage quality to change. For example, as the summer temperature increases the type and quantities of forages on nonirrigated rangelands change, so often animals move or are moved to higher elevations, where the temperature and conditions are still optimum for forage growth. In improved pastures, different combinations of grasses are planted so that as the climatic conditions change there is still one of the grasses that will grow well.

Table 6-2 shows the estimates of forages in the various regions of the world. The productivity of grasses is highly influenced by adequacy of water, solar energy and fertilization practices. When all these are present in their optimum amounts, then the grass will grow at its fastest rate. If one or more of these are not adequate, then grass growth rate will be depressed. Fertilizers will normally contain a combination of the following, depending on soil type: nitrogen (N), phosphorus (P), potassium (K), calcium (Ca), magnesium (Mg), sulfur (S), and other micronutrients. Many forage management regimes require that fertilizer be provided in order to sustain forage or animal productivity at a profit. Soil type and acidity of the soil often influence what needs to be provided by a fertilizer.

An adequate supply of nitrogen is especially important to grasses. Grasses use soil nitrogen in the form of nitrate or ammonium to satisfy their requirements. Growth will dramatically increase when nitrogen is raised from an inadequate level to an adequate level. Excessive use of nitrogen-containing fertilizers increases nitrogen runoff and percolation into the soil and possibly groundwater, which has a negative impact on the environment. Normally, the cost of fertilizer limits the amount of nitrogen that will be applied, so this does not present a major environmental problem. The environ-

TABLE 6-2

Estimates of total annual dry matter production of grasslands in the main climatic zones of the world (tons/hectare)

| Temperature | Water Supply | | | |
	Wet	Humid	Semi-arid	Arid
Subarctic	4	8	—	—
Temperate	25	15	9	4
Subtropical	120	40	10	4
Tropical	150	70	12	4

Source: Rodin et al (8).

mental problems have occurred more often in situations where animal numbers are highly concentrated and the nitrogen passing out in their manure starts to accumulate in the soil and finds its way into the groundwater or into surrounding water systems.

Other limiting nutrients (P, K, Ca, Mg, S) can also impair grass growth if not present in adequate amounts in the soils. Often soil tests are performed and then a customized fertilizer mixture is prepared and applied to the land. Soil type and acidity influence what needs to be provided.

FACTORS AFFECTING NUTRIENT CONTENT OF GRASSES The digestibility and feeding value of grasses are influenced dramatically by the stage of maturity at which they are harvested or consumed by an animal (Table 6-3). The digestibility of immature grasses is much higher than that of mature grasses. The reason for this is that as the grass matures it deposits more structural carbohydrates (cell wall components composed of cellulose and hemicellulose) and the lignin content increases. Lignin is almost completely indigestible, and it associates with the cell wall components, which reduces their digestibility. Cereal grain straw is a prime example of a mature grass that is very low in digestibility and feeding value.

Different varieties of grasses are often used for different growing conditions. For example, sorghum grass species are used in hot regions for forage production as well as pastures. Sudan and Johnson grasses are often used because they can be sown in early summer and will produce late summer and fall pastures. Timothy, on the other hand, is adapted to cool, humid climates, but not to drought conditions,

and grows better when the temperature is between 15° to 20°C.

Often cereal grains are used as cover crops in a crop rotation system. The cereal grain can then be harvested in an immature state as hay (oat hay, barley hay, etc.) or ensiled (oatlage, barlage, etc), which can then be fed to livestock. For example, in some regions oats or barley are planted as a cover crop for alfalfa and then harvested at an immature stage as hay or silage.

PROBLEM COMPONENTS ASSOCIATED WITH GRASSES Several compounds found in grasses can be detrimental to the animals consuming them. Sorghum species are prone to contain high levels of **glycosides,** which can be converted to **prussic acid** (a highly toxic compound) following drought or frost damage, so care must be taken when animals are grazing these species. Sorghum also contains **cyanogenic glycosides,** which can be converted to cyanide. **Ergot** is produced by fungi that infests the seed heads of grasses and produces **alkaloids** that are toxic to animals and humans. **Oxalate** increases in grasses under some growing conditions and complexes with the magnesium (Mg), causing a Mg deficiency referred to as **Mg tetany.** This normally occurs in the spring or early fall when the grass is growing rapidly under overcast conditions. Oxalate will also impair the utilization of calcium and other minerals.

LEGUMES

Legumes are second only to grasses with respect to number of species (over 15,000). Legumes are characterized by the fact that rhizobia organisms in the nodules attached to their root systems have the ability to transform atmospheric N into a form that can be used by the plant; therefore, nitrogen does not need to be provided. This symbiotic relationship benefits both the plant and the microorganisms and allows them both to grow. Grasses, on the other hand, must secure their nitrogen from fertilizer in order to grow at optimum rate. Often pastures are seeded with a combination of legumes and grasses so that the grasses can take advantage of the nitrogen-fixing ability of the legumes, thereby reducing the amount of nitrogen required as fertilizer and reducing the number of applications required, which reduces the cost. Often legume crops are plowed back into the soil as green manure, a practice that is desirable in a sustainable agriculture system.

TABLE 6-3

Effect of stage of maturity on digestibility of alfalfa

Stage	Percent Digestibility
Prebud	66.8
Bud	65.0
Early bloom	63.1
Mid-bloom	61.3
Full bloom	59.4
Late bloom	57.5
Mature	55.8

Source: Holland and Kezar (9).

The following are examples of legume species; alfalfa (lucerne), alsike clover, arrowleaf clover, red clover, sweet clover, white clover, bird's-foot trefoil, lespedezas, crimson clover, crownvetch, cicer milkvetch, kudzu, subterranean clover, lupines, soybean, cowpea, and peanut. There are legume species that will grow in a wide range of environments, ranging from very cool temperate to very hot tropical regions.

Alfalfa (lucerne) is the single most widely used forage legume in the world. Varieties of alfalfa have been developed that are disease resistant and will survive ($-25°$ to $60°C$) and grow under a wide range of environmental conditions. Approximately 55% of the hay produced in the United States is alfalfa or a mixture that contains some alfalfa. Alfalfa is best adapted to deep loam soils with porous subsoils. Good drainage is essential. It also does better on alkaline soils and requires large amounts of lime. Alfalfa harvested at an immature stage of development will have higher CP and feeding value than if it is harvested at a more mature stage. Continuous harvesting of alfalfa at an immature stage of development, such as the prebud or bud stage, reduces total tonnage and will reduce the longevity of the stand. Overgrazing also reduces the longevity of the stand. The reduction in longevity is caused in both cases by a reduction of carbohydrate (starch) reserves in the roots, which makes the alfalfa more susceptible to winter-kill. During the initial growth stages of alfalfa, root reserves of starches are depleted, and as the plant matures the root reserves are replenished. Intensive grazing systems can be effectively used with alfalfa without reducing the longevity of the stand. New cultivars are available with high persistencies that allow for early harvesting and production of hay with higher feeding value. The number of cuttings varies from one in the northern drier areas to as many as nine in the irrigated warmer areas.

As a forage matures, its yield increases and quality declines. This is because the stem portion increases and the leaf portion decreases. The changes that occur when a forage crop, such as alfalfa, is harvested at different maturities are shown in Table 6-4. Alfalfa leaves are higher in protein and lower in fiber than the stem portion. Thus hay with the highest nutrient content (CP and digestible energy) is produced when alfalfa is harvested at an immature stage of development. The feeding value of the hay declines as the plant matures, because the fiber content increases, which causes digestibility to decrease. However, when alfalfa is cut at an early stage of maturity, often the carbohydrate root reserved becomes depleted and the life of the alfalfa stand may be reduced.

PROBLEM COMPONENTS ASSOCIATED WITH LEGUMES Pastures that contain high amounts of legumes can cause **bloat** to occur. Gases build up in the rumen, which the animal is unable to expel, because of a stable foam that is produced as the legume is digested. Legumes contain a high amount of readily fermented carbohydrates that produce gas in the rumen. They also contain soluble proteins, which seem to increase the formation of stable foams in the rumen. The presence of these stable foams inhibits the eructation process, the animal cannot expel the gases, and bloat results. Plant breeders have developed varieties of alfalfa that reduce the occurrence of bloat. Legumes that are grazed have a higher potential to cause bloat than when they are harvested and fed as hay.

Pasture management practices can reduce the incidence and severity of bloat. Usually, the occurrence of bloat is highest when animals are first turned out on a legume pasture or when they overconsume because of a previous period of inadequate feed intake. This can be minimized by feeding a coarse roughage, such as straw or low-quality hay prior to allowing the animals to graze the alfalfa. Bloat can also be prevented by feeding an antifoaming agent, such as **poloxalene** (Bloat Guard) to the animals while they are grazing a legume pasture. A number of legumes do not cause bloat, such as arrowleaf, milkvetch, and bird's-foot trefoil.

Sweetclover disease (bleeding disease) is caused by **coumarin,** which is found in some clovers. The coumarin is converted to **dicoumarin,** which interferes with the blood-clotting process. Coumarin has a characteristic taste that if present in high concentrations, is not readily acceptable to animals, which reduces the potential of overconsumption to some extent. Other antiquality factors associated with some legumes and in some situations that can cause problems are **tannins** in lespedezas and trefoil, **alkaloids** in lupines, and **saponins** in alfalfa. Plant breeders have developed varieties that contain reduced levels of these antiquality factors. Various treatment methods have also been developed to reduce the concentration of these compounds so that they will not adversely affect the performance of animals consuming diets containing these compounds.

TABLE 6-4

Effect of different alfalfa cutting frequencies on yield, quality, weeds, and stand life

	Harvest		Dry Matter (%)			
Maturity	Interval (days)	Yield (tons/acre)	Protein	Leaves	Weeds	Stand Remaining[a]
Prebud	21	7.5	29.1	58	48	29
Mid-bud	25	8.8	25.2	56	54	38
10% Bloom	29	9.9	21.3	53	8	45
50% Bloom	33	11.4	18.0	50	0	50
100% Bloom	37	11.6	16.9	47	0	50

Source: Marble (10).

[a] percentage remaining after three years

CHEMICAL COMPOSITION OF HERBAGE

The chemical makeup of forages is complex and affected by many different factors. Some brief comments are in order so that the reader can obtain an understanding of the various nutrients contained in forages.

PROTEINS AND NONPROTEIN NITROGEN

The crude protein (CP) or nitrogen-containing fractions found in forages are composed of proteins and nonprotein nitrogen compounds. Proteins are long polymers of amino acids. Nonprotein nitrogen compounds are not composed of amino acids, but rather are compounds that contain nitrogen forms other than amino acids.

When the proteins are consumed by an animal, they can either be converted into amino acids and absorbed or converted into microbial protein in the rumen and then digested and absorbed. Proteins can be either soluble or insoluble under the conditions that exist in the GI tract. Soluble protein and nonprotein nitrogen are both readily converted into microbial protein in the rumen. The less soluble proteins are converted to a lesser extent into microbial protein in the rumen, and more of them are passed intact into the true stomach and the small intestine, where they are broken down into amino acids, which are then absorbed in the small intestine. Table 6-5 shows the amino acid profiles of two grasses and two legumes. The true protein is usually in the range of 75% to 85% of the N, but at times it may represent considerably less than this amount. The portion of the nitrogen found as protein increases as the plant matures.

The **nonprotein nitrogen** (NPN) fraction is composed of free amino acids (not associated with a protein molecule) and other nitrogen-containing compounds. Over 200 non-protein nitrogen compounds have been identified in forages. All these are not found in one forage, but this figure provides the reader with an idea of the diversity of the nitrogen compounds consumed by an animal when it consumes forage. Certain conditions influence the level of nonprotein nitrogen that is present in forage. Forages grown on saline soils have increased NPN levels. When mineral deficiencies occur in the soil, the free amino acid content increases in the herbage being produced. Nitrate levels of 0.07% may depress performance, and levels of 0.22% may be fatal to a ruminant animal. Ruminant animals can tolerate high nitrate levels in forages if they have been adapted to them and are fed them on a continual basis.

Other factors, such as duration and intensity of light, temperature, disease, and water stress, have been shown to affect NPN levels. Grasses, in general, contain fewer nonprotein amino acids than do legumes. Some amino acids, peptides, and amides are toxic. For nonruminants, the nonprotein amino acids have a lower nutritive value than the protein amino acids, perhaps only half the value of the latter. For ruminants, nonprotein amino acids probably have nearly the same value as protein amino acids, since they are converted to microbial protein, which provides the amino acid source for the animal.

CARBOHYDRATES

Forage carbohydrates are usually divided into nonstructural and structural types. In nutritional terminology, the nonstructural compounds are the readily available carbohydrates (sugars, starches, and fructosans). In herbage, glucose and fructose are the principal free simple sugars, being found in a 1:1 ratio and at a level of 1% to 3% of the dry matter. Sucrose is the only other sugar found in any appreciable

TABLE 6-5

Amino acid composition of wheat, ryegrass, red clover, alfalfa and soybean meal (grams of amino acids per 100 grams of crude protein)

Amino acid	Wheat[1]	Ryegrass[2]	Red Clover[2]	Alfalfa[2]	Soybean Meal[1]
Lysine[a]	4.68	5.78	6.14	6.35	6.85
Histindine[a]	1.47	2.26	2.47	2.49	2.09
Arginine[a]	4.88	6.06	5.97	5.91	7.78
Aspartic acid	7.84	9.53	11.17	11.26	11.20
Threonine[a]	4.34	4.78	4.81	4.89	4.17
Serine	3.95	4.30	4.36	4.60	5.46
Glutamic acid	10.88	12.35	11.88	11.77	18.68
Proline	5.86	5.28	5.07	5.13	5.00
Glycine	4.43	5.78	5.53	5.39	4.52
Alanine	6.91	6.95	5.97	6.02	4.49
Valine[a]	5.47	6.25	6.43	6.15	4.91
Isoleucine[a]	4.33	4.80	4.94	5.00	4.87
Leucine[a]	7.52	8.77	8.89	8.86	7.91
Tyrosine	3.34	4.01	4.48	4.31	4.08
Phenylalanine[a]	5.17	5.92	5.90	5.96	5.38
Cystine	0.54	1.29	0.98	1.25	1.18
Methionine[a]	2.16	2.15	1.76	1.84	1.78
Tryptophan[a]	ND	1.81	1.55	2.07	ND

[1]Source: Morey and Evans (11).
[2]Source: Eppendorfer (12).
[a]Essential amino acids.
ND = not determined.

amount; it is often present at about 4% to 5%. Other sugars are found only in trace amounts.

Starches and fructosans are the most commonly found nonstructural polysaccharides in forage. In plants, these compounds serve as a storage form of energy. They normally range between 1% and 6%, but higher levels are often seen in the leaf blades after a period of active photosynthesis following a cool night. Often forages are harvested early in the day at the time that their fermentable carbohydrate content is highest so that they will have a higher acid-generating capacity when they are being ensiled to prepare haylage or silage. Legumes accumulate sucrose and starch, with starch being the primary nonstructural polysaccharide. Grasses, on the other hand, fall into two distinct groups. Grasses of tropical and subtropical origin accumulate starches in their vegetative tissues, but grasses of temperate origin tend to accumulate greater amounts of fructosans. When temperate legumes and grasses are harvested at a comparable growth stage, it appears that there is relatively little difference in total nonstructural carbohydrates. Seasonal differences have been noted, but the main change seems to be some increase in fructosan content as a result of deposition in the stems. Likewise, N fertilization appears to result in a reduced content of fructosans in grasses, particularly temperate species. Cool weather tends to increase nonstructural carbohydrates, especially starch or fructosans. Temperature also influences the molecular size of fructosans; the molecule tends to decrease in size with warmer temperatures.

Forage structural carbohydrates, which are all polysaccharides, range from homogenous to highly varied molecules, which may be linear or highly branched and form amorphous to crystalline structures. Structural carbohydrates are grouped into three major groups: the pectic substances, which are believed to function as intercellular cement; the noncellulosic polymers (hemicellulose), which are composed primarily of five- or six-carbon sugars; and cellulose, a linear polymer composed of simple sugar units. The latter two are relatively insoluble and are resistant to digestion, because the sugar units are chemically linked in a different manner (beta 1–4 linkage) than are more available carbohydrates, such as the starches. The beta 1–4 linkage can only be broken by the action of microbial digestion.

Variations in the amount of these different fractions are observed within the plant cells, among plant parts (cellulose is especially higher in stems), among different species, in type of climate adaptation, and so on. With legumes, there appears to be little change in levels of leaf and stem hemicellulose or pectin during growth, but there is a marked rise in stem cellulose. With grasses, there is generally a rise in cellulose, with a lesser increase in hemicellulose in the stems. Less marked changes are seen in tropical grasses, but the cellulose level is usually considerably higher than in temperate species.

An example of changes in various plant components (primarily carbohydrates) is shown in Table 6-6 for four warm-season temperate grass species. There are noticeable differences in the neutral detergent fiber (NDF) values: high levels were recorded on the 7/23 sampling, and after that time regrowth tended to push the values down, particularly for the ryegrass and bromegrass. Acid detergent fiber (ADF), hemicellulose, and lignin values followed the same general trends, as did the NDF levels, with the reverse being the case for cell solubles. In this example the analyzed samples were obtained from pastures that had enough grazing

pressure to keep the grasses grazed closely most of the time.

The leaf fraction of grass contains between 15% and 30% cellulose, 10% and 25% hemicellulose, and 1% and 2% pectins. In legumes, leaf fractions contain 4% to 10% hemicellulose, 6% to 12% cellulose, and 4% to 8% pectins. Stems have a similar fiber content to that of grasses (5).

LIPIDS (FATS)

The lipid content of leaf tissues ranges from 3% to 10% and generally declines as the plant matures. Leaf lipids are composed of a variety of different components (Ch. 3), the bulk of which are galactolipids and phospholipids; most of these are found associated with the chloroplasts. Linolenic acid usually makes up 60% to 75% of the total fatty acids, with linoleic and palmitic found in next highest concentrations. Other lipids, such as the waxes on the surface of the leaf, are low in digestibility and thus have low nutritional value. Once the plant has been harvested and the outer protective surface has

TABLE 6-6

Seasonal changes in carbohydrate, lignin, and cell solubles of four grass species

Species and Chemical Fraction	Sampling Date				
	4/19	5/3	7/23	9/26	10/24
Tall fescue					
NDF[a]	45.6	56.4	65.7	56.4	54.6
ADF[b]	24.6	30.2	36.5	28.7	30.8
Hemicellulose					
Lignin					
Cell solubles[c]					
Perennial ryegrass					
NDF	42.2	51.0	55.8	53.0	38.3
ADF	22.3	28.3	32.9	24.6	20.4
Hemicellulose	19.9	22.8	22.9	28.4	18.0
Lignin	3.0	3.6	6.4	4.2	3.3
Cell solubles	57.8	49.0	44.2	47.0	61.7
Smooth bromegrass					
NDF	43.9	52.8	63.1	55.4	40.3
ADF	22.8	29.5	37.7	27.1	20.7
Hemicellulose	21.1	23.4	25.4	28.3	19.6
Lignin	3.0	3.5	6.9	4.9	2.8
Cell solubles	56.1	47.2	36.9	44.6	59.7
Orchardgrass					
NDF	46.6	56.3	64.3	54.4	47.9
ADF	23.6	30.0	39.8	28.0	24.8
Hemicellulose	23.3	26.4	25.0	26.4	23.1
Lignin	3.4	4.0	6.6	4.3	3.9
Cell solubles	53.4	43.7	35.7	45.6	52.1

Source: Powell et al. (13).

[a]NDF = neutral detergent fiber (cellulose + hemicellulose + lignin).

[b]ADF = acid detergent fiber (cellulose + lignin).

[c]Cell solubles = 100 − NDF

been disrupted, the unsaturated lipids can be altered by oxidative processes.

ORGANIC ACIDS

Generally, most reports on plant composition completely ignore the organic acids that are present in a plant. These acids, most of which are nonvolatile, are very important components of plants tissues. They have a primary role in respiration, amino acid synthesis, and cation–anion balance. Organic acids accumulate in plants in concentrations ranging from 2% to 8% of the dry weight. In the usual proximate analysis, they are a component of N-free extract (carbohydrate). If using the NDF and ADF methods, organic acids show up as cell solubles or contents.

Generally, malic, citric, guinic, shikimic, and aconitic acids are the predominant organic acids in common forage species (except for the higher fatty acids, which are classed as lipids). Others that have been identified in more than trace amounts include succinic, malonic, α-ketoglutaric and fumaric acids. The concentrations of quinic, malic, and total organic acids in top growth of perennial ryegrass decrease with maturity in direct proportion to the decrease in leaf tissue. Another study reported that malic and transaconitic acids did not change with the stage of growth of orchard grass and smooth bromegrass, but citric acid decreased and shikimic and quinic acids increased as grasses matured. The effect of fertilization varies with the plants species and type of fertilier. There are also rather marked differences in different genotypes of fescue grass species in acid content and composition (14). In addition to the important role these acids may have for the plant, there is evidence that they may be an important factor in taste and palatability for grazing animals. Other organic compounds present in only small amounts in plants tissue may also be quite important with respect to palatability for forages (15).

MINERALS

The mineral content of forages is highly variable and reflects soil levels, fertilization practices, plant species, and variation in climatic conditions. Table 6-7 shows the mineral composition of grasses and alfalfa. All forages have high levels of K, as shown in Table 6-7. Compared to grasses, legumes have characteristically higher concentrations of Ca, Mg, S, and frequently Cu. Legumes tend to be lower in Mn and Zn than do grasses.

Marked differences have been observed in the mineral content of many different grass species grown under similar conditions. The mineral content is usually highest in the leaves and normally decreases with stage of maturity, but there are many exceptions to this statement. Likewise, the influence of fertilizer applications on grasses at a vegetative stage (and other factors) are so diverse that it is difficult to make any general statements that cannot immediately be contradicted.

FACTORS AFFECTING FORAGE NUTRITIVE VALUE

A number of factors have already been mentioned that have some effect on the nutritive content of forages. In a given location with typi-

TABLE 6-7

Range and typical mineral concentration for pastures grasses and alfalfa

Mineral Element	Grasses			Alfalfa		
	Low	Typical	High	Low	Typical	High
Major elements, % of dry matter						
Ca	<0.3	0.4–0.8	>1.0	>0.60	1.2–2.3	>2.5
Mg	<0.1	0.12–0.26	>0.3	<0.1	0.3–0.4	>0.6
K	<1.0	1.2–2.8	>3.0	<0.4	1.5–2.2	>3.0
P	<0.2	0.2–0.3	>0.4	<0.15	0.2–0.3	>0.7
S	<0.1	0.15–0.25	>0.3	<0.2	0.3–0.4	>0.7
Trace elements, ppm of dry matter						
Fe	<45	50–100	>200	<30	50–200	>300
Co	<0.08	0.08–0.25	>0.30	<0.08	0.08–0.25	>0.3
Cu	<3	4–8	>10	<4	6–12	>15
Mn	<30	40–200	>250	<20	25–45	>100
Mo	<0.4	0.5–3	>5	<0.2	0.5–3.0	>5
Se	<0.04	0.08–0.10	>5	<0.04	0.08–0.1	>5
Zn	<15	20–80	>100	<10	12–35	>50

cal fertilization and management practices, the most important single factor is the maturity of the forage. Early in the growing season, grasses (especially cool-season species) have a very high water content, a high content of organic acids, and an excess of protein for most ruminant animals. The result is that animals may get diarrhea and, because of the low dry matter content, may have difficulty in obtaining a maximum intake of energy, even though energy digestibility at the immature stage may be high (70% to 85%).

As the plant matures the protein content decreases, structural carbohydrates increase along with lignin, readily available carbohydrates decrease, and digestibility of both protein and energy decreases (Table 6-8). Plant species and the environment in which the plant is growing affect these changes. For example, if the growing season progresses rather rapidly from cool spring weather to hot summer weather, changes in plant composition will be more rapid than when the weather remains cool and plants mature at a slower rate, especially in a cool-season grass. Alfalfa is very different from grasses; rapid changes take place as it matures and blooms. Crude protein values given by Dairy NRC publications indicate the following concentration (percent on dry basis) for alfalfa hay: early vegetative, 23%; late vegetative,

20%; early bloom, 18%; midbloom; 17% and full bloom, 15%. Corresponding changes in TDN range from 63% to 55%. Some of these differences are caused by loss of leaves as the plant matures and because leaves have a higher nutrient value than the remainder of the plant. There is also a decline in the concentrations of Ca, K, P, and most of the various trace minerals as forages mature. Data illustrating changes with maturity are shown in Tables 6-6 and 6-8.

The decline in crude protein in grasses is illustrated in Table 6-9. Note that it declined from 17.8% (dry basis) at a vegetative stage to 6.8% when seeds were at the dough stage. This decline may or may not be typical in all areas, but it illustrates the general changes that occur. In addition, the digestibility of energy declines rapidly, perhaps as much as 1% per day in some grasses in warm climates, although a more typical decline would be about 0.5% decrease per day. This rapid decline results in a reduction in animal performance of animals grazing these pastures within a time span of a week or less.

The decrease in digestibility in grasses is largely attributable to changes in the stem and leaf sheath, which decline in digestibility at a much more rapid rate than the leaf; there is also a decrease in the leaf to stem ratio with maturity. Similar changes occur in legumes, but

TABLE 6-8

Effect of stage of maturity on composition and digestibility of orchard grass

Parameter	Stage of Maturity			
	6 to 7 in. High, Cut 5/19, Pasture	8 to 10 in. High, Cut 5/31, Late Pasture	10 to 12 in. High, Cut 6/14, Early Hay	12 to 14 in. High, Cut 6/27, Mature Hay
Composition, % of DM				
Crude protein	24.8	15.8	13.0	12.4
Ash	9.3	6.8	7.1	7.2
Ether extract	4.0	3.5	3.9	4.2
Organic acids	6.3	6.0	5.4	5.0
Total carbohydrates[a]	49.9	63.0	64.4	63.1
Sugars	2.1	9.5	5.4	2.4
Starches	1.2	9.5	0.8	0.9
Alpha cellulose	19.5	19.8	19.1	27.7
Beta and gamma cellulose	3.4	5.4	3.8	2.5
Pentosans	15.1	15.8	16.8	18.1
Nitrogen-free extract	35.0	45.7	44.2	41.2
Lignin	5.7	5.0	6.2	8.1
Crude fiber	26.9	28.2	31.8	35.0
Digestibility, %				
Dry matter	73	74	69	66
Crude protein	67	63	59	59
Crude fiber	81	77	71	68

Source: Ely et al. (16).

[a]Total carbohydrates = (crude fiber + nitrogen free extract) − (lignin + organic acids).

TABLE 6-9

Effect of maturity on the crude protein content of grasses

Stage of Maturity	Crude Protein (% of dry weight)	
	Mean of 8 Grasses[a]	Brome-grass[b]
Vegetative	17.8	19.5
Heads half emerged	12.4	16.5
Flower parts fully expanded	9.5	14.5
One-fourth heads in bloom	8.6	10.2
Seeds at milk stage	7.4	8.8
Seeds at dough stage	6.8	7.4

[a]From Phillips et al. (17). Includes Alta fescue, bromegrass, Kentucky bluegrass, orchard grass, Reed canary grass, red top, timothy, and tall oat grass.
[b]From van Riper and Smith (18).

TABLE 6-11

Changes in composition of switchgrass hay (warm-season species) with advancing maturity

Cutting Date	Composition (% of DM)			
	Crude Protein	NDF	ADF	Lignin
6/23	9.4	74.9	43.7	4.7
7/16	7.5	78.1	42.7	5.2
8/9	6.2	82.5	50.3	4.9

Source: Vona et al. (20). First two cuttings were at an early head stage: the last was at an early bloom stage.

at a less rapid rate, because there is less lignification of the stems and also because the leaves change less. Legumes tend to lose their leaves more, so the leaf to stem ratio also changes with maturity.

The effect of advancing stage of maturity on legume–grass mixtures is illustrated in Table 6-10. Note that in a period of about 6 weeks dry matter digestibility declined about 11%. However, when grasses or legumes are maintained at a vegetative state, there is a less rapid change. Tall fescue clippings were frozen and then fed later in conventional digestion trials. Crude protein content and digestibility of switchgrass hay (warm-season species) declined during the season, but there was much less effect on digestible energy, as shown in Table 6-11. Cuttings were made over a shorter time span than for the fescue shown in Table 6-10, but the data indicate that there is not a marked change in composition when vegetation is maintained at about the same stage of matu-

rity, whether the grass be a warm- or cool-season species.

The major factor affecting the feeding value of forage is the reduction in digestibility coupled with a slower rate of passage, the combination of the two being reflected in lower voluntary consumption of the forage as it matures. In ruminant species, these factors effectively control feed consumption and thus the amount of energy that can be consumed. Voluntary intake can also be affected by a deficiency of some minerals and N. Normally, the critical level of crude protein required in a pasture before intake is reduced by a N deficiency is approximately 7.5% to 8%.

Unfortunately, there is not yet a good foolproof means of estimating voluntary intake and relating it to digestibility. Work some years ago indicated that voluntary intake of pastures was related to quality of dry matter soluble in acid pepsin, and a system of predicting a nutritive value index was suggested. This index was a combination of intake and digestibility. Other research involving pepsin solubility or digestion with fungal cellulase has been done (21). Even though digestibility can be predicted with

TABLE 6-10

Effect of stage of maturity of green chopped alfalfa–brome forage on digestibility, forage intake, and milk production

Stage of Maturity of Alfalfa	Harvest Date	DDM[a] (%)	DM Intake (lb/day)	DDM Intake (lb/day)	Milk Production (lb/day)
Prebud	5/17	66.8	34.0	23.0	42.5
Bud	5/24	65.0	33.2	21.6	39.5
Early bloom	5/31	63.1	32.0	20.2	31.4
Mild bloom	6/7	61.3	30.6	18.8	31.4
Full bloom	6/14	59.4	29.2	17.4	26.5
Late bloom	6/21	57.5	27.8	16.0	23.4
Mature	6/28	55.8	26.3	14.7	19.5

Source: Hibbs and Conrad (19).
[a]DDM = digestible dry matter.

reasonable accuracy using these laboratory methods, it is still difficult to estimate consumption. Further information is presented on this topic in the section on the evaluation of hay. Large differences have been observed in voluntary intake of other grasses when digestibility was the same. There is much variation among animals consuming the same forage.

Soil fertility and fertilization practices have been known to have a pronounced effect on the quantity of forage or crops being produced. In addition, it has been suggested that the composition of plants may be changed by the fertilization practices used. These changes are much less dramatic than those associated with increased maturity, and the results given in the literature show much variation in response. In pasture situations where mixed species are being grown, one obvious change that may occur as a result of fertilization is an alteration in the vegetative composition as some plants respond more to fertilizer than others. If a grass–legume mixture is fertilized with high levels of N, for instance, this practice is apt to kill out the legume or to stimulate the grass much more than the legume. Fertilization of grasses with N tends to increase total nonprotein and nitrate N of the forage. Potassium content and some other minerals may increase as a result of N fertilization.

Digestibility of protein is apt to be increased and, in some instances, palatability and dry matter intake may be increased when plants are fertilized with N, but the research is not all in agreement. The concentration of most minerals increases in the plant material when they are provided in the fertilizer. Palatability may be increased when P and N are applied together. Research indicates that soil fertility and/or fertilization practices may alter nutrient concentration and consumption of forage.

GRAZING SYSTEMS

A variety of different management practices for pastures have been developed to try to increase animal performance. These systems are generally based on continuous or some type of rotation grazing. Rotation grazing normally increases animal productivity, but requires a higher level of management and increased cost for fencing and watering systems.

For the output of a continuous grazing system to be maximized, the carrying capacity of the pasture must be matched with the appropriate number of animals. One problem is that pasture growth varies during the growing season, so there will be a surplus of forage during the spring and a deficiency during the summer (especially where it is dry), requiring that the number of animals must be varied over the season to match the production. In actual practice, there is usually too much pasture during the spring and too little in periods of limited growth later in the season. Often a portion of the pasture is set aside and put up as hay during these periods of excess production, which can then be fed during the winter months. During periods of excess, animals are much more selective, especially where several forage species are grown together. Animals tend to graze areas containing the more palatable species, while leaving the less palatable species to mature and become even more unpalatable. It is not uncommon for some areas to be grazed so intensely that the desirable species dies out, usually resulting in the invasion of weedy plants.

Continuous grazing has been shown to be just as productive during periods of rapid growth as other grazing systems. During less productive periods, some type of intermittent grazing usually gives greater animal production per unit of land. If feasible, the time periods when a pasture is grazed should not be the same each year. The sequence of grazing pastures should be rotated each year; if you have three pastures each year, a different pasture should be grazed first. This practice helps to maintain a broader range of forage species and is particularly important in dry areas.

From an animal productivity and pasture management point of view, some type of rotational grazing system is preferred. Continuous use is detrimental to some plants, such as alfalfa or other large erect, easily defoliated species. In addition, weedy plants tend to be unused, and there is a much higher percentage of herbage that is not utilized because of trampling or fouling with dung and urine. Short, semiprostrate species are more suitable for continuous grazing over extended periods of time.

In rotational grazing the stocking rate is high, and the animals are pastured for a relatively short period of time on a given piece of land. For optimal use, the time should be for no longer than 24 hours, but in practice many producers use considerably longer times. When the pasture is grazed down to a desirable height, the animals are then moved to another pasture, and the first pasture is allowed to rest and grow back to a desirable pasture height.

Rotational grazing in effect restores the plant to a physiologically younger stage of growth. If done at frequent enough intervals, the plant is maintained in an immature state and is much more digestible. Rotational grazing also allows time for the plants to recover and to build up root reserves. It may not result in higher forage production than continuous grazing, but digestible energy production will be considerably higher. This is because the plants are being consumed at a less mature state when digestibility is higher. When using the higher stocking rates (35 to 75 cows/ha in New Zealand), animals have little opportunity for selective grazing, and a higher percentage of the total forage is consumed, thus discouraging the growth of clumps of unused forage and weedy plants.

The frequency of grazing or harvesting and the effect on the pasture are a function of the species present and of various environmental factors. If overutilized, yield is greatly reduced. If underutilized, yield may increase, but digestibility is most likely reduced. In one experiment in which pasture grass was cut at frequent intervals, digestible energy was highest when cut at bi- or triweekly intervals, but total yield of digestible energy for the season was obtained when it was cut at 5-week intervals, even though the digestibility declined somewhat for maximum values. Some research indicates that maximal pasture production is achieved by close grazing followed by a relatively long interval (3 to 5 weeks) for regrowth, but this effect depends on the time of year and the species of forages involved. Close grazing is generally more appropriate in the spring. This tends to increase the relative amount of legume in mixed pastures, although some favored species such as ryegrass do not tolerate close grazing at frequent intervals. Another way to evaluate the effectiveness of a pasture management system is to assess the effect it has on the maximization of animal product production, for example, pounds of gain per unit of land.

It is always a problem to make optimal utilization of forage. With rotational grazing there is more opportunity to set aside some of the pasture for harvesting as hay during periods of rapid forage growth. However, even with rotational grazing there may be problems in getting maximal utilization of forage. High-producing animals react very promptly with a fall in production (especially dairy cows) if forced to consume overly mature, trampled, or contaminated forage. In some cases, herds may be split into high, medium or low groups and the high producers grazed first; or the milking cows can be followed by the dry cows and growing stock, which have lower demands. With sheep, sometimes lambs are grazed in advance of the main flock in order to allow them maximal selectivity.

Strip grazing is another variation of intermittent grazing. In practice, animals are given access to limited amounts of forage, which is consumed in a few hours. This can be controlled by the use of electric fences, which are moved every few hours. Stocking rates are usually very high so that the forage is consumed rapidly with less loss due to trampling and fouling. This method almost completely eliminates selectivity, but requires intensive use of labor. Bloat can be controlled easily by spraying the strips to be grazed with an oil just prior to the time that they will be grazed.

IRRIGATED PASTURES

In drier regions, irrigation has been used for many years to increase the productivity of pastures. The development of more sophisticated systems, such as the center pivot, has greatly improved and expanded the opportunity to irrigate land formerly not suitable because of uneven terrain, provided that water is available (Fig. 6-1). Another major benefit is that labor costs can be minimized.

As a general rule, grass–legume mixtures are favored for irrigated pastures, but there is good evidence that pure stands of legumes such as alfalfa or well-fertilized grasses such as orchard grass also allow a high level of production. Optimal combinations of grass and legume are more difficult to maintain, but the bloat problem is usually lessened when grass is

FIGURE 6-1 A central pivot irrigation system that can cover about 160 acres in one rotation.

mixed in with legumes such as alfalfa and ladino clover. Bird's-foot trefoil is sometimes used where bloat is especially severe (it does not cause bloat), but it is more difficult to maintain a good stand of trefoil than with other species. The most desirable grass species depend on numerous environmental factors. The ideal species would maintain a high level of production throughout the growing season.

Irrigated pastures are usually used to best advantage with high-producing animals (Fig. 6- 2), such as lactating dairy cattle. Yearling steers, weaned calves, lambs, cows and calf, and ewe and lamb combinations (Fig. 6-3) offer reasonable potential. It is not normally profitable to use good irrigated pastures for dry cows and ewes or other relatively nonproductive stock.

With yearling beef steers, production of 1,000 kg of gain/ha is not too uncommon over the total grazing season, provided proper fertilization and stocking rates are maintained. As much as 1,200 kg of gain/ha may be obtained under ideal conditions. Average daily gains of yearling cattle on good pasture should be on the order of 0.7 kg/day. Lactating dairy cows on good pasture alone should be able to produce on the order of 15 to 18 kg/day of milk. On good pasture, lambs should gain on the order of 0.2 to 0.3 kg/day.

The high protein content of pastures, particularly grass–legume or legume stands maintained in a vegetative state, results in an excess protein intake for most classes of livestock. If this type of diet is supplemented with energy in the form of grains, as would normally be done with dairy cattle, it is possible to improve the gain of growing cattle such as yearling steers and thus make more efficient use of the protein in the herbage.

PERMANENT PASTURES AND RANGELAND

Permanent pastures and rangeland account for about 405 million ha of the land in the United States, as compared to 142 million ha being used for all other agricultural production. This same relationship exists in other parts of the world, and in many countries pastures and rangeland account for a larger proportion of the agricultural land. Over half of the nutrients consumed by domestic livestock are provided by pastures and rangelands (Ch. 1). This makes them a substantial resource to the livestock industry.

Figure 6-4 provides a general indication as to the location of the different grasses of major importance in the United States. The native grasses are more important in the plains and

FIGURE 6-2 Cows grazing lush improved pasture.

FIGURE 6-3 Lambs grazing irrigated pasture.

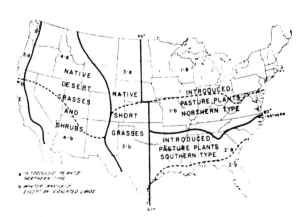

FIGURE 6-4 Five major pasture regions in the United States with type of forage that provides the majority of pasturage.

mountain states, while the introduced (improved varieties) species are of more importance in the central and eastern states, where rainfall is higher. Similar maps are available for other countries in the world.

NATIVE PERMANENT PASTURES

In the eastern states much of the permanent pastureland has at one time or other been used for crop production. Farther west, many areas of native grass remain on which many different grass species grow. Good examples of excellent pastureland still predominantly in the native prairie grasses are the Osage Hills of Oklahoma, the Flint Hills in Kansas, and the Sand Hills in Nebraska (Fig. 6-5).

Relatively little information is available on comparative production of native and introduced pasture grasses. That which is available indicates increased production of the introduced species; however, cultural and management practices with improved species are generally better than that applied to native grasses. Estimates indicate that approximately 70% of the improved species need treatment to reestablish or improve their productivity, and much of the land needs improved erosion control. Available information indicates that substantial improvement could be made in most humid areas. Approximately three-fourths of the pastures in humid regions are located on soils with great potential for improvement in the United States. Such practices as reseeding, liming, fertilizing, and weed control could improve productivity (22). Poor grazing management has, no doubt, contributed to the deterioration of some of this pastureland, resulting in the invasion of weedy species and loss

of some of the more palatable and productive native species.

RANGELANDS

A vast area of the western United States, as well as many other regions of the world, is dry enough so that forage production is limited and seasonal. In the United States these areas encompass deserts, arid woodlands, plateaus and grasslands covered with grasses, broad leaf herbs, shrubs, and noncommercial trees. These areas and the mountain meadows are the areas utilized by the livestock industry (Fig. 6-6). They are characterized by low precipitation, relatively low plant density, rough topography and soils that are usually shallow, rocky, and saline. Rangelands often contain many woody plants, such as sage species (*Artemisia* spp.), which are poorly utilized or not utilized at all by domestic livestock.

In addition, toxic broadleaf plants are often a problem in many areas. The grasses and other favored plants are easily subject to overgrazing because the ecology is more fragile than in more humid climates. In addition, the growing season is often of very short duration because of lack of moisture or late and early frosts.

Considerable research effort has gone toward improving forages for use on pastures and rangelands in drier regions. Many different species have been tried in various locations. As in other situations, both total and seasonal precipitation are very limiting factors. In areas with 8 to 10 in. of precipitation, crested wheatgrass (*Agropyron spicatum*) is extensively used. In areas receiving up to 15 in. of precipitation, western wheatgrass (*A. smithii*), intermediate

FIGURE 6-5 Cattle grazing native pasture in the Flint Hills of central Kansas.

FIGURE 6-6 A band of sheep grazing near a clear-cut in a forested area in the western United States.

wheatgrass (*A. intermedium*), and Siberian wheatgrass (*A. sibiricum*) have been used successfully. In mountain valleys, smooth brome (*Bromus inermis*) and orchard grass (*Dactylis glomerata*) find some use. Dryland alfalfas (*Medicago* spp.) also finds some limited usage.

In some areas in the west, overgrazing in the past has caused rangelands to be invaded by undesirable plants, or there has been a vast increase in poorly utilized plant species, such as sages, at the expense of more highly preferred species. The following control measures have been used in some areas: herbicides, mechanical removal, fire, and biological methods. Seeding of adapted grasses, particularly species such as creased wheatgrass, has been very successful, particularly when combined with control of undesired shrubs and browse plants. Fertilization has been practiced, but only on a limited scale. The development of dispersed water facilities has also resulted in improved range utilization. Insufficient water will greatly restrict these dry areas, especially during the hot summer months. Obviously, much more information is available on this topic. For further information, the reader is referred to reference 22.

MISCELLANEOUS FORAGE PLANTS

A variety of other plants is often utilized as pasture in specific situations. For example, rape and kale are often used for pasturing sheep. Beets and turnips tops are frequently used as forage for livestock. Feed turnips are grown in some areas for fall grazing by cattle and sheep, which learn rapidly to pull the turnips and eat them.

Cereal grains, such as oats and barley, may be used for winter pasture in many areas. Summer annuals such as Sudan grass may be used to extend the grazing season or to more nearly meet animals' demands when used to supplement the usual cool- and warm-season grasses. In addition to these, crop residues (straws, stover, etc.) are utilized heavily in some areas. In areas where corn is grown for grain production, cattle are often used to pasture the stalks remaining in the field after the grain is harvested. The same comment applies to straws and chaffs remaining after small grains have been harvested. Crop residues of this type can make up the major portion of the diet for beef cows after calves have been weaned, but some supplementation with protein and minerals is usually required to achieve optimal performance.

MULTIPLE USE OF GRAZING LAND

Often pastures and rangelands are used by a number of animal species. Different herbivore species, when given a choice, will select different plants or portions of the plant to consume. For example, cattle generally prefer grasses, although they will eat substantial amounts of forage legumes, some forbes, and a limited amount of browse. Cattle and elk are referred to as nonselective grazers. Sheep, deer, and goats, on the other hand, are called selective grazers and consume less grass and more forbes and browse. The net result is that there is less competition among animal species than one might think, and often utilization of the forage resources can be maximized by having more than one species utilize the forage resource. Preferences for a given forage species change with seasons of the year and with differences in plant maturity. In East Africa, where as many as 25 herbivorous species graze the same area, studies indicate that they seem to complement one another, provided overgrazing does not occur. The same type of studies in North America and Australia have indicated that optimum results can be achieved by having as many as four or more species grazing the same areas. When overgrazing of grass occurs, it tends to encourage the growth of shrubs. Likewise, if shrubs are overconsumed, it encourages the growth of grass.

In addition to natural preferences for different plant species, the terrain, temperature, and distance from water affect how effectively animals use the forage resources of a specific area. Cattle are rarely found on slopes of 45° or more, while sheep, deer, elk, and goats will graze on steeper slopes.

DETRIMENTAL SUBSTANCES IN FORAGE PLANTS

Many plants contain compounds that are toxic to grazing animals when the plant is ingested in normal amounts. With the exception of nitrates, which are normal metabolic products of plants, it is assumed by many scientists that the process of natural selection has resulted in the survival of plants with toxins that inhibit the consumption of the plants by animals. Fortunately, not many grasses are highly toxic to domestic animals.

Nitrate can be a problem in some grass species (such as corn, small cereal grains, sorghum species) following heavy fertilization and especially when accompanied by a drought. Sorghum species also are prone to produce exogenic glycosides, which are hydrolyzed to hydrocyanic acid in the rumen. This situation is apt to occur during a drought or following frost damage.

Perennial ryegrass produces a compound called perloline, which causes a condition called ryegrass staggers. *Phalaris* species produce toxic alkaloids, as do several other grass species. Other problems, such as fescue foot, occur from grazing tall fescue pastures infested with a toxic fungus, and a few species accumulate toxic levels of fluorine compounds.

In herbage other than grass, *Brassica* (cabbage family) often produce goitrogenic substances, and legumes often contain compounds (isoflavones) that have estrogenic effects. Many different species produce potent alkaloids, some accumulate toxic levels of oxalates, and a few contain toxic amino acids. Some plants, especially legumes, may contain toxic levels of copper, molybdenum, or selenium. Most plants contain compounds that may be toxic if ingested in sufficient amounts, but the level consumed is usually low enough and the diet varied enough that they rarely cause a problem. Acute toxicity is the exception rather than the rule in most cases, but it is much more common to have these materials impair the performance of an animal instead. Nevertheless, it is a topic that deserves careful consideration, especially in situations where herbage is scarce, because animals are then apt to consume plants that they usually do not eat.

HAY AND SILAGE MAKING

Forage crops are often preserved as hay or silage and are exclusively used as animal feeds. Hay and silage making has been practiced as a means of preserving a forage crop for many centuries in Europe, perhaps as far back as 750 B.C. (3). The following sections will describe the characteristics of both hays and silages. There is a great deal of research and practical information on the subject of hay and silage. Good overviews and review articles on this subject are available (3, 5, 9).

HAY AND HAYMAKING

In the United States each year in excess of 130 million tons of dried forage is produced from various forage crops and used primarily as a feed for wintering animals or for animals maintained in confinement facilities. This forage is preserved in a number of different forms, such as loose hay, baled hay, hay cubes, or hay pellets. The form selected depends on transportation and usage considerations. The primary dry form in which forage crops are preserved is as hay.

Preparation of hay is a means by which a forage crop can be harvested during periods of excess production (spring or early summer) and preserved and allows for it to be fed at a time that pastures are not producing adequate forage (winter). The cost of the nutrients contained in hay is higher than if the animal was just consuming the forage from a pasture, because of the labor and equipment costs associated with harvesting it. Fortunately, agricultural engineers have developed machinery that has also completely mechanized the process and made it financially feasible to harvest forage and put it up as hay. Continual improvements are being made in the machinery used for handling and feeding harvested roughages (see Fig. 6-7 and 6-8).

The intent in haymaking is to harvest the crop at or near its optimum nutrient content in order to provide maximal sustainable yield of digestible nutrients per unit of land. To make good hay, the moisture content of the forage must be reduced to a point low enough to allow it to be stored without marked nutritional changes occurring. The optimal moisture content is between 15% to 18%, which will minimize leaf loss. The moisture content of the forages at the time of harvest ranges from 60% to 85% moisture. At typical harvesting maturity, grasses contain 60% to 75% and legumes 70% to 75% moisture. This moisture level must then be reduced to between 15% and 18% prior to being baled. This process of lowering the moisture content is referred to as curing. If the moisture level is above 20% at the time of baling, excessive storage losses associated with heating and molding may occur.

FACTORS INFLUENCING HAY QUALITY   It is impossible to cut, cure (dry), and move hay into storage without some losses occurring in the process. However, it may be possible to harvest more units of nutrients per unit of land than can be obtained by grazing because of trampling and feed refusals resulting from contamination by dung and urine or selective grazing of some preferred species of plants. The quality and quantity of field-cured hay that can be harvested depends on the following factors: maturity of stand when cut, length of curing time, moisture content when baled, method of handling, and weather conditions during the

FIGURE 6-7 Newer type of machinery.

FIGURE 6-8 One example of equipment.

curing time. Losses that occur are a combination of physical loss of herbage, including leaves, and incomplete recovery of harvested forage. Other less visible losses are due to enzymatic activity of the plant tissues, oxidative losses during the drying process, and losses from leaching or weather damage.

Normal losses range from 20% to 40%, but can exceed 75% if conditions at harvest time are not optimal. When plant materials are first harvested, the moisture level is high enough to allow the plant enzyme to break down the components of the plant, which can result in a 4% to 15% loss due to the respiration process, which will continue until the moisture level is reduced by the curing process. Weather conditions have a significant effect on the magnitude of these losses. In dry, hot climates, these losses will be minimized, and under hot, humid conditions they will be maximized.

Leaf shatter is another loss that occurs, ranging from 2% to 5% of grass hay to 3% to 35% for legume hays. Leaf shatter losses are related to two factors: (1) dryness of the plant material and (2) roughness of equipment processing the hay. If a plant is very dry, the amount of leaf fraction lost increases, because leaves are more likely to shatter when they are being turned in the field or baled. Shatter losses increase if the equipment being used to harvest vibrates or shakes the material a lot and causes the leaves to be lost. Legumes are more fragile and therefore more susceptible to shatter loss. Leaching and losses associated with rain damage can range from 5% to 14%. The primary factor influencing these losses is weather conditions when the forage is being cured. If the conditions are dry and hot, leaching losses are minimized, but if it is rainy and cool, the losses are higher.

Weather damage to curing hay is generally considered to be the most destructive uncontrollable factor associated with making quality hay. It may vary from a slight loss of color from excess exposure to sunlight to extensive heating and molding, resulting in dark discolored hay that is almost worthless as a feed. Rain on freshly cut hay causes little damage, but damage becomes more severe as the curing time is extended. Consequently, hay that is harvested early in the season is more likely to have weather damage because of the longer drying time associated with cooler and wetter conditions (damper ground, higher humidity), resulting in more spoilage and leaching. Legumes normally have higher leaf losses under these conditions.

HAY MAKING PROCESS   The following steps are involved in the haymaking process: harvesting, curing (drying), and packaging for subsequent storage. The maturity at which the forage is harvested and the length of each of these steps influences the quality of the hay being produced. The following is an overview of the losses associated with each of these steps. For hay to be stored, it must be dried. Normally, it is harvested and allowed to dry prior to being baled, but in some areas this is not possible, and the hay is dried after it has been placed into storage. The shorter the time it takes to dry the forage prior to baling the better the quality of hay. Mechanical processing of forage can cause losses of up to 5% to 25%. The influence that each of these steps has on the quality of the hay produced will be discussed.

CHANGES OCCURRING DURING DRYING   Rapid drying, provided it is not accompanied by excessively hot temperatures, results in the least changes in chemical components of herbage. If drying is slow in the field, stack, or bale, appreciable changes may occur as a result of plant enzymes (respiration) and microbial growth (particularly molds) or oxidative changes. Thus machines such as crimpers (conditioners) have been developed to crush the stems of plants such as alfalfa to speed up the drying process. Evidence indicates that losses in the field are reduced by the use of this type of equipment.

Losses during drying are correlated with moisture content and ambient temperature. Plant enzymes continue to function until the moisture level falls below 40%. Under good drying conditions, losses associated with respiration can account for between 2% and 9% of dry matter. Under poor drying conditions, this loss can increase to as much as 15% to 18%. Usually, these losses go unnoticed, but several practices will reduce these losses. Cutting hay in the morning allows the hay to dry during the period when it is exposed to solar energy for the longest period of time. Not much drying occurs if the forage is cut at night or in the late afternoon, and respiration continues for a longer period of time. In one study it was shown that respiration losses in freshly cut immature alfalfa, which was dried at 27°C, amounted to 4.5% of the dry matter. Respiration rates of more mature and wilted herbage are reduced to about half of this value, but respiration losses are extremely variable. Rapid drying using artificial methods results in relatively minor changes in composition and nutritive value, but is normally too costly to feasibly use on a commercial operation.

The primary plant nutrients lost during respiration are the soluble carbohydrates, particularly glucose, fructose, and sucrose. Variable losses of starch and fructosans have been reported, and there may also be losses of organic acids. During drying there is generally some loss of N. This is due to the action of plant protease enzymes, which results in a decrease in protein N and an increase in soluble N. Total N losses are considerably less than those observed with readily available carbohydrates. Nitrates are affected very little by drying, but the cyanogenic glucosides (which cause prussic acid poisoning) of sorghum, white clover, and some other forages lose their toxicity when the plants are dried.

With regard to fat-soluble vitamins, slow drying results in a loss of as much as 80% of the carotene (provitamin A) when herbage is exposed to the sun. This is very obvious to the eye, as carotene losses more or less parallel the bleaching of cut herbage that occurs in the latter stages of prolonged drying. Rapid drying, particularly when the herbage is protected from sunlight, preserves most of the carotene content (for example, dehydrated hays). The vitamin D content of fresh herbage is very low. Curing in sunlight or ultraviolet light results in a marked increase in vitamin D activity as a result of synthesis of it from plant sterols, especially in the leaf, which may continue for as long as 6 to 8 days. Variable amounts of vitamin E normally occur during drying.

MECHANICAL LOSSES   The equipment used for harvesting and the moisture level affect the losses that occur, most of which are associated with leaf losses. If the harvesting equipment used tends to handle the hay roughly, the loss will be greater. Losses with the small rectangular bales range from 4% to 9%, while the large ton bales are slightly less. Losses in the large round bales can be as high as 10% to 15%. Losses associated with turning (raking) can be as high as 25%. New harvesting systems try to minimize the number of times that raking has to be done. Leaf loss in alfalfa is critical since leaves make up about 50% of the plant weight, but contain more than 70% of the protein, approximately 90% of the carotene, and over 60% of the available energy. Leaf loss increases as the moisture level at which it is baled decreases. New techniques are available that allow the forage to be baled at higher levels. Some forages are baled at high moisture levels (60% to 70%) and then placed in a bag and are allowed to ensile, which reduces the leaf loss.

HEAT DAMAGE   When hay is baled at a moisture content above 20%, microbial spoilage may occur. When microbes and fungi grow, they generate heat, and the heat causes changes, such as browning reaction, to occur. If temperatures rise above 440° to 460°F, spontaneous combustion will occur. If hay that has been put up containing too much moisture is placed in a storage unit, often the temperature is monitored to make sure that it does not heat to the point that it ignites.

STORAGE LOSSES   Further loss of nutritive value occurs primarily as a result of losses associated with exposure to moisture, which causes deterioration of the hay. Hay stored inside will lose 5% to 10%, while hay stored outside can lose as much as 30% to 35% of its dry matter. The conditions under which hay is stored affect the losses observed. The losses are greater for hays stored in regions where the temperature and humidity are high than for hay that is stored in a cool, dry region. The losses are primarily of sugars and other soluble carbohydrates.

A grower has little control over the weather. Our ability to project future weather has improved, but there is still a great deal of uncertainty associated with the weather. High-quality hay can only be made under ideal weather conditions. This is why in some areas forages are preserved as silage and not hay. The best hay is made when the humidity is low and there is no rain so that the forage can be dried in the shortest possible time. Unfavorable weather conditions extend the drying time and reduce the quality of the hay. If hay is rained on, the level of protein and carbohydrates is reduced. Rain leaches out the soluble proteins and carbohydrates, as well as other soluble nutrients. These components have the highest nutritive value, so the quality is directly related to the amount of rain that it is exposed to. Rain also increases the losses associated with plant enzyme losses. Plants contain enzymes that metabolize the nutrients (carbohydrates, protein, etc.) found in the plants until the moisture level drops below approximately 40%. These enzymes act on the more readily available nutrients first, which are also those that have the highest nutrient value for the animal. If the harvested forage is exposed to rain or high hu-midity, these enzymes continue to function and will reduce the amount of the readily available nutrients present in the hay. Table 6-12 shows losses that can result from forage being exposed to different amounts of rain.

Leaching losses increase from a low of 2% to a high of approximately 37% when hay is exposed to 2.5 in. of rain. The combined losses from leaf loss, leaching, and enzyme metabolism increases from less than 10% for hay exposed to no rain to 54% for hay exposed to 2.5 in. of rain. Methods are being evaluated that will increase the moisture level at which hay can be baled, which will reduce the losses associated with weather. Various methods have been developed, such as mechanical conditioning and chemical treatment, that increase the rate of moisture loss, thus reducing drying time and improving hay quality.

Excessive moisture can have a dramatic effect on hay quality. Research has shown that baled hay containing 16% moisture had a slight temperature rise and little microbial growth. Hay stored at 25% moisture heated spontaneously to about 45°C and became moldy. Bales containing more than 40% moisture became very hot (60° to 65°C) and contained high numbers of thermophilic (heat-loving) fungi (23). At elevated temperatures the proteins are complexed with the carbohydrates (browning reaction), which causes them to become less available to an animal (24). Excessive heating of stored hay results in the formation of brown or black hays, which have reduced palatability and nutritive value. The effect of feeding moldy hay to growing cattle is illustrated in Table 6-13. Note that the digestibility, daily gain, and feed conversion are lower for the cattle fed the moldy hay, even though it did not produce any clinical symptoms of toxicity. Severe toxicity can occur in horses that are fed moldy hay.

Hay protected from the weather changes very little in composition and can be stored for several years with little loss of nutritive value.

TABLE 6-12

Effect rain has on losses in alfalfa hay harvested at the bud stage of maturity

Loss	No Rain	1 in.	1.65 in.	2.5 in.
Leaf loss	7.6%	13.6%	16.6%	17.5%
Leaching and enzyme metabolism	2.0%	6.6%	30.1%	36.9%
Total	9.6%	20.2%	46.6%	54.4%

TABLE 6-13

Effect of feeding moldy hay on various parameters

Parameter	Good Hay	Moldy Hay
Dry matter intake, kg/day		
Hay	7.1	6.5
Total	8.8	8.2
Daily gain, kg	0.73	0.61
Feed-to-gain ratio	12.0	13.4
Total rumen VFA, μ mol/ml	88.0	72.5
Rumen ammonia, mg/dl	23.4	15.5
Digestibility, % dry matter	63.7	53.7
Crude protein	76.9	53.0
Energy	63.1	54.4

Source: Mohanty et al. (25).

If hay is stored outside and exposed to the weather, it may suffer external damage. The extent depends on the weather, type and amount of precipitation, and the physical nature of the hay. In dry regions, hay can be stored outside with little loss, but losses can be substantial in areas of high rainfall.

EFFECT OF MATURITY ON NUTRITIVE VALUE   The maturity of a forage crop harvested for hay production has a dramatic effect on its nutritive value and feeding characteristics. A marked decline in digestibility is observed with increasing maturity of a forage; this was illustrated previously in Tables 6-8 and 6-10. Table 6-14 shows that seasonal yield, TDN, and protein production of alfalfa were greatest when cutting was at one-tenth bloom. This is not the maximal point for digestibility of either TDN or protein, but it is the time at which maximal yield per unit of land is achieved. When determining optimal maturity for harvesting a forage, one must consider not only what effect harvesting has on nutrient yield, but also the effect it will have on animal performance. Forages harvested as hay at im-

mature stages of maturity will increase ruminant animal performance as compared to harvesting at more mature stages. The reason for this reduction in performance is related to changes in feed consumption associated with forage maturity. Nonruminant animals such as horses are able to eat larger quantities of poor-quality forages and pass them through their digestive tract faster to compensate for their lower quality. Unfortunately, in ruminant animals passage rate declines as the quality of the forage declines, because as a forage crop matures its fiber content increases. Thus it takes longer for the rumen microorganisms to digest the forage and increased retention in the rumen occurs, which causes a reduction in feed intake to occur, as shown in Table 6-15. Since digestion of the fibrous portion takes longer in a ruminant, intake declines. Nutritive intake also declines and causes the performance of the animal to decline as the quality of the forage declines. This is especially true in high-performing animals, such as dairy cattle.

Forage that is harvested at an immature stage of development has a high proportion of leaves as compared to stems. The leaves contain higher amounts of proteins and readily available carbohydrates as compared to the stems, which are more available to an animal. The digestibility of leaf constituents is higher than for stems; thus the feeding value of a forage harvested at an immature stage of maturity is higher than for one harvested at a mature stage. As a plant matures the stems become less digestible because the cellulose and lignin contents increase, as shown in Table 6-16.

The optimum time to harvest varies depending on whether you are a hay grower or a livestock producer. When forages are fed in their immature state, animal consumption and performance normally increases; therefore, it is

TABLE 6-14

Calculated yield of TDN and digestible protein from alfalfa cut at four stages of maturity in California

Item	Stage of Maturity			
	Prebud	Bud	One-tenth Bloom	One-half Bloom
Season yield, kg/ha	13,260	16,430	19,130	19,260
TDN, %	66.1	60.4	57.2	54.7
TDN yield, kg/ha	8,765	9,926	10,940	10,533
Crude protein yield, kg/ha	3,542	3,908	3,947	3,652
Crude protein digestibility, %	78.8	74.6	72.8	70.3
Digestible crude protein, kg/ha	2,791	2,916	2,873	2,567

Source: Wier et al. (26).

TABLE 6-15

Effect of neutral detergent fiber (NDF) content of forage on dry matter intake

Forage Quality	% NDF (dry matter basis)	Dry Matter Intake as a Percent of Body Weight
Excellent	38	3.16
	40	3.00
	42	2.86
	44	2.73
	46	2.61
	48	2.50
	50	2.40
	52	2.31
	54	2.22

Source: Van Soest and Mertens (27).

TABLE 6-16

Percent of changes in alfalfa composition at different maturities

Harvest Date	Leaf		Stem	
	Cellulose	Lignin	Cellulose	Lignin
April 22	7.1	2.43	11.0	1.80
28	7.0	2.51	10.2	2.10
May 5	6.9	2.83	15.2	3.76
13	7.1	2.37	16.6	4.73
22	7.1	2.85	22.5	6.77
June 4	7.6	2.82	23.5	8.79

Source: Burritt et al. (28).

beneficial for the livestock producer to purchase forages that have been harvested at an immature stage. Unfortunately, forage producers reduce their yield when they harvest their forage at an immature stage. The longevity of a legume stand, such as alfalfa, is decreased when it is cut continually at an immature stage of development. Therefore, the livestock producer must be willing to compensate the forage producer by paying more per ton for the less mature, higher-quality hay.

Relative feed value (RFV) is one method that has been proposed to compute the relative value of hay, which takes into account differences in consumption and digestibility as effected by maturity. This index provides a means to compare all types of forages. A forage that has an acid detergent fiber (ADF) content of 41% and a neutral detergent fiber (NDF) content of 53% would have an index of 100. The equation used to calculate this is

$$RFV = \frac{\% \ DDM \times \% \ DMI}{1.29}$$

where

$$\% \ DDM = 88.9 - (ADF\% \times 0.779) \quad \text{and}$$
$$\% \ DMI = \frac{120}{\% \ NDF}$$

The RFV has no units associated with it, but it does allow for comparison of similar forages. As the RFV increases above 100, quality increases. Values lower than 100 indicate that quality is lower, so by using this method forages can be compared with one another. Crude protein content does not assess the CP value of a forage. Table 6-17 shows the relationship between maturity and RFV for some typical forages.

ASSESSMENT OF NUTRITIVE VALUE OF HAYS

The nutritive values of hays are difficult to evaluate precisely, but several methods, both visual and analytical, can be used to provide reasonable estimates. These methods are discussed next.

VISUAL METHOD

The visual method can be used to provide some indication with respect to the following characteristics of a forage. The stage of maturity of the hay can be estimated with reasonable accuracy by visual appraisal, especially with alfalfa. The maturity can be estimated with some accuracy by assessing the relative numbers of buds, blossoms, or seed heads that are present. As a plant matures the less digestible carbohydrates increase and the digestibility of

TABLE 6-17

Relative feed values of various forages

Forage	CP	ADF	NDF	RFV
Alfalfa, prebud	23	28	38	164
Alfalfa, bud	20	30	40	152
Alfalfa, mid-bloom	17	35	46	125
Alfalfa, mature	15	41	53	100
Brome, late vegetative	14	35	63	91
Brome, late bloom	8	49	81	58
Bermuda grass, early	12	32	70	85
Bermuda grass, late	8	43	78	66
Fescue, late vegetative	12	36	64	88
Fescue, early bloom	10	39	72	76
Orchard grass, early vegetative	18	31	55	109
Orchard grass, early bloom	15	34	61	95
Wheat straw	4	54	85	51

Source: Holland and Kezar (9).

the nutrients decreases, which reduces the feeding value of the hay.

The available nutrient content of a hay can be partially judged by the leafiness and size of the stems. Since the leaf fraction contains the highest concentration of protein, there is a direct relationship with the CP content of the hay. A relationship also exists between the CP content and the digestibility of the other constituents contained in the hay. The CP content of a hay is highest when it is harvested in an immature stage of development. Also at this time a larger proportion of the nutrients are composed of those associated with the cell contents, which are higher in digestibility. The CP content is positively correlated with the digestibility of the hay. This is why hays that are prepared from immature forages have higher feeding value than those prepared from mature forages.

Quality can also be assessed by the color of the hay. If a hay has a bright color, it is a good indication that it has been cured (dried) properly and provides some indication as to what its carotene content will be. Carotene is the precursor in which plants store vitamin A. There are several reasons that can cause hay to become discolored, which results in a decline in its feeding value. If hay is rained on during the time that it is being dried, it will have a bleached out appearance. The rain leaches out the water-soluble components, such as sugars, proteins, and starches, thus reducing the more readily available components in the hay. When hay is first cut, enzymes present start to consume the more readily digestible components of the forage, but when the moisture level drops to a certain level, the enzymes are inactivated. Hay that is rained on takes longer to dry, which allows the intrinsic enzymes to break down a larger proportion of the more digestible components of the forage, thus reducing the feeding value of the hay. Excess exposure to sunlight can also affect the color. Color can also indicate if a hay has been baled too wet. If hay turns a dark color, this is often because the hay was put up too wet and has heated. Often when excess moisture is present the hay will heat, and this causes the proteins to react with the carbohydrates. This reaction is referred to as Mallard browning. This causes both the protein and carbohydrates involved in this reaction to become unavailable, which reduces the feeding value of the hay.

Odors associated with hay give some indication as to whether hay is moldy, musty, or putrefied. These will all reduce the **palatability** and feeding value of a hay. Odors often occur as a result of improper curing. Normally performed laboratory analysis cannot determine if a hay is moldy. When this type of hay is fed, often animals consume less, and in extreme instances the hay is toxic to the animals.

The presence of foreign materials, such as weeds or toxic plants, can only effectively be detected visually. In some cases, forage crops harvested for hay have large amounts of weeds present. These weeds can dramatically affect the palatability and performance of animals. In certain areas, toxic weeds invade fields that are used to prepare hays. The presence of weeds or toxic plants greatly reduces the value of the hay.

Visual methods can be used to estimate the quality of hay; indicators include leafiness, color, odor, and the presence of foreign materials or such things as mold and mildew. However, in practice hay is rarely sold or graded using these methods, because most hay never goes through any type of central marketing system. Some areas have hay grading stations, but these are limited to the regions that have large livestock numbers or grow large amounts of hay. Hays are quite variable and there has been a substantial effort to arrive at some method for quickly arriving at a quantitative measure of its value, which can be used in pricing. Many states have laboratories that perform hay analysis; these are used by both buyers and sellers to determine price. A voluntary certification program has been designed to assist the hay industry to assure that analyses performed in different laboratories are giving similar results. This program is administered by the National Forage Testing Association and provides assistance to cooperating members in maintaining acceptable laboratory performance. To be certified, a laboratory must correctly analyze three out of four reference samples that are sent to them annually. Laboratories do not have to be certified in order to analyze hay samples, but this certification process helps to standardize the results.

LABORATORY ANALYSIS METHODS

The analyses commonly performed on hay are summarized next. The **Proximate Analysis** procedure has been the industry standard for many years. Neutral Detergent Fiber (NDF) and Acid Detergent Fiber (ADF) determination are often used to assess the carbohydrate components of the hay. Use of the near infrared (NIR) analysis procedure has replaced the proximate analysis procedure in many areas. The NIR procedure can be performed at a hay testing center in a few minutes, so a driver can quickly obtain the results on a lot of hay being transported.

SAMPLING OF HAY   The first step is to collect a representative sample of hay to submit to the laboratory for analysis. Normally, only a couple hundred grams of sample is submitted for 20 to 100 tons or more of hay. Therefore, it is critical that the sample be taken correctly so that it accurately represents the lot of hay. A *lot* of hay is defined as hay taken from the same cutting, similar field, same stage of maturity, and the same species and variety. Hays that are similar can be further subdivided into different lots by the following: hay that contains higher amounts of grass and/or weeds; hay that has been rain damaged; hay quality that has been affected by soil differences within the same field; hay that has been handled differently after cutting or by additional raking or extended curing time; and others.

The use of a core sampling tube or probe is the most acceptable and accurate method for sampling hay. The results obtained from samples collected with a probe have been more accurate than those taken by a hand as a grab sample. Probes must be long enough to penetrate from 12 to 18 in. into small rectangular bales and to the center of large round or rectangular bales. The probe should be equipped with a cutting tip that cuts easily through the hay being sampled. Most accurate results are obtained when 20 samples are taken per lot of hay; fewer samples mean that the results will less accurately assess the lot of hay being sampled. Samples should be stored in a cool place in a sealed container so that the moisture content is retained. Figure 6-9 gives an example of a typical forage evaluation sheet.

PROXIMATE ANALYSIS   The proximate analysis procedure provides information on the following: moisture, crude protein, crude fiber, crude fat, ash, and nitrogen-free extract. This procedure uses wet chemical methods that were developed at the Weede Experiment Station in Germany over 100 years ago.

Moisture   The moisture content of a sample is a measure of the amount of free water associated with a sample. Moisture content is determined by weighing a sample and then heating the sample (50° to 100°C) until it comes to a constant weight and then determining the amount of water loss that occurred. This is important for two reasons: (1) the moisture will dilute the nutrient concentrations, and if more moisture than expected is present, it results in

higher costs per unit of nutrients being purchased; and (2) if the moisture level is excessive (>14% to 15%), the hay is likely to spoil.

Crude protein   The Kjeldahl procedure is used to estimate the amount of protein present in a hay. This is done by determining the nitrogen content of a sample and then, by using a conversion factor based on the typical nitrogen content of proteins found in hays, estimating the protein content. This is referred to as crude protein because it is not a direct measure of the protein.

The quality of forages is positively correlated with their crude protein contents. Hays vary greatly in their crude protein content. Protein is one of the most expensive feed nutrients, so often the price of hay is based on its crude protein content. For forages such as silages, it may be advisable to have additional analyses performed to assess the soluble protein, since the digestibility of the crude protein may be reduced drastically if the silage has been exposed to excessive heating during the ensiling process.

Crude fiber   This provides an estimate of the structural carbohydrates, which provides an estimate of the hemicellulose, cellulose, and lignin. The crude fiber (CF) fraction is composed of the less digestible carbohydrates contained in a hay and has an effect on performance and feed consumption. The CF is negatively correlated with the nutritive value of a hay. When the CF content increases, the nutritive value goes down; so the CF content is often used in conjunction with the CP content in computing the price of a hay. As a forage increases in maturity, its CF content increases; this provides a good estimate of maturity (Table 6-8).

Crude fat   This estimates the quantity of fats and lipids present in a hay. This is determined by extracting all the materials soluble in an organic solvent, such as ether or hexane. Fats are approximately 2.25 times higher in energy than carbohydrates and proteins and are normally very digestible, so this can have a major effect on the energy content of a feedstuff. Hays contain relatively small amounts of fats and the content does not vary, so the CF does not vary much among hays.

Ash   The ash content estimates the inorganic mineral present in a hay. Samples are ashed at high temperatures (500° to 550°C), and all the organic matter (carbon containing) is burned off. When extreme heat is applied, the carbon combines with oxygen to form carbon dioxide, which is volatilized off, leaving the inorganic minerals behind. The ash content does

FORAGE ANALYTICAL SERVICE
Department of Agricultural Chemistry
Oregon State University
Corvallis, Oregon 97331

Oregon State University

TRI-STATE ALFALFA COALITION

TRI-STATE REFERENCE ALFALFA HAY TEST:

- Sample No. ________
- Sampling Date ________
- Sampled by ________
- Cutting No. ________
- County ________

RESULTS SENT TO:

Name ________
Address ________
____
Telephone No. ________

LABORATORY APPRAISAL

Date Received ________ Laboratory No. ________
Percent Dry Matter ________ Percent Moisture ________

	SAMPLE 100% Dry Matter Basis	TRI-STATE REFERENCE 100% Dry Matter Basis
CHEMICAL ANALYSES:		
Crude Protein (CP)	________	18.5%
Acid Detergent Fiber (ADF)	________	32.0%
ESTIMATED VALUES:		
Quality Factor[1]	________	1.00
Reference Factor[2]	________	1.00
TDN[3]	________	58.7%

VISUAL APPRAISAL
(Provided by sampler)

- **COLOR** (Bright-green, light green, yellow-green, brown)
- ________ % **GRASS**
- ________ %**WEEDS**
- **DAMAGE** (Rain, mold, frozen, dirt)
- **AMOUNT** (Tons, bales, loaves)
- **LEAF HOLD** (Attached to stems, whole leaves, not attached, crumbled leaves)
- **PROCESSING** (wire, twine, round, loaves, chopped, cubed, pelleted, green chop, haylage)
- **IRRIGATION METHOD** (Flood, sprinkled, dry land)
- **STORAGE** (Covered, open)
- **SAMPLING METHOD** (Probe, broken bales, chopped pile, windrow, field clip)

[1]**Quality Factor**: The estimated feeding value of the hay as compared to the reference. QF=0.9199+0.0136 (CP)-0.0054 (ADF)

[2]**Reference Factor**: The estimated feeding value compared to the reference and adjusted for percent moisture. RF= (QF)(DM) 0.0114

[3]**TDN**: Total digestible nutrients estimated from CP and ADF and adjusted for 3X maintenance. TDN = 54.3208 + 0.7387 (CP) - 0.2915 (ADF)

OS J 3866

FIGURE 6-9 Example of forage evaluation sheet.

not provide any indication as to what specific minerals are present. Additional analyses must be performed to determine the specific mineral content of a hay sample. High ash values provide an indication if a hay is contaminated with high amounts of soil.

Nitrogen-free extract The nitrogen-free extract (NFE) provides an estimate of the soluble components contained in a hay; this content is not actually determined using chemical procedures, but rather calculated using the following equation:

$$\% \text{ NFE} = 100 - (\% \text{ CP} + \% \text{ CF} + \% \text{ C fat} + \% \text{ ash})$$

Thus one simply adds up all the other components that are determined using the proximate analysis method and subtracts the total from 100. If any of the analytical results are incorrect, then the NFE estimate will also be incorrect.

NEUTRAL DETERGENT AND ACID DETERGENT FIBER The CF method has had some problems in estimating performance in animals for feeds that contain high amounts of hemicellulose, because some of the hemicellulose is present in both the NFE and CF fractions. The digestibility of various components found in these fractions is quite variable; hemicellulose is 60% to 70% digestible, as compared to cellulose, which is 40% to 50%; lignin, which is 0% to 10%, and sugars, starches, and soluble proteins, which are 95% or more. Figure 6-10 shows which of the components are contained in the different fractions. There was a need for a more precise method to be developed. Van

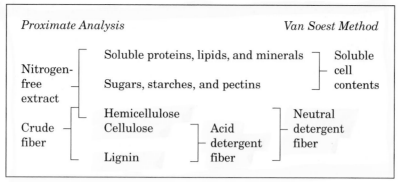

FIGURE 6-10 Assessment of components associated with feeds.

Soest and Goering developed the neutral and acid detergent fiber method to more accurately predict the digestibility of the components associated with feedstuffs. Neutral detergent fiber (NDF) provides an accurate estimate of cell wall components (hemicellulose, cellulose, lignin) (Figure 6-10). Acid detergent fiber (ADF) provides an estimate of the cellulose and lignin. The NDF and ADF method can be used to accurately predict the amount of hemicellulose present, thus providing a more accurate estimate of digestibility. For this reason, the NDF and ADF are more commonly used today to estimate the feeding value of hays.

NEAR INFRARED ANALYSIS Near infrared analysis (NIR) is a relatively new procedure that has been adopted by the feed industry to assess the nutritive content of feeds. Near infrared reflectance spectrophotometry is a rapid and low cost computerized method to analyze forages and other feedstuffs to determine their nutrient contents. Instead of using chemicals, as in conventional methods (proximate and neutral and acid detergent fiber), to determine protein, fiber, and minerals contents, NIR uses near infrared light. This procedure is very fast and no chemicals are required which makes it more compatible with the environment.

Near infrared analysis involves the drying and grinding of a sample, which is then exposed to an infrared light source in a spectrophotometer. The reflected infrared radiation is converted to electrical energy and measured and then fed into a computer for interpretation. Each major organic component of a feed will absorb and reflect near infrared light differently. By measuring these differences in reflected light, the NIR unit coupled with a computer can compute the quantity of each component in a feed. This procedure could be compared to that of a human's ability to visually distinguish color when light strikes a material that absorbs some wavelengths and reflects others. Reflected wavelengths are detected by the eye, and signals are sent to the brain to identify the color. The detection of specific nutrients is possible because reflectance spectra from forage samples of established nutrients values (wet chemistry methods) are programmed into the computer. When a similar feed sample is evaluated by NIR, the computer compares the wavelength reflections caused by the sample and matches them to the sample previously tested, thus determining its nutrient content.

DIFFERENT FORMS OF HIGH MOISTURE FORAGES

Greenchop (**soilage**) is herbage that has been cut and chopped in the field and then fed fresh to livestock, normally in confinement (Fig. 6-11). Crops used in this manner include the forage grasses, legumes, Sudan grass, corn, and forage sorghum, among others.

FIGURE 6-11 Greenchop harvesting equipment.

A major advantage of feeding forage crops as greenchop is that one can maximize digestible nutrient yield per unit of land with this method, as compared to pasturing, haymaking, or ensiling. This method has the advantage of using excess herbage growth for hay or silage before it becomes to mature for efficient use. In addition, weather is less of a factor than in haymaking. A big disadvantage is that feeding of greenchop (as opposed to grazing) requires daily harvesting and constant attention to changes in moisture content.

Currently, many successful large dairy operations and some feedlots utilize greenchop from different crops in their summer feeding programs. The greenchop is gradually phased in as it becomes available and then gradually phased out of the ration when it is no longer in adequate supply. This feeding method has worked out very successfully in many livestock operations and available machinery effectively harvests fresh cut or wilted herbage (Fig. 6-11).

Data, in general, indicate that beef steers gain essentially the same with greenchop as when pasture is intensively managed, such as short-term rotation or strip grazing. Note in Table 6-18 that beef production per hectare was considerably greater when fed as greenchop compared to pasture, and the gains were similar. Research with lactating dairy cows (Table 6-19) indicates slightly less milk production from alfalfa greenchop compared to pasturing. The data also show that greenchop had, on average, a higher content of crude protein and less fiber than pasture, possibly an indication of more intensive use of pasture than would result in maximal production because of reduced selectivity. Other research data show that the high level of crude protein in greenchop was used by cows more efficiently than that in hay or silage.

Silage has been used extensively for feeding livestock, primarily ruminants, for many years in Europe and in many other countries during the past 60 to 70 years. Development of improved machinery and ensiling systems has resulted in a large increase in ensiling of grasses, legumes, and other crops, primarily corn and sorghum, in the United States. Many different types of high-moisture crops and crop residues can be satisfactorily ensiled.

Good silage can most effectively maximize the utilization of the feed resources that can be produced on a unit of land. It is very palatable and little wastage occurs when it is fed. Excellent results can be obtained when fed to beef and dairy cattle. Ensiling is an excellent means of preserving high-moisture crops during periods when drying is not feasible (grasses or legumes) or for crops that deteriorate in quality if allowed to dry (corn or sorghum).

Ensiling does not result in improved nutritive value of the material being ensiled, because losses occur during the ensiling processes, which will be discussed in subsequent sections. However, the fermentation that occurs usually greatly reduces the nitrate content, if nitrate is present, and other toxic compounds, such as prussic acid, are reduced in amount. The bloat associated with legumes is less of a problem when they are ensiled. The biggest disadvantage of silage is that it is a bulky, high-moisture material and is not suitable for intermittent feeding. Therefore, for economical reasons, silage needs to be fed in very close proximity to where it is ensiled.

TYPES OF ENSILING FACILITIES   Silage is the material that is produced as the result of a controlled **anaerobic** fermentation of material high in moisture. Microbial fermentation produces organic acids, which subsequently inhibit the growth of microbes, and the preserved product is referred to as silage. The fermentation and subsequent storage must be in an

TABLE 6-18

Performance of beef steers given alfalfa herbage in different forms

Parameter	Pasture Rotation Grazing	Strip Grazing	Hay	Greenchop
Daily gain, kg	0.79	0.70	0.61	0.74
Beef production per hectare, kg	564.00	564.00	638.00	821.00
Daily feed consumed				
Alfalfa, kg	4.63	4.08	8.80	6.30
Oat hay, kg	2.17	2.27	—	1.72
Feed-to-gain ratio	8.6	9.1	14.5	10.8

Source: Meyers et al. (29).

TABLE 6-19

Production of lactating Holstein cows during a nine-week trial when fed alfalfa pasture, greenchop or hay

Parameter	Pasture	Greenchop	Hay
Milk, kg/day	24.5	23.3	22.1
Milk fat, %	2.9	3.2	3.3
Solids not fat, %	8.2	8.3	8.2
Dry matter consumed			
Forage, kg	13.6	13.6	14.7
Concentrate, kg	5.9	5.4	6.3
Composition of herbage			
Crude protein, %	18.2	21.0	22.7
Crude fiber, %	33.8	30.5	29.9

Source: Stiles et al. (30).

oxygen-limiting environment (anaerobic); otherwise, the material putrefies to an inedible and frequently toxic mess.

The oxygen supply is limited by the use of containers call silos, which can either be vertical (upright) (see Fig. 6-12) or horizontal (bunker or trench) (see Fig. 6-13). The open silos are normally covered with plastic or other type of material to prevent exposure to oxygen, which causes losses to occur. Large plastic bags are now being used, which have the advantage of lower cost, and they can be located where desired (see Fig. 6-14).

SILAGE MAKING The preparation of silage requires the following steps: harvesting, packing, and covering. The process is very labor and equipment intense. Often when silage making is started, the harvesting and filling of

FIGURE 6-13 Silage being loaded out of a large bunker silo. Silos of this type allow a high degree of mechanization in filling and unloading. (Courtesy of *Beef* magazine.)

FIGURE 6-12 Examples of oxygen-limiting upright (tower) silos often used for storing low-moisture silage. (Courtesy of *Beef* magazine.)

FIGURE 6-14 A plastic bag silo. This type, although more difficult to fill, has the advantage of flexibility in location.

the silo continue around the clock until all the material to be ensiled has been harvested. When silage has been properly prepared, it can be stored for an indefinite period of time without deteriorating.

The crop to be ensiled must be harvested at the proper stage of maturity in order to optimize nutrient yield per unit of land. It is also critical to have the moisture and fermentable carbohydrate contents at levels that are suitable for ensiling. The material being ensiled must provide the proper substrate for the microbes so that they will be able to produce adequate amounts of the organic acids that are essential to preserve the ensiled material. Materials that ensile best are relatively high in soluble carbohydrates, which can be used by the microorganisms to produce the organic acids. If the crop being ensiled has marginal levels of soluble carbohydrates, sources of soluble carbohydrates can be added, such as molasses or cereal grains, at the time of ensiling. The optimum harvesting maturity for various types of materials commonly ensiled is summarized in Table 6-20. The maturities listed are those that will optimize digestible nutrient yields of these crops. A number of these crops contain higher than the desired moisture level at the time they are harvested; therefore, it is necessary for them to be wilted (dried to remove a portion of their moisture). Crops requiring wilting are cut and allowed to dry while lying in the field prior to being chopped and ensiled. The moisture level needs to be reduced so that there is not excessive runoff.

The type of silo determines what the moisture level will need to be. If the material is placed in an airtight upright or a bag, then the moisture level can be lower (50% to 60%). If it is placed in a bunker or trench silo, the moisture content should be higher (65% to 75%). When ensiled materials contain higher moisture levels, they are easier to pack and therefore the oxygen is easier to expel, as compared to drier materials, and they can be ensiled in semi-airtight silos.

At the time of harvesting or after wilting, the material to be ensiled is chopped. This is done to allow for the material to be packed and the oxygen expelled more readily, so that an anaerobic environment can be established. Table 6-20 lists the optimal chop size for the various types of materials that are commonly ensiled. The chopping process requires additional mechanical power and thus slows output, so chopping finer than needed is costly both in time and fuel usage.

Packing is critical to good silage making. This is one of the primary ways in which the oxygen is eliminated from the silo. In trench or bunker silos, tractors are often driven over the stack to compact it. If the material being ensiled has been chopped to the proper size, this works very well. If the material has not been chopped well or if it is to dry, air pockets will occur in the silage, which will result in moldy spots.

It is critical to maintain an anaerobic environment. Ensiled materials deteriorate rapidly when exposed to oxygen. This is why a silo must be sealed and maintained in an anaerobic state. This is easy to do in an upright airtight silo, but more difficult to do in a bunker or trench silo. Both bunker and trench silos must be covered to prevent the exposed surface from deteriorating. This is done with plastic or other material that oxygen cannot penetrate.

ENSILING PROCESS Microorganisms start to metabolize the soluble carbohydrates. At first aerobic organisms start to grow, and then these are replaced by anaerobic organisms. Aerobic fermentation converts the organic (carbon-containing) materials to carbon dioxide and water, whereas anaerobic fermentation produces organic acids (butyric, acetic, and lactic acids) that increase the acidity to the point

TABLE 6-20

Harvest maturity for optimum ensiling characteristics

Crop	Maturity	Moisture (%)	Length of Cut (in.)
Alfalfa	Mid-bud to one-tenth bloom	50–70 wilt	1/4–3/8
Corn silage	One-half to two-thirds milk line	50–75 wilt	3/8–1/2
Sorghum	Medium to hard dough stage	50–77 wilt	3/8–1/2
Other cereal grains	Milk or soft dough	50–75 wilt	1/4–1/2
Clover	One-fourth to one-half bloom	50–72 wilt	1/4–1/2
Grasses	When first stem head appears	50–72	1/4–1/2

Source: Holland and Kezar (9).

that the microbes are killed off and the acidity preserves the ensiled material. The pH is reduced to 4 or lower, at which point acidity inhibits further microbial growth. The primary organic acid produced is lactic acid, which reaches concentrations of 4% to 8% in most cases, with less amounts of acetic and other acids such as formic, propionic, and butyric. Little butyric acid is present in well-preserved silages. The initial level of soluble carbohydrates is related to the amount of lactic acid that will ultimately be produced. Each 1% of soluble carbohydrate present in the material being ensiled, when fermented, yields approximately 0.3% of lactic acid.

From the point of view of energy conservation, lactic acid is the preferred acid to be produced, since only 3% of the glucose energy is lost in the conversion to lactic acid, compared to 22% if glucose is converted to butyric acid. In addition, at the high levels required for preservation, lactic acid is very palatable, especially as compared to butyric acid.

The action of the plant enzymes and the bacteria give rise to heat. Optimum temperatures during fermentation are said to be between 27° and 38°C (80° and 100°F). Excessive heat is primarily a problem in low-moisture-containing silages that do not compact well so that oxygen is excluded. Undesirable organisms, such as thermophilic (heat-loving) bacteria and some molds, grow in this condition. Excessive heating results in a reduction in the digestibility of the silages, particularly of the N-containing compounds.

After the plant material is harvested, there is a rapid and extensive hydrolysis of the protein, which is stopped by a low pH or a high dry-matter content. Even in well-preserved silage, the NPN content may be 40% or more of total N. Initially, a high percentage of NPN is present as amino acids, but continual breakdown converts them to such compounds as ammonia, amines, and other NPN compounds. While it has been shown that the water-soluble N found in silage is highly digestible, it does not seem to stimulate the digestion of cellulose in simulated rumens (in vitro) as readily as does urea, and there is some evidence that it is utilized poorly by animals.

Silage making involves an anaerobic fermentation of wet materials. So in addition to the detrimental effect of oxygen, the dry-matter content, pH, and availability of soluble carbohydrates are critical factors. If the silage material is very wet and if soluble carbohydrates are not present in sufficient supply or the plant materials are highly buffered (especially legumes), complications may arise and the material will not ensile properly. Any of the above conditions can result in the production of poor-quality silages.

For most crops a dry matter content of 25% to 35% is near optimal for silage making. High moisture content encourages the proliferation of clostridial bacteria, the production of relatively large amounts of butyric acid, and further fermentation of NPN compounds, resulting in the production of amines such as tryptamine and histamine. These amines, which may be toxic, result in a foul-smelling silage. This is particularly a problem with legumes, which have low soluble carbohydrate levels and are more highly buffered than grasses. Because of their greater buffering capacity, it takes more acid to lower the pH to a stable level.

Grasses can be ensiled successfully at rather low dry matter contents provided the soluble carbohydrate level is higher than 15% of the dry matter. Herbage with less that 10% soluble carbohydrate normally will not have enough fermentation capacity to ensile very well. In most areas where legumes or grass–legume mixtures are ensiled, it is a common practice to wilt the herbage before it is ensiled. Wilting raises the dry matter content and, as a result, less acid production is required for adequate preservation of the silage. A guide for estimating the dry matter content of green herbage is given in Table 6-21.

LOSSES ASSOCIATED WITH SILAGE MAKING Some losses are associated with harvesting and handling in the field, but since the materials being ensiled are normally fairly high in moisture, these losses are minimized. If the material is left in the field for extended periods of time prior to ensiling, plant enzymes and microbial fermentation can dramatically reduce the soluble nutrient content, especially

TABLE 6-21

Estimating forage moisture content

Condition of Forage Ball	Approximate Moisture Content, %
Holds shape, considerable juice	Over 75
Holds shape, very little juice	70–75
Falls apart slowly, no free juice	60–70
Falls apart rapidly	Below 60

Source: Shepherd et al. (31).

the carbohydrates. Another major loss can be surface and perimeter spoilage. This can amount to as much as 15% to 20% of the dry matter. This can be dramatically reduced by covering the surface with plastic. **Seepage** or runoff is another type of loss that can occur. This loss is quite variable, but is to some extent correlated with the moisture level of the material being ensiled. Seepage losses not only contain water, but also other soluble nutrients. If the moisture level is excessive, the amount of seepage loss increases; therefore, one way to reduce seepage losses is to ensile material at the correct moisture level. Moisture levels above 75% will increase seepage losses. Gaseous losses occur as a result of plant enzyme action and bacterial fermentation, both of which increase with prolongation of the ensiling process. To minimize gaseous losses, the crop must be harvested and ensiled as rapidly as possible.

Overall losses are quite variable. When field losses are included, total losses may be expected to be about 15% to 25% of the dry matter present in the field. This is illustrated in Fig. 6-15 and Table 6-22.

EVALUATION OF SILAGE QUALITY
No single test is available to evaluate silage quality accurately. For those who are thoroughly familiar with silages, smell and visual appearance are relatively good methods, but they are not very quantitative. Good silage should be free from moldy and musty smells and other objectionable odors, such as ammonia, butyric acid, and, especially in low-moisture silage, caramelized or tobacco odors. It should be green in color (forage silage) and not brown or black, and it should have a firm texture with no sliminess.

The pH of silage is used more commonly than any other chemical test, but by itself it is unreliable, because the optimal pH for forage silage depends on the dry-matter content. Generally, a good silage should have a pH of 4.2 or less, less than 0.1% butyric acid, and less than 11% volatile N as a percentage of total N.

NUTRITIVE PROPERTIES OF SILAGES

One problem associated with feeding high amounts of silage is that consumption of dry matter is always lower than when the same crop is fed as hay, and this seems to apply whether the crop in question is legume, grass, or other herbage. Animal consumption of silage made from grass or legumes is usually greater as the dry matter content increases, as illustrated in Table 6-23, but not necessarily so for corn or sorghum silages. In one study it was shown that silage consumption was positively related to digestibility for legumes, but negatively for grasses other than ryegrass (33). This information suggests that maximum intake should be achieved with silage containing just enough moisture to allow for preservation with minimum production of ammonia and acetic acid, which tend to increase at higher moisture contents.

GRASS AND GRASS–LEGUME SILAGE

If the herbage is harvested at an optimal stage of growth, grass or legume silages are normally high in crude protein (15% to 20%) and carotene, but only moderate N digestible energy. This combination may be too high in protein for efficient utilization for most ruminants, and more optimal results may be achieved by supplementing grass silage with some form of energy (grain) or by diluting the protein by feeding some other low-protein roughage. This is illustrated by the data presented in Table 6-24. The grass was harvested at an early head stage past the peak in protein content, probably resulting in some reduction in consumption and liveweight gain as compared to the grass–legume silage. The corn silage had urea added, which should have provided a more optimal protein-to-energy ratio, and the corn silage did result in greater and more efficient gains.

The most common procedure of harvesting grass or grass–legume silages is to cut the

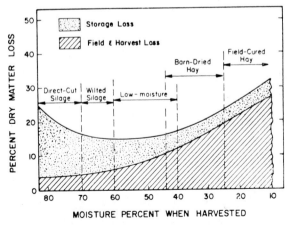

FIGURE 6-15 Estimated total field, harvest, and storage loss when legume–grass forages are harvested at varying moisture levels and by alternative harvesting methods. From Hoglund (43).

TABLE 6-22

A summary of harvesting methods for alfalfa

Method	Approximate Moisture Content (%)	Approximate Harvest Loss[a] (%)	Approximate Storage Loss[b] (%)	Total Loss (%)
Greenchop	75–80	3		3
Direct cut silage	75–80	3	18	21
Wilted silage	65–70	6	10	16
Haylage	50–60	9	4	13
Baled hay	85–90	19	8	27

Source: Fiez (32).

[a]Harvest loss is about average; management, weather, etc. influence this loss.

[b]About average loss; storage facilities, surface covers, etc., can reduce this loss.

TABLE 6-23

Effect of dry matter of alfalfa silage on intake and utilization by sheep

Parameter	Dry Matter of Silage, %			
	22	40	45	80
Silage pH	4.48	4.49	4.52	5.87
Acid content, % of DM				
Lactic	5.06	4.11	3.19	0.19
Acetic	4.91	3.21	2.88	0.62
Total	10.46	7.53	6.26	1.00
Dry matter intake, g/kg body wt$^{0.75}$	49.1	57.3	58.8	63.2
Dry matter digestibility, %	58.2	60.1	60.0	61.1
N intake, g/day	14.9	16.2	18.3	19.6
N retention, g/day	−0.9	1.0	2.0	2.9
N digestibility, %	70.5	72.2	72.0	70.0
Water intake, g/g DM	3.5	2.4	2.4	2.2

Source: Hawkins et al. (34).

TABLE 6-24

Comparison of silages made from grass, grass–legume, and corn when fed to beef steers for 126 days

Parameter	Silage Source		
	Grass	Grass–Legume	Corn with Urea
Forage DM, %	25.8	18.4	24.1
Silage DM, %	27.4	24.4	27.5
Crude protein, % of DM	10.9	15.5	13.1
DM lost as seepage, spoilage, respiration etc., %	18.4	22.6	5.1
DM digestibility, %	52.2	56.2	69.2
Silage consumption, kg/day	20.3	26.2	23.6
Silage DM consumption, kg/day	5.6	6.4	6.5
Gain, kg/day	0.56	0.89	1.05
Silage DM/kg gain, kg	10.0	7.2	6.2

Source: Calder et al. (35).

forage and let it wilt in the field so that the moisture content is reduced prior to ensiling. Excessive moisture would result in high seepage losses and detrimental effects on fermentation. A comparison of the feeding characteristics of fresh (frozen) grass, dried grass, direct cut (unwilted), and wilted silages is shown in Table 6-25. The drying or ensiling had little effect on digestibility, although there was a tendency for the unwilted silage to be more digestible than

TABLE 6-25

Composition and digestibility of green ryegrass and material from the same field that was dried artificially or made into wilted or unwilted silage

Parameters Measured	Fresh Grass	Dried Grass	Wilted Silage	Unwilted Silage
Composition of dry matter				
Organic matter, %	90.8	92.0	92.2	91.7
Total water-soluble carbohydrates, %	9.2	8.4	Trace	Trace
Cellulose, %	24.2	24.3	25.0	26.8
Hemicellulose, %	14.0	13.3	12.9	13.1
Crude protein, %	17.8	18.7	19.3	19.3
Gross energy, kcal/kg	4.59	4.55	4.46	4.89
Digestibility, %				
Energy	67.4	68.1	67.5	72.0
Cellulose	75.2	75.5	76.5	80.6
Hemicellulose	59.4	57.7	59.9	63.2
Crude protein	75.2	71.0	76.5	76.4

Source: Beever et al. (37).

In this experiment the fresh grass was quick-frozen so that it could be fed at the same time as the other forms.

the other preparations. More recent research has shown that the organic matter in a complete dairy diet was more digestible than when the forage was provided as silage, but there was no difference in the availability of protein between silages and hay rations (36).

A great deal of research effort has been put into improving ensiling methods for grass silage, especially in Europe, where corn and sorghum silages are less common than in North America. One development has been the use of formaldehyde or formic acid or combinations of similar chemicals. The effect of adding formic acid to grass silage is shown in Table 6-26. Note that ensiling resulted in a reduction in protein

N and marked increase in nonprotein N. Ensiling also resulted in a reduction in the consumption of dry matter, which was reversed by adding the formic acid. The formic acid also preserved some of the water-soluble carbohydrates, presumably because of a selective inhibition of microorganisms fermenting the silage (selective because there was no effect on the proteins). This has been an effective method of improving grass and grass–legume silages, provided that the addition of formic acid was not excessive.

Inoculation of forage crops being ensiled with microorganisms that produce lactic acid has become a common practice. This causes the fermentation to start more rapidly and the pH

TABLE 6-26

Comparison of silage made from fresh-cut grass, wilted grass, and wilted grass treated with formic acid

Parameter	Fresh Grass	Silage Fresh Grass	Silage Wilted Grass	Silage Wilted Grass with Formic Acid
Composition data, DM basis				
Crude protein, %	14.2	14.4	14.2	15.1
Total N, g/kg DM	22.7	23.0	22.8	24.2
Protein N, g/kg DM	14.7	5.4	6.6	7.6
Water-soluble N, g/kg DM	8.0	17.6	16.2	16.6
Volatile N, g/kg DM	0.9	1.8	1.8	1.6
Water-soluble carbohydrate, %	14.0	1.0	4.7	15.1
Digestibility, %				
Organic matter	79.7	80.9	76.8	78.8
Crude protein	75.2	78.2	72.3	78.4
Consumption, g/kg body weight	11.2	8.5	9.7	12.3

Source: Donaldson and Edwards (38).

of the ensiled material will drop quicker. This reduces the amounts of soluble carbohydrates that need to be present in order to ensile the forage crop.

Low-moisture silage (less than 50% moisture), often called haylage, is a very palatable feed, probably because it tends to have relatively less acetic acid and less N in the form of ammonia and other NPN compounds (see Tables 6-26 and 6-27). It is used to a great extent in the United States for feeding dairy cattle. This type of silage is best prepared in oxygen-limiting silos now in common use in many areas (Fig. 6-14), although other types, such as plastic silage bags, can be used successfully. Table 6-27 shows that low-moisture silage was not consumed as readily as either hay or high-moisture silage, and the digestibility of dry matter and N were lower than for hay. However, it would be expected that losses associated with seepage and plant enzyme action would be reduced for low-moisture silage.

If not prepared properly, low-moisture silage is more susceptible to heating, resulting in brown silage with a lower crude protein digestibility. Fires may also be a problem if heating is excessive.

Experimental results generally show that haylage feeding results in milk production slightly greater than when high-moisture silage is fed. Haylage is usually comparable to hay. Research has shown that maximum air exclusion (vacuum compression) resulted in a silage having the lowest pH and butyric acid and the highest lactic acid content. However, when the silage was fed as the sole diet to lactating dairy cows, intake was lower and butterfat was less than in silage having less air removed (33). Data in a number of the tables on silage are not complete enough to make evaluations of the methods to use, because information is frequently not given on the amount of digestible nutrients recovered after ensiling. Thus, even though one silage might be more digestible or have a higher crude protein content than another, if greater losses occur during harvesting or ensiling, the method may not be the preferred one.

CORN SILAGE

Corn silage is by far the most popular silage in the United States in areas where the corn plant grows well and where adequate water is available during the growing season. It is becoming more popular in other areas of the world because, except for sugar cane and cassava, the maximum yield of digestible nutrients per unit

TABLE 6-27

Composition and nutritive value of third cutting of alfalfa harvested as hay, wilted forage, greenchop, and low-moisture and high-moisture silages

Parameter	Hay	Wilted Forage	Greenchop	Low-moisture Silage	High-moisture Silage
Dry matter, %	92.6	58.1	28.8	59.0	28.1
Others on DM basis					
Crude protein, %	20.6	18.1	19.8	20.4	21.6
Cellulose, %	34.0	33.3	30.3	37.4	37.9
Soluble carbohydrates, %	7.8	5.2	3.8	2.5	3.7
Total N, %	3.3	2.9	3.2	3.3	3.4
Soluble N, % of total N	31.8	37.2	32.6	51.1	67.0
NPN, % of total N	26.0	28.4	22.6	44.6	62.0
Volatile fatty acids, %					
Acetic				0.4	4.2
Propionic				0.05	0.14
Butyric				0.002	0.11
pH				4.7	4.7
Dry matter intake, g/day/kg body weight	25.3			21.8	25.8
Digestibility, %					
Dry matter	64.5			59.0	59.5
Nitrogen	77.5			63.0	72.8
Celluose	58.8			60.4	64.3

Source: Sutton and Vetter (39).

Wilted herbage was used to make low-moisture silage, and greenchop was used to prepare high-moisture silage.

of land can be harvested from this crop. In addition, the corn plant can also be handled mechanically at a convenient time of the year and over a period of time.

Well-made corn silage is a very palatable product with a moderate to high content of digestible energy, but it is usually low to moderate in digestible protein, particularly for the amount of energy contained. On a dry basis, corn silage usually has 8% to 9% crude protein, 65% to 75% TDN, 0.33% Ca, and 0.2% P. Silage made from a well-eared stand may contain as much as 50% grain, particularly in silage made from mature plants, although average values are more on the order of 47% grain. High-yielding grain varieties of corn generally produce maximal yields of digestible nutrients. Even so, maximum growth rates or milk yields cannot be obtained from cattle without energy and protein supplementation.

Ensiling does not usually improve the palatability of herbage. Table 6-28 illustrates what happens when corn silage is acidified fresh or ensiled. Consumption of ensiled herbage is reduced (Table 6-28), as was shown when acids were added to frozen corn, indicating that acids, by themselves, have an effect on palatability. Likewise, ensiling results in a reduction of liveweight gain in this experiment.

In another research trial, corn with low or high dry matter was frozen or ensiled (Table 6-29); ensiling resulted in a reduction in consumption of low-dry-matter forage, but had no effect on consumption of high-dry-matter forage. In this case there was no depression in digestibility by ensiling the high-dry-matter corn.

Numerous studies have been conducted to determine the optimum stage of maturity at which corn should be harvested for silage. Table 6-30 gives the results of corn that was harvested for silage at the milk or dough stage or after one, two, or five frosts. The resultant silages were then fed to lactating dairy cows at a level equal to 70% of ration DM intake. Based on the data presented, the authors concluded that harvest after the second frost was optimal in this study. Dry matter was high enough to reduce seepage and runoff to minimal amounts; dry matter intake reached maximal levels at this date and milk production was as high as for any of the other silage-based rations. It did not result in maximal digestibilities of those items listed in the table, but that is not an uncommon finding. In general, other studies suggest that the optimum time to harvest corn for silage is when starch deposition in kernels is nearly completed. This can be determined best by the milk line location. The milk line is between the

TABLE 6-28

Effect of ensiling and adding acids on consumption of frozen corn or corn silage by heifers

	Frozen Corn		Ensiled Corn	
Parameter	+Sucrose	+Acids	+Sucrose	+Acids
Dry matter consumed per day, % of body weight	2.61	2.31	2.46	2.23
Gain per day	878	775	576	556

Source: Wilkinson et al. (40).

Sucrose was added to provide energy equivalent to the acids. Lactic and acetic acids were added at a rate of 6% and 2% of silage dry matter. Urea and a mineral–vitamin mixture were added to all diets at 2% and 5% of forage dry matter.

TABLE 6-29

Effect of ensiling on composition and consumption of corn

	Frozen		Ensiled Corn	
Parameter	Low DM	High DM	Low DM	High DM
Dry matter, %	26.3	38.9	27.4	38.1
pH	4.6	5.6	3.7	4.0
Lactic acid, % of DM	0.9	0.7	8.7	6.4
Total N, % of DM	1.3	1.4	1.3	1.4
DM intake, g/kg of body weight$^{0.75}$	88.2	72.1	84.2	72.0
DM digestibility, %	[a]	72.9	75.0	72.5
CP digestibility, %	[a]	68.4	69.8	67.8

Source: Phillip and Buchannan-Smith (41).

[a]Insufficient material to determine digestibility.

TABLE 6-30

Effect of stage of maturity of corn silage on composition and performance of lactating dairy cows

Item		Prefrost		No. of Frosts at Harvest		
Maturity: Harvest date:		Milk 8/30	Dough 9/7	First 9/18	Second 9/26	Fifth 10/17
Dry matter, %						
At ensiling		20.8	23.2	25.1	35.5	45.9
At feeding		23.2	25.7	28.5	34.1	45.0
Neutral detergent fiber, % of DM		59.0	58.8	59.4	61.6	65.9
Acid detergent fiber, % of DM		31.5	28.5	26.0	26.2	28.1
Acid detergent lignin, % of DM		2.92	2.65	2.68	2.57	3.48
DM intake, kg/day		14.5	14.3	15.4	16.5	14.9
4 % Fat-corrected milk, kg/day		19.6	18.7	18.7	19.7	17.6
Apparent digestibility, %						
Dry matter		64.7	64.7	64.3	63.7	61.2
Gross energy		64.9	64.9	63.7	62.6	60.6
Crude protein		52.7	49.1	53.3	52.5	48.4
Digestible energy, Mcal/kg		2.76	2.81	2.72	2.66	2.59
Net energy of lactation, Mcal/kg		1.55	1.59	1.52	1.46	1.41

Source: St. Pierre et al. (42).

Silage was fed to equal 70% of ration DM.

solid and liquid portion of the kernel (Fig. 6-16). The milk line will not appear until the corn is at the dent stage of maturity. Some hybrids do not show the milk line as readily as others; in these cases the kernel may have to be cut to determine its location. The milk line can also be determined by biting the kernel, starting at the bottom surface. At this stage (30% to 38% DM) the milk line is characterized by the appearance of a small black layer where the kernel attaches to the cob. Comparative values of corn silage at different dry-matter contents are

shown in Table 6-31, and Table 6-32 shows the equivalent dry matter and wet weights per acre for different grain yields.

As corn silage matures, the milk line moves down the kernel, and plant composition and energy values change. Table 6-33 shows how plant composition and energy values varied when harvested at three different maturities.

EFFECT OF CHOP SIZE Fine chopping of corn reduces the number of whole-grain kernels excreted in feces, but it is apt to result in a

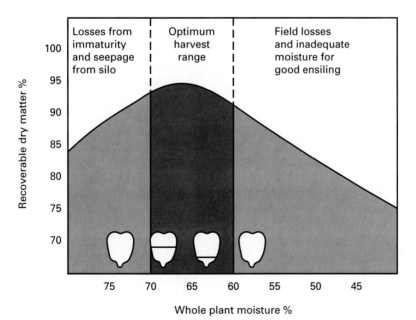

FIGURE 6-16 Kernel milk line (maturity) showing expected yields.

TABLE 6-31

Value of corn silage with different dry-matter content and different prices for bushel weight of corn

Price of No. 2 Corn, $/bu	Silage Value, $/ton			
	30% DM	35% DM	40% DM	45% DM
1.00	11.54	13.46	15.38	17.30
1.50	14.50	16.92	19.33	21.25
2.00	17.46	20.37	23.28	26.19
2.50	20.42	23.82	27.23	30.63
3.00	23.39	27.28	31.18	35.08
3.50	26.35	30.74	35.13	39.52
4.00	29.31	34.19	39.08	43.96

Source: Zimmerman (44).

Calculated on the basis that silage has 47% grain, with grain valued as indicated, and the nongrain DM valued at $30/ton, air dry basis.

TABLE 6-32

Silage yield compared to grain yields of corn with amounts of silage at different DM percentages

Corn Yield, bu/acre	Corn DM, lb/acre	Silage DM, lb/acre	Silage Yield, ton/acre			
			30% DM	35% DM	40% DM	45% DM
70	3,332	7,089	11.8	10.1	8.9	7.9
80	3,808	8,102	13.5	11.6	10.1	9.0
90	4,284	9,115	15.2	13.0	11.4	10.1
100	4,760	10,128	16.9	14.5	12.7	11.3
110	5,236	11,140	18.9	15.9	13.9	12.4
120	5,712	12,153	20.3	17.4	15.2	13.5
130	6,188	13,166	21.9	18.8	16.5	14.6
140	6,664	14,179	23.6	20.3	17.7	15.8
150	7,140	15,191	25.3	21.7	19.0	16.9

Source: Zimmerman (44).

TABLE 6-33

Plant composition and energy yields of whole plant corn silage harvested at three different maturities

Maturity	Dry Matter Basis				TDN (tons/acre)
	Grain (%)	Stover (%)	Sugar (%)	Starch (%)	
One-third milk line	32.4	59.1	9.8	22.2	7.2
Two-thirds milk line	41.8	50.2	7.1	28.4	7.8
Black line	46.1	45.8	6.6	31.0	7.7

Source: Holland and Kezar (9).

reduction in butterfat content in lactating dairy cows fed high levels of silage. Cows are also more prone to having displaced abomasums when they are fed finely chopped silage. There appears to be little effect on liveweight gain of cattle fed finely chopped corn silage.

SORGHUM SILAGE

Many different types of sorghum crops are grown, ranging from short-stalked crops such as milo to tall forage sorghums that contain little grain. Sorghum is very tolerant of water shortages and can be grown in areas where it is not feasible to grow corn. It seems likely that there will be an increase in sorghum usage in drier areas where irrigation water is becoming more expensive.

Sorghum plants contain less grain when harvested at similar maturities as corn, generally less than 40% grain for the harvested plant

TABLE 6-34

Effect of stage of maturity of forage sorghum silage on its feeding value

	Stage of Maturity					
	Early Bloom	Bloom	Milk	Late Milk to Early Dough	Dough	Hard Dough
Forage composition						
Dry matter, %	23.2	24.6	25.3	28.6	29.6	30.8
Crude protein, %	8.4	8.0	7.3	7.0	5.8	5.9
Acid detergent fiber, %	35.7	37.5	36.1	33.3	34.0	34.1
Silage composition						
Dry matter, %	19.8	23.2	23.7	27.4	27.6	32.3
Crude protein, %	9.2	8.3	7.7	7.2	7.1	7.4
Acid detergent fiber, %	35.3	36.9	38.5	34.4	37.4	34.4
Silage digestibility, %						
Dry matter	65.2	57.8	56.9	57.7	50.3	52.1
Total digestible nutrients (TDN)	64.9	57.9	57.1	57.0	50.3	53.5
Digestible energy, Mcal/ha	20,867	26,436	26,720	29,225	25,348	24,152

Source: Black et al. (45).

as it goes into the silo. For this reason, sorghum silages normally contain less energy than corn silage and normally the yield will be less. One example of a study on the effect the stage of maturity has on the nutrient values of forage sorghum is shown in Table 6-34. It was observed that as the forage dry matter increased, the crude protein decreased as the sorghum plant matures. Likewise, silage digestibility decreased with maturity. Maximum yield per hectare of digestible energy occurred at the late milk to early dough stage.

OTHER SILAGES

A wide variety of materials has been ensiled. Wastes from canneries processing food crops such as sweet corn, green beans, and green peas have been used to prepare silage. Various root or vegetable crop residues such as pea vines, beet tops and cull potatoes have been used suc-

cessfully. Residues of this type are often difficult to use on a fresh basis, because of the variable supply or because they are available only for a short period of time. Ensiling is advantageous in that it tends to result in a more uniform feed and a more constant supply.

Often silages are prepared from small cereal grain crops that are used as cover crops and must be harvested prior to maturing to the point that they can be harvested as grain. Table 6-35 compares several of these small-grain silages with corn silage. The crude protein content of silage from small-grain silages was higher than from corn, especially the oat silage. In the experiment shown on the left, dry matter of the small-grain silage was apparently higher than that of the corn, and this apparently stimulated greater dry matter consumption and higher daily gains, although less efficient gains.

TABLE 6-35

Response of growing cattle fed silages from small-grain forages as compared to corn silage

Parameter	Silage[a]				Silage[b]			
	Corn	Barley	Wheat	Triticale	Corn	Barley	Wheat	Oats
Silage composition								
Dry matter, %	30.3	42.7	46.4	58.2	37.2	35.7	39.2	30.1
Crude protein, % of DM	8.5	11.0	11.0	10.5	8.3	9.0	11.2	12.6
Acid detergent fiber, % of DM					28.9	32.3	36.5	42.3
Animal performance								
Daily gain, kg	0.82	0.98	0.92	0.92	1.15	1.06	0.94	0.50
Daily DM intake, kg	5.58	6.42	7.24	8.30	8.68	8.87	8.52	6.64
Feed-to-gain ratio	7.15	6.68	8.14	9.11	7.59	8.41	9.10	13.47

[a]Source: Hinman (47). Compared diets of 10% alfalfa hay, 15% barley grain, 5% protein–mineral supplement, and 70% silage (DM basis).
[b]Source: Oltjen and Bolsen (48). Cattle were fed a diet (DM basis) consisting of 84% silage, 12% of a milo–soybean meal mix, and 4% of a protein–mineral supplement.

TABLE 6-36

Effect of treating corn and wheat silage with formic acid

Parameters	Corn		Wheat	
	Untreated	Treated[a]	Untreated	Treated[a]
Forage as ensiled				
Dry matter, %	32.3	33.1	39.7	41.3
Crude protein, % of DM	9.0	9.4	9.3	8.9
Silage composition				
Dry matter, %	33.6	33.2	42.8	40.8
Crude protein, % of DM	8.4	8.7	8.9	9.0
pH	3.5	4.1	4.1	4.6
Lactic acid, % of DM	9.1	1.7	4.3	1.3
Animal performance				
DM digestibility, %	69.5	73.1	61.4	69.9
CP digestibility, %	57.6	64.4	57.7	65.7
DM intake, % of body weight	2.38	2.50	2.11	2.17
Fat-corrected milk, kg/day	16.8	16.6	15.0	15.5

Source: Baxter et al. (49).

[a]Silages were treated with 0.5% formic acid of the as-is weight of the silage.

In the experiment in Table 6-35, daily gains were lower with the small-grain silages, especially the oats. Although feed consumption was about the same for corn, barley, and wheat silages, feed efficiency was less for the cereal grain silages. Other comparisons have been made of silages from oats, peas, oats–peas, or barley–peas. These cool-season crops can be grown, harvested, and followed by warm-season crops such as pearl millet, sorghum, or Sudan grass (46).

A comparison between corn and wheat silage, both untreated and treated with formic acid, is presented in Table 6-36. In this study there was little difference in the protein content of forage or silage. Digestibility of dry matter and dry matter consumption were greater for the corn silage, as was milk production by lactating dairy cows. Treatment with formic acid inhibited production of lactic acid in both silages, but it increased protein digestibility, and tended to increase consumption, and had little effect on milk production.

Sunflowers are currently being used more extensively as a silage crop in areas where cooler temperatures prevail and that are relatively arid. It is primarily grown for its seed, but it can be made into acceptable silage. Table 6-37 compares sunflower silage to grass–legume silage as feed for dairy cows. The crude protein content of the sunflower silage was as high as the grass–legume silage and milk production was the same, but milk fat percentage was lower. Sunflower silage seems to be a viable alternative type of silage that can be used for high-producing animals.

TABLE 6-37

Comparison of grass–legume silage with sunflower silage for lactating cows

Parameter	Silage	
	Grass–Legume	Sunflower
Silage composition (DM basis)		
Dry matter, %	25.7	25.3
Crude protein, %	12.5	12.9
Acid detergent fiber, %	49.7	35.8
Calcium, %	1.2	1.6
Phosphorus, %	0.3	0.3
Cow performance		
Fat-corrected milk/day, kg	17.5	17.5
Milk fat, %	3.6	3.2

Source: Thomas et al. (50).

In tropical areas, sugar cane has been extolled as a high-potential crop for livestock feed because as much as 20 tons/ha of TDN can be produced from it. It has been suggested that either the whole plant or the derined stalk could be used as feed for ruminants (51). Since sugar cane is very high in sucrose, it would be expected to have excellent ensiling characteristics. However, one study showed that the digestibility of sugar cane silage was low, partially because large amounts of the sugars were fermented by yeast to ethanol (52).

Only limited studies have been reported of attempts to make silage from tropical grasses (53). Research (54) indicates that satisfactory silage can be made from various tropical grasses and that, where feasible, the product can probably be improved by the addition of a starch source such as cassava meal or a protein–starch source such as coconut meal. In general, the lactic acid content of tropical grass silages is low, and acetic acid is probably more important in preservation. The addition of a starch source will increase lactic acid to levels more comparable to that of silages made from temperate-zone grasses.

SILAGE ADDITIVES

A wide variety of additives has been used at one time or another to improve the ensiling process or nutrient content of the resultant silage. The general types of additives used are nutrients, preservatives, fermentation aids, and biological agents.

Several additives have been added to silage crops as they are ensiled, such as NPN and Ca sources to increase the nutrient content of the silage. Limestone or another Ca source has been added (0.5% to 1%) at the time of ensiling to increase the Ca level in crops that are deficient in Ca, such as, corn. The limestone also acts as a buffer and increases the amount of acids produced during the ensiling process, which normally results in improved feed consumption and animal performance.

The addition of urea or other NPN sources such as anhydrous ammonia has been made to increase the crude protein content of silages prepared from materials low in crude protein, but high in energy content. Some of the added NPN is converted to microbial protein during the ensiling process, but most of it is present as ammonia or, in lesser amounts, in NPN compounds such as amides; but the acidic pH prevents losses that might otherwise occur. Silage can also be put up with other sources of N; for example, manures of various types (broiler litter, cage layer manure, etc.) have been used in many experiments. Normally, feed consumption is reduced when manures are combined with materials being ensiled, but in some situations in developing countries this has proved to be a viable way of increasing the nitrogen content of a silage. With grass or grass–legume silages, the addition of molasses or grain has long been recommended to provide a source of fermentable carbohydrates, particularly when the herbage is ensiled without wilting.

Different chemicals have been used to inhibit fermentation in silage, particularly in direct-cut silage of high moisture content. This approach to preserving forage silages was pioneered in Europe. Organic acids, such as acetic, were used, which was very effective, but was corrosive, difficult to handle, and expensive. Subsequent research has used formic acid, formaldehyde, propionic acid, or mixtures of these. These compounds have proved to be very effective in inhibiting degradation of herbage protein and soluble carbohydrates; however, the results are rarely better than can be obtained by wilting (33, 55). Acid treatment inhibits growth of fungi (molds) and reduces heating caused by fermentation. Propionic acid has also been shown to be effective in inhibiting mold growth in haylage that was too dry (55).

Sodium metabisulfite is another relatively common additive. It is an effective inhibitor for most bacteria (and is used commonly in wine fermentation to kill wild yeasts and other microorganisms), but seems to have less effect on lactic-acid-producing bacteria. It would be interesting to see what effect its use might have when preparing sugar cane silage.

Bacterial cultures, often containing lactic-acid-producing species, are widely used as silage additives. Research evidence indicates that the ensiling process is improved and the resultant silage is more stable because of a more rapid establishment of a lactic-acid-producing bacterial population. These seem to be most effective on crops that are marginal in soluble carbohydrates and have limited fermentation capacity, such as alfalfa.

Microbial organisms are also being used that produce enzymes, such as cellulase, amylase, glucoamylase, or proteases. When added to silage, these enzymes break down the complex carbohydrates in the forage into simple sugars, which can then be readily fermented by the lactic-acid-producing bacteria. Enzymes have the greatest benefit on the ensiling process when a limited amount of plant sugars is available in the crop and an adequate level of lactic-acid-producing bacteria is present.

CROP RESIDUES

Vast quantities of crop residues, which are normally low-quality roughages, are available in many parts of the world. These residues include straws and chaff from the cereal grains: rice, wheat, barley, and others. Other residues include grass straws, stover from corn and sorghum, corn cobs, cottonseed hulls, cotton gin trash, hulls of various types, and other agriculture residues in lesser amounts. In addition, there are large amounts of standing winter feeds, such as grasses, forbes, and browse, especially in arid regions. In tropical and subtropical areas, residues like sugar-cane bagasses, pineapple stumps, and large quantities of coarse mature grasses may be found.

Table 6-38 summarizes the nutrient specifications for a number of these low-quality forages. A general characteristic of most low-quality forages is that their crude protein content is low. These materials are high in cellulose and other structural carbohydrates (crude fiber, ADF, NDF) and are often highly lignified. They have been exposed to excessive sun and rain, which has reduced their nutrient contents.

Low-quality roughages are generally lower in Ca than high-quality sources, especially legumes, and they are invariable low in P. The content of other minerals not listed would be expected to be low. Their minerals also seem to be less digestible, although there is very little information on this topic.

The carotene (provitamin A) content of plant tissues is highly related to the green color of the plant. Low-quality roughage, such as straw, is very low in carotene. Vitamins D and E and the B-complex vitamins are also extremely low.

FEEDING LOW-QUALITY ROUGHAGES

The low-quality roughages shown in Table 6-38 can normally only be used in feeding ruminants and to a lesser extent horses at near maintenance levels. Low-quality roughages normally require some type of supplementation in order to be effectively utilized. The major limitation of this type of feed is that its digestibility is low, because it contains high levels of cellulose and lignin. These feedstuffs are also bulky (low weight per unit of volume), which limits intake. The net result is that an animal simply cannot consume enough of these feedstuffs to maintain its body weight. One factor that limits the capacity of the ruminants digestive tract is that a feedstuff must be digested or chewed enough so that the rumen contents will pass through the retriculo-omasal orifice. Often these feedstuffs are chopped or pelleted to reduce practical size and increase passage rate. The effect of quality on digestibility, consumption, and rumen turnover is illustrated in Table 6-39. Note that as the digestibility of the roughages declined, consumption also declined and the rumen turnover rate decreased (which inhibits consumption).

Crude protein content and availability in low-quality roughages are other factors that reduce microbial digestion in the rumen. For microbes to digest the cell wall components (cellulose and hemicellulose), they require an adequate amount of an available nitrogen source. Depending on the solubility and digestibility of the proteins provided, at least 8% crude protein

TABLE 6-38

Compositional differences between high- and low-quality roughages

	Composition								
	CP	CF	ADF	NDF	TDN	NEm	NEg	Ca	P
High quality									
Orchard grass pasture, early vegetative	18	25	31	55	72	1.64	1.03	0.4	0.4
Alfalfa hay, early bloom	19	25	35	40	60	1.31	0.65	1.4	0.2
Sudan grass hay, mid-bloom	9	30	40	65	63	1.39	0.75	0.5	0.3
Low quality									
Grass hay, weathered	4	38	43	68	46	1.11	0.35	0.4	0.1
Grass straw	5	41	48	70	44	0.95	0.34	0.5	0.2
Oat straw	5	40	47	70	46	1.11	0.35	0.2	0.1
Wheat straw	4	42	56	85	40	0.86	0.0	0.2	0.08
Barley straw	4	42	57	82	46	0.99	0.11	0.3	0.05
Cottonseed hulls	4	48	67	86	50	1.03	0.22	0.2	0.07
Cotton gin trash	7	34	—	—	44	0.95	0.11	0.3	0.1
Corn cobs	3	36	39	88	48	1.04	0.20	0.1	0.04
Corn stover	6	35	58	87	50	1.02	0.22	0.5	0.1
Sugar-cane bagasse	2	49	59	86	48	1.03	0.19	0.5	0.3
Peanut hulls	7	63	65	74	22	0.77	0.0	0.2	0.07

Abbreviations: CP, crude protein; CF, crude fiber; ADF, acid detergent fiber; NDF, neutral detergent fiber; TDN, total digestible nutrients; NEm, net energy for maintenance; NEg, net energy for gain; Ca, calcium; and P, phosphorus.

TABLE 6-39

Effect of roughage quality on consumption, digestibility, and rumen turnover rate

| Parameters | Roughage | | |
	Orchard Grass Hay	Barley Straw	Corn Stover
Crude protein, %	10.1	4.2	4.1
Acid detergent fiber, %	40.7	57.0	58.6
Digestibility, %			
Dry matter	60.1	53.8	48.9
Crude protein	54.1	9.5	12.5
Consumption per day, kg			
Roughage dry matter	5.81	3.95	2.27
Rumen turnover rate per day	1.30	0.95	0.71

Source: Wheeler et al. (56).

Cattle were fed 1 kg/day of protein supplement.

appears to be necessary to sustain a reasonable level of activity. Ruminant animals have the ability to conserve nitrogen by recycling it via the saliva and across the rumen wall, so adequate levels can be maintained in the rumen. This recycling mechanism helps the animal during periods of inadequate N intake.

Supplementation with N is a common practice. This increases the digestion of the low-quality roughage and produces more digestive energy end products that can be used by the animal as a source of energy. NPN sources of N are often used in supplements, but performance can be improved when natural sources (containing amino acids) are fed (see Ch. 8 for further information on this topic).

Normally, low-quality roughages are limited to use in wintering or maintenance-type feeding programs and use in older animals (yearling or older for cattle). This is because of their bulkiness, low digestibility, and low protein content. Even with wintering beef cows, additional feed is required, since they will not be able to maintain their weight on most straws or similar feeds. However, these feedstuffs are of more value when fed to animals that are just being maintained, because they produce higher amounts of heat during rumen fermentation and body metabolism, which can be used to maintain body temperature in cool or cold environments.

Younger animals (weaned calves) simply cannot eat enough low-quality roughage, even when supplemented with proteins, vitamins, and minerals, to gain at recommended levels (1 to 1.25 lb/day). Limited amounts can be used for younger cattle, perhaps from 25% to 50%, depending on the type of low-quality roughage

and how it has been processed. Normally, when low-quality roughages are fed, the other dietary components need to be highly digestible to assure that adequate dietary nutrient intake is achieved.

Low-quality roughages are normally inexpensive and can be effectively used as a diluent in high-concentration rations, because a roughage source is required in nearly all ruminant and horse rations in order to maintain normal function of the digestive tract.

CHEMICAL TREATMENT OF LOW-QUALITY ROUGHAGES

Various methods of treatment of low-quality roughages to improve their nutrient content have been evaluated, such as heat, steam, various alkalis and acids, or a combination of steam and chemical treatment. The use of alkali treatment has resulted in the greatest improvement, especially when cost is considered. The use of NaOH (sodium hydroxide) seems to improve utilization to the greatest extent at the least cost. Methods have been developed to spray concentrated solutions of NaOH directly on straws or other low-quality roughages. The Australians spray NaOH on cereal straws prior to baling. Ensiling straws or corn stover with NaOH has also been shown to improve nutrient availability.

An example of the effectiveness of NaOH treatment is shown in Table 6-40. Note that feeding cattle diets with 15% to 25% straw that had been treated with NaOH resulted in improved daily gain and more efficient feed conversion, at least partially because the digestibility of the straw was increased from 43% to 62%. During the late 1970s and early

TABLE 6-40

Performance of steers fed rations containing ryegrass straw treated with NaOH

Item	Daily Gain, (kg)	Feed Conversion	Daily Straw Consumption (kg)	DE of Straw[a] (%)
Cubed with molasses	1.32	8.22	2.09	43.2
Sprayed with 4% dry weight of NaOH	1.43	7.49	2.13	61.9

Source: Author's unpublished research data. Growing rations contained 25% straw and finishing rations contained 15% straw.
[a]Digestible energy was determined by difference when fed with alfalfa hay to sheep.

1980s, some commercial mills utilized chemical processing methods for straws and other similar materials in Europe and other areas. However, the costs were relatively high. The cost of chemicals only for adding 4% NaOH to straw is about $14/ton. When the other costs for chopping and cubing are added, the method is not economically feasible in most areas.

Hydrogen peroxide has also been used, and Table 6-41 shows the results obtained when it was used to treat wheat straw. While it is evident that this treatment (soaking the straw in a chemical solution and then washing and drying it) is effective in increasing the digestibility of the ADF and cellulose, it is highly unlikely that such a process would be economically feasible.

Use of anhydrous ammonia seems to be a more feasible method to improve nutrient availability. This method requires that low-quality roughage be stacked and covered with plastic; then anhydrous ammonia is applied. The anhydrous ammonia changes from a liquid to a gas and penetrates the material being treated. The reaction time may require several weeks, depending on the ambient temperature and moisture content of the low-quality roughage. The

reaction proceeds faster at higher temperatures and higher moisture levels. Table 6-42 summarizes the results from one study that treated low-quality fescue with anhydrous ammonia. There was substantial N uptake (from 1.27% to 2.67%). The hemicellulose content of the fescue was reduced, as was the lignin. Digestibility of the fescue was increased substantially, but N retention was not improved much unless the treated fescue was fed in combination with corn grain. Ammonifying increased consumption by cattle more than it did with sheep.

Ammonifying treatment seems to be more feasible for an individual to use. Anhydrous ammonia is a common agricultural product used for fertilizing crops, is available in most areas, and is less hazardous to use than concentrated NaOH. Normally, it does not increase digestibility as much as NaOH treatment, but N is added, which increases the utilization of starch sources (grain) by ruminant animals. The ammonia is not completely bound, and some will be lost when the straw is processed or the stack is uncovered.

A considerable amount of research relating to treating low-quality roughages has been

TABLE 6-41

Performance of steers fed rations containing wheat straw treated with hydrogen peroxide

Parameter	Straw Level (% of diet)			
	Low		High	
	Treated	Untreated	Treated	Untreated
Diet composition, %				
Crude protein	11.0	11.4	10.6	11.8
Acid detergent fiber	35.2	25.4	52.3	42.8
Cellulose	29.8	18.1	42.4	21.8
Digestibility, %				
Crude protein	82.0	74.0	70.8	73.8
Acid detergent fiber	80.2	41.2	75.8	44.1
Cellulose	90.3	46.0	87.3	47.9

Source: Kerley et al. (57).

Low-level diets contained 33% straw and high-level diets contained 70% straw. The straw was soaked for about 5 hours in a solution of hydrogen peroxide, sodium hydroxide, and hydrochloric acid and then rinsed and dried.

TABLE 6-42

Effect of ammonifying treatment on utilization of low-quality roughages

	Roughage Treatment			
	Untreated Fescue, No Other Treatment	+SBM	Ammonified Fescue, No Other Treatment	+Corn
Total N, % DM	1.27		2.67	
Ammonia N, %	0.02		0.39	
Hemicellulose, %	30.7		25.5	
Lignin, %	8.1		6.7	
Digestibility, % dry matter	40.1	38.8	58.6	56.2
Nitrogen consumed, grams/head/day	7.92	12.84	20.55	24.36
Nitrogen absorbed, % of consumed	49.4	58.1	45.6	49.2
Nitrogen retained, grams/head/day	0.45	0.39	−1.61	1.12
Feed consumed, grams/body weight$^{0.75}$ kg				
Sheep	34.0	36.2	45.6	47.0
Cattle	39.9	49.1	65.4	68.6

Source: Buettner et al. (58).

done in recent years. For the reader that desires more information on this topic, consult references 59, 60, and 61.

SUMMARY

Roughages provide the majority of the feed consumed by herbivore species under natural conditions. Even in our highly developed agricultural feeding systems of today, roughages are still the most important feed ingredient for these species, but roughages must be supplemented with other nutrient sources in order to support high levels of production in finishing beef cattle, lactating dairy cattle, horse, and goats. Roughages are high in structural carbohydrates, which are less digestible than the starches found in cereal grains. The bulky nature of roughages, while essential for maintaining the health and well-being of herbivorous species, limits feed consumption and the level of production that can be achieved. High-quality roughages provide adequate levels of energy, protein, minerals, and vitamins for most ruminant species, while poor-quality roughages must be supplemented with additional nutrients if animals are forced to consume them for an extended time period. Young animals do not have the capability of consuming as much poor-quality roughage as do older animals. However, even poor-quality roughages can be used in moderate amounts in rations for fattening cattle or for feeding horses and in large amounts for maintaining wintering cattle, sheep, and horses.

REFERENCES

1. USDA. 1994. *Agricultural statistics* (Nov). Washington, DC: USDA.

2. NRC. 1982. *United States–Canadian tables of feed composition.* 3d. ed. Washington, DC: National Academy Press.

3. Barnes, R. F., D. A. Miller, and C. J. Nelson. 1995. *Forages.* 5th ed. Ames, IA: Iowa State Univ. Press.

4. Pathak, N. H. 1986. *Forage in ruminant nutrition.* New Delhi: Vikas, (distributed by Advent Books, New York).

5. Butler, G. W., and R. W. Bailey, eds. 1973. *Chemistry and biochemistry of herbage.* Vols. 1–3. London: Academic Press.

6. Fahey, G. C., Jr. ed. 1994. *Forage quality, evaluation, and utilization.* Amer. Soc. Agronomy, Crop Sci. Soc. Amer. and Soil Sci. Soc Amer.

7. Minson, D. J. 1990. *Forage in ruminant nutrition.* San Diego; CA: Academic Press.

8. Rodin, L. E., N. I. Bazilevich, and N. N. Rozov. 1975. In: *Productivity of world ecosystems.* Washington, DC: National Academy of Science.

9. Holland, C., and W. Kezar. 1990. *Pioneer forage manual—A nutrition guide.* Des Moines, IA: Pioneer Hi-Bred International, Inc.

10. Marble, V. L.. 1974. *Fourth California alfalfa symposium.* Univ. of California at Davis.

11. Morey, D. D., and J. J. Evans. 1983. *Cereal Chem.* 60:461.

12. Eppendorfer, W. H. 1977. *J. Sci. Food Agr.* 28:607.

13. Powell, K., R. L. Reid, and J. A. Balasko. 1977. *J. Animal Sci.* 46:1503.

14. Boland, R. L., et al. 1976. *Corp Sci.* 16:677.

15. Aderibigbe, A. O., and D. C. Church. 1982. *J. Animal Sci.* 54:164.

16. Ely, R. E., et al. 1953. *J. Dairy Sci.* 36:334.

17. Phillips, T. G., et al. 1954. *Argon. J.* 46:361.

18. Van Riper, G. E., and D. Smith. 1959. *Wisc. Expt Sta. Res. Rpt.* 4.

19. Hibbs, J. W., and H. R. Conrad. 1974. *Ohio Rpt.* 59(2):33.

20. Vona, L. C., et al. 1984. *J. Animal Sci.* 59:1582.

21. Wheeler, J. L., and R. C. Mochrie, eds. 1984. *Forage evaluation: Concepts and techniques.* CSIRO, Armidale, NSW, Australia and Amer. Forage and Grassland Council, Lexington, KY.

22. USDA. 1995. *Forage production handbook.* Washington, DC: USDA.

23. Gregory, P. H., et al. 1963. *J. Gen. Microbiol.* 33:147.

24. Weiss, W. P., H. R. Conrad, and W. I. Shockey. 1986. *J. Dairy Sci.* 69:1824.

25. Monhanty, G. P., et al. 1969. *J. Dairy Sci.* 52:79.

26. Wier, W. C., L. C. Jones, and J. H. Meyer. 1960. *J. Animal Sci.* 19:5.

27. Van Soest, P. J., and D. R. Mertens. 1985. *Agri-Practice.* Veterinary Publishing Co.

28. Burritt, E. A., et al. 1984. *J. Dairy Sci.* 67(6)1209.

29. Meyers, J. H., G. P. Lofgreen, and N. R. Ittner. 1956. *J. Animal Sci.* 15:64.

30. Stiles, D. A., et al. 1971. *J. Dairy Sci.* 54:65.

31. Shepherd, J. B., C. H. Gordon, and L. F. Campbell. 1953. *USDA Bureau Dairy Ind. Bull.* 149.

32. Fiez, E. 1976. In: *Beef* 12(7):20.

33. Waldo, D. R., J. E. Keys, and C. H. Gordan. 1973. *J. Dairy Sci.* 56:129.

34. Hawkins, D. R., H. E. Henderson, and D. B. Puser. 1970. *J. Animal Sci.* 31:617.

35. Calder, F. W., J. W. G. Nicholson, and J. E. Langille. 1976. *Can. J. Animal Sci.* 56:57.

36. Prange, R. W., et al. 1984. *J. Dairy Sci.* 67:2308.

37. Beever, D. E., et al. 1971. *Brit. J. Nutr.* 26:123.

38. Donaldson, E., and R. A. Edwards. 1976. *J. Sci. Food Agr.* 27:536.

39. Sutton, A. L., and R. L. Vetter. 1971. *J. Animal Sci.* 32:1256.

40. Wilkinson, J. M., J. T. Huber, and H. E. Henderson. 1976. *J. Animal Sci.* 42:208.

41. Phillip, L. E., and J. G. Buchannan-Smith. 1982. *Can. J. Animal Sci.* 62:259.

42. St. Pierre, N. R., et al. 1987. *J. Dairy Sci.* 70:108.

43. Hoglund, C. R. 1964. Mich. State Univ. *Agr. Econ.* Publ. 947.

44. Zimmerman, J. E. 1976. In: *Beef* 12(7):20.

45. Black, J. R., et al. 1980. *J. Animal Sci.* 50:617.

46. Jaster, E. H., C. M. Fisher, and D. A. Miller. 1985. *J. Dairy Sci.* 68:2914.

47. Hinman, D. D. 1982. *Univ. of Idaho Res. Rpt. 2.*

48. Oltjen, J. W., and K. K. Bolsen. 1980. *J. Animal Sci.* 51:958.

49. Baxter, H. D., M. J. Montgomery, and J. R. Owen. 1980. *J. Dairy Sci.* 63:1291.

50. Thomas, V. M., et al. 1982. *J. Dairy Sci.* 65:267.

51. Preston, T. R., and M. B. Willis. 1974. *Intensive beef production.* 2nd ed. New York: Pergamon Press.

52. King, L., and R. W. Stanley. 1982. *J. Animal Sci.* 54:689.

53. Wilkinson, J. M. 1983. *World Animal Rev.* 45:37.

54. Panditharatne, S., et al. 1986. *J. Animal Sci.* 63:197.

55. Waldo, D. R. 1977. *J. Dairy Sci.* 60:306.

56. Wheeler, W. E., D. A. Dinius, and J. B. Coombe. 1979. *J. Animal Sci.* 49:1357.

57. Kerley, M. S., et al. 1987. *J. Dairy Sci.* 70:2078.

58. Buettner, M. R., et al. 1982. *J. Animal Sci.* 54:173.

59. Males, J. R. 1987. *J. Animal Sci.* 65:1124.

60. Conner, M. C., and C. R. Richardson. 1987. *J. Animal Sci.* 65:1131.

61. Klopfenstein, T., et al. 1987. *J. Animal Sci.* 65:1139.

7

HIGH-ENERGY FEEDSTUFFS

D. C. Church and Richard O. Kellems

INTRODUCTION

Feedstuffs classed as high energy are those that are fed or added to a ration primarily for the purpose of increasing energy intake or increasing energy density of the ration. Included are the various cereal grains and many of their milling by-products, roots and tubers, liquid feeds such as molasses and mixtures in which molasses predominates, fats and oils, and other miscellaneous plant and animal products available in lesser amounts or in restricted geographical areas.

Energy from high-energy feedstuffs is supplied either by readily available carbohydrates (sugars and/or starches) or by fats and oils. Such feedstuffs are labeled high energy, in contrast to most roughages, because the available energy (digestible, metabolizable, or net) is much greater per unit of dry matter. In other words, the animal can obtain much more energy than from a typical roughage, even though there may be very little difference in gross energy (that resulting from burning in a bomb calorimeter) between straw and a cereal grain.

High-energy feedstuffs generally have low to moderate levels of protein, although several of the supplemental protein sources could be included on the basis of available energy. For our purposes here, only those feedstuffs with less than 20% crude protein and high in energy will be considered.

Depending on the type of diet and the class of animal involved, high-energy feedstuffs in this class may make up a substantial portion of the animal's total diet. In addition to providing a source of dietary energy, these feedstuffs also contain amino acids, minerals, and vitamins. The contribution of these nutrients needs to be accounted for when formulating rations.

CEREAL GRAINS

Cereal grains are produced by plants of the grass family (Gramineae) grown primarily for their seeds; consequently, they provide tremendous tonnages of harvested grains for animal feed and human food. Estimated world production of wheat, corn, and coarse grains (usually meaning oats, barley, and sorghum) for the 1991–1992 and 1995–1996 years is shown in Table 7-1. Note that cereal production has increased in the 5-year period shown in the table. Feed use for 1995–1996 accounted for 47% of grain production. If we include rice production for 1995–1996 (484 million T), of which very little was used for animal feed, then feed usage amounts to 37% of grain production during that year.

In the United States, estimated feed usage for 1986, 1990, and 1994 is shown in Table 7-2 for the cereal grains and a number of other energy sources, many of which are by-products of grain milling. Note that corn accounted for 119.8 out of

TABLE 7-1

World production and feed use of wheat, corn, and coarse grains
(in millions of tons)

Item	Crop Year 1991/92	Crop Year 1995/96	% of Production, 1995/96
Wheat			
Production	488.2	485.8	
Consumption	523.9	518.7	
Corn			
Production	487.3	527.8	
Consumption	486.5	549.1	
Coarse grains[a]			
Production	804.9	836.0	
Consumption	805.4	864.6	

Source: USDA (1).

[a]Primarily oats, barley, and sorghum.

TABLE 7-2

Grains and by-product feeds fed to livestock in the
United States (bushels, million)

Item	Crop Year 1986	Crop Year 1990	Crop Year 1994
Grains fed			
Corn	8,225	7,934	9,602
Sorghum	938	573	640
Oats	384	357	229
Barley	608	422	375
Total grains	10,155	9,286	10,846

Source: USDA (2)

159.9 million T of grain fed in 1986. This amounts to 74.9% of the total as compared to the world estimate of 60.5%. The relative amount of production from the other feed grains (sorghum, oats, barley) has declined, partly because yields of corn have increased more over the past 50 years than is the case for the other grains. In any case, in other areas of the world, much less of the total production of cereal grain production is used for animal feed. With increasing costs of fuel and fertilizer and increasing human populations, it may be anticipated that relatively less of the U.S. production will be used for animal feeds in future years.

Of course, some of these "feed" grains go into human food or for the production of industrial products of a wide variety. For example, corn may be consumed as popped corn, corn flakes, corn flour, corn starch, corn syrup, whiskey, and numerous other products, but the amount used in this manner is much less than currently goes into animal feeds. Wheat and rice are grown primarily for human consumption, although moderate amounts of wheat may go into animal feed in the United States when price and supply allow it. Barley and oats, although good feeds, are becoming relatively less important because they do not usually yield as well as some of the other feed grain crops. Barley is used extensively in the brewing industry for the production of malt. Only a very small amount of oats is used off the farm. Other grains, such as millet and rye, find only limited use in North America, although they are more widely used in Europe and Asia. Sorghums are mainly an animal feed in the United States. A wheat–rye cross, triticale, is grown in limited amounts for feed. Several of the most commonly feed cereal grains are shown in Fig. 7-1.

GRAIN MARKETING AND GRADING

In commercial practice in the United States, cereal grains, except for sorghums, are usually bought and sold on the archaic basis of bushel weights. This varies somewhat throughout the country, as it is more common in the West to sell by the ton, but the central markets quote prices on bushel weights.

Most of the cereal grains used for human food (and for export) pass through large terminal markets and most are sold on the basis of grades established by the U.S. Department of Agriculture, which also publishes information detailing the grade specifications (3). Grading standards, recently revised, are published for wheat, corn, barley, oats, rye, sorghum, triticale, flax, soybeans, and mixed grain. For animal feed, an appreciable percentage of the total grains fed is probably not graded. Nevertheless, a brief discussion of this topic is pertinent to this chapter.

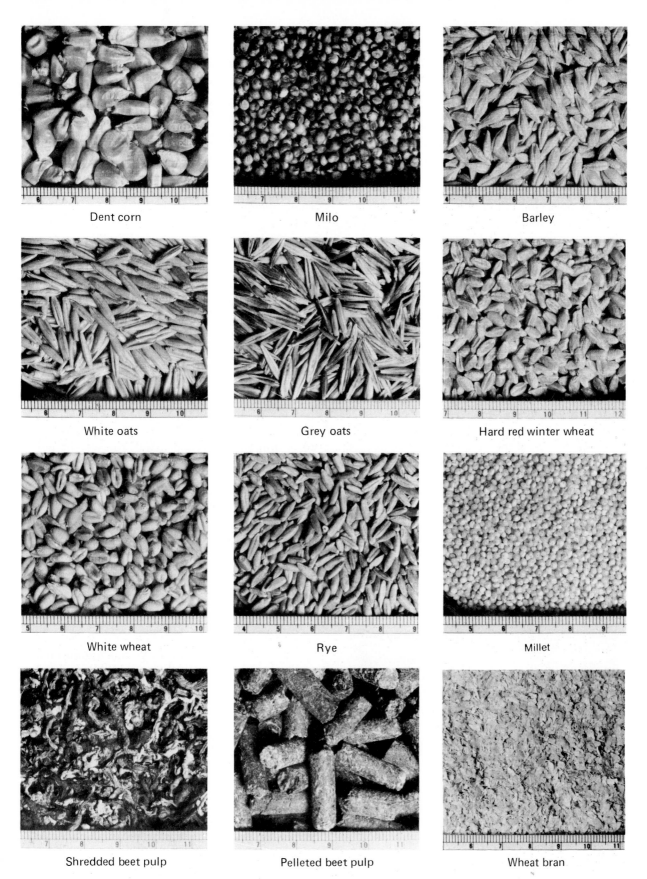

FIGURE 7-1 Cereal grains and by-product feeds.

Dent corn

Milo

Barley

White oats

Grey oats

Hard red winter wheat

White wheat

Rye

Millet

Shredded beet pulp

Pelleted beet pulp

Wheat bran

When a grain sample is obtained, the test weight is determined using the small device shown in Fig. 7-2. If the grain is to be graded, it is divided by the use of appropriate sieves into the sample and dockage; dockage consists of weed seeds and stems, chaff, straw, grain other than the principal one, sand, dirt, and any other material that can be removed readily from the sample. The sieved sample is then evaluated on the basis of various factors. For corn, factors include weight per bushel, broken kernels and foreign material, damaged kernels, and heat-damaged kernels. The minimum or maximum limits for the different U.S. grades of the common grains are shown in Table 7-3. In addition to the grades shown for corn in the table, there may be some special grades such as weevily corn or corn of "distinctly low quality," a term that indicates more than two crotalaria seeds (*Crotalaria* spp.) in 1 kg of grain. Similar comments apply to the other grains, although the details are not all listed in the table. As bushel weight increases, energy content increases, starch content increases, and fiber content decreases. For these reasons, pricing is often based on bushel weight.

As a result of the variability of feed grains, many large buyers, such as feedlots, may insist on discounts for light test weights, broken and damaged grain, presence of excessive foreign matter, and musty or sour odors. Moisture content will also vary greatly with high levels often found in freshly harvested grains. Most feedlots and grain elevators will discount grains having more than 15 percent moisture. This is, of course, a very legitimate practice, as the mois-

FIGURE 7-2 An apparatus used for measuring the bushel weight of grains.

ture is undesirable from the standpoint of keeping quality and the excess moisture is worthless unless the grain is to be stored and utilized as high-moisture grain.

NUTRIENT CONTENT OF CEREAL GRAINS

Average values for some of the nutrients in cereal grains are shown in Table 7-4. Note that relatively small differences occur between the grains, and identifying them on the basis of their chemical composition would be difficult at times.

Although grains are usually said to be less variable in composition than roughages, many factors influence nutrient composition and, thus, feeding value for a given grain. For example, factors such as soil fertility and fertilization, variety, closeness of planting, weather, rainfall, insects, and disease may all affect plant growth and seed production so that average book values may not be meaningful. With wheat, for example, if a hot, dry period occurs while the grain is ripening, shriveled, small kernels may result. Although they may have a relatively high protein content, the starch content is apt to be much lower than usual, the weight per bushel (bulk density) will be low, and the feeding value may be appreciably lower.

Although crude protein content of feed grains is relatively low, ranging from 8% to 14% for most grains, some may be much lower than this and, particularly with wheat, some may be higher, sometimes as high as 22% crude protein. Sorghum grains in particular are quite variable. Plant breeding practices are effective in increasing or decreasing the protein content.

Of the nitrogenous compounds in the seed, 85% to 90% is in the form of proteins, but the proteins, their solubility, and amino acid content vary from cereal to cereal. Most cereal grains are moderately low to deficient for monogastric species in lysine and often in tryptophan (corn) and threonine (sorghum and rice) and in methionine for poultry.

The fat content may vary greatly, ranging from less than 1% to greater than 6%, with oats usually having the most and wheat the least fat. Most of the fat is found in the seed embryo. The lipid content increases with maturity of the grain. Seed oils are high in palmitic acid and linoleic and oleic acids; the latter two are unsaturated acids that tend to become rancid quickly, particularly after the grain is processed.

The carbohydrates in grains, with the exception of the hulls, are primarily starch. Starch

TABLE 7-3

U.S. grade and grade requirements for the feed grains

Grain, Grade	Minimum Limits, %		Maximum Limits, %								
	Test wt/bushel lb[a]	Sound Grain	Damaged Kernels		Foreign Material	Broken Kernels	Thin Barley	Other Seeds[b]	Foreign Material Other Than Wheat	Total Defects
			Total	Heat Damaged						
Barley										
US No. 1	47.0	97.0	2.0	0.2	1.0	4.0	10.0	0.5		
2	45.0	94.0	4.0	0.3	2.0	8.0	15.0	1.0		
3	43.0	90.0	6.0	0.5	3.0	12.0	25.0	2.0		
4	40.0	85.0	8.0	1.0	4.0	18.0	35.0	5.0		
5	36.0	75.0	10.0	3.0	5.0	28.0	75.0	10.0		
Sample[c]	—	—	—	—	—	—	—	—		
Corn										
US No. 1	56.0		3.0	0.1		2.0				
2	54.0		5.0	0.2		3.0				
3	52.0		7.0	0.5		4.0				
4	49.0		10.0	1.0		5.0				
5	46.0		15.0	3.0		7.0				
Sample[c]	—		—	—		—				
Oats										
US No. 1	36.0	97.0		0.1	2.0			2.0		
2	33.0	94.0		0.3	3.0			3.0		
3	30.0	90.0		1.0	4.0			5.0		
4	27.0	80.0		3.0	5.0			10.0		
Sample[c]	—	—		—	—			—		

TABLE 7-3
Continued

| Grain, Grade | Minimum Limits, % | | Maximum Limits, % | | | | | | | |
| | Test wt/bushel lb[a] | Sound Grain | Damaged Kernels | | Foreign Material | Broken Kernels | Thin Barley | Other Seeds[b] | Foreign Material Other Than Wheat | Total Defects |
			Total	Heat Damaged						
Rye										
US No. 1	56.0		2.0	0.2	3.0				1.0	
2	54.0		4.0	0.2	6.0				2.0	
3	52.0		7.0	0.5	10.0				4.0	
4	49.0		15.0	3.0	10.0				6.0	
Sample[c]	—		—	—	—				—	
Sorghum, all classes										
US No. 1	57.0		2.0	0.2		4.0[d]				
2	55.0		5.0	0.5		8.0				
3	53.0		10.0	1.0		12.0				
4	51.0		15.0	3.0		15.0				
Sample[c]	—		—	—		—				
Triticale										
US No. 1	48.0		2.0	0.2	2.0	5.0[e]			1.0[f]	5.0
2	45.0		4.0	0.2	4.0	8.0			2.0	8.0
3	43.0		8.0	0.5	7.0	12.0			3.0	12.0
4	41.0		15.0	3.0	10.0	20.0			4.0	20.0
Sample[c]	—		—	—	—	—	—		—	—

Source: USDA (3).

[a] Note that the legal bushel weights are those given for No. 1 grades; regardless of the volume taken up by a lower grade, it still requires 56 lb of weight/bu of corn, etc.

[b] Black barley in barley samples; wild oats in oat samples.

[c] Sample grades include all grain that does not meet the listed standards for the other grades. In addition to foreign odors or grain that is musty, sour, or heating, other specifications may apply to specific grains. Refer to reference listed below for complete details.

[d] Foreign material, broken grains, plus other grains.

[e] Plus shrunken kernels.

[f] Other than wheat or rye.

TABLE 7-4

Average composition of the major cereal grains, dry basis

Item	Corn		Wheat		Rice with Hulls	Rye	Barley	Oats	Milo	Triticale
	Dent	Opaque-2	Hard Winter	Soft White						
Crude protein, %	10.4	12.6	14.2	11.7	8.0	13.4	13.3	12.8	12.4	14.0
Ether extract, %	4.6	5.4	1.7	1.8	1.7	1.8	2.0	4.7	3.2	4.6
Crude fiber, %	2.5	3.2	2.3	2.1	8.8	2.6	6.3	12.2	2.7	4.0
Ash, %	1.4	1.8	2.0	1.8	5.4	2.1	2.7	3.7	2.1	2.0
NFE, %	81.3	76.9	79.8	82.6	75.6	80.1	75.7	66.6	79.6	75.4
Total sugar, %	1.9		2.9	4.1		4.5	2.5	1.5	1.5	
Starch, %	72.2		63.4	67.2		63.8	64.6	41.2	70.8	
Essential amino acids, % of DM										
Arginine	0.45	0.86	0.76	0.64	0.63	0.6	0.6	0.8	0.4	0.4
Histidine	0.18	0.44	0.39	0.30	0.10	0.3	0.3	0.2	0.3	0.2
Isoleucine	0.45	0.40	0.67	0.44	0.35	0.6	0.6	0.6	0.6	0.3
Leucine	0.99	1.06	1.20	0.86	0.60	0.8	0.9	1.0	1.6	0.4
Lysine	0.18	0.53	0.43	0.37	0.31	0.5	0.6	0.4	0.3	0.3
Phenylalanine	0.45	0.56	0.92	0.57	0.35	0.7	0.7	0.7	0.5	0.3
Threonine	0.36	0.41	0.48	0.37	0.25	0.4	0.4	0.4	0.3	0.2
Tryptophan	0.09	0.16	0.20	0.12	0.12	0.1	0.2	0.2	0.1	0.2
Valine	0.36	0.62	0.79	0.56	0.50	0.7	0.7	0.7	0.6	0.3
Methionine	0.09	0.17	0.21	0.19	0.20	0.2	0.2	0.2	0.1	0.07
Cystine	0.09	0.22	0.29	0.34	0.11	0.2	0.2	0.2	0.2	0.2
Minerals, % of DM										
Calcium	0.02	0.05	0.06	0.09	0.06	0.07	0.06	0.07	0.04	0.1
Phosphorus	0.33	0.24	0.45	0.34	0.45	0.38	0.35	0.30	0.33	0.34
Potassium	0.33	0.30	0.56	0.44	0.25	0.52	0.63	0.42	0.39	0.4
Magnesium	0.12		0.11	0.11	0.11	0.13	0.14	0.19	0.22	

Source: NRC Publications.

accounts for about 72% of the corn kernel, but is as low as 41% in oats (Table 7-4). Starch is present in the endosperm (Fig. 7-3) in the form of granules that vary in size and shape in different plant species (Fig. 7-4). The starch granules increase in size but not in number with maturity. They are imbedded between a dense protein matrix and protein bodies that increase in

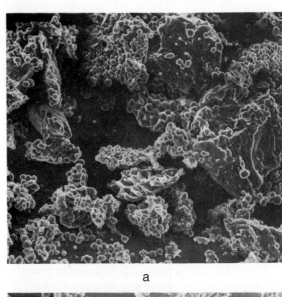

a

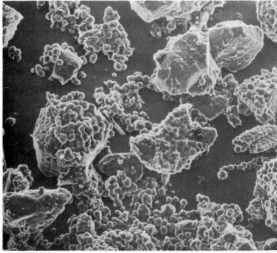

b

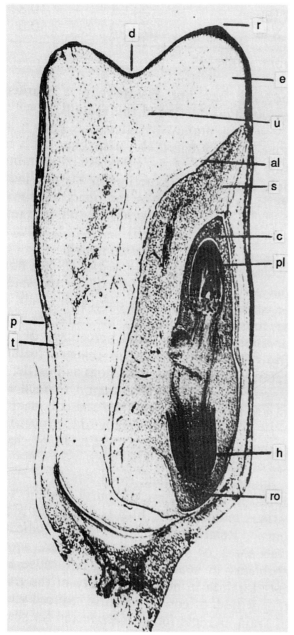

FIGURE 7-3 Half of a nearly mature dent corn kernel showing dent (d), remnant of style (r), endosperm (e), unfilled portion of immature endosperm (u), aleurone (al), scutellum (s), plumule (pl), hypocotyl and radicle (h), testa or true seed coat (t), and pericarp (p). The germ comprises the entire darker area, including the scutellum (s), coleoptile (c), and root cap (ro). (Courtesy of USDA.)

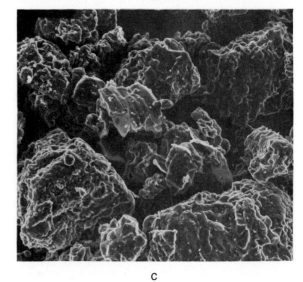

c

FIGURE 7-4 Electron micrographs of cereal starch from (a) ground corn, (b) milo, and (c) wheat. Note the lack of individual granules in (c). (Courtesy of L. H. Harbers, Kansas State University.)

thickness and size with maturity (4). The starch also differs somewhat in chemical composition among grains. Starch is a long-chain polymer made up of glucose molecules, and there are two primary types. Amylose is a linear, straight-chain molecule, while the other fraction, amylopectin, has many branched chains. The chain length of amylopectin varies among species and varieties of grains. These two fractions have somewhat different chemical properties, and it may be that relative differences in amounts in different grains have some effect on their feeding value. About 22% to 23% of the starch is amylose in corn, wheat, sorghum, barley and oats, about 27% in rice, and up to 100% in waxy corn and sorghum, although the usual level is 21% to 28%. Other carbohydrates in grains include a variety of simple sugars and other larger carbohydrate molecules, such as dextrins and pentosans, which together may account for 6% to 10% of the seed dry matter.

Cellulose is the principal constituent of the cell wall of the kernel and the hulls in grains such as barley and oats. While cellulose is also a glucose polymer, the linkage between the glucose molecules is much more resistant to bacterial or enzymic action (see Ch. 3). The lignin content of hulls is also a factor in reducing utilization, as is the waxy covering on grains such as corn.

Hulls of seeds have a substantial effect on feeding value. Most hulls or seed coats must be ruptured to some extent before feeding if efficient utilization is to be achieved, particularly for cattle, which do not chew grains as thoroughly as do some other species. Because of their heavy hulls, barley and oats are sometimes known as rough grains. Rice hulls are almost totally indigestible; this is related to a very high silica content.

With regard to minerals, cereal grains are universally low in Ca. Although the P content is reasonably high, much of it is present in the phytic acid P complex, which has a low biological availability for monogastric animals. Trace elements are generally present only in marginal amounts.

Of the vitamins, most cereal grains are fair sources of vitamin E and low in vitamin D and most of the B vitamins (see Appendix Table 4). Various milling by-products may have higher levels of the B vitamins and vitamin E. Except for yellow corn, cereal grains are also very low in carotene.

As a general rule, cereal grains are highly digestible, although the extent of digestibility varies among animal species and as the quality of the grain varies. More information is given on this subject in a section on the comparative evaluation of grains.

Additional information follows on specific cereal grains and other high-energy feeds. Cereal grain by-products that contain less than 20% crude protein are discussed in this chapter. Those containing more than 20% crude protein are discussed in Ch. 8.

CORN (ZEA MAYS)

Corn (Indian corn) is usually referred to as maize in countries other than North America. In the United States, corn is sometimes called the Golden Grain. This is because in areas where it grows well, corn will produce more digestible energy per unit of land than any other grain crop. Yields in excess of 400 bu/acre (11+ tons) of corn have been realized on small acreages. In addition to the high yield, corn is a very digestible and palatable feed, relished by all domestic animals, and it only rarely causes any nutritional problems when fed to animals.

Corn is commonly classified into seven groups based on kernel characteristics: dent, flint, flour, sweet, pop, waxy, and pod. Of these, the important commercial types are dent, sweet, and pop corn. Sweet corn, which has a high sugar content, is grown primarily for human consumption, as is pop corn. Dent corn is the main feed grain variety, and most of it is produced from double-cross hybrid seed varieties (Fig. 7-5). Many different hybrids are available that have been developed for specific areas or length of growing season, the result being that corn, as a major crop, is adaptable to a much wider geographical area than it used to be.

A diagram of a dent corn kernel is shown in Fig. 7-4. The endosperm, which is primarily starch, makes up about 85%; embryo and scutellum, 10%; and the pericarp and other parts, 5%. These various parts will be referred to in later sections dealing with milling by-products of the grains.

The chemical composition of corn has been studied in great detail; a very limited amount of data are presented in Table 7-4; other information may be found in sources such as reference 5. With regard to protein, about 73% of the total protein is found in the endosperm and most of the remainder, 24% is in the embryo. The protein in the embryo is of a better quality and is made up of a mixture of glutelin, globulins, albumins, and others. In the endosperm the principal protein is zein, a relatively insoluble

FIGURE 7-5 A hybrid corn field with a high yield of grain. (Courtesy of USDA.)

protein that makes up about half of the total kernel protein. This protein is low in several of the essential amino acids, particularly lysine and tryptophan; the total protein of corn is deficient in these amino acids for monogastric animal species and requires supplementation for adequate performance. This is illustrated clearly in Table 7-5. In addition, the low tryptophan (which is a precursor for niacin) plus the low niacin content lead, eventually, to a niacin deficiency and pellagra in simple-stomached animals depending on corn as a major dietary constituent. N fertilization has been shown to increase protein content (Table 7-6), but the increase is accompanied by a decrease in protein quality. This is a result of a relative increase in the zein fraction.

White and yellow corn have similar compositions, except that yellow corn has a fail content of cryptoxanthin, a precursor of vitamin A. Yellow corn also contains xanthophylls, yellow pigments that are of interest to the poultry industry because they contribute color to the skin of the bird and yolk of the egg. Yellow corn is generally preferred for feeding; consequently, very little white corn is now produced in the United States, although it is more widely used in other countries. White corn is often used for preparation of corn meal for human food.

The content of other nutrients in corn is similar to that in other cereal grains (see introduction), except that corn is particularly deficient in Ca.

Several genetic mutants of corn have been isolated and are being developed. One of these, known as opaque-2 (or high-lysine corn), is of particular interest because it has a high level of lysine and increased levels of most other essential amino acids. This change results from a reduction in zein and an increase in glutelin, a protein found in both the endosperm and the germ. The result is an improvement in the quality of the corn protein. Performance of monogastric species may be improved considerably by use of this mutant when livestock is not given supplementary protein (see Table 7-5). However, opaque-2 has not been shown to have added value for ruminants. Yields of opaque-2 are not equivalent to regular corn, nor are prices proportionately higher, so the interest of corn producers has not resulted in a marked increase in production of this mutant.

TABLE 7-5

Comparative performance of pigs fed corn, corn supplemented with adequate protein, or some of the experimental varieties of corn

Treatment	Protein, %	Daily Gain, g	Feed/Gain Ratio
Corn + soybean meal	16.0	520	2.16
Corn	9.6	91	8.94
Opaque-2 corn	9.4	310	3.32
Floury-2 corn	9.9	91	6.63
Corn + soybean meal	12.0	248	3.17
Opaque-2 + soybean meal	12.0	455	2.59
Floury-2 + soybean meal	12.0	323	3.14

Source: Pond and Maner (6).

TABLE 7-6

Effect of various fertilizer N levels on crude protein content (dry basis) of corn

Pounds N/Ac	Crude Protein, %	
	Farm 1	Farm 2
0	8.1	9.1
60	10.2	9.8
100	10.1	10.2
180	11.4	10.3
340	12.2	10.6

Source: Sunde (7).

FIGURE 7-6 A modern grain combine that, in this case, is harvesting and shelling corn in one continuous process. (Courtesy of Deere & Co., Moline, IL.)

A second mutant of interest is known as floury-2. Although it also has higher levels of lysine and some of the other essential amino acids, comparative experiments do not indicate it to be as good as the opaque-2 strain (Table 7-5). Other experimental mutants include those designated as high fat, high amylose, sugary-2 and brown midrib.

Harvesting of feed corn in the United States is now done with equipment that picks and shells the ear in the field (Fig. 7-6). This reduces the bulk of the finished product, and the cobs are harvested only on a small percentage of the crop. This has resulted in changed feeding practices, particularly with hogs and cattle, as it used to be relatively common to feed ear corn to hogs and ground corn and cob meal to cattle. Other information on processing methods will be discussed in Ch. 11 or in chapters on specific animals.

Spoiled and moldy feeds were mentioned in Ch. 4. Corn, of course, is subject to various molds, smuts, blights, and other problems. Relatively recent evidence indicates that some of the blights, which are severe problems in some years, do not cause any deterioration in the nutritive quality of the grain; they do result in reduced yields of grain and less forage if harvested for silage.

GRAIN SORGHUM (*SORGHUM VULGARE*)

Several different sorghum varieties (all *Sorghum vulgare*) are used for seed production. These include milo, various kafirs, sorgo, sumac, hegari, darso, feterita, and cane. Other nongrain types include forage sorghums, Sudan grass, and broomcorn. Milo (Fig. 7-7) is a favorite in the United States in drier areas because it is a short plant adaptable to harvesting with grain combines. The development of hybrid varieties, primarily milo and kafir crosses, which have higher yields, has increased production per unit of land greatly in recent years in the United States. On a worldwide basis, most of the sorghum seed production is in the United States, India, China, and Argentina, with lesser amounts in other countries, primarily in Africa.

The various sorghum varieties are able to withstand heat and drought better than most grain crops. In addition, sorghum is resistant to pests such as root worm and the corn borer and is adaptable to a wide variety of soil types. Consequently sorghum is grown in many areas where corn or other cereal crops do not do well. Sorghum yields less grain per unit of land than corn where corn thrives. The seed from all varieties of sorghum is small (2.5 to 3.5 × 4 mm) and relatively hard and usually requires some

FIGURE 7-7 Milo growing in western Kansas.

processing for optimum animal utilization (see Ch. 11).

Chemically, grain sorghums are similar to corn. Protein content averages about 11% (dry basis), but it is apt to be more variable than that of corn. Sorghum grains contain an alcohol-soluble protein, kafirin, which is similar to zein in corn. Lysine and threonine are said to be the most limiting amino acids, but some studies with turkeys and chickens indicate that methionine may be limiting (8). The content of other nutrients is similar to corn, but the starch content is somewhat lower and more variable than that of corn.

Different types of sorghum have been identified and developed by plant breeders following introduction of hybrids in the early 1960s. Those currently in use include normal, bird-resistant, waxy, and heteroyellow endosperm (hybrids) types. Some data on digestibility by sheep and cattle of the crude protein and starch from these different types is shown in Table 7-7. In nearly all cases where comparisons have been made, the brown, bird-resistant sorghums (which are high in Tannins) result in appreciably lower digestibility and lower animal performance than do other sorghums. The heteroyellow endosperm types generally show improved feeding values over the older, nonyellow cultivars. Waxy cultivars consistently give better feed efficiencies than do nonwaxy types. Waxy cereal starches are among the most digestible of all starches, and digestibility of a starch is generally inversely proportional to amylose content (9).

Most feeding studies indicate that sorghum grains are usually worth somewhat less than corn, although some data indicate an equal or higher value for swine and poultry. The presence of more protein bodies among the starch granules than in corn may affect utilization by some animal species. Fortunately, with a good job of processing, the feeding value of

most sorghum grains closely approaches that of corn grain.

Other studies of interest on sorghum grains have shown that lightweight grain has a lower feeding value (and more fiber) than that of heavy grains (weight per bushel). Other research on weathered grain with varying degrees of sprout damage, discoloration, or molds showed that weathered grain was less palatable to sheep, possibly because of dust, but there was little difference in digestibility between good-quality and weathered grain (10).

BARLEY (*HORDENUM VULGARE* OR *H. DISTICHON*)

Barley is widely grown in Europe and the cool, dry climates of North America and Asia. Although a small amount goes into human food and a very substantial percentage is used in the brewing industry in the form of malt, most of the barley is used for animal feeding. Even so, it accounted for only 4.4% of grain tonnage fed in the United States in 1996 (1).

Barley contains more total protein (11% to 16% dry basis) and higher levels of lysine, tryptophan, methionine, and cystine than corn (Table 7-4), but its feeding value for ruminants is appreciably less in most cases than that of corn or sorghum. This is in part because of its lower starch content (50% to 60%), somewhat higher fiber content, and lower digestible energy.

Barley is a very palatable feed for horses and ruminants, particularly when steam rolled before feeding, and few digestive problems result from its use, although it is more prone to cause bloat in feedlot cattle when used as a major portion of the ration than are the other cereal grains. Barley, as the only grain source, will not allow maximum gains or optimum feed efficiency for swine, and the fiber level is too high for use of more than small amounts in poultry rations. There are also appreciable differences in chick performance on different cultivars.

Hull-less varieties of barley are roughly equivalent to wheat or corn and are thus more suitable for swine and poultry feeding; however, not much hull-less barley is produced. Pearled barley, which has had the hull and most of the bran removed, has a high feeding value, but it is used primarily as a human food.

A considerable amount of malting barley is grown. Such barleys are generally higher in protein content with heavier bushel weights than feed grain varieties (Table 7-8). The comparison between malting barley and feed grain barley shown in the table does not indicate any

TABLE 7-7

Digestibility of crude protein and starch in different sorghum types by sheep and cattle (percent)

Sorghum Type	Sheep		Cattle	
	CP	Starch	CP	Starch
Normal	71.9	90.4	46.3	76.0
Bird resistant	61.4	82.6	23.9	60.0
Waxy	71.6	87.3	62.1	75.7
Heterowaxy	68.4	90.4	57.7	84.4
Heteroyellow	73.4	89.0	51.8	76.9

Source: Rooney and Pflugfelder (9).

TABLE 7-8

Composition and steer performance when fed two different types of barley and steer performance when fed light, medium, and heavy barley

| Item | Barley Type[a] | | | |
| | 2-Row Malting | | 6-Row Feed | |
	Irrigated	Dryland	Irrigated	Dryland
Barley composition				
Crude protein, %	12.7	12.5	10.6	12.4
Crude fiber, %	4.5	4.4	6.7	6.9
Calcium, %	0.05	0.07	0.06	0.04
Phosphorus, %	0.45	0.47	0.38	0.31
Weight, g/ℓ	702	698	625	634
Steer performance				
Daily feed consumed, kg DM	9.9	9.3	10.0	9.9
Daily gain, kg	1.15	1.12	1.13	1.09
Feed/gain ratio	8.61	8.30	8.85	9.08

| Item | Weight of Barley Fed[b] | | |
	Light	Medium	Heavy
Barley composition			
Crude protein, %	16.1	15.3	11.1
AD fiber, %	9.0	5.8	5.5
Weight, g/ℓ	469	554	666
Steer performance			
Dry matter intake, kg	9.34	9.15	8.89
Daily gain, kg	1.62	1.72	1.69
Feed/gain ratio	5.80	5.32	5.26

[a]From Hinman (11). The finishing ration was 81% steam-rolled barley, 10% chopped alfalfa hay, 7% corn silage, and 2% vitamin–mineral supplement.

[b]From Grimson et al. (43). Barley made up 85% of the ration DM. The remainder was made up of barley silage and a vitamin–mineral mix.

particular improvement in daily grain, although feed efficiency was improved slightly when the malting barley was fed to finishing cattle. Note also that there were minor differences in composition of the grain and in animal performance with barley grown under irrigated or dryland conditions. Other data from the same experiment station show that daily gain of fattening cattle decreased as the bushel weight of barley decreased. Feed efficiency was also less on the lightweight barley. This information was confirmed by another study, shown in Table 7-8.

A high-lysine barley containing 25% more lysine than common barley shows some promise as an energy source for nonruminant species. However, some of the high-lysine cultivars have been shown to be lower in some of the other essential amino acids (12), so the overall improvement may be less than anticipated from the higher lysine content.

OATS (AVENA SATIVA)

Although oats is the third most important cereal feed grain in the United States, it ac-counted for only 21% of feed grain use in 1994 (Table 7-2). Oats does not yield as much as the other grains and, considering the hull of the whole grain, the feeding value is relatively lower than for some of the other grains. Most of the world production is concentrated in northern Europe and North America. Only about 5% of the production goes into human food.

Three varieties of oats dominate the U.S. market. White oats account for most of the production and are grown primarily in the corn belt and northern plains. Red oats are grown in the south, and gray oats are grown in the Pacific Northwest.

The protein content of oats is relatively high (11% to 14%), and the amino acid distribution is the most favorable of any of the cereal grains, the ranking generally being in the order of oats first, followed by barley, wheat, corn, rice, rye, sorghum, and millet. Oats is still relatively deficient in methionine, histidine, tryptophan, and lysine. Oats is not widely used for feeding of poultry because of the hull, which makes up an average of 28% of the kernel. The hull may be as high as 45% of the total weight in lightweight oats. Because of this, oats with heavy test weights are more desirable for feeding, because there is an inverse relationship between test weight and percentage of hull.

In addition to the amount of hull, the hull itself is quite fibrous (31% fiber), and it is poorly digested. As a result, the fiber content of the whole grain is the highest of any of the cereal grains, and the starch content is the lowest. Even when the seed is ground, it results in a very bulky feed, and including a high percentage in the ration does not allow maximal feed intake for high-producing animals. Oats is of some value for young pigs in protecting them from stomach ulcers. The hulls contain an alcohol-soluble factor which provides the protective function. For older swine, oats is often used for finishing swine, particularly when it is desired to limit the consumption of energy so that the pig does not lay down too much back fat (see Ch. 19). For ruminants and horses, oats is a favored feed for breeding stock or in creep feeds for young animals. However, for feedlot cattle, oats do not supply sufficient energy for most rapid gains.

There are hull-less varieties of oats, but they are rarely grown. Oat groats (whole seed minus the hulls) are comparable or better than corn in feeding value, but the price is not usually attractive for animal feeding.

WHEAT (*TRITICUM* SPP.)

Wheat, like rice, is grown normally for the human food market, and nearly all commonly grown varieties were developed with flour milling qualities in mind rather than feed values. However, some feed grain varieties have been developed in recent years. Wheat is a versatile crop that is adapted to a wide variety of environmental conditions. The tonnage produced on a worldwide basis rivals that produced from rice, and wheat is grown in many different areas (Fig. 7-8).

Wheat is divided into hard and soft wheats. Within each type there are varieties classed as winter or spring wheats. Winter wheats are normally planted in the fall, while spring wheats are planted in the spring in areas where the winter is more severe. Soft wheats may be white or red. The major difference between soft and hard wheats is that hard wheats are higher in protein content (13% to 16%) and contain more gluten and gliaden proteins, which provide the sticky consistency important in bread making. White wheats contain less pro-

FIGURE 7-8 Grain combines harvesting wheat in the Pacific Northwest.

tein (8% to 11%) and are used more commonly for pastries, in breads that do not require high gluten levels, or in mixtures of flour with hard wheat (all-purpose flour).

The amino acid distribution of wheat is more favorable than that of corn, especially for lysine, tryptophan, methionine, cystine, and histidine. Soft wheats generally have lower levels of the essential amino acids (as percent of dry matter), but they are still a much better source than corn of those amino acids listed for hard wheats, except for tryptophan.

Wheat is a very palatable grain for most species. As an energy source, most comparative studies indicate that it does not produce quite as rapid a gain as corn does, but it very frequently produces more efficient gain than corn and, in some cases, the carcasses of fattened cattle are apt to have less subcutaneous fat but equally as good marbling as corn produces when wheat is fed as a high percentage of the diet (Table 7-9). If fed in large amounts, ground wheat tends to form a pasty mass on the beaks of birds. For ruminant animals, wheat is more prone to produce acidosis than any other cereal grain. Thus it is advisable to adapt animals to wheat over a period of time, and many nutritionists advise not feeding over 50% of the diet to fattening cattle and less than this to fattening lambs.

Feeding "new" wheat can cause problems in poultry and other species, if comments from producers and mill operators are correct. However, research data on the topic are not available. If such problems occur, they are presumably related to the solubility of the starch granules and/or the protein matrix surrounding the granules.

In more humid areas, weather conditions often result in lodging (plants fall over and mat on the ground). If the grain is not harvested quickly, the grain may sprout; this makes it unfit for bread making and, consequently, reduces the market value. However, there does not seem to be much effect, if any, on the feeding value of wheat that has a high percentage of sprouted kernels (Table 7-10). Molding is likely to occur under such conditions, however, and mold might reduce the feeding value in some conditions.

TRICALE

Triticale is a cereal grain derived from crosses between wheat and rye. Its nutrient content (Table 7-4) is similar to that of hard wheat or rye with, perhaps, more fiber. It does not have a particularly good distribution of amino acids such as lysine, methionine, and cystine. The feeding value is similar for nonruminant species to that of wheat (6), but with some supplemental lysine, some cultivars can be equal to a corn–soy diet for pigs, according to recent data. Note in Table 7-11 that gains, feed conversion, and back fat thickness were quite similar for pigs finished on triticale + lysine or corn–soy diets. In a study comparing corn and triticale, these two grains were fed in a partially purified diet in which the only source of protein was from the grain (16). In this instance (which may not be characteristic of a complete diet), the DE value for the corn was 4.08 kcal/g, and for each of two triticale varieties the DE values were 3.77 and 3.56 kcal/g. Respective ME values were 3.66, 3.19, and 3.12 kcal/g. The feeding value of triticale for finishing cattle appears to be somewhat lower than that of wheat, and the efficiency appears to be lower (Table 7-9). As with some other grains, triticale is subject to ergot infestation (ergot is a toxic fungus

TABLE 7-9

Comparison of triticale, corn and wheat in finishing rations for cattle

	Grain Source[a]		
Item	Corn	Wheat	Triticale
Daily feed consumed, kg	10.4	9.1	9.1
Daily gain, kg	1.33	1.25	1.13
Feed/gain ratio	7.81	7.24	8.02
Carcass marbling score	7.5	8.1	7.8
Adjusted fat thickness, cm	1.26	0.97	1.04

Source: Reddy et al. (13).

[a]Rations were 74% grain, 15% cottonseed hulls, 10% cottonseed meal, and 1% salt.

TABLE 7-10

Comparison of wheat and sprouted wheat with barley in finishing rations for cattle

	Grain Source[a]		
Item	Barley	Wheat	Sprouted Wheat
Feed consumption, kg/d	9.44	9.16	9.08
Daily gain, kg	1.32	1.30	1.29
Feed/gain ratio	7.15	7.05	7.05

Source: Preston et al. (14).

[a]Rations contained the following amounts of steam-rolled grain: barley, 77% barley; wheat, 50% wheat, 27% barley; sprouted wheat, 50% wheat that was 58% sprouted, 27% barley; the remainder was made up of 10% grass–legume hay, 10% supplement, and 3% cane molasses.

TABLE 7-11

Comparison of corn–soy rations with triticale for finishing swine

	Ration	
Item	Corn–Soy	Triticale + Lysine
Weight gain, kg	35.9	35.2
Average daily gain, kg	0.92	0.90
Feed/gain ratio	3.23	3.30
Average back fat thickness, cm	2.82	2.87

Source: Hale and Utley (15).

that replaces some of the kernels). At present, triticale is not grown widely as a feed grain.

RYE (*SECALE CEREALE*)

Rye is an important bread grain in northern Europe. It has a composition similar to that of hard wheat with, perhaps, a slightly higher protein content. Rye is generally said to be less palatable than most other cereal grains, and it is more susceptible to ergot than is wheat. Rye is not normally fed to poultry or, if so, in only small amounts. Studies with poultry indicate that rye contains at least two detrimental factors—an appetite-depressing factor in the bran and a growth-depressing factor in all fractions of the grain (16). For swine, data indicate a value of about 80% of corn for young animals (6).

Rye is often said to be unpalatable and difficult to masticate for ruminants when fed in large amounts. However, studies with lactating dairy cows (Table 7-12) do not indicate any problem in feeding up to 25% of the concentrate mix as rye (very little ergot was found), nor was there any effect on fat-corrected milk production when fed at this level. However, a study from the same laboratory showed that daily gain and feed consumption of calves were reduced when rye made up 60% of the starter mix and was fed from 6 to 18 weeks of age.

Roasting the rye grain restored performance to the level obtained with no rye. Other relatively recent studies with cattle suggest that rye should be restricted to perhaps 40% or less of the diet if depressing effects on performance are to be avoided.

RICE (*ORYZA SATIVA*)

Rice is grown in many areas of the world. Although most of the rice is grown in areas that can be flooded during active growth of the plant, dryland varieties of rice are also grown. Rice is always grown as a food grain (for human use) and, according to the Rice Council, more than 40,000 types of rice are grown in one place or another. As a food grain, rice is not normally cheap enough to use as a feed grain (for animal use). Occasionally, rough rice (unmilled) is available at attractive prices for animal feeding. Rough rice has about 8% crude protein, 9% fiber, and 1.7% ether extract (Table 7-4). The hulls make up about 25% of the kernel weight, and they greatly reduce the energy value and are, themselves, almost totally indigestible.

MILLET

Several different types of millet are grown, primarily in Asia and west Africa, for human food. They include a variety of small-grain cereal plants. In the United States, proso (*Panicum milliaceum*), a plant resembling some sorghums, is sometimes grown for a feed grain and is reported to be intermediate in feeding value between oats and corn. Others such as foxtail millet (*Setaria italica*), pearl millet (*Pennisetum glaucum* or *typhoideum*), Japanese millet (*Echinochloa crusgalli*), or finger millet (*Eleusine coracana*) are used for human food in most countries where they are grown or in other areas for forage or such things as bird seed or beer. Other than the uses mentioned, the millets are not an important source of animal feed.

TABLE 7-12

Comparative value of rye and barley for lactating dairy cows

	Grain Source[a]		
Item	Barley	25% Rye	50% Rye
Dry matter consumed, kg/d			
Concentrate mix	10.9	10.3	10.4
Total	20.6	19.4	19.1
FCM production, kg/d	21.6	21.3	19.8

Source: Sharma et al. (17).

[a]Concentrate contained 80% grain, of which 5% was oats and the remainder contained soybean meal, molasses, and minerals.

RELATIVE FEEDING VALUES OF CEREAL GRAINS

Although a considerable amount of information has been given on the nutrient values of the cereal grains, some additional information is in order. Cereal grains are primarily a source of energy; thus energy content is the most important basis of comparison, and corn is generally used as a standard against which the others are compared. If the relative value of corn is set at 100, the value of the other grains is usually lower, as illustrated in Table 7-13. This is a reflection, in part, of the differences in crude fiber and resultant lower digestibility of those grains having higher fiber levels. This can be confirmed by removing the hulls of grains such as oats and barley. The remaining material (oat groats or pearled barley) is highly digestible. The starch of corn is also highly digestible. In addition, as fiber increases, the content of starch and other readily available carbohydrates decreases (Table 7-4). The fat content has some effect on energy values, and it is also possible that there may be differences in utilization of the amylose and amylopectin in the different grains. Evidence also shows that starch granules in mature grains are more digestible by rumen microorganisms than are less mature kernels. Relative feeding values may be affected by various feed processing methods, because some grains are improved to a greater extent than others (see Ch. 11).

Corn and milo have less protein than the other feed grains and thus are not ranked as high on a relative basis as a protein source. The quality of corn protein is also low because it is low in lysine and lower than most grains in tryptophan.

TABLE 7-13

Relative value as compared to corn of the other cereal grains for crude protein content and metabolizable energy

Grain	Crude Protein	ME		
		Ruminants	Swine	Poultry
Corn	100	100	100	100
Barley	124	96.2	88.6	74.5
Millet, proso	118	96.2	89.0	86.5
Milo	114	98.8	96.3	96.7
Oats	122	87.1	80.9	75.0
Triticale	161	96.2	91.2	92.2
Rye	127	96.2	89.3	78.6
Wheat[a]	132	101.5	97.4	94.8

Source: NRC (18).

[a]Hard red winter wheat.

Research has gone on with feed grains for many years, but results are still reported rather frequently. Some data from three reports on swine are given in Table 7-14. These data are shown to illustrate the problem of comparative evaluation. In the first two experiments, pigs were fed the test grains in addition to a basal diet believed to have an excess of required nutrients other than energy. In the third experiment pigs were fed grain plus a mineral–vitamin supplement. The latter experiment would be a better estimate of the energy *and* amino acid contribution, while the former would be only a measure of energy contribution. With our current knowledge of the nutrient content of grains and of animal requirements, it is not logical to feed grain by itself; neither is it logical to evaluate grains without appropriate supplementation.

MILLING BY-PRODUCTS OF CEREAL GRAINS

The milling of cereal grains for production of flour and various other food or industrial products results in the production of a number of by-products used in the feed trade. For the United States, official descriptions (names, numbers, description of the product) are published by the American Feed Control Officials, Inc. (19). Essentially all these products are listed in Appendix Table 1 with data on nutrient content. Therefore, less information will be presented here than has been given previously on the cereal grains. In most cases the official description will be given as a minimum.

Grains are milled either by dry or wet milling processes (in addition to grinding, rolling, and other processes that may be done to feed grains). The dry milling methods are designed primarily to grind away the hulls (as with barley and rice) or to remove the outer layers of the seed to expose the starch endosperm for flour production. Wet milling methods are intended for the production of such products as starch, sugar, syrup, or oil for human food from corn and, to a much lesser extent, sorghum grains. The types of processes involved are illustrated in Fig. 7-9. Several different by-products result that are used for animal feeds.

WHEAT BY-PRODUCTS

With wheat, milling by-products account for about 28% of the intact kernel, the remainder being flour prepared for human food. Wheat millfeeds are usually classified and named on

TABLE 7-14

Comparative value of cereal grains for pigs

	Grains						Reference No.
Measurement	Corn	Oats	Wheat	Barley	Triticale	Milo	
Daily gain,[a] g	293	259					19
Feed conversion	1.17	1.36					
Digestible energy, kcal/g	3.43	2.84					
Daily gain,[a] g			251	230			20
Feed conversion			1.25	1.27			
Digestible energy, kcal/g			3.82	3.76			
Digestible energy,[b] kcal/g	3.80		3.71	3.38	3.60		21
Digestible energy, kcal/g	3.84	3.18	3.86	3.52		3.82	NRC

[a]Grains were fed at an average of 1.5% of body weight in addition to a basal diet fed at 3% of body weight per day. All nutrients except energy were considered to be fed in excess of requirements.

[b]Grains were fed with only a mineral and vitamin supplement.

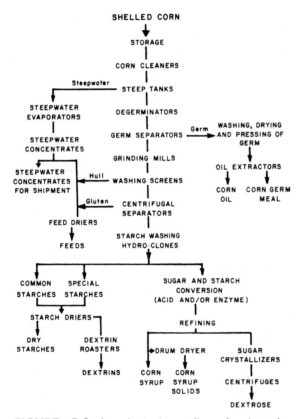

FIGURE 7-9 A schematic outline showing the processes involved in the wet milling of corn. As indicated, the wet-milling process is used primarily for starch, sugar, and syrup production. Other feed ingredients produced are by-products of the process.

the basis of decreasing fiber as bran, middlings, mill run, shorts, red dog, and wheat germ meal. The amount of each produced depends on the type of mill and the type of wheat being milled. Formula feed manufacturers use about 90% of the millfeed in the United States, with the largest users being the poultry industry. Typi-

cal specifications on wheat millfeeds are given in Table 7-15, and a brief description follows.

Wheat bran is the coarse outer covering of the wheat kernel as separated from cleaned and scoured wheat in the usual process of commercial milling. The appearance of bran is that of a flaky brown material. About 45% of total wheat millfeed is comprised of bran. **Wheat middlings** consist of the layer of the kernel just inside the outer bran covering (aleurone), endosperm, and bran particles. The appearance is that of a brownish, finely ground meal. Midds usually amount to about 40% of total millfeed. Lysine and threonine appear to be the most limiting amino acids in the midds of hard red spring wheat. **Wheat mill run** is a blend of bran and middlings. **Shorts** consist of fine particles of bran, germ, flour, and tailings. They contain somewhat more flour than midds and have the appearance of a finely ground meal containing somewhat less brownish material than midds. **Red dog** consists of mill tailings together with some fine particles of bran, germ, and flour. There are more floury particles than in any other millfeed. The appearance is much like grayish flour flecked with small brown

TABLE 7-15

Usual specifications on wheat mill feeds

Wheat Mill Feeds	Minimum Protein (%)	Minimum Fat (%)	Maximum Fiber (%)
Bran	13.5–15	2.5	12.0
Middlings	10–14	3	9.5
Mill run	14–16	2	9.5
Shorts	14–16	3.5	7
Red dog	13.5–15	2	4

Source: AFCO 1995 (19).

bran particles; red dog accounts for only 4% of total millfeed. **Wheat germ** is classed as a protein supplement and will be discussed in Ch. 9. Although these various milling by-products are supposed to be standardized, this is not always done, with the result that there may be more variability than indicated in Table 7-15.

Wheat millfeeds, particularly bran, are relatively bulky and laxative feeds, but they are quite palatable to animals except when used at a very high percentage (more than 40% to 50% in rations for fullfed cattle, when the fine particle size may decrease consumption). They generally contain more protein than the parent grain, and protein quality is usually somewhat improved, although wheat millfeeds are apt to be relatively deficient in lysine and methionine, as well as some other essential amino acids.

The bulky nature of wheat millfeeds tends to restrict intake by animals; thus maximal energy intake cannot be achieved on rations containing high levels. The energy values of wheat millfeeds have not been as well defined as might be desired. The available data indicate that ruminants utilize the energy to a greater extent than do monogastric species. Table 7-16 shows that pigs did well when wheat midds replaced up to 30% (by weight) of the corn in one experiment and up to 60% in finishing rations in a second experiment (not shown). Maximum protein utilization was at the 20% level and maximum ME at the 40% level. Overall ME was estimated to be 2.99 kcal/g, somewhat lower than the NRC (18) value of 3.25 kcal/g. In a recent report on lactating dairy cows, cows were fed rations with 60% concentrate. When the concentrate was altered to contain 20% or 40% wheat midds in one experiment or 40% to 60% in the second (with, by calculation, isocaloric diets), the cows consumed the rations with 40% of the concentrate as wheat midds without any change in production. However, 60% midds was accompanied by a reduction in milk production.

TABLE 7-16

Effect of increasing levels of wheat midds on performance of growing–finishing pigs

	Percentage of Midds Replacing Corn			
Item	0	10	20	30
Daily gain, kg	0.65	0.66	0.67	0.66
Daily feed, kg	1.82	1.85	1.85	2.00
Feed:gain	2.79	2.81	2.72	3.12
Back fat, cm	2.7	2.6	2.7	2.6

Source: Erickson et al. (20).

These millfeeds are also relatively good sources of most of the water-soluble vitamins, except for niacin, which is almost entirely unavailable. The Ca content is low, but the content of P, Mg, and Mn is relatively high. Most of the P is found as phytate P, which is only partially available to simple-stomached animals; some data indicate that heat treatment may increase the P availability. The content of trace minerals is also usually higher than in the parent kernel.

The bulk of wheat millfeeds is used for feeding poultry, swine, and dairy cattle. When available, bran is a favored feedstuff for all breeding classes of ruminants and for horses.

CORN MILLFEEDS

With dry milling methods, the primary products are corn meal, hominy grits, or flour (foodstuffs). Feedstuffs include hominy feed, corn flour, and corn bran. **Hominy feed** is a mixture of corn bran, corn germ, and part of the starch portion of either white or yellow corn kernels or a mixture thereof as produced in the manufacture of pearl hominy, hominy grits, or corn meal for table use, and it must contain no less than 4% crude fat. It is also referred to as **corn grits by-product.** Hominy feed is the most important feed on a volume basis. It may be sold also as solvent-extracted hominy feed, in which case the fat content is appreciably lower. Hominy feed is an excellent energy source for both monogastric and ruminant animals and is considered equal to or superior to whole corn as an energy source. **Corn flour** is the fine-sized, hard, flinty portions of ground corn containing little or none of the bran or germ. **Corn bran** is the outer coating of the corn kernel, with little or none of the starchy part or germ. **Gelatinized corn flour** is obtained from the endosperm of corn that has been gelatinized and reduced to a finely ground meal and must contain not more than 1% crude fiber. **Corn germ** is classified as a protein supplement because of its crude protein content.

As indicated in Fig 7-9, different products are available from the wet milling process. When 100 kg of corn is processed with wet milling, the average yields of products are as shown (kg): pearl starch, 62.5; oil, 2.86; animal feed, 34.6. About 71 kg of syrup or 49 kg of refined corn sugar may be obtained from the starch (22). Of course, the starch, sugar, syrup, and corn oil are used as human food. Except for starch molasses and liquefied corn product, which are discussed in a section on liquid feeds, the other feedstuffs from wet milling of corn—gluten feed, gluten meal, corn germ meal, condensed fermented corn extractives, and hydrolyzed corn protein—are classed as protein supplements and will be discussed in Ch. 8.

OTHER CEREAL MILLFEEDS

Some sorghum grain is wet milled in the same manner as corn. Feedstuffs classed as energy feeds from the process include **grain sorghum grits,** which consist of the hard flinty portions of sorghum containing little or no bran or germ. **Germ sorghum millfeed** is a mixture of grain sorghum bran, germ, part of the starch portion of the sorghum kernel, or a mixture thereof as produced in the manufacture of sorghum grits and refined meal and flour. It must contain not less than 5% crude fat and not more than 6% crude fiber.

By-products from barley include **barley hulls** (a roughage), **pearl barley by-product,** which is the entire by-product resulting from the manufacture of pearl barley, and **barley mill by-product,** which is the entire residue from the milling of barley flour and is composed of barley hulls and barley middlings.

As with barley, oats are processed primarily to separate the hull from the groat (inner portion of the kernel). **Oat groats** are defined as cleaned oats with the hulls removed. The groat is used both for food and feed products. **Feeding oat meal** is obtained in the manufacture of rolled oat groats or rolled oats and consists of broken rolled oat groats, oat groat chips, and floury portions of the groats, with only such quantity of finely ground oat hulls as is unavoidable in the usual process of commercial milling. It must not contain more than 4% crude fiber. **Oat hulls** consist primarily of the outer coverings of oats, obtained in the milling of table cereals or in the groating of oats. **Oat mill by-product** is the by-product obtained in the manufacture of oat groats, consisting of oat hulls and particles of the groat and containing not more than 22% crude fiber. **Clipped oat by-product** is obtained in the manufacture of clipped oats. It may contain the light chaffy material broken from the end of the hulls, empty hulls, light immature oats, and dust. It must not contain an excessive amount of oat hulls.

Rice is also milled primarily to remove the hull from the kernel. **Rice bran** is the pericarp or bran layer and germ of the rice with only such quantity of hull fragments, chipped, broken, or brewers rice, and calcium carbonate as is unavoidable in the regular milling. It must not contain more than 13% crude fiber. **Solvent ex-**tracted rice bran is also produced. **Rice polishing** is a by-product of rice obtained in the milling operation of brushing the grain to polish the kernel. Both of these two products are excellent feed materials comparatively high in fat (unless extracted), but the fat is oxidized easily unless it has been stabilized. Other by-products include **rice by-products fractions,** which are obtained by screening and aspirating ground rice hulls. They are used primarily as a pelleting aid and is composed of such fine particles of ground rice hulls, spongy parenchyma, and minute amounts of rice flour, rice germ, pericarp, and rice starch as will pass an 80-mesh screen. They contain not less than 5% crude protein, 1.5% crude fat, and not more than 25% crude fiber. **Rice hulls** are also a by-product, but they are of little if any value as a feed because of extremely low digestibility, unless they are processed with steam heat and added chemicals.

With regard to rye milling, by-products include **rye mill run,** which is obtained in the usual milling of rye flour and consists principally of the mill run of the outer covering of the rye kernel and the rye germ, with small quantities of rye flour and aleurone. Rye mill run must not contain more than 9.5% crude fiber. **Rye middlings** consist of rye feed and rye red dog combined in the proportions obtained in the usual process of milling rye flour and must not contain more than 8.5% crude fiber. Only very limited amounts of either are available.

Most of the millfeeds obtained from oats, barley, rice, and rye are relatively high in fiber or higher than the parent grain. The quality of the protein is not particularly appealing for monogastric species, although most millfeeds contain more protein than the original grain (hulls excluded).

Grain screenings are foreign materials obtained in the process of cleaning grains and seeds. They are likely to be extremely variable in composition and may contain broken seeds, numerous weed seeds, and other materials. When used in commerce, they must be identified as **grain screenings, mixed screenings,** and **chaff and/or dust.**

HIGH-CARBOHYDRATE LIQUID FEEDS

MOLASSES AND SIMILAR LIQUIDS

Molasses is a major by-product of sugar production, the bulk of it coming from sugar cane, but other sources include sugar beets, citrus fruits, starch, and wood. Various molasses types are standardized in terms of degrees Brix. This is determined by use of a refractometer and corresponds very closely to percentage of dry matter. In commercial use, molasses (cane) is usually adjusted to about 25% water content. Molasses may be dried for mixing into dry diets, although at appreciably higher costs.

New sugar extraction procedures are being adapted by many of the sugar manufacturers and the composition of the liquid by-products is changing. For example, many of the sugar beet plants are no longer producing beet molasses, but rather **concentrated separator by-product. Cane molasses** must contain not less than 43% sugars and have a density not less than 79.5° Brix, while **beet molasses** must contain not less than 48% total sugars, but with the same minimum density as for cane molasses. **Citrus molasses** is the partially dehydrated juices obtained from the manufacture of dried citrus pulp and must contain not less than 45% total sugars and have a density not less than 71° Brix. **Starch molasses** is a by-product of the manufacture of glucose from starch derived from corn or grain sorghums in which the starch is hydrolyzed by use of enzymes and/or acid. It must contain not less than 50% total sugars and not less than 73% total solids. **Hemicellulose extract** is a by-product of the manufacture of pressed wood. It is the concentrated soluble material obtained from the treatment of wood at elevated temperature and pressure without use of acids, alkalis, or salts. It contains pentose and hexose sugars and must have a total carbohydrate content of not less than 55%. **Lignin sulfonate** is produced when the lignin component of wood is solubilized by a combination of sulfonation and hydrolysis during the conversion of wood to wood pulp by the sulfite process. Depending on the method used, water-soluble salts of Ca, Na, or ammonium may be present. The liquor is sold as such or concentrated or dried, depending on intended usage.

Cane or blackstrap molasses is utilized widely as a feedstuff, particularly for ruminants. In the United States alone, more than 2.5 million T of cane and beet molasses is used annually, and large amounts are used in Europe and in other areas (semitropical or tropical) where it is produced. Worldwide about 10 million T of cane and beet molasses is produced each year. That produced contains less sugar than formerly because of more efficient extraction methods. Major exporters (1995) were Pakistan, Thailand, Indonesia, the Philippines, and Brazil. In 1979, Mexico was second and Cuba fourth in exports, but there has been a shift in production, particularly an increase by Pakistan.

Unfortunately, there is either little agreement of the typical analysis of molasses or the products are quite variable. No doubt, the composition varies from sugar mill to sugar mill. In a paper from the Netherlands (23), the authors quoted the following values from different molasses analyses: beet molasses, ash, 6.9% to 11.6%; crude protein, 4.7% to 14.0%; cane molasses, ash, 8.7% to 12.3%; crude protein, 4.1% to 6.5%. Digestibility values were: beet molasses, organic matter, 86% to 94%; crude protein, 34% to 71%; NFE, 91% to 98%; cane molasses, organic matter, 84% to 91%; crude protein, 0% to 60%; NFE, 87% to 93%. Comparable values can be found by comparing almost any source on nutrient content of molasses. For example, the NRC publication on beef gives TDN values of beet, cane, and citrus molasses of 79%, 72%, and 75%, respectively. The NRC on sheep gives values of 77% and 79% for beet and cane molasses, respectively.

Recent analyses from the Netherlands on cane and beet molasses are shown in Table 7-17. On an organic matter basis, the total sugar content was about 75% and 73% for beet and cane molasses, respectively. About 90% of total sugars in beet molasses was recovered as sucrose, and only small amounts were found as fructose or glucose. About 60% of total sugars in cane molasses was found to be sucrose and 30% fructose + glucose. However, in both products about 10% of total sugars were not identified. The authors state that part of this undetermined fraction is unfermentable material that

TABLE 7-17

Major components of cane and beet molasses

	Molasses Source	
Item, %	Beet	Cane
Dry matter	80.3	78.8
Ash	8.0	12.6
Crude protein	14.8	5.9
Betaine	5.2	0.1
Amino acids	7.0	2.2
Hexane extract	0.3	0.3
Crude fiber	0.1	0.3
NDF	—	—
Sugar	69.5	63.9
Sucrose	66.0	44.0
Fructose	1.0	13.0
Anhydro uronic acids	19.8	15.1
Ammonia	0.1	0.1

Source: Steg and Van Der Meer (23).

is not sugars, but reducing agents, possibly formed by the combination of normal reducing sugars with N compounds. For beet molasses, about 35% of the N was undetermined, and most of the N in cane molasses was not recovered in amino acids.

The ash content of molasses is variable, as indicated, and is largely made up of K, Ca, Cl, and sulfate salts. Cane molasses is usually a good source of most of the trace elements, but it has only moderate to low vitamin content. Part of the variability in ash (and other components) is, undoubtedly, caused by differences in sugar manufacturing procedures. Age, type and quality of sugar cane, soil fertility, and system of collection and processing also have a bearing on the composition of cane molasses.

The sweet taste of molasses, whatever the source, makes it appealing to most animal species. In addition, molasses is of value in reducing dust in feeds, as a pellet binder, as a vehicle for feeding medicants or other additives, and as a liquid protein supplement when fortified with an N source (see Ch. 8). The cost is often attractive as compared with that of grains.

Most molasses products are limited in use because of milling problems (sticky consistency) or because levels exceeding 15% to 25% of the ration are apt to result in digestive disturbances, diarrhea, and inefficient animal performance. The diarrhea is largely a result of the high level of various mineral salts in most molasses products. The problem is not caused by the sugar, because pigs or ruminants can utilize comparable amounts of sugar provided in other forms. High-test molasses, which has a lower ash content, can be fed at very high levels to either pigs or cattle without any particular problem, but not much high-test molasses is available for animal feeding. Under some conditions, feeding very high levels of molasses to cattle results in molasses toxicity, which resembles cerebrocortical necrosis, an induced thiamin deficiency. However, limited data on the topic suggest that the condition is caused by an abnormal metabolism of carbohydrates by brain tissue (24).

Most data indicate that cane and beet molasses, if fed in limited amounts (that is, less than 10% of the diet), are equivalent to a good-quality grain source for replacement of energy. When fed in increasing amounts (Table 7-18), animal performance tends to decrease in level of performance and efficiency of feed utilization. There is also a fair amount of information on hemicellulose extract (two examples are shown in Table 7-18) that indicates a compara-ble response to cane molasses when fed in restricted amounts. Much less information is available on starch and citrus molasses, but they appear to be comparable to other sources as an energy source.

Other liquid feeds containing moderate to high levels of sugars include **condensed soybean solubles,** which are a by-product of washing soy flour or flakes with water and acid. This product contains about 6% to 7% crude protein, 33% sucrose, and about 57% total soluble carbohydrates (dry basis). As shown in Table 7-18, it can be a satisfactory replacement for corn when fed at levels of up to 10% of the diet to lambs.

Liquefied corn product results from steam cooking and enzymatic treatment of the corn without removing any of the component parts. It contains not less than 30% solids. On a dry-matter basis it contains 8% to 9% crude protein. The author is not aware that much of this product is produced or used.

Other liquid products that contain relatively little soluble or fermentable carbohydrate include **extracted streptomyces solubles,** a by-product of streptomycin production. It contains about 17% to 19% crude protein (dry basis). **Condensed molasses solubles** (also called stillage), a by-product of rum or ethanol production, is available in limited amounts. It is a high-ash product with about 16% to 19% crude protein, (dry basis). Both of these products can be used in limited amounts with ruminant animals (29, 30). **Citrus condensed molasses solubles,** a residue from the fermentation of alcohol, contains about 10% to 12% crude protein (dry basis). Levels of up to 20% of the diet can be fed to cattle without depressing gain, although feed conversion appeared to decrease; with lambs, a level of 20% depressed gain as compared to 10% or 0%, and digestibility of crude protein was depressed at the 20% level (31).

LIQUID MILK BY-PRODUCTS

Several liquid by-products result from the production of cheese and/or recovery of products from whey. These include **fresh whey, acid whey, condensed whey,** and **dehydrated (dried) whey.** Whey is the liquid fraction of milk remaining after the removal of casein and butterfat in cheese making. Most of the lactose, minerals, and water-soluble protein present in milk remain in whey. When whey is passed through an evaporator so that only water is removed, the resulting product is defined as condensed whey (or whey concentrate). When sold

TABLE 7-18

Performance of animals fed molasses or other liquid energy sources

Treatment	Daily Feed Consumed, kg DM	Daily Gain, kg	Milk Production, kg/d	Feed/ Gain Ratio	Digestible Energy, %
Barley diet, finishing steers[a]					
Control (no molasses)	10.1	1.13		8.98	
9.6% molasses	10.4	1.12		9.08	
14.4% molasses	10.0	1.03		9.69	
High corn diet, finishing steers[b]					79.2
+10% cane molasses	9.3	1.36		6.83	83.8
+10% hemicellulose extract	9.1	1.30		7.00	79.6
+10% experimental wood molasses	9.2	1.35		6.80	84.1
Cottonseed hull roughage, lactating dairy cows[c]					
No molasses	20.0		20.8		
8% cane molasses	20.5		20.9		
8% hemicellulose extract	20.9		20.8		
Corn, oats diet, fattening lambs[d]					
Control	1.4	0.24		5.8	
5% condensed soy solubles	1.5	0.24		6.1	
10% condensed soy solubles	1.5	0.23		6.4	
15% condensed soy solubles	1.5	0.21		7.2	

[a]From Heinemann and Hanks (25). Basal diet was 53.4% barley and 22.9% beet pulp. For treatments with molasses, the barley–beet pulp fed was reduced by 10% and 20%, respectively.

[b]From Crawford et al. (26). The basal diet contained 81% corn. When liquids were fed, they were mixed with the basal diet prior to feeding.

[c]From Vernlund et al. (27). Molasses or hemicellulose extract replaced corn meal.

[d]From Perry et al. (28). Condensed soy solubles replaced ground corn.

as condensed whey, the minimum percentage of total whey solids must be declared on the label. A typical product with 40% to 50% solids will have 10% to 13% crude protein (DM basis) and 55% to 70% lactose. Dehydrated (dried) whey is dried to the point of less than 10% moisture. This results in a more expensive product, but one that costs less to transport long distances. Liquid and condensed wheys are subject to rapid spoilage unless steps are taken to prevent it.

Condensed whey has been used in liquid supplements for cattle and to a lesser extent in formulas for swine and poultry as a source of nutrients and to improve palatability and feed texture. Dried whey has long been used in starter feeds for poultry and baby pigs and in milk replacers for young ruminants. Liquid whey is more of a problem to utilize because of its low dry-matter content (4% to 5%). However, when it is available close at hand to livestock operations (so that transportation costs are minimal), it can be used very effectively. This is illustrated in Table 7-19. Data in the table show that high-producing cows consumed an average of 54 l/day, resulting in less concentrate consumption but somewhat more total dry-matter consumption and greater milk production. Schingoethe (33) has pointed out that ruminants can consume up to 30% of their dry matter from liquid whey without impaired performance, although

TABLE 7-19

Use of liquid (fresh) whey for high-producing dairy cows

	Dietary Treatments	
Item	Control	Fed Whey
Feed consumed per day		
Liquid whey, ℓ		54.1
Whey DM, kg		2.45
Concentrate, kg DM	16.4	14.7
Hay DM, kg	3.2	3.6
Total DM, kg	19.8	20.7
Milk yield, kg FCM/day	25.0	27.2
FCM milk/feed DM ratio	1.27	1.31

Source: Pihchasov et al. (32). The concentrate was made up primarily of sorghum, barley, corn, and wheat bran with supplemental protein, minerals, and vitamins. The concentrate, whey, and hay (medium quality vetch–oats) were fed ad libitum. Data shown are for 70 to 105 days of lactation.

swine may develop diarrhea when more than 20% of diet dry matter comes from liquid whey. Small amounts of whey or other whey by-products often increase weight gains and feed efficiency and improve nutrient utilization by cattle. Whey may also prevent milk fat depression in dairy cows without reducing concentrate consumption markedly.

Other by-products of whey include **condensed whey solubles,** which is obtained by evaporating whey residue from the manufacture of lactose after removal of milk albumin and partial removal of lactose. This product contains 40% to 50% solids with about 10% crude protein and 60% to 70% lactose (DM basis). **Condensed whey-product** is the residue obtained by evaporating whey from which a portion of the lactose has been removed. It will contain 44% to 48% dry matter with 15% to 16% crude protein and 60% to 70% lactose and lactic acid. These two products are used in a similar manner to condensed whey.

BY-PRODUCT DRY FEEDS

BEET PULP

Beet pulp is the residue remaining after extraction of sugar from sugar beets. In some localities, feeders near the processing plants may feed wet pulp. However, a high proportion of the pulp is dried, and frequently beet molasses is added to the pulp before drying. It may be sold in shredded or pelleted form. Beet pulp is highly favored in rations for lactating cows. Its physical texture resembles that of a roughage, but the pulp is much more digestible than roughages and it is also quite palatable. Although the crude fiber content is high for a concentrate (16%), the fiber is quite digestible, partly because the lignin content is low. Relatively recent digestion data with sheep and lactation studies with cows indicate that beet pulp provides as much energy as corn when fed in complete rations in amounts up to 70% of the total. Usual values given for digestible energy are appreciably lower than for most cereal grains.

CITRUS PULP AND MEAL

Citrus by-products are prepared from the residue resulting from the manufacture of citrus juices. The residue is shredded or ground, pressed to remove juices, and dried. Calcium hydroxide may be added before pressing. Dried citrus meal is composed of the finer particles obtained by screening dried citrus pulp. The protein content is low (5% to 8%), the fiber content is moderately high (11% to 12%), and the Ca content may be high. Data on ruminant animals indicate that citrus pulp is relatively palatable, and quantities approaching 50% to 60% can be used if desired. The fiber is quite digestible, and the energy value approaches that of some of the cereal grains. However, relatively large amounts (3 to 4 kg/day) have been shown to cause abnormal tastes in milk fat, with no effect when 2 kg/day were fed (34).

DRIED BAKERY PRODUCT

This is a feed produced form reclaimed (unused or stale) bakery products or other materials such as candy, inedible flours, unsalable nuts, and the like. The materials are blended and ground to produce a feedstuff that contains 9% to 12% crude protein, 10% to 15% fat, and a low level of crude fiber. It will be variable because the ingredients tend to vary from day to day and season to season. While relatively little is available, it is an excellent feed because the digestible energy is high, most of it being derived from starch, sucrose, and high-quality fats. It is well utilized by pigs and is a preferred ingredient in starter rations. It is also highly favored in rations for lactating dairy cows when available.

CASSAVA MEAL

A tropical root crop (*Manihot esculenta*), which goes by names such as cassava, yucca, manioc, tapioca, or mandioca, is of great potential importance as a livestock feed in tropical areas. In experimental plots it has yielded as much as 75 to 80 tons/ha/yr. This is much more than can be produced by rice, corn, or other grains adapted to the tropics.

Although it is strictly a tropical plant, significant amounts of dehydrated cassava meal are now used in the United States and Europe for feeding livestock. Cassava root contains about 65% water, 1% to 2% protein, 1.5% fiber, 0.3% fat, 1.4% ash, and 30% NFE. Thus its dry matter is largely readily available carbohydrates. Dried cassava is equal in energy value to other root crops and tubers and can be used to replace all of the grain portion of the diet for growing–finishing pigs if the amount of supplemental protein is increased to compensate for the very low protein content of cassava. It can also be used as the main energy source in diets of gestating and lactating swine. The stalk and leaf portions of the plant are well utilized by ruminants, but this portion is too high in fiber for monogastrics.

Freshly harvested cassava roots and leaves may be high in hydrocyanic acid (a very toxic material). Oven drying at 70° to 80° C, boiling in water, or sun drying are effective in reducing the HCN content. Principal sources of the dried product are Southeast Asian countries at this time.

ROOTS AND TUBERS

Root crops used for feeding animals, particularly in northern Europe, include turnips, mangolds, swedes, fodder beets, carrots, and parsnips. These crops frequently are dug up and left lying in the field to be consumed as desired when used as animal feed. The bulky nature of these feeds limits their use for swine or poultry, so most are fed to cattle or sheep.

Root crops are characterized by their high water content (75% to 90%), moderately low fiber (5% to 11%, DM basis), and crude protein (4% to 12%) content. These crops tend to be low in Ca and P and high in K. The carbohydrates range from 50% to 75% of the dry matter and are mainly sucrose, which is highly digestible by ruminants and nonruminants. Animals (sheep, cattle) not adapted to beets or mangolds (both *Beta vulgaris*) tend to be subject to digestive upsets, probably because of the high sucrose content.

Some root crops, turnips in particular, can be used in a double cropping system. They can be planted after harvest of small grain and still get substantial production for grazing in the fall and early winter in some areas. Both cattle and sheep do quite well when grazing on turnips with some supplementary feed. They learn rapidly to eat out the root portion of the plant, which is mostly below ground.

POTATOES

Surplus or cull white potatoes (*Solanum tuberosum*) are often used for feeding cattle or sheep in areas where commercial potato production occurs. Potatoes are high in digestible energy (dry basis), which is derived almost entirely from starch. Water content is 78% to 89%, crude protein content is low, and the quality of the protein is poor. The Ca content is usually low. Pigs and chickens do poorly on raw potatoes, but cooking improves digestibility of the starch so that it is comparable to corn starch Potatoes and, particularly, potato sprouts contain a toxic compound, solanin, which may cause problems if potatoes are fed raw or ensiled. In cattle finishing rations, cull potatoes are frequently fed at a level to provide about half of the dry-matter intake. With pigs, satisfactory performance may be obtained when growing pigs are fed cooked potatoes along with limited amounts of protein concentrates, but they are usually restricted to 30% or less of the total diet.

In North America the utilization of potatoes for processing has more than tripled in the past 20 years, and various by-products of potato processing are more available in some areas for feeding to livestock. It is estimated that about 35% of the preprocessed potato is discarded during processing. **Potato meal** is the dried raw meal of potato residue left from processing plants. **Potato slurry** is a high-moisture product remaining after processing for human food. It contains a high amount of peel. **Potato filter cake,** which represents about 20% of the residue from potato processing, is the residue recovered from the waste water by vacuum filtration. **Potato flakes** are residues remaining after cooking, mashing, and drying. **Potato pulp** is the by-product remaining after extraction of starch with cold water. The various raw meals have about the same relative nutritive values as raw cull potatoes. Recent studies with potato filter cake indicate that essentially all the barley (60% of total diet) could be replaced with filter cake without a marked reduction in digestibility of energy by cattle. When 15% of the diet barley was replaced with filter cake, it stimulated energy digestibility. However, at the higher levels (45% and 60%) there was an apparent reduction in feed consumption (35).

Data from two different experiments with potato by-products are shown in Table 7-20. In the experiment with potato meal, it was substituted for some of the corn and soybean meal in the concentrate mix for lactating cows. This particular product had a crude protein content of 8.45%, a 4.72% lipids content, and a 65% starch content. For some reason the 15% level resulted in less dry-matter intake, but it had no statistical effect on milk yield or milk fat. In the second experiment, potato processing waste (60% peel and sludge, 30% raw potato screenings, and 10% cooked packaging wastes) was fed. It had a dry-matter content of 24% to 28%, a crude protein content of 4.1% to 6.6% (dry basis), and a pH of 3.6 to 4.2. It was substituted for high-moisture corn on a dry-matter basis. Fat content of the milk tended to decrease with use of the processing wastes, but there were no statistical differences otherwise in milk production when using this level of potato waste. Thus these two studies tend to support earlier work that indicated that small to moderate amounts of potato wastes can be utilized very well by lactating or fattening cattle, but that using as much as 40% to 50% or more of the dry matter from potato waste will likely reduce performance.

Potato by-product meal (now called **dried potato products**) is produced in some areas. It contains residues of food production such as off-color french fries, whole potatoes, peelings, potato chips, and potato pulp. These are mixed, limestone is added, and the mixture is dried with heat. Generally, the value of the

TABLE 7-20

Effect of potato wastes on milk production by cows

Item	Potato Meal, % in Rations[a]			Processing Waste, % in Rations[b]			
	0	15	30	0	10	15	20
Dry matter intake, kg/day	19.8	18.6	19.5	18.2	19.0	18.3	18.7
Milk yield, kg/day	27.5	26.8	28.0	27.2	26.1	25.6	27.4
4% FCM/kg/day				24.2	22.8	22.8	23.4
Milk fat, %	3.33	3.41	3.30	3.31	3.20	3.20	3.01

[a]Data from Schneider et al. (36).
[b]Data from Onwubuemeli et al. (37).

various potato products is roughly comparable to that of raw or cooked cull potatoes, depending on how (or if) the product is dried. However, residues of the potato chip processing industry have much higher levels of fat, so the energy value is increased accordingly.

Many processing plants do not dry their wastes. It is a common practice for cattle feedlot operators to collect wastes and store the waste in pits. Because of the high moisture content, substantial fermentation occurs in warm and hot weather, often resulting in high losses (60% to 70%±) of the starch originally going into the pit unless the waste is fed soon after it is available. Potatoes have also been used successfully to make silage in combination with a variety of other feedstuffs.

FATS AND OILS

Although most animals need a dietary source of the essential fatty acids (see Ch. 3), these are usually supplied in sufficient amounts in natural feedstuffs, and supplementation is not required except when low-fat energy sources are fed. However, feeding fats are frequently used in commercial feed formulas. Fats are added to rations for several reasons. Nutritionally, fats are exclusively an energy source, because they contain very little, if any, protein, minerals, or vitamins. As a source of energy, fats are unequaled and are highly digestible (especially by simple-stomached animals). Digestible fat supplies about 2.25 times as much energy as digestible starch or sugar; thus fats can be used to increase the energy density of a ration. Fats often tend to improve rations by reducing dustiness and increasing palatability. Fats generally increase absorption of fat-soluble nutrients, such as the fat-soluble vitamins. However, they may form insoluble Ca or Mg soaps in the gut and reduce

absorption of P. There is some evidence that fats will reduce bloat in ruminants. From a manufacturing point of view, the lubrication of milling machinery by added fats is often of interest. There has been a very substantial increase in fat used for both hog and dairy feeds.

Feeding fats come from a variety of sources. Animal fats are primarily from slaughterhouses or other facilities that precut meat for the restaurant or grocery trade. Some also comes from rendering plants that process inedible animal tissues. Based on a survey of 40 rendering plants in the United States in 1978 (42), the raw material percentages used for producing feeding fats were composed of restaurant grease, 40; shop fat and bone, 21; packing house offal, 19; fallen animals (dead or sick), 9; poultry offal, 6; other materials, 4. However, the ingredients from any particular rendering plant may be appreciably different. In the feed trade, fat is supplied by renderers or by blenders or brokers. Blenders may purchase all their fats and then process and blend them, and they normally sell their finished products under brand names and/or in formula feeds. Brokers sell fats that they purchase from renderers or refiners but do not themselves process in any way, although many brokers deliver fat in their own trucks.

In addition to animal fats, restaurant fats or greases have become a major portion of fats recycled for feeding. These fats may be mixtures of a variety of animal or vegetable cooking fats. Another major vegetable component is acidulated vegetable soapstock. This material is primarily free fatty acids removed from crude vegetable oils as a first step in refining the oil. It is safe and well utilized, and approximately 100,000 T are used annually in feeds. Other fats such as lard, high-grade tallows, and high-quality seed oils (corn, safflower, soy, cottonseed, peanut, and others) are usually too expensive to use in feed but are used in the food trade or for other industrial purposes.

Descriptions and classification of feed-grade fat sources are not uniformly applied in the feeding industry. A proposed classification is given next (38) and further recommended specifications are given in Table 7-21.

1. **Animal fat.** Includes rendered fats from beef or pork by-products. This material is mainly packing house offal or supermarket trimmings from the packaging of meats. It can be identified as tallow if the titer (hardness measurement; temperature in °C at which a hydrolyzed fat solidifies) is 40 or higher or grease if under 40. A lower titer indicates higher unsaturated and/or polyunsaturated levels.
2. **Poultry fat.** Includes fats from 100% poultry offal.
3. **Blended feed-grade animal fat.** Includes blends of tallow, grease, poultry, and restaurant grease.
4. **Blended animal and vegetable fats.** Includes blends of feed-grade animal fat from category 3, plus vegetable fat.
5. **Feed-grade vegetable fat.** Includes vegetable oil, acidulated vegetable soapstocks, and other refinery by-products.

The official names of feeding fats (as of 1995) are **animal fat; vegetable fat or oil; hydrolyzed fat or oil, feed grade; fat product, feed grade; corn endosperm oil; esters of fats; calcium salts of long-chain fatty acids; hydrolyzed sucrose polyesters; vegetable oil refinery lipid, feed grade; and corn syrup insolubles, feed grade.**

Although the type of fat may vary considerably depending on the source, usually specifications state that feeding fats shall contain not less than 90% total fatty acids, not more than 2.5% unsaponifiable matter, and not more than 1% insoluble matter. Feeding fats must also be free from toxic or undesirable substances. Moisture is detrimental because it contributes to instability of the fat and is also an unneeded diluent. Occasionally, some residual solvent may be present, which may be an explosion hazard. Unsaponifiable substances are largely hydrocarbons (solvents used in some processing), waxes, and tars, which have little, if any, food value. Other unsaponifiable substances may include cholesterol, cholesterol esters, and some phospholipids. Of the toxic substances, the polychlorinated biphenyls (PCBs) have sometimes been a problem.

Fats are subject to oxidation with development of rancidity, which reduces palatability and may cause some digestive and nutritional

TABLE 7-21
Suggested quality specifications for feed fats

Feed Fat Categories	Quality Specifications, %					
	Minimum Total Fatty Acids	Maximum Free Fatty Acids	Maximum Moisture	Maximum Impurities	Maximum Unsaponifiable	Maximum Total MIU
Livestock	90	15	1	0.5	1	2
Poultry	90	15	1	0.5	1	2
Blended feed-grade animal	90	15	1	0.5	1	2
Blended animal and vegetable	90	30	1	0.5	3.5	5
Feed-grade vegetable	90	50	1.5	1.0	4.0	6

Source: Rouse (38). The following specifications apply to all fats:

- Fats must be stabilized with an acceptable feed or food-grade antioxidant added at levels recommended by the manufacturer. Fats should pass the AOM stability test at 20 h with less than 20 millequivalents (mEq) peroxide.
- No cottonseed soapstock or other cottonseed by-products should be included in fats for layer, breeder, or broiler rations.
- Blended fats shall include only tallow, grease, poultry fat, and acidulated vegetable soapstock. Any other by-products should be included only with the knowledge and consent of the buyer.
- Fats must be certified that any PCB and pesticide residues are within the allowable limits established by state and/or federal agencies.
- Fats for poultry rations should be certified as being negative for the chick edema factor as measured by the modified Liebermann–Burchard test.
- Fats shall not contain more than trace levels of any minerals, heavy metals, or other contaminants.
- The supplier should make every effort to provide a uniform fat structure in each delivery. A specification for minimum and/or maximum iodine values can be established for the type of fat purchased. Monitoring IVs can determine if product fat structure is uniform.
- Suppliers should furnish research data to support ME claims.

problems. Thus one of the first requirements for a feeding fat is that it be stable to oxidation. Feeding fats nearly always have antioxidants added, especially if the fat is held for any length of time or if the mixed feed is not fed immediately. The use of antioxidants also protects the feed sources against loss of some vitamins; vitamins E and A are particularly good examples.

If the proposed standards listed in Table 7-21 are finalized in this or some similar form, it will require quite a bit of policing by users or by some unbiased quality control laboratories to make them work as proposed. This is obvious when the results of the 40 samples from rendering plants are considered (Table 7-22). Some of these samples were probably quite good, but others must have fallen considerably outside the limits proposed in Table 7-21. Quality is also reflected in color and odor, as well as the chemical factors listed in Table 7-21. Fats that are rancid, off flavor, and unpalatable are not, of course, desirable feed ingredients.

Adding fat at low to moderate levels to animal rations can sometimes increase total energy intake through improved palatability, although animals usually consume enough energy to meet their demand when it is physically possible. In swine rations, 5% to 10% fat is often added to creep diets, but fat is usually used more sparingly for older market hogs. Relatively high levels may alter the character of body fat, particularly in nonruminant species because they tend to deposit dietary fatty acids relatively unchanged. In poultry rations, 2% to 5% fat is often added when fat is competitive as an energy source with the cereal grains. Amounts above 10% to 12% will usually cause a sharp reduction in feed consumption, so concentration of other nutrients may need to be increased in order to obtain the desired intake.

For ruminants, high levels of fats are used in milk replacers; depending on the purpose of the replacer, it may contain 10% to 30% added fat. Ruminants on dry feed are less tolerant of high fat levels than are monogastrics. Concentrations of more than 7% to 8% are apt to cause digestive disturbances, diarrhea, and greatly reduced feed intake. In practice, 2% to 4% added fat is an appropriate level for finishing rations. Some fat is occasionally added to rations for lactating dairy cows, but fat would only rarely be used in other situations.

At the present time the feeding of high-oil seeds such as soybean and cottonseed is a popular practice for dairy producers. The seeds are high in both fat and protein, and adding some to the diet generally increases milk fat percentage somewhat (see Ch. 15). Such seeds may also be processed through extruders for feeding to monogastric species. Some processing is necessary to destroy antiquality factors for monogastric species (see Ch. 8).

One problem in using fat (particularly in computer-formulated rations) is that there is considerable disagreement as to the feeding value (energy content) for ruminant animals (40). Tabular values for energy content do not appear to follow any logical thought patterns. For example, in the 1996 NRC publication on nutrient requirements of beef cattle, fat is considered to be 79% digestible and to have energy levels as shown: TDN, 177%; ME, 6.41 kcal/g; NEm and NEg values of 4.75 and 3.51 kcal/g, respectively. The author's opinion is that different values should be used for preruminants or other suckling mammals than for older animals, because the young animals generally digest fats more completely and/or they are fed fats that are more digestible. One report provides data on fattening cattle fed 0% or 4% added yellow grease (41). The results of two comparative slaughter trials and a digestion trial indicated that this fat product had the following energy values: NEm, 6.20, and NEg, 4.53 kcal/g. These appear to be more realistic values than some others quoted in tables of feedstuff composition.

GARBAGE

On a commercial scale, garbage fed to animals is primarily food waste from restaurants, hotels, food markets, and other institutions handling large amounts of food. In the United States, garbage feeding is restricted to feeding of swine near large metropolitan centers where collections can be made daily. Most of the states in the United States have laws requiring

TABLE 7-22

Analyses of feed-grade fat from 40 rendering plants

Component	Mean	Range
Moisture and volatiles, %	0.44	0.01–1.99
Insolubles, %	0.21	0.01–2.97
Unsaponifiables, %	0.68	0.24–3.48
Total M, I, U, %	1.33	0.48–7.33
Free fatty acids, %	6.50	0.70–36.81
Capillary melting point, °C	39.8	27.8–45.3
FAC color	25	11–45
Peroxide value at 20 h	105.2	1.5–340

Source: Boehme (39).

that garbage be cooked to some minimum temperature in order to prevent the spread of diseases such as salmonellesis, trichinosis, and tuberculosis. Nevertheless, occasional cases of trichinosis occur from humans eating pork that has been fed on garbage, usually from animals that were not processed through an inspected slaughterhouse.

Research data indicate that garbage can be fed successfully to market pigs if the garbage is supplemented with energy and protein sources. The high moisture content of garbage would, otherwise, result in some reduction in performance. Garbage tends to produce pork with softer fat than when cereal grains are fed, so it is a common practice to remove the garbage from the ration during the final finishing stage.

OTHER ENERGY SOURCES

Many other materials that have not been discussed in this chapter are fed to animals. However, most of them are considered to be by-products of the food industry in one form or other or nonfood-related products produced in small amounts. Many of the products are listed in a publication on by-product and unusual feedstuffs published by the California Experiment Station (42). It should be noted that good animal data are not available on many of the products often available only in localized areas.

SUMMARY

Feedgrains (corn, barley, sorghum, oats, wheat triticale, millet) are the primary sources of high-energy feed for farm livestock. The grains and various milling by-products of the grains provide a readily digested source of starch with lesser amounts of other carbohydrates. In general, the feedgrains are moderate in protein content, low in Ca, moderate in P, and variable in vitamins and trace minerals. Molasses of various types are also available, as are other liquid by-products in lesser amounts. These are widely used in commercial feed formulas or in liquid supplements.

Although there is some competition between animals and humans when grains are fed to animals, the use of such high-energy feeds allows animals to produce at considerably higher levels and with greater efficiency than they could otherwise do if dependent on forage or other high-fiber feeds for their energy supply, which increases the availability of animal products for human consumption.

Other energy sources available in substantial amounts include the various milling by-products of flour milling, primarily from wheat, but with lesser amounts from rice, barley, sorghum, and rye. Corn milling by-products are also available in substantial amounts, though some are classed as protein supplements. Although the by-product feedstuffs are sometimes standardized for protein and/or fiber content, they are still rather variable. Molasses, primarily a by-product of sugar production, is used in relatively large amounts. It is also available from citrus juice production as well as from some wood processing. Liquids such as molasses have multiple uses in formula feeds or as liquid protein supplements when fortified with N. Other by-product feeds include beet pulp, citrus pulp and meal, and dried bakery product. Some roots and tubers are fed directly to animals. In addition, wastes from food production with potatoes are very useful energy sources. Inedible (for humans) fats and oils are used in many modern finishing rations for swine, poultry, and cattle. Garbage and a host of other by-products of food processing are available in limited amounts in localized areas.

REFERENCES

1. USDA. 1996. *World grain situation and outlook.* Foreign Agr. Ser. Cir. Series FG-2-87 (November). Washington, DC: USDA.

2. USDA. 1994. *Feed situation and outlook report.* Econ. Res. Ser. FDS-304 (August). Washington, DC: USDA.

3. USDA. 1987. *The official United States standards for grain.* Washington, DC: Federal Grain Inspection Service.

4. Subramanyam, M., C. W. Deyoe, and L. H. Harbers. 1980. *Nutr. Rept. Internat.* 22:657;667.

5. Kent, N. 1983. *Technology of cereals.* Oxford, UK: Pergamon Press.

6. Pond, W. G., and J. H. Maner. 1984. *Swine production and nutrition.* Westport, CT: AVI.

7. Sunde, M. L. 1971. *Hatch* (February). Madison: University of Wisconsin Poultry Dept.

8. Luis, E., and T. Sullivan. 1984. *Poultry Sci.* 61:321.

9. Rooney, L. W., and R. L. Pflugfelder. 1987. *J. Animal Sci.* 63:1607.

10. Lichtenwalner, R. E., et al. 1979. *J. Animal Sci.* 49:183.

11. Hinman, D. D. 1979. *Proc. West. Sec. Amer. Soc. Animal Sci.* 30:49.

12. Misir, R., W. C. Sauer, and R. Cichon. 1984. *J. Animal Sci.* 59:1011.

13. Reddy, S. G., M. L. Chen, and D. R. Rao. 1975. *J. Animal Sci.* 40:940.

14. Preston, R. L., D. C. Rule, and W. E. McReynolds. 1980. *Proc. West. Sec. Amer. Soc. Animal Sci.* 31:269.

15. Hale, O. H., and P. R. Utley. 1985. *J. Animal Sci.* 60:1272.

16. Misir, R., and R. R. Marquardt. 1978. *Can. J. Animal Sci.* 58:717.

17. Sharma, H. R., et al. 1981. *J. Dairy Sci.* 64:441.

18. NRC. 1982. *United States–Canadian tables of feed composition.* 3d ed. Washington, DC: National Academy Press.

19. AFCO. 1995. *Official Publication. Amer. Feed Control Officials.* Washington, DC: AFCO.

20. Erickson, J. P., et al. 1985. *J. Animal Sci.* 60:1012.

21. Adeola, O., et al. 1986. *J. Animal Sci.* 63:1854.

22. Anon. 1972. *Millfeed manual.* Chicago: Miller's National Federation.

23. Steg, A., and J. M. Van Der Meer. 1985. *Animal Feed Sci. Tech.* 13:83.

24. Lora, J., et al. 1978. *Tropical Animal Prod.* 3:19.

25. Heinemann, W. W., and E. M. Hanks. 1977. *J. Animal Sci.* 45:13.

26. Crawford, D. F., W. B. Anthony, and R. R. Harris. 1978. *J. Animal Sci.* 46:32.

27. Vernlund, D. S., et al. 1980. *J. Dairy Sci.* 63:2037.

28. Perry, T. W., et al. 1976. *J. Animal Sci.* 42:1104.

29. Randel, P. F., and B. Vallejo. 1982. *J. Agr. Univ. P.R.* 66:11; Vallejo, B., and P. F. Randel. 1982. *J. Agr. Univ. P.R.* 66:44.

30. Potter, S. G., et al. 1985. *J. Animal Sci.* 60:839.

31. Chen, M. C., et al. 1981. *J. Animal Sci.* 53:253.

32. Pihchasov, Y., et al. 1982. *J. Dairy Sci.* 65:28.

33. Schingoethe, D. J. 1976. *J. Dairy Sci.* 59:556.

34. Bartsch, D. B., and R. B. Wickes. 1979. *Australian J. Exp. Agr. Animal Husb.* 19:658.

35. Stanhope, D. L., et al. 1980. *J. Animal Sci.* 51:202.

36. Schneider, P. L., et al. 1985. *J. Dairy Sci.* 68:1738.

37. Onwubuemeli, C., et al. 1985. *J. Dairy Sci.* 68:1207.

38. Rouse, R. H. 1986. In: *Feed Mgmt.* 38(2):18.

39. Boehme, W. R. 1978. Personal communication. Des Plaines, IL: Fats & Proteins Res. Foundation.

40. Church, D. C., ed. 1979. *Digestive physiology and nutrition of ruminants. Vol. 2: Nutrition.* 2d ed. Corvallis, OR: O & B Books.

41. Zinn, R. A. 1988. *J. Animal Sci.* 66:213.

42. Bath, D. L., et al. 1980. *By-products and unusual feedstuffs in livestock rations.* West. Regional Ext. Pub. 39, Davis, CA: Univ. of California.

43. Grimson, R. E., et al. 1987. *Can. J. Animal Sci.* 67:43.

8

SUPPLEMENTAL PROTEIN SOURCES

D. C. Church and Richard O. Kellems

INTRODUCTION

Protein is a critical nutrient in the diet (i.e., one likely to be low or deficient), particularly for young, rapidly growing animals and for mature animals such as high-producing dairy cows. Optimal use of protein is a must in any practical feeding system, because protein supplements are usually much more expensive than energy feeds, and wasteful usage increases the cost of production in almost all instances. Another consideration is that when excess protein is fed there will be increased elimination of nitrogen in the feces and urine, which has environmental implications.

As pointed out in Ch. 3, the need for protein differs with different species. For monogastric species and young suckling ruminants (preruminants), a diet must supply the essential amino acids; thus quality is important because protein quality is a measure of the ability of a protein to supply needed amino acids in the diet. For ruminant species the dietary need is a combination of needs to nourish the microorganisms and needs for an adequate supply of digestible essential amino acids in the gut. High-producing ruminants pass appreciable amounts of some ingested proteins into the intestines without it being metabolized in the rumen; protein quality is more important under these circumstances than for animals producing at low levels and consuming much less feed. This topic is discussed in more detail later in the chapter.

Most energy supplements (except for fat, starch, or refined sugar) supply some protein, but usually not enough to meet total needs except for adult animals in a maintenance situation. Thus supplementary protein sources are often needed in rations for all species of animals.

Protein supplements are arbitrarily defined (by NRC) as those feedstuffs that have 20% or more crude protein (dry basis). Many proteins from animal, marine, plant, or microbial sources are available as well as nonprotein N sources such as urea and biuret, which come from chemical manufacturing processes. Some commentary will be presented on most of the more common supplemental protein sources.

The selection of a given protein to use in a feed formula is affected by several different considerations. Principal items are availability and cost in an area. Many protein sources are available in limited supply in some localities, while others are available on a nationwide or worldwide basis.

Another important factor for monogastric species is the content and availability of amino acids. The content is reasonably well defined, but much less information has been published on availability. Data on digestibility of N are available on all common protein supplements. Such information is not complete, because it does not provide information on availability of critical amino acids. This type of information must be collected using other techniques to measure absorption from the small intestine with the use of animals with intestinal cannulas and/or catheters implanted in blood vessels draining the small intestine. Obviously, not much information of this type is available, especially under practical conditions. Once absorbed, amino acids are often metabolized in the tissues prior to being transported to the various tissues.

Likewise, digestibility data on N sources such as urea are of little value. This is so because such compounds are highly soluble and are probably completed absorbed, thus yielding high digestibility values. However, excretion of urea via the urine will also be high, showing that it is not retained by the animal as well as digestibility data indicate.

At the present time most nutritionists (and computer programs) simply match up analytical data on critical amino acids with amino acid needs as listed in feeding standards. If information were available on true availability of amino acids, listed requirements for various amino acids would undoubtedly be lower than now shown.

Another important factor is the presence of undesirable or toxic compounds in protein supplements. This is particularly a problem with plant sources, but it may also be a problem with some animal proteins. This topic will be discussed in more detail in a later section.

Last, but not always least, is the content of other nutrients. Many protein sources are an excellent source of P, which often needs to be added to rations in some form or other.

Relative nutrient value is a summation of all the nutrients that are provided by a feed.

PROTEIN SOURCES

The major plant protein sources manufactured and utilized in the feed trade or sold directly to livestock enterprises are listed in Table 8-1. These data show clearly that soybean meal is the predominant plant protein source, providing 64.4% of the world tonnage in 1995/96 of those sources listed in this table. It is obvious that cottonseed meal use has declined. Part of

TABLE 8-1

Processed protein sources used as feed in the world (millions of metric tons)

Source	Crop Year 1991/92	Crop Year 1995/96[a]
Oilseed meals		
Soybean[b]	73.2	87.1
Cottonseed	13.3	12.2
Copra	1.6	1.8
Peanut	4.8	5.6
Sunflower	8.6	10.1
Canola	15.6	18.3
Total of oilseed	117.1	135.1

Source: USDA (1).
[a]Partially forecast.
[b]Includes use in edible soy products and shipments to U.S. - territories.

the decline may be due to production of less cotton, but part of it is also due to a change in usage, because whole cottonseed has become a favored feed ingredient in dairy cow rations. Peanut meal use has declined, while sunflower has increased tenfold in the past 10 years. The oilseed meal not listed in this table (statistics not readily available) is safflower meal. Gluten feed and meal production has increased appreciably, reflecting a greater use of corn for wet milling use. Likewise, the use of distillers dried grains has increased markedly. No doubt part of the increase reflects increased production of ethanol; however, in the past it was common for feedlots to be located close to distilleries so that the wet distillers grains could be used with only a short haul to the feedlot. Thus increased use may partly represent a shift in usage. Alfalfa meal (most of it is dehydrated) has declined to about half of that produced 10 years ago.

Animal protein sources include feather and hair meals and several products produced from poultry slaughtering plants and hatcheries. In addition, there are many other minor plant and animal products that will be described to some extent later in this chapter. Overall, we are producing about 13.3% more tonnage of the plant protein sources listed in Table 8-1 than was the case in 1991.

PLANT PROTEIN SOURCES

OILSEED MEALS

Oilseed meals are produced from a variety of crops (see Table 8-1) that have seeds that are high in oil. The oils all have important nutritional or industrial uses. Soybeans, peanuts, and sunflowers are grown primarily for their seed, and all produce edible oils used in the human food trade. Cottonseed is strictly a by-product of cotton production, but its oil is widely used in food and for other uses. Flax used to be grown to provide the fibers for linen cloth production and for the oil from the seed, which is used as a drying oil in paint and for other industrial uses. The demand for linen cloth is much less than it was before the days of synthetic fibers, and production of other types of paints and varnishes has reduced the need for linseed oil.

As a group, the oilseed meals are high in crude protein (Table 8-2); protein levels of those meals shown in the table are all 40% or more except safflower meal with hulls. The crude protein content is standardized before marketing by dilution with hulls or other material. A high percentage of the N is present as true protein ($\pm$90%), which is usually highly digestible and of moderate to good biological value, although of usually lower value than good animal protein sources. As an average about 9% of the crude protein of oilseed meals is from nucleic acid protein as compared to 8.8% in fish meal, 3.5% in meat and bone meal, and 14% to 20% in yeast. This source of N is of questionable value even for ruminant species.

TABLE 8-2

Data on different types of protein sources (dry-matter basis)[a]

Source	Typical Dry Matter (%)	Crude Protein (%)	Ether Extract (%)	ADF (%)	TDN (%)	Minerals			
						Ash (%)	Ca (%)	P (%)	Mg (%)
Plant sources									
20% to 30% CP range									
Beans, cull navy	90	25.0	1.6	6	83	4.6	0.17	0.60	0.15
Brewers dried grains	92	26.0	7.2	23	66	4.1	0.29	0.54	0.15
Coconut meal, solvent extracted	92	23.1	2.7	24	74	7.3	0.18	0.66	0.39
Corn distillers solubles, dried	93	28.9	5.7	6	88	7.2	0.38	1.47	0.69
Corn gluten feed	90	27.5	2.8	10	82	8.6	0.45	0.89	0.32
Malt sprouts	92	28.0	1.6	20	68	6.7	0.26	0.84	0.23
Safflower meal, mechanical extraction (w/hulls)	91	22.8	7.6	45	57	4.2	0.28	0.79	0.36
Wheat germ meal	90	28.1	10.2	5	95	5.8	0.06	1.16	0.28
30% + CP									
Alfalfa seed screenings	90	34.4	10.9	15	86	5.6			
Brewers dried yeast	93	48.3	0.8	4	78	7.7	0.14	1.54	0.25
Corn gluten meal	90	48.0	2.4	5	87	3.9	0.15	0.45	0.05
Cottonseed meal, 41% solvent extracted	92	44.8	2.3	20	75	6.9	0.17	1.31	0.61
Linseed meal, solvent extracted	90	40.7	1.1	13	82	6.4	0.43	0.95	0.66
Rapeseed meal, solvent extracted	92	44.0	1.2	13	71	7.8	0.72	1.01	0.50
Soybean meal, solvent extracted, 44%	89	49.6	1.4	10	81	6.8	0.36	0.75	0.30
Sunflower meal, solvent extracted	93	50.3	1.2	30	65	6.3	0.40	1.10	0.81
Animal sources									
Blood meal	89	89.6	1.1	—	68	4.9	0.31	0.25	0.25
Feather meal	90	87.4	2.9	1	63	3.8	0.20	0.75	0.21
Meat meal, 55%	93	59.3	7.8	2	73	26.9	8.19	4.31	0.29
Meat and bone meal, 50%	93	54.0	9.2	2	70	35.6	9.93	4.75	1.22
Marine sources									
Crab meal	95	31.6	2.3	9	27	31.0	18.95	1.57	0.92
Fish meal, herring	93	77.4	10.7	1	75	11.2	2.15	1.07	0.19
Fish solubles, dried	94	69.9	9.9	1	79	15.8	1.36	1.80	0.29
Milk sources									
Buttermilk, dried	93	34.2	5.6	—	86	10.8	1.07	0.73	0.10
Skim milk, dried	94	36.0	1.1	—	86	8.5	1.25	1.03	0.11

[a]Values are from NRC publications

Most meals are low in cystine and methionine and have a variable and usually low lysine content (Table 8-3; Appendix Table 3); soybean meal is an exception in lysine content. The energy content varies greatly. Note in Table 8-2 that the Ca content is usually low. Most meals are high in P content, although half or more is present as phytin P, a form poorly utilized by monogastrics. These meals contain low to moderate levels of the B vitamins and are low in carotene and vitamin E.

The oilseed meals are processed to remove the oil by three primary methods at present. The methods are designated as screw press (or expeller), direct solvent, and prepress solvent. In the expeller process the seed, after cracking and drying, is cooked for 15 to 20 minutes, then extruded through dies with the use of a variable-pitch screw. This results in rather high temperatures, which may cause reduced solubility as well as reduced biological value of the protein. Usually, only a moderate amount of heating for short periods of time is necessary to inactivate some of the antinutritional factors (see later section). If longer or excessive levels of heat are involved, reactions involving carbohydrates (glucose) and amino acids may occur (browning reaction), which result in the formation of linkages between glucose and some amino acids. When this reaction occurs, the amino acid(s) becomes less available to the animal because the linkage cannot be digested completely in the intestinal tract. The result is that the biological value of the protein decreases

markedly. Amino acids of concern are lysine and, to a lesser degree, arginine, histidine, and tryptophan. A similar reaction occurs with gossypol (in cottonseed) and lysine. Fortunately, these problems are well recognized by the oilseed processors, with the result that most meals are of higher or more uniform quality than formerly.

Solvent-extracted meals are extracted with hexane or other solvents, usually at low temperatures. When low temperatures are used, usually the meal will be heated in the solvent recovery step. The heating is required to inactivate antiquality factors. A combination method, called prepress-solvent extraction, is often used. The seed oils are partially removed with a modifier expeller process and then extracted with solvents. The maximum amount of fat can be removed with the prepress solvent method, while the least amount of fat is removed with the expeller process. Protein N solubility is also higher with the solvent method than with the other two methods.

In addition to the high-protein meals (40% ± CP), a number of protein supplements are available in the 20% to 30% CP range (Table 8-2; Appendix Tables 2,3). Feedstuffs in this range include grain legumes (peas, beans), several milling by-products such as corn gluten feed, germ meals, and whole cottonseed, distillery and brewery by-products, and other feeds such as coconut meal (or copra), an important source in some tropical countries. As a group, these lower-protein meals tend to have

TABLE 8-3

Crude protein and essential amino acid content of several important protein supplements[a]

Item	Dried Skim Milk	Meat Meal	Fish Meal, Herring	Corn Gluten Meal, Solvent Extracted	Cottonseed Meal, Solvent Extracted	Safflower Meal without Hulls, Solvent Extracted	Soybean Meal, Solvent Extracted	Yeast, Dried Brewers
Dry matter, %	94.3	88.5	93.0	90.0	91.0	90.0	89.1	93.7
Crude protein	36.0	55.0	77.4	48.0	45.5	46.5	52.4	47.8
Essential amino acids								
Arginine	1.23	3.0	4.5	1.6	4.6	4.1	3.8	2.3
Cystine	0.48	0.4	0.9	0.7	0.7	0.8	0.8	0.5
Histidine	0.96	0.9	1.6	1.0	1.1	1.1	1.4	1.2
Isoleucine	2.45	1.7	3.5	2.6	1.3	1.9	2.8	2.2
Leucine	3.51	3.2	5.7	7.3	2.4	2.8	4.3	3.4
Lysine	2.73	2.6	6.2	0.9	1.7	1.4	3.4	3.2
Methionine	0.96	0.8	2.3	1.1	0.5	0.8	0.7	0.7
Phenylalanine	1.60	1.8	3.1	3.2	2.2	2.1	2.8	1.9
Threonine	1.49	1.8	3.2	1.6	1.3	1.5	2.2	2.2
Tryptophan	0.45	0.5	0.9	0.2	0.5	0.7	0.7	0.5
Valine	2.34	2.2	4.1	2.4	1.9	2.6	2.8	2.5

[a]Composition on dry basis. Data from NRC publications.

protein that is usually less digestible (Table 8-2) and of lower biological value than the high-protein meals. In addition, the fiber content is usually higher and the energy value is lower than for the oilseed meals. Photos of some of the common meals are shown in Fig. 8-1.

SOYBEAN MEAL

Whole soybeans (*Glycine max*) contain 15% to 21% oil, which is usually removed by solvent extraction during preparation of the meal. The meal is toasted, a process that improves the biological value of its protein; the protein content is standardized at 44% or 50% (as fed basis) by dilution with soybean hulls. Soybean meal is produced in large and ever increasing amounts in the United States and is a highly favored feed ingredient because it is quite palatable, is highly digestible, has high energy value (Table 8-2), and results in excellent performance when used for different animal species. Methionine is the most limiting amino acid for monogastric species (Table 8-3), and the B vitamin content is low. In overall value, soybean meal is considered to be the best plant protein source available in any quantity, and it is the standard protein source in many rations used for broilers and swine.

As with most other oilseeds (and plant high-protein seeds), soybeans have a number of toxic, stimulatory, or inhibitory substances. Therefore, raw soybeans have lower nutritional value than heat-treated soybeans or soybean meal. For example, chicks show a growth depression on a ration with raw beans for the first 8 to 12 weeks of life, but eventually they gain weight at a rate equal to those fed heated beans. Sheep, swine, and calves are also affected by inhibitors in raw beans. Hens produced eggs just as fertile as those they produce when fed treated beans, but the eggs have more blood spots in the yolks. Raw soybeans cause the pancreas of affected animals to become enlarged, and fat absorption is reduced. Heat treatment, particularly pressure cooking, is effective in removing most of the inhibitory effect. The inhibitory factor(s) depresses utilization of methionine and cystine or both, but addition of these amino acids to the ration does not restore performance compared to that of birds fed heat-treated beans. Addition of antibiotics enhances the performance of animals fed raw soybeans.

Soybeans also contain at least four compounds that inhibit trypsin (or chymotrypsin) activity (antitrypsin factor). The presence of these factors (not restricted to soybeans) reduces protein digestibility, which is accompanied by increased excretion of N and S. Heat treatment inactivates these factors. A goitrogenic factor is found in soybean meal, and its long-term use at high levels may result in goiter in some animal species, particularly if the iodine content of the ration is low. Other anti-quality factors include saponins and proteins that cause agglutination of red blood cells in the laboratory. The latter is readily inactivated by pepsin, and both are inactivated by heat treatment. Soybeans also contain genistein, a plant estrogen, which may account, in some cases, for part of the high growth-inducing properties of the meal. Some meals also contain relatively high levels of phytic acid, which may interfere with Zn utilization.

Dehulled, solvent-extracted soybean meal is sometimes used in animal feeds. The dehulling process results in a higher protein content and reduces the fiber content. Some unextracted soybeans are fed after appropriate heat processing (110°C for 3 min) and grinding. The product is known in the feed trade as **full-fat soybean meal** and officially as **ground extruded whole soybeans.** It contains about 38% CP, 18% fat, and 5% fiber and has a higher energy value because of the high oil content. Heating-extruding equipment has been developed for on-the-farm processing, and its use appears feasible in relatively small operations. Such meal has found some favor in dairy cow rations and, in moderate amounts, in rations for swine and poultry. Heat-treated soybeans can be used to replace all the soybean meal in corn–soy rations for growing–finishing pigs.

Soy flour is the finely powdered material resulting after screening the ground, dehulled, extracted meal. It is often used as a partial replacement for milk proteins in milk replacers. For the food trade, **soy protein concentrate** is prepared from dehulled beans that have been fat extracted and leached with water to remove water-soluble nonprotein constituents. It must have not less than 70% CP (dry basis). It is used as a protein extender in food products and to produce texturized products resembling meat because it can be spun into fibers.

Examples of recent feeding trials comparing soybean meal to other proteins are shown in Table 8-4 for swine and poultry and in Table 8-5 for ruminants. Note that soybean meal always produced very acceptable performance, although some of the data indicate that combinations with other proteins might, in some

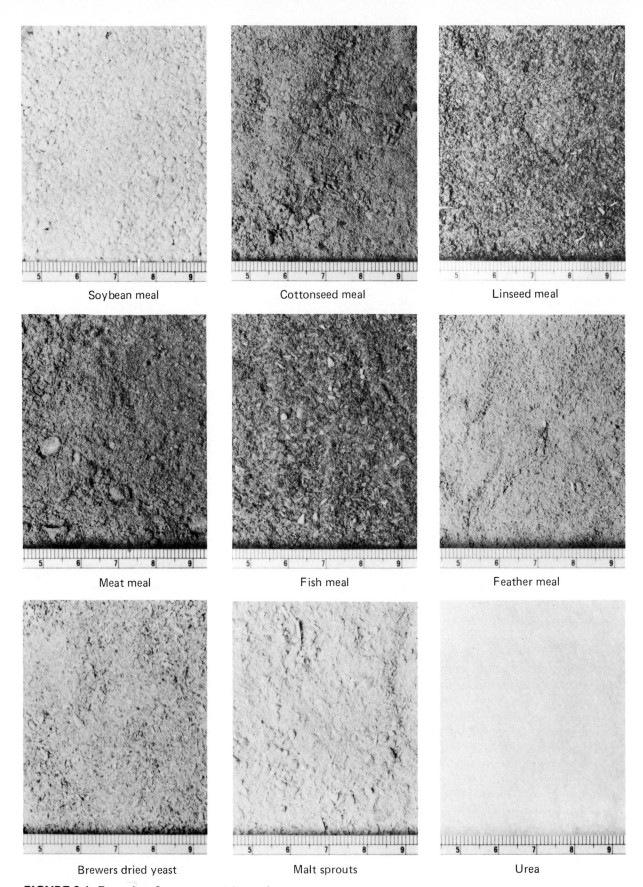

FIGURE 8-1 Examples of common protein supplements.

TABLE 8-4

Recent examples of comparative feeding trials with swine and poultry fed different protein sources

Comparison	Daily Gain	Feed Consumed per Day	Feed to Gain Ratio	Reference No.
Growing–finishing pigs, 25–95 kg				
Soybean meal[a]	0.68 kg	2.11 kg	3.08	2
4% blood meal	0.71	2.11	2.98	
6% blood meal	0.69	2.05	2.96	
8% blood meal	0.60	1.90	3.21	
Soybean meal	0.74	2.55	3.40	2
Blood–meat meal	0.72	2.40	3.33	
Soybean–blood meal	0.73	2.33	3.16	
Soybean–meat meal	0.76	2.45	3.22	
Growing–finishing pigs, 18–98 kg				
Soybean meal	0.85	2.35	2.79	3
SBM + glandless cottonseed				
Ground	0.82	2.27	2.27	
Extruded	0.82	2.79	2.78	
Young pigs (25 days old, fed for 4 weeks)				
Soybean meal	408 g	639 g	1.57	4
1/3 yeast[b]	418	630	1.51	
2/3 yeast	399	609	1.53	
All yeast	373	599	1.61	
Young chicks, fed for 2 weeks[c]			Gain/feed	
Dehulled soybean meal	226 g/2 wk		0.75	5
Extruded soybeans	241		0.80	
Extruded red beans	178		0.68	
1/3 beans, 2/3 soybeans	221		0.78	
2/3 beans, 1/3 soybeans	228		0.75	
Turkey poults, fed from 7 to 42 days of age				
Soybean meal	1.79 kg total gain		0.66	6
Soybean–fish meal[d]	1.87		0.65	
Soybean–rapeseed meal[d]	1.79		0.63	
Rapeseed–fish meal[d]	1.89		0.63	

[a]Basal diet contained 17.5% soybean meal; blood meal (new process) at the designated levels replaced an equivalent amount of protein, and corn was increased accordingly.

[b]Soybean meal was replaced by a combination of yeast and corn.

[c]The red beans (*Phaseolus vulgaris*) were extruded, as were the mixtures of beans and soybeans. For the data shown here, methionine was added to all diets.

[d]Herring meal was fed at 0, 60, or 120 g/kg and rapeseed meal at levels of 0, 150, 300, or 450 g/kg in all possible combinations. Only the overall means are shown.

cases, allow slightly improved production (gain, efficiency, or lactation).

COTTONSEED MEAL

The cotton plant (*Gossypium* spp.) grows in hot areas; thus cottonseed meal (CSM) is available in many areas where soybeans do not grow, particularly in some areas in South America, northern Africa, and Asia. The CSM protein is of good, although variable, quality as a result of variations in processing procedures. Most meals are standardized (in the United States) at 41% CP (as fed), but meals may be found, at times, with 44% and 48% CP.

The protein of CSM is low in cystine, methionine, and lysine, and the meal is low in Ca and carotene. Although palatable for ruminants, CSM is less palatable for swine and poultry. Nevertheless, it finds widespread use in animal feeds, although its use is more limited by various antinutritional factors than that of soybean meal.

The cotton seed contains a yellow pigment, gossypol, which is relatively toxic to monogastric species, particularly young pigs, chicks, and calves. A high percentage of the gossypol in the seed is free (readily removed), but heat processing results in the formation of various complexes, including a gossypol–lysine complex. Prepress solvent meals tend to have the highest levels of bound gossypol, while the screw press meals tend to have the highest levels of free gossypol. Research evidence indicates that free gossypol is the toxic form. It is generally agreed that free gossypol levels (total

TABLE 8-5

Examples of comparative feeding trials with lambs, beef cattle, and lactating dairy cows fed different protein sources

Comparison	Daily Gain	Feed Consumed per Day	Feed to Gain Ratio	Reference No.
Finishing lambs, high sorghum ratio				
Cottonseed meal	289 g	1.48 kg	5.29	7
Soybean meal	290	1.38	4.76	
Guar meal	200	1.22	6.10	
Blood meal	250	1.37	5.48	
Feather meal	250	1.39	5.56	
Urea	260	1.41	5.42	
Finishing lambs, sorghum–cottonseed hull ration				
Cottonseed meal	227	1.84	8.11	8
Sunflower meal	256	1.91	7.46	
Cottonseed–sunflower	246	1.96	7.97	
Finishing lambs, high roughage pellet				
Alfalfa–barley pellet	210	1.95	9.29	9
+10% cull beans	200	1.89	9.45	
+20% cull beans	160	1.79	11.19	
Finishing lambs, high corn–corncob ration				
Soybean meal	289	1.86	6.6	10
Blood meal	250	1.90	7.6	
Meat and bone meal	210	1.65	7.9	
Dehydrated alfalfa	250	2.06	8.2	
Finishing lambs, high-roughage pellet				
Cottonseed meal	390	2.35	6.0	11
Liquefied fish–urea	370	2.26	6.1	
Urea	380	2.29	5.9	
Growing cattle, high corn silage ration				
Urea	0.79 kg	5.93 kg	7.6	12
Soybean meal	0.94	6.33	6.8	
Blood meal	0.95	6.35	6.7	
Meat and bone meal	0.89	6.20	7.0	
Dehydrated alfalfa	0.94	6.69	7.1	
Finishing steers, high corn ration				
Urea	1.27	7.76	6.13	10
Soybean meal	1.33	8.13	6.13	
Wet corn gluten feed	1.38	8.80	6.37	
Dry corn gluten feed	1.35	9.46	7.01	
Finishing cattle, high corn ration[a]				
Soybean meal	1.25	7.97	6.4	10
Blood meal	1.33	8.29	6.2	
Meat and bone meal	1.36	8.41	6.2	
Finishing cattle, medium grain ration				
Cottonseed, meal–urea	1.72	9.9	5.8	13
Cottonseed–feather meal	1.72	10.1	5.8	
Feather meal–urea	1.72	9.4	5.5	
Finishing cattle, high-moisture corn, corn silage[b]				
Soybean meal	0.98	6.32	6.4	14
Soybean–feather meal	0.99	6.29	6.4	
Soybean–hair meal	0.99	6.67	6.7	
Finishing cattle, ground shelled corn, corn cobs				
Soybean meal–urea	1.05	7.2	6.9	15
Crambe meal–urea	1.03	7.1	6.9	

ration) of less than 40, 80, and 100 ppm are not problems for layers, broilers, and swine, respectively. Toxicity for swine and poultry can be reduced or eliminated by addition of iron salts such as ferrous sulfate. Recommended levels are 1 to 2 ppm of additional Fe for each ppm of free gossypol for broilers and about 4:1 for layers. Older research indicated that gossypol might be inactivated in the rumen of ruminant animals. However, more recent research (25) indicates that levels of 24 g/day of free gossypol for high-producing dairy cows resulted in toxic-

TABLE 8-5 (*cont.*)

Comparison	Daily Gain	Feed Consumed per Day	Feed to Gain Ratio	Reference No.
Lactating dairy cows	DM consumed per day (kg)	FCM/day (kg)	Milk fat (%)	
Corn silage, alfalfa–grass hay, concentrate				
Soybean meal[c]	17.8	20.7	3.96	16
Sunflower meal[c]	17.9	20.5	3.87	
Corn silage, alfalfa hay, concentrate				
Basal ration[d]	17.3	22.3	3.52	17
+ Whole cottonseed[d] (1.9 kg/day)	18.1	25.1	3.57	
Control[e]	18.0	24.1	3.66	17
Whole cottonseed[e]	18.4	25.4	3.70	
Energy equivalent[e]	18.5	24.4	3.60	
Urea-treated corn silage, alfalfa–grass hay, concentrate				
Soybean meal[f]	19.3	21.2	3.8	18
Tower rapeseed meal[f]	19.1	23.1	3.9	
Turret rapeseed meal[f]	18.7	20.7	3.9	
Hay–concentrate diet				
Cottonseed meal	21.0	36.7	3.52	19
Rapeseed meal (low glucosinolate)	20.8	38.5	3.54	
Corn silage, alfalfa silage, ensiled grain corn				
Soybean meal	20.1	26.2	3.45	20
Wet brewers grains	20.1	28.9	3.22	
Dried brewers grains	22.3	29.4	3.09	
Corn silage, alfalfa cubes, concentrate				
Soybean meal	19.7	33.3	3.74	21
75% ESB, 25% CGM[g]	19.2	29.4	3.58	
50% ESB, 50% CGM[g]	18.8	29.3	3.59	
25% ESB, 75% CGM[g]	19.6	32.0	3.81	
Corn silage, alfalfa hay, concentrate (complete diet)[h]				
Soybean meal	19.3	27.2	2.98	22
Heated soybean meal	20.1	29.0	2.89	
Extruded soybeans	20.5	28.5	2.63	
Corn silage, corn grain rations				
Soybean meal	24.2	31.6	3.48	23
+ Dehydrated alfalfa	23.8	31.0	3.58	
+ Dehydrated alfalfa + urea	24.0	30.7	3.63	
Corn silage, alfalfa hay, concentrate				
Grain corn, soybean meal	20.1	29.9	3.06	24
Corn, soybean meal + chickpeas	20.0	30.4	3.09	
Chickpeas	20.6	31.1	3.28	

[a]All rations contained 0.2% urea.

[b]Feather or hair meal supplied half the supplemental protein in rations where fed.

[c]When sunflower meal was fed, the concentrate contained more rolled oats and less corn to equalize the fiber content.

[d]When fed, whole cottonseed replaced an equivalent amount of concentrate. Trial was for 56 days.

[e]Cows were fed in a Latin square arrangement of treatments so that each cow was fed each ration for a 4-week period. The control ration was the same as in the previous experiment (d). When whole cottonseed was fed, it replaced 20% of the concentrate fed. For the energy equivalent ration, additional concentrate was fed so that the energy (ME) was equivalent to the whole cottonseed ration.

[f]Rapeseed meal replaced soybean meal and straw. Tower RSM is a newer variety low in glucosinolates (20.5 vs. 36.2 μmol/g for Turret). Cows were fed for an 8-week period.

[g]A different concentrate was fed to those cows getting extruded soybeans (ESB) and corn gluten meal (CGM); it had added fat to compensate for the higher level of fat in the ESB.

[h]All protein supplements were fed with and without methionine. Data shown here are without methionine.

ity (reduced blood hemoglobin, erythrocyte fragility, increased total protein of plasma, and elevated respiration rates at hot ambient temperatures). Although gossypol was detected in liver and plasma, it was not found in milk from the cows fed high levels of CSM. Reduced feed intake also occurred. Feeding large amounts of whole cottonseed also results in small amounts of gossypol in serum and liver tissue and small amounts of cyclopropene fatty acids in tissue lipids and milk fat (26). These experiments suggest that some of the gossypol must pass through the rumen without being metabolized by rumen microorganisms, as was previously thought to be the case.

Gossypol is produced by glands in the seed that can be reduced in size or removed by genetic changes in the plant. Meals from glandless seeds have resulted in good performance with poultry, although not with young pigs. Low gossypol and/or biologically tested meals are available in some areas. The biologically tested meals are tested by feeding to hens, but this, obviously, increases the cost.

CSM also creates problems with egg quality if fed in high levels. The gossypol tends to produce green egg yolks. In addition, a fatty acid (sterculic acid) found in CSM can cause egg whites to turn pink during storage. Obviously, either condition would be undesirable. If fed to poultry, direct solvent or prepress solvent meals are recommended because of these factors. As a result of the presence of the antiquality factors mentioned, the feeding of CSM to monogastric species is usually limited to less than 25% of total protein supplement fed.

The feeding of whole cottonseed to lactating dairy cows has become popular in recent years. Evidence (see Table 8-5) indicates that it may be somewhat stimulatory for milk production, usually for milk fat percentage, but also for FCM production. Other examples where CSM has been compared to other protein supplements for ruminants are shown in Table 8-5.

SUNFLOWER MEAL

Sunflowers (*Helianthus annuus*) are produced for oil and seeds, primarily in northern Europe and Russia and, in recent years, in Canada and the northern states of the United States. This plant will grow in cooler and drier climates than will soybean or cotton plants. The meal, although high in protein (±50% if dehulled, dry basis), is deficient in lysine, but otherwise the quality of the meal is comparable to that of soybean meal. However, the relatively high fiber content (Table 8-2) discourages use in poultry and swine feeds; when used for these species, the amount fed should be restricted because of the fiber, particularly for younger animals, or the dehulled meal should be used.

It has also been suggested that whole sunflower seeds are an efficient way of providing nutrients for lactating dairy goats without the risk of producing acidosis and enterotoxemia (because of the high fiber content).

PEANUT MEAL

Peanut meal (called groundnut meal in England) is available in substantial amounts in many countries because peanuts (*Arachis hypogaea*) are produced for human food in many warmer areas. The protein content varies from 45% to 55%, and the quality is influenced by the processing as well as the amount of hulls in the meal. Peanut meals are quite deficient in lysine, and the protein is low in digestibility, possibly because of tannins found in the skins. In addition, peanuts may frequently be contaminated with molds such as *Aspergillus flavus,* a fungus that produces a potent toxin of particular concern with young animals. Peanut meal is quite palatable to swine, and research with swine has indicated a value equivalent to soybean meal if fed with feeds not low in lysine or methionine.

SAFFLOWER MEAL

Safflower (*Carthamus tictorius*) is a plant grown in increasing (but limited) amounts for its oil. The plant also has the advantage that it does not require as much water as many other oilseed plants. The meal is high in fiber and low in protein unless the hulls are removed. The hulls make up about 40% of the seed and 60% of the meal, which contains 18% to 22% CP. Partial removal of hulls results in a meal with about 46% CP and 21% ADF. The protein is deficient in S-containing amino acids and lysine. Research studies with both swine and poultry indicate that its use should be restricted to provide only part of the supplementary protein. For ruminants, the high fiber is no problem, and it is utilized well by both sheep and cattle (Table 8-5).

LINSEED MEAL

Linseed meal is made from flax seed (*Linum usitatissimum*), which is now produced primarily for the drying oils it contains, although the flax plant was used extensively in the past to produce fibers used to weave linen cloth. Linseed meal accounts for only a small part of the total plant proteins produced in North America (Table 8-1). The CP content is relatively low (35%) and it is deficient in lysine, but it contains selenium in higher amounts than most feeds owing to the fact that most of the flax in the United States is grown on soils relatively high in Se. The meal may contain a cyanogenic glycoside (produces toxin hydrogen cyanide) in small amounts, as well as an antipyridoxine factor. Linseed meal is favored in rations for ruminants, horses, and, sometimes, for sow diets, but it is rarely used for poultry because of its poor amino acid distribution, its high fiber, and its laxative nature. In addition, it is usually priced higher per unit of protein than some of the other oilseed meals.

CANOLA MEAL

Canola meal may be produced from a Polish-type rapeseed (*Brassica campestris*) or the Argentine-type (*B. napus*). Rape is a crop that will grow in colder climates than soybean or cotton plants; thus it is of interest in areas such as northern Europe and Asia, Canada, the northern states in the United States, and in colder areas of South America (27).

Solvent-extracted meals run about 41% to 43% CP (dry basis) and have a good distribution of amino acids. Canola meal contains less lysine than soybean meal but more methionine and, in general, is probably less palatable than soybean meal. Some meals contain a high level of tannic acid, which may depress the growth of young animals.

In common with other members of the *Brassica species*, rapeseed contains goitrogenic compounds, which are attributed to a group of compounds called glucosinolates. Although the glucosinolates themselves are biologically inactive, they can be hydrolyzed by enzymes in the seed to produce isothiocyanates, thiocyanates, nitriles, and various derivatives of these chemicals. Fortunately, selected cultivars of both species have been developed that contain lower levels of the glucosinolates. Recent feeding trials with dairy cows and calves indicate that the low-glucosinolate meals do not cause any problems other than a slight increase in thiocyanate in milk (18, 28). One example of data from a trial with turkey poults is shown in Table 8-4, and two from lactation trials are shown in Table 8-5. Thus, if canola meal is to be used in substantial amounts to replace other protein supplements, the low-glucosinolate meals should be used.

SESAME MEAL

Sesame meal is produced as a by-product of extracting oil from sesame seed (*Sesame indicum*). The plant is grown primarily in India and China, although production is increasing in California. The meal contains 38% to 48% CP (dry basis) that is low in lysine but that has good levels of methionine. It can be used in limited amounts in diets for simple-stomached animals. The meal apparently contains enough phytic acid to interfere with Zn utilization under some conditions.

MISCELLANEOUS OILSEED MEALS

A few other oilseed meals are available in very limited amounts in some areas. These include **babassu meal, castor seed meal, crambe meal** (one example given on crambe meal is in Table 8-5), **hempseed meal, meadowfoam meal,** and **mustard seed meal.** On the basis of quantity produced, these meals are not important at this time, although several of them show some promise in particular climatic and soil conditions. Several of them, such as castor meal, crambe meal, and meadowfoam meal, contain antinutritional factors that either require special processing or feeding methods that restrict their use for most animal species.

SUMMARY

Some of the good points and limitations of the various oilseed meals have been discussed. These protein sources can be used in moderation in rations for nearly all animals. As a general rule, one-quarter to one-third of a basal protein concentrate can often be replaced with less common meals without any marked effect on animal production and often at a reduced cost of production.

MILLFEED PROTEIN SOURCES

Corn millfeeds with 20% or more CP include corn gluten feed, corn gluten meal, and condensed fermented corn extractives (corn solubles) produced from wet milling methods and corn germ meal that comes from either wet or dry methods. Similar products are produced from milling sorghum grain, although in much smaller quantities. Small amounts of germ meal are produced from wheat or other grains milled for flour.

Corn gluten meal is marketed at levels of about 41% or 60% CP (as fed). It is the residue remaining after removal of the larger part of the starch and germ and separation of the bran. **Gluten feed** contains gluten meal (corn bran with or without the fermented extractives) and usually has a protein level of 21% to 23% (as fed). The **fermented extractives** are condensed material derived from steeping corn in water prior to milling. The water content is about 50% and the protein 21% to 23%. The **germ meal** has the best overall balance of amino acids.

The protein in these different products resembles somewhat that in the parent grain in that lysine or tryptophan are usually the most limiting amino acids but, on the other hand, the S-containing amino acids are found in higher concentrations (Table 8-3). Gluten meal is used in rations for all farm livestock, and the 60% meal with added xanthophylls is used in poultry feed as a source of protein, but also for the pigments (for skin or yolk color). The gluten feed, because of its high fiber content, is not often used for poultry or growing pigs. The water content of the solubles limits its use in dry feeds.

A comparative trial is given with finishing steers and lactating cows in Table 8-5. Two other papers might be mentioned (29, 30), both of which present data that show that corn gluten feed makes a satisfactory supplement to either fescue hay or native grass hay. The protein in corn gluten feed or meal is not as soluble as the protein in some products, such as soybean meal.

Similar products produced from sorghum grains include **grain sorghum gluten feed, grain sorghum gluten meal,** and **grain sorghum germ cake** or **germ meal.**

DISTILLERY AND BREWERY BY-PRODUCTS

Over 1% of the corn crop in the United States is used for manufacturing beers and distilled liquors, and most of the residue, minus part of the starch, is returned as animal feed. In the manufacturing of bourbon whiskey, a mash must contain a minimum of 51% corn, but more typical ingredients include 75% corn, 12% rye, and 13% malted barley (31). The grains are cooked with water and cooled, and the barley malt is added to provide a source of enzymes to produce maltose and dextrin. Yeast is then added and the mixture is fermented. Following fermentation and distillation, the solids are recovered in different ways and dried. Feedstuffs produced include condensed and dried distillers solubles, distillers dried grains, and distillers dried grains with solubles. **Distillers dried grains** account for the largest volume of distillery by-products. If official terminology is used in naming the grains, they may be listed as corn distillers grains, rye distillers grains, and so on, according to the predominant grain in the mix. Other protein supplements may also be available from the distilling industry. These include **molasses distillers condensed** or **dried solubles** and **potato distillers dried residue.**

Analytical data on some of these products are given in Table 8-2 and in Appendix Table 1. The distillers dried grains are mainly differentiated by the relatively high fiber content (11% to 13%), but all have a protein content on the order of 27% to 29% due to the presence of yeast. The energy value is medium to high, and the B vitamin content is high. The grains are relatively deficient in lysine and phenylalanine, but the solubles have a more balanced amino acid distribution. With regard to the minerals, the P content is relatively high, as is the S content. Some trace minerals, such as Se, are found in relatively high concentrations.

Protein sources from the beer industry include brewers dried grains, dried spent hops, malt sprouts, and brewers yeast. The **malt sprouts** are dried rootlets that are removed in the preparation of malted barley. **Brewers dried grains** are the dried residue of barley malt and other grains that have been used to provide maltose and dextrins for fermenting. **Brewers yeast** is surplus yeast used to ferment the wort, the liquid drawn off from the malt–grain mixture. **Dried spent hops** are obtained by drying the material filtered from hopped wort. **Brewers grains meal** is scalped brewers dried grains containing not less than 35% Cp and not more than 10% moisture. **Converted cereal extractives** may be produced from effluents resulting from the production of malt, wort, ale, or beer.

Protein levels of the grains and malt sprouts range from 26% to 29%, and the quality is good. Lysine appears to be the most limiting amino acid in the dried grains and methionine for malt sprouts; tryptophan is relatively high in all brewery by-products. Brewers yeast has about 45% CP, with high levels of lysine and tryptophan. The B vitamin content is high, as it is in most fermentation by-products.

Use of brewers grains and malt sprouts is limited in monogastric rations because of the relatively high fiber level (18% to 19% and 14% to 16%, respectively). Brewers grains and distillers grains, as well as other fermentation by-products, have been touted as a good source of unidentified growth factors. Whether there are, indeed, unidentified factors or an optimal combination of different nutrients remains to be seen, but they do produce favorable production responses in many circumstances (Table 8-5). The dried grains, in particular, have proteins that are resistant to degradation by rumen microorganisms. Research results show a good response when the dried grains are fed to animals needing higher levels of some amino acids. Wet brewers and distillers grains are often fed to cattle, but high moisture content requires that they be used regularly and close to the source of supply. Researchers have shown that ammonia can be used as a preservative for wet distillers grains without any detrimental effect on milk production by cows (32). This would, no doubt, be cheaper than drying.

OTHER PLANT PROTEIN SOURCES

Coconut or **copra meal,** the residue remaining after extraction and drying of the meat of the coconut (*Cocus nucifera*), is available in many tropical areas of the world. The protein content is low (20% to 26%, dry basis), it is relatively deficient in lysine and methionine, and it may be quite variable in digestibility and quality because of variations in processing methods. It is also subject to molding while drying. Coconut meal can be used to supply some of the supplementary protein for swine and poultry, but results with both poultry and swine have indicated a drop in performance when the meal has been used as a major protein source.

Palm kernel meal, a meal remaining after extraction of palm oil from seeds of *Elaeis guineenis,* apparently has some promise as a low-protein meal. It should be noted that the better-quality palm oil used as a feedstuff is extracted from the husk of the fruit rather than from the seeds.

Cull beans and **peas** of a number of different species are sometimes available for animal feeding, although they are usually grown primarily for human food. Those available include kidney beans, pinto beans, navy beans (*Phaseolus vulgaris*), broad or horse beans (*Vici faba*), lima beans (*P. lunatus*), and various types of peas, such as the common green pea (*Pisum sativum*), chick-pea or garbanzo (*Cicer arietinum*), cowpea or black-eyed pea (*Vigna sinensis*), pigeon pea (*Cajanus indicus*), winged pea (*Lotus tetragonolobus*), and Austrian field pea (*Pisum* spp.).

Pea and bean seeds generally contain 20% to 28% CP that tends to be deficient in S-containing amino acids as well as tryptophan. Pea seeds that have had adequate testing with animals generally are quite satisfactory when fed without processing (other than grinding) and allow very satisfactory performance with all species. One example of the use of chickpeas with lactating dairy cows is shown in Table 8-5.

Bean seeds, on the other hand, often contain antinutritional factors such as toxins or trypsin inhibitors. Because of the toxins and poor protein quality, use in feed for monogastric species should be limited, and some type of heat processing is highly desirable. Recent data (62) indicate that steaming at 100°C for 75 min was adequate when cull beans were fed to pigs. Those steamed only 45 min resulted in reduced feed intake, diarrhea, and lower digestibility. Most beans can be used efficiently in ruminant rations, provided they are fed in moderate amounts (Table 8-5). Some of these grain legumes hold promise as complete energy–protein feeds for swine and poultry in tropical areas of the world. Considerable research is being carried out to identify high-yield varieties and to develop economical methods of removing or inactivating inhibitors and toxins.

Legume seed screenings (from seed production of alfalfa, clover, and the like) are often available in limited amounts in restricted areas. Limited research indicates that they are quite satisfactory as feeds for ruminants. Little research has been reported for other species. Crude protein content is usually on the order of 22% to 25%.

PROTEIN SUPPLEMENTS OF ANIMAL ORIGIN

Protein concentrates derived from animal tissues have been used for many years and are highly prized for supplemental sources. Because these products come from animal sources, the amino acid distribution is generally similar to dietary needs. However, it must be recognized that quality may be quite variable because products with essentially the same label may come from many different sources. In addition, quality may be affected markedly by processing methods and the use of appropriate temperatures in cooking and drying the meals.

MEAT MEAL, MEAT AND BONE MEAL

AFCO listings for animal products include **meat meal, meat and bone meal, meat meal tankage,** and **meat and bone meal tankage.** The only difference in the specifications between meal and tankage is that the meals shall not contain added dried blood. Meat meal is differentiated from meat and bone meal (or tankage) on the basis of P content. If the product contains more than 4.4% P, it shall be labeled as meat and bone. For all products (mentioned above) there are specifications that not more than 14% of the material shall be indigestible by pepsin and that not more than 11% of the CP shall be pepsin indigestible. These products are made from carcass trimmings, condemned carcasses, condemned livers, inedible offal (such as lungs), and bones. They are not supposed to include hair, hoof, horn, hide trimmings, manure, and stomach contents except in such amounts as may occur unavoidably in good factory practice.

In practice, tankage is the unground material produced by dry rendering, and meat meal is the material remaining after grinding (33). Nearly all the tankage and other similar products go to blenders, who then incorporate the different ingredients into meals that are standardized on the basis of protein and ash content. This results in much more uniform products to the feeder. These meals are usually standardized to have 45%, 50%, or 53% to 55% CP (as fed), usually by the addition of dried blood or dried meat solubles.

Most large meat packers process and produce meat meal and meat and bone meal. In addition, such products may be produced by rendering plants that utilize dead animals (from farm and feedlots), meat and bone wastage from wholesale meat houses, grocery stores, hotels, and so on. Products from rendering plants of this type are apt to be more variable in quality and in protein and ash content.

Data on animal experiments indicate that the protein is 81% to 87% digestible and that the limiting amino acids for swine with cereal diets are lysine, methionine, and threonine (34) and methionine and cystine for poultry (35). In meals with higher protein content, isoleucine may be limiting also, because this amino acid is quite deficient in blood meal. The protein quality is generally considered to be lower than that of fish meals or soybean meal (35).

In addition to protein, these meals have about 8% fat, which is, of course, high in digestible energy. The ash content generally ranges from about 28% to 36% in blended meals, and about 7% to 10% of the meal will be Ca and 3.8% to 5% will be P; K, Mg, and Na also are present in substantial amounts. Consequently, meat meals may be an important dietary source of these minerals. On the other hand, the mineral content may limit usage in some cases. Meat meals (as with most animal products), fish meals, and fermentation products are a good to excellent source of vitamin B_{12}.

BLOOD MEAL

Blood meal is produced from dried, ground blood. Drying is done by methods referred to as drum, spray, ring, and flash. Flash drying, a relatively new process, results in a more uniform product than the other methods and a meal that has a high available lysine content. The protein content of blood meal is about 85% and, except for isoleucine, it is an excellent source of other amino acids. Data on young pigs indicate that the better methods of drying produce a blood meal with about 7% available lysine (34). The meal is low in minerals (except Fe) and fiber as well.

MISCELLANEOUS MAMMALIAN PRODUCTS

Other products available include **dried meat solubles,** which are obtained by drying the defatted water extract of the clean, wholesome parts of slaughtered animals prepared by steaming or hot water extraction. It must contain no less than 70% CP. **Dried liver meal** is sometimes available. A high proportion is derived from livers condemned because of abscesses, infection with liver flukes or worms of various types, or for other reasons. **Glandular meal** is produced by drying liver and other glandular tissues from slaughtered mammals. If a significant portion of the water-soluble material has been removed, it may be called **extracted glandular meal. Fleshings hydrolysate** is obtained by acid hydrolysis of the flesh from fresh or salted hides. It is defatted, strained, and neutralized. If evaporated to 50% solids, it is referred to as **condensed fleshings hydrolysate. Hydrolyzed hair** is produced in limited amounts. It must be cooked under pressure for relatively long periods to hydrolyze the hair protein. When prepared in this manner, it contains a minimum of 80% CP, which is well utilized by ruminants (Table 8-5). **Animal by-product meal** is the dry product from animal tissues prepared for feeding when processed by steam or dry rendering. It must be designated according to its protein content. **Hydrolyzed leather meal** is also available in some areas. A considerable amount of steam cooking is required to produce a digestible product.

MISCELLANEOUS POULTRY PRODUCTS

Hydrolyzed poultry feathers are produced in relatively large amounts. They must be cooked under steam pressure for 30 to 45 min to produce a digestible product in which the CP must be 75% digestible by pepsin. Feather meal may contain 85% to 90% CP, but it is deficient in cystine, methionine, lysine, histidine, and tryptophan for poultry (35), so care must be used in selecting other protein sources to complement the amino acid content. It is a satisfactory protein source for ruminant animals (Table 8-5).

Poultry by-product meal is made from ground, rendered, clean parts of the carcass (except feathers) of slaughtered poultry, such as heads, feet, undeveloped eggs, and intestines. It must contain not more than 16% ash and not more than 4% acid-insoluble ash. The CP content is on the order of 58% (as fed). It has proved to be a very satisfactory protein source for chickens.

Poultry hatchery by-product is a mixture of egg shells, infertile and unhatched eggs, and culled chicks that have been cooked, dried, and ground with or without removal of part of the fat. **Hydrolyzed poultry by-products aggregate** is the product resulting from heat treatment of all by-products of slaughter. The product may, if acid-treated, be neutralized. It provides a satisfactory protein source for broilers if supplemented with methionine and lysine. **Dried poultry waste** is the dried excreta collected from caged layers. It contains 25% to 28% CP (dry basis), of which about one-third is true protein and the remainder is nonprotein compounds such as uric acid (excreted by birds in urine). This product is not suitable for monogastric species because of the NPN and the high ash content (25% to 30%) but it can be used in some situations for ruminants, particularly in wintering or maintenance rations. **Poultry litter,** primarily from broiler operations, usually has enough CP to qualify as a protein source. There is a considerable amount of information on its use in ruminant diets for both sheep and cattle. It has been fed as is, ensiled with or without other products (such as corn forage), or fed after sun drying or dehydration. Overall, it is a rather cheap source of N for growing calves or for maintenance feeding of ewes or wintering beef cows. Neither dried poultry waste nor litter is permitted to be fed to lactating dairy cows.

MILK PRODUCTS

Dried whole milk, although considered to be almost the perfect food for young suckling mammals, is nearly always too expensive for use as an animal feed. When used as animal feed, **dried skim milk** is used primarily in milk replacers and, to a lesser degree, in solid starter rations for young pigs and ruminants and in some pet foods.

The quality of dried milk can be impaired by overheating during the drying process (drum drying); therefore spray dried milk is preferred. Poor-quality milk, when used in milk replacers, is apt to lead to diarrhea and digestive disturbances.

From a nutritional point of view, dried skim milk is apt to be deficient in fat-soluble vitamins and, depending on the animal species, in Mg, Mn, Fe, and Cu. Normally, other minerals and vitamins are supplied in adequate amounts for young animals.

Condensed or **dried buttermilk** is often available, and it qualifies as a protein concentrate, since the CP content is about 34% (dry basis). It has a comparable feeding value to skim milk except for having a slightly higher energy content. **Cheese rind** (cooked cheese trimmings) is available in some areas and is said to have a good feeding value. It has about 60% CP. **Dried whey protein concentrate** (25% minimum CP) is also available in some areas.

Other milk products are available but usually are not economical for feeding animals, except for whey, which does not qualify as a protein concentrate because its CP content is below 20%. **Casein,** the solid residue obtained by acid or rennet coagulation of defatted milk, is available, as are **dried hydrolyzed casein, dried milk albumin,** and **dried milk protein.**

MARINE PROTEIN SOURCES

FISH MEALS

Fish protein sources are primarily of two types: those from fish caught for making meal and those made from fish residues remaining after processing for human food or industrial purposes. Anchovy (*Engraulis* spp.), herring (*Clupea* spp.), and menhaden (*Brevootia tyrannus*) provide a majority of the meals made from whole fish. These fish have a high body oil content, much of which is removed in preparation of the finished meal. In addition to these sources, residues from any species processed for human food may be used to make fish meal. Some of the processing residue may be fed directly as is (or ground) to mink, foxes, and other species raised for fur and is often included when making up diets for pets.

Fish meal is defined as the clean, dried, ground tissue of undecomposed whole fish or fish cuttings, with or without extraction of part of the oil. It must not contain more than 10% moisture and, if it contains more than 3% salt, the amount of salt must be specified. In no case must the salt content exceed 7%. Fish meal may be processed using several different methods. The oil content is objectionable if fed in relatively large amounts to poultry or swine (because of fishy flavor in the meat); thus most of it should be removed. In addition, fish oils oxidize readily; thus it is common for most processors to add antioxidants to prevent oxidation, overheating, and molding. Unfortunately, the quality of the meal may still be variable if it is not well processed.

Good-quality fish meals are excellent sources of proteins and essential amino acids (Tables 8-2 and 8-3). The protein content is high and highly digestible. Fish meals are especially high in essential amino acids, including lysine, that are deficient in the cereal grains. In addition, fish meal is usually a good source of the B vitamins and most of the mineral elements. As a result, fish meal is a highly favored ingredient for swine and poultry feeds (Table 8-4), although the cost is usually higher than for other protein sources except milk. Some fish meal is used in ruminant rations because they are not digested in the rumen by the microorganism. Some use of fat-extracted meals appears feasible in milk replacers.

OTHER MARINE PROTEIN SOURCES

Although fish meals make up the majority of marine protein feeds, a variety of other products may be found on the market in some places. **Fish residue meal** is prepared from the residue from the manufacture of glue from nonoily fish. **Fish protein concentrate, feed grade,** is prepared from the residues used to prepare fish protein concentrate for human food. **Fish liver and glandular meal** is made up of viscera of the fish; at least 50% of the dry weight must be from fish liver. **Condensed fish solubles** are obtained by condensing the stickwater, that is produced during the canning process, so that the product contains at least 30% CP (as fed), while **dried fish solubles** must be dried and contain at least 60% CP. Fish solubles are considered to contain protein of high quality, to be a good to excellent source of water-soluble B vitamins, and to contain "unidentified growth factors."

Fish by-products are comprised of the nonrendered portions of fish. **Dried fish protein digest** is the dried enzymatic digest of fish or fish cuttings using the enzyme hydrolysis process. It must be free of bones, scales, and undigested solids with at least 80% CP and not more than 10% moisture. **Condensed fish protein digest** is the condensed enzymatic digest of fish or fish cuttings with at least 30% CP. **Fish digest residue** is the undecomposed residue (bones, scales, and the like) of the enzymatic digest.

In addition to the products mentioned, the Scandinavians have developed a product labeled **fish silage.** Fish are ground and treated with acids (mineral acids and/or formic acid). The combination of the body enzymes and acids reduces the material to a relatively liquid product, which will keep at ambient temperatures for some period of time. Although this product is probably more suitable for nonruminants, it can be used as a feed for ruminants (36, 37). **Liquefied fish,** prepared from ground fish or fish residues that are liquefied by fish enzymes, pasteurized briefly, strained to remove the bones, and acidified to a pH of about 4, is another promising way to use fish without the need for dehydration. Liquefied fish has been used in liquid protein supplements or in complete diets for ruminants with no problems (Table 8-5). Both methods (ensiling and liquefication) offer a way to utilize nutritious products without going to the high expense of drying.

Shrimp and **crab meals** prepared from processing residues are also on the market at times. Although shrimp meal is said to be about equivalent to tankage for swine, crab meal is unpalatable to swine. Crab meal does not show much promise for ruminants. Both of these meals contain a fair amount of chitin (a major component making up the animal's exoskeleton), which has a CP content of about 40%. Chitin is composed of a glucose polymer with N–acetyl groups on the second carbon atom of the glucose. Although the glucose molecules are linked in the same manner as in cellulose, rumen microbes do not appear to produce chitinase, the enzyme required to hydrolyze chitin (38). We would expect to find this enzyme primarily in the GI tract of species with exoskeletons. The relative values of these meals probably depend on the amount of nonchitinous residue in the meal.

MISCELLANEOUS PROTEIN SOURCES

As the world protein supply becomes more critical, efforts are underway to identify and develop additional sources for use in livestock feeding.

Potential sources include animal wastes, plant extracts, and single-cell organisms, such as algae, bacteria, fungi, and yeasts. Some discussion of these sources follows.

Poultry wastes have been mentioned previously. In areas with a high poultry population, **broiler litters** have fairly widespread usage for wintering adult cattle or as a supplement or in silage (see Ch. 6) for growing calves or for feeding to ewes during some stages of their reproductive cycle. **Cage layer wastes** have also been used (after air drying) to feed sheep or goats, and the commercially dried product **dried poultry waste** is marketed in a number of areas. Wastes from cattle do not contain enough N to qualify as a protein source. However, they show some promise as a N source when making silage. They have also been mixed fresh with other feed ingredients and fed to cattle, they have been included in silage (see Ch. 6), and they have been screened to remove solids and the fluid portion has sometimes been mixed with other ingredients and fed to cattle. There is less interest in feeding wastes from swine, although some experimental data are available.

There is interest in preparing **leaf protein concentrates** from crops such as alfalfa because of its high yield of protein per unit of land. Such concentrates (when dried) contain 40% to 47% CP of high quality, even though some concentrates have antitrypsin activity. The dehydrated product has been shown to be an effective protein substitute for soybean meal for fattening lambs, and the presscake (residue after extracting leaf juices) is said to be equal to alfalfa hay when used in limited amounts for fattening cattle (39).

SINGLE-CELL PROTEINS

Single-cell protein (SCP) could be developed into a very large source of supplemental protein to be used in animal feeding. Some individuals have gone so far as to suggest that SCP sources will provide the principal protein source for domestic animals (worldwide basis), depending on world population growth and the availability of feed proteins from plant sources. This could develop because microbes can be used to ferment some of the vast amounts of waste materials, such as straws; wood and wood processing wastes; food, cannery, and food processing wastes; and residues from alcohol production or from human and animal excreta.

Producing and harvesting microbial proteins is not without costs, unfortunately. In nearly all instances where a high rate of production would be achieved, the SCP will be found in rather dilute solutions, usually less than 5% solids. Methods used for concentration include filtration, precipitation, coagulation, centrifugation, and the use of semipermeable membranes. Such materials must either be dried to about 10% moisture, or condensed and acidified to inhibit spoilage, or be fed in the fresh state. Removal of the amount of water necessary for storage would, in most instances, not be economical.

Algae are a potential feed source. Preliminary results with cultivated freshwater algae indicate a potential for about 10 times as much protein per unit of land area as from soybeans. Algae contain about 50% protein, 6% to 7% fiber, 4% to 6% fat, and 6% ash. Because of its bitter taste and protein of low biological value, it should not be used at levels exceeding 10% of the diet of growing swine. A major problem is how to harvest and dry a product of this type.

Yeast products have been available to the feed trade for many years. They are described as dried yeast (nonfermentative, that is, inactivated with heat), live yeast, and irradiated yeast. Those available include **brewers dried yeast** and **brewers liquid yeast** (both *Saccharomyces* spp.), which must contain a minimum of 35% CP (dry basis). Other nonfermentative yeasts include **grain distillers dried yeast, molasses distillers dried yeast,** and **primary dried yeast** or **dried yeast** (all *Saccharomyces* spp.). **Torula dried yeast** (*Torulopsis* spp.) is also available as a feed ingredient. The yeast products just mentioned must contain a minimum of 40% CP. Yeasts contain protein of high quality and are high in most B vitamins, although there are differences between brewers and Torula yeast in vitamin content. The costs are also relatively high, so not much yeast is used in most animal feeds. In moderate amounts it is a very satisfactory protein supplement (Table 8-4).

Live yeast (active) products include **active dry yeast, yeast culture** (a dried product), and **molasses yeast condensed solubles,** which are a condensate of broth remaining after removal of bakers yeast cells propagated on molasses.

Irradiated yeast or **irradiated dried yeast** is prepared from yeast that has been subjected to ultraviolet light to increase the concentration of vitamin D_2, which is a satisfactory form of the vitamin for domestic animals except poultry, which utilize D_3 much more efficiently than D_2. This is a common source of the vitamin in many feedstuffs.

Bacterial SCP sources (which may include fungi) are only just beginning to come on the market. In Europe, a long-researched microbial product known as **Pruteen** is available for commercial use in the European Economic Community (40). This SCP source is derived from microorganisms selected to utilize methanol (wood alcohol). It is said to contain 71% CP and 13% fat. Numerous research reports have shown that it can be used to replace a high proportion of the dried skim milk normally used in milk replacers for young calves or lambs (41). In the United States, Coor's Brewery markets a product grown on effluent from their brewery. It is designated as **Brewers SCP.** ITT Rayonier, for a time, marketed a dried SCP product (with added fat) designated as **Raypro**™. It was produced from aerobic microbes grown on effluent from a paper-pulp mill operation. It contained about 50% CP (dry) and about 10% to 13% fat. Limited data on this product show that it was a satisfactory replacement for some of the standard protein sources in complete rations for fattening beef cattle or lactating dairy cows (42). Other preparations from pulp mills, but dried in different ways, have been shown to be satisfactory feeds for sheep or cattle (43). Microbes can be used to ferment solubles in processing waters, thus reducing microbial growth in streams, lakes, or oceans; consequently, with only costs for drying, a valuable source of protein can be brought on the market.

NONPROTEIN NITROGEN

Nonprotein nitrogen (NPN) includes any compounds that contain N but are not present in the polypeptide form of protein that can be precipitated from a solution. Organic NPN compounds include ammonia, urea, amides, amines, amino acids, and some peptides. Inorganic NPN compounds include a variety of salts, such as ammonium chloride, ammonium phosphates, and ammonium sulfate.

Although some feedstuffs, particularly some forages and silages, contain substantial amounts of organic NPN, from a practical point of view, NPN in formula feeds refers to added materials, primarily urea or, to a lesser extent, such compounds as biuret and ammonium phosphate. There is adequate research evidence to show that a number of other compounds could be used, but they are too costly to use for feeding to animals at the present time. In most areas, urea is the least costly source of crude protein.

This simple fact accounts for the tremendous interest in its use in feeds for ruminants.

NPN, especially urea, is primarily of interest for feeding to animals with a functioning rumen. The reason for this is that urea is hydrolyzed rapidly to ammonia and carbon dioxide, and the ammonia is then incorporated into amino acids and microbial proteins by rumen bacteria, which are utilized later by the host. Thus the animal itself does not utilize urea directly. In simple-stomached species the only microorganisms that can synthesize protein from urea are found in the lower intestinal tract at a point where absorption of amino acids, peptides, and proteins is believed to be rather low or nonexistent. Research with pigs, poultry, horses, and other species indicates only slight utilization of N from urea.

LIMITATIONS ON UREA USAGE

A variety of factors must be considered in utilizing urea in feeds. Urea can only be used as a source of N for ruminant animals. One of the more important facts is that urea is not a satisfactory source of N to feed to animals fed primarily on poor-quality roughage. This is illustrated by data in Table 8-6. Note that when steers were fed cottonseed hulls they responded fairly well to urea supplementation, but rate of gain was increased by adding fish meal or soybean meal. When fed barley straw or corn stalks, performance was quite low. With wintering beef cows, winter losses were not reduced by adding urea to a basal supplement with 15% CP. Neither did biuret or the addition of methionine hydroxy analog (a substitute for methionine) improve performance. Many other examples of a similar nature could be gleaned from the literature to illustrate this point. Although some divergence in responses can be found in the literature, the writer believes that the illustrations presented are representative of typical responses. Other examples were also given in Table 8-5, in which cattle were fed corn cobs or corn silage as the principal roughage. Urea did not allow the same level of production as supplements with natural proteins.

There have been hundreds of research reports dealing with the utilization of urea since it was first tried in the early 1940s. While it is true that most cellulose-digesting rumen microorganisms require ammonia and also that in vitro (laboratory) rumen studies show that urea as the only added N source will stimulate very adequate cellulose digestion, this is not the case for the live animal. Under laboratory con-

TABLE 8-6

Response of cattle fed nonprotein N with low-quality roughage

Item	Daily Gain (kg)	Feed Consumed (kg)		Reference No.
		Roughage	Supplement	
Steers fed cottonseed hulls				
Urea	0.51	7.92	0.36	44
80% Urea, 20% fish meal	0.59	7.94	0.48	
80% Urea, 20% soybean meal	0.71	8.44	0.96	
Steers fed chopped barley straw				
80% Urea, 20% SBM	0.20	4.72	0.21	
Steers fed corn stalks				
80% Urea, 20% soybean meal	−0.08	2.98	0.24	
	Winter Loss (kg)			
Beef cows wintered on native pasture				
15% CP, natural protein	90.0		1.34	45
30% CP, natural protein	76.0		1.31	
Biuret[a]	86.1		1.40	
Urea[a]	91.2		1.24	
Biuret + MHA[a]	88.5		1.32	
Urea + MHA[a]	98.5		1.02	

[a]Supplements formulated to be 30% CP with half of the CP from the source identified in the table. MHA = methionine hydroxide analog. Cows were fed for an 88 day period.

ditions where fermentations are usually carried out in nonpermeable containers (glass), the urea and/or ammonia remain in solution and are available to the microorganisms. In the rumen the ammonia can be absorbed through the rumen wall or pass into the lower stomach. In either case it is no longer available to the bacteria, although some of it eventually is recycled back to the rumen through saliva or, under some conditions, through the rumen wall. Ammonia is absorbed from the rumen more rapidly as the pH rises toward 7 or higher. At acid pHs of 6 or lower, absorption is slow or nil. If enough starch is fed to the animal to reduce the pH to this range, then utilization of urea by the live animal is very satisfactory. Sugars, such as those in molasses, do not support maximum urea utilization.

UREA TOXICITY

Urea may also be toxic or lethal, depending on the size and timing of the dose (feed consumption). Clinical symptoms may show up as soon as 30 min after ingestion of the feed. They include uneasiness, staggering, and kicking at the flank, at which point affected animals tend to go down. Labored breathing, incoordination, tetany, slobbering, and bloating have been observed. Fatally affected animals are prone to regurgitate rumen contents just before death. The reaction tends to be an all or none type, that is, the animal either dies or recovers with little if

any aftereffect. Fatal levels of urea are affected by adaptation of the animal to urea, how long it has been without food, the type of diet, and other things. Generally, an intake in a period of about 30 min of about 40 to 50 g/100 kg of body weight may be lethal. The toxicity occurs because urea is hydrolyzed rapidly in the rumen to ammonia and carbon dioxide. If the rumen fluid is alkaline (urea and ammonia make it more alkaline), ammonia will be absorbed rapidly, resulting in an overload on the liver, which would normally remove most of it and convert it to urea, which is less toxic. Ammonia builds up in the peripheral blood, resulting in the symptoms observed. There is really no feasible means of treatment. Provided animals are observed before the late stages of toxicity, they can be treated with vinegar (or other acids). One recommendation is to administer 1 gal of vinegar and 1 gal of water/100 lb of body weight. The point is that if several animals are affected not enough vinegar would be available soon enough to prevent death if it would otherwise occur. Prevention of urea (or ammonia) toxicity amounts to good feeding practices, to careful mixing of feeds containing urea, and to adapting animals gradually to feeds containing large amounts (i.e., liquid supplements or dry protein supplements with high percentages of urea).

Where livestock management is good and feed is formulated and mixed properly, urea can provide a substantial amount of the supplemental N required for feedlot animals, dairy cows, and other ruminants. In practical rations, current recommendations are that not more than one-third of the total N be supplied by urea or other NPN compounds. In complete feeds, urea should be restricted to 2% or less. More than this may be unpalatable and cause reduced feed intake. Some recent data indicate that feeding urea-based rations may result in meals of shorter duration, although the cows tended to eat more frequently than when fed rations without urea. Note in Ch. 5 that most states require labeling of feed tags with maximal amounts of urea or other NPN compounds as a protective measure for the buyer.

METHODS FOR IMPROVING UREA UTILIZATION

A considerable amount of research effort has been expended to improve utilization of urea when used as a protein concentrate or to increase its versatility in feeding. Several different examples of the use of urea or anhydrous ammonia in silage were given in Ch. 6. This appears to be one of the most cost-effective means of increasing the CP content of forages without increasing the likelihood of any problems with feeding NPN.

At least two extruded urea-grain mixtures are on the market, **Starea** and **Golden-Pro.** If extrusion is carried out under the correct conditions of heat, moisture, and pressure, the process results in a product that releases ammonia at a slower rate than urea in the rumen. Most studies indicate satisfactory results when such products are used for feeding dairy cows, young calves, or wintering beef cows. Results are normally better than with grain–urea mixes and approach or equal the value of soybean meal (in most but not all experimental studies).

Urea has been used also with other feed ingredients. A mixture of dehydrated alfalfa meal, urea, dicalcium phosphate, sodium sulfate, and sodium propionate formulated to contain 100% CP has produced very satisfactory results with high-producing dairy cows. It is called **Dehy-100.** Urea solutions absorbed on **flaked soybean hulls** also appear to be a satisfactory means of feeding dairy cows (46). Pressure cooking **urea with cassava meal** is a possible means of expanding the use of urea in tropical areas (47). In Europe, it is apparently fairly common to absorb **urea solutions** (50% urea–water) **on whole grains** for feeding to cattle or sheep. In addition, there have been ef-

forts to produce treated urea products that release ammonia at a slower rate than untreated urea (**Slo-release**). Although some of the products do release urea at a slower rate, animal performance on low-quality roughage has not been improved much, if at all. Studies from the same laboratory indicate that release rate is not the problem. In one experiment when urea was administered directly into the rumen over a 24-h period or during a 6-h or 1-h period, there was no effect on N retention by the animals. Thus it may be that other factors have more influence on utilization of ammonia from urea than the speed of urea hydrolysis in the rumen.

AMMONIATED LIQUID FEEDS

Two liquids containing ammonia are worthy of note for use with ruminants. **Ammonium lignin sulfonate** is produced by the paper manufacturing industry when wood is reacted under conditions of heat and pressure with sulfur dioxide and ammonium bisulfite (other sulfonates are produced using Na or Ca bisulfite). The liquids are partially evaporated, resulting in a product with about 50% to 55% solids, 25% to 30% sugar or sugar acids, and 15% to 25% CP. When used at levels approved by FDA (4% of finished feed or 11% of a sulfonate–molasses mixture), it appears to be a satisfactory feed ingredient.

Fermented ammoniated condensed whey is another product of interest. It is produced from whey by fermenting with *Lactobacillus bulgaricus* (which metabolizes lactose to lactic acid) accompanied by continual neutralization with ammonia. The resulting product contains about 68% ammonium lactate with a crude protein equivalent of 69% (dry basis) with about 64% DM. Research results from various experiment stations indicate that the product is utilized more effectively than urea and that it appears to be quite comparable to soybean meal for dairy cattle. Certainly, it also provides another means of utilizing large amounts of surplus whey.

OTHER NPN COMPOUNDS

Biuret, a condensation product of urea, has been used to some extent as an NPN feed ingredient (Table 8-6). It usually gives a better response when fed with low-quality roughage than does urea. However, it is more costly and not as readily available on the market. Biuret is also less soluble and is much less likely to cause toxicity. It is not approved for use with lactating dairy cows because biuret will show up in the milk.

A wide variety of other N-containing salts could be used as NPN sources in place of urea. **Urea phosphate** and **isobutylidendiurea** appear to be well utilized. In addition, many different ammonium salts have been used experimentally. These include the **chloride, bicarbonate, acetate, propionate, lactate,** and **sulfate** salts as well as the **mono-** and **diammonium phosphates** (the latter two are not very palatable). Available evidence indicates that most of these compounds are less likely to be toxic than urea, but most of them are not used because of cost. Ammonium chloride is frequently used as a preventative or treatment for urinary calculi, and the mono- and diammonium phosphates are used at times to supply both N and P.

The only other NPN compounds of practical interest for ruminants are **methionine hydroxy analog** and protected methionine. They are used as a source of methionine for poultry because the compound can be converted to methionine by the liver. There has been limited use in dairy rations, because some evidence indicates that they have been effective to a limited extent in increasing milk fat percentage in early lactation and they are less likely to be degraded in the rumen than is free methionine. **Lysine** is, of course, often added to poultry diets, since it can be produced commercially at a reasonable cost.

LIQUID SUPPLEMENTS

Liquid supplements (LS) have been used for some time (±50 years), primarily for cattle, to a much lesser extent for sheep, and with very limited use for horses. In the early days, LS were primarily molasses with minor amounts of additives. Most manufacturers still use some molasses, but many other liquids are available from the production of sugar, paper, and cheese and from many different products involving fermentations of one kind or another. The increased use of these many different liquids may have been caused by several factors, including the increasing cost of cane molasses and reduced availability of beet molasses, poor results with cattle fed low-quality roughage and LS primarily of molasses–urea, and, last but not necessarily least, tougher environmental regulations prohibiting indiscriminate dumping of many wastes down the sewer and into the nation's water supply.

If they are consumed regularly and used with reasonable efficiency, LS offer a great deal of convenience for supplying needed nutrients to cattle and sheep on winter ranges. Lickwheel feeders have been developed to restrict consumption (one example is shown in Fig. 8-2). Generally, LS are formulated on the basis of a desired consumption of 0.23 to 2.27 kg (0.5 to 5 lb) for cattle and 50 to 100 g/day (2 to 4 oz) for sheep. Unfortunately, it is difficult to obtain a uniform daily consumption because it will vary among animals with the weather, palatability of the liquid, availability and kind of feed, and other factors. Sheep are slower to learn to use lickwheel feeders than are cattle, possibly because they have less of a preference for sweets. Studies in Australia involving the use of tritium-labeled water in molasses indicated that about half of the sheep never did consume LS (tritium can be detected in very low concentrations in the blood), and consumption by those that did consume the supplement was quite variable. Methods for injecting LS directly into low-quality roughages have also been utilized and have proved quite effective.

LS are also used for feeding in drylot situations to beef and dairy cattle and to sheep. As a result, formulas may vary depending on the intended purpose. Most companies manufacturing LS provide custom mixes with additional ingredients or modification of their basic formula if desired by the user.

The other major use of LS is for addition to complete feeds for finishing beef cattle or for dairy cattle or for use as a top dressing for all or part of the remainder of the ration. In these circumstances the amount consumed is controlled by the feeder, so it is then a matter of cost or convenience if LS are used. LS serve as a very

FIGURE 8-2 One example of a lickwheel feeder for liquid supplements.

useful vehicle for including small amounts of additives in rations in these circumstances, because more uniform mixing is always likely when micro ingredients are diluted before addition to the major part of a ration. LS can also improve palatability of the ration.

LIQUID SUPPLEMENTS AS AN ENERGY SOURCE

LS are usually based on use of one or more of the various types of molasses (usually cane or beet in the United States) diluted with water and other liquids and with various materials in suspension. In addition to molasses, other liquid energy sources (discussed in Ch. 7) that find increasing use in LS include hemicellulose extract, lignin sulfonate, hydrolyzed grain, whey, condensed whey, condensed whey product, concentrated separator by-product, condensed whey solubles, condensed distillers or brewers solubles, liquid streptomyces solubles, blending feeding fats, and, in some cases, propylene glycol; the latter compound is metabolized for energy but also acts as a preservative and depresses the freezing point.

NITROGEN SOURCES

Most commonly, some urea is used in most LS, but its use is lower and less frequent than was the case several years ago. Urea is usually added as a saturated urea solution (ca. 50% urea in water). Protein equivalents in LS may range up to 80% or more, although most advertised LS are in the range of 30% to 35% CP. Other liquid N sources that qualify as protein supplements include fermented ammoniated whey, ammonium lignin sulfonate, condensed fermented corn extractives, condensed extracted glutamic acid fermentation product (also listed sometimes as condensed beet fermentation solubles, Dynaferm or Manaferm), condensed molasses fermentation solubles, liquid brewers yeast, condensed fish solubles, or liquefied fish waste. In addition to these various liquids, dry products can be suspended in limited amounts in the supplement with the aid of suspending agents such as bentonite, xanthan gum, or alginates. Dry protein sources that have been used include dry poultry waste, feather or hair meal, fish meal, blood meal, corn gluten meal, and meat and bone meal.

OTHER ADDITIVES IN LIQUID SUPPLEMENTS

Phosphoric acid has been used extensively in LS. In addition to supplying P, the acid reduces viscosity, inhibits microbial growth, an important factor in cold climates, and limits intake as well because of the lower pH of the mixture. Ammonium polyphosphates are also used as P sources (as well as a source of NPN); some companies also use monosodium phosphate and dicalcium phosphate as P sources (48).

Other minerals such as Mg and S may often be added. Although it may be desirable to add Ca, most Ca salts have a low solubility in aqueous solutions; if large amounts are added, they must be in suspension. Mineral compounds added by some companies include limestone, salt, potassium chloride, magnesium oxide, ammonium sulfate, copper sulfate, sodium bicarbonate, sodium sesquicarbonate, sodium carbonate, and calcium phosphate (48).

Depending on the intended use, average vitamin A fortification is about 30,000 IU/lb (10,000 to 100,000 range). Vitamin D is added at average levels of about 9,000 IU/lb (1,500 to 40,000 range) and vitamin E at levels of about 12 IU/lb (4 to 67 range). A variety of trace minerals may be added, depending on local needs.

Nonnutritive additives added to LS may include various antibiotics, ionophores, hormones such as melengestrol acetate, antioxidants, ethanol, and flavoring agents. Emulsifiers and other agents such as clays or gums to aid in keeping ingredients in suspension are normally used if dry ingredients are added.

Research Reports on LS Although LS have been (and probably still are) used widely for supplementing cattle consuming poor-quality forage (winter range, crop residues, or harvested roughage), urea as the principal N source does not give the response of a dry supplement such as soybean meal in stimulating added consumption or improved digestibility of the total ration. A fairly typical response between the two types of CP sources is illustrated in Table 8-7. Note that SBM stimulated greater consumption of wheat straw and greater digestibility than the urea-based LS. Likewise, when fed to cows on winter range (Table 8-8), urea in either dry or liquid form and biuret in a LS resulted in greater winter weight loss than did a dry supplement with no urea. These results are reasonably typical of most of the reports in the literature. Animals do respond when fed most NPN sources as a supplement to poor-quality roughage, but not to the extent that they will if fed a dry supplement. A response is more likely with medium-quality roughages such as cottonseed

TABLE 8-7

Comparison of straw consumption and energy digestibility when cattle were fed soybean meal or a molasses–urea LS

| | Supplement | | | |
| | Soybean Meal | | Molasses–Urea | |
Supplement Level[a]	Straw Consumed[a]	Energy Digestibility (%)	Straw Consumed[a]	Energy Digestibility (%)
0	43.3	38	43.3	38
1	63.1	55	44.3	43
2	69.6	48	48.4	41
3	72.6	49	50.8	43
4	70.4	52	51.4	38

Source: Church and Santos (49).

[a]Expressed as g/kg body weight$^{0.75}$.

TABLE 8-8

Supplement consumption and winter weight loss of cows fed various nitrogen supplements on winter range

Supplement	Supplement Consumed (kg/day)	Winter Weight Loss (kg)
Dry supplements		
SBM–sorghum–grain	1.18	−72
SBM–sorghum–urea	1.33	−92
SBM–sorghum–biuret	1.23	−92
Liquid supplements		
Molasses–urea	1.55	−96
Molasses–biuret	3.08	−117

Source: Rush and Totusek (50).

hulls or grass hay than with the very poor quality roughages.

The amount of urea in a LS will influence consumption. This is illustrated by two experiments shown in Table 8-9. In the first experiment, increasing urea from 1.35% to 4% increased consumption of both supplement and straw. A further increase to 9.8% urea decreased supplement but not straw consumption. In the second experiment, steers were first fed a supplement with 6% urea. When urea was subsequently increased to 9.8%, consumption of supplement decreased to about half. This is probably one reason most commercial supplements are formulated to contain about 30% CP when using a substantial amount of urea.

The addition of preformed (natural) protein sources to LS has, in a limited number of cases, resulted in some improvement in animal performance when low-quality roughages were fed. Some data on this topic are shown in Table 8-9 when part of the CP was made up of liquefied fish, feather meal, or pulp mill single-cell protein. Note in each experiment shown that

there was some modest improvement in daily gain of young cattle fed ryegrass straw and LS with the preformed protein.

Two other examples are shown in Table 8-10 in which cows with calves or yearling heifers were fed supplements on winter range. They received a dry (NC) supplement with a low CP level or a dry supplement with 29% CP compared to liquids containing part of the N from corn steep liquor or fermented ammoniated condensed whey. In this experiment the LS with the two liquid N sources allowed performance at least equal to the 29% CP dry supplement, thus providing additional data showing that preformed proteins are needed for adequate performance of cattle consuming low-quality roughage.

A number of reports are available in which a variety of different liquid N sources has been incorporated into complete diets. For example, in one case cattle were fed corn silage, shelled corn, and 10% LS (54). The liquids fed were ammonium lignin sulfonate–molasses, condensed cane solubles–molasses–urea, condensed beet solubles–molasses–urea, or molasses–urea. No differences were found in daily gain of the cattle. Dry-matter consumption was lower by cattle fed the condensed beet solubles and highest for those fed the molasses–urea; however, the feed to gain ratio was the best for those fed the beet solubles and cane solubles. In another case the addition of fish solubles increased the growth rate of steers fed a mixture of ground ear corn and LS (55).

The addition of preformed proteins may affect palatability of LS; certainly, it will affect consumption over a short period of time. It has been demonstrated that small additions of SCP from pulp mills, feather meal, hair meal, and liquefied fish waste all stimulated greater

TABLE 8-9

Effect of level of urea or addition of native proteins on performance of cattle self-fed liquid supplements

	Feed Consumption		
Item	Liquid (kg/day)	Straw (kg/day)	Average Daily Gain (kg)
10-day trial, individually fed steers[a]			
1.35% urea, 7.5% CP	1.69	3.1	
4.03% urea, 15% CP	2.24	3.5	
9.8% urea, 30% CP	1.1	3.5	
60-day trial, group-fed steers, 20% CP[a]			
6% urea	1.45	3.2	0.18
5.3% urea, 13.3% LF[b]	1.81	3.1	0.23
60-day trial, group-fed steers, 30% CP[a]			
9.8% urea	0.69	3.2	0.23
8.5% urea, 20% LF[b]	0.91	3.5	0.37
88-day trial, group-fed steers, 30% CP[c]			
9.8% urea	0.85	5.2	0.55
8.6% urea, 3.9% FM[d]	1.10	5.3	0.61
8.6% urea, 7.9% SCP[e]	0.60	5.6	0.59

[a]From Kellems et al. (51).
[b]LF = liquefied fish.
[c]From Kellems (52). The basal roughage was ryegrass straw in each experiment.
[d]FM = feather meal.
[e]SCP = pulp mill single-cell protein.

consumption than urea when fed to cattle given LS and ryegrass straw (56). No doubt other additives such as phosphoric acid may also have some effect on consumption.

CURRENT USAGE OF LS   It is obvious from the preceding discussions that a wide variety of materials could be used to satisfy nutrient specifications in LS. There is some indication that more and more commercial products are routinely formulated to contain preformed proteins of one type or another. This is an encouraging step and one that will without doubt result in more efficient use of LS by livestock feeders using them to supplement poor-quality roughage of one kind or another.

TABLE 8-10

Response of cattle to liquid supplements based on corn steep liquor (CLS) or fermented ammoniated condensed whey (FACW)

	Supplement[a]			
Item	NC	PC	CSL	FACW
Cows (fall calving) with calves, 85 days				
CP content of supplement, %	12.9	29.0	16.8	16.0
Supplement intake, kg/head[b]	2.4	2.3	2.1	2.2
CSM intake, kg/hd	—	—	0.8	0.8
Total CP intake, kg/hd	0.31	0.67	0.68	0.68
Change in cow (weight, kg)	−54.4	−20.8	−25.0	−22.5
Change in calf (weight, kg)	31.7	37.7	34.9	36.0
Heifers, yearlings, 85 days				
Supplement intake, kg/hd	0.8	0.7	0.7	0.8
CSM intake, kg/hd	—	—	0.2	0.2
Total CP, kg/hd	0.10	0.20	0.20	0.21
Weight change, kg	−22.4	−4.8	8.0	4.2

Source: Wagner et al. (53).

[a]NC = negative control; PC = positive control. NC supplement was made up of 11.5% CSM, 41.8% grain corn, 41.8% grain sorghum, plus minerals and vitamins. PC supplement was 66.7% CSM, 30.6% grain corn, plus minerals and vitamins, CSL was 69.9% CSL, 27.4% cane molasses, 1.1% H_2SO_4, plus minerals and vitamins; FACW was 37.2% FACW, 60% cane molasses, 1% H_2SO_4, plus minerals and vitamins. Cattle were fed individually 6 days/week.
[b]CSM was fed at a level so that the CP intake of cattle on the LS was equal to that in the NC.

A recent innovation with LS is the use of *pumpable dry* supplements. This term is used to describe a product with about 70% solids that appears as a thick slurry. Some pumpable dry products have suspending agents, giving them thixotropic characteristics (they tend to gel when at rest but take on liquid characteristics when agitated or stirred). Some that are not thixotropic products use suspending agents such as xanthan gum. According to trade sources, pumpable dry products are currently being used to mix with other feed ingredients in complete rations rather than being used for self-feeding, as is often done with typical LS.

PROTEIN SOLUBILITY AND DEGRADABILITY BY MICROORGANISMS AS RELATED TO RUMINANTS

Proteins can be evaluated (and rated) for monogastric species on the basis of digestibility of the protein and its ability to supply specific amino acids for the animal's needs. With ruminant animals, if optimal utilization is desired, the needs can be more complex because of microbial fermentation taking place in the forestomach. For optimal protein efficiency, it would be desirable to be able to quantitate and express in feeding standards the needs for rumen microorganisms *and* for the host animal separately for all the likely production situations. Although several systems have been proposed, there appear to be too many variables and unknowns to make them workable except in very specific situations.

Proteins provided by rumen microorganisms provide an adequate supply to the tissues in most instances for animals producing at moderate levels. However, for young, rapidly growing animals and for high-producing dairy cows, this does not appear to be the case. That is, these two classes of ruminants probably require a greater supply of some amino acids than are supplied by the microorganisms. This has been shown by infusing either intact proteins or different amino acids into the abomasum or duodenum. In such conditions it can be shown that wool growth, milk production, or growth of young animals can be increased under experimental conditions.

For animals that have a higher than normal need for some of the limiting amino acids, the optimal situation for highly efficient N utilization would be to supply adequate N for the rumen microbes (primarily ammonia for rumen bacteria) and to have any excess dietary protein digested and absorbed in the gut. If this is to be accomplished, some of the protein in the feed must escape degradation by the microorganisms but be digested in the gut. Proteins that are digestible yet not degraded in the rumen are sometimes referred to as bypass proteins (rumen inert). This term should, more logically, be applied only to proteins that go through the esophageal groove (primarily milk) and do, in truth, bypass the rumen. Proteins escape degradation in the rumen if (1) they pass out too rapidly or (2) rumen microorganisms cannot metabolize them. More recently the term *escape proteins* has been applied to those proteins that are not degraded in the rumen.

Protein reaching the lower GI tract is a combination of microbial protein synthesized in the rumen and that from the feed that escapes degradation. The latter is highly affected by a number of factors, which include solubility of the various N fractions in the feed, the particle size of ingested material (small particles pass out of the rumen at a faster rate, thus allowing less digestion in the rumen), the speed of digestion in the rumen (affected by particle size and resistance to digestion), and the level of feeding (a high level of feeding pushes feed through the system at a faster rate, thus reducing potential digestion).

There is considerable interest by nutritionists, feeders, and milling companies in applying information on protein utilization to improve protein efficiency by ruminant animals and to allow more efficient use of added NPN. However, it must be realized that there is a great variation in solubility and utilization of different N fractions within and among various feedstuffs. With forages and silages, about $31 \pm 15\%$ of the N will be in NPN form, which is metabolized very rapidly by rumen microorganisms. About $32 \pm 16\%$ is insoluble leaf protein, which is metabolized at a slower rate. Insoluble protein from other plant parts accounts for $12 \pm 5\%$ that is slowly metabolized, and about $18 \pm 4\%$ is unavailable to the microorganisms, a high proportion being bound to lignin-fiber complexes (57). In some animal products, proteins such as keratins (high in hair, feathers, and connective tissues) are almost totally indigestible unless they are cooked enough with moist heat to partially hydrolyze these proteins and break the sulfur–sulfur bonding.

Although a variety of methods has been tried to measure solubility (extraction with hot or cold water, rumen fluid, buffers, alkaline or acid solutions; pepsin digestibility; suspension of the product in nylon or Dacron bags in the rumen of rumen-fistulated animals), from the preceding information it is obvious that one of the first criteria is that NPN compounds should be quantitated separately from proteins. This is so because most ruminant nutritionists believe that essentially all the soluble NPN is metabolized in the rumen. Note in Table 8-12 that many common feedstuffs have large proportions of the total N present as NPN. Most feedstuffs contain rapidly degradable protein, more slowly degradable protein, unavailable protein, and other N-containing compounds that are not protein. Furthermore, evidence shows that some proteins insoluble in buffers may be degraded rapidly and that some soluble proteins show a slow rate of degradation. Also, protein solubility is quite different in different types of solutions. Consequently, methods of defining solubility and relating it to rumen utilization need further improvement.

The solubility of proteins in forages is generally greatest in lush, young plants and decreases with age. Drying, particularly with heat, generally results in a reduction in solubility of cytoplasmic proteins, which is associated with denaturation of the proteins. Solubility may be reduced intentionally by heating feed ingredients, resulting in coagulation or denaturation. This is routinely done to soybean to increase its bypass potential. Heat treatment appears to reduce rumen degradation partly by blocking reactive sites for microbial enzymes and partly by reducing solubility. If excessive heat is applied to the point of causing browning (Maillard reaction), solubility is reduced by the formation of complexes between amino groups and carbohydrates. Such products may be totally indigestible, although not necessarily insoluble. Such products can be detected by increasing amounts of N in the acid detergent fiber fraction. Solubility may also be reduced intentionally by treating feedstuffs with formic acid, formaldehyde, tannins, acids, bases, or ethanol. If the amount added is not overdone, these chemicals inhibit rumen degradation, yet allow an adequate level of digestion in the small intestine. Solubility, as determined using buffer solutions, of N in a wide range of feedstuffs is shown in Table 8-11. Note that it ranges from 3% in beet pulp to a high of 63% in a corn solubles–germ meal–bran mix. If urea were included, it would be 100%. Note also that the amount of N bound to ADF tends to increase as the percentage solubility decreases. Either value may be quite variable depending on previous treatment of the feedstuff, especially when it has been subjected to heat. Another example of solubility of feed proteins when done with a bicarbonate–phosphate buffer is shown in Table 8-12. In this case, soluble material was classed as soluble N ranged from a low of 4.1% for brewers dried grains to a high of 53.1% for oats. However, except for oats and peanut meal, most of the soluble material was classed as soluble NPN. Except for oats and dried corn silage, the majority of the N was insoluble in this buffer system.

A major reason for discussing this topic is that there is much interest in learning under what conditions soluble N sources such as urea can best be utilized and when proteins undegradable in the rumen should be fed. There is a reasonable amount of literature on the topic, but it does not present a clear picture at this time because results with diets of differing N solubility have been contradictory. In some cases, animals fed diets with low N solubility have given greater yields of milk or improved gain or improved N retention, but in others there has been no response. One example with growing calves is shown in Table 8-13. In this instance it was observed that inclusion of blood meal or meat meal (both rather insoluble) resulted in somewhat improved feed efficiency and more gain per unit of N than did a urea or soybean meal–urea based ration. Other examples are given in Chs. 13, 14, and 15.

Information on proteins that escape rumen degradation but that are digested in the small intestine is even more scarce. French work, based on a number of studies with cattle and sheep, suggests that about 65% of insoluble N escapes rumen degradation. Their calculated values for digestibility of undegraded protein that was digestible in the gut ranged from 8% to 46% of dietary N with a mean of 28% for 31 feedstuffs, most of which were forages (61).

SUMMARY

Concentrated sources of protein are necessary in modern-day agriculture to supplement other ration ingredients for high-producing animals. A wide variety of sources is available from plants, milling by-products, the brewing and distilling industries, animal and marine sources, and various miscellaneous sources. For ruminants, a number of nonprotein N sources are available that can be utilized with good efficiency in some situations. Protein sources vary greatly in content of total N, essential amino acids, digestibility, and solubility. Many plant sources

TABLE 8-11

Soluble proteins and acid detergent fiber-bound protein in various feedstuffs

Feedstuff	Typical Crude Protein Content (%)	% of Crude Protein	
		Soluble in Buffer Solution	Bound in ADF
High-solubility feeds			
Corn solubles + germ meal and bran	29	63	3
Corn gluten feed	22	55	3
Rye middlings	18	48	2
Wheat middlings	18	37	2
Wheat flour	15	40	0.2
Intermediate-solubility feeds			
Oats	13	31	5
Corn gluten feed with corn germ	24	32	3
Cotton waste product	22	24	2
Hominy	11	24	3
Soybean meal	52	24	2
Soybean mill feed	15	22	14–20
Distillers dried grains with solvents	29	19	15
Low-solubility feeds			
Cottonseed meal	44	12	3
Corn	10	15	5
Brewers dried grains	29	6	13
Corn gluten meal	66	4	11
Beet pulp	8	3	11
Distillers dried grains	27	6	19
Forages and silages			
Alfalfa hay	15–25	30	10
Alfalfa, dehydrated	17–25	25–30	10–30
Alfalfa silage[a]	17–25	30–60	15–40
Corn silage[a]	9	30–40	10–30

Source: Sniffen et al. (58).

[a]Higher-moisture silages tend to have greater soluble NPN, while low DM silages may have greater heat damage.

TABLE 8-12

Nitrogen fractions in various feedstuffs as determined by solubility in bicarbonate–phosphate buffer

Feedstuffs	Total Crude Protein (% of DM)	Crude Protein Fractions (% of Total N)			
		Insoluble	Soluble	Soluble True Protein	Soluble NPN
Corn grain	9.9	88.9	11.1	3.4	7.7
Brewers dried grains	27.9	95.9	4.1	1.2	2.9
Corn gluten feed	24.0	60.8	39.2	0.2	39.0
Beet pulp	10.2	73.5	26.5	0.7	25.8
Corn gluten meal	69.0	95.8	4.2	0.5	3.8
Oat grain	14.0	46.9	53.1	43.3	9.8
Distillers grains with solvents	25.1	88.8	11.3	1.1	10.2
Soybean meal, solvent extracted	54.8	79.7	20.4	9.1	11.3
Rapeseed meal, solvent extracted	42.3	67.6	32.4	9.4	23.0
Peanut meal, solvent extracted	47.5	67.1	32.9	24.3	8.6
Dried corn silage	7.9	47.8	52.2	4.0	48.2
Guinea grass hay, mature	7.1	61.2	38.8	3.3	35.5
Timothy hay, mature	8.1	74.2	25.8	0.8	25.1
Tall fescue hay, mature	11.3	74.1	25.9	0.5	25.4
Rice straw	3.9	59.2	40.8	6.6	34.2
Corn stover	4.4	57.1	42.9	3.0	39.9

Source: Krishnamoorthy et al. (59).

TABLE 8-13

Effect of feeding growing calves diets with different nitrogen solubility

| Item | N Source[a] | | | |
	Urea	SBM–Urea	BM–Urea	MM–Urea
Daily gain, kg	0.72	0.81	0.91	0.85
Daily feed, kg	6.20	6.32	6.47	6.06
Feed-to-gain ratio	8.36	7.82	7.12	7.10
Gain-to-protein ratio		0.64	1.37	1.38

Source: Stock et al. (60).

[a]SBM = soybean meal, BM = blood meal, MM = meat meal.

contain antinutritional factors, most of which can be inactivated with heat or by use of various solutions or solvents. Ruminant animals are, generally, more flexible in that they can utilize moderate to large amounts of most protein sources. Use is more restricted with swine and poultry because of fiber or antinutritional factors and because essential amino acids are more critical. Liquid supplements have been devel-oped to the point that they can be used in many situations, but recent evidence indicates that they can be improved when fed with low-quality roughage if some native protein is added, rather than depending on NPN sources. Solubility and rumen degradability of proteins appear to be important factors in efficient N utilization by ruminants, but further research is required for utilization in most practical situations.

REFERENCES

1. USDA. 1996. *Feed situation and outlook report.* Econ. Res. Serv. FDS-304 (August). Washington, DC: USDA.

2. Wahlstrom, R. C., and G. W. Libal. 1977. *J. Animal Sci.* 44:778.

3. La Rue, D. C., et al. 1987. *J. Animal Sci.* 64:1051.

4. Slagle, S. P., and D. R. Zimmerman. 1979. *J. Animal Sci.* 49:1252.

5. Myer, R. O., C. N. Coon, and J. A. Froseth. 1982. *Poultry Sci.* 61:2117.

6. Salmon, R. E. 1982. *Poultry Sci.* 61:2126.

7. Huston, J. E., and M. Shelton. 1971. *J. Animal Sci.* 32:334.

8. Richardson, C. R., et al. 1981. *J. Animal Sci.* 53:557.

9. Doyle, J. J., and C. V. Hulet. 1978. *Proc. West. Sec. Amer. Soc. Animal Sci.* 32:96.

10. Loerch, S. C., and L. L. Berger. 1981 *J. Animal Sci.* 53:1198.

11. Shqueir, A. A., R. L. Kellems, and D. C. Church. 1981. *Proc. West Sec. Amer. Soc. Animal Sci.* 32:96.

12. Firkins, J. L., L. L. Berger, and G. C. Fahey, Jr. 1985. *J. Animal Sci.* 60:847.

13. Church, D. C., D. A. Daugherty, and W. H. Kennick. 1982. *J. Animal Sci.* 54:337.

14. Wray, M. I., et al. 1979. *J. Animal Sci.* 48:748.

15. Perry, T. W., et al. 1979. *J. Animal Sci.* 48:758.

16. Schingoethe, D. J., et al. 1977. *J. Dairy Sci.* 60:591.

17. Anderson, M. J., et al. 1979. *J. Dairy Sci.* 62:1098.

18. Papas, A., J. R. Ingalls, and L. D. Campbell. 1979. © *J. Nutr.* 109:1129, American Institute of Nutrition.

19. DePeters, E. U., and D. L. Bath. 1986. *J. Dairy Sci.* 69:148.

20. Polan, C. E., et al. 1985. *J. Dairy Sci.* 68:2016.

21. Annextad, R. J., et al. 1987. *J. Dairy Sci.* 70:814.

22. Schingoethe, D. J., et al. 1988. *J. Dairy Sci.* 71:173.

23. Price, S. G., L. D. Satter, and N. A. Jorgensen. 1988. *J. Dairy Sci.* 71:727.

24. Hadsell, D. L., and J. L. Sommerfeldt. 1988. *J. Dairy Sci.* 71:762.

25. Lindsey, T. O., G. E. Hawkins, and L. D. Guthrie. 1980. *J. Dairy Sci.* 63:562.

26. Hawkins, G. E., et al. 1985. *J. Dairy Sci.* 68:2608.

27. Bell, J. M. 1984. *J. Animal Sci.* 58:996.

28. Laarveld, B., R. P. Brockman, and D. A. Christensen. 1981. *Can. J. Animal Sci.* 62:131.

29. Cordes, C. S., et al. 1988. *J. Animal Sci.* 66:522.

30. Fleck, A. T., et al. 1988. *J. Animal Sci.* 66:750.

31. Anon. Undated. *Distillers feeds research.* Cincinnati, OH: Distillers Feed Research Council.

32. Johnson, C. O. L. E., and J. T. Huber. 1987. *J. Dairy Sci.* 70:1417.

33. Boehme, W. R. 1975. Personal communication. Des Plaines, IL: Fats and Proteins Research Foundation, Inc.

34. Parsons, M. J., P. K. Ku, and E. R. Miller. 1985. *J. Animal Sci.* 60:1447.

35. Church, D. C., and W. G. Pond. 1988. *Basic animal nutrition and feeding.* 3d ed. New York: Wiley; Scott, M. L., M. C. Nesheim, and R. J. Young. 1982. *Nutrition of the chicken.* 3d ed. Ithaca, NY: M. L. Scott & Assoc.

36. Raa, J., and A. Gildberg. 1982. *CRC Crit. Rev., Food Sci. Nutr.* 16:383.

37. Chirase, N. K., M. Kolopita, and J. R. Males. 1985. *J. Animal Sci.* 61:661.

38. Ortega, E., and D. C. Church. 1979. *Proc. West. Sec. Amer. Soc. Animal Sci.* 30:302.

39. Zinn, R. A. 1988. *J. Animal Sci.* 66:151.

40. Phelps, A. 1982. *Feedstuffs* 54(2):17.

41. Hinks, C. E. 1978. *J. Sci. Fd. Agric.* 29:99.

42. Church, D. C., J. C. Steinberg, and B. N. L. Khaw. 1982. *Feedstuffs* 54(3):30.

43. Kellems, R. O., M. S. Aseltine, and D. C. Church. 1981. *J. Animal Sci.* 53:1601.

44. Oltjen, R. R., D. A. Dinius, and H. K. Goering. 1977. *J. Animal Sci.* 45:1442.

45. Rush, I. G., R. R. Johnson, and R. Totusek. 1976. *J. Animal Sci.* 42:1297.

46. Peyton, S. C., and H. R. Conrad. 1979. *J. Dairy Sci.* 61:1742.

47. Schultz, T. A., E. Schultz, and C. F. Chicco. 1972. *J. Animal Sci.* 35:865.

48. Jimenez, A. A. 1986. *Feedstuffs.* 58(8):12.

49. Church, D. C., and A. Santos. 1981. *J. Animal Sci.* 53:1609.

50. Rush, I. G., and R. Totusek. 1976. *J. Animal Sci.* 42:497.

51. Kellems, R. O., E. Ortega-Rivas, and D. C. Church 1981. Unpublished data. Corvallis, OR: Dept. of Animal Sci., Oregon State Univ.

52. Kellems, R. O. 1982. Unpublished data. Corvallis, OR: Oregon State Univ. Dept. of Animal Sci.

53. Wagner, J. J., K. A. Lusby, and G. W. Horn. 1983. *J. Animal Sci.* 57:542.

54. Wahlberg, M. L., and E. H. Cash. 1979. *J. Animal Sci.* 49:1431.

55. Velloso, L., et al. 1971. *J. Animal Sci.* 32:764.

56. Ortega-Rivas, E., and D. C. Church. 1981. Unpublished data. Corvallis, OR: Oregon State Univ. Dept. of Animal Sci.

57. Van Soest, P. J. 1982. *Nutritional ecology of the ruminant.* Ithaca, NY: Cornell Univ. Press.

58. Sniffen, C. J., et al. 1980. *Feedstuffs* 52(20):25.

59. Krishnamoorthy, U., et al. 1982. *J. Dairy Sci.* 65:217.

60. Stock, R., et al. 1981. *J. Animal Sci.* 53:1109.

61. Verite, R., M. Journet, and R. Jarrige. 1979. *Livestock Prod. Sci.* 6:349.

62. Rodriguez, J. P., and H. S. Bayley. 1987. *Can. J. Animal Sci.* 67:803.

9

MINERAL AND VITAMIN SUPPLEMENTS

D. C. Church and Richard O. Kellems

MINERAL SUPPLEMENTS

Minerals are the inorganic components and make up only a relatively small amount of the total diet of animals. Nevertheless, they are vital to the animal, and in most situations some diet supplementation is required to balance the various minerals and satisfy the requirements of high-producing animals. All the required mineral elements are needed in an animal's diet (or water supply), but needed supplementary minerals vary according to the animal species, age, type, and level of production and diet, which, in turn, is affected by the mineral content of water, soils, and crops in the area where grown and of the fertilization practices used.

Generally, the minerals are classified as either macrominerals or microminerals. Macrominerals are found present in concentrations greater than 100 ppm in an animal's body, while microminerals are found in concentrations less than 100 ppm. Macrominerals of concern include common salt (NaCl), Ca (calcium), P (phosphorus), Mg (magnesium), and, sometimes K (potassium), and S (sulfur). With regard to the microminerals (trace elements), Cu (copper), Fe (iron), I (iodine), Mn (manganese), Se (selenium), and Zn (zinc) are often deficient for domestic animals, and Co (cobalt) may be defi-

cient for ruminant species. With rare exceptions, the other required mineral elements (see Ch. 3) are not normally a problem.

Adequate mineral concentrations and balance between minerals are also important items. Thus excessive amounts of one mineral may interfere with utilization of one or more other elements. For example, excessive Ca may cause problems with P, Mg, and Zn, and excessive aluminum may interfere with P. Other similar interactions are known, and there are, without doubt, interactions among other mineral elements that have not been identified. Even if no interaction occurs, minerals may be outright toxic, so maximal levels are of some concern in many situations (1). This is one reason why state and federal regulations often specify maximum concentrations in labeling laws for formula feeds (Ch. 5).

Mineral supplements must be provided by sources that are available to the animal. The amount that may be dissolved and absorbed may vary with different compounds. For example, ferric oxide (Fe_2O_3) is almost insoluble and, therefore, of very little use to the animal. An appreciable amount of information is available on this topic (2, 3).

Whenever possible or practicable, the mineral requirements should be met by selection or combination of available feedstuffs that supply energy or protein. Most energy and N sources, fat and urea being two marked exceptions, provide minerals in addition to the basic organic nutrients. Although considerable opportunity exists to provide most of the mineral needs from basal feedstuffs, the flexibility needed in formulating rations often requires concentrated sources of one or more mineral elements, particularly for high-producing animals. Some of these sources are discussed in subsequent sections. Mineral sources used commercially are listed in Table 9-1, with those used most commonly identified by an asterisk.

The choice of a mineral supplement is governed by the cost per unit of the element or elements required; the chemical form in which the mineral is combined, which affects its biological availability; its physical form, especially its fineness of division; and its freedom from harmful impurities. These factors are generally of more importance for minerals required in large amounts and usually of less consequence for the trace minerals. Note also that the mineral concentration is not always constant in any given source unless the source is highly purified. Figure 9-1 is included to provide some idea of the appearance of some of the mineral supplements.

SALT (NaCl)

Common salt, which is practically pure sodium chloride, is the most common mineral supplement added to any livestock ration. It is unique in that it is quite palatable and attractive to animals, and they can be relied on to consume enough and usually more than enough to meet their needs. In addition, salt is often used as a carrier for other required elements or other materials, such as pesticides, medicines, or antibloating drugs. Mixtures that contain about 40% salt are usually consumed at adequate levels to achieve satisfactory consumption of other elements.

Salt is usually required for good animal production, particularly for lactating animals and species that sweat profusely. Requirements increase as fluid losses increase, for example, when sweating or milk production increases. It is a common practice to add 0.5% to 1% salt to most commercial formulas for ruminants and horses, and some supplemental concentrates may contain 1% to 3% salt. Poultry and pig feeds usually contain from 0.25% to 0.5% salt. In most cases this amount of salt is more than required but less than will be consumed free choice.

Salt is often fed ad libitum, particularly to ruminants and horses, because their requirements are probably higher than that of swine or poultry, and different feeding methods lend themselves to this practice. Salt may be fed in loose form or as compressed blocks (Fig. 9-2). The blocks tend to restrict consumption if access is limited, but they are often convenient to use and are more resistant to losses from rain and moisture condensation. Where rainfall is high, loose salt needs protection (Fig. 9-3).

Excess salt may be a problem for all species, but it is particularly so for swine and poultry, because they are much more susceptible to toxicity than are other domestic species, especially if water consumption is restricted in some manner. In areas where the soil and/or water are quite saline, it may not be necessary to feed salt, nor will the consumption be very high if it is available.

When salt is mixed in feed, it should be fairly fine in texture, noncaking, and free flowing. Iodized salt is regular feed-grade salt to which iodine has been added to supply a minimum of 0.007% iodine. If KI is used, a stabilizer is usually added to preserve the iodine content. If other compounds such as potassium iodate (KIO_3) or diiodosalicylic acid are used, then a stabilizer is unnecessary.

Trace-mineralized salt is available in most areas and is usually compounded so that sources of Co, Cu, Fe, I, Mn, and Zn are added. Sulfur and Mg may sometimes be added, as may be Se, but the latter mineral is closely restricted by the Food and Drug Administration (FDA) (discussed later). In most instances there are sound reasons for feeding trace-mineralized salts. In rare instances it might cause a problem if local soil and forage are particularly high in some element such as Cu.

In addition to the trace minerals, various medicines or drugs are frequently added to salt. Ethylenediamine dihydroiodide (EDDI) is often added to mineralized salt. It is a therapeutic agent for prevention and treatment of footrot and lumpy jaw in cattle, as well as a source (poor) of iodine. Antibloating compounds may be combined with salt or other mixtures. Phenothiazine or other drugs used to control stomach worms are often added to salt fed to sheep.

CALCIUM AND PHOSPHORUS

Supplementary Ca and P are generally needed in many animal diets because the demands for them are higher than demands for most minerals for skeletal growth, lactation, or egg

TABLE 9-1

Sources of mineral supplements used in feed supplements or complete feeds or in mineralized salt

Name, Principal Element	Name, Principal Element
Macrominerals	Microminerals[a]
Calcium	Cobalt
Calcium carbonate*	Cobalt acetate
Calcium chloride	Cobalt carbonate*
Calcium gluconate	Cobalt chloride
Calcium hydroxide	Cobalt oxide
Calcium oxide	Cobalt sulfate*
Chalk, precipitated	Cobalt–choline–citrate complex
Dolomitic limestone*	
Limestone*	Copper
Oyster or clam shells*	Copper carbonate
	Copper chloride
Calcium–Phosphorus	Copper gluconate
Bone charcoal	Copper hydroxide
Bone charcoal, spent	Copper orthophosphate
Bone meal, steamed*	Copper oxide*
Dicalcium phosphate*	Copper pyrophosphate
Monocalcium phosphate	Copper sulfate*
Rock phosphate	
Rock phosphate, defluorinated*	Iodine
Rock phosphate, soft	Calcium iodobenhenate
Tricalcium phosphate	Calcium periodate*
	Diiodosalicylic acid
Magnesium	Ethylenediamine dihydroiodide*
Dolomitic limestone*	Iodized salt*
Magnesium carbonate	Potassium iodate*
Magnesium hydroxide	Potassium iodide*
Magnesium oxide*	Sodium iodate
Magnesium sulfate*	Sodium iodide
	Thymol iodide
Phosphorus	
Diammonium phosphate*	Iron
Monoammonium phosphate	Ferric ammonium citrate
Phosphoric acid, feed grade*	Ferric chloride
Sodium phosphate, dibasic	Ferric phosphate
Sodium phosphate, monobasic*	Ferric pyrophosphate
Sodium phosphate, tribasic	Ferrous chloride
Sodium tripolyphosphate*	Ferrous fumarate
Triammonium phosphate	Ferrous gluconate
	Ferrous sulfate*
Potassium	Iron carbonate*
Potassium bicarbonate	Iron oxide (coloring agent)*
Potassium carbonate	Reduced Iron
Potassium chloride*	
Potassium sulfate	Manganese
	Manganese acetate
Sodium	Manganese carbonate
Sodium bicarbonate*	Manganese chloride
Sodium chloride*	Manganese citrate
Sodium phosphates	Manganese gluconate
Sodium sulfate	Manganese orthophosphate
	Manganese oxide*
Chlorine	Manganese phosphate (dibasic)
Potassium chloride	Manganese sulfate*
Sodium chloride*	

TABLE 9-1 *(cont.)*

Name, Principal Element	*Name, Principal Element*
Sulfur	Selenium
Ammonium sulfate	Sodium selenate*
Elemental sulfur (ruminants)*	Sodium selenite
Macromineral sulfates*	
Organic sulfur in proteins*	Zinc
	Zinc acetate
	Zinc carbonate
	Zinc chloride
	Zinc oxide*
	Zinc sulfate*

*Sources used most commonly in commercial feeds. All these materials are generally recognized as safe except for the two selenium salts.

[a]Trace minerals such as fluorine, molybdenum, nickel, or others are not commonly needed in feed supplements.

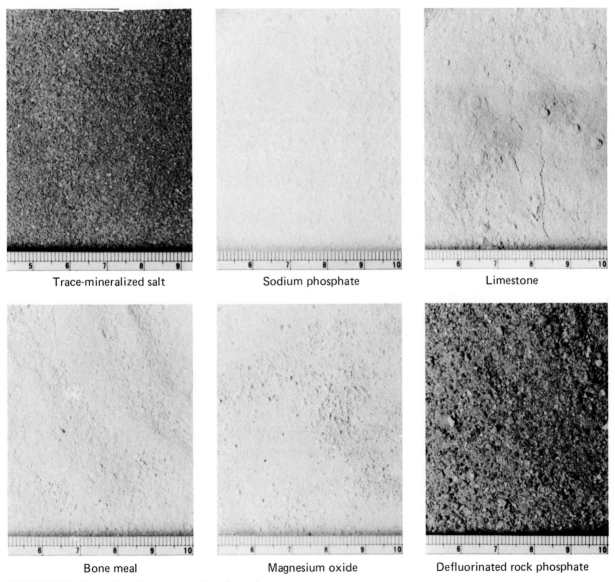

Trace-mineralized salt   Sodium phosphate   Limestone

Bone meal   Magnesium oxide   Defluorinated rock phosphate

FIGURE 9-1 Examples of common mineral supplements.

FIGURE 9-2 Cattle consuming salt from a compressed block.

production and because many feedstuffs are borderline to deficient in these minerals. Phosphorus is usually more apt to be deficient in the diets of herbivorous animals, since more forage is relatively lower in P than in Ca. Animals on high-cereal diets are more likely to be deficient in Ca because grains and similar feedstuffs are usually a relatively good source of P but quite low in Ca.

Most nonplant sources of Ca are well utilized by different animal species. Although net

FIGURE 9-3 One sample of a salt-mineral feeder that turns with the wind and thus offers some protection from the weather.

digestibility may be low, particularly in older animals, there is little difference between Ca sources provided particle size and other physical factors are similar. However, this situation does not apply to P because sources may differ greatly in availability. In plants, about half of the P is bound to phytic acid and phytin P is utilized poorly by nonruminant species. The usual recommendation is to consider only half of plant P available for nonruminant species, although ruminants utilize phytin P very well because of metabolism of phytin by rumen microorganisms.

Marked differences also exist in the biological availability of some inorganic P sources. Phosphoric acid and the mono-, di-, and tricalcium phosphates are well utilized, but sources such as Curacao Island and colloidal (soft) phosphates are utilized less well by most animals. Some sources derived from rock phosphates must be defluorinated; otherwise, long-term consumption may produce chronic fluorine toxicity. Calcium and P sources are listed in Table 9-1, and some descriptions of a few of the common sources are given in following paragraphs. Analytical data of some common mineral supplements are given in Appendix Table 5.

Dicalcium phosphate is one of the more common supplements for both Ca and P. The commercial sources are generally prepared from bones that have been treated first in a caustic solution, then in a HCl solution, and then precipitated with lime and dried. The other form is manufactured by adding Ca compounds to phosphoric acid and precipitating the dicalcium phosphate. These forms of dicalcium phosphate are excellent sources of both Ca and P, as they are readily soluble in the digestive tract and are low in fluorine. They contain 18% to 21% P and 25% to 28% Ca. Although a considerable amount of work has been done to determine the availability of P from some of these sources, the information is still inadequate (4).

Other products prepared from bone include steamed bone meal, bone charcoal, spent bone charcoal, and bone ash. The steamed bone meal is prepared from bones by cooking under steam, grinding, and drying. Bone charcoal (bone black) is obtained by charring bones in closed retorts. Spent bone charcoal is the product resulting from repeated charring of bone charcoal after use in clarifying sugar solutions in the manufacturing of table sugar, and bone ash is the ash obtained by burning with free access to air. The Ca and P content vary somewhat in these products, but they are all good sources of these elements. Much smaller amounts of other required elements are also present in most bone products.

Other Ca and P sources include monocalcium and tricalcium phosphates, salts that are prepared from mixtures of Ca oxide and phosphoric acid. The mono salt contains about 20% Ca and 21% P and tricalcium salt about 38% Ca and 18% P. Defluorinated phosphate is also a low-fluorine product prepared from Ca phosphates and must have less than 1 part of F to 100 parts of P.

Feed-grade phosphoric acid is a solution of phosphoric acid and water and is used extensively in liquid supplements for ruminants. In addition to being a good source of P, it has the property of reducing the viscosity of molasses. It must not contain more than 100 ppm of F, 3.2 ppm of arsenic, and 1.3 ppm of heavy metals.

Ammonium phosphate is prepared by treating phosphoric acid with ammonia. The monoammonium form must contain at least 9% N and 25% P with limits on F, arsenic, and heavy metals such as lead. It is more palatable than the diammonium form, which must have at least 17% N and 20% P. These products are restricted to ruminant feeds (when used as a source of N) in an amount that supplies not more than 2% of crude protein equivalent in the total daily ration. Rumen microorganisms can utilize the N as well as the P. In some situations it may be desirable to add P without Ca or ammonia. When it is not feasible to use phosphoric acid, mono- or disodium phosphate may be useful sources.

Rock phosphates of a variety of different types are also available. Those of domestic origin are usually relatively high in F, so their use must be restricted to prevent F toxicity. Soft rock phosphate, also known as colloidal phosphate, is not utilized well by some species.

Other Ca sources used extensively are usually some form of Ca carbonate, which is the principal Ca salt in limestone, oyster shells, calcite, chalk, or dolomitic limestone. Calcium sulfate (gypsum) is sometimes added to ruminant rations containing nonprotein N to provide needed S. These different Ca sources are well utilized, and the Ca content of the carbonates runs from about 33% to 38% except for dolomitic limestone, which must contain at least 10% Mg.

OTHER MAJOR MINERALS

A supplementary source of Mg is frequently required in animal rations. Dolomitic limestone must have 10% or more Mg; however, it is not highly available to some species and Mg oxide is usually the preferred source. Magnesium carbonate or sulfate are also used. Potassium is required for ruminants on high-concentrate rations as well as in some other situations. The chloride, bicarbonate, or carbonate are the usual sources used. Sulfur may be needed in some situations for ruminants. It can be supplied as elemental S or as any of a variety of sulfate salts.

MICROMINERALS (TRACE MINERALS)

The microminerals most commonly added to animal diets include cobalt, copper, iodine, iron, manganese, selenium, and zinc; various sources are listed in Table 9-1. With regard to Fe, a considerable amount of iron oxide is used primarily as a coloring agent in things such as trace-mineralized salt, but the Fe is only slightly soluble and, as a result, it is a poor nutritional source of Fe. Other compounds commonly added include iron carbonate and ferrous sulfate. Copper is probably most commonly added as the sulfate, but the oxide, carbonate, and hydroxide are used sometimes. Experimental evidence indicates that small Cu "needles" can be given to cattle in a capsule and that they will lodge in the abomasum and provide a relatively long term supply of Cu. Manganese is usually required in poultry diets. The oxide and the sulfate are most often used in feeds. Cobalt is required only by rumen microorganisms (to synthesize vitamin B_{12}). When added to feed, it is most often added as the carbonate or sulfate. Cobalt "bullets" (small heat-treated cylinders of various metallic compounds) are utilized extensively in Co-deficient areas in Australia and New Zealand. The bullet is administered with a balling gun and, if successfully administered, gets positioned in the ventral sac of the rumen, where it will last for a year or more and provide a steady supply of the metal. Iodine is most commonly fed in the form of iodized salt, but it is also fed in various forms as indicated previously. Zinc is needed frequently in swine, poultry, and ruminant diets. The oxide is the most common compound used.

If any of the trace minerals are added to premixes, the minimum amount must be specified according to most regulations. Selenium (Se) is the only trace mineral closely regulated in the United States by the FDA. This is so because there is evidence that high levels can be carcinogenic; there is also evidence that low levels may be anticarcinogenic. At any rate, although evidence has been available since the late 1950s that Se is a required nutrient, the FDA did not approve it in some animal feeds until 1974 and for others it was much later. The current (early 1990) regulations are as given. Selenium may be added at a level of 0.1 ppm in complete diets for broilers, for chicken layers and breeders, and for ducks. A level of 0.2 ppm is approved for turkeys. For swine, a level of 0.3 ppm is permitted for pigs up to 50 lb in weight and 0.1 ppm for swine over this weight. Selenium can be added to sheep, beef, and dairy complete feeds at a level of 0.3 ppm. For horses, 1 to 2 mg/head (hd/d) is approved with a maximum of 3 mg/hd/d for up to 3 months and then a level of 1 mg/hd/d, with the specification that meat from such horses is not to be used for human food. In all cases, Se must be provided in a premix that is produced by a registered manufacturer, and all premixes must carry this statement: "Caution: Follow label directions. The addition to feed of higher levels of this premix containing selenium is not permitted." If Se is added to a salt–mineral mix, the FDA limits are 90 and 120 ppm, respectively, for sheep and cattle.

Some interest remains in utilization of chelated trace elements. Chelates are compounds in which the mineral atom is bound to an organic complex (hemoglobin for example). Feeding of chelated minerals has been promoted based on the contention that chelates will prevent the formation of insoluble complexes in the GI tract and reduce the amount of the particular mineral required in the diet. Limited evidence indicates that this works sometimes (with Zn in poultry diets), but not in all cases. Usually, the cost per unit of mineral is appreciably higher than that of using more common forms. Chelates are also promoted for use in premixes containing vitamins or other compounds that may oxidize readily. Chelation reduces this type of reaction. One method of chelation is with soluble proteins. **Metal proteinate** is the product resulting from the chelation of a soluble salt with amino acids and/or partially hydrolyzed protein. It must be declared as an ingredient as the specific metal proteinate, for example, copper proteinate. **Metal polysaccharide complex** is the product resulting from complexing of a soluble salt with a polysaccharide solution declared as an ingredient as the specific metal complex. Several different products are on the market, but the author is not familiar with the amount used in the feed trade, and little research evidence has been forthcoming.

VARIATION IN CONTENT AND UTILIZATION OF MINERALS IN NATURAL FEEDSTUFFS

Data were given earlier on the variation in chemical composition of forage samples coming into an analytical laboratory (Table 4-3). Another example is presented on minerals in corn and soybean meal in Table 9-2. In this example, values quoted in the NRC on beef (6) are listed and the analytical values and range in values are shown. With regard to corn, the NRC and the mean analytical values for the macrominerals were reasonably close, although P was lower than might be expected. With the trace minerals, Cu analytical values were lower than indicated, and Zn values were quite a bit higher than suggested by NRC. With soybean meal, analytical means for Ca, Mg, Fe, and Zn were appreciably higher than NRC, and K and Cu were appreciably lower. These data show clearly that trying to get by on minimal levels for several months might cause some problems to livestock feeders if they depend only on book values that may not pertain to their area or to the crops they feed.

In addition to the problem of using incorrect values for the mineral content of feedstuffs, there is a great scarcity of information on how much of the minerals in the feed are absorbed by the animal. Certainly, it will not be 100% from natural feedstuffs, although it might or might not approach that level for some of the manufactured mineral sources. A random check of the literature shows that Ca in legumes is generally less available than that from a good supplement, such as bone meal or dicalcium phosphate (7). Data of Ward et al. (8) indicated that 20% to 33% of Ca in alfalfa is in the form of oxalate and is apparently unavailable to ruminants. Other information shows that Cu is more available from silage than from hay and that feeding forages highly fertilized with N results in reduced utilization of Mg and increased use of Ca by sheep. Unfortunately, there is just not much information on this topic. Even in studies with mineral supplements, it is difficult to obtain information showing how much of a given source was absorbed unless the use of radioactive isotopes can be incorporated into the study. One reason for this is that there is a great deal of recycling of the mineral elements within the body, particularly of those excreted primarily in the

TABLE 9-2

Variation in mineral content of corn grain and soybean meal
(dry-matter basis) samples from Iowa

| Mineral | NRC[a] | Iowa Samples[b] | |
		Mean	Range
Corn grain			
Ca, %	0.02	<0.01	<0.01–0.18
K, %	0.37	0.34	0.27–0.42
Mg, %	0.13	0.17	0.14–0.21
Na, %	0.02	<0.01	<0.01–0.03
P, %	0.35	0.30	0.18–0.41
Cu, ppm	4	2.3	0.3–8.0
Fe, ppm	26	33.1	21.9–94.8
Mn, ppm	6	5.2	1.7–21.1
Se, ppm	—	0.21	0.04–0.66
Zn, ppm	16	58.7	16.9–147.7
Soybean meal, 44%			
Ca, %	0.33	0.55	0.33–0.82
K, %	2.14	1.92	1.21–2.37
Mg, %	0.30	0.48	0.45–0.53
Na, %	0.03	<0.01	<0.01–0.03
P, %	0.71	0.75	0.63–0.85
Cu, ppm	30	16.2	11.8–21.6
Fe, ppm	142	213.6	132.3–326.3
Mn, ppm	32	36.3	30.6–41.2
Se, ppm	0.14	0.71	0.28–0.93
Zn, ppm	61	91.5	59.8–168.3

Source: Ewan (5).

[a]From NRC beef (6).

[b]Values represent data on 49 corn and 16 soybean meal samples collected on farms in Iowa.

feces. Oral administration of isotopes shows that the majority of Cl, K, Na, S, and I are excreted via the urine. A majority of the (normal) dose of all other minerals is excreted via the feces. Depending on the mineral, excretion may occur in the rumen, abomasum, or glandular stomach, via bile or pancreatic juice, or through the intestinal wall (small and/or large). In any case, unless samples can be obtained at various sites in the GI tract, it is difficult to determine how much of an element is truly absorbed.

PROBLEMS WITH COMMERCIAL MINERAL SOURCES

There is often a problem that mineral mixes, particularly those intended to supply the trace minerals, may not be satisfactory supplements to the usual feedstuffs. There is also evidence that special premixes (prepared for specific feedlot use) may not always be as close to the target as they should be. Data illustrating these problems are presented in Table 9-3. These data were put together by Owens (9), who used a base diet with the major components of corn,

milo, or wheat, alfalfa hay, and corn silage, supplemented as needed by soybean meal and cottonseed meal (except for the wheat ration). Based on the NRC requirements for 600-lb steers (6) (these specifications are not necessarily exactly right for all animals) and on NRC mineral composition for these feed ingredients (6), Owens calculated what the maximum shortage of the various trace minerals might be (shown in the table) and then showed how specialized mixes prepared by nutritionists might meet these requirements or how commercial trace-mineralized salts fill the need when fed at levels recommended by the manufacturers. As is evident from the table, some of the mineral mixes were below and some were above the suggested target amounts. One of the mixes put out by nutritionists was off only very slightly on its I content and one had no Se, but the others were off on two or three minerals. With the commercial trace-mineralized salts, all five mixes were on or above the Co target; two out of five were adequate for Cu and I, one was low in Fe, all five were low in Mn and Se, and only one

TABLE 9-3

Mineral concentrations of typical feedlot rations, deficits for 600-pound steers, and amounts provided by some premixes and trace-mineralized salts

Item	Mineral (ppm)						
	Co	Cu	I	Fe	Mn	Se	Zn
Diet base[a]							
Corn	0.08	6.20	0.0	45	9	0.07	19
Milo	0.19	11.80	0.03	65	19	0.40	21
Wheat	0.15	5.70	0.0	43	31	0.39	39
Requirement							
NRC 1984 Beef	0.1	8.0	0.5	50	40	0.2	30
Maximum shortage	0.02	2.3	0.5	7	31	0.13	11
Mineral content of mixes prepared by nutritionists							
A	2.1	11.30	1.0	19	10*	0*	43
B	6.0	0.6	1.5	9	15*	0*	30
C	0.3	5.3	0.4*	0*	6*	0*	24
D	1.2	18.7	1.2	73	37	0*	88
E	0.6	7.0	0.8	30	8*	0*	52
F	0.1	6.0	0.4*	10	40	0.2	60
G	0.2	7.8	0.2*	25	25*	0*	67
Mineral content of commercial trace-mineralized salts[b]							
H	0.2	3.3	0.7	20	25*	0*	1*
I	0.5	2.5	0.5	13	20*	0*	25
J	0.5	1.0*	0.4*	33	3*	0*	1*
K	0.1	0.5*	0.1*	4*	5*	0*	6*
L	0.02	0.1*	0.3*	10	0*	0*	0*

Source: Owens (9).

[a]Diets calculated on the basis that alfalfa and corn silage made up 5% each of the dry matter, and cottonseed meal and soybean meal were added at a level of 3% of the dry matter to the corn and milo diets, but not to the wheat diet (which is already high enough in CP).

[b]The amount of minerals that should be present according to the manufacturer's recommendations to add levels of 0.5% to 1%.

*Potentially deficient ration.

of five was above the target for Zn. If this situation is representative of what exists around the country, it shows that livestock feeders must be more careful in the future to be reasonably certain that the mineral needs of their animals are taken care of. The cost is so small in comparison to the likely benefits that feeders should consider careful selection of mineral sources as potentially critical to success.

VITAMIN SOURCES

Almost all feedstuffs contain some of the various vitamins. However, vitamin concentration in plant or animal tissues varies tremendously. In plant tissues, vitamin concentration is affected by harvesting, processing, and storage conditions, as well as by plant species and plant part (seed, leaf, stalk). In animal tissues, the liver and kidney are generally good sources of most of the vitamins. Yeasts and other microorganisms are also excellent sources, particularly of the B vitamins.

As a rule, vitamins are easily destroyed by heat (especially in conjunction with exposure to air), sunlight, oxidizing conditions, or storage conditions that allow mold growth. Thus, if any question arises of adequacy of a diet, it is frequently better to err on the positive side than to have a diet that is deficient. It is now a common practice to provide premixes in diets of poultry and swine to be sure that deficiencies do not occur.

As a general rule, vitamins that are likely to be limiting in natural diets include vitamins A, D, E, riboflavin, pantothenic acid, niacin, choline, and cobalamin (B_{12}), depending on the species and class of animal with which we are concerned. Biotin may also be a problem with both swine and poultry under some conditions. Vitamin K would normally be synthesized in adequate amounts by microbes present in the GI tract, but it may also be needed, because synthesis may be inhibited by some feed additives.

With regard to ruminant species, normally, vitamin A is most likely to be deficient, particularly for animals that have not had access to green feed for some period of time. There is also a fair amount of evidence indicating that betacarotene may be an important factor affecting fertility in high-producing dairy cows, a situation that would not be remedied by normal vitamin A supplementation. It is also a fairly standard practice to supplement dairy cows and, sometimes, feedlot cattle with vitamins D and E. There is increasing evidence that niacin may be a factor in ketosis. In some situations an induced thiamin deficiency (polioencephalomalacia) develops, most often in feedlot cattle. It apparently is brought about by either excessive levels of thiaminase enzymes (which inactivate thiamin) or by thiamin analogs (which compete with thiamin). Thus it is a common practice to supplement ruminant diets with more vitamins than used in earlier years with less productive animals.

Sources used for supplementing rations with vitamins are listed in Table 9-4, and the vitamin content of a limited list of feedstuffs is shown in Appendix Table 4. Appendix Table 5A lists the various vitamins used commonly in feed formulation and the typical concentrations available from manufacturers and in premixes.

FAT-SOLUBLE VITAMINS

The best feed sources of carotene are green and yellow plants; generally, the more intense the green color, the higher the carotene content. Commercially, dehydrated alfalfa leaf and alfalfa meals or sun-cured alfalfa are normally used as sources of carotene (as well as other nutrients). More concentrated sources of carotene, such as carrot oil or alfalfa extracts, are also available, as are dry products in which carotene has been absorbed on a millfeed product for use in feeds. Carotene products in vegetable or animal oils are also available.

Vitamin A, itself, is not found in plants, but only in animal tissues. In animal products the liver, kidney, and liver oils from fish such as cod and shark are highly concentrated sources. Although fish liver oils are still used commercially, they have been replaced to a large extent with vitamin A that is produced synthetically or concentrated from fish liver oil by molecular distillation. For addition to feed, vitamin A is normally sold in a dry, gelatin-coated form to which antioxidants have been added. The chemical form normally sold is the ester (a combination of an alcohol and an acid), usually as vitamin A acetate, propionate, or palmitate.

Vitamin A preparations normally are quite stable and can be added to most feed mixes or liquid supplements without much loss of vitamin activity during normal storage periods. However, there is probably more loss during processing than normally realized. A common practice such as grinding a feed makes it susceptible to

loss of vitamin activity because of the heat involved and because ground feed is more exposed to oxygen. One example of the effect of pelleting is shown in Table 9-5. Note that pelleting reduced vitamin A content in this particular study by about 30% to 40%. Other factors that may reduce vitamin A content of stored feeds include time in storage, temperature, the presence of ultraviolet light, and trace mineral content of the diet. In addition, moisture or hygroscopic compounds such as choline chloride or urea or the initial concentration of the microingredients may reduce vitamin content. It is now a common practice to add antioxidants to premixes to avoid vitamin destruction. Modern methods of preparation of vitamin A (use of emulsifying agents, antioxidants, gelatin, and sugar in spray-dried, beaded, or prilled products) all help to reduce losses in storage (10). At any rate, the cost of adding vitamin A to rations is low enough that there is no economic reason not to add it when any likelihood of a need exists.

Important dietary sources of vitamin D are sun-cured forages, fish liver oils, and synthetic vitamin D produced by irradiating yeast, plant, or animal sterols with ultraviolet light. Although animals may obtain all the vitamin D they need as a result of ultraviolet light acting on sterols in the skin, most if not all commercial feeds have vitamin D added.

Four-footed animals are able to convert vitamin D_2 to D_3 (see Ch. 3), whereas poultry utilize D_2 very inefficiently. When intended for use with poultry, vitamin D is standardized in International Chick Units (based on preparations containing crystalline D_3). Use of D_3 (rather than D_2) in feeds for animals other than poultry is more efficient, resulting in lower dietary requirements. Vitamin D_3 is found naturally in animal products, but vitamin D produced from irradiation of plant or yeast products is in the D_2 form. D_3 not originating from fish products is usually derived from irradiated animal sterols.

In addition to the fish liver oils, which may be fortified or concentrated, D-activated animal sterol is available; it may be dissolved in oil or absorbed by flour or other fine powders suitable for feeding. Vitamin D_2 supplements may also be purchased, as may irradiated yeast products.

Vitamin D is relatively stable in mixed feeds. However, when mixed directly with such materials as limestone, oxidizing compounds, and some organic ingredients, it is subject to rapid losses.

Vitamin E (primarily α-tocopherol) is present in most common feedstuffs, but it is found in

TABLE 9-4

Recognized supplementary sources of vitamins for use in feedstuffs

Name and IFN Number	Source of	Status under Food Additives Amendment
Ascorbic acid, 7-00-433	Crystalline ascorbic acid, commercial feed grade	Reg. 582.5013
Betaine hydrochloride, 8-00-722	Crystalline chloride of betaine, commercial feed grade	Effective date extended
Biotin 7-00-723	Biotin, commercial feed grade	Reg. 582.5159
Calcium pantothenate, 7-01-079	Crystalline calcium pantothenate, commercial feed grade	Reg. 582.5212
Carotene, 7-01-134	The refined crystalline carotene fraction of plants	Reg. 582.5245
Choline chloride, 7-01-228	Choline chloride, commercial feed grade	Reg. 582.5252
Choline pantothenate, 7-01-229	Crystalline choline pantothenate, commercial feed grade	GRAS[a]
Choline xanthate, 7-01-230	Choline xanthate, commercial feed grade	Reg. 573.300
Erythorbic acid, 7-09-823	Acid or sodium salt	GRAS[a]
Folic acid, 7-02-066	Crystalline folic acid, commercial feed grade	GRAS[a]
Inositol, 7-09-354	Vitamin B complex vitamin; lipotropic. Chemical name: cyclohexanehexol. Also referred to as *i*-inositol or meso-inositol.	Reg. 582.5370
Menadione dimethylpyrimidinal bisulfite, 7-08-102	Crystalline menadione, dimethyl pyrimidinal bisulfite, commercial feed grade	Reg. 573.620
Menadione sodium bisulfite, 7-03-077	The addition product of menadione and sodium bisulfite containing not less than 50% menadione	GRAS[a]
Menadione sodium bisulfite complex, 7-03-078	The addition product of menadione and sodium bisulfite containing not less than 30% menadione	GRAS[a]
Niacin; nicotinic acid, 7-03-219	Crystalline nicotinic acid, commercial feed grade	Reg. 582.5530
Niacinamide; nicotinamide, 7-03-215	Crystalline amide of nicotinic acid, commercial feed grade	Reg. 582.5535
Pyridoxine hydrochloride, 7-03-822	Crystalline chloride of pyridoxine, commercial feed grade	Reg. 582.5676
p-Aminobenzoic acid, 7-03-513	*p*-Aminobenzoic acid, commercial feed grade	GRAS[a]
Riboflavin, 7-03-920	Crystalline riboflavin, commercial feed grade	Reg. 582.6595
Salmon liver oil, 7-02-013	The oil extracted from salmon livers	Not a food additive
Shark liver oil, 7-02-019	The oil extracted from shark liver	Not a food additive
Thiamin, thiamin hydrochloride, 7-04-834	Crystalline chloride of thiamin, commercial feed grade	Reg. 582.5875
Thiamin mononitrate 7-04-829	Crystalline mononitrate of thiamin, commercial feed grade	Reg. 582.5878
α-Tocopherol, 7-00-001	α-Tocopheral, commercial feed grade	Reg. 582.5890
Vitamin A acetate, 7-05-142	Vitamin A acetate, commercial feed grade	Reg. 582.5933
Vitamin A palmitate, 7-05-143	Vitamin A palmitate, commercial feed grade	Reg. 582.5936
Vitamin A propionate, 7-26-311	Retinol or esters of retinol formed from edible fatty acids	GRAS[a]
Wheat germ oil, 7-05-207	The oil extracted or expressed from wheat germ	Not a food additive

[a]GRAS is an abbreviation for the phrase "generally recognized as safe by experts qualified to evaluate the safety of the substance for its intended use."

TABLE 9-5

Influence of pelleting, initial vitamin A concentration, and urea content on the supplement vitamin A content prior to storage

Initial Vitamin A Concentration (IU/lb)	Feed Form		Loss by Pelleting (%)
	Meal	Pellet	
3000			
0% Urea	4.0 ± 0.2	3.0 ± 0.1	25
3.5% Urea	9.0 ± 0.3	7.0 ± 0.1	22
7.0% Urea	12.0 ± 0.1	7.0 ± 0.2	42
30,000			
0% Urea	23.8 ± 0.2	15.5 ± 0.3	35
3.5% Urea	24.2 ± 0.5	14.5 ± 0.5	40
7.0% Urea	24.5 ± 1.5	14.8 ± 0.2	40
300,000			
0% Urea	260 ± 4.0	164 ± 5.0	37
3.5% Urea	293 ± 0.5	177 ± 5.0	40
7.0% Urea	235 ± 2.5	154 ± 4.5	34

Source: Shields et al. (10).

Values indicate vitamin A content expressed in thousands of IU/lb on a 90% DM basis.

highest concentrations in the germ or germ oil of plants and in moderate concentrations in green plants or hays or in dehydrated alfalfa meal. Commercially, the germ oil is used extensively, although synthetically produced concentrates of vitamin E are available. Vitamin E is an antioxidant; thus it is lost rapidly in any situation resulting in oxidizing conditions (heat, light, high trace mineral content of feed, etc.).

Vitamin K is widely distributed in green plant material and is produced by the microbes in the GI tract. A number of different compounds have vitamin K activity, but menadione, a naturally occurring compound, is usually the normal reference standard. It is fat soluble and may be stored in relatively high concentrations in animal tissues or in seeds such as soybeans. Two common water-soluble forms, menadione sodium bisulfite and menadione dimethylpyrimidinol bisulfite, are often used as feed supplements. Both have good stability in feeds.

WATER-SOLUBLE VITAMINS

Animal and fish by-products, green forages, yeast, fermentation products, milk by-products, oilseed meals, and some seed parts are usually good sources of the water-soluble vitamins. The bran layers of cereal grains are fair to moderate sources, and roots and tubers are poor to fair sources. Cobalamin (B_{12}) is the only required vitamin that is not found in plants. It is produced exclusively by microorganisms. Thus good sources are yeast or similar products. Animal and fish products, particularly liver, are good sources. Animal manures contain B_{12}, also.

Commercially, some of the crystalline vitamins or mixtures used for humans are prepared from liver, yeast, or other fermentation products, but a number are produced at competitive prices from synthetic processes. Water-soluble vitamins produced synthetically include thiamin hydrochloride, riboflavin, nicotinic acid or nicotinamide, pyridoxine, ascorbic acid, choline chloride, and pantothenic acid. These sources may be used when especially high vitamin content is needed in some particular situation. Supplementary sources of the vitamins are listed in Table 9-4, and some of the synthetic products and their vitamin content are listed in Appendix Table 6.

SUMMARY

Mineral supplements, which are concentrated sources of needed minerals, are available for all of the required elements needed for domestic animals. Of the Ca supplements, it is generally agreed that those normally used are more completely available than are natural feed sources such as the legume hays. Calcium sources include a variety of materials, and most of them are inexpensive. Phosphorus is usually a much more costly mineral to provide and often is found in association with Ca (e.g., in bone meal and other bone products and a number of the Ca–P salts prepared from rock phosphates). Most P sources are utilized efficiently, with the exception of some of the colloidal clays. Rock phosphates may be too high in fluorine to be used as a mineral supplement because of potential fluorine toxicity, unless they are defluorinated.

Defluorinated phosphate and dicalcium phosphate are the most common supplements. Potassium is easily provided from carbonates, bicarbonates, chlorides, or sulfates. Magnesium can be provided easily by the oxide, sulfate, or dolomitic limestone. Sodium and Cl can be provided easily and cheaply by common salt. Sulfur can be provided to ruminants in any of many different forms, including elemental S, but nonruminants need an organic source such as methionine. Numerous different salts are available as sources of the different trace minerals. Iodine presents a problem in that ordinary KI is partially volatile and, if used, stabilizers must be added.

With regard to the vitamins, natural sources such as germ meals, brans, high-quality dehydrated grass or legume products, and various other products such as yeast and liver meals are often used as supplementary sources. In addition, relatively pure sources of a number of the vitamins are available in feed-grade purity and at prices that enable them to be used routinely where needed. A number of these vitamins are produced synthetically. Some natural sources such as fish liver oils are still used in moderate amounts as sources of vitamins A, D, and E, but vitamin K is primarily added, where needed, as a synthetic product.

REFERENCES

1. Ammerman, C. B., et al. 1980. *Mineral tolerance of domestic animals.* Washington, DC: Nat. Acad. Sci.

2. Church, D C., ed. 1988. *The ruminant animal.* Englewood Cliffs, NJ: Prentice Hall.

3. Underwood, E. J. 1981. *The mineral nutrition of livestock.* 2d ed. Farnham Royal, England: Commonwealth Agr. Bureaux.

4. Miller, W. J., et al. 1987. *J. Dairy Sci.* 70:1885.

5. Ewan, R. C. 1986. *Report* AS-580-K. Ames, IA: Iowa State Univ. Dept. Animal Sci.

6. NRC. 1984. *Nutrient requirements of beef.* 6th rev. ed. Washington, DC: National Academy Press.

7. Church, D. C., ed. 1979. *Digestive physiology and nutrition of ruminants.* Vol. 2: Nutrition. 2d ed. Corvallis, OR: O & B Books.

8. Ward, G., L. H. Harbers, and J. J. Blaha. 1979. *J. Dairy Sci.* 62:715.

9. Owens, F. N. 1988. *Beef* 24(7):29.

10. Shields, R. G., et al. 1982. *Feedstuffs* 54(47):22.

10

FEED ADDITIVES

D. C. Church and Richard O. Kellems

INTRODUCTION

A feed additive is defined by AFCO (American Feed Control Officials) as "an ingredient or combination of ingredients added to the basic feed mix or parts thereof to fulfill a specific need. Usually used in micro quantities and requires careful handling and mixing" (1). In practice, feed additives are defined as feed ingredients of a nonnutritive nature that stimulate growth or other types of performance or improve the efficiency of feed utilization or that may be beneficial in some manner to the health or metabolism of the animal. The official federal definition is somewhat different (and ambiguous); it is outlined in Public Law 65.929 for those interested in further detail.

Many of the commonly used feed additives are classified as drugs. Drugs are defined by the FDA as "A substance (a) intended for use in the diagnosis, cure, mitigation, treatment or prevention of disease in man or other animals or (b) a substance other than food intended to affect the structure or any function of the body of man or other animals" (3). It is obvious that this is a very broad definition.

Of the various groups of additives classed as drugs, the major groups include many different compounds; such as, antibiotics, nitrofurans and sulfa compounds (synthetic antibacterial compounds), coccidiostats, wormers (antihelmintics as well as other types), and hormonelike compounds. These are discussed in some detail in a later section in this chapter.

Feed additives have been used extensively in the United States and in many other countries following the discovery and commercial production of antibiotics and sulfa drugs in the late 1940s and thereafter. The simple reasons for use were early observations that low levels of the additives, fed more or less continuously, often resulted in a marked improvement in the health and growth or feed efficiency of young animals such as chicks, baby pigs, or calves. In more recent years a trend has developed (particularly in Europe) for government regulatory agencies and lawmaking bodies to restrict usage of additives for a variety of reasons. For example, it has been claimed but not proved that usage of some additives with animals results in the development of resistant strains of microorganisms that might be more pathogenic to humans or that might be resistant to antimicrobial agents used in treating diseases in humans.

While it is undoubtedly true that feed additives are misused from time to time, it is also true that they have been extremely beneficial to livestock producers under our modern methods of production. Development of more intense systems of management and concentration of animals such as broilers, laying hens, growing–finishing pigs, and fattening cattle and sheep has been possible only with the use of a number of the additives to help control various diseases and/or parasites. Further information is given on this in later sections.

USE OF FEED ADDITIVES CLASSED AS DRUGS

In the United States the use and regulation of additives classed as drugs are controlled by the FDA (or more specifically, within the FDA, the Center for Veterinary Medicine). The responsibility of the FDA is to determine that drugs and medicated feed are properly labeled for their intended use and that animal feeds and food derived from animals are safe to eat. The federal law plainly states that no animal drug can be used in feed until there has been extensive research proving to the FDA that the drug is both safe and effective. In the process of developing a new drug for use with animals, manufacturers must go through extensive testing. They must establish the safety for the target species, effectiveness of the drug, environmental impact, safety to humans, and chemistry and manufacturing specifications for labeling of the drug. Some of these data may be collected while using a permit for investigational new animal drugs (INAD), which allows some compounds to be tested under practical conditions and, providing there are no significant problems with tissue residues, for the animals to be marketed. General marketing of a new drug can be done only after approval of a new animal drug application (NADA).

MEDICATED FEED REQUIREMENTS

The FDA's medicated feed requirements focus on those medicated feed mixers who use human-risk drug sources. A human-risk drug source is a drug premix that requires a form FD-1900 for its use. Medicated feed mixers who choose not to use human-risk drug sources are subject to less demanding regulation and are exempt from mandatory FDA inspections, establishment registration, and drug assay requirements. Obviously, there is some incentive not to use the human-risk drugs as defined by the FDA.

All animal drugs used in medicated feeds are divided into one of two classes, called Category I or Category II. These drugs are listed in Table 10-1. As of February 1989, there were 31 drugs or drug combinations in Category I and 74 drugs or drug combinations in Category II. Category I drugs are the safest to use, have the least potential for unsafe residues, and require no withdrawal period before slaughter of meat-producing animals. Category II drugs are those

(1) for which a withdrawal period is required at the lowest use level for at least one species for which they are approved, or (2) that are regulated on a "no-residue" basis or with a "zero" tolerance because of a carcinogenic concern, regardless of whether a withdrawal period is required.

Manufacturers normally provide drugs (or other microingredients) to feed mixers in rather concentrated mixes because it is impossible to add only a few grams per ton of feed and have it distributed with any accuracy. Thus drugs, vitamins, and other additives are diluted in packages that might range from 5 to 100 lb for addition to other feeds. This practice ensures more safety and/or accuracy for the feed mixer.

With regard to medicated feeds, they are classified as Type A, B, or C. A type A medicated product contains a drug(s) at a potency higher than permitted in Type B feed levels (see Category I and II levels, Table 10-1). In other words, it is a concentrated premix of the product(s) that is added to other feed, with resultant dilution of the drug(s). Type B is a medicated feed for further mixing that contains drug(s) and that is intended solely for manufacture into another Type B or Type C medicated feed. It is produced from a drug component, a Type A medicated article, or another Type B medicated feed. A Type C medicated feed is one not intended for further mixing. It may be fed as a complete feed or it may be fed top dressed or offered free choice in conjunction with other animal feed to supplement the animals' total daily ration. It is produced by substantially diluting a drug component, a Type A medicated article, or a Type B or C medicated feed with ingredients to a level of use that is covered by an approved new animal drug application.

Any feed mixer must be registered with the FDA before mixing any Type A feeds containing Category II animal drugs into animal feed. Feed mills that use any Category I animal drug source or a Category II, Type B, animal drug source are not required to register. Individuals mixing medicated feed for feeding only to their own animals are subject to the same rules as commercial feed mills. Labeling of medicated feeds is under strict control. It is discussed briefly in Ch. 5.

ANTIBIOTICS

Antibiotics are compounds produced by microorganisms; these compounds have the properties of inhibiting the growth or metabolism of some (not all) other microorganisms and, in some instances, they may be toxic to warm-blooded

TABLE 10-1

Category I and II animal drugs for use in Type B feeds

Drug	Maximum Potency[a] g/lb	%	Drug	Maximum Potency[a] g/lb	%
Category I Drugs			**Category II Drugs (cont.)**		
Aklomide	22.75	5.0	Levamisole	113.5	25.0
Ammonium chloride	113.4	25.0	Lincomycin	10.0	2.2
Amprolium with			Melengestrol acetate	2g/T	0.00022
ethopabate	22.75	5	Morantel tartrate	66.0	14.52
Bacitracin methylene			Neomycin	7.0	1.54
disalicylate	25.0	5.5	Neomycin	7.0	1.54
Bacitracin zinc	5.0	1.1	Oxytetracycline	10.0	2.2
Bambermycins	0.4	0.09	Nicarbazin	5.675	1.25
Buquinolate	9.8	2.2	Nitarsone	8.5	1.87
Chlortetracycline	40.0	8.8	Nitrofurazone	10.0	2.2
Coumaphos	6.0	1.3	Nitromide	11.35	2.5
Decoquinate	2.72	0.6	Sulfanitran	13.6	3.0
Dichlorvos	33.0	7.3	Nitromide	11.35	2.5
Erythromycin			Sulfanitran	5.65	1.24
(thiocyanate salt)	9.25	2.04	Roxarsone	2.275	0.5
Fenbendazole	4.54	1.0	Novobiocin	17.5	3.85
Iodinated casein	20.0	4.4	Phenothiazine	66.5	14.6
Lasalocid	40.0	8.8	Piperazine	165	40.25
Monensin	40.0	8.8	Pyrantel tartrate	4.8	1.1
Narasin	7.2	1.6	Robenidine	1.5	0.33
Nequinate	1.83	0.4	Ronnel	27.2	6.0
Niclosamide	225	49.5	Roxarsone	2.275	0.5
Nystatin	5.0	1.1	Roxarsone	2.275	0.5
Oleandomycin	1.125	0.25	Aklomide	11.35	2.5
Oxytetracycline	20.0	4.4	Roxarsone	2.275	0.5
Penicillin	10.0	2.2	Clopidol	11.35	2.5
Penicillin	1.5	0.33	Bacitracin	5.0	1.1
Streptomycin	7.5	1.65	methylene		
Polaxalene	54.48	12.0	disalicylate		
Salinomycin	6.0	1.3	Roxarsone	2.275	0.5
Tiamulin	3.5	0.8	Monensin	5.5	1.2
Tylosin	10	2.2	Sulfadimethoxine	5.675	1.25
Virginiamycin	10	2.2	Ormetoprim (5/3)	3.405	0.75
Zoalene	11.35	2.5	Sulfadimethoxine	85.1	18.75
			Ormetoprim (5/1)	17.0	3.75
Category II Drugs			Sulfaethoxypyridazine	50.0	11.0
Amprolium	11.35	2.5	Sulfamerazine	18.6	4.0
Apramycin	7.5	1.65	Sulfamethazine	10.0	2.2
Arsanilate sodium	4.5	1.0	Chlortetracycline	10.0	2.2
Arsanilic acid	4.5	1.0	Penicillin	5.0	1.1
Butynorate	17.0	3.74	Sulfamethazine	10.0	2.2
Butynorate	63.45	14.0	Chlortetracycline	10.0	2.2
Piperazine	49.85	11.0	Sulfamethazine	10.0	2.2
Phenothiazine	263.32	58.0	Tylosin	10.0	2.2
Carbadox	2.5	0.55	Sulfanitran	13.6	3.0
Carbarsone	16.0	3.74	Aklomide	11.2	2.5
Clopidol	11.4	2.5	Sulfanitran	13.6	3.0
Dimetridazole	9.1	2.0	Aklomide	11.2	2.5
Famphur	5.5	1.21	Roxarsone	2.715	0.6
Fenbendazole	8.87	1.96	Sulfaquinoxaline	11.2	2.5
Furazolidone	10.0	2.2	Sulfathiazole	10.0	2.2
Halofuginone			Chlortetracycline	10.0	2.2
hydrobromide	0.136	0.3	Penicillin	5.0	1.1
Hygromycin B	0.6	0.13	Thiabendazole	45.5	10.0
Ipronidazole	2.84	0.63			

Source: Published in the Federal Register (2) and compiled by Anon. (3).

[a]The maximum amount of Category I drugs that may be used in Type B feeds in 200 times the amount permitted for continuous use in a complete feed. That for Category II drugs is 100 times that approved for continuous use.

animals. Most antibiotic names end in "cin" or "mycin." All antibiotics used commercially for growth promotion are produced by fermentation processes using fungi or bacteria. Those compounds currently approved for use with farm animals are listed in Table 10-2 along with data on which compound is approved for a given species, the range of antibiotic concentration approved, claims by the manufacturer, and, if required, any withdrawal time before slaughter.

Antibiotics have, in general, been effective as production improvers when fed at low levels to young, growing animals, as is illustrated in Table 10-3. Their use tends to result in an increased feed intake, which may account for some of the improvement usually found in feed efficiency. Growth is nearly always increased, particularly when the animal is exposed to adverse environmental conditions. Weekly intramuscular injections (cattle) may work just as well as daily oral intake, but it is not a feasible method of administration except for dairy animals.

The response in growth and feed efficiency is apt to be variable among animal species and from time to time and place to place. For example, there is little or no response in new animal facilities, in very clean surroundings, or in germ-free animals raised under aseptic conditions. Generally, the most pronounced responses in young animals occur in situations where organisms such as *E. coli* or other organisms, causing diarrhea are in high concentrations, because antibiotics, as a rule, will help to reduce the incidence or severity of several types of diarrhea. Antibiotic-fed animals are less apt to go off feed. In addition, evidence indicates that antibiotics may have a sparing effect on dietary needs for some amino acids and B-complex vitamins for young chicks, pigs, or rats, the beneficial response being greater when diets contain submarginal or minimal levels of these nutrients.

Antibiotics may be useful for a number of other purposes, for example, in the prevention and control of a wide variety of animal and poultry diseases. Various antibiotics are approved at low levels of continuous use for reducing the incidence of enterotoxemia in lambs, liver abscesses in fattening cattle, or diarrhea in young mammals deprived of colostrum. With poultry, some of the claims include reduction in respiratory disease, nonspecific enteritis (blue comb), and infectious sinusitis, as well as improved egg production and hatchability.

Antibiotics are often used at therapeutic levels in the treatment or control of many common diseases or control of gastrointestinal worms. When used at higher levels for therapeutic treatments (usually for only a few days), antibiotics have been very useful for treating or preventing stresses associated with transportation and adjustment to new conditions with cattle; for treatment of diseases such as anaplasmosis in cattle, and bacterial enteritis in swine; and respiratory diseases, diarrhea, fowl cholera, fowl typhoid, and breast blisters in poultry. In most instances the higher levels are not approved for long-term usage.

Two antibiotics approved in recent years for cattle, monensin and lasalocid, are unusual in that they give a good response in both growing and mature animals. Technically, they are classed as rumen additives. Although the mode of action is not completely worked out, these antibiotics result in a shift in the rumen volatile fatty acid production to relatively more propionic acid and a reduction in methane production, which in turn results in more efficient gain of growing animals and some improved efficiency and increased growth or gain in cattle (both young and adult) on pasture or forage. An example of the performance of growing–finishing cattle fed monensin and lasalocid is given in Table 10-4.

In the case of monensin and lasalocid, approval was first received for use as coccidiostats with poultry (they can also serve the same function in cattle or sheep). Although a few antibiotics do act on coccidia, most of the coccidiostats shown in Table 10-2 are synthetic compounds. Presumably, this normal difference accounts for the way these two compounds were named. Both of these antibiotics are quite toxic to horses.

Obtaining approval for new drugs to be used as feed additives has been more difficult in recent years. More investigative effort and much more expense has been involved, the result being that not many have been approved. For example, with antibiotics, only apramycin and tiamulin have been approved for use with swine, nystatin for use with chickens and turkeys (for treating crop mycosis and mycotic diarrhea), and salinomycin (a coccidiostat) for use with broilers. There have been some other changes in approvals of different species for compounds other than antibiotics, some alterations in approved levels, removal of some drugs from the list, but the addition of only a few compounds.

The reader may note (Table 10-2) that very few additives have been approved for horses, rabbits, sheep, goats, and species such as ducks, pheasants, and quail. No approvals are given for geese or pets such as cats and dogs. The primary

TABLE 10-2

Drugs approved for use as feed additives in the United States

Drug Name	Animal	Approved Dose Level[a]		Claims[b]	Withdrawal Time
		Grams per Ton of Feed	Other Measure of Dose		
Antibiotics					
Apramycin	Swine	150	For 14 d	DR	
Bacitracin methylene disalicylate	Feedlot cattle		70 mg/hd/d or 250 mg/hd/d for 5 d	LA	
	Swine	10–250		GP, FE, DR	
	Chickens, turkeys	4–200		GP, FE, EP, NE	
	Pheasants	4–50		GP, FE	
	Quail	5–20		GP, FE	
Bacitracin zinc	Feedlot cattle	35–70		GP, FE	
	Swine	10–50		GP, FE	
	Chickens, turkeys	4–50		GP, FE	
	Pheasants	4–50		GP, FE	
	Quail	5–20		GP, FE	
Banbermycins	Swine growing, finishing	2–4		GP, FE	
	Broilers, growing turkeys	1–4		GP, FE	
Chlortetracycline	Calves		0.1–0.5 mg/lb BW	GP, FE, DR	
	Beef cattle, nonlactating dairy cattle		70–750 mg/hd/d	GP, FE, LA, DR, FR, RD, AP	48 h at 350 mg/hd/d or higher
	Beef		5 mg/lb BW for 60 d	AP	
	Dairy cows		0.1 mg/lb BW/d	DR, FR, RD	
	Sheep	20–50		GP, FE	
	Breeding sheep		80 mg/hd/d	Vibrionic abortion	
	Swine	10–400		GP, FE, BE, leptospirosis	
	Horses to 1 yr of age		85 mg/hd/d	GP, FE	
	Mink	20–50		GP, FE, pelt size	
	Chickens, turkeys	10–500		GP, FE, RD, EH, BC	
	Ducks	200–400		FC	
Erythromycin	Chickens, turkeys, layer and breeder	92.5–185		RD	24–48 h
Hygromycin B	Swine	12		GW	15 d
	Chickens	8–12		GW	3 d
Lasalocid sodium	Cattle Feedlot	10–30		GP, FE	
	Pasture except for lactating dairy cows		60–200 mg/hd/d	GP	
	Sheep	20–30	15–70 mg/hd/d	GP	
	Broilers, fryers	68–113		CC	

TABLE 10-2 (*cont.*)

Drug Name	Animal	Approved Dose Level[a] — Grams per Ton of Feed	Approved Dose Level[a] — Other Measure of Dose	Claims[b]	Withdrawal Time
Lincomycin	Swine	20–200		DR, GP	6 d
	Broilers	2–4		GP, FE, NE	5 d
Monensin	Cattle, feedlot pasture	5–30 5–30	with complete feed or 25–400 with supplement	GP GP	
Neomycin	Chick, broilers and replacements	90–110		CC	
	Turkeys	54–90		CC	
	Cattle, sheep, goats	70–140		DR, BE, ET (sheep)	Cattle 30 d, sheep 20 d
	Young animals on milk replacer		200–400 mg/gal MR	BE, DR	20 d
	Swine, horses, mink	70–140		BE, DR	
	Poultry	70–140		BE	14 d
Novobiocin	Mink	200–350	200 mg/T wet feed	BP, FE	
	Chickens, turkeys	350	For 5–7 d	Breast blisters	4 d
	Ducks		For 5–7 d	Fowl cholera	
Nystatin	Chickens, turkeys	50	Continuously or 100 g/T for 7–10 d	Crop mycosis and mycotic diarrhea	
Oxytetracycline	Cattle Calves, 0–12 wk starter feeds or milk replacer	50–100	0.5–1 mg/lb BW/d 0.5–5 mg/lb BW/d or 25–75 mg/hd/d	GP, FE	5 d if fed in MR or 2 g/hd in dry feed
	Feedlot Beef		75 mg/hd/d	DR, FE GP, FE LA, DR, RD MP, BT, DR	
	Dairy cows		75–100 mg/hd/d	GP, FE DR, ET	
	Sheep	10–20 20–100	75–100 mg/hd/d	GP, FE, BE GP, FE, DR	
	Swine	7.5–500		GP, FE	5 d at 500 g/T
	Mink	25–100		FE, RD, GP, CC, BC	
	Rabbits	10			
	Chickens	5–500			3 d at 200 g/T
	Turkeys	5–200		EP, FE, BC	3 d at 200 g/T

Drug	Animal	Amount (g/ton)	Purpose	Withdrawal
Penicillin, procane	Swine	10–50	GP, FE	
	Chickens, turkeys	2.4–100	GP, FE, RD, BE, BC	
	Pheasants, quail	2.4–50	GP, FE	
Salinomycin	Broilers	40–60	CC	
Streptomycin	Swine	7.5–75	GP, FE, DR	
	Chickens, turkeys	12–180	EP, BC, RD, IS	2 d
Tiamulin	Swine	10–35	GP, FE, DR	2 d
Tylosin	Beef cattle	8–10	LA	
	Swine			
	Starter feed	10–100	GP, FE, DR	
	Grower feed	10–40	GP, FE, DR	
	Chickens	4–50 or 800–1,000 for 5 d	GP, FE; Chronic RD	
Virginiamycin	Swine	5–100	GP, FE, DR	
	Poultry			
	Broilers, replacements	5–20	GP, FE, NE, CC	
	Turkeys	10–20	GP, FE	
Arsenicals				
Arsanilic acid or sodium arsanilate	Swine	45–90	GP, FE, DR	5 d
	Chickens, turkeys	90	GP, FE	5 d
Nitarsone	Chickens, turkeys	170	BH	5 d
Roxarsone	Swine	22.5–67.5	GP, FE, DR	5 d
Coccidiostats				
Amprolium	Calves	5 mg/kg BW for 21 d or 10 mg/kg for 5 d	CC	24 h
	Chickens	72.5–113.5	CC	
	Turkeys	113.5–227	CC	
	Pheasants	159	CC	
Butynorate	Chickens, turkeys	340–680	CC	
Clopidol	Broilers, replacements	113.5–227	CC	
Decoquinate	Cattle, goats	0.5/kg BW/d	CC	28 d
	Broilers	27.25	CC	
Halofuginone hydrobromide	Broilers	2.72	CC	4 d
Nicarbazin	Chickens	90.5–181	CC	4 d
Robenidine HCl	Broilers	30	CC	5 d

TABLE 10-2 (cont.)

		Approved Dose Level[a]			
Drug Name	Animal	Grams per Ton of Feed	Other Measure of Dose	Claims[b]	Withdrawal Time
Nitrofurans					
Furazolidone	Sows, pigs	100–300		DR, BE, GP	5 d
	Chickens, turkeys	7.5–200		GP, FE, CC, BC, TT	5 d
Nitarsone	Chickens, turkeys	170		BH	
Nitrofurazone	Broilers, replacements and turkeys	50		CC	5 d
	Swine	500		BE	
	Mink	100		DR	5 d
Sulfa drugs					
Sulfamethazine	Swine	100		BE	15 d
Sulfadimethoxine	Chickens, turkeys	72.5		FC, bacterial infection	
Sulfaquinoxaline	Rabbits	159		CC	10 d
Sulfathiazole	Swine	100		BE	7 d
Other antimicrobials					
Carbadox	Swine	10–50		GP, FE, DR, BE	10 wk
Dimetridazole	Turkeys	136–725		GP, FE, BH	5 d
Ipronidazole	Turkeys	56.7–227		GP, FE, BH	4–5 d
Drugs for gastrointestinal, lung, or kidney worms					
Coumaphos	Cattle	27.2–36.3	0.091 g/100 lb BW for 6 d	GW	
	Chicken over 8 wk of age			GW	
Dichlorovos	Swine	348–499		GW	
Fenbendazole	Swine		3 mg/kg BW/d for 3 d	GW	
	Cattle		5 mg/kg BW/d	GW	
Levamisole HCl	Beef cattle, young dairy cattle	72.5–725		Internal worms	13 d
				GW, lung worms	48 h
	Swine	725		GW, lung worms	72 h
Morantel tartrate	Cattle		0.44 g/100 lb BW	GW	14 d

Drug	Animal	Amount (g)[a]	Use level	Claims[b]	Withdrawal
Phenothiazine	Cattle, except lactating dairy cows		0.25 g/100 lb BW or 2 g/hd/d in feed or salt	GW	
	Sheep, goats		12.5–25 g/d or 1 g/hd/d in feed or salt	GW	
	Swine		5–30 g for 1 d	GW	
	Horses, mules		2 g/hd/d for 21 d	GW	
Piperizine	Chickens, turkeys		0.5%–1 g/bird for 1 d	Cecal worms	
	Swine		0.1%–0.2% in water or 0.2%–0.4% in feed for 1 d	GW	
	Chickens, turkeys		0.2%–0.4% in feed or 0.1%–0.2% in water for 1 d	GW	
Pyrantel tartrate	Swine	96–800		GW	
Thiabendazole	Cattle		3–5 g/100 lb BW	GW	3 d
	Sheep, goats		2–3 g/100 lb BW	GW	30 d
	Swine	45–908		GW	30 d
	Pheasants	454	For two weeks	GW	21 d
Miscellaneous drugs					
Larvadex	Chickens			FP	
Methoprene	Cattle	454	22.7–45.4 mg/100 lb BW/month	FP	3 d
Poloxalene	Cattle		1–2 g/100 lb BW/d	BT	
Propylene glycol	Dairy cows		0.25–0.5 lb/hd/d prior to or after calving	KT	
Rabon	Cattle		0.07 g/100 lb BW/d	FP	
	Swine		0.05 g/100 lb BW/d	FP	
	Horses		0.07 g/100 lb BW/d	FP	
	Mink		3 mg/kg BW/d	FP	
Hormonelike compounds					
Melengestrol acetate	Beef heifers		0.025–0.50 mg/hd/d	GP, FE, ES	48 h

Source: Original data are from the Federal Register (2) and were excerpted from Anon. (3). *Note:* Many compounds are approved for use in various combinations that are not listed here. Approved use levels vary according to the specific purposes. The claims listed may not be complete in all cases, nor are many of the warnings listed here.

[a]Quantities are given in grams (g) per ton (T) of feed or in amounts per unit of body weight (BW).

[b]The claims are for improvement of such things as growth, GP; feed efficiency, FE; egg production, EP; egg hatchability, EH; improved pigmentation, IP; milk production, MP; or for control or treatment of diseases or problems as listed: AP, anaplasmosis; BC, blue comb (nonspecific infectious enteritis); BE, bacterial enteritis; BH, blackhead; CC, coccidiosis; DR, diarrhea; ES, estrous suppression; ET, enterotoxemia; FP, fly prevention; FR, foot rot; FT, fowl typhoid; GW, gastrointestinal worms; IS, infectious sinusitis; KT, ketosis; LA, liver abscesses; NE, necrotic enteritis; RD, respiratory diseases.

TABLE 10-3

Effect of feeding low levels of antibiotics to beef cattle

Item	Growing–Wintering		Finishing	
	Controls	+Antibiotic	Controls	+Antibiotic
Daily gain, kg	0.56	0.61	1.16	1.24
Feed per unit gain	12.70	12.02	9.85	9.37
Number of animals	5353		2354	
Number of days on feed	112		117	
Improvement in gain, %	8.9		6.7	
Improvement in efficiency, %	5.4		4.9	

Source: Wallace (4). Antibiotics were fed continuously, most at 70 mg/hd/d. Antibiotics involved were chlortetracycline, oxytetracycline, and bacitracin.

TABLE 10-4

Effect of feeding of monensin and lasalocid on performance of cattle

Treatment	Measure of Performance		
	Feed Consumed (kg/d)	Daily Gain (kg/d)	Feed-to-Gain Ratio
Feedlot cattle[a]			
Controls	8.27	1.09	8.09
Monensin	7.73	1.10	7.43
Controls	8.89	1.14	8.01
Monensin	8.65	1.18	7.43
Implant	9.61	1.28	7.63
Monensin + implant	8.97	1.32	6.85
Pasture cattle[a]			
Controls		0.61	
Monensin		0.69	
Feedlot cattle[b]			
Controls	8.68	0.99	8.75
Lasalocid, 30 g/T	8.27	1.02	8.09
Lasalocid, 45 g/T	8.09	1.04	7.79
Monensin, 30 g/T	8.09	1.03	7.93

[a]Data excerpted from a review (5) that contains a compilation of many different experiments done in many different places.

[b]Data from Berger et al. (6) from cattle (48 per treatment) fed a 60% high-moisture corn, 30% corn silage diet (dry basis).

reason for this is the cost of obtaining approval in relation to potential sales volumes. The approval procedure must be done with each species and at different levels that are recommended. Thus, if projected sales do not indicate sufficient volume to justify the costs, no approval will be requested. Many of the antibiotics approved for cattle have been tested with sheep, and they may be quite effective for a specific purpose, but few ever get approved for sheep and even fewer for species other than cattle, swine, chickens, and turkeys.

With cattle, most of the antibiotics (in feed) are used for baby calves and growing–finishing animals. Although there is approved use of chlortetracycline and oxytetracycline for lactating dairy cows, experimental data have indicated little if any advantage from long-term feeding of low levels.

With regard to poultry, the trend is to use one or more antibiotics in nearly all broiler feeds. Most of the antibiotics (Table 10-2) can be used for layers, with the exception of high levels of chlortetracycline and erythromycin and normal levels of lincomycin, novobiocin, and virginiamycin. Note that approval must be obtained for using different combinations of antibiotics or antibiotics and other controlled drugs by the manufacturers of at least one of the drugs. Far more drug combinations have been approved for use with chickens and turkeys than for all other animals combined. Also note that it is illegal to feed antibiotics at different levels or in different combinations (with other antibiotics and other controlled drugs) than those that have been approved.

With swine (Table 10-2) there are restrictions based on age or weight, but probably less so than for poultry. A fairly large number of antibiotics have been approved for use with swine.

ARSENICALS

Arsenicals, which are all synthetic compounds, include a number of drugs used in turkey, chicken, and swine rations (see Table 10-2). These drugs were developed as a means of controlling parasites, but it was soon discovered that some of the compounds stimulated growth in the same manner as do antibiotics, and in fact sometimes it was additive to that of antibiotic stimulation. Note (Table 10-2) that several of the arsenicals have claims of improved growth production as well as improved feed efficiency for chickens, turkeys, or swine and for control of blackhead in poultry and diarrhea in swine. There have been other claims in earlier years, but they have been withdrawn as data on most of the older additives have been reworked and updated with added research.

Arsenicals have the disadvantage that they may accumulate in body tissues, particularly the liver. At the levels fed, they are not considered to be toxic, but all have a minimum 5-day withdrawal period before animals are to be slaughtered for human food.

COCCIDIOSTATS

Coccidiostats include a wide variety of compounds, ranging from a number of synthetic drugs to several of the antibiotics. These drugs are of considerable importance to the poultry producer because close confinement methods used in modern facilities accentuate the possibility of coccidiosis outbreaks. In addition, there is evidence that coccidiosis is becoming more of a problem with sheep and cattle kept under close confinement. In addition to the drugs listed under the heading of coccidiostats in Table 10-2, several antibiotics—lasalocid, monensin, oxytetracycline, salinomycin, and virginiamycin—have claims for prevention or treatment of coccidiosis. Also, some of the nitrofurans and sulfa drugs are effective against coccidiosis.

NITROFURANS

The nitrofurans are antibacterial compounds and are effective against a relatively large number of microbial diseases. Continued use of nitrofurans has not as yet developed bacterial resistance, as is the case for some antibiotics. The nitrofurans are often used in combination with other drugs, especially with swine and poultry.

SULFAS

The first sulfa drug was synthesized in the 1930s, and some of the early ones were used extensively against some human diseases that were very difficult to treat at the time. Unfortunately, most sulfas present problems with tissue residues, and some of the injectibles are even worse, resulting in tissue residues in edible cuts of meat. The result is that there has been a gradual withdrawal of sulfa drugs as feed additives; fortunately, most of the problems alleviated by sulfas can be treated successfully with other drugs that cause fewer problems.

DRUGS FOR CONTROL OF GASTROINTESTINAL PARASITES

Gastrointestinal and other types of worms (lung, kidney, and others) are a problem with all domestic species. Fortunately, a number of compounds have been developed to control these pests. In most cases these additives are used in the feed or water for only a few days.

HORMONELIKE PRODUCTION IMPROVERS

Only one hormonelike production improver remains on the approved list. Melengestrol acetate (a synthetic progestogen) has found rather extensive use with beef heifers. It acts to suppress estrus, resulting in more efficient and more rapid gain.

Although not feed additives, several products are available for use as subcutaneous implants (in the ear). These include hexestrol (used outside the United States), zeranol (Ralgro™), said to be an anabolic agent, Synovex™, a combination of estrogen and progesterone, Rapid Gain™, a combination of testosterone and estrogen, and Steer-oid™, a combination of progesterone and estradiol. Most of these products also appear to work well for feedlot lambs, although much lower dosages are required. A high percentage of growing–finishing cattle is treated with one or another of these implants.

In ruminants the various natural or synthetic hormones appear to produce a response that results from increased N retention accompanied by an increased intake of feed. The result is, usually, an increased growth rate, an improvement in feed efficiency, and, frequently, a reduced deposition of body fat, which may, at times, result in a lower carcass grade for animals fed to the same weight as nontreated animals.

There is ample evidence at this time (1996) that injections of natural growth hormone (somatotropin) can be used to increase milk production of dairy cows substantially. The hormone, which would normally be in very short supply, can be produced from bacteria by modern gene-splicing techniques (as are insulin and some other drugs or "natural" products).

Somatotropin (bst) is given as an intramuscular injection every 14 days.

OTHER SPECIAL-PURPOSE ADDITIVES

In addition to the many different compounds discussed as drugs in Category I or II (and the brief discussion of implants), many others are used in manufactured feeds, supplements, liquids, mineral or vitamin premixes or other special-purpose products (Table 10-5). Many of these products are used to improve the handling or processing properties of feedstuffs or ration ingredients, to alter their flavor, to

TABLE 10-5

Additional special-purpose additives

Name	Classification under Food Additives Amendment	Limitations or Restrictions
Aluminum sulfate	Antigelling agent for molasses	None
Aniseed[a]	Spice, seasoning, essential oils	None[b]
Attapulgite clay[a]	Anticaking agent and pelleting aid; suspension aid in liquid feed supplement	Not to exceed 2% in finished feed or 1.5% in supplement
Ball clay	Anticaking agent and pelleting aid	Not to exceed 2.5% in finished feed
Calcium silicate	Anticaking agent	Not to exceed 2%
Calcium stearate	Anticaking agent	In accordance with good manufacturing practices
Capsicum; red pepper[a]	Spice, seasoning	None[b]
Diatomaceous earth	Inert carrier and anticaking agent	Not to exceed 2% of total ration
Disodium EDTA	To solubilize trace minerals in aqueous solutions	Not to exceed 0.024% in finished feed
Ethyl cellulose	Binder or filler in dry vitamin preparations	Specified in Reg. 121.230
Fennel[a]	Spice, seasoning, essential oils, etc.	None[b]
Fenugreek seed[a]	Spice, seasoning, essential oils, etc.	None[b]
Ginger[a]	Spice, seasoning, essential oils, etc.	None[b]
Glycyrrhizin ammoniated	Spice, seasoning, essential oils, etc.	None[b]
Iron ammonium citrate	Anticaking agent in salt	Not to exceed 0.0025% in finished salt
Kaolin[a]	Anticaking agent	Not to exceed 2.5% in finished feed
Lecithin[a]	Stabilizer	None[b]
Methyl glucoside coconut oil ester	Surfactant in molasses	Not to exceed 0.032% in molasses
Mineral oil	To reduce dustiness of feed or mineral supplements, to serve as lubricant in the preparation of pellets, cubes, or blocks, to improve resistance to moisture of such pellets, cubes, and blocks, and to prevent segregation of trace minerals in mineralized salt	Not to exceed 3% in mineral supplements
Monosodium glutamate[a]	Spices, seasoning	None[b]
Montmorillonite clays[a]	Anticaking aid, pelleting aid, and nonnutritive carrier	Not to exceed 2% of the finished material
Petrolatum or a combination of mineral oil and petrolatum		Not to exceed 0.06% in finished feeds

Substance	Purpose	Restrictions
Paraffin	Dust control agent	Not to exceed 3% in mineral supplements or 0.06% in finished feeds
Petroleum jelly	Dust control agent in mineral mixes	Not to exceed 3% in mineral mixes or 0.06% in finished feed
Phosphoric acid	Miscellaneous and/or general purpose	None[b]
Polyethylene glycol (400) mono and dioleates	Processing aid when present as a result of its addition to molasses	Not to exceed 0.025% in the molasses
Polyoxyethylene glycol (400) mono & dioleate	Emulsifier	Calf milk replacers
Polysorbate 80	Emulsifier	Calf milk replacers
Polysorbate 60	Emulsifier	Calf milk replacers, mineral premixes
Propylene glycol alginate	Emulsifier, stabilizer, or thickener	None except in those specific foods listed in 121.1015[b]
Pyrophyllite	Anticaking aid, blending agent, pelleting aid, or carrier	Not to exceed 2% in finished feed
Saccharin sodium[a]	Nonnutritive sweetener	None[b]
Sodium carboxymethyl cellulose[a]	Stabilizer	Not to exceed 2% in finished feed
Sodium silico aluminate[a]	Anticaking agent	Not to exceed 2% in finished feed
Sodium stearoyl-2-lactylate	Emulsifier and stabilizer	None[b]
Sorbitan monostearate with or without polysorbate 60	Emulsifier in mineral premixes and dietary supplements for animal feed	None
Tagetes (Aztec marigold) meal and extract	To enhance the yellow color of chicken skin and eggs	Sufficiently supplemented with xanthophyll and other carotenoids to accomplish the desired effect
Tetrasodium pyrophosphate	Dispersant	Not to exceed 1% of the finished product
Titanium dioxide	Color additive	
Urea formaldehyde[a]	Coating for feed-grade urea for ruminant animal feed	Not to exceed 1% of the finished material
Yellow prussiate of soda	Anticaking agent in salt	Not to exceed 0.0013%

Source: *Feed Industry Red Book* (7).

[a]Generally recognized as safe.

[b]No quantitative restrictions, although use must conform to good manufacturing practices.

prevent some ingredient from settling out of a liquid, to control dust, and so forth. Because these compounds are not considered to be drugs, they are not regulated at the feed mill level by the FDA. There are some limits to how much can be used and in what types of feed they can be used, and all users are expected to follow accepted "good manufacturing practices," which apply to feed mills or other manufacturers producing and using these chemicals.

SUMMARY

Feed additives are a most important part of modern-day animal production, especially in any situation where animals are housed in large numbers in limited spaces. Many of the additives used are classed as drugs, and all drugs used in animal production are under some degree of control by the Food and Drug Administration (in the United States), which must approve a feed additive for use before it can be used at the commercial level on a routine basis.

Antibiotics are among the most widely used feed additives, and many of these antibiotics result in more rapid growth, improved feed efficiency, and improved general health, primarily in young animals, when fed continuously at low levels. They may be used at intermediate levels for some types of disease control or at high levels for therapeutic use for short periods of time. Other antibiotics may be effective against problem organisms such as coccidia (a protozoan) or gastrointestinal parasites. Arsenicals, nitrofurans, coccidiostats, sulfa drugs, and other special-purpose additives have been developed for use with domestic animals. In some cases there may be problems with tissue residues (particularly with arsenicals or sulfas), and the animal producer may be required to withdraw the drug for a period of time before the animal is slaughtered for human food. Two hormonelike products are approved for use with some species. Other than use to suppress estrus in beef heifers, the primary use of hormones or hormonelike compounds is as subcutaneous implants (not as a feed additive) with finishing cattle or sheep. However, it is likely that growth hormone, produced from bacterial sources, will be approved for use as an injectible with lactating dairy cows in the near future.

REFERENCES

1. AFCO. 1995. *Official publication.* Washington, DC: Assoc. Amer. Feed Control Officials.

2. Superintendent of Documents. *Federal register.* Washington, DC: Government Printing Office.

3. Anon. 1996. *Feed additive compendium.* Minnetonka, MN: Miller.

4. Wallace, H. D. 1970, *J. Animal Sci.* 31:1118.

5. Goodrich, R. D., et al. 1984. *J. Animal Sci.* 58:1484.

6. Berger, L. L., S. C. Ricke, and G. C. Fahey, Jr. 1981. *J. Animal Sci.* 53:1440.

7. Anon. 1996. *Feed industry red book.* Eden Prairie, MN: Communications Marketing.

11

FEED PREPARATION AND PROCESSING

D. C. Church and Richard O. Kellems

INTRODUCTION

Feed represents a major cost in any intensive system of animal production. Even with sheep, which typically consume more forage (as a percentage of their diet) than do other domestic species, feed may represent 55% or more of total production costs. A value of 75% to 80% might be more appropriate for poultry. Thus it is imperative to supply a diet adequate (in terms of nutrient content) and in a form that will encourage consumption without excessive feed wastage.

Feed-processing methods may involve mechanical, chemical, and/or thermal methods or a combination of these methods. Microbial fermentation may also be involved. Feeds may be processed to alter the physical form or particle size, to prevent spoilage, to isolate specific parts of a seed or plant, to improve palatability, or to inactivate toxins or antinutritional factors of one type or another. In some cases feed may be processed primarily to improve the capability of machinery to handle it, for example, chopping or grinding of baled hay.

Generally, feed preparatory methods become more important as the level of production and feeding increases and when maximum efficiency is desired. This is so because heavily fed animals become more selective and are more in-clined to sort out less palatable ingredients or to refuse and/or waste feed if a ration is not to their liking.

With ruminants, digestibility generally decreases as the level of feeding increases. This occurs primarily because feed does not remain in the GI tract long enough for maximum digestion to occur. Appropriate processing methods may, then, be used to partially counteract the normal decline in digestibility.

Some large feedlots must purchase and store large amounts of grain (Fig. 11-1) so that they will have an adequate supply. It is then processed as needed. Feed preparation becomes more important as animal production units become more concentrated, larger in size, and more mechanized. This applies particularly to roughage, because many roughages must be processed to some degree if they are to be used with feeding equipment that is now available, and (where complete rations are fed) chopped, pelleted, or ground roughages are much easier to mix with other ingredients.

The methods discussed in this chapter are those used for processing feed grains or roughages or for preserving high-moisture grains. Methods used in milling grains for flour, processing oil seeds, or processing pet foods are not discussed.

FIGURE 11-1 Milo grain stored outside at a large feed-lot in Arizona. It is not obvious in the picture, but this large pile of grain is covered with wire netting to keep birds away from it.

GRAIN-PROCESSING METHODS

Grain-processing methods may be divided conveniently into dry and wet methods or into cold and hot methods. Heat is an essential part of some of the methods, but it is not utilized at all in others. Likewise, added moisture is essential in some methods but may even be detrimental in others. Examples of grains processed in different ways are shown in Fig. 11-2, and various methods are listed and discussed in following sections.

COLD PROCESSING METHODS

ROLLER MILL CRACKING AND GRINDING   Roller mills act on grain by compressing it between two corrugated rolls that can be screwed together to produce smaller and smaller particles (Figs. 11-3 and 11-4). With grains like corn, wheat, or milo, the product can range in size from cracked grain to a rather fine powder. With the coarse grains (barley and oats) the product may range in size from a flattened kernel to a relatively fine ground product, but the hulls will not be ground as well as with other types of grinding mills. Roller mills produce a less dusty feed than a hammer mill. If not ground too finely, the physical texture is very acceptable to most species. Roller mills are not used with roughage.

GRINDING   Grinding is by far the most common method of feed processing and, other than soaking, is the cheapest and most simple process. A variety of equipment is available on the market, and all of it allows some control of the particle size of the finished product. The hammer mill is probably the most common food-processing equipment used in North America. Hammer mills (Fig. 11-5) process feed

with the aid of rotating metal bars (hammers) that blow the ground product through a metal screen. The size of the product is controlled by changing the screen size. These mills grind anything from a coarse roughage to any type of grain, and the product size varies from particles similar to cracked grain to a fine powder. There may be quite a bit of dust lost in the process, and the finished product is usually more dusty than grain ground with a roller mill or other type of grinding equipment.

Grinding generally improves digestibility of all small, hard seeds. Coarsely ground grains are preferred for ruminants because they dislike finely ground meals, particularly when the meals are dusty. Finer grinding is more common for poultry and swine. Grinding is just as satisfactory as other more expensive methods when grain intake is relatively low.

SOAKED GRAIN   Grain soaked for 12 to 24 h in water has long been used by livestock feeders, particularly when they wish to pamper prize animals a bit. The soaking, sometimes with heat, softens the grain, which swells during the process, making a palatable product that should be rolled before it is used in finishing rations. Most research results do not show any marked improvement in feed efficiency as compared to other methods of processing. Space requirements, problems in handling, and potential souring (during warm weather) have discouraged large-scale use.

RECONSTITUTION   Reconstitution is a process somewhat similar to soaking; it involves adding water to mature, dry grain to raise the moisture content to 25% to 30% and storage of the wet grain in an oxygen-limiting silo for 14 to 21 days prior to feeding. This procedure has worked well with sorghum and corn grains, resulting in improved gain and feed conversion by beef cattle fed high-concentrate rations when whole grain was used, but it does not work well if grain is ground prior to reconstitution. Some fermentation takes place during the holding period, obviously. A major disadvantage is that storage is needed for a considerable amount of feed and, if sorghum grain is used, it should be rolled before feeding.

HIGH-MOISTURE GRAIN   This term refers to grain harvested at a high moisture content (ca. 20% to 35%) and stored in a silo (or under plastic) to preserve the grain. Unless stored in such a manner or treated with chemicals, the grain will heat and mold if the weather is not cold. High-moisture grain may be ground

Steam-rolled corn   Coarsely cracked corn   Finely cracked corn

Popped, rolled barley   Steam-rolled barley   Ground Barley

Popped, rolled corn   Popped, rolled wheat   Popped, rolled milo

FIGURE 11-2 Cereal grains processed in different ways.

(a)

(b)

FIGURE 11-3 (a) One type of a roller mill used for processing grain. (b) The large corrugated rolls that physically crush the grain as shown. (Courtesy of Automatic Equipment Mfg. Co., Pender, NB.)

FIGURE 11-4 One example of feed-processing equipment that will roll grain, mix it with other feed ingredients, and deliver the final mix into a bin or delivery vehicle. (Courtesy of Automatic Equipment Mfg. Co., Pender, NB.)

or rolled before ensiling or before feeding. This is a particularly useful procedure when weather conditions do not allow normal drying in the field, and it obviates the need to dry the grain artificially with expensive fuel. Storage costs may be relatively high, but high-moisture grain produces good feedlot results, comparable to some of the better processing methods discussed. Feed conversion, particularly, is improved over that of dry grain. Wet grains are more difficult to dispose of on the market, of course, than are dry grains.

ACID PRESERVATION OF HIGH-MOISTURE GRAINS With higher fuel costs, increased interest has developed in recent years in eliminating artificial drying of newly harvested cereal grains. Data with barley or corn for pigs and research with corn or sorghum grains for beef cattle show promise with the use of acids to preserve high-moisture grains. Thorough mixing of 1% to 1.5% propionic acid, mixtures of acetic and propionic acids, or formic and propionic acids into high-moisture whole corn or other cereal grains retards molding and spoilage without affecting animal performance appreciably, compared with that obtained with dried grains.

HOT PROCESSING METHODS

Methods used in recent years for heat processing grains and other products (such as oilseed meals, pet food, and so on) include steam rolling and flaking, extruding, pelleting, popping, micronizing, and roasting. Pressure cooking and exploding are methods that have been tried at the feedlot level; however, the cost of equipment and maintenance problems have discouraged continued use. Information available to the author indicates that micronizing and roasting are not being used at the present time, and popping is on its way out. Comments on some of these methods follow.

STEAM-ROLLED AND STEAM-FLAKED GRAINS Steam rolling is a process that has been used for many years, partly to kill weed seeds (in the early days). The steaming is accomplished by passing steam up through a chamber that holds the grain above the roller mill (Fig. 11-6). Grains are subjected to steam for only a short time in the usual procedure (3 to 5 min) prior to rolling, usually just enough to soften the seed, but not long enough to modify the starch granules to any degree. Results indicate improvement in animal performance as

FIGURE 11-5 One example of a portable grinder (hammermill)–mixer that is available for on-the-farm use. (Courtesy of Ford New Holland, Inc., New Holland, PA.)

FIGURE 11-6 A high-capacity roller mill used in a large cattle feeder.

compared to dry rolling and steam does allow production of larger flakes and fewer fines, thus resulting in an improved physical texture (as compared to the result of dry rolling). This is an advantage when feeding very high levels of grains.

Steam-flaked grains are prepared in a similar manner except that the grain is subjected to high-moisture steam for a longer time (15 to 30 min), usually sufficient to raise the moisture content to 18% to 20% and the grain is then rolled between corrugated rolls to produce a rather flat flake. Feedlot data with cattle indicate that the best response is produced with thin flakes, which allow more efficient rupture of starch granules and produce a more desirable physical texture in the finished product. Corn, barley, and sorghum usually give a good response in terms of increased gain; although feed efficiency is improved with corn and sorghum, steam-flaked barley shows little improvement over steam-rolled barley.

PELLETING Pelleting is accomplished by grinding the feed and then forcing it through a thick, spinning die with the use of rollers, which compress the feed into the holes in the pellet die. Feedstuffs are usually, but not always, steamed to some extent prior to pelleting. Pellets can be made in different diameters, lengths, and hardnesses, and they have been available commercially for more than 50 years. All domestic animals generally like the physical nature of pellets, particularly as compared to meals, and a high percentage of poultry and swine feeds are pelleted. However, results with ruminants on high-grain diets have not been particularly favorable because of decreased feed intake, even though feed efficiency is usually improved over other methods. Pelleting the ration fines frequently is desirable because the fines are often refused otherwise. Pellets mix well with rolled grains in a complete or textured feed. Supplemental feeds such as protein concentrates are often pelleted so that they can then be fed on the ground or in windy areas with much less loss. An example of a pellet mill is shown in Fig. 11-7.

EXTRUDING   Extruded grains or other feeds are prepared by passing the feed through a machine with a spiral screw that forces the feed through a tapered head (Fig. 11-8). In the process the feed is ground, heated, and extended, producing a ribbonlike product. Results with cattle fed high-grain rations are similar to those with other processing methods. Some extruders are being used to process whole soybean seeds or other oilseeds for feeding to domestic animals. The heating is sufficient to get rid of most of the antinutritional factors found in soybeans or other types of field beans (see Ch. 8). Extruders are also used extensively for processing pet and human foods and for making fat-extracted oilseeds into oilseed meals.

POPPING, MICRONIZING, ROASTING Most readers are familiar with popped corn, which is produced by action of dry heat, causing a sudden expansion (caused by water going

FIGURE 11-8 The spiral screws that are the unique part of an extruder used for processing pet feeds, some fish foods, oilseed meals, and numerous human foods. (Courtesy of Wenger Mfg. Inc., Sabetha, KS.)

from a liquid to a gas state) that ruptures the endosperm of the grain. This process increases gut and rumen starch utilization, but results in a low-density feed. Consequently, the popped feed usually is rolled before feeding to reduce its bulkiness. Micronizing is essentially the same as popping except that heat is provided in the form of infrared energy. Roasting is accomplished by passing the grain through a flame, resulting in heating and some expansion of the grain, which produces a palatable product, but the process has not been adopted to any extent by livestock feeders.

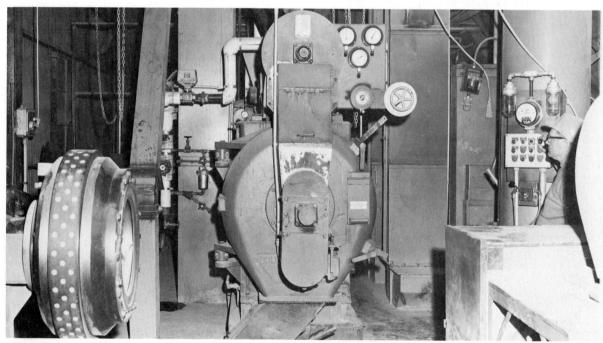

FIGURE 11-7 One type of pellet mill widely used for pelleting concentrates. A quick-change dye is shown at the left. Roughages can be pelleted, but production is reduced and costs are appreciably higher than for other feedstuffs.

ROUGHAGE PROCESSING

BALED ROUGHAGE

Baling is still one of the most common methods of handling roughage, particularly where it is apt to be sold or transported some distance. Baling has a considerable advantage over loose hay stacked in the field or roughage in other less dense forms insofar as transportation is concerned. Large bales are becoming more common. Although baled hay can now be handled mechanically for the most part, it still requires considerably more hand labor than many other feedstuffs. Furthermore, considerable waste may occur in feeding, depending on how it is fed (feed bunks, on the ground) and on the level of feeding. Heavily fed animals such as dairy cows may be quite selective so that coarse stems will not be consumed. Thus increased loss nearly always occurs in feeding baled hays.

CHOPPED AND GROUND ROUGHAGE

Chopping or grinding puts roughage in a physical form that can be handled readily by mechanical equipment (Figs. 11-9 and 11-10). Processing of this type also tends to provide a more uniform product and usually reduces feed refusal and waste. However, additional expense is incurred by grinding, and loss of dust may be appreciable from grinding with a hammer mill. The dust loss is sometimes reduced in commercial mills or feedlots by spraying fat or molasses on bales before they are ground. Ground hays are, as a rule, quite dusty and may not be consumed readily. Adding molasses, fat, or water or combining with wet feeds, such as silages, usually improves intake. Chopping pro-

FIGURE 11-10 An example of a tub grinder, a type that is very useful for grinding roughages coarsely.

duces a physical texture of a more desirable nature for ruminants or horses than does grinding, but chopped hay does not lend itself as well to incorporation into mixed feeds as does ground hay.

PELLETING

Pelleted roughages are usually consumed readily by ruminants, horses, and rabbits because the particle size and physical texture are of a desirable nature, provided very large pellets are not fed to small animals. Roughages such as long hay must be ground before pelleting, a slow, costly process compared to similar treatment of grains. Thus the cost of processing is a bigger item than for most other feed-processing methods. Pelleting usually gives the greatest relative increase in performance for low-quality roughages. This appears to result from an increase in density with more rapid passage through the GI tract and not to any great improvement in digestibility. Pelleted roughages are also digested somewhat differently by ruminants; as a result of more rapid passage out of the rumen, less cellulose is digested and relatively less acetic acid is produced, with relatively greater digestion in the intestines.

Pelleted high-quality roughages produce performance (gain in weight) in young cattle or lambs that is almost comparable to high-grain feeding. However, it should be noted that feeding finely ground, pelleted rations for long periods of time may be detrimental to the rumen. Animals fed in this manner will develop hyperkeratosis of the rumen papillae, which reduces absorption and performance. An example is

FIGURE 11-9 An example of a simple hay chopping machine.

FIGURE 11-11 The effect of baling (left), grinding in a hammermill (center), or grinding and pelleting (right) on the volume of alfalfa hay after processing. Each pile contains 5 lb of hay.

shown in Fig. 11-11 of the change in density achieved with grinding and pelleting of hay.

CUBED ROUGHAGES

Cubing is a process in which hay is forced through dies that produce a square product (about 3 cm in size) of varying lengths and hardnesses. Grinding before cubing is not required with field-cubed hays, but water is usually sprayed on dry hay as it is cubed. Field cubers have been developed and stationary cubers are also used to cube hay from stacks or bales, usually after grinding. Alfalfa hay produces good cubes that are less likely to break up than are cubes of grass hays.

Research data indicate that cubes produce satisfactory performance in cattle provided that the cubes are not too hard. Cubes are often used for dairy cattle.

DRIED, DEHYDRATED ROUGHAGES

Discussion of this topic will be very limited. However, it should be pointed out that a substantial amount of dehydrated alfalfa meal is produced in the United States (see Table 9-1), and some dehydrated alfalfa is produced in Europe. In northern Europe, where alfalfa does not thrive, grass forage is dried in limited volumes. Equipment such as that illustrated in Fig. 11-12 is used for such purposes.

Dehydrated alfalfa, Bermuda grass, or others are harvested at an early stage of growth (prebud for alfalfa) when the protein content is high and the fiber content relatively low. Such products also have a high content of carotene and xanthophylls. The cost is also relatively high; thus dehydrated forage is usually used in limited amounts in poultry or swine rations as a source of carotene, vitamins, or unidentified

growth factors. Only limited use is made of dehydrated forage for other species. A substantial portion of the dehydrated alfalfa produced in the United States is exported.

EFFECT OF PROCESSING ON NUTRITIVE VALUE

A high percentage of feedstuffs used for nongrazing animals is processed to some degree by use of heat and grinding and, to a lesser extent, by pelleting. These methods may result in substantial alteration of the nutritive value of feeds, and, because there is quite a volume of literature on the subject, particular attention will be focused on these methods.

HEATING

Heat may be used to dry plant or animal materials to a point that will allow storage without refrigeration, use of preservatives, or ensiling; to sterilize some products; to alter the chemistry of proteins or carbohydrates; or to detoxify some plant toxins.

With regard to proteins (see Ch. 8), it has been known for many years that excessive heating in the presence of sugars may result in browning (Maillard reaction). As a consequence of this reaction, lysine (and possibly some other amino acids) reacts with the sugars and, as a result, the lysine becomes partially unavailable to the animal. This reaction is mainly a problem with plant proteins or milk and milk byproducts (most other animal sources have little or no carbohydrate present).

Data generally indicate that heating of fish or animal proteins makes them less efficient in promoting growth. With plant sources, heating of the cereal grains to a moderate degree and for short periods of time may result in a slight improvement in protein utilization for ruminants but little if any improvement for nonruminants. However, with legume seeds, heating results in an improvement in protein quality, partly by degrading antiquality factors that are present (see Ch. 8). Heating of soybeans also increases the metabolizable energy value, and it will reduce solubility of the soy protein, resulting in less degradation of protein in the rumen.

With cereal grains, ample evidence exists to show that heat, especially in the presence of moisture, results in partial gelatinization of the starch. Especially for ruminants on high-grain diets, this results in more efficient utilization of the cereal grains, particularly if the heating

FIGURE II-12 One example of a commercial dryer used for drying and dehydrating feeds.

and processing results in mechanical rupture of the starch granules (1).

With regard to vitamins, any treatment that increases exposure to air (oxygen) or results in prolonged exposure to heat or light normally results in some deterioration of most of the vitamins. The fat-soluble vitamins and thiamin, pantothenic acid, folic acid, and biotin are particularly susceptible to destruction by heat. In studies with dehydrated alfalfa, it has been shown that high temperatures result in the conversion of *trans-β*-carotene to compounds with much less vitamin A activity. Xanthophylls, which are important as skin pigments for poultry, are also much reduced if temperatures are excessive. Additional information shows that natural plant antioxidants in alfalfa (in addition to vitamin E) are reduced with high-temperature drying (2).

With regard to minerals, there is some evidence that heat processing may slightly alter the availability of some trace minerals, probably as a result of changes in natural chelating agents in the original feedstuffs. With fats, it is well known that excessive heating results in production of acreolins, which are toxic to most animals. In most instances, excessive heating is also likely to be conducive to increased rancidity.

GRINDING AND PELLETING

Grinding results in a substantial reduction in particle size and exposure of much more surface area to the action of chemicals, as well as to the digestive juices. Digestibility is usually increased, but storage of ground grains or other feedstuffs results in destruction of those nutrients that are readily oxidized, such as unsaturated fats, especially in the presence of trace minerals such as iron and manganese.

Pelleting has been shown to be quite advantageous for swine and poultry in many instances (see later sections). With some feedstuffs the improved performance may be due to the increased density of the diet, thus allowing greater consumption. When unpalatable ingredients are a factor, pelleting tends to mask the flavor and promote greater consumption. In addition, the heat involved in pelleting is believed to be a factor in inactivating heat-labile toxins in feeds. With corn there may be some improvement in utilization of amino acids by swine. Various studies have shown that larger amounts of feedstuffs, such as dehydrated alfalfa, rye, wheat bran, wheat germ meal, rapeseed meal, or field peas, can be used in poultry rations when they are pelleted without any appreciable effect on performance (3).

Even regrinding of the pellets still allows the same response with feedstuffs such as wheat middlings, while the effect is less apparent with the cereal grains. Some of the improvement with wheat middlings from pelleting is believed to be due to improved protein availability and, possibly, enhanced availability of phosphorus, as well as the reduction in bulk. However, pelleting corn–soy diets appears to be of relatively little value for poultry. It may improve performance if protein is adequate, but it may depress production in some cases with marginal levels of protein or lysine, suggesting reduced availability of protein as a result of heat and pelleting. With some feeds the reduction in bulk may be less of a factor than chemical changes brought about by the heat of pelleting or rupturing of cells by mechanical means. Poultry can, apparently, eat reground pelleted mash more easily than the original mash (3).

FEED PROCESSING FOR NONRUMINANT SPECIES

A discussion of feed processing for different species is given in this section. In addition, some comparative research data have been cited. Additional information will be presented by most authors in Chs. 13 through 25 when discussing specific classes or species of domestic animals.

SWINE

With respect to cereal grains or roughages such as alfalfa hay, grinding and pelleting are the most common feed-processing methods used for swine. As with other animals, grinding usually increases digestibility; the improvement is probably greater with older pigs or hand-fed pigs, which masticate food less well than do young pigs or self-fed pigs.

Rather fine grinding (0.16-cm hammer mill screen) results in an improvement in feed conversion compared to coarse grinding (Table 11-1), but pigs fed finely ground rations are prone to develop stomach and esophageal ulcers. Thus most swine nutritionists recommend a medium degree of fineness. Rate of gain is not affected appreciably by fineness of grinding.

Pelleting generally results in a very slight improvement in gain and an improvement in feed efficiency with corn-based diets (Table 11-2). As a rule, feed intake may decrease slightly and feed wastage is normally reduced

TABLE 11-1

Effect of fineness of grinding on pig performance

	Hammer Mill Screen Size (cm)		
Item	0.16	1.27	2.54
Experiment I			
Average daily gain, kg	0.65	0.63	0.63
Feed conversion	3.19	3.56	3.67
Experiment II			
Average daily gain, kg	0.70	0.71	
Feed conversion	3.52	3.74	

Source: Pickett et al. (4).

as compared to meal diets. With the high-fiber grains (oats, barley) there is generally less improvement from pelleting than with other cereal grains. As with most animals, pigs usually show a preference for rations in pelleted form. Pelleting is of less value for older hogs, which do not need to produce at maximal rates.

Other feed preparatory methods have been used for swine but have not found much acceptance. With barley, soaking and treatment with amylase enzymes results in some improved efficiency, but the improvement usually does not justify the expense. Treating whole barley by spraying with an aqueous NaOH solution to avoid milling is not an effective method for swine (8). Steam flaking and reconstitution may improve digestibility for pigs, but they have little effect on gain. Feeding diets in liquid or paste form is also used (see Ch. 21), but not to the extent predicted a few years ago.

TABLE 11-2

Effect of pelleting corn-based diets on pig performance

Item	Daily Gain (kg)	Daily Feed (kg)	Feed Conversion
Growing pigs[a]			
Meal	0.725	2.02	2.78
Pellets	0.745	1.86	2.50
Advantage of pellets, %	+2.8	−7.9	+10.1
Finishing pigs[a]			
Meal	0.845	3.10	3.68
Pellets	0.905	3.00	3.32
Advantage of pellets, %	+7.1	−3.2	+9.8
Total period[b]			
Meal	0.69	2.52	3.71
Pellets	0.72	2.43	3.42
Advantage of pellets, %	+4.3	−3.6	+7.8

[a]Data from Becker (5) and Perry (6).
[b]Data from Baird (7).

Roughage processing for swine is very simple because alfalfa is generally the only roughage fed in North America. Except in instances where high levels might be fed to restrict caloric intake, only small amounts of dehydrated alfalfa or ground, sun-cured hay are fed, usually incorporated into the complete diet, which is often pelleted.

HORSES

Relatively little recent information is available on this topic as related to horses (9). The few reports available suggest that ground or rolled grains are probably utilized more efficiently by young animals and working horses than are whole grains. Rolled or cracked grains have the advantage of being less dusty and are favored in this respect.

With respect to roughages, limited research data show that horses eat more alfalfa hay when cubed or pelleted as compared to baled hay. Chopping appears to have no particular advantage. Grass hays and other products such as grass seed screenings are often fed to horses. Pellets, in general, are well accepted by horses, and this is a feasible way to feed some finely ground feedstuffs, which are not particularly palatable to horses. Limited data indicate that horses can be fed complete diets in pelleted form.

POULTRY

Poultry feeds are nearly all fed as a rather finely ground mash, pellets, or crumbles, a product that has been pelleted and then put through rollers to partially break up the pellets. In producing crumbles, the feed can be pelleted in a larger size than is used normally for poultry, resulting in a savings in the cost of pelleting as compared to the production of very small pellets.

It has been estimated that as much as 70% of poultry feeds is fed in pelleted or crumbilized form. Some of the advantages and disadvantages of feeding pelleted, steam-heated feed have been mentioned previously. In addition, with poultry, there tends to be an increased incidence of cannibalism and feather picking and an increased consumption of water when pelleted feeds are used. If wet litters are a problem in broiler production, pelleting would appear to worsen the condition.

Data illustrated in Tables 11-3, 11-4, and 11-5 show that broiler chicks or turkey poults gain more and have improved feed conversions when fed pelleted feed, crumbles, or reground crumbles as compared to performance of birds on mash diets. This type of improvement has resulted in a widespread use of pellets or crumbles for broilers and turkeys. Fortunately, the cost of pelleting rations of this type is low as compared

TABLE 11-3

Performance of broiler chicks fed mash, crumbles, and reground crumbles

Item	Mash	Crumbles	Reground Crumbles
Corn–soy diet			
Average weight, g	407	471	461
Feed/gain	2.06	1.90	1.96
Corn–soy + 30% wheat bran			
Average weight, g	388	417	413
Feed/gain	2.16	2.00	1.97

Source: Summers (10).

TABLE 11-4

Influence of amount of corn or barley in mash or pellet form on broiler performance

Grain Component (%)				
Corn	Barley	Form	Body Weight (lb)	Feed Conversion	Water Consumption (lb)
---	---	---	---	---	---
100	0	Mash	2.76	2.73	10.8
100	0	Pellets	2.92	2.56	12.0
50	50	Mash	2.70	2.88	11.5
50	50	Pellets	2.82	2.60	13.5
0	100	Mash	2.50	3.27	12.9
0	100	Pellets	2.77	2.82	15.8

Source: Arscott et al. (11).

TABLE 11-5

Influence of amount of corn or barley in mash or pellet form on turkey poults (0–8 Weeks)

Grain Component (%)		Form	Body Weight (lb)	Feed Conversion
Corn	Barley			
100	0	Mash	4.20	2.07
100	0	Pellets	4.40	2.02
50	50	Mash	3.94	2.14
50	50	Pellets	4.38	2.11
0	100	Mash	3.94	2.33
0	100	Pellets	4.31	2.24

Source: Harper (12).

to costs of pelleting high-fiber feedstuffs. Australian data (56) show that feeding crumbled pellets, as opposed to mash, resulted in an increased digestibility by chicks of 2.1% for a barley-based diet versus 4.5% for corn or wheat.

If the data in Table 11-6 are representative of most diets for laying hens, there is no great advantage in pelleting rations containing high levels of cereal grains. For layers the advantages are more likely in the use of higher levels of fibrous grains such as barley, less dense feedstuffs such as wheat bran or shorts, or other feedstuffs that are not particularly palatable for poultry.

With regard to roughage, the same comments apply as to swine because very little roughage is included in poultry diets. Small amounts of dehydrated alfalfa or ground, sun-cured alfalfa are used in most instances. Ground, good-quality dehydrated grass may be used in some areas.

GRAIN PROCESSING FOR BEEF CATTLE

FEEDLOT CATTLE

Processing of cereal grains for feedlot cattle is done primarily to improve efficiency of utilization by improved digestibility and/or greater

consumption because grain is already in a physical form that can be handled easily with mechanical equipment. Improvement in utilization can usually be obtained by a number of methods discussed previously, provided that the hull or waxy seed coat is broken up, thus allowing easier and more rapid access of rumen microorganisms and digestive juices to the interior of the kernel. Some physical disruption of the starch granules of the endosperm and, normally, partial gelatinization of the starch by heat treatment, soaking, or the like is also required for maximal efficiency. Heat treatment often improves protein utilization of cereal grains for ruminants.

Some processing methods may also provide a more favorable particle size, particularly for the smaller grains. Optimal particle size and density facilitate more timely passage through the rumen. In addition, particle size and freedom from dust are important factors affecting palatability.

Grain (or roughage) processing is expected to give greater returns per unit of cost when intake of grain is high. Usually, as the percentage of roughage in the diet increases, the physical nature of the concentrate becomes less important, although processing may still influence digestibility and efficiency. Animals on a main-

TABLE 11-6

Effect of pelleting corn or barley rations for White Leghorn layers

Gain Component	Form	Egg Production (%)	Feed Intake (lb)	Feed per Dozen Eggs (lb)	Body Weight Gain (lb)
Corn	Mash	67.3	68.1	4.00	0.65
	Pellets	70.8	74.2	4.15	0.82
Barley	Mash	70.8	74.1	4.14	0.48
	Pellets	69.6	77.2	4.38	0.75

Source: Arscott et al. (13).

tenance diet are not likely to return the added costs. In beef feedlot rations typically in use at this time in the United States, grain and other concentrates in finishing rations for cattle may account for 70% to 90% of total intake. With this type of ration the need and benefits of grain processing can be shown in most instances.

Data from about 50 different feeding trials with cattle fed corn or sorghum processed with different methods were summarized by Hale (14) on research done before 1980. Although the data were lumped together for many different experiments (not always an informative process), the data indicate that various methods (dry rolled, steam flaked, high moisture and ensiled, and reconstituted) had little effect on daily gain of cattle fed corn. It did appear that flaking resulted in some reduction in feed consumed and an improved efficiency as compared to the other methods. Whole corn was used as efficiently as rolled, but less efficiently than flaked, early harvested (high moisture), or reconstituted. With regard to sorghum, any of the methods resulted in improvement in daily gain as compared to dry-rolled grain, with some reduction in feed consumption and improved feed efficiency.

Data on processed grain are shown in Table 11-7, although it should be pointed out that not much work has been reported in the past decade compared to the previous one. It is rather surprising that very little information is available on steam-flaked barley in comparison with other methods. Available data indicate that steam flaking improves consumption and early gain with no appreciable effect on efficiency as compared to dry-rolled grain.

The example shown in Table 11-7 on wheat is fairly typical in that processing did not have any effect on daily gain of cattle fed wheat, but dry rolling resulted in efficiency as high as any of the other methods, and it was considerably better than feeding whole wheat. The one comparison of sorghum grain that was reconstituted shows that this method improved daily gain and efficiency, information typical of that reported earlier. Four examples are shown where corn was processed in different ways. In each case it is apparent that some type of processing improved the feeding value of whole corn both in gain per day and feed efficiency and with rations in which the grain content ranged from 50% to 80%+. The evidence suggests that almost any method of breaking up the kernel will result in some improvement. Most of the papers listed in Table 11-7 give data on starch or organic matter digestibility. In each case where it was given, starch digestion was improved by heat treatments or flaking of corn and, for sorghum, it was improved by using high-moisture or reconstituted grain.

OTHER FEEDING SITUATIONS WITH BEEF CATTLE

Varying amounts of grain may be fed to growing beef cattle (stockers, feeders), but only limited amounts are fed to adult cattle. In either case, grinding or dry rolling appear to be satisfactory methods, and there is little justification for resorting to more expensive processing (see Chs. 13 and 14). Some processing is normally justified when grain is fed, because cattle do not masticate most grains efficiently. In one experiment cattle excreted in feces 7%, 48%, and 40%, respectively, of the oats, barley, and wheat fed in rations containing about two-thirds grain; dry rolling increased grain digestibility from 77% to 81% for oats, 52% to 85% for barley, and 63% to 88% for wheat (21).

ROUGHAGE PROCESSING FOR BEEF CATTLE

Roughage is frequently chopped or coarsely ground to facilitate mixing with other ration ingredients. Roughage may also be processed in this manner to reduce wastage or selectivity. Other than information on chemical treatments or addition of ammonia, there has been very little material published on roughage processing in the past 10 years.

A limited amount of older data are shown in Table 11-8. These data generally show that chopping and grinding do not have much effect on daily gain or feed efficiency (unless waste is reduced). Pelleting or cubing will almost always result in increased gain and feed consumption along with improved efficiency, but feed efficiency values may be biased, as pointed out in the footnote in Table 11-8. Unless waste is reduced, normally, cubing or pelleting may not be enough more efficient to pay the added costs for growing cattle. Cubed or pelleted roughage is sometimes used as supplementary feed for wintering animals. When it is necessary to feed on the ground, wastage will be reduced considerably.

Pellet size is probably not as critical with cattle as with some other species. However, young animals may have difficulty in consuming cubes (usually at least 1 in.²), especially if they are quite hard. In one study with calves 7 to 14 weeks of age, there was little effect of size (9 to 18 mm in diameter) or density of grass

TABLE 11-7

Performance of feedlot cattle fed grains processed in different ways

Grain, Method	Feed Consumed per Day, Dry Basis	Average Daily Gain (kg)	Feed:Gain Ratio	Reference No.
Barley, dry rolled	9.44	1.31	7.22	14
Steam flaked	10.31	1.41	7.32	
Barley, dry rolled	8.96	1.67	5.41	57
Steam flaked	9.29	1.69	5.51	
Wheat (hard, red winter)[a]				15
Whole	11.53	1.35	8.50	
Dry rolled	9.72	1.32	7.34	
High moisture	9.81	1.32	7.39	
Steam flaked	9.90	1.33	7.45	
Extruded	9.99	1.30	7.71	
Corn vs. sorghum (80% grain)				16
Corn, dry rolled	8.87	1.48	5.99	
Sorghum				
Dry rolled	9.00	1.42	6.34	
High moisture, ensiled	9.87	1.42	6.95	
Reconstituted	9.00	1.52	5.92	
Sorghum, reconstituted (80% grain)				16
At 23.5% moisture	9.39	1.39	6.76	
At 31.5% moisture	8.81	1.37	6.43	
Corn (5% to 15% hay)				17
Whole	7.37	1.22	6.06	
Cracked	7.81	1.31	5.97	
Fine ground	7.87	1.33	6.00	
50:50 Whole cracked	7.54	1.33	5.75	
50:50 Whole, fine ground	7.82	1.33	5.95	
Corn (66.8% to 68.4% corn)				18
Whole	7.20	1.18	6.09	
75:25 Whole flaked	7.35	1.23	5.95	
50:50 Whole flaked	6.95	1.12	6.17	
25:75 Whole flaked	7.01	1.19	5.89	
Steam flaked	6.30	1.05	5.97	
Corn (67% of ration)				19
Whole, shelled	7.01	1.25	5.62	
Whole, steamed	7.59	1.31	5.79	
Steam flaked	6.71	1.33	5.06	
Corn (50.6% of ration)				20
Dry rolled	6.77	1.19	5.71	
Steam flaked	6.41	1.21	5.32	

[a]Data expressed on an air-dry basis.

pellets when calves could also consume hay. Consumption of high-density pellets was reduced, but the calves ate more hay to compensate (28).

FEED PROCESSING FOR DAIRY COWS

Feed processing may result in somewhat different responses in dairy cows than for growing or finishing cattle or lambs. Generally, feeding lactating cows high-grain rations (more than 60% grain), particularly heat-treated grains or all of their roughage in ground, pelleted, or cubed form, results in reduced rumen acetate production and lower milk fat percentages (Table 11-9).

Total milk fat production may not be decreased because milk production may be increased. However, when milk is sold on the basis of its fat percentage (a wasteful method), the use of heat-treated grains or pelleted roughages may need to be restricted.

As a rule it is generally accepted that processing of grain by grinding or rolling is about all that needs to be done for dairy cows, although pelleting the concentrate speeds up consumption considerably when cows are fed some portion of their concentrate in the milking parlor, particularly when eating time is limited. In one study (29), eating rate was in the order (highest to lowest) of pelleted, coarse cracked

TABLE 11-8

Effect of roughage processing on performance of growing-finishing cattle

Item	Gain (kg/d)	Dry Matter Intake (kg/d)	Feed Conversion[a]	Reference No.
Coastal Bermuda grass				
Long	0.33	4.67[b]	17.0[b]	22
Ground	0.50	6.07[b]	13.9	
Pelleted	0.66	6.53	11.3	
Meadow hay				
Chopped	0.18	4.49	26.8	23
Wafered	0.14	4.58	33.7	
Pelleted	0.32	5.58	17.3	
Alfalfa hay				
Baled	0.29	4.31	15.1	24
Chopped	0.28	4.22	15.1	
Pelleted	0.78	6.49	8.3	
Alfalfa hay				
Baled	0.67	6.14	9.1	25
Cubed	0.86	6.65	7.8	
Haylage	0.77	6.79	8.9	
Concentrate–wheat straw				
15% Straw, loose	1.05	10.4	9.9	26
15% Straw, pelleted	1.16	10.4	9.0	
30% Straw, loose	1.05	12.1	11.5	
30% Straw, pelleted	0.98	9.9	10.1	
Concentrate, alfalfa cubes				
Chopped and cubed hay	1.15	5.85C, 3.27B[b,c]	8.6[b]	27
Field-cubed hay	1.15	6.12C, 3.22B	8.8	
0.5%[d] Barley	1.14	8.35C, 1.72B	9.5	
1.0%[d] Barley	1.15	6.08C, 3.31B	8.8	
1.5%[d] Barley	1.16	3.49C, 4.72B	7.8	

[a]Feed conversion values for animals fed cubes or pellets with all roughage or high roughage tend to bias the results because rumen and/or gut fill will nearly always be greater in animals fed roughage compressed in such a manner. More realistic values on comparative weight gains would be obtained by adjusting live weights using the carcass weight divided by an average dressing percentage or by using empty body weight.

[b]As fed basis.

[c]C = daily consumption of cubes, B = daily consumption of barley which was fed at about 0.5%, 1%, or 1.5% of body weight per day.

[d]Cattle were also fed limited amounts of corn silage.

corn with pelleted premix, crumbled pellet, and concentrate in meal form.

Pelleting complete diets for dairy cows is generally advantageous only when the roughage level is high and when the quality of roughage is moderate to low. Cubing or pelleting roughage often results in somewhat greater consumption of feed, but milk production may or may not be improved. Some data on the effect of feed processing are shown in Table 11-9. Further comments will be found in Ch. 15.

FEED PROCESSING FOR SHEEP AND GOATS

It is not feasible to spend much money on processing feed for mature sheep or goats. The convenience and reduced waste related to use of chopped hay may make chopping feasible, but the cost of pelleting or cubing roughage is usu-

ally prohibitive. Pelleted feeds are sometimes used where ewe bands are fed on the range or in winter lambing quarters for short periods of time. The savings in feed plus the fact that feed bunks are not needed allow it to be done. When grains are fed, anything more than coarse grinding, cracking, or rolling is of doubtful benefit. Very limited studies with lactating dairy goats also show that ground or pelleted alfalfa did not result in any greater milk production than did feeding long hay.

In the feedlot, not much if any benefit is obtained by processing cereal grains (Table 11-10) for lambs. However, it may be feasible to pellet complete rations when they contain appreciable amounts of roughage (see Ch. 19). Lambs do very well on high-roughage pellets and they will, in fact, finish to a satisfactory degree (USDA grades) when fed nothing but alfalfa pellets. Maximal gains usually are

Ration, Process	Feed per Day (kg)	Milk per Day		Milk Fat (%)	Reference No.
		Total	4% FCM		
Processed concentrates					
Alfalfa, 40%, concentrate, 60%					
Whole corn	14.6	14.1	13.7	3.81	30
vs. corn meal	14.2	16.0	13.8	3.09	
Corn meal	13.4	15.2	14.5	3.68	
vs. cracked corn	13.3	14.3	14.1	3.90	
53% to 55% Concentrate (80% sorghum)					
Finely ground grain	20.0	28.4	25.6	3.42	31
Reconstituted grain	20.3	28.0	25.1	3.34	
Concentrates as					
Meal	13.2	17.8	15.0	2.98	32
Pelleted without steam	12.9	17.9	13.7	2.47	
Pelleted with steam	13.0	17.4	13.4	2.53	
Processed concentrates and roughage					
Ration form, % concentrate					
Milled, 60%	15.7	24.5	22.7	3.51	33
Cubed, 60%	17.2	21.5	17.3	2.72	
Cubed, 50%	19.4	23.8	20.0	2.90	
Cubed, 40%	19.2	22.6	19.4	3.06	
Alfalfa, high-barley concentrate					
Pelleted concentrate, baled hay	25.9	30.1	31.0	4.19	34
Cubed ration[a]	24.7	28.6	25.1	3.28	
Pelleted concentrate, cubed hay[b]	22.9	29.3	26.5	3.40	
Processed roughage					
Alfalfa hay fed as					
Baled	17.6	30.4	28.2	3.48	35
Baled + corn silage	17.2	27.3	24.4	3.32	
Cubes	20.3	32.7	30.7	3.60	
Cubes + corn silage	21.3	31.8	29.8	3.54	
Alfalfa hay fed as					
Baled		20.2	18.7	3.7	36
Ground		19.9	18.2	3.6	
Pelleted		19.1	16.1	3.1	

[a]Alfalfa hay was baled in the field and then ground and cubed in a stationary cuber.
[b]Alfalfa hay was field cubed and fed with pelleted concentrate.

achieved by using 15% to 25% grain along with alfalfa hay; feed conversion will be improved proportionately as the grain is increased.

When lambs are fed pelleted roughages (good-quality grass hay, legume hays), several research studies have shown clearly that consumption increases considerably as compared to when they are fed ground or chopped hay. In one study, lambs ate 40% more of a grass pellet than comparable chopped grass. When given a choice of chopped and pelleted grass hay, the lambs ate about 90% pellets (40). Pelleting (with sheep as well as other ruminants) of roughage results in more rapid passage out of the rumen and proportionately more digestion in the intestines. This may result in somewhat lower digestibility of energy, but it also results in less degradation of protein in the rumen and greater protein absorption (41). The overall re-

sult of feeding pelleted forage to lambs is that energy is utilized more efficiently when it is pelleted than when fed in other forms (long, ground, chopped) (42), but this situation is applicable (in economic terms) to young, fast-growing animals rather than to mature sheep, which do not need the higher level of energy intake. It should be pointed out that lambs fed pelleted or heat-processed feeds are apt to have softer fat than if fed otherwise. Presumably, this is a reflection of differences in volatile acids produced in the rumen as a result of processing.

HIGH-MOISTURE GRAIN

As pointed out in Ch. 7, grains must be dried to a moisture content of less than 15% to avoid spoilage in storage. In some areas it is not possible to get grains dry enough to store without

TABLE 11-10

Effect of feed processing on feedlot performance of lambs

Rations	Gain (g/d)	Feed Intake (kg/d)	Feed Conversion	Reference No.
Chopped alfalfa	136	1.41	10.31	37
Pelleted alfalfa	177	1.68	9.43	
Chopped alfalfa + 30% barley	141	1.27	9.16	
Pelleted alfalfa + 30% barley	163	1.45	8.97	
Alfalfa hay–corn (50:50)				
Long hay				
Ground corn	141	1.30	9.18	38
Pelleted corn	145	1.25	8.54	
Ground hay				
Ground corn	145	1.34	9.38	
Pelleted corn	154	1.20	7.79	
Pelleted hay				
Ground corn	177	1.42	8.02	
Pelleted corn	177	1.37	7.83	
Pelleted complete ration	186	1.43	7.88	
Pelleted complete, then reground	168	1.39	8.36	
Shelled corn	241	1.48	6.24	39
Extruded corn	241	1.41	5.91	

resorting to drying with a source of heat. In addition, harvesting grains at an earlier date may avoid losses to bad weather or, in the case of sorghum grains, large amounts of damage from birds in some areas.

Early-harvested corn or sorghum may be expected to have a moisture content of 25% to 30%. While corn or sorghum can be harvested without difficulty at this moisture content, it is then necessary to dry, ensile, or treat it with chemicals if the grains are stored. Early harvesting has other advantages in that it requires less energy to grind or roll grains and they are, generally, quite palatable to livestock. The protein and starch fractions are more soluble and more completely digested or are digested at a more rapid rate by ruminant animals. A major disadvantage is that high-moisture grains cannot be sold and transported any distance without spoilage, and large amounts of storage facilities are required if a season's supply of grain is laid in at harvest time.

In some sorghum-growing areas, producers harvest the whole sorghum head, chop it, and ensile it. While there is essentially no information in the scientific literature on the product (sorghum head silage), information in field day reports and trade magazines indicates that it is a useful method of harvesting and storing sorghum grains.

Examples of research data using high-moisture grains for feeding beef cattle are shown in Tables 11-7 and 11-11 and for dairy cattle in Table 11-11. In general, the data indicate that high-moisture grains (sometimes called early-harvested grains) are not likely to increase the rate of gain of beef cattle, but efficiency generally is improved *provided* that the grains are ground or rolled prior to storage or before feeding. Results with dairy cattle indicate that high-moisture barley and corn were about equivalent to the dry-rolled grains (on an equal DM basis).

Although use of high-moisture corn or sorghum grains may be somewhat more efficient in some cases, it must be remembered that any fermentation will result in some energy losses, as well as some change in protein solubility; it may enhance the B vitamin content. One example of losses associated with ensiling corn grain is shown in Table 11-12. Note that the dry-matter loss amounted to about 4%, with an additional loss of 3% of the grain that was considered to be spoiled and not usable as feed. In another example (50), it was calculated that 7.5% to 13.6% of dry matter from high-moisture corn was lost during storage. In most experiments, only the weight of the feed going onto the scales is used to calculate consumption and efficiency; thus, if losses even approached these levels, the data would not look as good for the high-moisture feeds. The amount of fermentation, leaching, and spoilage loss would vary considerably with the type of silo (or plastic bag) used, its size, location, and so on. Whether these values are typical is not known, but losses are undoubtedly more than most feeders realize. One example

TABLE 11-11

Utilization of high-moisture (HM) or chemically treated grains for feeding ruminants

Item	Feed Consumed per Day (kg DM)	Daily Gain (kg)	Feed Conversion Poststorage DM	Feed Conversion Prestorage DM	Reference No.
Beef finishing trials					
Corn, corn silage, supplement[a,b]					
Dried corn	6.73	1.15	6.39	5.83	43
vs. acid-treated HM corn	6.72	1.16	5.97	5.77	
Dried corn	6.56	1.15	6.17	5.70	
vs. HM corn treated with ammonium isobutyrate	6.24	1.17	5.50	5.33	
HM corn, high corn silage					
Ensiled, ground corn[c]	6.36	1.08		5.93	44
Urea-treated whole corn	6.24	1.02		6.14	
Corn, corn silage, supplement					
Whole shelled	9.62	1.07		9.0	45
Whole HM	9.03	1.00		9.1	
Rolled HM	9.08	1.14		8.0	
NaOH treated[d]					
Whole shelled	9.40	0.95		9.9	
Whole HM	9.40	0.93		10.1	
Corn, supplemental					
Whole, shelled	6.8	1.41		4.8	46
HM corn					
Whole, acid treated	6.7	1.38		4.8	
Rolled, acid treated	6.4	1.33		4.7	
HM corn:dry-rolled grain sorghum (80% grain)					
100:0	10.65	1.64		6.49	47
75:25	10.61	1.66		6.39	
50:50	10.84	1.66		6.53	
0:100	11.07	1.58		7.01	

Item	DM Consumed (kg/d) Grain	DM Consumed (kg/d) Total	Milk Production kg/d	Milk Production FCM/d	Butterfat (%)	Reference No.
Dairy feeding trials						
Barley trials						
Dry barley	11.3	18.7	23.2	20.8	3.37	48
Dry barley, acid treated	11.2	18.7	22.8	21.1	3.53	
HM barley						
Acid treated	11.1	18.0	21.8	19.8	3.49	49
Ensiled	10.9	17.7	22.1	20.4	3.53	
Corn, corn silage						
Dried corn	9.8	17.1	25.2	19.6	2.64	49
HM corn, ground, ensiled	9.3	17.2	24.1	20.6	3.09	
Alfalfa, concentrate, 70% to 30%±						
Rolled rye grain		19.4	25.0	22.2	3.29	50
NaOH-treated rye		20.9	26.0	23.6	3.38	
Corn, corn silage, alfalfa hay						
Dry-rolled corn		24.0	22.6	21.8	3.76	51
vs. HM corn, rolled, ensiled		23.4	22.2	21.3	3.74	
Dry-rolled corn		19.9	27.4	26.0	3.67	
vs. HM corn, rolled +1% acid		18.9	27.8	26.1	3.59	
Dry-rolled corn		21.0	23.2	22.3	3.75	
vs. HM ear corn, ground + acid		21.0	23.1	22.1	3.75	

[a]These data represent six trials on acid-treated corn and four on AIB-treated corn.
[b]Values on feed consumption are expressed as grain DM consumed per day.
[c]When fed, urea was added to equalize that added to the other treatment (3.7%).
[d]NaOH added equivalent to 3% of DM of corn.

TABLE 11-12

Storage loss from ensiling high-moisture corn

Nutrient	Recovery after Ensiling (%)[a]
Dry matter	96.0 ± 1.9
Crude protein	97.2 ± 2.1
Crude fiber	95.5 ± 4.1
Ether extract	100.5 ± 2.6
N-free extract	95.9 ± 1.9
Ash	82.4 ± 2.0

Source: Chandler et al. (49).

[a]In addition to these losses, 3% of the ensiled corn was classified as spoiled grain.

of a large trench silo used on a commercial beef feedlot is shown in Fig. 11-13.

In addition to ensiling, grains may be preserved by various chemicals. Propionic acid or mixtures of propionic acid with formaldehyde or other similar compounds, ammonium isobutyrate, or the like have proved to be effective means of preventing mold and/or bacterial growth. Data in Table 11-10 show that acid-treated grains produce performance in beef or dairy cattle that is comparable to other methods of preservation. There has also been quite a bit of interest (particularly in Europe) in using NaOH or urea solutions on wet grains. Urea used at a level of about 1% of the grain has proved effective in inhibiting bacterial and fungal growth in grains with enough moisture to cause spoilage. Urea also has the advantage of providing a source of N that rumen microorganisms can utilize. NaOH also inhibits spoilage of high-moisture feed grains, and several reports show that it increases digestibility of whole grains fed to cattle and, at the same time, increases somewhat the digestibility of the fibrous components of the diet (presumably because of higher rumen pH). However, other reports indicate that increases in digestibility were no greater than if the grains were cracked or rolled and that the cost of mechanical processing was less. Two examples of using NaOH-treated grains are shown in Table 11-11. With beef cattle, feeding NaOH-treated grains did

FIGURE 11-13 Loading out ensiled high-moisture corn. Many feedlots are making extensive use of high-moisture grains. In the photo, the corn is covered with a layer of silage, which reduces moisture loss and spoilage.

not result in improved performance, while feeding NaOH-treated rye resulted in some apparent increase in milk production of dairy cows.

A limited number of feeding trials have been done using high-moisture or reconstituted corn or sorghum for feeding swine. Results indicate a slight improvement in the use of high-tannin sorghums, but otherwise not much effect.

ECONOMICS OF FEED PROCESSING AND GRAIN PRESERVATION

The relatively higher energy costs experienced in the past 15 years or so have increased interest in minimizing feed processing costs or in obtaining the maximum return per unit of cost for processing or preservation. Current fuel costs are higher in proportion to grain costs than when most of the various methods discussed previously were developed. Costs in megacalories of fossil fuel (oil, coal, gas) per ton of processed feed have been estimated and are shown in Table 11-13. It is obvious that grinding is the cheapest method available for minimal processing when measuring fuel costs. Note that the greatest use of fuel is required to dry very wet materials such as green forage (alfalfa) or raw beet pulp.

Data presented in Table 11-14 confirm that any grain processing more elaborate than grinding or rolling normally does not result in any marked increase in daily gain of cattle and that the primary benefit may be some improvement in feed efficiency. Of course, good processing can stimulate feed consumption just as poor processing may depress feed consumption, and many reports show appreciable differences in gain (see earlier tables). Wastage

TABLE 11-13

Fossil fuel energy use for some feed-processing methods

Process	Fuel Energy Used per Ton (Mcal)
Drying corn	240
Adding propionic acid	200
Grinding corn	40
Pelleting corn	80
Steamflaking corn	120
Dehydrating and pelleting alfalfa	1,800
Dried beet pulp	1,750

Source: Ward (52).

may be reduced by good processing and feed bunk management.

The relative values of different grain-processing methods for feedlot cattle have been calculated by Schake and Bull (53) based on daily gains and efficiencies shown in Table 11-14, which was compiled from a number of experimental trials done in Texas. Flaking, reconstitution, or high-moisture storage all resulted in some improvement in feed efficiency when compared to feeding dry-ground or rolled grain. However, because of higher costs for equipment, power, storage (high-moisture grain requires that all grain be purchased at one time), and the like, the most efficient methods did not necessarily result in the greatest net value per ton of grain. Under their specific conditions, reconstitution or steam flaking had the greatest net values.

EFFECT OF PROCESSING ON DENSITY

Frequent references have been made to the effect processing methods have on the density of feedstuffs. Density, which is a measure of weight per unit of volume, is sometimes referred to as bulk density in contrast to caloric density, which is a measure of usable energy per unit of weight. In view of the importance of density, it is desirable to present some data and a few comments. Values on selected feedstuffs are shown in Table 11-15.

In feedstuffs such as unprocessed roughage and grains, density is inversely related to fiber content. Hays, which have higher fiber content than most other feedstuffs, have a much lower density than the grains and other concentrates. This is partly because of the physical nature of the hay and not entirely because of the fiber content. Grinding will always increase the density of hay and usually that of other roughages. Pelleting and cubing result in a marked increase in the density of hay, although there should be little if any change in fiber content.

With grains, grinding is more apt to result in a decrease in density. Pelleting and cubing grains increases density, but not nearly as much as they do with roughage. Other processing methods such as flaking, popping, and roasting result in reduced density of feed grains.

Some increase in density of feedstuffs, particularly roughages, is advantageous in transportation, milling, and handling during feeding operations. Nutritional benefits are less clear

TABLE 11-14

Daily gain and feed conversion values of cattle fed corn or sorghum processed in different ways

Method	Corn			Sorghum		
	Gain (lb)	Feed:Gain lb	Feed:Gain %[a]	Gain (lb)	Feed:Gain lb	Feed:Gain %[a]
Dry ground or rolled	2.62	5.79	—	2.50	6.09	—
Steam flaked	2.68	5.36	7.46	2.65	5.60	8.11
Reconstituted	2.61	5.33	7.97	2.54	5.34	12.32
High moisture						
Ground, ensiled	2.57	5.14	11.14	2.64	5.52	9.35
Acid treated	2.68	5.56	3.97	2.45	5.60	8.11

Source: Schake and Bull (53).

[a]Percent improvement in feed conversion of grain dry matter compared to dry processing.

except for the marked increase in consumption of pelleted and cubed roughages by ruminants. With processing methods that result in a marked expansion and reduced density of grains (such as popping), data on feedlot cattle indicate that consumption is reduced unless these processed grains are rolled (and made more dense) before feeding. However, digestibility appears to be altered very little, if any.

Caloric density is used extensively in ration formulation. There are appropriate values beyond which feed intake may either decrease (high values) or energy intake decrease (low values). However, the relationships between caloric density and bulk density have not been well defined. For example, the digestibility of pelleted alfalfa may actually be less than that for ground or long hay, but the intake by ruminants is nearly always considerably higher when the hay is pelleted, and the gain increases considerably.

When dealing with complete rations, it seems obvious that there is probably an optimum density range for a given feeding situation. It seems reasonable to this author that density probably could be substituted for maximum and/or minimum fiber specifications when formulating least-cost rations with computerized methods (see Ch. 12).

OXIDIZED FEED

A chapter on feed processing is probably a good spot to mention the subject of oxidized feed, because processed feed is more subject to oxidation and deterioration than grains (or roughage) that have not been ground, rolled, chopped, pelleted, or otherwise disturbed. Feeds that have been ground or otherwise broken up have more unprotected surfaces exposed to oxygen or other chemicals, which may cause problems. Feeds, particularly complete feeds, with added fats or metal ions are more subject to oxidation than a ground grain or hay might be. Thus the sooner a mixed feed is used, the better off the feeder is likely to be, because the feed will not improve with age.

In the oxidation process, free radicals formed by unsaturated fatty acids react with oxygen to form peroxides that serve as the entry point into a multitude of reactions producing numerous by-products, which may result in decomposition of the feed. Aldehydes, ketones, acids, esters, and polymerized fats are direct products of the oxidation process, and they result in reduced energy values, off-tastes, and off-odors. Peroxides can initiate the oxidation and destruction of the fat-soluble vitamins or other susceptible feed constituents such as xanthophylls or aroma and flavor constituents. In situations where animals may be forced to consume oxidized feed for some period of time, they may develop such problems as steatitis in swine and cats, exudative diathesis, muscular dystrophy, and necrotic tissues in various species (sheep, swine, cattle, poultry) or poor fertility and reduced hatchability in poultry. Less severe problems, which may not be readily detectable, includes such things as reduced rate of gain, lower body weight, and poor feed efficiency, which have been observed in studies with broilers and growing pigs, or inadequate egg yolk pigmentation in hens' eggs because of oxidation of xanthophylls in the feed (55).

Normally, feeding fats have antioxidants added to them (see Ch. 7). Other feed ingredients such as dehydrated alfalfa may also have added antioxidants, as may some complete

TABLE 11-15

Density of some selected feedstuffs

Feedstuff	lb/cu ft	g/ℓ[a]	Feedstuff	lb/cu ft	g/ℓ[a]
Alfalfa			**Cereal by-products** (continued)		
Long hay	2-3	40	Rice grits	42-45	695
Chopped hay	4-5	70	Rye bran	15-20	280
Ground hay	8-10	145	Rye middlings	42	670
Dehydrated meal	18-22	320	Rye shorts	32-33	520
Pelleted hay	41-43	670	Wheat bran	11-16	215
			Wheat midds	18-25	345
Other roughage			Wheat red dog	22-28	400
Bagasse	7-10	135	Wheat flour	32-42	590
Corn cobs, ground	17	270			
Cottonseed hulls	12	190	**Protein supplements**		
Oat hulls, ground	11-12	185	Blood flour	30	480
Soybean hulls, ground	20	320	Blood meal	39	625
			Corn gluten feed	26-33	470
Cereal grains			Corn gluten meal	32-43	600
Barley, whole	38-43	650	Cottonseed meal	37-40	615
Barley, ground	24-26	400	Feather meal	34	645
Barley, rolled	21-24	360	Fish meal	30-34	510
Corn, whole, shelled	45	720	Flax seed	33-45	625
Coarsely cracked	35	560	Meat meal	37	590
Finely cracked	37-38	600	Shrimp meal	25	400
Roasted	39	625	Soybeans, whole	46-48	750
Steam flaked	34	545	Soybeans, ground	25-34	470
Popped, rolled	19	305	Soybean meal, expeller	36-40	610
Corn and cob meal	36	575	Soybean meal, solvent 44%	35-38	585
Kafir	40-46	690	Soybean meal, solvent 50%	41-42	665
Millet, whole	38-40	625	Sunflower seed	26-38	510
Milo, whole	40-45	680	Tankage	49	785
Finely ground	37	590			
Dry rolled	31	495	**Miscellaneous**		
Steam flaked	18	290	Beans, dry, lima	45	720
Whole, reconstituted	40	640	Beet pulp, dried	11-16	215
Popped, rolled	16	255	Brewers dried grains	14-15	230
Oats, whole	25-35	480	Buttermilk, condensed	31	495
Rolled	19-24	345	Citrus pulp, dried	20-21	330
Crimped	19-25	350	Corn distillers, dried grains	18-19	295
Ground	20-25	360	Corn distillers, dried solubles	25-26	410
Rice, rough	32-36	545	Malt sprouts	13-16	230
Polished	30	480	Molasses, 79.5° Brix	88	1,410
Rye, whole	43-45	705	Sugar	54-56	880
Ground	39	625	Peas, dried	48-50	785
Wheat, whole	45-52	775	Tallow, melted	54	865
Ground	38-39	615	Urea	34-42	610
Coarse cracked	35-38	585	Whey, dried	35-46	650
Finely cracked	38-41	630	Yeast, dried	41	655
Steam flaked	31-33	510			
			Mineral supplements		
Cereal by-products			Bentonite	51	815
Corn bran	13	210	Bone meal	50-53	830
Corn hominy feed	25-28	425	Ca carbonate	75	1,200
Grain screen, light	22-24	370	Limestone	68	1,090
Oat groats	46-47	745	Oyster shell, ground	53	850
Oat middlings	38	610	Salt, coarse	62-70	1,055
Rice bran	20-21	345	Salt, fine	70-80	1,200

Source: Most values taken from Anonymous (54); other values are from various research papers and the author's unpublished data.

[a]Values given as grams per liter; where a range in weight is given in pounds per cubic foot, the mean is used to calculate density in grams per liter, and values have been rounded off to the nearest 5 units.

feeds that the manufacturer anticipates might not be used immediately. Antioxidants often used in the feed trade include ethoxyquin and butylated hydroxy tolalene.

SUMMARY

Feed processing is an important aspect of animal production for confined animals, especially those expected to produce at high levels. Various methods are available for processing feed grains. At the present time, grinding and pelleting are used extensively for preparing feeds for broilers, turkeys, growing–finishing swine, rabbits, and to a lesser extent other classes or species of animals. Dairy cows may be fed some pelleted concentrates while being milked, some portion of the ration may be pelleted for finishing beef cattle, and protein supplements may be pelleted or cubed when fed to range cattle. Only minimal amounts of pelleted feeds are normally fed to other animals.

Grinding or rolling of grains are satisfactory methods for lactating dairy cows or beef cattle fed moderate to low levels of grain. Likewise, such methods are satisfactory for adult swine or horses. Although whole corn is used with reasonable efficiency by finishing beef cattle, other grains should be processed in some manner to improve efficiency of use. Utilization of reconstituted corn or sorghum or high-moisture (early-harvested) grains usually results in improved efficiency when fed to finishing beef cattle, but the economics of using a particular process depends on relative costs of fuel and feed and on the size of the operation. High-moisture feeds are not, generally, used much by large beef feedlots.

Hay processing is less complex. Grinding or chopping usually reduces wastage and selectivity, but may not increase consumption. Feeding large amounts of finely ground, pelleted, or cubed roughage may be expected to result in reduced butterfat percentage of milk from lactating dairy cows. Pelleting or cubing of roughage increases consumption, particularly of lower-quality roughage, but it is a relatively expensive process because of the power and time requirements for processing.

REFERENCES

1. Theurer, C. B. 1987. *J. Animal Sci.* 63:1649.
2. Kohler, G. O., A. L. Livingston, and R. M. Saunders. 1973. In: *Effect of processing on the nutritional value of feeds.* Washington, DC: Nat. Acad. Sci.
3. Slinger, S. L. 1973. In: *Effect of processing on the nutritional value of feeds.* Washington, DC: Nat. Acad. Sci.
4. Pickett, R. A., et al. 1969. *J. Animal Sci.* 28:837.
5. Becker, D. E. 1966. *Merck Agr. Memo.* Vol. 11, No. 1. Rahway, NJ: Merck & Co.
6. Perry, T. W. 1973. In: *Effect of processing on the nutritional value of feeds.* Washington, DC: Nat. Acad. Sci.
7. Baird, D. M. 1973. *J. Animal Sci.* 36:516.
8. Patterson, D. C. 1984. *Animal Prod.* 38:271.
9. Ott, E. A. 1973. In: *Effect of processing on the nutritional value of feeds.* Washington, DC: Nat. Acad. Sci.
10. Summers, J. C. 1974. *Can. Poult. Rev.* 93:44.
11. Arscott, G. H., et al. 1958. *Poultry Sci.* 37:117.
12. Harper, J. A. 1959. *Proc. 17th Oregon Animal Ind. Conf.* Corvallis, OR: Oregon State Univ.
13. Arscott, G. H., et al. 1962. *Oregon Agr. Expt. Sta. Tech. Bul.* 64.
14. Hale, W. H. 1980. In: *Digestive physiology and nutrition of ruminants. Vol. 3: Practical nutrition.* 2d ed. Corvallis, OR: O & B Books.
15. Prasad, D. A., et al. 1975. *J. Animal Sci.* 41:578.
16. Stock, R. A., et al. 1987. *J. Animal Sci.* 65:548.
17. Turgeon, O. A., Jr., D. R. Brink, and R. A. Britton. 1983. *J. Animal Sci.* 57:739.
18. Lee, R. W., M. L. Galyean, and G. P. Lofgreen. 1982. *J. Animal Sci.* 55:475.
19. Ramirez, R. G., et al. 1985. *J. Animal Sci.* 61:1.
20. Zinn, R. A. 1987. *J. Animal Sci.* 65:256.
21. Toland, P. C. 1976. *Australian J. Expt. Agr. & Animal Husb.* 16:71.
22. Cullison, A. E. 1961. *J. Animal Sci.* 20:478.
23. Wallace, J. D., R. J. Raleigh, and W. A. Sawyer. 1961. *J. Animal Sci.* 20:778.
24. Webb, R. J., and C. F. Cmarik. 1957. *Univ. of Illinois Rpt. 15-40-329,* Dixon Springs Sta., IL.
25. Kercher, C. J., W. Smith, and L. Paules. 1971. *Proc. West. Sec. Amer. Soc. Animal Sci.* 22:33.
26. Levy, D., et al. 1972. *Animal Prod.* 15:157.
27. Kercher, C. J., et al. 1978. *Proc. West. Sec. Amer. Soc. Animal Sci.* 29:410.
28. Tetlow, R. M., and R. J. Wilkins. 1978. *Animal Prod.* 27:293.
29. Kertz, A. F., B. L. Darcy, and I. R. Prewitt. 1981. *J. Dairy Sci.* 64:2388.
30. Moe, P. W., J. F. Tyrrell, and N. W. Hooven. 1973. *J. Dairy Sci.* 56:1298.
31. Bush, L. J., et al. 1979. *J. Dairy Sci.* 62:1094.
32. Carpenter, J. R., R. W. Stanley, and K. Morita. 1972. *J. Dairy Sci.* 55:1750.

33. Dobie, J. B., et al. 1974. *Feedstuffs* 46(6):30.

34. Murdock, R. F., and A. S. Hodgson. 1977. *J. Dairy Sci.* 60:1921.

35. Anderson, M. J., et al. 1975. *J. Dairy Sci.* 58:72.

36. NRC. 1973. *Effect of processing on the nutritional value of feeds.* Washington, DC: Nat. Acad. Sci.

37. Weir, W. C., et al. 1959. *J. Animal Sci.* 18:805.

38. Fontenot, J. P., and H. A. Hopkins. 1965. *J. Animal Sci.* 24:62.

39. Jordan, R. M. 1965. *J. Animal Sci.* 24:754.

40. Greenhalgh, J. F. D., and G. W. Reid. 1974. *Animal Prod.* 19:77.

41. Beever, D. E., et al. 1981. *Brit. J. Nutr.* 46:357.

42. Thomson, D. J., and S. B. Cammell. 1979. *Brit. J. Nutr.* 41:297.

43. Ware, D. R., H. L. Self, and M. P. Hoffman. 1977. *J. Animal Sci.* 44:722.

44. Mowat, D. N., P. McCaughey, and G. K. Macleod. 1981. *Can. J. Animal Sci.* 61:703.

45. Anderson, G. D., L. L. Berger, and G. C. Fahey, Jr. 1981. *J. Animal Sci.* 52:144.

46. Macleod, G. K., D. N. Mowat, and R. A. Curtis. 1976. *Can. J. Animal Sci.* 56:43.

47. Stock, R. A., et al. 1987. *J. Animal Sci.* 65:290.

48. Ingalls, J. R., K. W. Clark, and H. R. Sharma. 1974. *Can. J. Animal Sci.* 54:205.

49. Chandler, P. T., C. N. Miller, and E. Jahn. 1975. *J. Dairy Sci.* 58:682.

50. Sharma, H. R., J. R. Ingalls, and J. A. McKirdy. 1983. *Animal Feed Sci. Tech.* 10:77.

51. Voelker, H. H., et al. 1985. *J. Dairy Sci.* 68:2602.

52. Ward, G. W. 1981. Unpublished data. Ft. Collins, CO: Colorado State Univ.

53. Schake, L. M., and K. L. Bull. 1980. *Texas Agr. Expt. Sta. Tech. Rpt.* 81-1.

54. Anon. 1982. *Feedstuffs* 54(30):136.

55. Calabotta, D. F., and W. D. Shermer. 1985. *Feedstuffs* 57(48):24; Arnold, R. L. 1985. *Feedstuffs* 57(42):30.

56. Farrell, D. J., E. Thomson, and A. Choice. 1983. *Animal Feed Sci. Tech.* 9:99.

57. Grimson, R. E., et al. 1987. *Can. J. Animal Sci.* 67:43.

12

RATION FORMULATION

Hugo Varela-Alvarez and D. C. Church

INTRODUCTION

Ration (or diet) formulation is a topic that should be mastered to some degree by anyone concerned with feeding livestock. For a nutritionist or feed formulator, it is an absolute necessity. The previous statement may require some qualification with the increasing availability of personal and office computers that have the necessary programs to do all the manipulations without the operator needing to know much about the mathematical problems involved.

In ration formulation the objective is to utilize knowledge about nutrients, feedstuffs, and animals in the development of nutritionally adequate rations that will be eaten in sufficient amounts to provide the level of production desired at a reasonable cost. Obviously, a blend of knowledge is required for optimum results. Fortunately, the mathematics required are not complex, and the reader will be led through a series of processes that will enable him or her to formulate rather complex rations by the time this chapter is completed. It should be emphasized that it is necessary to master each step as presented.

INFORMATION NEEDED

Before any mathematical manipulations can begin, several different types of information are needed for an organized approach to ration formulation for any given situation. These are discussed in the following sections.

NUTRIENT REQUIREMENTS OF THE ANIMAL

To develop a satisfactory diet, we must have some knowledge of the limiting nutrients needed by an animal in a specific situation and of the amounts of nutrients required in the diet. In the United States, the National Research Council (NRC) publications are the generally accepted standards used for domestic animals (and for others, as well). These publications represent the opinions of different committees that have organized the publications. The NRC publications provide a generally satisfactory starting point, and tables delineating these requirements are given in the Appendix. Similar feeding standards have been developed in other countries.

Assuming that the NRC standards are to be used as is, the first step in formulating a ration is to look up and tabulate the nutrient levels needed in the particular situation. Depending on the situation, it is not usually necessary to be concerned with all known nutrients, because some are always more critical than others; that is, some nutrients are more likely to be deficient in the usual feed ingredients than are other nutrients. For example, if it were desired to formulate a protein supplement for cows on the range, protein would be the first consideration, with other considerations being the energy content and one or two of the macrominerals, such as P and Ca. Most other nutrients would probably be disregarded, with the possible exception of vitamin A. This can be done because past experience indicates that other vitamins and minerals are not likely to be as limiting as those mentioned; in addition, the animals in question will be consuming an appreciable percentage of their total diet from roughage sources. However, if a broiler ration is to be formulated where the chick has access only to the feed the ration provides, then it is necessary to be concerned with a much more detailed list of nutrients, because all required nutrients must be contained in the diet formulated and because the nutrient requirements of the chick are more demanding (numerous vitamins, amino acids, some minerals) than those of the range cow. In any case, the various nutrients of concern should be tabulated on a form such as that shown in Fig. 12-1.

FEEDSTUFFS

The next step is to list available feedstuffs that are suitable for the particular animal in question. In many cases only a limited number of locally grown feeds are available at competitive prices. Although a long list of by-product and supplementary feeds might be contrived, this is usually unnecessary when we are concerned with ration formulation at the farm or ranch level and are primarily concerned with utilization of homegrown products. Commercial feed mills would normally have a much wider variety of available feedstuffs. If a relatively long list is required, use of a form similar to Fig. 12-1 will facilitate summarization of the needed data.

The contribution of critical nutrients by each feed should be listed. Analytical data on roughages and grains are preferred where they are available from a local laboratory on the actual feedstuffs to be used. If no information of this type is available, the average composition data can be used from NRC tables or from other sources.

The list of feeds it is necessary to consider can be reduced greatly by careful consideration. Is this feed suitable and economical for its intended use? For each species and class of animal, some feeds are more useful than others. For adult ruminants, we would usually want to include urea where protein is of concern; but on the basis of present regulations, urea is not permitted in rations for any of the monogastric species or for milk replacer formulas. Meat and fish meals, on the other hand, may be favored feeds for poultry and pigs, but they are usually too high in price for ruminant feeds. Remember, also, that a given feed may have different feeding values for different animal species.

Furthermore, we must consider whether the feed should be processed and, if so, in what manner and at what cost. What effect will the processing have on animal production? Is it a palatable feed or will the mixture be palatable? Does it present problems in handling, mixing, or storage? Are feed additives required and, if so, what additives and at what concentrations?

Obviously, it takes a fund of knowledge and experience to answer these questions. The ability to answer them correctly is a must for a practicing nutritionist. For our purposes here, the beginner can ignore such questions as posed in the last paragraph, as time and experience will help provide the answers that cannot be covered in this chapter. Other chapters on the different animal species will illustrate and answer some of these questions.

TYPE OF RATION

The type of ration has a great deal to do with its needed composition and nutrient content. That is, is it a complete feed, is it a finishing grain mix to feed along with roughage, or is it a supplemental feed formulated primarily for its protein, vitamin, or mineral content? If a complete feed, is it intended to be fed on a restricted or free-choice basis? If the feed is for herbivorous animals such as ruminants, it would be normal to first evaluate roughage as the base feed and then determine what nutrients are needed to supplement the roughage. Sometimes, roughage may be added only as a diluent to rations for fattening animals to control intake or produce a desired physical texture for the ration.

EXPECTED FEED CONSUMPTION

Complete rations should be designed so that animals will consume a desired amount, because the required concentration of a nutrient in a ration depends on consumption. For example, if we want a steer to consume 500 g of protein per day, the feed needs to contain only 10% protein if the steer eats 5 kg of feed; if it will eat only 4 kg, then 12.5% protein is needed to achieve the desired protein intake.

Energy concentration greatly affects feed intake, as do other factors, such as physical (bulk) density, deficiency of some nutrients, or

Feed Formulation Worksheet

Date _____

Prepared by _____

Animal: (species, weight, purpose, expected performance) _____

Animal Requirements:

Dry matter (as g, kg or lb/day) _____

Protein (as g, kg or lb/day or as % of ration) _____

Energy (as g, kg or lb TDN/day or as % of ration or as Mcal of DE, ME, NE or NEm, NEg, NEℓ per day or per unit of ration) _____

Ca (as g/day or as % of ration) _____

P (as g/day or as % of ration) _____

Nutrient Composition of Feedstuffs   ☐ as fed   ☐ dry basis

Feed ingredient	DM, %	Protein,% Crude Digest.	Energy	CA, %	P, %	Other*	Cost lb ton

Ration ingredients	Lb/ ton	% dry basis	Lb, as fed basis	% as fed	Protein	Energy	Ca	P	Others	Cost

*Fat, fiber and ash values are usually required for commercial mills because of legal restrictions (see Ch. 4); depending on the species and age, we might wish to include Mg, S, and K of the macrominerals and several of the trace minerals in some areas; vitamins A, D, E, and K and riboflavin, niacin, pantothenic acid, folic acid, and B12; and amino acids such as lysine, arginine, methionine, cystine, glycine, and tryptophan.

FIGURE 12-1 One type of ration formulation worksheet.

presence of unpalatable ingredients. Pelleted hay, for example, will usually be consumed in much larger amounts by ruminants than will long hay; thus the concentration of some nutrients can be reduced when pelleted hay is fed.

Some information on this topic is presented in the various appendix tables. Additional information is also presented in the chapters on individual species.

GUIDELINES AND RULES OF THUMB FOR RATION FORMULATION

Note in the tables on feed composition (Appendix Tables 1–5) that feed data may be given either on the basis of dry feed (oven-dry basis) or on an as-fed (air dry) basis, depending on which publication the information was taken from. When it is desirable to formulate rations to rather exact specifications or when part of the ration ingredients contain a considerable amount of water (silage, for example), then it is preferable to formulate rations on a dry basis. For the feedstuffs listed in the appendix tables for swine and poultry, bear in mind that these are given on an as-fed basis. Therefore, if formulation is to be done on a dry basis, some recalculation will be required before ration formulation is commenced. If laboratory data are available on dry matter, then these should be used; if not, then use appropriate values from the appendix tables or other sources. For example, if the usual dry-matter content is 90% and it is desired to convert the nutrient concentration from an as-fed (or air dry) to a dry basis, divide the nutrient content by 0.9. If there is difficulty in remembering whether to multiply or divide by dry-matter percentage, just keep in mind, when converting from as-fed to dry basis, that the resulting values must be higher than those started with. The reason for this, of course, is that removal of water increases the concentration of the remaining ingredients. However, after the ration is formulated, be sure to convert it to an as-fed basis. Feed mills mix rations in the amounts given in formulas, and the percentage composition on a dry basis will nearly always be different than on an as-fed basis. This simple procedure is illustrated later in this chapter. It might be noted that some people (in research or commercial formulators) prefer to formulate all rations (even those containing wet feeds) on a 90% DM basis. The logic of this practice escapes the writer.

Formulation can be done on the basis of daily needs, although in practice this is seldom done for most animals. Rather, it is more common to formulate on the basis of a given weight unit: 100 lb, 100 kg, 1000 lb, or 1 ton. Use of percentage units is the simplest means, because the final values can be converted easily to any final weight unit.

If formulas are worked out to exact specifications, fractional units (g, lb, kg) may result, which are undesirable in commercial feed mill usage. Where pound units are used to make up ton batches, these should be rounded off to at least 10-lb increments, except for ingredients included in the ration in very small amounts (urea, minerals, antibiotics, and the like). As a rule, the errors caused by this procedure will be small.

An important rule of thumb is that **simple nutrient needs can be met adequately by simple feed formulas.** Thus, when nutrient needs are simple, complex formulas do not guarantee improved performance. Also, the more complex the nutrient specifications, the more complex a formula is required to meet all specifications without having an excess of some nutrient(s). One individual feedstuff rarely will suffice to supply all needed nutrients without one or more being in excess. For example, if it is desired to feed a roughage to cows to supply 12% CP, the feeding of alfalfa hay, which usually has 15% or more protein, will be wasteful of protein, although it may have other nutrient properties of considerable value. Therefore, the alfalfa could be diluted with some other roughage that has a lower protein content, such as grass straw, without having excess protein consumption. In some instances it may be cheaper to use the feed with excess nutrients rather than dilute it with another ingredient. Suppose that alfalfa is the only hay available in an isolating farming area. It might actually cost less to feed the alfalfa hay, under some price situations, rather than to bring in another roughage over long distances.

Commercial concentrate mixes frequently contain a wide variety of feedstuffs, even in situations where nutrient requirements are relatively simple. There are two reasons why a variety may be included. First, it is felt that a variety of feedstuffs in a mixture may be more palatable for heavily fed animals; perhaps this is true where the same mix is fed for a long period of time. The second reason is that a mixture of several energy or protein sources may provide some insurance against trace nutrient deficiencies. This may or may not be true, depending on the feed ingredients chosen for the mixture.

As seasonal changes occur in availability or costs of feeds, many times it is desirable to alter formulas, particularly for the feed grains and protein concentrates. Some substitution can take place with relatively little effect; for example, corn, sorghum, and wheat have about the same energy value for most animals. For ruminants, barley and oats can be substituted more liberally than for most monogastric species. If drastic changes are in order, then formulas should be recalculated.

With respect to protein sources, most nutritionists suggest using protein (either crude or digestible) for ruminants on the basis of least cost per unit of protein. Factors other than protein content may affect value (plant hormones, Se in linseed meal), but these are poorly quantified. Some meals may not be very palatable, so some judgment and knowledge of the specific feedstuff are required when liberal substitution is practiced. For monogastric species, the amino acid content of protein substitutes must be evaluated, as well as other factors such as gossypol content of cottonseed meal.

Attention should be called to a few other simple guidelines. Salt (NaCl) can be supplied for herbivorous species by providing it ad libitum in a separate container from other feed. In complete feeds or concentrate mixtures, usually 0.25% to 0.5% salt is included for poultry and swine and 0.5% to 1% for ruminants. Likewise, trace-mineralized salt can be used as a supplementary source of some trace minerals, and mineral supplements are commonly used to supply sources of additional Ca, P, and Mg. When fat is added, it is the usual practice to add not more than 5% for swine and poultry and more on the order of 2% to 3% in finishing rations for cattle and, sometimes, in rations for dairy cattle. Feeds with high levels of fat do not store well because they are apt to become rancid unless antioxidants have been added to the fat. With molasses or other similar liquid feeds, the amount commonly added in mixed feeds is usually restricted to 7% to 8% because of handling and mixing problems and because, in stored feed, more than this is apt to set up into a firm mass that does not lend itself to good feeding practices.

When a tentative formula is finished, check it over to determine if needed nutrients are present in desired concentrations. Evaluate the ration with respect to any excesses and specify what (if any) nutrients or feedstuffs will be needed in addition (e.g., salt or salt-minerals fed free choice) and compute the cost for the ration.

MATHEMATICS OF RATION FORMULATION

Unless we are concerned with least-cost formulation, nothing more than very simple algebra is required to put together some rather complicated diets. Most beginners have trouble with it, but it is not difficult provided the various steps are learned thoroughly. Various techniques that are useful for hand formulation are illustrated in the following sections.

PEARSON'S SQUARE

Pearson's square is a simple procedure that was originally devised for use in blending milk products to a known fat percentage. Use of the square allows blending of two feedstuffs (or two mixtures) with different nutrient concentrations into a mixture with a desired concentration. To solve a problem with Pearson's square, the desired solution is placed in the center (X) and feed source A, with its usual level of protein or energy (or any other nutrient needing a solution), is then added. If we are working for a solution of crude protein (CP), the CP level must be either higher or lower than X. Then a second feed source (or a mixture of several) with its typical CP level is placed in the B position. To solve,

the difference ($+$ or $-$) between X and A goes in the D position, and, likewise, the difference between B and X goes in the C position. The answer is expressed as parts. Add up the total parts, and the amounts of A and B needed to provide the desired solution (X) can be expressed as parts of the whole or as percentages, as illustrated in the following example. Suppose we have a protein concentrate, such as cottonseed meal (CSM) with 40% CP, and a grain with 10% CP, and we wish to have a blend with 18% CP.

Using this square,

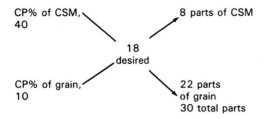

the answer is

8 parts of CSM and 22 parts of grain

or, if expressed as percent,

CSM in mix = 8 ÷ 30 × 100 = 26.67%
Grain in mix = 22 ÷ 30 × 100 = 73.33%

Check for CP:

26.67 (CSM) × 40% = 10.67
73.33 (grain) × 10% = 7.33
Total   18.00

As shown in the illustration, compare the CP percentage of each feed on the left with the desired percentage in the middle of the square. The **lesser value is subtracted from the greater value and the answer,** in parts of a mixture (rather than in percent), **is recorded diagonally.** It would usually be best to calculate the percentage of the final mixture. Although this is not necessary in this illustration, working with percentages is easier than working with fractions. The same procedure can be used for energy, minerals, and so on, but remember the **one feed** (or mixture) **must have a value higher and the other must be lower than the desired solution.** Furthermore, calories or ppm can be used as well as percentage. Further illustrations of the use of Pearson's square will be given later in the chapter.

As a matter of information, the check run on CP in the example illustrates the use of a weighted average. When there are two or more ration ingredients with different nutrient concentrations, the total is obtained by multiplying the amount of each ration ingredient by its nutrient concentration and summing the answers for all ingredients.

ALGEBRAIC SOLUTION

Some individuals prefer to solve the simple problems of ration formulation using algebraic solutions from equations with two unknowns.

For the same problem as with the Pearson's square, the approach would be as follows:

X = lb of CSM in mix
Y = lb of grain in mix
$X + Y = 100$ lb of mix
$0.40X + 0.10Y = 18$ (lb of CP in final mix)

To solve this problem, it is necessary to multiply the equation ($X + Y = 100$) by a unit that will allow one of the unknowns in the second equation to factor out. Thus, if we multiply by 0.1, we have $0.10X + 0.10Y = 10$. The problem is then solved as shown:

Unknown equation:	$0.40X + 0.10Y = 18$
Subtract from it.	$0.10X + 0.10Y = 10,$
The answer is	$0.30X = 8$
So one can compute	$X = 8 ÷ 0.3 = 26.67$
Thus	$Y = 100 - 26.67$
	$= 73.33$

The answer, obviously, is the same as with Pearson's square.

DOUBLE PEARSON'S SQUARE

In many situations we might want to have exact amounts of two major nutrients, such as CP and energy. We can accomplish this by going through three squares, as shown below. Suppose we want a final mix with 12 percent CP and 74% TDN. We have corn with 10% and 80%, CSM with 40% and 68%, and alfalfa hay with 15% and 55% CP and TDN, respectively. We must first go through two squares and get a mix exact for one of the nutrients: in this example we will do CP first, but it makes no difference which goes first. We must have one mix with 12% CP and greater than 74% TDN and one mix with 12% CP and less than 74% TDN. For this we must have a minimum of three feedstuffs, and we could use four if desired. Proceed as shown.

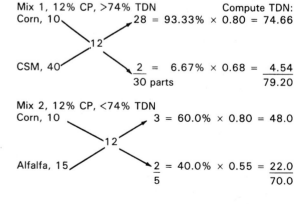

Then solve for TDN:

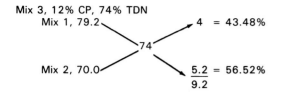

Mix 3, 12% CP, 74% TDN

Mix 1, 79.2 → 4 = 43.48%

74

Mix 2, 70.0 → 5.2 = 56.52%

9.2

Calculate the ingredient composition as follows:

Corn in mix 1, 93.33
    (43.48% of mix 1 in mix 3) = 40.58 ⎱
Corn in mix 2, 60.00   74.49
    (53.52% of mix 2 in mix 3) = 33.91 ⎰
CSM in mix 1, 6.67
    (43.48% of mix 1 in mix 3) =   2.90
Alfalfa in mix 2, 40.0
    (56.52% of mix 2 in mix 3) =   22.61
                                100.00

Check

$$TDN = (0.7449 \times 80 + 0.029 \times 68 + 0.2261 \times 55)$$
$$= 59.59 + 1.97 + 12.44$$
$$= 74.00$$

As the reader can see, this procedure works very well to produce a mix with exact specifications on two nutrients. However, if we want to add a third nutrient, it could require as many as nine squares, although it would not necessarily need this many. Therefore, diets that have exact requirements for more than two major nutrients are tedious to formulate by hand. Minor nutrients can be handled with slack space (see later section).

If we want to use algebra to solve this problem with corn, alfalfa, and CSM, the solution would be the same. We can arrive at a solution for two nutrients with only two feeds (sometimes), but the answer will be in weight units rather than percent. For example, if we want a mix to supply 12 lb of protein and 74 lb of TDN using only corn and CSM, the algebraic solution would be as shown (where X is corn and Y is CSM):

For CP, $0.1X + 0.40Y = 12$
For TDN, $0.8X + 0.68Y = 74$

Now, divide all elements of the first equation by 0.1 and the second by 0.8, and the equations become

CP, $X + 4Y \quad = 120$
TDN, $X + 0.85Y = \;\;92.5$

Subtract the second equation from the first and the answer is

$$3.15Y = 27.5$$
$$Y = \;\;8.73$$

X can be solved by substituting the value of Y in one of the original equations; thus, with the first equation,

$$0.1X + (0.4 \times 8.73) = 12$$
$$0.1X + 3.492 = 12$$
$$0.1X = \;\;8.508$$
$$X = 85.08$$

Thus the answer is 8.73 lb of CSM and 85.08 lb of corn. Unless there are some coincidental values, it is not possible to formulate a mixture with exact values for two nutrients using only two feeds. In this example, with the amounts of CSM and corn shown, we have 12 lb of protein and 74 lb of TDN, but the percentage of protein in the mix is 12.79 ($12.0 \div 93.81$) and that for TDN is 78.88 ($74.0 \div 93.81$). In other illustrations the Pearson's square will be used because it is the authors' opinion that fewer mistakes are made by most students using the square, and it certainly is just as fast, if not faster.

REQUIRED INGREDIENTS

In a number of situations there is some need to specify exact or minimal amounts of specific feedstuffs or quantities of supplements, additives, or the like. Suppose, for example, that we want a pig ration with 3% fish meal and 10% ground wheat grain in a final mix with 18% CP and 3500 kcal of DE/kg of feed. Corresponding values are 70.6 and 3650 (as-fed basis) for fish meal (herring) and 12.7 and 3520 for wheat. Now, what we must do is calculate how much of the required nutrients is provided by the required ingredients and what concentration is needed in the remaining portion of the mix, as shown below:

Ingredient	Fish Meal	Wheat Grain
Amount required, %	3.0	10.0
Composition		
CP, %	70.6	12.7
DE, kcal/kg	3650	3520
Nutrients supplied/ 100 kg		
CP, kg	2.12	1.27
DE, kcal	10,950	35,200
Total CP, kg	3.39	
Total DE, kcal	46,150	

To calculate the amount needed in the remaining part of the diet, proceed as follows:

	DP, kg	DE, kcal
Amount required/100 kg	18	350,000
Supplied by fish and wheat	3.39	46,150
Still needed in 87 kg	14.61	303,850

Thus the concentration of CP required is $14.60 \div 87 = 16.78\%$, and for DE it is $303,850 \div 87 = 3492.5$ kcal/kg. Knowing the concentrations required for these two nutrients, we could proceed to formulate the remainder of the ration.

SLACK SPACE

The use of slack or reserved space is quite convenient and, indeed, necessary if we want to adjust amounts of nutrients, such as some of the minerals (Ca, P) or amino acids, without changing specifications for other nutrients or required feed ingredients. Usually, 1% to 2% of the total is quite adequate for slack. If all of it is not needed for adjustment, then a normal ingredient such as salt could be added. One brief example will be given here and another in a later example of a complete ration being formulated.

If, for example, we have reserved 1% slack and we have formulated a ration to exact amounts of CP and DE, but we want 0.6% Ca and 0.35% P, and we only have 0.4% Ca and 0.30% P, we could solve this problem in the following manner. The amount of Ca still needed is 0.2% (= 0.2 lb/100) and for P, it is 0.05% (= 0.05 lb/100). If we pick dicalcium phosphate to supply the P, it contains 23.3% Ca and 18.2% P. The amount needed to supply 0.05 lb of P is equal to 0.27 lb of dical ($0.05 \div 0.182$). Now, this amount will also supply 0.06 lb of Ca ($0.27 \times 0.233 = 0.06$). We needed 0.2 lb of Ca; with the addition of the dical, we still need 0.14 lb ($0.2 - 0.06 = 0.14$). We could supply this with limestone (35.8% Ca). It would require about 0.39 lb ($0.14 \div 0.358$) of limestone.

Thus we have added 0.27 lb of dicalcium phosphate and 0.39 lb of limestone for a total of 0.66 lb. We could use up the remaining slack space by adding salt if we want the other ingredients to remain unchanged.

Slack space is easy to use and allows adjustment to exact amounts of nutrients required in relatively small quantities. We haven't altered the composition otherwise. Granted that replacement of a small amount of a major ingredient with less than 1% of the total diet would not alter nutrient content greatly. Nonetheless, formulation to exact levels should be the desired objective, and, in commerce, certain minimums or maximums may be legal requirements.

RATION EXAMPLE

We will now go through the steps of formulating a ration for growing pigs (30 to 120 lb) using the various procedures already illustrated.

Step 1 Determine dietary requirements on a dry-matter basis from NRC (given in the appendix tables). For our example, they are 16% CP, 1670 kcal DE/lb, 0.60% Ca, 0.50% P, 0.74% lysine, and 0.50% methionine. There are specifications on other nutrients, but these are all that we will be concerned with in this example.

Step 2 Determine any other ration restrictions. It is a common practice with both swine and poultry rations to add a vitamin–trace mineral package (premix) as a needed supplement or in the form of insurance. We will add this at the rate of 10 lb/ton or 0.5%. We will also specify the addition of 5% dehydrated alfalfa (15% CP), 10% wheat (hard red spring), and 1.5% fish meal (herring). For the remainder of the diet, we will choose from corn, wheat middlings, soybean meal, and tallow. Thus our required ingredients and their nutrient compositions are shown in the following table:

Required	%	DM, %	Composition, DM basis, % or kcal[a]					
			CP	DE[a]	Ca	P	Ly	Meth
Slack	1.0	—	—	—	—	—	—	—
Vit–min premix	0.5	—	—	—	—	—	—	—
Alfalfa meal	5	93.1	16.3	736	1.32	0.24	0.64	0.21
Fish meal	1.5	92.0	76.7	1730	3.20	2.39	7.94	2.17
Wheat	10.0	89.1	14.6	1735	0.06	0.47	0.40	0.20
Total of nutrients from required, lb or kcal/100 lb								
	18.5		3.42	23,625[a]	0.124	0.095	0.191	0.128

Step 3 Calculate the nutrient concentration needed for the remainder (81.5%) of the ration:

| | Composition, DM basis, % or kcal[a] | | | | | |
	CP	DE[a]	Ca	P	Ly	Meth
Required in diet	16.0	167,000	0.60	0.50	0.74	0.50
Still needed in 81.5 lb	12.58	143,375	0.476	0.405	0.549	0.372
Concentration required/lb	15.44	1759.3	0.584	0.497	0.674	0.456

Step 4 Assemble the needed data on other feed ingredients:

| | Composition of other ingredients, % or kcal,[a] dry basis | | | | | | |
	DM	CP	DE[a]	Ca	P	Ly	Meth
Corn	89.0	10.0	1684	0.02	0.35	0.20	0.10
Wheat middlings	89.0	20.2	1575	0.09	0.58	0.67	0.11
Soybean meal (44%)	90.0	44.0	1871	0.30	0.70	3.00	0.89
Tallow	98.0	—	3690	—	—	—	—

Step 5 Use the square (or algebra) to formulate a final mix that is exact for needed CP and DE.

Mix 1, 15.44% CP, <1759 kcal DE

Calculate DE.

$$\begin{aligned} \text{Corn} &= 84.0\% \times 1684 = 1415 \text{ (rounded)} \\ \text{Soybean} &= 16.0\% \times 1871 = \underline{299} \\ & 1714 \end{aligned}$$

Mix 2, 15.44% CP, >1759 kcal DE

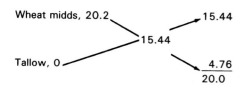

Calculate DE:

$$\begin{aligned} \text{Wheat midds} &= 76.44\% \times 1575 = 1204 \\ \text{Tallow} &= 23.56\% \times 3690 = \underline{869} \\ & 2073 \end{aligned}$$

Mix 3, 1759 kcal DE

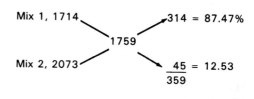

Step 6 Calculate the ingredient composition:

Required	%
Slack	1.5
Premix	0.5
Alfalfa meal	5.0
Fish meal	1.5
Wheat	10.0
Corn	59.88 = 81.5(0.840 × 0.8747)*
Wheat midds	7.80 = 81.5(0.7644 × 0.1253)
Soybean meal	11.41 = 81.5(0.16 × 0.8747)
Tallow	2.41 = 81.5(0.2356 × 0.1253)

*Values are lb of total diet supplied by these four ingredients: % of mix 1 × % of mix 1 in Mix 3, etc.

Step 7 Check to see that CP (16.01) and DE (1669.7) are correct. The slight discrepancies between the amounts required and the final values are caused by rounding errors.

Step 8 Compute the amounts of other nutrients specified (lb/100):

	Ca	P	Ly	Meth
In required ingredients	0.124	0.095	0.191	0.128
In remaining 81.5 lb	0.053	0.329	0.514	0.171
Total	0.177	0.424	0.705	0.299
Still needed	0.323	0.046	0.055	0.261

Step 9 Determine how to meet the remaining specifications. We can meet P needs by adding 0.25% of dicalcium phosphate (0.046 ÷ 0.182, or the pound of P needed divided by the amount supplied by dical). The dical supplied 0.058 lb of Ca, but we still need 0.265 lb of Ca (0.323 − 0.058), which can be provided by 0.74 lb of limestone (0.265 ÷ 0.358).

Because rather pure sources of lysine and methionine are available commercially, we could complete the lysine by using the commercial product (76.9% L-lysine). This would require 0.07 lb (0.055 ÷ 0.769). For methionine, we could use methionine hydroxy analog. The product is at 93% purity, and it is generally rated at about 80% the value of methionine, so this material has a replacement value of 74.4% (0.8 × 0.93). It would require about 0.35 lb (0.261 ÷ 0.744) for our diet. In place of using purified amino acid sources, we could go back and revise the diet by increasing feedstuffs that are higher in these amino acids; a logical place to start would be to eliminate most if not all of the alfalfa meal and/or to increase the fish meal to perhaps 3% or more. As you have seen, normally there is more than one way to get a final solution on most problems.

Step 10 Convert final formula to an as-fed basis. The final diet is shown in a dry basis (in the following table). Of course, we do not mix feed or feed it to animals on a dry-weight basis. Therefore, we must convert it to the as-fed basis. This requires that the amounts of ingredients be divided by the decimal of the dry-matter content, summing up the total weights and calculating a final percentage for the as-fed ration.

COST FACTORS IN RATION FORMULATION

In the examples presented so far in this chapter, cost has not been considered. However, in the normal course of events, cost of feedstuffs is an important factor that must be considered in ration formulation. Only in a few instances—show animals, some hobby situations, or for racing animals—is cost not a primary factor.

EFFECT OF DIFFERENCES IN DM

Feeds, of course, are purchased on an as-fed basis, so some calculations are always involved to get comparative costs on a dry basis. If the

Final pig grower ration

Ingredient	%	DM, %	As-Fed Weight[a]	As Fed, %[b]
Alfalfa meal	5	93.1	5.37	4.82
Fish meal	1.5	92.0	1.63	1.46
Wheat	10.0	89.1	11.24	10.09
Corn	59.88	89.0	67.28	60.38
Wheat middlings	7.80	89.0	8.76	7.86
Soybean meal	11.41	90.0	12.68	11.38
Tallow	2.41	98.0	2.46	2.21
Vit–min premix	0.50	100	0.50	0.45
Limestone	0.74	100	0.74	0.66
Dicalcium phosphate	0.25	100	0.25	0.22
Methionine hydroxy analog	0.35	100	0.35	0.31
Lysine	0.07	100	0.07	0.06
Remainder	0.09[c]	100	0.09	0.08
			111.42	100.00

[a]For alfalfa = 5 ÷ 0.931 = 5.37.
[b]For alfalfa = 5.37 ÷ 1.1142 = 4.82.
[c]Add salt.

buyer ignores the moisture content of feed, water in feed may be an expensive purchase. It is always best to calculate costs on a dry-weight basis or relative to a standard DM content, such as 90% DM.

Suppose that corn with 89% DM costs 7 cents/lb. What is corn of 75% DM worth, assuming utilization of nutrients on a dry basis is comparable? This is easily calculated by the method shown:

$$(0.75 \div 0.89) \times 7¢ = 5.9¢ \quad \text{or} \quad 8.4\% \text{ less}$$

Although it is often not a common practice to buy feed grains or other products such as soybeans on a standard moisture content, it would be a fair practice that would not be difficult to do. Various devices are on the market that can determine moisture content quickly. As now practiced, not buying on moisture content may encourage some unscrupulous producers to add water before selling. Likewise, it certainly does not pay producers to overdry grain in areas where drying is necessary.

Data shown in Table 12-1 could be used for a standardized bushel weight and moisture content. The values for equivalent bushel weights shown in Table 12-2 show quite clearly that one bushel is not always equal to another bushel! Assuming that the grain is of the same grade (for variables other than moisture content), grain with less moisture has a greater value than grain with more moisture. As it stands now, the old maxim of "let the buyer beware" is very applicable to buying and selling grains. Note that grains are bought on a weight basis. However, except for sorghum, central markets quote prices on a bushel weight basis, an archaic practice.

EVALUATION ON BASIS OF REPLACEMENT COSTS

One way to evaluate feedstuffs with different CP or energy concentrations is to calculate how

TABLE 12-1

Base data for a standardized bushel for feed grains in the United States

	Legal Weight, lb	Base Moisture,[a] %	Dry Matter, %	Dry Matter, lb
Corn	56	15.5	84.5	47.32
Soybeans	60	13.0	87.0	52.20
Sorghum	56	14.0	86.0	48.16
Wheat	60	13.5	85.5	51.90

Source: Hill (1).
[a]These are the accepted values used by the grain trade when buying and selling these products.

TABLE 12-2

Equivalent bushels at base moisture from 1000 bushels of grain at selected moisture contents

Moisture Content, %	Feedstuff[a]			
	Soybeans	Wheat	Sorghum	Corn
10.0	1034	1040	1047	1065
11.0	1023	1029	1035	1053
12.0	1011	1017	1023	1041
12.5	1006	1012	1017	1036
13.0	**1000**	1006	1012	1030
13.5	994	**1000**	1006	1024
14.0	989	994	**1000**	1018
14.5	983	988	994	1012
15.0	977	983	988	1006
15.5	971	977	983	**1000**
16.0	966	971	977	994
16.5	960	965	971	988
17.0	954	960	965	982
17.5	948	954	959	976
18.0	943	948	953	970
19.0	931	936	942	959
20.0	920	925	930	947

Source: Hill (1).
[a]Boldfacing denotes base moisture content generally used for pricing in the market. Refer to note at bottom of Table 12-1.

much money would be saved (or spent) by using a similar feed with higher or lower nutrient concentration. If, for example, a 7% CP corn costs 6 cents/lb and you want a 12% CP ration with corn and cottonseed meal and the CSM is valued at 9 cents/lb, what is 9% CP corn worth? The answer can be obtained by putting these ingredients through Pearson's square, as follows:

Calculate cost

Corn, 7% CP   29 = 85% × 6¢ = 5.10
      12
CSM, 41% CP   5 = 15% × 9¢ = 1.35
     34   6.45¢/lb for mix

Corn, 9% CP   29 = 91% × ?
      12
CSM, 41% CP   3 = 9% × 9¢ = 0.81
     32

Thus, if 9% corn is substituted, 54 cents is saved per 100 lb by using less CSM (1.35 − 0.81). The second mixture should be worth the same as the first, and the corn should be worth 6.45 cents − 0.81 or 5.64 cents. Because more corn was used in the second mix, the price per pound is 5.64 ÷ 0.91 = 6.198 cents/lb. On a ton basis, the 7% corn would be worth $120 and the 9% corn worth $124 when CSM sells for $180. In this particular example, each additional 1% CP in corn is worth $2/ton.

With the use of these simple procedures, it is obvious that some feed ingredients can be selected that have a lower cost than others. Evaluating feeds on the basis of cost of only CP or energy is not a very complete method, although feeds can be ranked easily on cost of one nutrient. A combined value is not obtained for both protein and energy, nor is there a good means of evaluating other nutrients such as P, which may be relatively expensive to provide from concentrated supplements; thus other methods can be more informative.

MATHEMATICAL PROGRAMMING*

Mathematical programming is widely used in the United States, by both private and public institutions, to formulate rations for livestock. The rapid development and distribution of desktop computers with the appropriate programs has greatly improved the opportunity for even modest-sized livestock producers or small feed mills to do their own programming.

With the Pearson's square and the simultaneous equations methods, the mix obtained is of a predetermined CP percentage or a predetermined energy level. The amounts of the ingredients considered for the mix are therefore fixed. In reality, the percentage of protein required or the amount of energy in a diet is considered a minimum, and therefore the final mix should have "at least" the required amount of the nutrient. Sometimes the specified nutrient is required within a range, that is, at least some minimum quantity, but less than a maximum quantity. These two methods, the Pearson's square and the simultaneous equations, cannot handle inequalities or ranges, and both methods are independent of price. When price is considered, optimization is done by trial and error.

In cases of multiple nutrient requirements, multiple feed sources, price consideration, and requirements being greater than and/or less than some level, diet formulation should be done by mathematical programming. This technique, when properly used, helps to achieve both a nutritionally balanced diet and economic optimization, because it allows for simultaneous consideration of economical and nutritional parameters.

Nutritionists should have a good knowledge of diet specifications, should be familiar with formulation and interpretation of results, and should think of the solution process, that is, the mathematical manipulation, as a "black box." Nutritionists (or students) need not be concerned with the mechanics of the mathematical solution of the linear programming matrix. Diet formulation by mathematical programming should be treated as an interactive process, in which the nutritionist should verify, interpret, and reformulate, if necessary, all diet formulas.

With the availability of microcomputers, mathematical programming can be performed quite easily if the programming is done properly. We use the term "programming" as the planning of "economic" activities for the sake of optimization, subject to some constraints. There exist a large number of computer programs for the solution of mathematical programming problems, so a detailed explanation of the mathematics involved is not given here. Further information may be found in other sources (2–5).

Not all computer ration formulating programs utilize proper methods for linear programming. When mathematical programming is solved by not recognized algorithms, it will generate incorrect answers leading to improper recommendations (5). The student should keep in mind that some good software is improperly utilized, resulting in erroneous answers. It is important to be "mathematically correct" if we are to utilize computers as a tool in ration formulation.

LEAST-COST FORMULATION

A great percentage of production cost is due to feed; thus diet formulation using least-cost techniques has been used extensively during the past 20 years. For any of the mathematical programming techniques to work, formulation of the problem should be done properly. Once the formulation of the problem is stated, then we can use any linear programming package to solve the diet formulation. Once we have a solution, it is one of the functions of the nutritionist to verify whether the solution conforms with nutritional knowledge.

EXAMPLE 1

To illustrate the problem statement for a least-cost formulation, the following steps of this

---

*Part of this section has been taken from Chapter 22, Diet Formulation, by H. Varela Alvarez, from *Basic Animal Nutrition and Feeding*, 3rd ed., by D. C. Church and W. G. Pond. It is reproduced here by permission of the publisher, John Wiley & Sons, Inc., New York, NY. Copyright 1988 by John Wiley & Sons, Inc.

technique are presented. Let us formulate a diet for growing pigs (20 to 35 kg).

Step 1 DETERMINE THE ANIMAL'S DIETARY REQUIREMENTS from the NRC tables (given in the Appendix). They are 16% CP, 3380 kcal DE/kg, 0.60% Ca, 0.50% P, 0.70% lysine, 0.45% methionine. There are other nutrient specifications, but for this example only these requirements will be utilized.

Step 2 DETERMINE ANY OTHER RATION RESTRICTIONS. It is a common practice with both swine and poultry rations to add a vitamin–trace mineral package (premix) as a needed supplement. We will add the vitamin premix at a rate of 5 kg/ton or 0.5%. We will also specify the addition of 5% dehydrated alfalfa (17.5% CP), 10% wheat (hard red spring), and 1.5% fish meal (herring). For the remainder of the diet, we will choose from corn, wheat middlings, soybean meal, tallow, dicalcium phosphate, limestone, L-lysine, methionine hydroxy analog, and salt. Our required ingredients, price per ton, and their nutrient compositions are presented in Table 12-3. Note that in the case of amino acid supplements the amount of protein content is equal to the amount of the amino acid (or sum of amino acids) supplied.

Step 3 STATE THE PROBLEM IN EQUATION FORM, as follows: Let $X_1, X_2, X_3, X_4, X_5, X_6, X_7, X_8, X_9, X_{10}, X_{11}, X_{12}$, and X_{13} be the nonnegative quantities of the vitamin–mineral premix, alfalfa meal, fish meal, wheat, corn, wheat middlings, soybean meal, tallow, dicalcium phosphate, limestone, L-lysine, methionine hydroxy analog, and salt, respectively. They will be mixed to yield 100 units of a minimum-cost diet that will satisfy all the specified nutritional requirements. The mix will also consider the constraints on the first four ingredients.

Using the crude protein content of the different feedstuffs under consideration, we can say that there are $0X_1$ units of protein in X_1 units of vitamin–mineral premix, $0.175X_2$ units in X_2 units of alfalfa meal, $0.723X_3$ units in X_3 units of fish meal, $0.141X_4$ units in X_4 units of wheat, $0.088X_5$ units in X_5 units of corn, $0.16X_6$ units in X_6 units of wheat middlings, $0.44X_7$ units in X_7 units of soybean meal, $0X_8$ units in X_8 units of tallow, $0X_9$ units in X_9 units of dicalcium phosphate, $0X_{10}$ units in X_{10} units of limestone, $0.769X_{11}$ units in X_{11} units of L-lysine, $0.744X_{12}$ units in X_{12} units of methionine hydroxy analog, and $0X_{13}$ units in X_{13} units of salt. Expressing it in an equation form we have

$$0X_1 + 0.175X_2 + 0.723X_3 + 0.141X_4 \\ + 0.088X_5 + 0.16X_6 + 0.44X_7 + 0X_8 + 0X_9 \\ + 0X_{10} + 0.769X_{11} + 0.744X_{12} + 0X_{13} \geq 16$$

This equation states that the sum of the contribution of protein by each ingredient should be greater than or equal to 16, which is the minimum requirement of CP in this diet. This also implies that

$$X_1 + X_2 + X_3 + X_4 + X_5 + X_6 + X_7 + X_8 \\ + X_9 + X_{10} + X_{11} + X_{12} + X_{13} = 100$$

That is, that the sum of all ingredients should be equal to 100 units, thus having a diet with a

TABLE 12-3

Feedstuff composition on an as-fed basis

Ingredient	Feed Reference Number	DM, %	$/Tᵃ	CP, %	DE, Mcal/ kg	Ca, %	P, %	LYS, %	METH, %
Vit–min premix		100	900						
Alfalfa meal	1-00-023	92	137	17.5	2270	1.44	0.22	0.73	0.20
Fish meal	5-02-000	93	310	72.3	2500	2.29	1.70	5.70	2.10
Wheat	4-05-268	87	120	14.1	3220	0.05	0.37	0.31	0.20
Corn	4-02-935	89	77	8.8	3325	0.02	0.28	0.24	0.20
Wheat middlings	4-05-205	88	83	16.0	2940	0.12	0.90	0.69	0.20
Soybean meal	5-04-604	89	209	44.0	3090	0.29	0.65	2.93	0.70
Tallow	4-00-409	99	220	—	7900	—	—	—	—
Dicalcium phosphate	6-01-080	100	292	—	—	23.70	18.84	—	—
Limestone	6-02-632	100	72	—	—	36.07	0.02	—	—
L-Lysine		100	4080	76.9	—	—	—	76.9	
Methionine hydroxy analog		100	3420	74.4	—	—	—	—	74.4
Salt		100	66	—	—	—	—	—	—

ᵃ1 ton = 1000 kg.

minimum of 16% CP. The equation that includes the sum of all ingredients would be called the "amount" equation.

The difference between mathematical programming and simultaneous equations is that in mathematical programming we work with equalities and inequalities at the same time, while in simultaneous equations all are equalities. By handling inequalities, mathematical programming allows us to handle ranges for ingredients or nutritional specifications.

The rest of the equations (constraints) for the nutritional requirements are as follows:

Energy: $0X_1 + 2270X_2 + 2500X_3 + 3220X_4$
$+ 3325X_5 + 2940X_6 + 3090X_7$
$+ 7900X_8 + 0X_9 + 0X_{10} + 0X_{11}$
$+ 0X_{12} + 0X_{13} \geq 338,000$

The figure 338,000 comes from the fact that the specified requirement for energy is 3380 kcal DE/kg. Because we are formulating a 100-kg diet, then the total content of energy will be at least $3380 \times 100 = 338,000$ kcal DE.

Ca: $0X_1 + 0.0144X_2 + 0.0229X_3 + 0.0005X_4$
$+ 0.0002X_5 + 0.0012X_6 + 0.0029X_7$
$+ 0X_8 + 0.237X_9 + 0.3507X_{10} + 0X_{11}$
$+ 0X_{12} + 0X_{13} \geq 0.6$

P: $0X_1 + 0.0022X_2 + 0.017X_3 + 0.0037X_4$
$+ 0.0028X_5 + 0.009X_6 + 0.0065X_7 + 0X_8$
$+ 0.1884X_9 + 0.002X_{10} + 0X_{11} + 0X_{12}$
$+ 0X_{13} \geq 0.5$

Lysine: $0X_1 + 0.0073X_2 + 0.057X_3$
$+ 0.0031X_4 + 0.0024X_5 + 0.0069X_6$
$+ 0.0293X_7 + 0X_8 + 0X_9 + 0X_{10}$
$+ 0.769X_{11} + 0X_{12} + 0X_{13} \geq 0.7$

Methionine: $0X_1 + 0.002X_2 + 0.021X_3$
$+ 0.002X_4 + 0.002X_5 + 0.002X_6$
$+ 0.007X_7 + 0X_8 + 0X_9 + 0X_{10}$
$+ 0X_{11} + 0.744X_{12} + 0X_{13} \geq 0.45$

Step 4 SET THE RESTRICTIONS FOR INDIVIDUAL FEEDSTUFFS.

Vitamin–mineral premix	$X_1 =$	0.5
Alfalfa meal	$X_2 =$	5
Fish meal	$X_3 =$	1.5
Wheat	$X_4 =$	10

Step 5 SET UP THE OBJECTIVE FUNCTION. Construct the equation dealing with the price of the diet. Price in dollars per ton or any other unit does not affect the results. As long as the prices for each ingredient are in the same

units ($/kg, $/ton, $/cwt), the cost of the resulting diet formula will be correct. We add all the prices and state the minimum-cost diet as the objective function in the following way:

$$900X_1 + 137X_2 + 310X_3 + 120X_4 + 77X_5$$
$$+ 83X_6 + 209X_7 + 220X_8 + 292X_9 + 72X_{10}$$
$$+ 4080X_{11} + 3420X_{12} + 66X_{13}$$
$$= \text{minimum price}$$

Step 6 UTILIZING ANY APPROPRIATE LINEAR PROGRAMMING SOFTWARE, RUN THE DIET FORMULATED. The final setup will be as follows:

Minimize price: $900X_1 + 137X_2 + 310X_3$
$+ 120X_4 + 77X_5 + 83X_6$
$+ 209X_7 + 220X_8 + 292X_9$
$+ 72X_{10} + 4080X_{11}$
$+ 3420X_{12} + 66X_{13}$

Thus

$$X_1 + X_2 + X_3 + X_4 + X_5 + X_6 + X_7 + X_8$$
$$+ X_9 + X_{10} + X_{11} + X_{12} + X_{13}$$
$$= 100 \quad \text{(amount)}$$

$0X_1 + 0.175X_2 + 0.723X_3 + 0.141X_4$
$+ 0.088X_5 + 0.16X_6 + 0.44X_7 + 0X_8 + 0X_9$
$+ 0X_{10} + 0.769X_{11} + 0.744X_{12} + 0X_{13}$
$\geq 16 \quad \text{(protein)}$

$0X_1 + 2270X_2 + 2500X_3 + 3220X_4 + 3325X_5$
$+ 2940X_6 + 3090X_7 + 7900X_8 + 0X_9 + 0X_{10}$
$+ 0X_{11} + 0X_{12} + 0X_{13} \geq 338,000 \quad \text{(energy)}$

$0X_1 + 0.0144X_2 + 0.0229X_3 + 0.0005X_4$
$+ 0.0002X_5 + 0.0012X_6 + 0.0029X_7 + 0X_8$
$+ 0.237X_9 + 0.3607X_{10} + 0X_{11} + 0X_{12} + 0X_{13}$
$\geq 0.6 \quad \text{(Ca)}$

$0X_1 + 0.0022X_2 + 0.017X_3 + 0.0037X_4$
$+ 0.0028X_5 + 0.009X_6 + 0.0065X_7 + 0X_8$
$+ 0.1884X_9 + 0.002X_{10} + 0X_{11} + 0X_{12} + 0X_{13}$
$\geq 0.5 \quad \text{(P)}$

$0X_1 + 0.0073X_2 + 0.057X_3 + 0.0031X_4$
$+ 0.0024X_5 + 0.0069X_6 + 0.0293X_7 + 0X_8$
$+ 0X_9 + 0X_{10} + 0.769X_{11} + 0X_{12} + 0X_{13}$
$\geq 0.7 \quad \text{(lysine)}$

$0X_1 + 0.002X_2 + 0.021X_3 + 0.002X_4$
$+ 0.002X_5 + 0.002X_6 + 0.007X_7 + 0X_8 + 0X_9$
$+ 0X_{10} + 0X_{11} + 0.744X_{12} + 0X_{13}$
$\geq 0.45 \quad \text{(methionine)}$

$X_1 = 0.5$	(vitamin–mineral premix)
$X_2 = 5$	(alfalfa meal)
$X_3 = 1.5$	(fish meal)
$X_4 = 10$	(wheat)

Step 7 FROM THE RESULT OF THIS RUN (run 1 in Table 12-4), VERIFY WHETHER THE DIET IS APPROPRIATE. We can see that in this case we have too much wheat middlings and no salt. The reader should note that this was done on purpose to illustrate the need for verification. We should specify an upper limit of 20% wheat middlings and include salt within a range of 0.2% to 0.5% in our diet. Therefore, we have to include the following additional equations (constraints) to our initial formulation:

$X_6 \leq 20$	upper limit for wheat middlings
$X_{13} \geq 0.2$	lower limit for salt
$X_{13} \leq 0.5$	upper limit for salt

Then, we run the program again.

After verifying the new diet (run 2, Table 12-4), we may decide that this ration is acceptable. Notice that because of the new constraints this new diet has a higher price than the first one ($127.8827 versus $125.2653), but it will be *the lowest-cost diet given these constraints*. Also, in the second run we obtained a lower tallow and limestone content, as well as an increase in soybean meal. With the *proper mathematical formulation and interactive interpretation of the results,* we can arrive at a satisfactory formulated diet that is lowest in cost for a particular set of requirements and constraints.

MAXIMUM-PROFIT FORMULATION

With maximum-profit formulation, both the nutritional requirements and the animal performance are considered. The formulation contains all known feeding and nutrition inputs and animal production outputs. In this method we utilize feedstuffs based on cost and composition, animal performance (kilogram of milk or daily gain) as a function of nutrients, and total animal product output. The animal product output is treated as an income and price of feed as an expense. The objective is to maximize profit. This type of formulation requires knowledge of the maximum daily dry-matter intake (DMI) of the animal, production response to nutrient intake, and daily nutrient requirements for maintenance.

EXAMPLE 2

To illustrate this technique, a diet will be formulated to achieve maximum profit while supporting an optimum level of milk production in a cow with a good dairy potential. The cow has a weight of 600 kg and is in her first lactation.

In this case, minimum daily requirements for maintenance and growth for the lactating cow are obtained from the NRC tables. These requirements are net energy for lactation (NEℓ), 12.36 Mcal/day; crude protein, 0.881 kg/day;

TABLE 12-4

Diet composition and analysis (as fed)

		Run 1	Run 2
X_1	Vitamin–mineral premix	0.5000	0.5000
X_2	Alfalfa meal	5.0000	5.0000
X_3	Fish meal (herring)	1.5000	1.5000
X_4	Wheat (hard red spring)	10.0000	10.0000
X_5	Corn	—	41.8823
X_6	Wheat middlings	67.9875	20.0000
X_7	Soybean meal (44%)	3.4641	12.6562
X_8	Tallow	10.1408	6.7764
X_9	Dicalcium phosphate	—	0.2482
X_{10}	Limestone	1.1007	1.0001
X_{11}	L-Lysine	—	—
X_{12}	Methionine hydroxy analog	0.3068	0.2368
X_{13}	Salt	—	0.2000
	Amount, kg	100.0000	100.0000
	Price/ton	125.2653	127.8827
	DE, kcal/kg	3380.0000	3380.0000
	Protein, %	16.0000	16.0000
	Ca, %	0.6000	0.6000
	P, %	0.7081	0.5000
	Lysine, %	0.7236	0.7623
	Methionine, %	0.4500	0.4500

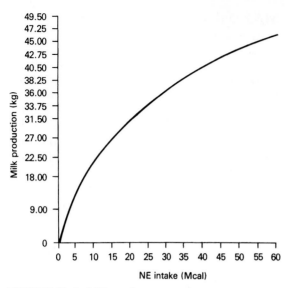

FIGURE 12-2 Milk production response to NEℓ intake.

Ca, 0.0264 kg/day; and P, 0.0204 kg/day. We will also specify salt within a range of 0.05 to 0.1 kg/day.

The optimum level of milk production and maximization of returns are interdependent; we want to maximize the amount of money that is left after we subtract feed price from milk sales, so the solution will include the amounts of each feedstuff that should enter into the formula and also the optimal amount of milk to be produced. Requirements for production depend on amount of milk produced and milk butterfat. In this case we will be working with fat corrected milk (FCM) at 3.5% butterfat.

In dairy cows, milk production response to NEℓ intake is quadratic as described in Fig. 12-2. The procedure for incorporating that selected curvilinear milk production response into our mathematical programming setup is to linearize portions of that curve and then specify them as inequalities. These linearized values are presented in Table 12-5.

For the formulation of this diet, the objective function will be to maximize the difference between the income from milk and the expense of feed. Because the diet is being formulated on a dry-matter basis, we have to convert the prices of the feedstuffs from an as-fed to a dry-matter basis. The price per kilogram of X_1 will be $90/ton divided by dry-matter percent of X_1 and then divided by 1000 (1000 kg/ton); that is, $(90 \div 0.89) \div 1000 = 0.1011$. Using the prices (dry-matter basis) presented in Table 12-6 and using $0.25 per kilogram of milk as the selling price, the objective function will look like this:

$$
\begin{aligned}
\text{Maximize profit:} \quad & 0.25\text{MILK} - 0.1056X_1 \\
& - 0.0865X_2 - 0.292X_3 \\
& - 0.072X_4 - 0.2348X_5 \\
& - 0.1419X_6 - 0.1348X_7 \\
& - 0.1254X_8 - 0.066X_9
\end{aligned}
$$

In this equation, MILK is the optimum amount of milk to be produced per cow per day.

The amount of energy supplied by the feed minus the energy required for production must be greater than or equal to the energy required for maintenance and growth. The equation for energy (NEℓ) will be as follows:

$$
\begin{aligned}
1.89X_1 &+ 2.42X_2 + 2.07X_5 + 2.66X_6 + 1.21X_7 \\
&+ 1.70X_8 - 0.33P_1 - 0.41P_2 - 0.77P_3 \\
&- 1.1P_4 - 1.21P_5 - 1.34P_6 - 1.7P_7 - 2.38P_8 \\
&- 3.22P_9 \geq 12.36
\end{aligned}
$$

The next point to consider will be to estimate total milk production. Total milk production will be the variable MILK. Therefore, if from total milk production we subtract the milk from each portion of production, the results must be equal to zero:

$$
\begin{aligned}
\text{MILK} - P_1 &- P_2 - P_3 - P_4 - P_5 - P_6 - P_7 \\
&- P_8 - P_9 = 0
\end{aligned}
$$

TABLE 12-5

Linear representation of milk production response to net energy for lactation (NEℓ)

Portion	Milk Production Portion, kg	Amount of Milk, kg	NE Lactation /kg Milk, Mcal
P_1	0.00–9.00	9.00	0.33
P_2	9.00–18.00	9.00	0.51
P_3	18.00–22.50	4.50	0.77
P_4	22.50–27.00	4.50	1.10
P_5	27.00–31.50	4.50	1.21
P_6	31.50–33.75	2.25	1.34
P_7	33.75–36.00	2.25	1.70
P_8	36.00–38.25	2.25	2.38
P_9	38.25–48.25	10.00	3.22

TABLE 12-6

Feed composition on a dry-matter basis

| | | As Fed | | | Dry-Matter Basis | | | | | |
| | Reference | DM, | $/ | | NEℓ, Mcal | MEm, Mcal | NEg, Mcal | CP, | Ca, | P, |
Feedstuff	Number	%	ton[a]	$/kg	/kg	/kg	/kg	%	%	%
X_1 Barley	4-07-939	89.0	94	0.1056	1.89	2.12	1.45	10.7	0.05	0.36
X_2 Corn	4-02-931	89.0	77	0.0865	2.42	2.24	1.55	10.0	0.02	0.35
X_3 Dical Phosphate	6-01-080	100.0	292	0.2920	—	—	—	—	23.70	18.84
X_4 Limestone	6-02-632	100.0	72	0.0720	—	—	—	—	36.07	0.02
X_5 Soybean meal	5-04-604	89.0	209	0.2348	2.07	2.09	1.43	51.5	0.36	0.75
X_6 Cottonseed	5-01-608	93.0	132	0.1419	2.66	2.41	1.69	24.9	0.15	0.73
X_7 Alfalfa hay	1-00-063	89.0	120	0.1348	1.21	1.24	0.68	16.0	1.35	0.22
X_8 Corn silage	3-08-154	27.9	35	0.1254	1.70	1.56	0.99	8.0	0.27	0.20
X_9 Salt		100.0	66	0.0660	—	—	—	—	—	—

[a]1 ton = 1000 kg.

This equation ensures that the amount MILK is restricted by the milk produced under each portion of production; otherwise, the amount could go very high because the object is to maximize profit.

To set the limits on each portion of production, as presented in Table 12-5, the following equation must be set:

$$P_1 \le 9.00 \quad \text{1st portion}$$
$$P_2 \le 9.00 \quad \text{2nd portion}$$
$$P_3 \le 4.50 \quad \text{3rd portion}$$
$$P_4 \le 4.50 \quad \text{4th portion}$$
$$P_5 \le 4.50 \quad \text{5th portion}$$
$$P_6 \le 2.25 \quad \text{6th portion}$$
$$P_7 \le 2.25 \quad \text{7th portion}$$
$$P_8 \le 2.25 \quad \text{8th portion}$$
$$P_9 \le 10.00 \quad \text{9th portion}$$

If we assume that the production requirements for protein, Ca, and P are linear, that is, directly proportional to the amount of milk, then these equations can be written in such a way that the amount of protein (or Ca or P) supplied by the feedstuffs minus the amount of protein required for production (protein per kg of milk times MILK) is greater than or equal to the requirement of protein for cow maintenance and growth. According to the NRC tables, a cow needs 0.074, 0.0026, and 0.0019 kg of CP, Ca, and P per kilogram of milk (3.5% FCM), respectively.

Maximum dry-matter intake equations can be set directly, that is, the sum of all feedstuffs being less than certain maximum dry-matter intake, or we can set the equation dependent on production, similarly to the way we have set the energy equation. Here we will set it in the former way.

$$X_1 + X_2 + X_3 + X_4 + X_5 + X_6 + X_7 + X_8 + X_9 \le \text{MDMI}$$

In this equation, MDMI is the maximum dry-matter intake. The figure used here is 3.7% of the body weight of the animal; that is, $600 \times 0.037 = 22.2$ kg dry matter.

The rest of the equations are set in the same way as in the least-cost formulation problem. The final maximum-profit formulation will be as follows:

Maximize profit: $0.25\text{MILK} - 0.1056X_1$
$- 0.0865X_2 - 0.292X_3$
$- 0.072X_4 - 0.2348X_5$
$- 0.1419X_6 - 0.1348X_7$
$- 0.1254X_8 - 0.066X_9$

Thus

$$X_1 + X_2 + X_3 + X_4 + X_5 + X_6 + X_7 + X_8 + X_9 \le 22.2 \quad \text{(dry-matter intake)}$$

$$1.89X_1 + 2.42X_2 + 2.07X_5 + 2.66X_6 + 1.21X_7 + 1.70X_8 - 0.33P_1 - 0.51P_2 - 0.77P_3 - 1.1P_4 - 1.21P_5 - 1.34P_6 - 1.7P_7 - 2.38P_8 - 3.22P_9 \ge 12.36 \quad \text{(energy)}$$

$$0.107X_1 + 0.1X_2 + 0X_3 + 0X_4 + 0.515X_5 + 0.249X_6 + 0.16X_7 + 0.08X_8 + 0X_9 - 0.074\text{MILK} \ge 0.881 \quad \text{(protein)}$$

$$0.0005X_1 + 0.0002X_2 + 0.237X_3 + 0.3607X_4 + 0.0036X_5 + 0.0015X_6 + 0.0135X_7 + 0.0027X_8 + 0X_9 - 0.0026\text{MILK} \ge 0.0264 \quad \text{(Ca)}$$

$$0.0036X_1 + 0.0031X_2 + 0.1884X_3 + 0.002X_4 + 0.0075X_5 + 0.0073X_6 + 0.0022X_7 + 0.002X_8 + 0X_9 - 0.0019\text{MILK} \ge 0.0204 \quad \text{(P)}$$

$$MILK - P_1 - P_2 - P_3 - P_4 - P_5 - P_6 - P_7$$
$$- P_8 - P_9 = 0$$

$P_1 \leq$	9.00	(1st portion)
$P_2 \leq$	9.00	(2nd portion)
$P_3 \leq$	4.50	(3rd portion)
$P_4 \leq$	4.50	(4th portion)
$P_5 \leq$	4.50	(5th portion)
$P_6 \leq$	2.25	(6th portion)
$P_7 \leq$	2.25	(7th portion)
$P_8 \leq$	2.25	(8th portion)
$P_9 \leq$	10.00	(9th portion)
$X_9 \geq$	0.05	(lower limit of salt)
$X_9 \leq$	0.10	(upper limit of salt)

As with the least-cost formulation, maximum-profit formulation has to be run with a linear programming software package, and the result should be interpreted and modified accordingly. After running this formulation for the first time, the diet does not contain any forage (run 1, Table 12-7). A correction is needed in order to produce a diet that is technically acceptable, that is, a diet that the animal can eat and that will create no problems. We have done this on purpose in order to illustrate the iterative procedure of ration formulation. As it is, this diet may not be consumed by the animal, or it will cause acidosis. We will explain how to correct this in the next section.

The reader should note that when diets are formulated on a dry-matter basis the for-mulation also should include the prices on a dry-matter basis. Failure to use prices on a dry-matter basis will give erroneous results that do not satisfy the maximum-profit property. When the result is obtained, it should be converted to an as-fed basis.

Under a maximum-profit formulation, we obtain a ration together with the optimum level of production for maximum profit. The reader should note that the *mathematical statement of the diet and the revision of the results* are important components for an appropriate diet formulation.

PROPORTIONS IN MATHEMATICAL PROGRAMMING FORMULATION

In diet formulations the nutritionist wants to fulfill not only the quantity of a given nutritional requirement of an animal, but also to specify the proportion in which some feedstuff must be in the diet or the proportions in which two nutrients should be present in the diet. Examples of these are the proportion of Ca to P in dairy rations.

EXAMPLE 3

We shall reformulate the previous diet in such a way that the forage-to-concentrate ratio is between 40:60 and 50:50; also, we shall include an upper limit for whole cottonseed such that it cannot exceed 25% of the concentrate.

TABLE 12-7

Diet composition and analysis

	Feedstuff	Run 1 Dry Matter	Run 1 As Fed	Run 2 Dry Matter	Run 2 As Fed
X_1	Barley, kg	0.000	0.000	0.000	0.000
X_2	Corn, kg	10.085	11.331	6.772	7.610
X_3	Dicalcium phosphate, kg	0.000	0.000	0.030	0.030
X_4	Limestone, kg	0.316	0.316	0.217	0.217
X_5	Soybean meal, kg	0.000	0.000	2.921	3.282
X_6	Cottonseed, kg	11.749	12.634	3.330	3.581
X_7	Alfalfa hay, kg	0.000	0.000	0.000	0.000
X_8	Corn silage, kg	0.000	0.000	8.880	31.828
X_9	Salt, kg	0.050	0.050	0.050	0.050
	Profit, $/cow/day	7.748		6.709	
	Optimum milk, kg/day	41.257		38.379	
	Feed, kg/cow/day	22.200	24.331	22.200	46.600
	Energy, NEℓ, Mcal	55.658		46.389	
	CP, kg	3.934		3.721	
	Ca, kg	0.134		0.126	
	P, kg	0.121		0.093	
	Cottonseed in concentrate, %	52.925		25.000	
	Forage-to-concentrate ratio	0:100		40:60	

To specify that the diet requires a forage-to-concentrate ratio between 40:60 and 50:50, we first write an expression that specifies the *forage,* that is $X_7 + X_8$; then, we represent the *total diet* by the expression

$$X_1 + X_2 + X_3 + X_4 + X_5 + X_6 + X_7 + X_8 + X_9$$

The ratio between the first and second expressions (forage and total diet) gives us the proportion of forage in the diet. The lower ratio (40:60) can be represented as follows:

$$\frac{X_7 + X_8}{X_1 + X_2 + X_3 + X_4 + X_5 + X_6 + X_7 + X_8 + X_9} > 0.4$$

Multiplying both sides by $(X_1 + X_2 + X_3 + X_4 + X_5 + X_6 + X_7 + X_8 + X_9)$, we get

$$X_7 + X_8 \geq 0.4(X_1 + X_2 + X_3 + X_4 + X_5 + X_6 + X_7 + X_8 + X_9)$$

Performing the multiplication on the right side and then subtracting from the left side, we obtain

$$-0.4X_1 - 0.4X_2 - 0.4X_3 - 0.4X_4 - 0.4X_5 - 0.4X_6 + 0.6X_7 + 0.6X_8 - 0.4X_9 \geq 0 \quad \text{(forage} \geq 40\%)$$

This linear equation then can be used in mathematical programming to specify the lower limit. With the following equation for the upper limit, the diet will have the range of 40% to 50% forage in the final form.

$$-0.5X_1 - 0.5X_2 - 0.5X_3 - 0.5X_4 - 0.5X_5 - 0.5X_6 + 0.5X_7 + 0.5X_8 - 0.5X_9 \leq 0 \quad \text{(forage} \leq 50\%)$$

With these manipulations, we can set specified percentages of any nutrient in a diet whose total amount is determined by the formulation; that is, it is a part of the optimal solution. This formulation is used to set ratios of an individual or a group of feedstuffs with respect to the total or part of the diet and to set ratios of individual nutrients with respect to the diet or any other nutrient. These manipulations are necessary because ratios (proportions) and percentages are not additive and, therefore, not linear. Linearity is essential for solving linear programming problems.

To specify that the whole cottonseed (X_6) must be less then 25% of the concentrate, we take the ratio of X_6 and the total concentrate ($X_1 + X_2 + X_3 + X_4 + X_5 + X_6 + X_9$) and make it less than 0.25:

$$\frac{X_6}{X_1 + X_2 + X_3 + X_4 + X_5 + X_6 + X_9} \leq 0.25$$

Multiplying both sides of this expression by the total concentrate $(X_1 + X_2 + X_3 + X_4 + X_5 + X_6 + X_9)$ and then subtracting the right side from X_6, we obtain

$$-0.25X_1 - 0.25X_2 - 0.25X_3 - 0.25X_4 - 0.25X_5 + 0.75X_6 - 0.25X_9 \leq 0$$

After adding this equation and the two forage-to-concentrate equations to our original maximum-profit formulation, we obtain a new result (run 2, Table 12-7). The diet in run 2 complies with the constraints that the forage-to-concentrate ratio should be between 40:60 and 50:50 and that whole cottonseed should be less than 25% of the concentrate.

This procedure can be used with the simultaneous equations method. Two things should be taken into consideration when using the simultaneous equations procedure: (1) the equations should have the = sign; (2) ranges are not possible, and therefore the ratio should be set at some fixed point.

FORMULATING WITH NEm AND NEg VALUES

The NRC uses a system for expressing net energy requirements (NE) and feed values for beef cattle that separates the NE requirements for maintenance (NEm) from those for gain (NEg) and gives different energy values for feedstuffs used for these two functions (7). NEg is applied only if the total energy intake is above that required for maintenance. NEm and NEg are not independent of each other. To handle this dependency with linear equations, some mathematical manipulations are necessary. Let us illustrate how to formulate the energy requirement for beef cattle with the NEm–NEg system, using the following example.

EXAMPLE 4

Using the feedstuffs in Table 12-6, let us formulate a minimum-cost diet for a growing steer with the following daily requirement: a minimum 9.4-kg dry-matter intake, 6.89 Mcal NEm, 5.33 Mcal NEg, 0.87 kg CP, 0.021 to 0.042 kg of Ca, 0.02 kg of P, and a minimum of 45% forage. We would also specify that salt must be included within a range of 0.03 to 0.06 kg.

The equation to specify crude protein is as follows:

$$0.107X_1 + 0.1X_2 + 0.515X_5 + 0.249X_6$$
$$+ 0.16X_7 + 0.08X_8 \geq 0.87$$

Dividing both sides by 0.87, we obtain

$$\frac{0.107}{0.87}X_1 + \frac{0.10}{0.87}X_2 + \frac{0.515}{0.87}X_5 + \frac{0.249}{0.87}X_6$$
$$+ \frac{0.16}{0.87}X_7 + \frac{0.08}{0.87}X_8 \geq \frac{0.87}{0.87}$$

Let us examine the coefficient of each feedstuff: $0.107 \div 0.87$ for X_1 is the inverse of the amount of X_1 necessary to supply the 0.87 kg of protein required; that is, $0.87 \div 0.107 = 8.1308$ kg of X_1 is necessary to supply **all** the required protein; $0.44 \div 0.87$ is the inverse of the amount of X_2 necessary to supply the 0.87 kg of protein required; that is, $0.87 \div 0.10 = 8.7$ kg of X_2 is necessary to supply **all** the required protein; and so on with X_5, X_6, X_7, and X_8.

Based on this, we can construct an equation to specify NE requirement in which $X_1, X_2, X_3, X_4, X_5, X_6, X_7, X_8$, and X_9 have coefficients equal to the inverse of the total amount of each feedstuff required to supply **both** NEm and NEg.

The amount of X_1 necessary to supply all the NE requirement for maintenance is the amount of NEm required divided by the amount of NEm/kg in X_1 ($6.89 \div 2.12 = 3.25$ kg); the amount of X_1 necessary to supply all the NE requirement for gain is the amount of NEg required divided by the amount of NEg/kg in X_1

($5.33 \div 1.45 = 3.6759$ kg). Therefore, the total amount of X_1 required to supply all the requirement of NEm and NEg will be $3.25 + 3.6759 = 6.9259$ kg. The inverse of this number will be $1 \div 6.9259 = 0.144586$. If a feedstuff does not have a NEm and NEg value, such as dicalcium phosphate, the inverse value is taken as zero (0). Table 12-8 presents the amounts of each ingredient necessary to supply NEm and NEg and the inverse value that will be used as the coefficient for the equation to specify the NE (NEm and NEg) requirement. The equation to specify the requirement of both NEm and NEg will be as follows:

$$0.144586X_1 + 0.153501X_2 + 0.142371X_5$$
$$+ 0.166313X_6 + 0.074656X_7 + 0.102036X_8 \geq 1$$

This procedure is tedious if done by hand, but fast and accurate if done by computer. Formulating a diet under the Lofgreen–Garrett system with this approach is a simple and accurate method.

For our example, the equation to specify dry-matter intake will be as follows:

$$X_1 + X_2 + X_3 + X_4 + X_5 + X_6 + X_7 + X_8$$
$$+ X_9 \geq 9.4$$

To specify the minimum 45% forage, we will divide the total forage ($X_7 + X_8$) by the total intake ($X_1 + X_2 + X_3 + X_4 + X_5 + X_6 + X_7 + X_8 + X_9$) and make it greater than or equal to 0.45:

$$(X_7 + X_8)/(X_1 + X_2 + X_3 + X_4 + X_5 + X_6 + X_7 + X_8 + X_9) \geq 0.45$$

Multiplying both sides of this expression by total dry-matter intake and subtracting the right side from ($X_7 + X_8$), we get

TABLE 12-8

Amounts required to supply NEm and NEg and inverse of total NEm + NEg

| Feedstuff | Amount Required to Supply all | | | Inverse |
	NEm	NEg	NEm + NEg	NEm + NEg
X_1	3.2500	3.6759	6.9259	0.144586
X_2	3.0759	3.4387	6.5146	0.153501
X_3	—	—	—	0.000000
X_4	—	—	—	0.000000
X_5	3.2967	3.7173	7.0240	0.142371
X_6	2.8589	3.1538	6.0127	0.166313
X_7	5.5564	7.8382	13.3946	0.074656
X_8	4.4167	5.3838	9.8005	0.102036
X_9	—	—	—	0.000000

$$-0.45X_1 - 0.45X_2 - 0.45X_3 - 0.45X_4$$
$$-0.45X_5 - 0.45X_6 + 0.55X_7 + 0.55X_8$$
$$-0.45X_9 \geq 0$$

The final formulation to produce a diet that is minimum cost for our example will be as follows:

$$\text{Minimum price} = 0.1056X_1 + 0.0865X_2 +$$
$$0.292X_3 + 0.072X_4 + 0.2348X_5 + 0.1419X_6 +$$
$$0.1348X_7 + 0.1254X_8 + 0.066X_9$$

such that

$$X_1 + X_2 + X_3 + X_4 + X_5 + X_6 + X_7 + X_8 + X_9$$
$$\geq 9.4 \quad \text{(dry-matter intake)}$$

$$0.107X_1 + 0.1X_2 + 0.515X_5 + 0.249X_6 + 0.16X_7$$
$$+ 0.08X_8 \geq 0.87 \quad \text{(protein)}$$

$$0.144586X_1 + 0.153501X_2 + 0.142371X_5$$
$$+ 0.166313X_6 + 0.074656X_7 + 0.102036X_8$$
$$\geq 1 \quad \text{(NE)}$$

$$-0.45X_1 - 0.45X_2 - 0.45X_3 - 0.45X_4 -$$
$$-0.45X_5 - 0.45X_6 + 0.55X_7 + 0.55X_8 - 0.45X_9$$
$$\geq 0 \quad \text{(minimum 40\% forage)}$$

$$0.0005X_1 + 0.0002X_2 + 0.237X_3 + 0.3607X_4$$
$$+ 0.0036X_5 + 0.0015X_6 + 0.0135X_7 + 0.0027X_8$$
$$\geq 0.021 \quad \text{(lower limit of Ca)}$$

$$0.0005X_1 + 0.0002X_2 + 0.237X_3 + 0.3607X_4$$
$$+ 0.0036X_5 + 0.0015X_6 + 0.0135X_7 + 0.0027X_8$$
$$\leq 0.042 \quad \text{(upper limit of Ca)}$$

$$0.0036X_1 + 0.0031X_2 + 0.1884X_3 + 0.002X_4$$
$$+ 0.0075X_5 + 0.0073X_6 + 0.0022X_7 + 0.002X_8$$
$$\geq 0.02 \quad \text{(P)}$$

$$X_9 \geq 0.03 \quad \text{(lower limit of salt)}$$
$$X_9 \leq 0.06 \quad \text{(upper limit of salt)}$$

The results of this formulation after running it with a linear programming program are presented in Table 12-9.

SENSITIVITY ANALYSIS

The final result of mathematical programming is more than a feed formula. In addition to the quantities of each ingredient that will comprise a given ration, mathematical programming yields a series of decision-making indicators. With these, the nutritionist can make management decisions and future ration adjustments that will maximize economic return. Information is given for ingredients that are not included in the ration, as well as those that are part of the ration.

For the growing steer ration presented in Table 12-9, information on all ingredients submitted for consideration is presented in the following tables. Table 12-10 has information regarding upper and lower limit prices for all ingredients, while Tables 12-11 and 12-12 have additional information regarding substitutions or requirements changes of the diet when prices reach upper or lower limits. All this information is derived from mathematical programming.

INGREDIENT PRICE RANGES

Table 12-10 presents ingredient price ranges for our beef example (Table 12-9). For those

TABLE 12-9

Diet composition and analysis

	Feedstuff	Dry Matter	As Fed
X_1	Barley, kg	0.000	0.000
X_2	Corn, kg	5.038	5.661
X_3	Dicalcium phosphate, kg	0.000	0.000
X_4	Limestone, kg	0.072	0.072
X_5	Soybean meal, kg	0.000	0.000
X_6	Cottonseed, kg	0.000	0.000
X_7	Alfalfa hay, kg	0.347	0.390
X_8	Corn silage, kg	3.883	13.918
X_9	Salt, kg	0.060	0.060
	Minimum price $	0.979	
	Feed, kg/day	9.400	20.100
	Energy, total NEm, Mcal	17.774	
	total NEg, Mcal	11.890	
	available NEg, Mcal	7.280	
	CP, kg	0.870	
	Ca, kg	0.042	
	P, kg	0.024	
	Forage, %	45.000	

TABLE 12-10

Ingredient price ranges

Ingredient	Allowable Decrease	Allowable Increase	Price ($/kg)	Lower Limit	Upper Limit
Barley	0.016265	Infinity	0.094000	0.077735	Infinity
Corn	0.002481	0.016237	0.077000	0.074519	0.093237
Dicalcium phosphate	0.219078	Infinity	0.292000	0.072922	Infinity
Limestone	0.244321	0.002788	0.072000	−0.172321	0.074788
Soybean meal	0.088329	Infinity	0.209000	0.120671	Infinity
Cottonseed	0.035156	Infinity	0.132000	0.096844	Infinity
Alfalfa hay	0.008738	0.001985	0.120000	0.111262	0.121985
Corn silage	0.000622	0.002739	0.035000	0.034378	0.037739
Salt	Infinity	0.008688	0.066000	−Infinity	0.074688

ingredients that are part of the solution, price ranges indicate the lower and upper limit at which these ingredients can fluctuate in price without affecting the levels of ingredient inclusion in the present ration formula. For ingredients not included in the ration, the lower limit is the price at which the given ingredient will be considered for inclusion in the diet. These prices are referred to as *opportunity prices* and can be interpreted as the price at which it is economical to pay if the ingredient is to be included in the ration.

For those ingredients included in the ration, if the price increases (within the range), then the ration price will increase, and if the price decreases within that range, then the ration price will decrease. However, the amounts and proportions of the ingredients included in the final answer will remain the same. If the ingredient price is outside this range, then the ration composition will change; the change could be one ingredient replacing another or that the proportions in the ration will be different, in which case nutrient levels could be different.

Regardless of the price change, the mathematical programming will try to find a solution to the formulation problem. However, it is im-

portant to verify every intermediate answer to make sure that the ration is technically correct; that is, the ration not only fulfills all nutritional requirements, but the diet can be consumed by the animal without causing any metabolic problem. If the diet is not technically correct, then it should be modified and formulated again.

From Table 12-10, for example, corn (one of the ingredients included in the ration), can fluctuate between $0.074519 and $0.93237/kg without changing its level of 5.038 kg (see Table 12-9) in the diet. However, barley is an ingredient not included in the final answer, but will enter the final diet only when the price is below $0.077735/kg. Barley's current price is $0.094/kg (Table 12-6).

INGREDIENTS NOT INCLUDED IN THE RATION

For those ingredients submitted for consideration but not included in the ration, additional information is presented in Table 12-11. The prices referred to in this table are based on the price ranges presented in Table 12-10. Ingredients submitted for consideration may not be included in the ration because they are "expensive." We observe that soybean meal

TABLE 12-11

Ingredients not included in ration

Ingredient	If Price Is Reduced		If Ingredient Is Forced	
	Replaces Ingredient	Constraint at Limit	Penalty Price	Maximum Can Force
Barley		Phosphorous	16.2650	1.9534
Dicalcium phosphate	Limestone		219.0780	0.1070
Soybean meal	Alfalfa hay		88.3288	0.0692
Cottonseed	Alfalfa hay		35.1556	0.1928

is not included in the ration. If the price of soybean meal is reduced from $0.209 (current price) to $0.1206712/kg (Table 12-10, the lower limit) or lower, then soybean meal will replace alfalfa hay and the price of the ration will be lower.

The student can visualize that the information relates not only at what price the ingredient will enter but also what ingredient will be replaced. This information is useful when considering feedstuff procurement. In some cases, there is no replacement but a change in the level of a nutrient, in which case a decision should be made as to whether the change in level is appropriate in relation to animal performance.

Part of the results for ingredients not included in the final diet is the *penalty price*. If we decide to force an ingredient that has not been included, then for every kilogram that we force, the ration price will increase by the penalty price; but this penalty is applicable only for the quantity specified as *maximum can force*. For example, barley is not included in the ration. If we force barley, then the ration price will increase by $16.265/ton (Table 12-11, penalty price) per kilogram of barley included, but this penalty price holds only for the first 1.9534 kg (Table 12-11, maximum can force) of barley forced. If we force more than 1.9534 kg, then two things can happen: the penalty price will increase, or there may not be a solution to the formulation problem.

INGREDIENTS INCLUDED IN THE RATION

Information referring to the ingredients submitted for consideration and included in the ration for the beef example (Table 12-9) is presented in Table 12-12. Increases refer to increases above the upper limit of that range and decreases to those below the lower limit of the range (Table 12-10).

If the price of an ingredient in the ration falls below the lower limit, it becomes "inexpensive," and we may want to increase its amount in the ration. If the price of an ingredient in the ration goes above the upper limit, it becomes "expensive" and we may want to decrease its amount in the ration or eliminate it completely. In both cases we are unbalancing the ration. Mathematical programming may correct this condition by including a new ingredient. If the change in the diet composition is within the specified requirements, then no new ingredient will enter.

We observe in Table 12-12 that, if the price of alfalfa hay is reduced below the lower limit (Table 12-10, $0.1112617), then protein in the diet will be above the specified minimum requirement (Example 4, minimum of 0.87 kg).

REQUIREMENT ECONOMIC VALUATION

Mathematical programming gives a value to nutritional requirements, more specifically the diet's constraints, similar to the way that ingredients are priced. Even though the general ration formulation assumes that constraints (nutritional requirements) are fixed, in practice there is a degree of flexibility as some requirements are not strictly fixed. Constraints valuation is quantified by what the economists define as *dual prices*. The dual price is the economic valuation of the diet's requirements and is expressed as dollars per unit ($/Mcal, $/kg, $/ppm). These requirement economic values quantify how much is the value per unit of each constraint, representing the value of an additional unit of a given constraint.

Dual prices for the beef ration are presented in Table 12-13. Dual prices exist only for those constraints for which the ration level is equal to the specification. For example, the dual price for protein level is $0.118299/kg (Table 12-13) and not zero, because the level of protein in

TABLE 12-12

Ingredients included in ration

	If Price Is Reduced		If Price Is Increased	
Ingredient	Ingredient Entering	Constraint in Excess	Ingredient Entering	Constraint in Excess
Corn		Calcium	Barley	
Limestone		Salt		Calcium
Alfalfa hay		Protein		Calcium
Corn silage		Calcium		Protein
Salt				Salt

TABLE 12-13

Dual prices and constraint slack or surplus

Constraint	Slack or Surplus	Dual Prices	Constraint Value	Value (%)
Amount	0.0000	0.093281	0.876837	89.4192
NEm and NEg	0.1955	0.000000	0.000000	0.0000
Protein	0.0000	0.118299	0.102920	10.4957
Calcium, <	0.0210	0.000000	0.000000	0.0000
Calcium, >	0.0000	−0.007453	0.000313	0.0319
Phosphorus	0.0062	0.000000	0.000000	0.0000
Forage	0.0000	0.041316	0.000000	0.0000
Salt, <	0.0300	0.000000	0.000000	0.0000
Salt, >	0.0000	−0.008688	0.000521	0.0532

the final ration meets, exactly, the requirement specified. However, the dual price of energy (NEm and NEg) is zero because the level of NEg is not exactly as specified, but above the minimum required. If a nutritional requirement in the final ration is in excess, then it is not a limiting resource; therefore, it will have no value. For the protein requirement the dual price is $0.118299/kg, meaning that if the protein requirement is increased by 1 unit (kilogram of protein) then the ration's price will increase by $0.118299. The unit here is 1 kg because the constraint was specified as "more than 0.87 kg of protein." If we increase the requirement by 10 g, then the final price of the ration will change from $0.979 (Table 12-9) to $0.980183 [(0.01 × 0.118299) × 0.979].

Dual prices have signs. A positive sign indicates that as the constraint increases the ration price will increase, and if the constraint decreases, the ration's price will decrease. A negative sign indicates that as the constraint increases the ration price will decrease, and if the constraint decreases, the ration's price will increase. Multiplying the dual price by the specified constraint level will yield the constraint value, and then this value can be converted to percentage. For example, if we multiply the dual price of the protein requirement (0.118299) by its specified level (minimum 0.87 kg of protein) we will obtain 0.118299 × 0.87 = 0.10292. We can express the constraint value as a percentage of the ration total price. In the case of protein, the percent value is calculated as follows: 0.10292 (constraint value) divided by 0.979 (ration price), all multiplied by 100 (0.10292 × 100/0.979 = 10.4957). The percent value will indicate which constraint contributes the most (or the least) to the total price of the ration. Dual prices can help you decide if a given nutritional requirement could be changed. Atten-

tion should be paid to the decrease in ration price as compared to decrease in performance if a requirement is changed.

EXACTNESS IN RATION FORMULATION

In practice, most rations for farm animals are milled and mixed with little or no analysis of individual batches of ingredients, even though the nutrient content of a new batch may differ from that of previous batches. The result is that the finished ration is not likely to have precisely its expected nutrient content, and it may be higher or lower in quality than it is supposed to be. It might also be noted that the long detailed lists of nutrients that may be obtained from computer printouts, sometimes expressed to the 1/1000 of a pound, tend to give the impression of far greater precision than actually exists.

The commercial feed mill that sells mixed diets with this uncertainty in composition may have a substantial percentage of batches that fall outside legal limits, thus rendering the miller liable to prosecution by regulatory agencies. The miller can respond by formulating to a higher average standard than is claimed or gain greater control by analyzing ingredients before use.

As a general rule, milling by-products of cereals and many commercial protein supplements are standardized by the processors, so they tend to vary less in crude protein content than feedstuffs such as cereal grains and roughage sources. Consequently, if variation in nutrient content is to be reduced, analyses must be done on the major ration ingredients that are most likely to vary from batch to batch.

Least-cost ration formulation implies that a ration will be produced that will result in least cost to the feeder; however, this concept is misleading for several reasons. As illustrated previously, there are problems in knowing what the nutrient concentration is in the diverse variety of feedstuffs utilized today, not to mention the problems in evaluating animal utilization of nutrients from different sources of feedstuffs.

It is true that least-cost formulas can be obtained from computers if feedstuff composition is accurately known, but there is no guarantee that such formulas will result in least-cost animal production. A diet that maximizes profit must consider the effects of the nutrient levels on animal performance, costs of the nutrients, costs of the livestock operation, and returns from production.

From the nutritionist's point of view, it is not a simple matter to establish a fixed nutrient requirement for animals, because of differences in breeds, variation between animals of similar genetic background, differences in utilization of nutrients from different sources, and many other environmental factors. Theoretically, as a limiting nutrient is added to a ration, animal performance will increase up to a point where the animal has all the nutrient it can use and further additions will not improve response at all.

FORMULATING PREMIXES

In nutritional use a premix refers to a small amount of a total mixture. Commercial premixes are made of antibiotics, vitamins, trace minerals, and the various drugs and medicines that may be used in livestock feed. In nearly all instances the amount of material to be added is very small, perhaps as little as a few grams per ton of finished feed. Quantities as small as this are difficult to mix uniformly into large batches, thus the reason for the larger premixes. In addition, many small feed mills do not wish to be bothered with weighing out micro amounts of these different ingredients and they would prefer to use the premixes; fewer errors probably result, also, from use of the premixes.

Premixes are often made up so that they come in packages of 5 to 50 lb, which will be added to one or more tons of feed. They are mixtures of the microingredients mixed with some type of carrier, which may be soybean meal, ground grain, wheat middlings, or other mill feeds. Sometimes about 2% stabilized fat will be added to reduce loss of the microingredients as dust in any future mill operations. Normally, the vitamins are not mixed in with the mineral premixes because a high concentration of mineral elements is apt to result in oxidation and destruction of vitamin activity.

PREPARATION OF A VITAMIN A–ANTIBIOTIC PREMIX FOR CATTLE

Suppose that we want to formulate a premix to be used at 10 lb/ton in a complete finishing ration for cattle that should have 1000 IU of vitamin A/lb and 5 mg/lb of antibiotic. This means we need 10 g of antibiotic (5 mg × 2000 = 10 g) and 2 million IU of vitamin A (1000 IU × 2000 lb). Antibiotics may come in different concentrations, varying from pure sources (usually expensive) to much more dilute concentrations (see Appendix Table 7). For our purposes here it will be specified that the product has 50 g/lb. Likewise, vitamin A may be purchased in different concentrations. In this example, a source with 650,000 IU/g will be used. Consequently, for the premix 3.08 g (rounded to 3.1) of vitamin A concentrate and 0.20 lb of antibiotic (or 0.2 × 453.6 g/lb = 90.7 g) for a total of 93.8 g will be needed; the remainder of the 10 lb (4536 g), or 4442.2 g, will be made up of soybean meal. In practice, probably enough of this for several tons of feed would be made up at one time. If 100 lb of premix is wanted, enough for 10 tons, it is merely a matter of multiplying the different ingredients by 10 in this instance.

VITAMIN AND MINERAL PREMIXES FOR BROILERS

For many poultry rations, it is sometimes common to ignore most of the vitamins likely to be present in feedstuffs or, in some instances, to formulate premixes on the assumption that about half of the vitamins in the feed will be available at the time of feeding. An example of a premix used for broilers getting typical corn–soy rations is shown. It is made up in a concentration so that 5 lb will be added to each ton of finished feed. Since this total is to be in 5 lb (or 2268 g), we need, in grams, 2268 − 456.21 = 1811.79 g of carrier. This amounts to 79.88% of the 5 lb of premix. If premix is needed for more than 1 ton, it is simply a matter of increasing the amount.

With mineral premixes, which are usually used for the trace minerals, the procedure is similar. A mixture that might be used for broilers is shown, along with the amount, source, concentration in the source, and amount needed for a 5-lb premix.

The total amount of these mineral salts, 296.9 g, would then be diluted to 5 lb (2268 g) with some appropriate carrier.

Vitamin	Amount per 5 lb	Concentration in Source	Amount Source Needed in Premix
Vitamin A	3 mil IU	650,000 IU/g	4.62 g
D	1 mil ICU[a]	200,000 ICU/g	5.0 g
E	1000 IU	275 IU/g	3.64 g
K	500 mg	Pure	0.50 g
Riboflavin	3 g	0.50 g/g	6.0 g
d-Pantothenic acid	5 g	0.41 g/g	12.2 g
Niacin	20 g	Pure	20.0 g
Choline	173.6 g	0.434 g/g	400.0 g
B_{12}	5 mg	1.32 mg/g	3.8 g
Folacin	200 mg	0.45 g/g	0.45 g
			456.21 g

[a]International chick units.

Mineral Element	Amount per 5 lb (g)	Source	Element in Feed-Grade Salts, %	Amount of Source Needed in Premix, g
Mn	54.4	$MnSO_4$	28	194.6
Fe	18.16	$FeSO_4$	21	86.5
Cu	1.82	$CuSO_4$	25	7.3
I	1.09	KI	69	1.6
Zn	2.50	$ZnSO_4$	36	6.9
				296.9

USE OF FIBER OR PHYSICAL DENSITY IN FORMULATION

Limitations on fiber are often used in formulation, particularly for dairy cows (minimums). Maximum values may be used for other species. For cows, this is done because of the well-known effect of low-fiber rations on milk fat depression (see Ch. 15). In the opinion of the writer, this is not a very effective nor versatile means of achieving the objective of presenting a ration having an optimal bulk density to the animal—an important factor in maintaining good rumen function over a prolonged period of time. For example, if we have hay available from the same source as long (baled or stacked), chopped, ground, cubed, or pelleted, then the fiber concentration should be the same in each product, provided no losses occur in harvesting or processing. Obviously, pelleted hay does not produce the same production response as long hay because of much greater consumption and more rapid passage through the rumen (see Ch. 11). Furthermore, the fiber restrictions do not always give the same type of response when used with dry roughages as opposed to wet roughages (e.g., silage) or with different types of roughage. Thus the fiber limitation is limited to a restricted usage rather than as a versatile means of defining ration characteristics. In addition, processing of grains will alter bulk density (see Ch. 11), although it may not always affect digestibility.

In the opinion of the author, nutritionists who use fiber (crude or ADF) and concentrate limitations for dairy cows to help define a ration are really trying to come up with a measure of physical density. Consequently, it seems logical that utilization of some measure of bulk density, particularly in computer formulation, should allow a more complete characterization of a ration.

Physical density data have been included for some feedstuffs listed in Table 11-15. Such data are not readily available on many roughages, partly because such measurements are more tedious to collect and more variable than for concentrates.

Mertens (8) has presented data that suggest that NDF (neutral detergent fiber) is a reasonably good measure of density when the roughage in the ration was ground. In contrast to the other fiber values, NDF contains all the normal fibrous components (lignin, cellulose, hemicellulose). NDF is also highly correlated to digestibility (negatively), rumination (posi-

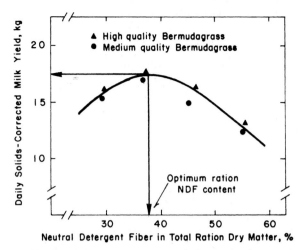

FIGURE 12-3 Determination of the optimal neutral detergent fiber content of the total ration resulting in maximum solids-corrected milk production using the quadratic regression equation. Courtesy of D. R. Mertens (8).

tively), and intake (negatively). A graph showing response of dairy cows to different NDF levels is shown in Fig. 12-3. However, this procedure may not resolve the problem of changes in density that occur with different types of feed processing, with roughages fed in different physical forms, or when different roughages are substituted in the ration.

The effect of feeding beef cattle diets with different densities is shown in Table 12-14. In this instance the concentrate portion of the ration remained the same, and density was altered by varying the proportions of cottonseed hulls and ryegrass straw, each presumably containing about the same level of TDN. It is obvious from the table that decreasing the density resulted in less DM consumption, consumption of a greater volume of feed, and less intake of TDN. These data are too limited to extrapolate

TABLE 12-14

Consumption by beef cattle of diets varying in bulk density

		Consumption by Cattle		
Diet	Density of Diet, kg/bu[a]	DM, % of BW	Volume, bu/1000 lb of BW	TDN, kg/1000 lb of BW
1	12.8	3.52	1.25	11.4
2	10.7	3.42	1.46	11.2
3	9.2	3.21	1.59	10.5
4	8.0	2.91	1.65	9.5
5	7.1	2.56	1.63	8.4

Source: Kellems and Church (9).

[a]Density is expressed as kg/bushel. It was altered by changing the proportions of cottonseed hulls and chopped ryegrass straw, presumably without altering caloric density of the diets. Cattle were fed each diet individually for a two-week period.

to other situations, but the method seems to be one that could be used with minimal effort in computer programs.

FACTORS AFFECTING NUTRIENT NEEDS OR FEED UTILIZATION

Now that the reader has been exposed to the methods used to formulate rations by hand, it is appropriate to call attention to various factors that may alter nutrient needs or utilization of feed. Unfortunately, a number of these are not, at this time, built into feeding standards. Nevertheless, the reader should be aware of them.

ANIMAL VARIATION

Animals of similar breeding do not always have the same nutrient needs, even though it is economically feasible to treat them as if they do. Previous exposure to various stresses (disease, injury) may have altered their capabilities. Hormone stimulus (thyroid, growth, and sex hormones) can easily alter metabolism as well as produce differences in physical activity. In addition, some individuals are more susceptible to a given stress than others. In animal research, many documented instances show marked differences in nutrient requirements. Thus it is well to remember that closely related animals may differ in nutrient requirements just as they differ in taste preferences, hair color, or temperament. The consequence (for ration formulation) is that our objectives should be to satisfy the needs of most of the animals. A few will receive more of one or more nutrients than needed; most should get what they need; and a few may be underfed on one or more nutrients. Only in the case of outstanding animals (such as superior dairy cows) is it feasible to feed on an individual basis so that none is underfed.

MISCELLANEOUS FACTORS

In many instances experienced nutritionists may modify specified requirements in a given situation on the basis of experience under similar conditions with a given class and species of animal. This is often needed, particularly for animals (especially ruminants) grown under nonstandard conditions and fed markedly different ration ingredients. In contrast, poultry (and many swine) are produced under rather standard conditions regardless of the area and, in addition, basic rations are rather similar regardless of where the birds are grown. Consequently, it is much more difficult to refine the dietary requirements of ruminants and horses to the degree that can be accomplished with poultry or swine.

Nutrient requirements of healthy animals may be altered by a considerable number of genetic, management, and environmental factors (7, 10, 11, 12). Some that have been identified include

Genetic: species, breed, strain.

Production related: age, sex, pregnancy, lactation (or egg production) and level of lactation, growth and rate of growth, desired carcass fatness, and composition of milk produced.

Management related: production system (total confinement, grazing, caged, etc.), additives use, distance to milking parlor, and groupings (7).

Other environmental: disease, nutrient deficiencies, climatic factors such as temperature and humidity, muddy lots, wind, rain, and other miscellaneous stresses.

Although they may not always alter the requirement for absorbed nutrients, factors such as level of feed consumption, energetic and physical density of the diet, feed additives, growth stimulators, and feed processing methods may greatly affect efficiency or completeness of digestion and metabolism of absorbed nutrients. Thus, if we are basing animal needs on chemical composition or digestibility, these factors usually alter efficiency and have the same final effect as if they altered nutrient requirements. For example, if we feed lambs a pelleted hay–grain diet, consumption will be increased greatly, as compared to a non-pelleted diet, and the lambs will gain more per day even if they are on a diet with a lower concentration of digestible protein and energy.

The NRC publications generally are developed to make allowances for some differences related to species, occasionally for breeds, and for body size, sex, age, pregnancy, lactation and milk fat percentage, level of egg production, and rate of growth. Information is not, at this time, sufficient to include the marked effect of the environmental factors in the nutrient requirement tables.

SUMMARY

Formulation of a satisfactory ration for a given situation represents the summations and utilization of information on animal requirements and nutritive value and palatability of feedstuffs. While a variety of different methods can be used to formulate acceptable rations for livestock, the result should be a ration that is acceptable to the animal, one that will produce the desired results (level of production), and one that will be economical for the livestock feeder.

Hand methods of ration formulation are quite satisfactory for formulating rations in which exact amounts are required for two major nutrients. More complex formulas can be put together using reserved feedstuffs or slack space for minor ingredients. Very complex rations involving minimums of a number of different nutrients are, generally, better done with linear programming methods using computers. The reader should remember that exact mathematical methods do not guarantee exactness in actual formulas because tables of nutrient composition of feedstuffs may not always be representative of the feedstuffs that will be combined into a formula that will be fed to the animals.

REFERENCES

1. Hill, L. B. 1982. *Feedstuffs* 54(23): 30.

2. Cooper, L., and D. Steinberg. 1974. *Methods and applications of linear programming.* Philadelphia: W. B. Saunders Co.

3. Hadley, G. 1962. *Linear programming.* Reading, MA: Addison–Wesley.

4. Varela-Alvarez, H. 1978. Description of an introductory course in Operations Research for Animal Science students. M. A. Paper, University Park, PA. Department of Statistics, Pennsylvania, State University.

5. Varela-Alvarez, H., and R. Grapes. 1991. *J. Animal Sci.* Suppl. 1, 69:563–564. Abs. 885.

6. Dean, G. W., et al. 1972. Giannini Foundation Monograph 31, Berkeley, CA: Calif. Agr. Exp. Sta.

7. NRC. 1996. *Nutrient requirements of beef cattle.* 7th rev. ed. Washington, DC: National Academy Press.

8. Mertens, D. R. 1980. *Proc. Distillers Feed Conf.* 35:35.

9. Kellems, R. O., and D. C. Church. 1981. *Proc. West. Sec. Amer. Soc. Animal Sci.,* 32:26.

10. Fox, D. G., C. J. Sniffen, J. D. O'Connor, J. B. Russell, and P. J. Van Soest. 1990. *Search: Agriculture.* Ithaca, NY: Cornell Univ. Agr. Exp. Sta. No. 34.

11. NRC. 1981. Effect of environment on nutrient requirements of domestic animals. Washington, DC: National Academy Press.

12. Varela-Alvarez, H., and R. Grapes. 1994. *FEED. Ration formulation program.* Manual and Reference Guide. Association Consultants International.

13

FEEDING THE BEEF COW HERD

Jan Bowman and Bok Sowell

INTRODUCTION

Cow–calf production relies heavily on grazed forages to supply nutrients for both cows and calves. For much of the cow–calf production cycle, requirements for protein and energy can be met with low- to medium-quality forages. The forages that form the basis of beef cattle production can be native rangelands and introduced pastures. In addition, conserved forages (hay) are fed when weather conditions limit grazing or forage availability. Dormant range forage is usually high in fiber and may be deficient in both protein and energy, especially for cows during late gestation and lactation (Fig. 13-1). Limited forage quantity or quality may require supplemental feeding.

The most important factor influencing the performance of beef cattle consuming forage diets is dry-matter intake. In these types of diets, the actual physical volume that the forage occupies in the rumen can limit intake. The clearance of feed residues from the rumen is the primary process determining forage intake and nutritive value. Forage removal from the rumen depends on two processes, digestion by ruminal microorganisms and passage of the digesta out of the rumen. It is necessary for the forage to be colonized by ruminal bacterial, digested and

physically reduced in size before it can pass out of the rumen and provide room for additional forage intake. Bacterial colonization is important both in supplying nutrients from forage digestion to the animal and in subsequent reduction of particle size and passage from the rumen. The kinetics of digestion and passage in beef cattle affect intake by determining the speed with which forages are broken down, nutrients made available, and indigestible residues leave the rumen.

BIOLOGICAL CYCLE AND REQUIREMENTS

A cow's biological cycle can be divided into four periods: the first, second and third trimesters of gestation, and the postpartum period (Table 13-1). Cows should be fed for optimum production, not necessarily maximum production, and therefore the requirements should be expressed and considered on an amount per day basis. The breeding herd is fed primarily roughages, with supplemental feed used as necessary to meet specific requirements.

The first trimester of gestation begins on the day that conception occurs. During the first trimester, the cow requires nutrients for maintenance and lactation if the cow has a calf at

FIGURE 13-1 Hereford cow–calf pairs grazing summer range.

side. Maintenance requirements are greater for cows with greater body weight and may vary by breed type, with breeds that have a larger mature size requiring relatively more and breeds with a small mature size requiring relatively less. Nutrient requirements for lactation increase as the level of milk production and milk fat content increases. However, during the first trimester, milk production is declining, and spring-born calves are consuming more forage. Fetal growth is a relatively minor proportion of nutrient requirements in the first trimester.

During the second trimester, this year's calf is weaned and lactation requirements end. From weaning to the start of the third trimester is the time of lowest nutrient requirements for the beef cow. A nonlactating cow in adequate body condition can be fed minimally during this time, and it is during this period that it is easiest and most economical to increase body condition in thin cows.

During the third trimester, the cow's nutrient requirements are increasing rapidly due to rapid fetal growth. A cow can be expected to gain approximately 0.9 lb/day during the last trimester just to support fetal growth. Cow body condition must be watched closely during this time. Cows that are too thin at calving have a greater incidence of dystocia, weak calves, sick calves, and decreased milk production.

TABLE 13-1

Biological cycle of the cow

Period		Length (days)
First trimester gestation		95
Second trimester gestation		95
Third trimester gestation		95
Postpartum		80
	Total	365

Cows should not lose more than 15% of their body weight during the winter and through calving. Forage quality becomes more critical during this time due to the increased nutrient requirements. The best-quality hay should be saved for the 60 days prepartum and the 90 days postpartum.

The postpartum interval is a critical period nutritionally for the cow. During this time lactation requirements are high, and the cow's reproductive system is recovering from parturition. Feed intake by lactating cows is 35% to 50% greater than for nonlactating cows. A cow must be in adequate body condition to begin cycling and to conceive during the 80-day postpartum period. If cows are nutritionally stressed during this time, reproductive performance is the first thing to suffer. It is difficult and expensive to increase cow body condition during this time period. Requirements are approximately 50% higher during this period of the biological cycle (Fig. 13-2).

ENERGY REQUIREMENTS

The maintenance energy requirement is the amount of dietary energy needed to maintain an animal with no loss or gain in body energy. The amount of energy required for maintenance in beef cattle has been estimated as 77 kcal/$W^{0.75}$, where W is body weight in kilograms (1). Maintenance energy requirements, therefore, increase as the cow's body weight increases. Larger mature size breeds may have greater maintenance energy requirements due to having a greater metabolic mass to support and to being physiologically younger at a given age than breeds with a smaller mature size. Physical activity increases the energy requirement, and free-grazing cows that range long

FIGURE 13-2 Lactating cow during postpartum interval when nutrient requirements are high.

distances may have a 30% to 50% greater energy requirement than cows under more confined conditions. Energy requirements are also increased if the effective environmental temperature is above or below the cow's thermoneutral zone (the temperature zone above or below which energy must be expended to maintain a constant body temperature; Table 13-2).

Energy is considered first in balancing diets to meet nutrient requirements. Carbohydrates, lipids, and proteins can supply energy to cows, and the feed ingredients used to supply energy make up the largest part of the ration in terms of weight and cost. Energy supply and utilization determine to a large extent the ability of the cow to utilize other nutrients; therefore, energy intake is important in determining the requirement for other nutrients.

PROTEIN REQUIREMENTS

Protein from microbial protein synthesis supplies on average 50% of the protein and amino acids needed by cattle. However, protein deficiency is the most common nutrient deficiency for cows grazing mature forage or consuming low-quality hays or straws. When forages contain less than 7% crude protein, there may not be sufficient ruminal ammonia for the growth and activity of the microbes. This reduces digestion of forage and thereby limits forage intake. Positive responses in digestibility and intake of low-quality forages are often seen when protein supplements are supplied.

MINERAL REQUIREMENTS

Young cattle have relatively higher Ca and P requirements due to their rapid rate of bone growth. High levels of milk production and pregnancy also increase the requirements for Ca and P. Calcium or P deficiency results in rickets in growing animals, and adult cattle can suffer from osteomalacia, or a demineralization of bone tissue, resulting in brittle bones that may break easily. An acute Ca deficiency called parturient paresis, or milk fever, may occur in high-milk-producing cows shortly after the start of lactation. Forage legumes contain relatively high levels of Ca and low levels of P, while grains are low in Ca and higher in P. Sources of Ca and P with high availability include dicalcium phosphate and monocalcium phosphate.

Cobalt is required by the rumen microorganisms to synthesize Vitamin B_{12}. The Co requirement for all classes of cattle is estimated to be 0.1 ppm (Table 13-3). Forages vary considerably in their Co content; cereal grains are generally low in Co, and oilseed meals are good sources. Cobalt from cobalt carbonate is highly available, while the oxide, chloride, and sulfate forms have medium availability.

Dietary levels of 10 ppm of Cu are considered to be adequate for beef cattle (Table 13-3). However, Simmental and Charolais cattle have been shown to have higher Cu requirements than Angus cattle (2) and may require up to 15 ppm. High concentrations of sulfur and/or molybdenum interfere with Cu utilization and can induce a Cu deficiency. To prevent Mo interfering with Cu utilization, the Cu:Mo ratio should be 4.5 or above. Symptoms of Cu deficiency include lack of pigmentation in the hair, joint and leg weakness or abnormality in newborn calves, delayed estrus, and embryonic death. The hair on the edge of the ears in black cattle may turn reddish and in red cattle, yellowish. Serum Cu levels below 0.6 ppm may indicate a potential deficiency. Liver biopsy is a good indicator of Cu status, with levels below 25 ppm of liver tissue (75 to 90 ppm on a dry-matter basis) considered deficient. Copper has a relatively narrow range of requirement and toxicity. Adding copper sulfate to ponds to reduce algae growth can result in consumption of toxic levels of Cu. Forages are relatively good sources of Cu, while grains are generally low. Copper carbonate has a high Cu availability, while the sulfate form is intermediate in its availability.

Deficiencies of I are found in the Northwest and Great Lakes regions due to low levels of I in the soil. Symptoms of I deficiency include weak calves, hairless calves, goiter, reduced growth rate, early embryonic mortality, abortion, silent estrus, retained placenta, and decreased conception rate. The use of iodized salt is the most convenient method of supplying adequate iodine.

Iron is important in the transport of oxygen in the body, and apparently 50 ppm is adequate for adult cattle (Table 13-3). Iron toxicity can result with levels above 1,000 ppm in the diet. Forages contain higher levels of Fe than do

TABLE 13-2

Critical temperatures for beef cows

Conditions	Critical Temperature(°F)
Summer hair coat or wet coat	59
Fall hair coat	45
Winter hair coat	32
Heavy winter hair coat	17

TABLE 13-3

Mineral requirements for beef cows

Mineral	Requirement	Range	Toxic Level
Cobalt	0.1 ppm	0.07–0.11 ppm	>5 ppm
Copper	8.0 ppm	4–10 ppm	>100 ppm
Iodine	0.5 ppm	0.2–2.0 ppm	>50 ppm
Iron	50 ppm	50–100 ppm	>1000 ppm
Magnesium	0.1%	0.05–0.25%	>0.4%
Manganese	40 ppm	20–50 ppm	>1000 ppm
Potassium	0.65%	0.5–0.7%	>3%
Selenium	0.2 ppm	0.05–0.3 ppm	>2 ppm
Sodium	0.08%	0.06–0.1%	>10%
Sulfur	0.1%	0.08–0.15%	>0.4%
Zinc	30 ppm	20–40 ppm	>500 ppm

Source: NRC (1).

grains, while oilseeds are intermediate. Iron from ferrous sulfate is highly available, whereas ferric oxide from it is almost unavailable.

Milk production increases the Mg requirement, as do high levels of K, P, or Ca, which interfere with Mg utilization. Hypomagnesemic tetany, or grass tetany, can result from Mg deficiency.

Manganese functions in the synthesis of reproductive hormones, and a deficiency of Mn can result in reproductive failure in both males and females, including impaired spermatogenesis, delayed estrus, decreased ovulation, and delayed conception. Manganese blood levels below 0.005 ppm or liver levels below 9 to 15 ppm (dry-matter basis) indicate a potential Mn deficiency. Manganese requirements are increased by high levels of Ca and P in the diet. Forages contain relatively high levels of Mn, while feed grains are lower.

Potassium deficiency is not common in cattle, although it can occur with high grain diets or in cows grazing low-quality standing winter forage. Potassium chloride is the most commonly used source of supplemental K.

Selenium deficiency can result in retained placenta, abortion, weak calves at birth, premature births, erratic estrus, and poor fertility. A dietary level of 0.2 ppm is considered adequate (Table 13-3). Liver biopsy is the best indicator of Se status, with levels below 0.2 ppm (wet basis) considered deficient. Blood Se levels below 0.05 to 0.08 ppm are considered indicative of deficiency. Selenium toxicity can be a problem, especially in regions with Se-accumulating plants (locoweeds, *Astragalus* spp.) or high Se soils, and 2 ppm or above is considered toxic.

Supplementation with Na and Cl is necessary for cattle because most plants contain very low levels of Na. Providing salt free choice is the easiest method of supplying adequate Na and Cl, and daily intake will range from 1 to 4 oz. Some regions of the United States, including much of the West, have soils and/or waters that are high in salt. Levels of 7,000 ppm dissolved salts are considered detrimental.

Cattle fed nonprotein nitrogen may benefit from supplemental S by increasing the rumen microbial synthesis of protein. Sulfur interferes with the utilization of Cu and Se.

Dietary requirement for Zn appears to be 20 to 40 ppm. Zinc deficiency can result in low conception rates, dermatitis, and reduced performance. Levels below 0.4 μg/ml in blood or 20 to 40 ppm (80 to 100 ppm on a dry-matter basis) in the liver are considered deficient.

Recent research has indicated that some of the trace minerals, such as Cu, Zn, and Mn, play an important role in the body's immune function. Deficiencies of these minerals may impair the immune system's ability to respond to vaccinations and to disease organisms (Fig. 13.3). These minerals can also be toxic (Table 13-4).

VITAMIN REQUIREMENTS

The microorganisms in the ruminant digestive system synthesize vitamin K and the B vitamins, meeting most of the requirements for these nutrients. Vitamin D is synthesized in the skin when cattle are exposed to direct sunlight, and sun-cured forages contain large quantities of vitamin D, vitamin E, and β-carotene. Requirements for vitamins A, D, and E are given in Table 13-5.

The liver can store enough vitamin A to meet the cow's need for a period of 2 to 4 months; however, liver stores can be depleted rapidly if the diet contains little vitamin A. De-

FIGURE 13-3 Feeder to supply salt and minerals to cows on pasture.

TABLE 13-4

Toxic minerals for beef cattle

Mineral	Toxic Level (ppm)
Aluminum	1,000
Arsenic	50–100
Bromine	200
Cadmium	0.5
Fluorine	20–100
Lead	30
Mercury	2
Molybdenum	5–6

Source: NRC (1).

TABLE 13-5

Vitamin requirements for beef cattle

Vitamin	Animal	Requirement (IU/kg DM)
Vitamin A	Feedlot	2,200
	Pregnancy	2,800
	Lactation and bulls	3,900
Vitamin D	All	275
Vitamin E	Calves	15-6

Source: NRC (1).

ficiency signs include rough hair coat, diarrhea, excessive tearing or lacrimation, abortion, stillbirth, low conception rates, and increased susceptibility to respiratory diseases. High-quality forages are generally good sources of vitamin A, while most feed grains contain relatively low amounts. Dietary conditions that can lead to vitamin A deficiency include high-concentrate diets; mature, dry pasture; weathered hay; processed feeds; and stored feeds.

If cattle are fed sun-cured forages and are exposed to sunlight, vitamin D deficiency rarely develops.

Symptoms of vitamin E deficiency are similar to those of Se deficiency and can include white muscle disease in calves, retained placenta, and lengthened calving-to-conception interval in cows. Most practical diets contain adequate levels of vitamin E.

Vitamin K is synthesized by the ruminal microbes and rarely needs to be supplemented to cattle. However, the formation of dicoumarol in moldy sweet clover hay and its antagonistic action on vitamin K can lead to hemorrhages. In this situation, supplemental vitamin K can be very effective.

The B vitamins have many important metabolic roles; however, due to extensive ruminal microbial synthesis, deficiencies are not very common. Vitamin B_{12} deficiency occurs as a result of a deficiency of cobalt, where inadequate cobalt is available for microbial synthesis of Vitamin B_{12}. Polioencephalomalacia can occur in grain-fed cattle in response to thiaminase activity in the rumen.

WATER REQUIREMENTS

Nonlactating cows consume approximately 3 parts water for each 1 part dry-matter intake. Lactating cows need an additional 0.1 gallon (0.87 lb) water per day for each 1 lb of milk produced. A restriction in water intake reduces total dry-matter intake and results in decreased production. A diet high in protein or salt content or high environmental temperatures increases the need for water intake (Table 13-6).

TABLE 13-6

Expected daily water intake by a 1,200 pound cow

Temperature (°F)	Water Intake (gal)
<40	8
50	8–9
60	10
70	11–12
80	13–14
90	15–19

Add 0.1 gal. of water for each 1 lb of milk produced.

GROWTH EFFECTS ON NUTRIENT REQUIREMENTS

Growth rate by cattle (measured as an increase in body weight) is very rapid in young calves, declines gradually until puberty, and then becomes even slower until mature body size is reached. Nutrient requirements per unit body weight or per unit metabolic body size are greatest for very young calves and then decline as they approach maturity. Dry-matter intake per unit body weight is greater for younger animals and declines as physiological age increases. Nutrient requirements per unit gain are lowest and efficiency of nutrient utilization highest when growth rates are rapid.

REPRODUCTION EFFECTS ON NUTRIENT REQUIREMENTS

During pregnancy, energy is required for maintenance, deposition of protein and fat in maternal tissue, and deposition of protein and fat in the fetus. Nutrient deficiencies prior to breeding may result in a longer interval from calving to first estrus, low fertility, silent estrus, or failure to conceive. Underfeeding energy and/or protein during growth results in delayed sexual maturity. Underfeeding, especially after calving, or overfeeding energy results in reduced fertility. Fetal tissues have priority over maternal tissues for nutrients; thus a cow's body reserves may be depleted if nutrition is not adequate during gestation.

LACTATION EFFECTS ON REQUIREMENTS

Heavy lactation requires more nutrients than any other production state. All nutrient requirements increase because milk components must either be supplied directly via the blood or synthesized in the mammary gland. Limiting water or energy intake decreases milk production, with a restriction in protein intake having less of an effect.

BODY CONDITION SCORING

Body condition scoring is a management tool used to estimate the body fat reserves of a cow and to monitor the effectiveness of nutritional programs (Table 13-7). Under practical management conditions, a cow may lose body weight and condition at calving and during early lactation and gain body weight and condition during late lactation and the first or second trimester of gestation. In general, this cyclic loss and gain of weight does not reduce productivity, as long as cows are not losing

TABLE 13-7

Body condition scoring in beef cattle

BCS	Description
1	Emaciated: extremely emaciated, bone structure of shoulder, ribs, backbone, hips and pelvic bones is easily visible and sharp to the touch, tailhead and ribs are very prominent, no detectable fat deposits, little muscling
2	Poor: somewhat emaciated, tailhead and ribs less prominent, backbone still sharp to the touch, some evidence of muscling in the hindquarters
3	Thin: ribs are individually identifiable but not quite as sharp to the touch, some palpable fat along spine and over tailhead, some tissue cover over rear portion of ribs
4	Borderline: individual ribs are no longer visually obvious, backbone can be identified individually by touch but feels rounded rather than sharp, some fat covers ribs and hip bones
5	Moderate: good overall appearance, fat cover over ribs feels spongy, and areas on either side of tailhead have some fat cover
6	High moderate: firm pressure now needs to be applied to feel backbone, fat deposits in brisket, over ribs and around tailhead, back appears rounded
7	Good: fleshy appearance, thick and spongy fat cover over ribs and around tailhead, some fat around vulva and in pelvis
8	Fat: very fleshy and overconditioned, backbone difficult to palpate, large fat deposits over ribs, around tailhead, and below vulva
9	Extremely fat: smooth, blocklike appearance, tailhead and hips buried in fatty tissue, bone structure no longer visible and barely palpable.

Source: Richards et al. (3)

weight during late gestation and the post-partum period. However, there is a clear relationship between body condition score and reproductive performance (Table 13-8, Fig. 13-4). Compared with cows in moderate body condition, thin cows have a longer postpartum interval (calving to rebreeding) and lower pregnancy rates (Figs. 13-5 and Fig. 13.6). The feedstuffs and daily gain necessary to change the body condition score (BCS) of a 1,100-lb pregnant cow from BCS 4 to 5 and from BCS 3 to 5 are given in Table 13-9. The closer the cow is to calving the more difficult and expensive it becomes to change body condition. This reinforces the need to monitor body condition of cows prior to the last trimester of gestation.

ENVIRONMENTAL EFFECTS ON REQUIREMENTS

Environment affects nutrient requirements by changing both requirements for maintenance and feed intake (Table 13-10). High environmental temperatures depress intake by up to 30% depending on the severity of the temperature. Low environmental temperatures increase intake by up to 30%. Maintenance energy requirements increase, especially with cold environmental temperatures (Table 13-11, Fig. 13-7).

FORAGES FOR BEEF CATTLE

Beef cattle usually depend on forage during some stage of their life. The majority of free-ranging cattle diets consist of grasses (9). There are periods, especially in the spring, when cattle may consume 30% to 45% of their diets in forbs (10). Shrubs may constitute 25% to 40% of cattle diets (11) depending on location. Beef cow producers in the more humid regions of the United States usually rely on introduced forages (brome grass, timothy, ryegrass, tall fescue) to supply nutrients for beef cow herds. Cow–calf producers in arid and semi-arid western United States rely on a mixture of native grasses, forbs, and shrubs to provide forage for their herds. Many cattle producers in the

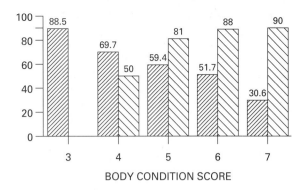

POSTPARTUM INTERVAL, DAYS

PREGNANCY RATE, %

FIGURE 13-4 Effect of body condition score at calving on postpartum interval and pregnancy percentage.

FIGURE 13-5 Cow in body condition score 3.

FIGURE 13-6 Cow in body condition score 5.

TABLE 13-8

Relationship of body condition score at weaning and pregnancy rate[a]

| | Body Condition Score | | | | |
	<3	4	5	6	>6
Number of cows	3,415	23,811	37,970	26,213	9,654
Percent pregnant	75.5	85.4	93.8	95.6	95.6

Source: Cherni (4).
[a]Nine-year summary.

TABLE 13-9

Feedstuffs and daily gain necessary to change body condition score of a 1,100-pound pregnant beef cow

| | Days to Calving | | | | | |
	120	105	90	75	60	45
Change from 4 to 5 BCS						
Gain required, lb/day	0.62	0.71	0.82	0.98	1.23	1.63
Feedstuffs, % of diet						
Low-quality forage	64	62	—	—	—	—
Medium-quality forage	—	—	61	55	48	37
Grain	21	21	18	20	22	25
Wheat middlings	15	17	21	25	30	39
Change from 3 to 5 BCS						
Gain required, lb/day	1.24	1.40	1.63	1.96	2.44	[a]
Feedstuffs, % of diet						
Low-quality forage	52	47	—	—	—	
Medium-quality forage	—	—	37	25	—	
Grain	24	23	29	33	47	
Wheat middlings	24	30	34	42	53	

Source: Buskirk et al. (7).

[a]Not possible.

TABLE 13-10

Intake adjustments for environmental conditions[a]

Temperature (°F)	Intake Adjustment (%)
>95, with no night cooling	−35
>95, with night cooling	−10
77–95	−10
59–77	None
41–59	3
23–41	5
5–23	7
<5	16

Source: NRC (8).

[a]Adjustments assume that cattle are not exposed to wind and storms.

FIGURE 13-7 Cow on winter range during second trimester of gestation.

TABLE 13-11

Estimated increase in energy intake due to cold temperature for 1,000-pound pregnant cow

°F below Critical Temperature	Additional Mcal ME Needed Per Day
0	0
5	0.9
10	2.0
15	3.2
20	3.6
25	4.5
30	5.4
35	6.4
40	7.3

northern latitudes of the United States extend the grazing season by moving cattle from cool- to warm-season-dominated pastures.

FORAGE QUALITY AND INTAKE

Forage intake is modified by forage quality in addition to energy demands. Intake of ruminants grazing low-quality native or introduced forages (below 50% digestibility) appears to be limited by bulk fill. When NDF levels of the forage are above 50%, digestibility decreases, which decreases rate of passage and subsequently decreases forage intake. When NDF content is lower, digestibility and rate of passage increase, which allows more forage to be

TABLE 13-12

Expected dry-matter intake of various forages by beef cows

Forage Type	DM Intake (% BW)
Low-quality hay	1.0–2.0
Average-quality hay	1.5–2.5
High-quality hay	2.2–3.3
Green pasture	2.0–3.0
Spring rangeland	2.7–2.9
Late spring, early summer rangeland	2.4–2.9
Late summer rangeland	1.8–2.4
Fall rangeland	1.7–2.2
Winter rangeland	1.4–1.8

consumed. When beef cattle are grazing low-quality forages, the increased cell wall content (NDF) limits forage intake. Expected intake of various forage by beef cows is presented in Table 13-12.

Grazing cattle usually select live over dead plant material. Likewise they have been shown to select leaf over stem and young versus old plants. Selection of forages influences animal performance by affecting the intake of nutrients.

GRAZING MANAGEMENT

The greatest challenge facing the beef cow producer is to reduce purchased feed costs and increase profitability per cow. One of the first steps in that direction is to maximize the use of inexpensive forages when animal demands are greatest. Unfortunately, this does not occur under most growing conditions in North America. Use of improved pastures early or later in the growing season is one of the best methods to match forage quality and animal demands. Determining the optimum stocking intensity for native or introduced pastures is the next step toward long-term profitability. Experience is the only proven method to determine the proper stocking intensity for any given pasture. There are a number of specialized grazing systems designed to increase animal gains per acre, but many reduce individual animal performance due to increased stocking intensity.

IMPROVED PASTURES

Pastures comprised of introduced forage species are used throughout the United States to supply beef cow herds with higher-quality forage than native forages. In the South, perennial ryegrass (cool season) has been effective in meeting the winter nutritional needs of graz-ing cattle and reducing the use of protein supplements. A 1:3 ratio of introduced forages to native range for use in Kansas beef cow operations has been suggested (12). Using sorghum–Sudan grass hybrid pastures from June to July, utilizing range in fall and early winter, and then switching to cereal crop grazing in the late winter has been suggested. Use of crested wheatgrass and Russian wildrye in addition to native range reduced the land required to support an animal unit for the 7.5-month grazing season from 25 to 11 acres. Crested wheatgrass has been used extensively throughout the western United States to provide forage to meet the nutritional requirements of lactating beef cows in the early spring. The use of crested wheatgrass in the spring prevents early spring use of native ranges, which typically depletes cool-season grasses. Cow–calf producers in Nebraska have used introduced cool-season grasses in the spring and late fall. Cornstalks can be used as late fall and winter grazing in place of native range.

Irrigated pastures provide high-quality forage to beef cows following calving and improve conception rates. Mixtures of grasses and legumes have been used successfully to improve livestock performance compared to poor grass stands. Incorporating legumes into pastures also eliminates the need for nitrogen fertilizer, since the legume provides nitrogen for both plant species.

STOCKING INTENSITY

Many studies have demonstrated the influence of grazing intensity on cow–calf productivity per unit area and per animal unit. As stocking rate is increased, productivity per animal declines, but productivity per acre is increased up to a certain point (Fig. 13-8). Both individual animal performance and gain per acre cannot be maximized at the same time regardless of the grazing system. Long-term studies demonstrate maximum monetary returns per acre are realized at moderate stocking levels.

Reduced individual animal performance with increased grazing intensity is due to reduced animal intake and forage quality. Beef cows are selective grazers. When a large number of animals is present in the same pasture, the best-quality forage is removed early and only poorer-quality forage is left behind.

Consequences of high grazing intensity include lower gain per head, higher gains per acre, reduced dry-matter intake, more time and

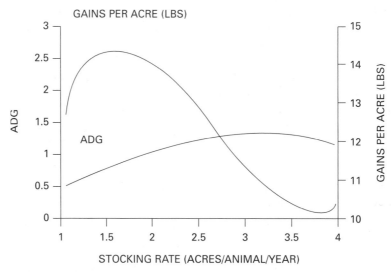

FIGURE 13-8 Stocking rates and cattle gains.

GRAZING SYSTEMS

energy spent grazing, lower nutrient quality of forage, increased nutrient deficiencies, lower pregnancy rates, lower calf crops, no drought emergency forage, and increased losses from poisonous plants (13).

GRAZING SYSTEMS

Most grazing systems should manipulate beef cattle to maintain or improve range conditions. The premise is that gradual improvement in forage conditions eventually increases carrying capacity, and any depression in animal performance is offset by the increased carrying capacity. This premise is supported more by testimonials than by specific research data (12, 14).

Continuous grazing is the most common type of grazing scheme used by beef producers (Fig. 13-9). Animals are left to graze one area for the entire grazing season or sometimes the entire year. Total production per acre is lower compared to other systems, but maintenance

costs are low. It is usually difficult to maintain preferred plant species, especially when stocking intensity is excessive. However, when stocking intensity is controlled, continuous grazing can maintain range condition and meet livestock nutritional needs better than most specialized grazing systems (12).

Deferred rotational grazing systems are used to allow desirable plants to achieve some phenological stage, such as flowering or seed maturity, without being grazed. In a four-pasture deferred rotation, one pasture would not be grazed from spring to mid-summer. The following year, another pasture would not be grazed for the same period. After 4 years, the deferment would be rotated through all pastures. This grazing system was developed in the Pacific Northwest to prevent bunchgrass losses due to selective cattle grazing. It has proved to be superior to continuous grazing to improve range condition. A problem with deferred grazing is that it reduces individual animal performance and increases supplementation costs during the deferred period.

Rest rotation grazing systems usually contain three to five pastures. One pasture is not grazed for an entire year while the single herd utilizes the other pastures. The 12-month rest is rotated across all pastures in different years. This system was developed in foothill rangelands of California and has been used extensively in the western United States. Vegetation response has usually been superior on rest rotation pastures compared to continuously grazed pastures; however, animal performance has usually been decreased. Animal selectivity is reduced due to increased stocking intensity

FIGURE 13-9 Cows and calves on range during the summer.

in grazed pastures. Forage quality of the rested pasture is also reduced due to an increased quantity of mature stems and leaves.

Short duration grazing was developed in France with dairy cattle. Allan Savory used this system in Africa and promoted its use in the United States. The system involves a wagon wheel arrangement of pastures with a central watering point and livestock-handling facilities. It usually has at least 8 pastures, but some systems have had as many as 40 paddocks. Each pasture is grazed intensively for 2 to 3 days and then not grazed for a period of several weeks. Stocking intensities are usually increased substantially (two- to three-fold). The high stocking density is supposed to remove grazing selectivity and force animals to uniformly graze all plants in the pasture. The long rest periods are supposed to allow plant recovery from grazing. Some pastures may be grazed two or three times per growing season, depending on growing season length. This grazing system works best on highly productive soils with grazing-resistant plants that have very similar palatabilites. The greatest drawback with this system has been a large reduction in individual animal performance on pastures that were already moderately stocked. There have been problems in arid and semiarid regions because this system leaves no emergency forage for drought conditions. If regrowth does not occur during the nonuse period, no forage is available for the next grazing period.

ANIMAL REQUIREMENTS, FORAGE RESOURCES, AND COSTS

The greatest challenge to cow–calf producers is to increase long-term profitability. This goal must be accomplished when forage production can vary two- to five-fold from year to year. Annual and seasonal cattle prices also fluctuate widely. One of the best methods to monitor profitability is to evaluate production costs per cow. When total feed costs per cow are reduced without drastically altering performance, profitability should increase. Reductions in feed costs are usually the result of matching the annual forage availability curve with the animal requirement curve.

Feed costs represent the largest single cow expense to beef cow–calf producers. Feeding costs may represent 60% to 80% of the total costs per cow to the average cow–calf producer. Most producers in northern latitudes spend large amounts of money on purchased or produced hay (Fig. 13-10). Some producers spend more to produce hay than it would cost to purchase. Cattle can be allowed to graze directly on hay meadows or feed on loose windrowed forage with a portable electric fence. Crested wheatgrass can be utilized in the early spring, reducing feed costs. Some hayfields can be reirrigated and grazed in the fall and winter, reducing total hay demand and winter feed costs.

NUTRITIONAL MANAGEMENT

Nutritional management of the cow herd involves matching the available feed resources to animal requirements. To make the most economical use of available forage resources, producers should know the nutritional characteristics or quality of their forages. Forage quality includes nutrient content and availability, digestibility, and intake, all factors that influence animal performance.

FORAGE ANALYSIS

It is difficult to find one chemical analysis that indicates forage quality, so generally forages need to be analyzed for several components. These components are as follows:

CRUDE PROTEIN   Crude protein content is positively related to quality.

FIBER   Fiber is negatively related to forage quality. Dry-matter intake (as a percent of body weight) and forage digestibility for beef cows can be estimated by the following equations:

$$\text{Dry matter intake} = \frac{120}{\text{percent NDF}}$$

$$\text{Dry matter digestibility} = \frac{88.9}{-(0.779 \times \text{percent ADF})}$$

FIGURE 13-10 Wintering cows consuming hay during last period of pregnancy.

For example, a cow consuming a high-quality alfalfa hay with 35% NDF and 25% ADF would be expected to eat (120/35) = 3.5% of body weight per day, and the digestibility of that hay would be [88.9 − (0.779 × 25)] = 69.4%. However, a cow consuming a low-quality alfalfa hay with 55% NDF and 35% ADF would be expected to eat (120/55) = 2.2% of body weight per day, and the digestibility of that hay would be [88.9 − (0.779 × 35)] = 61.6%.

AVAILABLE ENERGY   Most laboratories use fiber content (either ADF or NDF) to estimate available energy in forages. The digestibility value estimated above can be used as a rough estimate of the energy or TDN value of the forage for beef cows.

Other equations can be used to predict the energy value of forages:

Legumes $NE_m = 1.044$
                        $− (0.0123 × \text{percent ADF})$

Grasses $NE_m = 1.085$
                        $− (0.015 × \text{percent ADF})$

Grass− $NE_m = 1.044$
legume mix $− (0.0131 × \text{percent ADF})$

Corn silage $NE_m = 1.044$
                        $− (0.0132 × \text{percent ADF})$

Once the expected intake and digestibility (energy content) of the forage are calculated, a comparison can be made with the cow's nutrient requirements. Consider the following example. A producer is feeding grass hay (an example of forage analysis is given in Table 13-13) to 1,200-lb cows during the last trimester of gestation. Using the laboratory analysis, the hay dry-matter intake by a cow can be calculated.

$$\text{DM intake, percent of body weight} = \left(\frac{120}{\text{percent NDF in forage}} \right)$$

$$= \frac{120}{65}$$

$$= 1.85\% \text{ of body weight}$$

TABLE 13-13

Example forage analysis of grass hay fed to 1,200-pound cows during the last trimester of gestation

Laboratory Analysis	% on a DM basis
Dry matter	88.0
Crude protein	8.5
NDF	65.0
ADF	36.0

This means that a 1,200-lb cow would on average consume 1,200 lb × 0.0185 or 22.2 lb of hay DM per day.

The estimated energy value (percent TDN) of the forage can also be calculated.

$$\text{Energy value, percent TDN} = \frac{88.9 − (0.779 × \text{percent ADF in forage})}{}$$

$$= 88.9 − (.779 × 36)$$

$$= 60.9\% \text{ TDN}$$

If the cow consumes 22.2 lb of this hay per day, she is consuming 22.2 lb × 0.609 = 13.5 lb TDN per day. The crude protein intake can also be calculated: 22.2 lb × 0.085 = 1.9 lb crude protein. Now the nutrient intake by the cow can be compared with her nutrient requirements.

Table 13-14 shows that the forage quality as predicted by laboratory analysis is adequate to meet the nutrient requirements for a cow during the third trimester of gestation. However, when the cow calves, her nutrient requirements go up, and Table 13-15 shows that now the forage quality does not meet the cow's requirements during the postpartum period. Either a higher-quality forage must be fed during this time or supplemental nutrients must be supplied to the cow (Fig. 13-11).

SUPPLEMENTATION

The goal of supplementation is to provide nutrients to the cow that are not available in sufficient quantity or adequate balance and to do this in an economical way. Dormant range forage is high in fiber and deficient in both crude protein and energy for cows during late gestation and early lactation. When forage is limited in quantity and/or quality, supplementation may be necessary to maintain the desired level of productivity (Fig. 13-12).

Supplements are classified broadly as protein supplements or energy supplements. Both these supplements contain both protein and energy. The difference between the two is the relative concentration of protein. Protein supplements contain high concentrations of protein, generally 25% crude protein or greater. Energy supplements contain lower concentrations of protein, usually less than 18% crude protein.

For most cows grazing range forages, protein is the first limiting nutrient. Energy may be deficient as well, however, but usually is not as deficient as protein. The energy available in range forage is of little use without protein available to support microbial digestion. For

TABLE 13-14

Comparison of nutrient intake estimated by forage analysis with requirements of a 1,200-pound cow in third trimester of gestation

Item	Requirement (lb)	Estimated Intake (lb)	Comments
Dry matter	20.4	22.1	OK
TDN	11.8	13.5	OK
Crude protein	1.7	1.9	OK

Source: NRC (1).

TABLE 13-15

Comparison of nutrient intake estimated by forage analysis with requirements of a 1,200-pound cow during postpartum interval

Item	Requirement (lb)	Estimated Intake (lb)	Comments
Dry matter	27.6	22.1	Deficient by 5.5 lb
TDN	15.2	13.5	Deficient by 1.7 lb
Crude protein	2.1	1.9	Deficient by 0.2 lb

Source: NRC (1).

FIGURE 13-11 Low-quality hay fed to cows after weaning.

low-quality forages, small quantities of high-protein (25% to 40%) supplements should be emphasized.

Voluntary intake of mature range forages by cattle is approximately 1.5% to 1.8% of the

FIGURE 13-12 Supplementation of cows on pasture.

cow's body weight per day (16.5 to 19.8 lb/day for a 1,100-lb cow). This level of intake may be inadequate to meet the cow's nutrient requirements. When forage crude protein falls below 6% to 7%, intake can be stimulated by supplementation with protein or nonprotein nitrogen (NPN). Supplementation of cows grazing native range has not always been cost effective, because subsequent reproduction or calf growth has not been consistently improved. Supplementation may become cost effective if it can increase intake and utilization of low-quality range forage and allow ranchers to enhance efficient use of available forage resources.

Protein supplements are generally expensive, and supplying supplemental N via less expensive NPN sources such as urea is a possibility. Forage intake has been increased by feeding urea, but often intake is increased only if molasses or other readily available energy sources are applied. This is probably due to a mismatch in the supply of N and the available energy found in the digestion of low-quality forage. Ruminal microbes are able to use urea to synthesize protein if adequate amounts of energy are available. Molasses not only supplies a source of energy, but also sulfur and branched-chain volatile fatty acids that are required nutrients for cellulolytic ruminal microbes.

Traditional hand feeding of supplements to grazing cattle increases labor costs and has the potential to alter grazing behavior and reduce forage utilization. Liquid and/or block supplements have been used to reduce labor costs and decrease possible interference with normal grazing behavior. Liquid or block supplements containing molasses, NPN, and natural protein can be fed free choice and may be more profitable than conventional supplements under range conditions. Variable animal performance in studies with liquid supplements may be a result of differences in supplement formulations. Poor response by grazing cows to liquid supplements has been seen, especially when the protein concentration of the supplement is low. Levels of P, S, and trace minerals and whether additional ingredients such as phosphoric acid are added may affect animal performance.

A major limitation to providing additional nutrients to grazing cattle is the inability to control individual animal consumption of supplement. This inability to regulate supplement intake results in a mismatch between animal nutrient requirement and nutrient supply. Several techniques have been used with limited success to control supplement intake. Inclusion of salt, additives that alter palatability, or changing the physical characteristics of supplements has given inconsistent results.

SUPPLEMENTATION WITH NONSTRUCTURAL CARBOHYDRATES

Supplementation of forage diets with concentrates containing nonstructural carbohydrates (NSC) often decreases fiber digestion and forage intake. Starch has been shown to have a negative effect on fiber digestion. Supplementation with feed grains has decreased fiber digestion and forage intake. Differences exist in the source and amount of NSC, as barley has been shown to have a greater negative effect on fiber digestion compared with corn. This depression in fiber digestion may result from the effects of low ruminal pH from increasing volatile fatty acid production on the growth of fibrolytic bacteria, from an increase in lag time for fiber digestion due to preferential use of NSC by fibrolytic bacteria, or from competition for nutrients between fibrolytic and amylolytic microorganisms. When fiber digestion is depressed, passage rate may be decreased as well, reducing forage intake. The decrease in forage digestion with NSC supplementation becomes more severe as forage quality decreases. However, effects of NSC supplementation on forage intake are greater with high-quality forages than with low-quality forages.

Despite the negative effects on forage digestion and intake that can result, supplementation of forage diets can increase animal performance due to an increase in digestible organic matter intake, a shift in rumen fermentation pattern to a greater proportion of propionate, and an increase in the flow of undigested feed and microbial protein to the intestine. In addition, reducing forage intake with starch-containing supplements or substituting supplement for forage can be beneficial if forage supplies are limited.

BY-PRODUCT FEEDS AS SUPPLEMENTS

By-products of concentrate feeds, such as wheat middlings, soybean hulls, and corn gluten feed, generally contain more fiber and less starch than the parent feedstuffs. This results in a lower NSC to structural carbohydrate ratio. In addition, by-product fiber or structural carbohydrate is highly digestible compared with fiber in most forages.

Supplementation with fibrous by-product feedstuffs, such as soybean hulls and wheat middlings, has been shown to increase intake and utilization of low-quality forages by cattle. Wheat middlings supplement increased forage intake and fiber digestibility when fed to beef cattle consuming dormant range forage.

FEEDING BEHAVIOR

Cattle typically graze early in the morning and late in the evening. This pattern can be altered by temperature and other factors. Disrupting this behavior by feeding supplements may reduce forage intake and increase supplement intake. Therefore, if supplements are handfed, it should be done at time (midday) when cattle are not actively grazing (Fig. 13-13).

Cattle are social animals and form linear dominance hierarchies. Larger, older, more

FIGURE 13-13 Cattle grazing summer range may need to travel some distance to water.

dominant animals usually consume more supplement than smaller, younger, less dominant animals. This results in inefficient use of self-fed supplements because larger animals in better body condition receive more supplement than the animals who need the supplement to maintain or improve body condition. Therefore, young cows should be separated from the rest of the herd when feeding supplements.

CALF SURVIVAL AND COW NUTRITION

Providing the proper nutrition for calves begins before the animal is born. Deaths of newborn calves have been attributed to protein deficiencies in the dam 60 days prepartum. Results from the western region of the United States found that low protein levels in the dam caused weak calf syndrome and resulted in the deaths of young calves. Cow herds that consumed an average of 2 lb crude protein per day during the prepartum period did not have weak calves. Herds consuming hay with at least 10% crude protein had no problems, while herds that were fed hay containing less than 10% crude protein had serious weak calf problems (15).

CREEP FEEDING

Creep feeding is the practice of providing suckling calves with supplemental feed. Creep feeds can be comprised of hay, grain, or mixed rations. Research has shown that creep feed can add 25 to 50 lb to a calf's weaning weight. Many studies have also demonstrated that creep feeding is not always profitable, because calves usually require 7 to 10 lb of creep feed for each pound of gain. It is important for producers to calculate the break-even costs of creep feeding.

Milk production by the lactating dam is usually adequate in the first 100 days postpartum to maximize calf growth rates. Milk production declines after this period, and the calf must rely on grazed forages to meet its nutrient requirements. This is when most producers practice creep feeding. Creep feeding may be profitable when calf prices are high relative to feed prices. It may benefit calves of first- or second-calf heifers and when cows and calves are kept in confinement. Creep feeding has the greatest effect on calf weight when forage quality is low and milk production of the dam is reduced.

Creep feeding may not be profitable if pasture forage is high in quality and calves are going to continue to graze as yearlings. It may be detrimental to long-term heifer productivity if they become too fat. Excessive fat tissue can inhibit the formation of milk-secreting tissue in the udder and increase calving difficulty. Small-framed calves may experience overfattening when creep feeding is unrestricted. Overfattened calves gain more slowly the first 2 or 3 months in the feedlot and finish at lighter weights. Therefore, some cattle feeders prefer calves that have not been creep fed.

Creep feeders should be located where cows bed down and where shade and water are available. If creep feeders are located large distances from water or bedding grounds, calves will not use them very frequently. Multiple feeders may be necessary if cows and calves are in large pastures.

Cow–calf producers should analyze forage quality of the pasture before selecting a ration for creep feeding. Cost of products and palatability should also be considered. Creep feeding may be profitable when the price of calves per pound is greater than 10 times the price of the creep feed since it takes an average of 10 lb of feed to add an additional pound of weaning weight to a calf (16).

A variety of commercial products is available. They are usually very palatable and well formulated to match the nutritional needs of the calf. However, they should be compared with the costs of producer rations. Grains such as shelled corn and wheat should be cracked and barley rolled to increase palatability. Cane molasses has been added to creep rations to reduce dust and increase palatability. An example creep feed might include 30% cracked corn, 30% cracked oats, 30% cracked barley, 5% molasses, and 5% soybean meal.

WEANING

Calves are normally weaned from 6 to 9 months of age. When weaned, calves are kept out of sight and sound of the dams if possible. Clean water and high- to medium-quality hay should be made easily accessible, and eventually calves should be fed from bunks to minimize adaptation time when they reach the feedlot. Calves should be consuming 2% of body weight before the feed or water is medicated. Calves that have been creep fed should continue to receive creep feed. Calves usually shrink 3% to 5% due to weaning, but should gain 30 to 50 lb during a month-long preconditioning period.

Early weaning is a management technique designed to improve rebreeding of first and second calf heifers, since suckling delays the onset of estrus in cattle. After early weaning, calves may continue to graze forages or may be fed in drylot. Most calves are functional ruminants at 45 to 60 days of age and theoretically capable of maintaining themselves on an all forage diet. Unfortunately, if forage quality is low at the time of early weaning, intake may be limited and calf performance reduced compared to normal weaning. Calves should not be weaned prior to 5 months of age unless there is an emergency.

HEIFER DEVELOPMENT

It is generally not recommended to creep feed heifers that will join the breeding herd as replacement females. Although creep feeding allows more growth, it will mask the effects of cow milk production, a criterion that is generally used to select females. In addition, the increased gain from creep feeding is not marketed and does not increase the value of replacement heifers.

If heifers are expected to calve for the first time as 2-year olds, they need to reach approximately 60% of their mature body weight by 15 months of age. The appropriate postweaning gain depends on the breed type and environmental conditions, but usually the target rate of gain is 0.75 to 1.0 lb/day. A slower rate of gain may delay puberty, and a faster rate of gain may impair subsequent reproduction and lactation. Heifers should be managed separately from the main cow herd because they have increased requirements due to the additional nutrient needs for growth (Fig. 13-14).

The first winter, when spring-born heifers are 7 to 15 months old and are not pregnant, they should gain 100 to 200 lb depending on weaning weight and breed type. Pregnant heifers should enter their second winter at approximately 750 to 850 lb and should weigh 900 to 1000 lb at calving. Nutritional management

FIGURE 13-14 Replacement heifers grazing lush pasture.

during the postpartum period is critical for first-calf heifers because they need nutrients for lactation, rebreeding, and growth. During their third winter, pregnant 3-year-olds should not lose more than 5% to 10% of their fall weight. Maintaining adequate body condition is essential for pregnant heifers. Feeding ionophores can reduce the age at puberty and enhance heifer reproductive performance.

NUTRITIONAL DISORDERS

PASTURE BLOAT

Pasture bloat results from consuming lush legumes such as alfalfa and Ladino or red clover. Factors that increase the incidence of pasture bloat include a predisposition to bloat (genetic) and the amount and rate of feed intake.

The most effective method to relieve bloat is to insert a stomach tube (a rubber hose case, ¾ to 1 in. in diameter) or in extreme cases to puncture the rumen with a trocar to relieve the pressure. At least 1 pint of defoaming material, such as vegetable oil, should be administered into the rumen immediately.

The incidence of pasture bloat can be reduced by maintaining pastures that do not exceed 50% legumes, filling cattle on dry roughage before grazing legume pasture, feeding straw or dry hay in the legume pasture while cattle are grazing, feeding poloxalene once or twice daily following heaviest legume consumption (early morning or late evening), not grazing legume pasture wet with dew or rain, not removing cattle from alfalfa pasture unless bloat is severe, and feeding the ionophore monensin while cattle are grazing bloat-producing pastures (or administering a controlled-release capsule containing monensin).

GRASS TETANY

Grass tetany (hypomagnesemia) is usually most pronounced in the spring and is caused by low blood Mg levels. Levels of 0.2% Mg or greater in forages reduce the risk of grass tetany. Tetany occurs most frequently in cows that are nursing calves under 2 months of age. High levels of N fertilization or high levels of K in the soil, along with cool, rainy weather, increase the incidence of grass tetany. This condition usually occurs when cows are grazing lush green grass pastures and less frequently occurs in legume–grass mixtures. Clinical signs are not often observed until death. Affected animals may become excitable and appear to be blind. If tetany is suspected, a blood sample should be obtained.

Normal blood Mg levels are 2.25 mg/100 of serum. Affected animals may have below 1.0 mg Mg/100 ml of serum. Cows that develop tetany once are more likely to develop symptoms the following year. Death losses can be 100% if the condition is not treated rapidly.

To prevent grass tetany, free-choice mineral supplement containing Mg should be made available during the early grazing season. Adequate Mg can be supplied by consumption of 2 oz/day of magnesium oxide. However, MgO is unpalatable, and a mixture of 30% MgO), 32% dicalcium phosphate, 30% trace mineralized salt, and 8% grain products will increase consumption.

NITRATE TOXICITY

When nitrate levels are consumed in excess of the rumen's ability to convert them to ammonia, nitrites accumulate in the rumen and are absorbed into the bloodstream. High blood nitrate causes hemoglobin to be converted to methemoglobin, which cannot transport oxygen, and death can result from asphyxiation. Oat, barley, wheat, sorghum, corn, Sudan grass, Johnson grass, sweetclover, pigweed, lamb's quarters, Russian thistle, sunflower, kochia, bindweed, sugar beets, potatoes, and carrots are forage plants that can commonly accumulate nitrates. Other conditions that increase the potential for nitrate accumulation in plants include stresses such as drought, frost, low temperatures, hail damage, and cloudy weather. Nitrogen fertilization increases the potential nitrate level in plants. Young plants usually have higher nitrate levels, and the nitrate level decreases as the plant matures. Nitrate levels are highest in the lower portions of the plant, so harvesting at a higher level or leaving a 12-in. stubble reduces the level of nitrate in the forage. Table 13-16 gives toxic levels of nitrate in forages. Laboratories may express the nitrate level as nitrate-nitrogen or as nitrate. To convert from nitrate to nitrate-nitrogen, multiply the nitrate level by 0.23. Symptoms of acute nitrate poisoning include staggered gait, tremors, rapid pulse, and dark mucous membranes due to lack of oxygen. Abortion may result from asphyxiation of the fetus. Water can also be a source of nitrates and can contribute to the total nitrate load that the cow is receiving (Table 13-17). Total nitrate loads of from 15 to 45 g/100 lb body weight are considered toxic. A veterinarian should be consulted immediately if nitrate poisoning is suspected. Methylene blue injection intravenously (2 g/500 lb body weight) will convert methemoglobin to hemoglobin.

FESCUE TOXICOSIS

Cattle grazing or consuming harvested hay from tall fescue pastures, a common grass species in the Midwest and southern parts of the United States, can exhibit toxic symptoms or reduced performance (Fig. 13-15). Symptoms of fescue toxicosis can include decreased intake, reduced growth rate, rough hair coat, excessive salivation and urination, increased body temperature and respiration, decreased milk production, and impaired reproductive performance. Hot temperatures in the summer may aggravate these symptoms, and cattle will generally spend more time standing in ponds and shade, resulting in less time spent grazing. Animals may exhibit soreness in one or both hind limbs (fescue foot), and hooves and tails may be sloughed off. The toxicity symptoms have been linked to the presence of an endophytic fungus growing between the cells of tall fescue plants. Results of testing tall fescue samples throughout the United States indicate that close to 90% of tall fescue pastures have some level of endophyte infection.

TABLE 13-16

Nitrate levels in forages for cattle

Nitrate-Nitrogen (ppm)	Recommendations
<1,000 (< 0.1%)	Safe to feed to all cattle
1,000–1,500 (0.1–0.15%)	Safe to feed to nonpregnant cattle. Limit use for pregnant cattle to 50% of total ration dry matter.
1,500–2,000 (0.15–0.20%)	Safe to feed if limited to 50% of total ration dry matter
2,000–3,500 (0.20–0.35%)	Should not be fed to pregnant cattle. Can be fed to nonpregnant cattle if limited to 35% to .40% of total ration dry matter.
3,500–4,000 (0.35–0.40%)	Should not be fed to pregnant cattle. Can be fed to nonpregnent cattle if limited to 25% of total ration dry matter.
>4.000 (>0.40%)	Do not feed to any class of cattle.

To convert nitrate-nitrogen: Nitrate-N = (nitrate × 0.23).

TABLE 13-17

Nitrate levels in water for beef cattle

Nitrate-Nitrogen (ppm)	Comments
<100	Not harmful for any class of cattle.
100 to 300	This level not itself harmful, but may add significantly to the total nitrate intake.
>300	Can cause nitrate toxicity.

FIGURE 13-15 Calf exhibiting symptoms of fescue toxicosis.

Fescue toxicosis can be reduced by planting endophyte-free fescue cultivars, replacing fescue with other cool-season grass species, keeping fescue in a vegetative state as long as possible by grazing or clipping pastures, not grazing infected fescue when seed heads are present, grazing fescue pastures in the spring then moving cattle to other grass pastures, and grazing tall fescue regrowth in the fall, cutting fescue hay early, feeding fescue hay in combination with other feedstuffs such as alfalfa or orchard grass hay to dilute the effects of the endophyte, and incorporating legumes into tall fescue pastures.

ACUTE PULMONARY EMPHYSEMA

This disease, commonly called asthma, occurs in western regions of the United States when cattle are moved from dry rangelands to lush meadow pastures in late summer or early fall. It can occur on many types of forages, including rape, kale, alfalfa, turnips, small grains, introduced grasses, and mixed meadow grasses. Abrupt dietary changes from dry to lush forage may trigger the onset of this condition. It is commonly noticed 4 or 5 days after cows have been grazing a lush meadow or irrigated pasture. The first indication is labored breathing. Moderate exercise, such as moving sick cows, may precipitate severe respiratory distress. Afflicted animals extend their necks, breathe with an open mouth, and grunt on expiration. Physical exertion may cause death. It is believed that the amino acid tryptophan in lush forage is converted to a compound that is highly toxic to lung tissues.

Prevention of death loss from pulmonary emphysema appears to be a combination of grazing management and ionophore use. Prevention is critical, because there is no effective treatment for cattle with pulmonary emphysema. Cattle should be introduced to green lush forages gradually over a period of several days and not when they are exceptionally hungry. Feeding dry hay along with the lush pasture can help to prevent the problem. A daily dose of 200 mg of monensin has been shown to prevent pulmonary emphysema. Cattle should be checked twice daily when they are moved abruptly from dry forage diets to green meadow pastures. Sick cattle should be treated and moved with care to avoid excess oxygen demand.

SUMMARY

Nutritional management of the beef cow herd requires a thorough knowledge of the factors that influence animal requirements, such as breed type, mature size, milk production, biological cycle, and environmental conditions, as well as factors that affect forage quality and grazing management. Matching the cow's nutrient requirements with forage production is the key to profitable nutritional management.

REFERENCES

1. NRC. 1984. *Nutrient requirements of beef cattle.* 6th rev. ed. Washington, DC: National Academy Press.

2. Ward, J. D., J. W. Spears, and G. P. Gengelbach. 1995. *J. Animal Sci.* 73:570.

3. Richards, M. W., J. C. Spitzer, and M. B. Warner. 1986. *J. Animal Sci.* 62:300.

4. Cherni, M. 1995. Personal communication.

5. Houghton, P. L., et al. 1990. *J. Animal Sci.* 68:1438.

6. Selk, G. E., et al. 1986. *Okla. Animal Sci. Res. Rept.* MP-118.

7. Buskirk, D. D., R. P. Lemenager, and L. A. Horstman. 1992. *J. Animal Sci.* 70:3867.

8. NRC. 1987. *Predicting feed intake of food-producing animals.* Washington, DC: National Academy Press.

9. Sowell, B. F., et al. 1985. *Cattle nutrition in Wyoming's Red Desert.* Univ. Wyo. Agr. Exp. Stn. Sci. Mono. 45.

10. Streeter, C. L., D. C. Clanton, and O. E. Hoehen. 1968. *Univ. Nebr. Agr. Exp. Stn. Res. Bull.* 227.

11. Rosiere, R. E., R. F. Beck, and J. D. Wallace. 1975. *J. Range Manage.* 28:94.

12. Launchbaugh, J. L., et al. 1978. Grazing management to meet nutritional and functional needs of livestock. In: D. N. Hyder (ed.) *Proc. First Intern. Rangeland Cong.,* Denver, CO.

13. Vallentine, J. F. 1990. *Grazing management.* New York: Academic Press.

14. Heady, H. F. 1975. *Rangeland management.* New York: McGraw-Hill Book Co.

15. Bull, R. C., et al. 1983. Nutrition and weak calf syndrome in beef cattle. CL645. In: *Cow–calf management guide.* Cattleman's Library. Univ. Wyo. Agr. Ext. Serv., Laramie, WY.

16. Lusby, K. S., and D. R. Gill. 1992. Creepfeeding. GPE 1550. In: *Great Plains beef cattle handbook.* South Dakota State Univ. Agr. Ext. Serv., Brookings, SD.

17. Thomas, V. M. 1986. *Beef cattle production, an integrated approach.* Philadelphia: Lea and Febiger.

14

FEEDING GROWING–FINISHING BEEF CATTLE

M. L. Galyean and G. C. Duff

INTRODUCTION

Cattle feeding and allied concerns are very progressive and competitive industries. New and recently introduced products make decision making more difficult, and cattle feeders need to have a thorough understanding of both basic and applied nutrition. Approximately 88% of larger feedlot operations (1,000 + head, one-time capacity) use a consulting nutritionist to help in nutritional decision making (1).

The cattle feeding industry is concentrated in the Great Plains region of the United States, with some cattle fed in the midwestern and western states. Arizona, California, Colorado, Idaho, Illinois, Iowa, Kansas, Minnesota, Nebraska, Oklahoma, South Dakota, Texas, and Washington accounted for approximately 86% of the U.S. cattle-on-feed inventory as of January 1, 1994 (1). Moreover, estimated gross receipts from all cattle amounted to more than $40 billion in 1993 (Table 14-1). The cattle business is responsible for 20% of all cash receipts from farm marketings and is more than three times larger than any other sector of meat-animal agriculture (2). An estimated additional $5 in related economic activity is generated for each cattle dollar when the meat processing and marketing and agricultural supply and service industries are taken into account (2).

CURRENT CATTLE TYPES AND CATTLE GROWTH

Today's feeder calves come from diversified cow herds across the United States and Mexico. In addition, many new and exotic breeds of cattle have entered the United States in recent years, resulting in a considerable variation in size, body type, and potential growth response among feeder cattle. Because differences in size and rate of maturing have an effect on nutritional management, the NRC (3) included medium- and large-frame sizes in the prediction equations for estimating nutrient requirements and feed intake. As a rule of thumb, large-framed steers are deemed as those that will produce Choice carcasses at liveweights of more than 1,200 lb; medium-framed steers, from 1,000 to 1,200 lb; with small-framed steers producing Choice carcasses at liveweights of less than 1,000 (4). Muscle thickness standards denote differences in muscle-to-bone ratio among

TABLE 14-1

Estimated cattle of feed and gross income by state for 1993

State	Cattle on Feed[a] (Thousands)	Gross Income[b] (1,000 dollars)	State	Cattle on Feed[a] (Thousands)	Gross Income[b] (1000 dollars)
Alabama	—	377,098	Montana	19.0	813,130
Alaska	—	977	Nebraska	6,617.4	4,715,933
Arizona	368.3	559,345	Nevada	1.3	136,212
Arkansas	22.8	380,467	New Hampshire	—[c]	9,199
California	857.1	1,543,114	New Jersey	18.4	19,900
Colorado	2,441.0	2,426,227	New Mexico	52.4	749,175
Connecticut	42.0[c]	18,641	New York	63.9	252,040
Delaware	40.4[d]	10,767	North Carolina	152.4	205,274
Florida	—	364,577	North Dakota	—	514,307
Georgia	—	316,936	Ohio	185.4	393,202
Hawaii	20.8	27,659	Oklahoma	41.9	2,141,967
Idaho	—	698,647	Oregon	27.7	376,346
Illinois	—	724,555	Pennsylvania	993.8	541,323
Indiana	80.3	357,705	Rhode Island	—[c]	1,112
Iowa	1,668.0	2,217,380	South Carolina	—	150,444
Kansas	6,232.2	4,375,292	South Dakota	241.3	1,514,095
Kentucky	65.8	775,780	Tennessee	—	424,463
Louisiana	26.9	218,138	Texas	6,052.8	6,369,370
Maine	—[c]	26,929	Utah	—	321,276
Maryland	—[d]	81,143	Vermont	—[c]	59,880
Massachusetts	—[c]	13,572	Virginia	21.4	343,644
Michigan	—	314,283	Washington	817.5	710,693
Minnesota	1,050.3	1,103,627	West Virginia	15.1	131,628
Mississippi	—	259,835	Wisconsin	1,393.3	804,223
Missouri	200.3	886,863	Wyoming	5.4	596,200
			Total U.S.	33,324.1	40,374,593

Source: USDA, National Agricultural Statistics Service (83).

[a]Includes slaughter in federally inspected and other slaughter plants; excludes animals slaughtered on farms. States with no data printed are included in the U.S. total but data are not printed to avoid disclosing individual operations.

[b]Preliminary values. Includes cash receipts from sales of cattle, calves, beef, and veal, plus value of cattle slaughtered for home consumption.

[c]Connecticut, Maine, Massachusetts, New Hampshire, Rhode Island, and Vermont are summed.

[d]Delaware and Maryland are summed.

cattle of similar fatness with No. 1 = thick, No. 2 = moderate, and No. 3 = thin.

The average slaughter weight of cattle has increased from 1,072 lb in 1984 to 1,161 in 1993 (Table 14-2). Part of the increase in weight most likely is related to the selection of larger-framed cattle. However, management of the cattle by grazing for an extended length of time before placement in the feedlot and decreasing margins and the resultant holding of cattle for longer periods of time may be partly responsible. As finishing cattle become more mature, fat deposition increases, whereas protein and water deposition decrease. More NE_g is required per unit of fat gain than per unit of protein gain (3). As a result, as animals deposit more fat they become less efficient. Along with decreasing returns resulting from lowered efficiency, there are additional yardage and interest costs associated with holding cattle for extended periods after they are finished.

TABLE 14-2

Number of cattle slaughtered under federal inspection and average liveweight from 1984 to 1993

Year	No. Slaughtered (thousands)	Average Liveweight (lb)
1984	35,880	1,072
1985	34,765	1,103
1986	35,913	1,105
1987	34,468	1,109
1988	34,048	1,124
1989	33,010	1,138
1990	33,242	1,136
1991	32,690	1,163
1992	32,874	1,169
1993	33,324	1,161

Source: USDA, National Agricultural Statistics Service (83).

Three age classes and nine frame-by-muscle thickness subclasses fed to a constant 0.53 in. of fat thickness were used (5) to evaluate the effects of time on feed of large-, medium, and small-framed feeder steers with No. 1, No. 2, and No. 3 muscle thickness for calves, yearlings, and long yearlings. Increasing age and decreasing frame size resulted in a significant decrease in days on feed (Table 14-3). Slaughter weights and carcass weights were heavier for long yearlings than for calves and yearlings (Table 14-3). Increasing frame size also resulted in heavier slaughter weights and hot carcass weights. Dressing percent of large- and medium-framed steers was higher than dressing percent of small-framed steers. However, the USDA cattle muscling categories did not consistently identify differences in muscle:bone ratio, and some cattle may be incorrectly perceived as thinly muscled (5). There were no differences in time on feed required to reach .53 inches of fat thickness for No. 1 and No. 2 steers, but No. 3 steers required more time on feed than the other two muscle thickness groups. Muscle thickness increased slaughter weights and hot carcass weight for No. 3 steers compared with No. 1 steers, with intermediate hot carcass weight for No. 2 steers (Table 14-3). Days on feed required to reach .53 inches of fat thickness for the nine frame size x muscle thickness subclasses within each age group are shown in Table 14-4. These data agree with Table 14-3 in that as frame size increases, days on feed required to reach .53 inches of fat thickness increase, and as animal age increases, fewer days on feed are required.

The study by Dolezal et al. (5) made no attempt to standardize the breeds of cattle used. Intramuscular fat deposition has been shown to be a function of the number of days cattle are fed a high-concentrate for certain breeds. For yearling British-cross steers (Angus × Hereford, approximately 16 months of age), approximately 112 days on high-concentrate diet were needed to reach Choice quality grades (6). Feeding beyond 112 days did not increase quality grade or enhance the palatability of the steaks, but increased waste fat. Choice carcasses can be produced from moderate-framed Angus steers with only 45 days in the feedlot following a growing period of 150 days at 1.3 lb/day (7); however, more time may be required to achieve acceptable tenderness. For some large breeds of cattle (Chiaina), quality grades were increased up to 128 days on feed, after which they reached a plateau (8). Feeding the cattle for 182 days did not result in these cattle reaching Choice quality grade. Instead, subcutaneous fat was increased with no improvement in quality grades.

With the large number of heifers that are marketed for slaughter each year, producers need to understand the effects of chronological age on carcass maturity score. The effects of carcass maturity on meat palatability were studied (9) using concentrate-feed cattle with known history to determine the relationship of chronological age to carcass maturity scores. Results indicated that the magnitude of differences of tenderness between yearling heifers and 2-year old cows was not large enough to justify the price of discrimination in the marketplace. As a guide to managing and marketing

TABLE 14-3

Effects of age class, frame size, and muscle thickness on time on feed, slaughter weight, hot carcass weight, and dressing percentage for cattle fed to 0.53 inches fat thickness

Item	Days on Feed	Slaughter Weight (lb)	Hot Carcass Weight (lb)	Dressing (%)
Age class				
Calf	251[a]	1,134[b]	712[b]	62.6
Yearling	166[b]	1,181[b]	737[b]	62.2
Long yearling	98[c]	1,301[a]	816[a]	62.7
Frame size				
Large	214[a]	1,421[a]	897[a]	63.2[a]
Medium	163[b]	1,176[b]	739[b]	62.8[a]
Small	139[c]	1,024[c]	631[c]	61.5[b]
Muscle thickness				
No. 1	139[b]	1,181[b]	741[b]	62.7
No. 2	156[b]	1,197[ab]	749[ab]	62.4
No. 3	220[a]	1,239[a]	774[a]	62.5

Source: Dolezal et al. (5).

[abc]Least squares means in the same column and for the same item without common superscripts differ ($P < 0.05$).

TABLE 14-4

Number of days on feed required to reach 0.53-inch fat thickness by steers

Frame Size and Muscle Thickness	Calf	Yearling (days on feed)	Long Yearling
Large No. 1	235	162	105
Large No. 2	247	194	135
Large No. 3	368	286	189
Medium No. 1	196	148	84
Medium No. 2	254	142	91
Medium No. 3	272	198	78
Small No. 1	185	76	55
Small No. 2	213	92	42
Small No. 3	287	198	100

Source: Data from Dolezal et al. (5) as reported by Ritchie and Rust (4).

cull cows, the following chronological age groups were recommended to reflect each USDA maturity class: A, 9 to 24 months, B, 24 to 36 months; C, 36 to 48 months; D, 48 to 60 months; and E, > 60 months.

Matching feeding programs with cattle types is important (10). With interest in producing a value-added product for the consumer, various companies have developed programs and/or systems to sort cattle into outcome (slaughter) groups. Advantages of these endeavors may be increased profitability by the producers and presumably increased market share for the beef industry. For more information on these programs and systems, readers are referred to reference (4).

SYSTEMS USED FOR GROWING AND FINISHING BEEF CATTLE

Cattle production systems in the United States vary by locality and may be affected by feed sources and climate. Generally, the eastern and southeastern United States support most of the cow–calf production, with calves sold at weaning to go either directly to the feedlot or to a backgrounding (growing phase) operation. There has been increasing interest by cow–calf producers in retaining ownership of the calf through the growing period or through slaughter.

In the United States most cattle go through a backgrounding period after weaning before placement in a feedlot on a finishing diet. In particular, it may be necessary to grow small-to medium-framed cattle (400 to 500 lb) for a period of time (1.75 to 2.5 lb/day) to allow for frame growth in order to produce a desirable carcass weight. Most often, steers are grown to a body weight (BW) of approximately 700 to 750 lb before placement on a finishing diet.

A vast number of systems is available for growing cattle, and to classify all the systems would be a formidable task. Hence, we will group growing systems into two general areas: intensive systems, in which cattle are placed directly in the feedlot (see Fig. 14-1); and extensive systems, in which cattle are grazed on forages for varied lengths of times or fed milled diets in drylot. With heightened interest in retained ownership and producing a value-added product, it is increasingly important for producers to know what effects growing programs have on feedlot performance.

INTENSIVE SYSTEMS

Besides the obvious advantage of decreased cost of gain as a result of less expensive sources of energy, feeding cereal gains can decrease manure production compared with ad libitum con-

FIGURE 14-1 Cattle using shades in a large commercial feedlot in the southwestern United States. (Courtesy of D. C. Church.)

sumption of roughage-based diets. Decreased costs are also associated with mixing and handling of concentrates versus roughages. A further advantage of intensive systems is that intensively fed calves are generally more efficient than calves subjected to extensive production systems (11). However, these animals also will put on more fat and finish at lighter weights than cattle grown for a specified length of time before finishing (11). Thus, a major disadvantage with this system is that with certain cattle types (small-framed cattle) intensive systems may produce light carcasses. However, with large-framed cattle intensive systems will produce carcasses within an acceptable range for the packing industry.

LIMIT OR PROGRAMMED FEEDING
One alternative to ad libitum feeding of a finishing diet, as in typical intensive systems, is for producers to feed a high-concentrate diet to provide a specified or programmed rate of gain. Limit feeding an all-concentrate diet to achieve gains similar to those from a corn-silage-based diet improved feed conversion dramatically in the growing phase as a result of increasing diet digestibility (12). In addition, limit feeding the all-concentrate diet during the growing phase did not have detrimental effects on finishing performance; however, incidence of liver abscesses increased when all-concentrate diets were fed in both the growing and finishing phases. Concentrate or roughage level may be important in limit or programmed feeding systems. A New Mexico study (13) examined the effects of percentage of concentrate in limit-fed growing on feed intake, average daily gain, and feed efficiency during the finishing phase. Cattle were limit fed 60%, 75% or 90% concentrate diets to gain approximately 2.2 to 2.25 lb/day. Results suggested that percentage of concentrate during the growing phase did not greatly affect finishing-phase performance or carcass characteristics. In addition, large-framed steers did not need a growing phase to produce acceptable carcasses.

EXTENSIVE SYSTEMS

The second general option for cattle producers is to grow cattle at low to moderate rates of gain (1.75 to 2.5 lb/day) using roughage-based diets (extensive systems). Cattle raised in extensive-based systems are generally older and yield heavier carcasses than cattle given a finishing diet after weaning. Roughage and yardage costs can be a major disadvantage to these programs; however, when cattle graze forages, costs can be decreased.

FESCUE PASTURES A number of cattle entering commercial feedlots have previously grazed tall fescue pastures. These cattle often exhibit signs of heat stress and display rough hair coats. Cattle that are grazed on endophyte-infected fescue pastures are more likely to exhibit these symptoms when grazed during periods of high ambient temperatures or humidity. Generally, when cattle are received in the feedlot later in the fall when these conditions are not prevalent, no sacrifice in performance should be noted. Using anabolic agents (estradiol 17-β) may improve the performance by steers grazing endophyte-infected fescue (14); however, performance was still compromised by the infected pastures when compared with performance from noninfected varieties and low-endophyte-infected fescue. Steers compensated for poorer performance during grazing of endophyte-infected fescue when placed in the feedlot.

STOCKPILED FORAGES One method to decrease labor and equipment cost is to graze stockpiled forages during the winter. Stocker cattle grazing stockpiled tall fescue from November to April in Virginia gained 0.75 lb/day and required only 527 lb of hay–stocker fed during 36 days (15). Interseeding legumes for stockpiled fescue eliminated the need for nitrogen fertilization but decreased the quantity of grazed forage, thereby requiring more hay to be fed. Performance was similar between feeding orchard grass–alfalfa hay versus stockpiled fescue–alfalfa, but more stored feed was required.

WINTER WHEAT PASTURE Cattle that have grazed annual winter wheat pasture (Fig. 14-2) generally perform well in the feedlot. However, each pasture is inherently different, with cattle generally gaining approximately 2 lb/day. Wheat varieties are generally selected based on factors other than animal performance. Research in Oklahoma evaluated the effects of wheat varieties on cattle performance (16). In general, selection of wheat varieties for gain yield will decrease beef gain per acre. Stocking densities of 1.7 acres/steer seemed to be the optimum level for maximum production. Feedlot performance was not affected by difference in gain during the wheat pasture growing phase.

In certain years, cattle may need to be fed in a drylot before being placed on the wheat

FIGURE 14-2 A heifer showing stiff forelegs and foundered front feet, both conditions typical aftereffects of acidosis. (Courtesy of D. C. Church.)

pasture, and it is advantageous to feed newly weaned/or received cattle a receiving diet for a period of time to prevent outbreaks of disease (see later discussion). When adequate wheat pasture is available, cattle will compensate for decreased short-term gains in the drylot before placement on wheat (17).

Another advantage of wheat pasture is that there is no need for supplementation of the cattle to achieve adequate gains, other than a mineral supplement. A disadvantage of wheat pasture, however, is the possibility of bloat. Feed additives like poloxalene (fed at 1 to 2 g/100 lb of BW daily; 18) and the ionophore monensin help to prevent bloat in cattle grazing wheat pasture. It is preferable to include these products in some type of supplement, and not supply them in the mineral, so that the animals receive the daily recommended levels. Supplemental feeding also may increase performance by cattle grazing wheat pasture. Supplementing either a high-fiber (soybean hulls and wheat middlings-based energy supplement) or a high-starch (corn-based energy supplement) at approximately 0.65% of BW to growing cattle on wheat pasture increased stocking rates by approximately one-third. In addition, daily gains were increased by 0.33 lb by supplementation but were not influenced by supplement type (19).

EFFECTS OF SYSTEMS ON PERFORMANCE   As with all segments of the cattle industry, buying cattle for finishing is a very competitive business. Some buyers may pay a premium for cattle that seem to have gone through a period of restricted growth. These cattle in turn often exhibit compensatory gain in which the animal displays rapid and, in some instances, more efficient growth following the period of feed restriction. However, the severity,

nature, and duration of growth restriction are all important aspects to consider. During the growing phase, growth restriction has its greatest impact on fat accretion without impairing muscle development (20). Increased dry-matter (DM) intake was responsible for compensatory growth upon refeeding for steers that were fed a high-roughage diet and for steers limit fed on a high-concentrate diet, but feeding a low-concentrate diet during the growing phase may impair overall feedlot performance (20).

Reports from Oklahoma (21, 22) evaluated the effects of different management schemes on performance during the finishing phase. The management schemes employed were early weaning and placement in the feedlot at 3.5 months of age; normal weaning and placement in the feedlot at 7.9 months of age; weaning at normal age and grazing on wheat pasture for 112 days before placement in the feedlot at 11.6 months; weaning at normal age followed by wintering on native range and grazing on managed native range for 68 days before placement in the feedlot at 15.4 months (short grazed); and treated the same as short-grazed cattle but grazed for 122 days before placement in the feedlot at 17.4 months. Starting weight in the feedlot was 314, 540, 765, 848, and 918 lb for the five management schemes, respectively. Younger steers were fed for longer periods of time in the feedlot to reach a similar fat thickness endpoint (0.5 in.; 287, 198, 134, 123, and 101 days, respectively). Early weaned calves tended to have lower daily gains than the other management schemes (2.93, 3.22, 3.70, 3.36, and 3.02 lb/day, respectively). Slaughter weights were 1,154, 1,178, 1,259, 1,259, and 1,222 lb when adjusted to a 64% dressing percent. Average daily feed intakes were 15.59, 18.20, 23.35, 25.36, and 25.11 lb, respectively. The feed-to-gain ratios were 5.33, 5.66, 6.32, 7.55, and 8.36, respectively. Based on yield grade or carcass specific gravity, younger cattle were fatter at slaughter. In addition, carcasses of younger cattle were worth more because these cattle reached slaughter weight earlier in the year.

After cattle have gone through a growing program, producers need to know how long to feed the cattle to obtain optimum performance. The effects of feeding for 105, 119, 133, or 147 days on performance by British × Continental beef steers were evaluated in another Oklahoma study (23). Carcass-weight-adjusted daily gains were highest for steers for 119 days, whereas feed intake tended to increase linearly as the cattle were fed longer. Feed conversion was superior for steers fed for 119 days compared with steers fed for 147 days.

As is evident from the previous discussion, U.S. beef cattle producers have almost limitless means of feeding cattle from weaning to slaughter. Researchers in Nebraska used a computer model to simulate the influence of postweaning production systems on performance and carcass composition of different biological types of cattle (24). The model used steers from F_1 crosses of 16 sire breeds mated to Hereford and Angus dams, growing under nine backgrounding systems [high average daily gain (1.98 lb) for 111, 167, or 222 days; medium average daily gain (1.1 lb) for 200, 300, or 400 days; a low average daily gain (0.55 lb) for 300 or 400 days; and 0 days backgrounding] and finished at either a low (2.2 lb) or high (3.0 lb) average daily gain. Results suggested that considerable flexibility exists in the choice of postweaning production systems for several genotypes of steers to produce acceptable carcass composition and retail product and that carcasses with a specified composition, retail product, or quality can be produced from a mixed group of steers fed and managed similarly from weaning to slaughter.

MANAGING NEWLY RECEIVED CATTLE

Losses associated with morbidity and mortality from bovine respiratory disease (BRD) in newly weaned or received cattle are a significant eco-

FIGURE 14-3 A typical feed bunk arrangement used by many large commercial feedlots. (Courtesy of D. C. Church.)

nomic problem for the beef cattle industry (See Fig. 14-3). In feedlots with 100 to 1,000 animals marketed annually (25), death losses ranged from 1.5 to 2.7 per 100 animals marketed. Two-thirds to three-quarters of these deaths were attributed to respiratory disease (25). Two factors contribute to the high incidence of BRD in newly received, lightweight (e.g., less than 400 to 500 lb) cattle. First, stresses associated with weaning and normal marketing procedures have a negative effect on the immune system (26) at a time when the animal is often challenged by infectious agents. Second, feed intake by stressed calves is usually low (27), averaging approximately 1.5% of BW during the first 2 weeks after arrival of lightweight feeder cattle (28). Low nutrient intake may further impair immune functions (27). Practices that have been used to offset these negative factors that affect the health of newly weaned or received cattle include preconditioning, on-ranch vaccination programs, nutritional management after arrival at the feedlot, and prophylactic medication.

ENERGY

Stressed calves have low feed intake. They also tend to prefer and will consume greater quantities of a high-concentrate than a high-roughage diet. When given a choice among feeds varying on concentrate level, stressed calves selected diets with 72% concentrate during the first week after arrival (29). Recovery of pay weight is typically faster and cost of gain less with higher than lower concentrate diets, but higher concentrate receiving diets can increase the severity of morbidity (days of medical treatment per calf purchased). Feeding alfalfa or good-quality grass hays in addition to a 75% concentrate diet during the first week of the receiving period offset the negative effect of higher concentrate receiving diets on morbidity (30). An effective receiving program for newly received, stressed cattle can be based on diets with $\geq$ 60% concentrate, with either free-choice or limited quantities of hay (approximately 2 lb/animal daily) during the first week of the receiving period. If feed milling facilities are limited, feeding good-quality hay plus protein supplement can be sufficient (27), but calves may not fully compensate for lower gains during the receiving period often observed with such programs (30). A typical concentrate-based receiving diet is shown in Table 14-5.

PROTEIN

Newly weaned or received calves have a low capacity for protein deposition during the first

TABLE 14-5

A concentrate-based diet for newly received cattle typical of those fed in the Great Plains region[a]

Ingredient	% of DM
Sudan grass hay	10
Alfalfa hay	20
Corn or milo grain	56
Soybean or cottonseed meal	5
Molasses	5
Limestone	0.7
Dicalcium phosphate	0.5
Salt	0.3
Urea	0.5
Premix[b]	2

[a]Approximate composition using corn and soybean meal: NE_m (Mcal/100 lb) = 83.9; NE_g (Mcal/100 lb) = 55.3; CP = 14.1%; Ca = 0.76%; P = 0.42%; K = 1.05%.

[b]Supplies all vitamins and trace minerals. See Table 14.7 for suggested levels.

week or two after arrival because of low feed intake relative to subsequent periods when feed intake is normal. Generally, a supplemental source of crude protein (CP) does not seem to be a major factor in receiving diets as long as a natural source is used. One exception may be corn-silage-based receiving diets, which would likely supply a fairly large amount of ruminally degraded nitrogen (N) from corn silage. Improved gain and gain efficiency were reported (31) when blood meal was the source of supplemental protein in corn-silage-based receiving diets compared with soybean meal. New Mexico workers (32; Table 14-6) compared diets with 12%, 14%, or 16% CP for stressed calves (19.5 h in transit, 6.8% shrink from pay weight). Daily gain and DM intake increased linearly with increasing CP concentration for the 42-day receiving period. After this period, all calves were fed a 14% CP, 85% concentrate diet. Those fed the 12% CP diet during the receiving period compensated for decreased gain during the subsequent 42-day period, such that dietary CP concentration fed during the receiving period did not affect performance for the 84-day trial. Practical receiving diets based on hay and other dry roughages should probably contain at least 14% CP.

MINERALS AND VITAMINS

Although potassium requirements may be increased in stressed calves, requirements for other minerals do not seem to differ greatly from those of nonstressed calves (33). Because of low feed intake, however, concentrations of most minerals need to be increased in receiving diets. Among the trace minerals, Zn, Cu, Cr, and Se have potential effects on immune function (see reference 34), but further practical experiments are needed to adequately define fortification levels. Effects of B vitamin supplementation on the performance and health of newly weaned or received cattle have been variable (33, 27), and there seems to be little economic justification for B vitamin supplementation to practical receiving diets. Vitamin E has effects on the immune system, seemingly stimulating the immune response when given before an infectious challenge, but having little or no effect when given after the challenge (33). Feeding 400 to 800 IU of vitamin E per animal daily has increased performance and decreased morbidity in some field studies (35). Suggested ranges for vitamin and mineral concentrations in receiving diets are shown in Table 14-7.

FEED ADDITIVES

Ionophores are often added to receiving diets, frequently as a means of controlling coccidiosis. In one trial (36), 250 calves (initial BW = 391 lb) purchased from auction barns in southern Arkansas were shipped to New Mexico and fed one of four 70% concentrate diets: control (no ionophore); lasalocid (Bovatec®, Roche Animal Nutrition and Health) at 30 g/ton; monensin (Rumensin®, Elanco Animal Health) at 20 g/ton; and monensin at 30 g/ton of the dietary DM. Ionophores decreased feed intake compared with the control diet for the 28-day trial, but daily gain and feed efficiency were not significantly altered by treatments. Numerically, monensin at 30 g/ton resulted in a lower feed intake than lasalocid at 30 g/ton. All three ionophore treatments decreased the presence of coccidial oocysts. Although ionophores have negative effects on the intake of receiving diets, these effects can be minimized by the choice of ionophore or, with monensin, by decreasing its dietary concentration.

PROPHYLACTIC MEDICATION

Antibiotics are often added to feed or water of newly received cattle as a means of decreasing morbidity. Chlortetracycline and oxytetracycline included in the feed or water of new cattle are effective as a means of controlling BRD, particularly when morbidity and mortality levels are low (33). The treatment of individual animals with antibiotics on a prophylactic or preventive basis is another approach that has been successful in decreasing the incidence of BRD. Animals are typically mass medicated at

TABLE 14-6

Influence of protein concentration on performance by calves during a 42-day receiving period

Item	Dietary CP Concentration (%)			Contrast[a]	SE[b]
	12	14	16		
Receiving period performance					
No. of calves	40	40	40	—	—
Initial BW, lb	412.4	409.2	403.3	—	2.6
Day BW, lb	520.1	536.6	537.2	—	5.5
Daily gain, lb					
0 to 21 days	1.50	1.85	1.97	NS	.34
21 to 42 days	3.52	4.23	4.40	L[d]	.21
0.42 days	2.51	3.04	3.19	L[d]	.11
Daily DM intake, lb/steer					
0 to 21 days					
Hay	1.34	1.36	1.34	NS	.05
Concentrate	5.84	6.36	6.67	NS	.34
Hay + concentrate	7.17	7.72	8.01	NS	.35
21 to 42 days	12.55	12.40	13.30	NS	.39
0 to 42 days	9.86	10.06	10.635	L[e]	.27
Feed to gain ratio					
0 to 21 days	5.64	4.72	4.35	NS	.94
21 to 42 days	3.61	2.95	3.04	L[d]	.17
0 to 42 days	3.95	3.32	3.35	Q[d]	.08
Calves treated for BRD,%[c]	37.5	22.5	47.5	—	—
Mortality (no.)	2	1	0	—	—
Postreceiving performance					
Daily gain, lb	3.49	3.40	3.21	NS	.18
Daily DMI, lb/steer	15.37	15.76	15.49	NS	.47
Feed-to-gain ratio	4.44	4.66	4.83	NS	.16
Overall performance					
Daily gain, lb	3.00	3.22	3.20	NS	.12
Daily DMI, lb/steer	12.62	12.91	13.07	NS	.33
Feed-to-gain ratio	4.23	4.01	4.09	NS	.09

Source: Galyean et al. (32).

[a]Orthogonal contrasts: L = linear, Q = quadratic effect of protein concentration.

[b]Standard error of means, n = four pens per treatment.

[c]Distribution differs; for 12% versus 14% CP, $P < 0.15$; for 14% versus 16% CP, $P < 0.03$.

[d]$P < 0.05$.

[e]$P < 0.10$.

TABLE 14-7

Typical vitamin and mineral fortification levels in diets for newly received cattle[a]

Item	Suggested Level in the Final Diet (DM Basis)
Vitamin A	1,000–1,200 IU/lb
Vitamin E	30–50 IU/lb
Cobalt	1–2
Copper	10–15
Iodine	.5–2
Iron	50–70
Manganese	40–60
Selenium	0.1
Zinc	60–120

[a]See NRC (3) for possible sources of these vitamins and minerals.

or near the time of arrival in the feedlot. High-risk cattle that undergo extensive stress and shrink during marketing and transport or cattle from locations with a history of health problems are most likely to benefit from such programs. A combination of long-acting oxytetracycline and sustained-release sulfadimethoxine as a mass-medication treatment for newly received, stressed calves decreased morbidity from 63.3% in control calves to 7.1% in mass-medicated calves (37). Three trials were conducted (38) in which tilmicosin phosphate was given at the time of processing to newly received, lightweight calves. Mass medication with tilmicosin phosphate did not affect daily gain, but the percentage of calves treated for BRD was decreased from 46.4% to 0% in one trial and from 32.8% to 12.1% in a second trial. In a third trial, treatments included no arrival medication (control), mass medication with tilmicosin phosphate, or medication with tilmicosin phosphate if a calf's rectal temperature at arrival processing was ≥ 103.5°F. Treatment of calves with tilmicosin phosphate, either by mass medication or based on arrival rectal temperature, increased daily gain and DM intake during a 28-day receiving period and a subsequent 28-day feeding period. The percentage of calves treated for BRD was decreased from 43.6% in control calves to 11.9% in mass-treated calves and 12.9% in calves treated on the basis of arrival temperature. Hence, individual prophylactic medication of newly received, stressed cattle can be used to decrease the incidence of BRD.

YEARLING CATTLE

Older (e.g., yearling) cattle or preconditioned cattle that have undergone some type of backgrounding or growing problem before shipment to the feedlot typically have greater intake and less health problems than lightweight cattle subjected to shipping stress, although outbreaks of BRD can also be a problem with such cattle. Generally, the nutritional and management principles discussed for receiving programs with lightweight cattle are effective guidelines for older or preconditioned cattle. One notable exception is that because of greater feed intake the length of the receiving period for yearling-type cattle typically does not need to be as long as for lightweight, newly weaned or received cattle (e.g., the transition from the receiving diet to the finishing diet will likely be more rapid with yearling cattle). As discussed previously, small- to medium-framed lightweight cattle may need to be grown at moderate rates of gain. Growing programs should generally not be initiated until cattle have been through a receiving period in which the emphasis was placed on controlling health problems.

STARTING CATTLE ON FEED

The term "starting cattle on feed" refers to the transition between the receiving or growing diet and the final, high-grain finishing diet. The nature of this transition period depends on the type of receiving and/or growing programs to which cattle have been exposed. For cattle received or grown on roughage-based diets, the transition involves gradually changing the diet from one of primarily roughage to one of primarily concentrate. For cattle that have been grown on an 85% to 90% concentrate limit or

programmed feeding program, little or no step-up in diet concentrate level is required; rather, the quantity of diet can be increased gradually until the animals have reached the point of ad libitum consumption.

For cattle undergoing the transition from a grazing situation or a roughage-based growing diet, the transition is typically accomplished by a series of step-up diets that contain an increasing concentrate level. The objective in this situation is to provide a smooth changeover that avoids excessive consumption of concentrate, leading to acidosis and decreased feed intake. Acute acidosis has long-term consequences on cattle health and performance and must be avoided (10). The traditional approach to starting cattle on feed is to feed a relatively low concentrate diet (40% to 50%) as the starter diet. In some cases, free-choice hay may be provided for the first few days in the feedlot. In a survey of operations of 1,000 head or more capacity, 5.3% of operations fed 0% concentrates, 70.1% fed from 1% to 55% concentrates, 13.6% fed from 56% to 74% concentrates, and 11% of operations fed 75% + concentrates to cattle on arrival (1). The transition to the final high-concentrate diet (90% + concentrate) is typically made in steps of 10% to 20% concentrate over a period of 21 to 28 days. Each diet of increasing concentrate level is fed for a period of 3 to 7 days, depending on the concentrate increment in the step and the intake pattern of the cattle. Changes that are too abrupt are likely to cause cattle to decrease feed intake because of acidosis, sometimes necessitating a step back to the previous diet.

Because most commercial feedlots feed cattle two to three times daily, another approach that seems to work effectively is to make diet concentrate changes gradually by feeding the next diet in the step-up program as the second or third feeding of the day. On the following day, the next step-up diet might comprise two of the three daily feedings, with a complete transition made over a 3-day period. Depending on intake, cattle can be held at any step for a few days or the transition can continue. Another approach is to allow cattle to reach ad libitum intake of a moderate concentrate (approximately 60%) starter diet, followed by a gradual substitution, 10% daily for example, of the final high-concentrate (90% + concentrate) diet. Both methods theoretically lead to a smoother transition curve in dietary concentrate level as opposed to a stair-step curve in the more traditional approach of abrupt changes in dietary concentrate level.

Sound bunk management is critical to any approach to starting cattle on feed. During transition to higher concentrate levels, it is generally beneficial to have some feed in the bunks to prevent overconsumption. Feed intake should be monitored closely to determine that it meets with expectations based on experience. Alternatively, intake prediction equations, such as those provided by NRC (58), or more recently derived equations (26) can be used to determine whether feed intake is meeting appropriate targets.

FEED BUNK MANAGEMENT

Feed bunk management is one of the most critical jobs in the feedlot (see Fig. 14-4). Nutritional consultants often devote a large share of their time with feedlot clients dealing with bunk management issues. Once cattle have been started on feed, the approach to bunk management changes from that used during diet step-up periods. Continued close monitoring of feed intake to ensure that cattle are consuming expected quantities of feed is important, but once cattle are on the final high-concentrate diet, a further objective of bunk management is to achieve a relatively constant intake across time. The traditional goal of bunk reading (10) has generally been considered to have "the last mouthful of feed consumed as the feed truck is dumping additional feed in the bunk." Hence, the effort is to ensure that feed bunks are basically clear or "slick" within a 24-h feeding cycle. Obviously, careful attention to bunk reading is critical to ensure that the cattle are not short on feed, and thereby hungry, or fed too much so that excessive feed remains in the bunk and becomes stale. Recently, many nutritional consultants have slightly modified this traditional goal to achieve a clean bunk in somewhat less that a 24-h cycle, which requires that the feed bunks be read during the late night hours. This approach also involves constantly challenging the cattle to increase their feed intake. For example, if a pen of cattle has a clean bunk at 12:00 A.M., the next day's feed call might be increased by 0.5 to 1 lb/animal. If the pen fails to clean up the new allotment of feed, the allotment would be decreased back to the original level. However, if over a period of 2 to 3 days the pen cleans up this new level of feed, the cattle are once again challenged with a higher level of feed. Obviously, the bunk management approach requires consistent bunk reading practices and typically involves the use of computerized records for each pen to assist the bunk reader. The net effect of this type of intensive bunk management is to allow the cattle to achieve maximum feed intake with minimal day-to-day variance.

FINISHING DIETS

ROUGHAGE IN FINISHING DIETS

High-concentrate finishing diets typically contain small amounts (3% to 15%) of roughage, but all-concentrate diets are used by feedlots. On an energy basis, roughage can be one of the most expensive ingredients in finishing diets. However, roughages are an important component of feedlot diets and have a large influence on ruminal function Low dietary roughage levels have been associated with digestive upsets, including acidosis and liver abscesses. Common roughage sources include alfalfa hay, grass hays, silages (corn, wheat, and grasses), and by-product feeds (cottonseed hulls).

LEVEL Roughage level in finishing diets can affect cattle performance. Kansas workers (40) reported faster, more efficient gains by cattle fed either 5% or 10% roughage compared to those fed 0% or 15% roughage in steam-rolled wheat diets. With roughage levels of 0%, 3%, 6%, and 9% (50:50 mixture of alfalfa and corn silage) in high-moisture corn and dry-rolled sorghum diets fed to yearling steers, a quadratic effect on DM intake and daily gain was reported (41); gain was maximized at 9% roughage, but steers were less efficient as roughage level increased from 0% to 9%. Similarly, with diets based on a combination of dry-rolled wheat and dry-rolled corn (75:25), increasing roughages level (0%, 3.75%, and

FIGURE 14-4 Newly received, stressed calves require careful monitoring for outbreaks of respiratory disease.

7.5%) increased DM intake but linearly decreased the gain-to-feed ratio. Adding roughage to dry-rolled wheat, but not to dry-rolled corn or sorghum diets (0% versus 7.5%), was beneficial to gain efficiency (41). When whole-shelled corn diets were fed to growing–finishing heifers (42), daily gain did not differ with roughage levels of 7.5% versus 15%, but DM intake was increased with 15% roughage diets. Also, with whole-shelled corn diets, daily DM intake increased linearly as alfalfa level increased from 0% to 8% (43), but daily gain did not increase sufficiently to compensate for increased feed intake, leading to decreased gain efficiency with increased roughage level. Nonetheless, cattle fed 8% roughage had greater carcass weights than those fed 0% or 4% roughage, leading to greater profitability with the higher roughage level.

SOURCE   Roughage source is often as important as roughage level, and both source and level can interact with grain processing. The effects of roughage source on performance by cattle fed high-concentrate diets may depend on the corn-processing method (44). In one trial, steers fed ground high-moisture or dry-rolled corn diets with 9.2% alfalfa silage had greater DM intake and an improved feed-to-gain ratio compared with those fed 11.5% corn silage diets. In a second trial, DM intake and feed-to gain ratio were increased with 9.2% alfalfa hay versus 11.5% corn silage as roughage sources in ground high-moisture corn diets, but roughage source did not markedly affect performance with dry-rolled corn diets. In two other trials, roughage source did not affect performance by cattle fed whole, dry corn diets, but the feed-to-gain ration was improved with alfalfa silage versus hay in whole, high-moisture corn diets. In growing–finishing heifers (42), use of alfalfa as the roughage source resulted in consumption of 10.5% less feed than use of sorghum–Sudan grass hay, with intermediate intake by heifers fed cottonseed hulls; daily gains reflected differences in feed intake. Matching roughage sources and levels with grains for optimal utilization in high-concentrate diets is desirable, but data on which to base such pairings are limited.

EFFECTS OF ROUGHAGE ON DIGESTION AND PASSAGE   Effects attributable to added roughage in high-concentrate diets may be related to changes in fiber digestion and digesta passage with added roughage. Increasing dietary roughage decreases residence time of grain in the rumen (45, 46). Roughage source affects digestion and passage (47), but these effects could vary with the type of grain in the concentrate. Variations in roughage level and source could be used to modify flows of particular dietary components from the rumen, and provide a means to alter the site of nutrient digestion, but further research on roughage–concentrate interactions is needed before such an approach can be used successfully in practice.

High-grain (starch) diets often decrease digestion in the rumen, shifting it to the large intestine (48), but it is not known whether the large intestine can compensate fully for the increased fiber load. Lowered fiber utilization may partially explain decreased feed efficiency with increasing roughage level. With highly processed grains that are digested extensively in the rumen, little starch should reach the large intestine, and compensatory digestion of fiber in the large intestine could proceed without negative associative effects of starch. With unprocessed grains, more starch would reach the large intestine, which might have negative effects on fiber digestion in that organ (48). Roughage level and source also might influence small intestinal digestion of starch. Increased pancreatic α-amylase activity in forage-fed compared with concentrate-fed calves at equal energy intakes has been reported (49).

GRAINS, GRAIN PROCESSING, AND OTHER FEEDSTUFFS

SOURCES   Grain sources used by the feedlot industry depend on many factors, including location of the feedlot, feed availability, commodity cost, equipment, and palatability. In the midwestern and upper Great Plains states, corn is readily available for a source of energy, whereas in the southern Great Plains states, sorghum grain is common. During certain periods of the year, wheat or barley may become available and, in certain locations, oats may be available and will be least cost into the diets of feedlot cattle.

In addition to the traditional sources of cereal grains, recent advances in plant genetics and the resulting hybrids may affect cattle performance. High-lysine corn may be a viable alternative to normal corn hybrids for livestock feeding (44). Enhanced cattle performance may be a result of the indirect selection for improved grain utilization.

STARCH IN GRAINS AND GRAIN PROCESSING   Starch exists in grains as granules in the endosperm, and the properties of starch granules depend on the particular grain (51). A highly organized crystalline region (mostly amylopectin) in the granule is surrounded by a less dense, amorphous region that has a high percentage of amylose (51). Granules also are embedded in a protein matrix, which, particularly in corn and sorghum, can decrease the potential for enzymatic attack and digestion. If sufficient energy is applied to break bonds in the crystalline region of the granule, starch gelatinization occurs (51). Mechanical, thermal, and chemical agents can instigate gelatinization, but water is necessary for the process to occur (51). Processing methods that employ heat and moisture (e.g., steam flaking) cause extensive gelatinization and rupture of starch granules. Such heating also may denature grain proteins, which could affect starch digestion.

Gelatinized grains are typically digested to a greater extent in the rumen and total tract than unprocessed grains (52). The benefits from processing are related inversely to the digestibility of the unprocessed grain, with the magnitude of increase being sorghum > corn > barley (52). Even without processing, barley and wheat are digested extensively in the rumen, but some degree of processing is necessary for sorghum to be used efficiently by cattle. Bulk density is a practical means of determining the degree of processing. Decreasing bulk density of steam-flaked sorghum from 35 to 18 lb/bu increased both enzymatic and ruminal (in situ) rates of starch breakdown (53). Dry-matter intake by steers decreased as bulk density of steam-flaked sorghum grain decreased from 34 to 22 lb/bu; feed efficiency was optimized at the least bulk density (54). Bulk density may have less effect on processed corn than on sorghum. Steam flaking corn to bulk densities of either 28, 24, or 20 lb/bu did not affect feedlot performance (55). Ruminal starch digestion was increased by decreasing bulk density, and ruminal pH was decreased with the least bulk density. For corn and sorghum, steam flaking, and presumably similar processing methods, should improve feed efficiency by 7% to 10% compared with unprocessed or minimally processed grain. Although response to grain processing is usually limited with grains like barley and wheat, some degree of processing (e.g., steam rolling) may be desirable to improve palatability and the physical characteristics of the grain.

Rate of digestion can be affected by particle size of grains. As particle size decreased, in situ DM and starch digestion for steam-flaked, dry-rolled, and high-moisture corn increased (56). Nonetheless, passage from the rumen is faster for smaller particles (57), so the effects of grain particle size on the site and extent of digestion seem limited in vivo (58).

Mixing different grains in finishing diets or mixing the same grain processed by different methods may be an effective way to take advantage of different digestion characteristics among grains and grain-processing methods and a means of decreasing grain-processing costs. In particular, complementary effects of mixing high-moisture corn with dry corn and sorghum have been reported (10), and up to 25% whole corn can be substituted for steam-flaked corn in finishing diets without effects on daily gain and feed efficiency (59).

BY-PRODUCTS AND OTHER FEEDSTUFFS Because of lower costs in some instances, by-product feeds have become important feedstuffs for feedlot diets. Some commonly used by-product feeds include whole cottonseed, brewers grains, distillers grains, beet pulp, citrus pulp, soybean hulls, wheat middlings, cull potatoes, and potato manufacturing products. Large quantities of whole cottonseed have been fed to feedlot cattle in the southwestern United States and northwestern Mexico and on occasion are priced competitively with more common concentrates (60).

Recent increases in ethanol production from grain have increased the availability of grain by-products to feedlots. Distillers grain, both in wet and dry forms, is often available at attractive prices throughout the midwestern United States and Great Plains feeding regions. Dry milling yields distillers dried grains with solubles, whereas wet milling yields corn gluten feed and meal (61). Depending on manufacturing processes and grains used, the composition of these products can be variable, making it important to determine the composition and establish links with potential sources. Wet corn gluten feed can effectively replace up to 25% of the corn in finishing diets without adversely affecting performance (62). Research at the University of Nebraska (63) has shown that wet corn gluten feed contained more NE_g than dry-rolled corn when fed in growing diets for cattle. In addition, the grain by-products typically have higher protein contents than the original grain, and their lower starch content may make them particularly suitable for use in starting diets because of a decreased potential for acidosis (64). Additional by-products of the corn milling industry are wet distiller grains and thin stillage that result from yeast fermentation of corn for ethanol production (65). In ruminant finishing diets, these wet distillers by-products are efficiently utilized as a protein and energy source (65).

Another by-product feed that has gained popularity in recent years is soybean hulls. When priced competitively, soybean hulls can be utilized efficiently by feedlot cattle. Soybean hulls contain approximately 50% acid detergent fiber and 12.1% CP and are low in lignin (2%; 3). A study at Purdue University evaluated soybean hulls as a replacement for corn in concentrate diets for beef cattle (66). The feeding value of soybean hulls was estimated at 75% to 80% of that of corn when included in moderate to high amounts in cracked-corn-based finishing diets. The researchers further suggested that using > 20% soybean hulls in concentrate-based diets should be evaluated based on cost of the product relative to cereal grains, given the decreased gain, increased feed intake, and poorer feed conversion.

Various liquid feeds are included in diets for growing and finishing cattle. Fat (tallow, animal and vegetable blends, and soap stocks) is commonly added as 2% to 4% of finishing diets, with level of addition depending on price. Molasses and liquid by-products of the corn milling industry also are common ingredients in feedlot diets. Addition of liquid feeds to diets generally decreases dustiness and often improves palatability.

Examples of finishing diets based on various feed ingredients are shown in Table 14-8. The use of particular ingredients in finishing diets depends on cost, availability, and feeding and milling facilities.

PROTEIN AND OTHER NUTRIENTS IN FINISHING DIETS

PROTEIN The feedlot industry has tended to increase dietary CP levels over the past 5 to 10 years, with current levels exceeding the values suggested by NRC (3). In a survey of six feedlot nutritional consultants (67), CP levels in finishing diets ranged from 12.5% to 14.4%. Added urea ranged from 0.5% t o 1.5% of the dietary DM. Diets fed by these consultants were low in roughage (3% to 11% of DM) and based on highly processed grains. In contrast, the factorial approach to calculation of CP requirements for beef cattle (3) suggests that dietary CP levels for finishing beef steers started

on feed at 700 lb and gaining 3 lb/day range from 8.9% to 11.7% depending on BW during the feeding period. Research reviewed by Galyean (67) indicated improved performance with CP levels in excess of those suggested by NRC (3), particularly with cattle subjected to aggressive implant programs involving estrogen + trenbolone acetate (see below). Improvements in performance with added protein were more consistent when added CP was derived from ruminally degraded sources of CP (e.g., urea and soybean meal), which may explain the fairly high urea levels used by consultants (67). Generally, limited advantages in performance by finishing cattle have been shown for supplemental protein sources that escape ruminal degradation. The reasons for responses to these relatively high CP level in current use by the feedlot industry are not known, but highly processed grains that are rapidly fermentable in the rumen may increase microbial needs for ruminally degraded N, and N in excess of requirements might provide ammonia for maintenance of acid–base balance by the kidney (67).

Galyean (unpublished)

VITAMINS Vitamin A is typically added to finishing diets at levels recommended by NRC (3; approximately 1,000 IU/lb of dietary DM). A common practice is to supply the required amount as supplemental vitamin A, ignoring contributions from dietary ingredients. Vitamin E requirements of feedlot cattle are not fully established (3), but vitamin E is commonly added to receiving diets and often added to finishing diets at levels of approximately 6 to 7 IU/lb of dietary DM. Recent research suggests that high levels of vitamin E (300+ IU/animal daily) added to finishing diets may have beneficial effects on the color stability of beef products (68).

MINERALS Generally, finishing diets for beef cattle are formulated to follow NRC (3) guidelines for required major minerals (Ca, P, K, S, Mg, Na, and Cl) and trace minerals (Co, I, Fe, Mn, Se, and Zn). Guidelines for potentially toxic or maximum tolerable levels are also available (3). Considerable debate exists relative to the merits of organic complexes of trace minerals versus inorganic sources, with some research suggesting greater bioavailability of organic complexes and other research suggesting no major difference between sources. Some inorganic sources of trace minerals (e.g., copper oxide) are relatively unavailable, but further research is needed before definitive

TABLE 14-8

TABLE 14-8

Composition of typical beef cattle finishing diets

Ingredient	Processed Corn and Dry Roughage	Whole Corn and Corn Silage (% of DM)	Dry-rolled Corn and Wet Corn Gluten Feed
Roughages			
Sudan grass hay	4	—	8
Alfalfa hay	6	—	—
Corn silage	—	10	—
Grain and grain by-products			
Steam-flaked corn	74.5	—	—
Dry-rolled corn	—	—	52.5
Whole shelled corn	—	71	—
Wet corn gluten feed	—	—	35
Liquid feeds			
Molasses	5	—	—
Condensed distillers solubles	—	4	—
Fat	3	—	—
Supplement[a]	7.5	15	4.5

[a]Supplement supplies calcium and phosphorus sources, urea and/or natural protein, trace minerals, vitamins, and feed additives. See Table 14-9 for the composition of a typical commercial supplement.

recommendations regarding organic versus inorganic sources can be made.

ANABOLIC AGENTS AND FEED ADDITIVES

ANABOLIC AGENTS

A variety of growth-promoting implants are available for use in growing–finishing cattle. Currently approved implants are of three general types: estrogen-based, androgen (trenbolone acetate)-based, and estrogen + trenbolone acetate-based. Implants increase rate of gain, improve feed efficiency, and increase carcass weight at Choice grade by 55 to 99 lb (69). Implants typically increase daily protein and fat gain in the carcass (70, 69), and the effects of estrogen + trenbolone acetate-based implants may be greater than estrogen-based implants (70). Growth-stimulating implants also usually increase feed intake. In two trials with beef steers fed a 60% concentrate diet, an estradiol benzoate–progesterone implant increased DM intake from 4% to 16%, depending on the time the implant was given relative to slaughter (71). Cattle that are not implanted typically consume 6% less DM than implanted cattle (72). Yearling steers that are on feed for 140 to 150 days typically receive an implant at the start of the feeding period and are reimplanted after 60 to 80 days. One common procedure is to use an estrogen-based implant initially and an estrogen + trenbolone acetate-based implant as the second implant. Finishing heifers often receive an initial estrogen-based implant and a final androgen-based implant. The use of such aggressive implant programs may increase dietary protein requirements, presumably because of an increased rate of daily protein gain (73).

IONOPHORES

Ionophores are fed to the majority of finishing beef cattle. At present, three of these compounds are approved for use in confined beef cattle diets: monensin, lasalocid, and laidlomycin propionate (Cattlyst®, Roche Animal Nutrition and Health). The effects of ionophores on performance by growing–finishing cattle have been summarized (74, 75). Generally, monensin decreases feed intake, with little change in daily gain, resulting in improved feed efficiency. Feed intake with lasalocid-based diets is typically equal to or greater than intake with diets that do not contain an ionophore, and daily gain is often increased, resulting in improved feed efficiency. Similarly, laidlomycin propionate tends to have little effect on feed intake, but increased daily gain results in improved feed efficiency. Diet composition (e.g., protein and minerals) may modify response to ionophores.

OTHER FEED ADDITIVES

A variety of other antibiotic feed additives are used by the feedlot industry. The antibiotic Virginiamycin (V-Max™, Pfizer Animal Health) is approved for use in confined cattle fed for slaughter to increase weight gain, improve feed efficiency, and decrease the incidence of liver abscesses. Bambermycins (Gainpro™, Hoechst Roussell Agri-Vet) is a feed-grade antibiotic approved for increased weight gain and improved feed efficiency in confined cattle fed to slaughter and for increased weight gain in pasture cattle. Tylosin (Tylan®, Elanco Animal Health) is an antibiotic that is commonly fed to decrease the incidence of liver abscesses in confined cattle. Tylosin is approved for combination feeding with monensin for steers and heifers and can also be fed in combination with melengesterol acetate (see below) alone or with melengesterol acetate plus monensin or lasalocid for confined beef heifers. The antibiotic chlortetracycline (CTC) is approved for a variety of purposes in various classes of beef cattle. In general, CTC is used to prevent BRD and anaplasmosis and for growth promotion and improvement of feed efficiency. Oxytetracycline is another antibiotic that is approved for purposes similar to CTC. Oxytetracycline is also approved for combination feeding with lasalocid for confined cattle to decrease the incidence of liver abscesses. Decoquinate is a compound approved for feeding to cattle for prevention of coccidiosis in cattle, a practice that is common for cattle during periods of exposure to coccidiosis or with newly received cattle suspected to be infected. Melengestrol acetate (MGA) is used to inhibit estrus in finishing beef heifers, resulting in improved gain and feed efficiency (10). Response to MGA can be variable, depending on heifer age, pen space per animal, and implant effects (10). Implant programs for heifers often involve the use of a trenbolone acetate-based implant as the final implant when MGA is fed. A complete listing of feed additives approved for use with beef cattle is avaiable in the *Feed Additive Compendium* (18).

Probiotics (microbial preparations and growth-media extracts) are used widely in ruminant production. Yeast cultures and live cultures of *Lactobacillus acidophilus* and *Streptococcus faecium* are the most common members among this group. Microbial cultures have been used primarily for either food preservation, as an aid to restoring gut function, or an agent to enhance feed utilization by ruminants (76). Because results are often conflicting, with some experiments showing benefits from using probiotics and other studies showing no benefit, further research is needed to more clearly de-

fine the effects of various probiotics on feed intake, digestion, and ruminal function.

Ionophores and other feed additives, as well as vitamins and trace minerals, are often added to finishing diets in the form of a pelleted supplement. Such supplements are also a convenient means of including supplemental protein and/or urea to finishing diets. An example of this type of supplement is shown in Table 14-9.

METABOLIC DISORDERS

ACIDOSIS

Lactic acidosis (Fig. 14-5) is a practical consequence of the expanded use of highly processed grain in beef cattle finishing diets. Cattle with acute acidosis experience a drastic decrease in ruminal pH (4.5 or less), with absorption of acids into the bloodstream leading to systemic acidosis, stiffness in the legs, founder, severe ruminal lesions, and sometimes death. In suba-

TABLE 14-9

Ingredient and chemical composition of a commercial supplement for beef cattle finishing diets[a]

Item	%, As-fed Basis
Ingredient composition	
Wheat middlings	33.882
Sunflower meal, 32% CP	30.000
Feather meal	6.846
Urea	6.968
Ammonium sulfate	1.500
Salt	5.500
Calcium carbonate	14.006
Bio-phos	.537
Trace mineral premix[b]	.350
Vitamin A/D$_3$ premix	.021
Vitamin E premix	0.13
Rumensin 80	.287
Tylan 100	0.90
Chemical composition	
Dry matter	92.91
Organic matter	67.65
Crude protein, %	42.00
NPN, %	22.00
Fat, %	1.82
Crude fiber, %	9.26
Calcium, %	5.80
Phosphorus, %	.80
Magnesium, %	.31
Salt, %	5.59
NE$_m$, Mcal/100 lb	49.92
NE$_g$, Mcal/100 lb	32.65
Added vitamin A, IU/lb	25,000
Added vitamin E, IU/lb	15
Monensin, g/ton	460
Tylosin, g/ton	180

[a]Formulated to be included at 5% of complete diet (as-fed basis).
[b]See Table 14-7 for suggested levels of trace minerals.

FIGURE 14-5 Winter wheat pasture is used throughout the Great Plains as a growing program for calves.

cute acidosis, ruminal pH drops substantially (5.0 to 5.5), with decreased feed intake following the acid load. To decrease acidosis, cattle are adapted gradually to higher concentrate levels. Roughages are typically much lower in rapidly fermentable carbohydrate than grains are, so added roughage decreases the chances for disastrous ruminal acid loads and proliferation of lactate-producing bacteria. Inherent buffering capacity of many roughages also may help to moderate ruminal pH. Careful bunk management designed to provide cattle with relatively constant day-to-day feed intake is critical in preventing episodes of acute and subacute acidosis. Diet ingredients (e.g., roughage or digestible fiber sources like wet corn gluten feed) and feed additives also may affect acidosis. Monensin has been shown to decrease day-to-day variance in intake, thereby decreasing the potential for acidosis (77).

LIVER ABSCESSES

A continual high acid load in the rumen can lead to parakeratosis, which in turn leads to clumping and necrosis of ruminal papillae (78). Such conditions, along with the occurrence of hairs embedded in the ruminal epithelium, can provide route of entry into the portal system for microbes like *Fusobacterium necrophorum* and *Actinomyces pyogenes* that cause liver abscesses (78). Incidence of liver abscesses can be high in certain geographical regions of the United States, especially in the Midwest feeding belt and northern Great Plains region. Cattle with severely abscessed livers have decreased gain and are less efficient than cattle with either no abscesses or less severely abscessed livers (7a). Affected livers are condemned at slaughter.

BLOAT

Feedlot bloat is typically of the frothy type, with formation of a stable foam. Free-gas bloat, which occurs secondary to factors that prevent eructation, is observed in some cases. Highly fermentable grain-based diets and(or) fine particle size presumably promote frothiness and high levels of gas production, and slime production by certain microorganisms may be involved (80). Cattle that die from acute bloat usually show no previous indication of illness. Feeding monensin decreases the incidence of frothy bloat on irrigated wheat pasture (81), and it is likely to be beneficial for feedlot bloat, presumably because of the effect of monensin on gas production and eating patterns of cattle.

OTHER METABOLIC DISORDERS

Outbreaks of polioencephalomalacia (PEM) occur sporadically in feedlots. The disease is characterized by initial blindness and muscle tremors, followed by recumbency and death; characteristic brain lesions are evident at postmortem examination (80). If animals are treated at the onset of signs with intravenous thiamin hydrochloride, the chances of recovery are good. The cause of PEM seems to be a thiamin deficiency induced by production of thiaminase or thiamine antimetabolites in the rumen (80).

Foot rot, caused by *Fusobacterium,* spp., is a contagious disease that causes inflammation of the foot and lameness (80). Outbreaks of foot rot are more likely with wet, humid weather and wet or muddy conditions in the feedlot. Antibiotic therapy and treatment of foot lesions are the usual course of action for foot rot. Prevention involves maintaining pens so as to decrease muddy and damp conditions. Supplementation of pasture cattle with zinc methionine has been effective in decreasing the incidence of foot rot (82), and many feedlot consultants feed higher levels of Zn or organic complexes of Zn in feedlots that frequently experience foot rot problems.

SUMMARY

Considerable diversity exists in size, body type, and growth potential of feeder cattle. As a result of this diversity, a variety of growing and finishing systems are used by beef cattle producers. Once cattle are placed in the feedlot, the time required to reach slaughter condition varies with breed type and previous management system. Newly weaned cattle and/or cattle subjected to marketing and transportation stress require special feeding and management to ensure adequate feed intake and to decrease the incidence of respiratory disease before placement in growing and finishing programs. Once cattle are started on feed, care must be taken to provide a smooth transition from higher-roughage starting diets to high-concentrate finishing diets. Bunk management is critical to successful cattle feeding, and the objective of any bunk management program should be to ensure consistency in daily feed intake, thereby decreasing the incidence of acidosis and other metabolic disorders and optimizing performance. Most finishing diets contain only small amounts of roughage, with the bulk of such diets made up of rapidly fermented grain. Composition of finishing diets varies with ingredient costs, availability of ingredients, and feed milling facilities. A variety of feed additives that influence efficiency of feed utilization and incidence of metabolic disorders is available for use in finishing diets.

REFERENCES

1. USDA-APHIS. 1995. Cattle on feed evaluation. Part I. Feedlot management practices. Fort Collins, CO: USDA-APHIS.

2. *The Beef Brief,* R. Price, ed. 1995. December issue. Englewood, CO.

3. NRC. 1984. *Nutrient requirements of beef cattle.* 6th ed. Washington, DC: National Academy Press.

4. Ritchie, H. D., and S. R. Rust. 1995. *Feedstuffs* 67(43):12.

5. Dolezal, H. G., J. D. Tatum, and F. L. Williams, Jr. 1993. *J. Animal Sci.* 71:2975.

6. Duckett, S. K., et al. 1993. *J. Animal Sci.* 71:2079.

7. Coleman, S. W., et al. 1995. *J. Animal Sci.* 73:2609.

8. Wheeler, T. L., et al. 1989. *J. Animal Sci.* 67:142.

9. Shackelford, S. D., M. Koohmaraie, and T. L. Wheeler. 1995. *J. Animal Sci.* 73:3304.

10. Klopfenstein, T. J., R. Stock, and J. K. Ward. 1991. Feeding, growing, and finishing beef cattle. In: D. C. Church, ed. *Livestock feeds and feeding* (3rd ed.), pp. 258–277. Englewood Cliffs, NJ: Prentice Hall.

11. Lewis, J. M., et al. 1990. *J. Animal Sci.* 68:2517.

12. Loerch, S. C. 1990. *J. Animal Sci.* 68:3086

13. Gunter, S. A., M. L. Galyean, and K. J. Malcolm-Callis. 1996. *Prof. Animal Sci.,* July–September.

14. Beconi, M. G., et al. 1995. *J. Animal Sci.* 73:1576.

15. Allen, V. G., J. P. Fontenot, and D. R. Notter. 1992. *J. Animal Sci.* 70:588.

16. Horn, G., G. Krenzer, D. Bernardo, and B. Mc-Daniel. 1994. Okla. Agr. Exp. Sta., P-939:151.

17. Highfill, G. A., et al. 1994. Okla. Agr. Exp. Sta., P-939:169.

18. *Feed Additive Compendium.* 1995. Minnetonka, MN: Miller.

19. Horn, G. W., et al. 1995. *J. Animal Sci.* 73:45.

20. Sainz, R. D., F. De la Torre, and J. W. Oltjen. 1995. *J. Animal Sci.* 73:2971

21. Gill, D. R., et al. 1993. Okla. Agr. Exp. Sta., P-933:197.

22. Gill, D. R., et al. 1993. Okla. Agr. Exp. Sta., P-933:204.

23. Van Koevering, M. T., et al. 1995. *J. Animal Sci.* 73:21.

24. Williams, C. B., G. L. Bennett, and J. W. Keele. 1995. *J. Animal Sci.* 73:674.

25. USDA-APHIS. 1994. Report N134.594. Fort Collins, CO: USDA-APHIS.

26. Blecha F., S. L. Boyles, and J. G. Riley. 1984. *J. Animal Sci.* 59:576.

27. Cole, N. A. 1996. Review of bovine respiratory disease: Nutrition and disease interactions. In: R. A. Smith, ed. *Bovine respiratory disease: Sourcebook for the veterinary professional,* pp. 57–84. Trenton, NJ: Veterinary Learning Systems.

28. Galyean, M. L., and M. E. Hubbert. 1995. Okla. Agr. Exp. Sta., P-942:226.

29. Lofgreen, G. P. 1983. *Veterinary clinics of North America: Large animal practice,* pp. 87–101, Vol. 5, No. 1, March, 1983.

30. Lofgreen, G. P. 1988. *Veterinary clinics of North America: Food animal practice,* pp. 509–522, Vol. 4, No. 3, November, 1988.

31. Fluharty, F. L., and S. C. Loerch. 1995. *J. Animal Sci.* 73:1585.

32. Galyean, M. L., et al. 1993. *Clayton Livestock Res. Ctr. Prog. Rep. No. 88,* N.M. Agr. Exp. Sta., Las Cruces.

33. Cole, N. A. 1993. *Proc. southwest nutr. and mgmt. conf.,* pp. 1–9, University of Arizona, Tucson.

34. Cole, N. A. 1995. Publ. No. TAMUS-AREC-95-2, pp. 33–45, Texas A & M Univ. Res. and Ext. Center, Amarillo.

35. Hays, V. S., et al. 1987. Okla. Agr. Exp. Sta., MP-119:198.

36. Duff, G. C., et al. 1995. *Clayton Livestock Res. Ctr. Prog. Rep. No. 96,* N.M. Agric. Exp. Sta., Las Cruces.

37. Lofgreen, G. P. 1983. *J. Animal Sci.* 56:529.

38. Galyean, M. L., S. A. Gunter, and K. J. Malcolm-Callis. 1995. *J. Animal Sci.* 73:1219.

39. Galyean, M. L., M. E. Hubbert, and S. A. Gunter. 1994. *Proc. Calif. Animal Nutr. Conf.*, pp. 25–34, Sacramento: California Grain and Feed Assoc.

40. Kreikemeier, K. K., et al. 1990a. *J. Animal Sci.* 68:2310.

41. Stock, R. A., et al. 1990. *J. Animal Sci.* 68:3441.

42. Galyean, M. L., et al. 1991. *Clayton Livestock Res. Center Prog. Rep. No. 69,* N.M. Agr. Exp. Sta., Las Cruces.

43. Milton, C. T., R. T. Brandt, Jr., and S. A. Shuey. 1994. *J. Animal Sci.* 72(Suppl. 2):70 (Abstr.).

44. Mader, T. L., J. M. Dahlquist, and L. D. Schmidt. 1991. *J. Animal Sci.* 69:462. High L Corn

45. Owens, F. N., and A. L. Goetsch. 1986. Digesta passage and microbial protein synthesis. In: L. P. Milligan, W. L. Grovum, and A. Dobson, eds. *Control of digestion and metabolism in ruminants,* pp. 196–223. Englewood Cliffs, NJ: Prentice Hall.

46. Wylie, M. J., et al. 1990. *J. Animal Sci.* 68:3843.

47. Moore, J. A., M. H. Poore, and R. S. Swingle. 1990. *J. Animal Sci.* 68:3412.

48. Owens, F. N., R. A. Zinn, and Y. K. Yim. 1986. *J. Animal Sci.* 63:1634.

49. Kreikemeier, K. K., et al. 1990b. *J. Animal Sci.* 68:2916.

50. Ladely, S. R., et al. 1995. *J. Animal Sci.* 73:228.

51. Rooney, L. W., and R. L. Pflugfelder. 1986. *J. Animal Sci.* 63:1607. Starch

52. Theurer, C. B. 1986. *J. Animal Sci.* 63:1649. Processing

53. Swingle, R. S., J. Moore, M. Poore, and T. Eck. 1990. *Proc. Southwest nutrition and Mgmt. Conf.,* pp. 52–64. University of Arizona, Tucson.

54. Xiong, Y., S. J. Bartle, and R. L. Preston. 1991. *J. Animal Sci.* 69:1707.

55. Zinn, R. A. 1990. *J. Animal Sci.* 68:767.

56. Galyean, M. L., D. G. Wagner, and F. N. Owens. 1981. *J. Dairy Sci.* 64:1804.

57. Ewing, D. L., D. E. Johnson, and W. V. Rumpler. 1986. *J. Animal Sci.* 63:1509.

58. Galyean, M. L., D. G. Wagner, and F. N. Owens. 1979. *J. Animal Sci.* 49:204.

59. Lee, R. W., M. L. Galyean, and G. P. Lofgreen. 1982. *J. Animal Sci.* 55:475.

60. Zinn, R. A. and A. Plascencia. 1993. *J. Animal Sci.* 71:11.

61. Weigel, J. C. 1995. Proceedings of the 56th Minnesota Nutrition Conference and Alltech, Inc. Technical Symposium, pp. 93–95, Minnesota Extension Service, Univ. of Minnesota, St. Paul.

62. Hussein, H. S., and L. L. Berger. 1995. *J. Animal Sci.* 73:3246.

63. Ham, G. A., et al. 1995. *J. Animal Sci.* 73:353.

64. Krehbiel, C. R., et al. 1995. *J. Animal Sci.* 73:2931.

65. Larson, E. M., et al. 1993. *J. animal Sci.* 71:2228.

66. Ludden, P. A., M. J. Cecava, and K. S. Hendrix. 1995. *J. Animal Sci.* 73:2706.

67. Galyean, M. L. 1995. (Unpublished.)

68. Arnold, R. N., et al. 1992. *J. Animal Sci.* 70:3055

69. Fox, D. G., and T. C. Perry. 1995. Publ. No. TAMUS-AREC-95-1, pp. 38–47, Texas A & M Univ. Res. and Ext. Center, Amarillo.

70. Byers, F. M., et al. 1994. *J. Animal Sci.* 72(Suppl. 1):325 (Abstr.).

71. Rumsey, T. S., A. C. Hammond, and J. P. McMurtry. 1992. *J. Animal Sci.* 70:995.

72. Fox, D. G., et al. 1992. *J. Animal Sci.* 70:3578.

73. Trenkle, A. 1992. *A.S. Leaflet R919, Beef and Sheep Res. Rep.,* pp. 102–107. Ames: Iowa

74. Galyean, M. L., and M. E. Hubbert. 1989. *Proc. of the southwest nutr and mgmt. conf.,* pp. 64–81. Tucson: Univ. of Arizona.

75. Minson, D. J. 1990. *Forage in ruminant nutrition.* San Diego, CA: Academic Press.

76. Males, J. R. 1990. *J. Animal Sci.* 68(Suppl. 1):504 (Abstr.).

77. Stock, R. A., et al. 1995. *J. Animal Sci.* 73:39.

78. Orskov, E. R. 1986. *J. Animal Sci.* 63:1624. State Univ.

79. Brink, D. R., et al. 1990. *J. Animal Sci.* 68:1201.

80. Blood, D. C., and O. M. Radostits. 1989. *Veterinary medicine* (7th Ed.), *A Textbook of the Diseases of Cattle, Sheep, Pigs, Goats and Horses.* 7th ed. London: Baillere Tindall.

81. Branine, M. E., and M. L. Galyean. 1990. *J. Animal Sci.* 68:1139.

82. Brazle, F. K. 1994. *Prof. Animal Sci.* 10:169.

83. USDA. *National Agricultural Statistics Service.* 1994. Washington, DC: U.S. Government Printing Office.

15

FEEDING DAIRY COWS

David J. Schingoethe

INTRODUCTION

Milk and milk products are an important part of the American diet, with annual per capita consumption of dairy products requiring about 280 kg of milk. Dairy products supply about 75% of our dietary Ca and are also an important dietary source of protein, vitamins, other minerals, and energy. In some countries, per capita consumption of dairy products is 50% to 100% higher than in the United States. Even when one averages in those countries that consume much less milk, world consumption of dairy products is more than 100 kg per capita.

To meet these needs for human food, the approximately 9.5 million U.S. dairy cows each produce an average of 7,500 kg/year, most of it in dairies that are highly mechanized (Fig. 15-1). A number of herds average more than 11,000 kg/cow, and some cows have produced more than 25,000 kg of milk annually. Approximately 381 million metric tons of milk are produced annually in the world, mostly in the temperate zones.

The sale of milk and milk products accounts for approximately 11% of all farm cash receipts in the United States, while sales of veal calves, dairy beef, and other dairy cattle account for another 3% of the nation's farm income. On the expenditure side of the dairy farm ledger, feed cost is an important consideration, because feeds usually account for 50% of the cost of operating the dairy farm enterprise. To achieve profitable milk production, it is important to feed dairy cows sufficient amounts of nutritionally balanced diets at reasonable costs.

Dairy cows need to consume a lot of feed to achieve the levels of production expected today; however, the nutrient needs of dairy cows vary tremendously throughout the lactation and dry period cycle, as illustrated by the example in Table 15-1. At peak production, a dairy cow may require 3 to 10 times as much protein and energy as she required during late gestation. This is further complicated by the fact that the cow's appetite usually lags behind her nutrient requirements. The challenge for a dairy feeding program is to meet the cow's nutrient needs while minimizing body weight loss, minimizing digestive upsets, and maintaining health.

THE LACTATION AND GESTATION CYCLE

Figure 15-2 illustrates the relationships between milk production, dry-matter intake, and body weight changes typically observed during the normal lactation and gestation cycle. Milk production increases rapidly and reaches peak (maximum) production 6 to 8 weeks after calving. However, appetite lags behind production such that maximum daily dry-matter intake often does not occur until 12 to 15 weeks

FIGURE 15-1 The interior of a modern milking parlor.

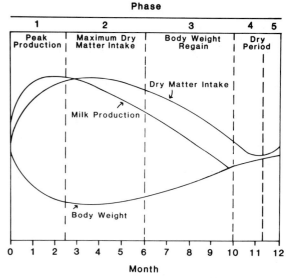

FIGURE 15-2 Milk yield, dry-matter intake, and body weights of cows during phases of the lactation–dry period cycle.

postpartum. As a result, most cows are in negative energy balance for 8 to 10 weeks and possibly as long as 20 or more weeks for some high producers. The cow makes up these nutrient deficits by borrowing from her body stores. Cows in good condition often lose 90 to 135 kg of body weight during early lactation, which is sufficient to support 700 to 900 kg of milk production. If that source of nutrients is not available, peak production and total lactational production will likely be less than optimal.

After optimal dry-matter intake is achieved, intake tends to follow production requirements and decreases as production decreases. There is still a lag in intake, only now the cow tends to consume more than she needs during later lactation. This allows her to regain the body weight lost in early lactation. The cow should regain most of this body weight during late lactation for optimal energetic efficiency, with most of the weight gain during the dry period being accounted for by fetal growth.

A PHASE FEEDING PROGRAM

The lactation and gestation cycle can be divided into five phases, as illustrated in Fig. 15-2, based on particular nutritional considerations

at various times. Phase 1, which is approximately the first 10 weeks of lactation, is when peak production occurs, but body stores are being utilized to make up for deficits in nutrient intake. Phase 2, which starts approximately 10 weeks postpartum for most cows and may continue for 10 to 20 weeks, corresponds with maximum dry-matter intake and is the period in which intake is in balance with requirements. In later lactation, phase 3 is entered as intake exceeds nutrient requirements for production. This is the main period for restoring body reserves for the next lactation.

The exact timing and relative length of these three phases of lactation may be altered if one is injecting cows with bovine somatotropin (bst) or other similar compounds that may alter the lactation curve. Production increases rapidly after the initiation of bst injections regardless of when they are started in the lac-

TABLE 15-1

Daily protein, energy, calcium, and phosphorus needs of a 650-kg cow at maintenance, late gestation, and when producing 50 kg of milk daily

Nutrient	Status		
	Maintenance	Late Gestation	Peak Lactation[a]
Protein, kg	0.428	1.120	4.628
Neℓ, Mcal	10.30	13.39	44.8
Calcium, g	26	43	174
Phosphorus, g	19	26	110

Source: Based on NRC for daily cattle (1).

[a]Requirements for maintenance plus production of 50 kg of milk containing 3.5% fat

tation cycle, but appetite lags behind the increased production and may not increase until 5 to 8 weeks after starting bst administration.

Phases 4 and 5 are in the dry period. This time may be considered as only one phase by some authors, but this author chooses to divide it into two phases. Phase 4, which may be most of the dry period, is a period for any final regaining of body weight and for regeneration of secretory tissue in the udder for the next lactation. Phase 5 is the last 1 to 3 weeks prepartum. During this time segment, one should start increasing grain intake as a means of preparing the rumen for the increased nutritional intakes that will soon follow upon parturition.

Feeding programs for the various phases of the lactation and gestation cycle are discussed in following sections. The discussion starts with the dry period, or late gestation for bred heifers, and then proceeds through lactation.

DRY PERIOD AND BRED HEIFERS

Cows need a short dry period as rest while preparing for the next lactation. The optimum dry period is 6 to 8 weeks (Table 15-2). Dry periods shorter than 40 days do not allow enough time for udder regeneration, which may cause a decrease in production during the next lactation. Dry periods longer than 60 days do not promote increased production, may result in excess body condition and calving complications, and are costly to the dairy producer. The economic loss due to long dry periods is estimated to be about $3 per cow daily for dry periods longer than 60 days and $2 for each day dry under 40 days.

For most cows, an abrupt cessation of milking is the best way for drying off. When cows that need to be dried off are still milking quite heavily (e.g., more than 23 kg/day), re-move grain mix from the ration for a week or more before the planned dry-off date to reduce production; then abruptly quit milking.

The dry period is a time for cows to regenerate new secretory tissue in the udder as well as to replace lost body condition. Several changes occur in the udder during the dry period: active involution, steady-state involution, and lactogenesis plus colostrogenesis. Milk continues to be secreted for several days after drying off before active involution takes over. During active involution, which is completed by 30 days into the dry period, milk-secreting tissue is reabsorbed. The second stage, steady-state involution, can exist indefinitely, and the mammary gland remains in a collapsed state. The third stage, lactogenesis plus colostrogenesis, begins 15 to 20 days prepartum and involves the onset of lactation and the secretion of colostrum (15). The first and third stages suggest that the dry period should be at least 45 to 50 days in length.

Dry-cow feeding programs emphasize maintaining the body condition that a cow carries when dried off. This is because USDA research (3) demonstrated that cows convert feed energy to body tissue more efficiently in late lactation than during the dry period. A body condition score of 3.5 to 4 at calving (on a scale of 1 to 5, with 1 very thin and 5 excessively fat) is ideal for high milk yield, fat test, and reproductive performance. Thus, the cow should be in ideal calving condition when she goes dry. Then weight gain during the dry period is mostly accounted for by growth of the fetus. Approximately two-thirds of the fetal growth occurs during the last 2 months of gestation, which could account for 25 to 35 kg of weight gain. Of course, if a cow is still thin at drying off, one will want to replenish body stores as well as provide for fetal growth.

FAT COWS   A cow should be fed enough during the dry period to maintain or get into good condition but not become excessively fat (Fig. 15-3). Cows consuming excess energy from grain and/or corn silage are more likely to develop a disorder called fat cow syndrome, which is characterized by high blood lipids and fatty livers. Such cows are more likely than the average to have calving difficulties, displaced abomasums, ketosis, and other health problems. Cows fed hay and/or haylage are less likely to have problems than cows receiving free-choice corn silage. If corn silage must be fed, limit dry cows to 9.5 to 11 kg/day of corn silage, plus

TABLE 15-2

Effect of days dry on milk production in the next lactation

Days Dry	Milk (kg/lactation)
<39	7,998
40–49	8,316
50–59	8,375
60–69	8,106
70–79	7,852
80–89	7,473
>90	7,197

Results from a study of 123,181 cows in 808 herds (4).

FIGURE 15-3 An excessively fat cow shown just prior to parturition.

FIGURE 15-4 Healthy heifers being group fed in drylot.

0.5 kg/day of protein supplement such as soybean meal, and provide a Ca–P supplement free choice. In all cases, dry cows should be separated from the milking herd to avoid excess energy intake.

BRED HEIFERS Nutrient requirements of bred heifers during late gestation are slightly greater than are requirements of dry cows of similar size, because they are still growing. Bred heifers will likely need some grain along with forages during the last 3 to 4 months of gestation in order to continue growing, provide nutrients for the fetus, and not become thin. Good-quality forages can provide all the nutrient needs of bred heifers during early gestation. But if forages are not of good quality or if heifers are exposed to severe weather conditions, additional grain may be needed to maintain optimal growth rates. As with dry cows, heifers should be in good condition but not excessively fat at calving. Heifers are usually fed in groups of similar ages or sizes (Fig. 15-4).

DRY COW RATIONS The pregnant dry cow requires more protein, energy, Ca, and P than when not pregnant or lactating, but not nearly as many nutrients as when lactating (Table 15-1). More detailed nutrient requirements and ration formulation guidelines are in Appendix Tables 14 and 17. A dry cow's nutrient requirements can often be met with only forages and no grain. Neither legume–grass hay nor corn silage alone meet all the requirements; however, a combination of the two needs only vitamins and a small amount of P to meet requirements.

The dry cow ration can be quite simple but should include the following considerations. Include a minimum of 1% of body weight as long-stem, dry forage. Preferably this should be grass hay, because legumes contain excessive amounts of Ca and are low in P, a dietary combination that may increase the incidence of milk fever. Free-choice feeding of corn silage should be avoided, because it leads to excessive energy intake and increases the likelihood of displaced abomasum and fat cow syndrome. Grain should be limited to the amount needed to meet energy and protein needs. Calcium intake should be kept under 100 g/day while providing adequate amounts of P (35 to 140 g/day for larger breeds). Higher amounts of Ca, especially if the diet is deficient in P, will increase the incidence of milk fever. Under cold weather conditions, cows housed with minimal protection from the cold may need additional energy because of increased maintenance requirements.

Additional Se (selenium) (3 to 5 mg) may be needed in areas where crops are deficient in Se. A Se deficiency, especially if vitamin E intake is not large, may result in retained placenta. Most stored feeds contain adequate but not excessive amounts of vitamin E, but usually only green forages contain greater amounts. Injecting Se and vitamin E or feeding supplemental amounts of these two nutrients during the last 3 weeks of the dry period may effectively provide this need and reduce the incidence of retained placentas. Most dairy farmers provide supplemental amounts of vitamins A and D as well as trace minerals to all their cattle as insurance against any possible deficiencies. But high mineral intake should be avoided, especially Na-based buffer mixtures, and salt intake should be limited to 28 g daily. High salt intake can aggravate the fluid retention problems of edema, which is a problem with some cows, especially first-calf heifers.

APPROACHING PARTURITION (PHASE 5)

The last couple of weeks before parturition is the time to make several nutritional changes to help the cow to prepare for parturition and the ensuing lactation. Most of these nutritional changes are aimed at adapting the ruminal microflora to the higher-energy diets that will be needed postpartum to meet the greatly increased nutrient requirements. This adjustment is often best achieved by including small amounts of all ingredients of the lactating ration and gradually increasing the amounts of grain fed so that by parturition the cow is consuming 0.5% to 1% of her body weight as grain mix. Such a "steaming up" approach may also minimize the chance for milk fever, because most grain mixes have a more desirable Ca to P ratio than do legume forages, and this approach may minimize the chance for ketosis during lactation by helping the cow to adapt to higher-energy diets more rapidly in early lactation.

It is advisable to start switching to some forage that the cow will receive after calving, but continue to provide at least 0.5% to 1% of her body weight as long-stem hay. This will help provide sufficient bulk in the diet to minimize the likelihood of displaced abomasum. It is important to make these ration changes as gradually as possible, because abrupt changes are likely to cause digestive upsets and throw cows off-feed. Most cows experience a sharp decrease in total dry-matter intake 24 to 48 h before calving, so stabilizing the rumen is important to avoid displaced abomasum, acidosis, and off-feed.

This late prepartum period is a time to take steps to minimize several problems likely to occur around parturition. In milk-fever-prone cows, one may consider lowering the Ca intake to 14 to 18 g daily for several days prepartum. This helps to activate the Ca-mobilizing hormonal system to increase Ca absorption from the gut and Ca mobilization from the bones.

Some research (5) has shown a reduction in milk fever by feeding diets with a negative dietary cation-anion difference (CAD) to close-up cows. The primary minerals considered in these rations are Na, K, S, and Ca. If using this approach, one should aim for a CAD of −10 to −15 mEq/100 g of diet dry matter. One should know the K content of the forages, because this seems to be a critical factor. When feeding anionic salts, increase dietary Ca to 150 g/day (1.3% to 1.5% Ca), balance the diet for P, Mg, and S, and avoid Na buffers, because the Na works against the anionic salts.

For cows that are fat (body scores of 4 and 5), ketotic prone, and high producers, one may consider feeding 6 g/day of supplemental niacin, starting the last couple of weeks prepartum and continuing during the last 8 weeks of lactation. Increasing the amount of grain fed to the close-up cow, as recommended above, helps to reduce the incidence of both milk fever and ketosis. The milk fever reduction is due to supplying adequate P, while the ketosis reduction occurs because of supplying more energy in the diet, which reduces the demand for body fat mobilization. In Se-deficient areas, one should increase vitamin E intake to 500 to 1,000 IU/day and Se to 0.3 ppm or inject Se and vitamin E the last 2 to 3 weeks prepartum.

PEAK MILK PRODUCTION

The most critical period for a dairy cow is from parturition until peak milk production. Each 1-kg increase in milk production usually means an additional 200 kg of milk production during lactation. However, since appetite lags behind nutritional requirements, phase 1 (peak milk production) is a period of negative nutrient balance.

The objective during this phase of lactation is to increase feed intake as rapidly as possible so as to minimize the nutritional deficit, but not to introduce ration changes so rapidly as to cause digestive upsets and off-feed. Once the stress of calving has passed, grain intake can usually increase 0.5 to 0.7 kg/day. Or, if total mixed rations are fed, the concentrate-to-forage ratio can be gradually increased to a maximum of 60:40. At higher concentrate ratios, it may be difficult to maintain a minimum of 18% to 19% acid detergent fiber.

A successful phase 1 feeding program maximizes peak milk yield, utilizes body weight as an energy source, minimizes ketosis, and returns cows to a positive energy balance by 8 to 10 weeks postpartum. This is the period of lactation that requires the best in nutritional management. During this phase, positive responses to feed additives, special feed treatments, and special ration formulations are most likely to occur. Some examples of nutritional considerations are listed in following sections.

Cows can compensate for much of their deficit in energy intake by borrowing the remaining needed energy from body fat; however, they cannot borrow very much protein. Thus, most of their protein must be supplied in the diet. When energy intake equals energy requirements, a diet containing 16% to 17% crude protein will likely meet protein requirements for most cows. However, in early lactation, 18% to 20% crude protein may be needed to meet protein requirements when energy intake is not supplying all the cow's energy needs.

Early lactation cows will also likely benefit from ruminal bypass (escape) proteins (see Chs. 8 and 14). The protein requirement of cows producing up to 5 kg milk/100 kg body weight can usually be met by rumen microbial protein synthesis plus normal amounts of bypass protein. Cows producing more than this amount of milk will likely benefit from bypass proteins and/or ruminally protected amino acids. Nonprotein nitrogen (NPN) supplements will not be efficiently used by these cows, although NPN utilization can be improved by increasing the amount of fermentable carbohydrates in the diet.

Increasing the energy density also helps the early lactation cow more nearly to meet her energy requirements. This can be achieved to a certain extent by increasing the concentrate to forage ratio. However, higher starch–lower fiber diets are more apt to cause acidosis, digestive upsets, and milk fat depression. Supplemental dietary fat may allow one to increase energy density of the diet while maintaining adequate fiber intake. There are limits to how much fat can be fed (see later section), but cows can easily consume an additional 0.5 to 0.7 kg fat, which will increase energy density of the diet. Calcium content of the diet should be increased to more than 0.9% and Mg to about 0.3% of ration dry matter when feeding added fat.

Buffers such as $NaHCO_3$ (sodium bicarbonate) alone or in combination with MgO (magnesium oxide) may be helpful during early lactation. Cows fed ensiled forages, especially if of small particle size, and high amounts of soluble carbohydrates will likely benefit from 100 to 200 g/day of $NaHCO_3$ or its equivalent. The primary benefit is from maintaining ruminal pH, which minimizes acidosis, reduces digestive upsets, and results in increased dry-matter intake.

Niacin supplementation, which may have already started during the late dry period, should be continued with high-producing cows. Feed intake will likely increase faster than without niacin supplementation, and the chances of ketosis are reduced.

Providing at least 2.25 kg of dry hay in the daily ration helps to maintain normal rumination and digestion, especially during early lactation. If chopped or ground forages are used, strive to maintain a particle length of at least 2.5 to 4 cm. Cornell research demonstrated few cases of displaced abomasums in cows fed long-stem hay, a slight increase when fed legume silages, and the most cases of displaced abomasums in cows fed corn silage as the only forage (6).

MAXIMUM DRY-MATTER INTAKE

To maintain peak milk production, maximum dry-matter intake should be achieved as early in lactation as possible. This will minimize the negative nutrient balance experienced in early lactation and shift cows from negative to positive energy balance earlier. Conception rates are greater for cows in positive energy balance than for cows in negative energy balance, which is an important consideration, because cows are usually being bred during this phase of lactation. Body weights should stabilize and weight gains should actually start occurring during this phase.

Maximum dry-matter intake will likely reach 3.5% to 4% of body weight for most cows but can vary with production and individual cow appetites (Appendix Table 16). Dry-matter intakes are usually higher for higher-producing cows, and some cows may consume more than 5% of their body weight.

The percent of protein needed in the diet during this period may likely be lower than what was needed in early lactation, because the cow is now getting all her needed energy as well as her needed protein supplied by the diet. Increased microbial protein synthesis stimulated by the greater dry-matter intake, coupled with lower dietary protein percentages, means that these cows are less likely to benefit from bypass protein than they would have benefited in early lactation. However, one should strive to maintain a balance between ruminally degradable crude protein and carbohydrates for optimal nutrient utilization.

While a cow is less likely to go off-feed during this phase than in early lactation, one should still strive to prevent conditions that may precipitate digestive upsets or result in less than optimal intake. Adequate fiber intake is still important, and grain intake should not exceed 2.5% of body weight. Frequent feeding, especially if fed as total mixed rations, minimizes digestive upsets and maximizes dry-matter intake. If all or some grain is fed through electronic grain feeders (Fig. 15-5), limit the grain available per meal to 2.25 to 3.25 kg. Dry-matter content of the diet may limit dry-matter intake if the total ration is too wet. Dry-matter intake may be limited by gut fill if the feeds are more than 45% to 50% moisture.

LATE LACTATION

Late lactation should be the easiest phase to manage because the cow is pregnant, nutrient intake normally exceeds requirements, and milk production is declining. This is the time to replace the weight lost during early lactation so that the cow is in good condition at drying off, but one should also maintain milk persistency as much as possible. Keep in mind that young cows are still growing and thus need additional nutrients for growth as well as for weight regain. The usual guidelines for estimating nutrient requirements for growth are 20% of maintenance requirements for 2-year-olds and 10% maintenance for 3-year-olds.

During this phase, one has an opportunity to minimize feed costs by increasing the forage-to-concentrate ratio to match nutrient needs based on milk production and body condition and by utilizing NPN. A lower protein content is likely needed because the protein-to-energy ratio needed for weight gain is less than the ratio needed for milk production. NPN sources may be well utilized for a portion of the crude protein needs of these cows, while bypass proteins will be less cost effective than in earlier lactation when production was higher.

FIGURE 15-5 A magnet feeder that allows individual animals to be fed additional concentrate when in pens with other cows.

SPECIAL NUTRIENT CONSIDERATIONS

There are certain nutritional factors to consider when feeding dairy cows. Some of these factors apply in all cases, whereas others may be of concern only during certain phases, such as during early lactation.

ENERGY

One of the greatest challenges to a dairy producer is to get cows to consume sufficient amounts of energy, especially during early lactation. Energy intake may be increased by increasing the energy density of the diet (by providing a greater proportion in the form of fat), by increasing the amounts of readily fermentable carbohydrates in the diet (by increasing the concentrate-to-forage ratio in the total diet), or, in some instances, by using individual feeders to supplement intake.

ADDED FAT One kilogram of fat contains approximately 2.25 times as much energy as contained in 1 kg of carbohydrates; thus, the energy density of the diet can be increased by replacing portions of the carbohydrates (grains) in the diet with fat. This may allow an increase in energy intake while avoiding excessive starch or deficient fiber intakes.

Cows can consume more fat than is usually present in forages and grain mixes, but they cannot consume an unlimited amount. Most forages and grains contain 2% to 4% fat. The amount of fat can be increased to 5% to 7% of total diet dry matter without adversely affecting feed intake or nutrient utilization. However, diets containing more than 8% to 10% fat may reduce feed intake, reduce fiber digestibility, and cause digestive upsets (7). This is because many fatty acids are inhibitory to rumen microorganisms, especially fiber digesters, and ruminants cannot digest as much fat in the intestinal tract as nonruminants can. Unsaturated fatty acids are generally a greater problem than are saturated fatty acids.

The optimal energetic efficiency for milk production is when fat provides 15% to 17% of the dietary metabolizable energy, which can be supplied by 5% to 8% of the total ration dry matter as fat (7). Adding 0.5 to 0.7 kg of fat to a cow's daily diet already containing 2% to 4% fat from forages and concentrates will provide this level. The energy in this amount of added fat is sufficient to produce 3.5 to 4.9 kg of milk, or 2.4

to 3.3 kg more milk than would have been produced from the 0.5 to 0.7 kg of carbohydrates the fat may replace. Researchers have observed increases in milk production of 2% to 12% from feeding added fat.

High-producing cows, especially during the first 2 to 5 months of lactation, benefit most from added fat. This generally means cows producing more than 30 kg of milk daily. Herds with averages higher than 7500 kg/cow/y will likely benefit from added fat for the high-producing cows. Low producers and cows in the last half of lactation are not likely to respond to additional energy in their diet unless they are presently receiving a poor-quality diet in restricted amounts. Cows under heat stress may also benefit from added dietary fat.

Production increases from feeding added fat are not always as great as some would expect, but the added energy intake may result in other benefits. Cows fed added fat usually attain peak daily milk yields a couple of weeks later than do other cows, but they then maintain their production with greater persistency. Field reports also indicate that cows fed added fat have less weight loss in early lactation, which may result in improved reproductive efficiency and fewer cases of ketosis; however, research data have not given definite support to such claims. With some unpalatable fat sources, dry-matter intake may be reduced such that the intended increase in energy intake does not occur.

FAT SOURCES Several acceptable fat sources for feeding lactating cows are listed in Table 15-3. Not all sources of fat are suitable feeds for milking cows. For instance, while soybeans, sunflower seeds, and cottonseeds are acceptable fat sources, free oils such as soybean oil, sunflower oil, cottonseed oil, corn oil, and fish oil should not be used because they affect rumen fermentation adversely, resulting in reduced fiber digestion and lowered milk fat tests. Fat sources listed in Table 15-3 are essentially inert in the rumen in that they do not significantly alter rumen fermentation. Oilseeds such as cottonseed, soybeans, and sunflower seeds also provide some protein and fiber, which can be considered when formulating diets.

Cottonseed, especially if it still includes the lint, contains relatively high amounts of protein and digestible fiber to go along with the fat. Thus, it is a popular feed in areas where it is available at reasonable cost. Gossypol can be a potential problem only if cattle are consuming more than 3 kg of cottonseed daily. Feed handling may be a consideration with cottonseed because it does not flow well, which creates problems with some feed-handling equipment. It is best stored in a dry, flat area and fed through some type of total mixed ration feeding system.

Soybeans can be fed whole without processing, although they are often rolled prior to feeding. Soybeans usually benefit from heat treatment, while heating is not necessary with most other oilseeds. Soybeans contain several antinutritional factors that are destroyed by heat (see Ch. 8). These factors, such as soybean trypsin inhibitors, may interfere with intestinal digestion of proteins in cows consuming more than 1.8 to 2.3 kg of unheated soybeans daily. Heated soybeans may be stored longer than ground unheated soybeans because they are less likely to develop rancidity. Heated soybeans are often more palatable and may have greater protein bypass properties than unheated soybeans.

Sunflower seeds can be fed without processing, although rolling prior to feeding may slightly improve utilization by animals. Most sunflowers in the United States are oilseed varieties that contain more than 40% fat; however, about 5% of sunflowers are confection varieties that contain about 30% fat. The fibrous hull in sunflower seeds is poorly digested. Canola seeds should be ground or crushed prior to feeding to improve utilization by animals.

Some fat supplements, such as tallow and some animal and vegetable fat products, are more difficult to handle on the farm because they are not dry and free flowing at normal handling temperatures. Calcium salts (soaps) and prilled fats are dry free-flowing products that can be handled conveniently in most feeding systems. These products are formulated for maximum rumen bypass and utilization for milk production, although research data indicate that these products are not any more ruminally inert than are oilseeds.

Some nutritionists have recommended that supplemental fat be provided by two or more sources. However, this author is unaware of research that supports multiple fat sources as being more effective than the same total amount of fat from a single source.

Fatty acid composition of the fat source should be a consideration also. Highly unsaturated fat sources such as vegetable oils are more likely to interfere with ruminal fermentation and cause milk fat depression. Hydrogenated vegetable fats and animal sources of

TABLE 15-3

Fat sources for lactating cows

Source	Composition (% of dry matter)			Recommended Amounts to Feed (kg/cow/day)
	Fat	Crude Protein	Acid Detergent Fiber	
Oilseeds				
Canola	40	20	12	1.5–2.2
Cottonseeds	23	23	35	2–3
Soybeans	19	41	6	2–3.6[a]
Sunflower seeds	30–45	19	16	1.5–2.2
Fat supplements				
Tallow	99	—	—	0.5–0.7
Hydrolyzed animal–vegetable blend	99	—	—	0.5–0.7
Calcium salts of fatty acids	85	—	—	0.6–0.8
Prilled fat	99	—	—	0.5–0.7

[a]Greater than 2.3 kg only if the soybeans are heated.

fats are more saturated and thus less likely to cause problems with cows.

Extensive hydrogenation of unsaturated fatty acids normally occurs in the rumen, explaining why the fatty acid composition of milk fats and carcass fats of cattle remain virtually unaffected by the fatty acid composition of dietary fats. However, when cows are fed highly unsaturated fats such as soybeans and sunflower seeds, the milk fat becomes slightly more unsaturated (8). This slight alteration of the milk fat composition may offer some marketing advantages for dairy products, such as allowing the manufacture of butter that is more spreadable at refrigerated temperatures than butter from normal milk. However, if the unsaturated fatty acids completely escape ruminal hydrogenation, the milk may develop undesirable oxidized flavors.

CONSIDERATIONS WITH ADDED FAT
There are several possible concerns when feeding added fat. Milk protein percentages are usually reduced 0.1 to 0.2 percentage units, a problem that researchers have not yet solved (9). Milk fat percentages are usually unaffected, but they may be reduced slightly, especially when consuming highly unsaturated fat sources. Palatability is sometimes a problem with some fat sources, and feed handling presents problems with some products.

Several other nutritional adjustments should be made when feeding added fat. Boost the Ca content to more than a 0.9% of the diet dry matter. Calcium complexes with fatty acids, forming soaps. This helps to minimize the undesirable effects of fatty acids on fiber digestion in the rumen, but also means that some Ca may be lost in the feces as soaps. The Mg content of the diet should be increased slightly to more than 0.3% of dry matter to maintain a balance between Ca and Mg. Crude protein content may need to be increased slightly to maintain the proper protein to energy balance for optimal milk production. As a guideline, for each 0.5 kg of supplemental fat fed, the crude protein content of the diet should be increased by 1%. Providing the extra protein in a bypass form rather than in a highly soluble form may be a good way to ensure that the protein is utilized efficiently. This is because the fat replaces nonstructural carbohydrates, which may result in a loss of energy to the rumen microbes and reduce animal microbial protein synthesis.

CARBOHYDRATES Most of the cow's energy is provided by carbohydrates, but the form and digestibility of carbohydrates vary substantially among various feed sources. Structural (fibrous) carbohydrates such as cellulose and hemicellulose are digested in the rumen, whereas the noncarbohydrate lignin is virtually undigestible. Nonstructural carbohydrates such as starches and sugars are readily digested in the rumen and in the intestinal tract, although some grain sources of starch are fermented more rapidly in the rumen than are other sources. It is necessary to maintain a sufficient amount of structural carbohydrates in the diet for normal rumen function, and fine-tuning the ratio of ruminally degradable carbohydrates and crude proteins can increase ruminal microbial protein synthesis and energy utilization by the cow.

Some researchers have suggested that neutral detergent fiber (NDF) is related to dry-matter intake and that one should formulate diets for an optimal amount of NDF in order to optimize dry-matter intake (1). Theoretically, this principle is probably correct; however, the digestibility of NDF, which contains all the fibrous components, varies substantially among various types of feed sources. Thus, one cannot be confident that a certain percent of NDF will always allow maximum dry-matter intake. In general, at higher proportions of NDF (probably greater than 36% NDF with most forages), energy intake will likely be limited by gut fill, while at lower proportions (24% to 26% NDF) chemostatic factors may limit intake and insufficient fiber may be available for normal rumen function. By-product fiber sources such as soy hulls can be fed but should not provide more than 20% to 30% of the dietary NDF.

Acid detergent fiber (ADF) should generally be at least 19% of dry matter to avoid milk fat depression, although some research studies with 16% to 17% ADF observed no milk fat depression. This amount of fiber can usually be supplied by a minimum consumption of 1% to 1.5% of body weight as forage dry matter. Fiber physical form is also a factor because ground, finely chopped, or pelleted forages are more likely to cause ruminal upsets and milk fat depression than are longer feed particles. From a practical standpoint, the above items generally mean that the concentrate-to-forage ratio should not be greater than 60:40. If corn silage comprises all or most of the forage, 50% to 55% concentrate may be maximum because of the higher nonstructural carbohydrate content of the corn silage.

Feed grains are the primary dietary source of nonstructural carbohydrates, so increasing the amount of grain in the diet generally increases the amount of readily digestible carbohydrates and, hence, available energy for the cow. However, not all grain sources are equal in providing fermentable energy in the rumen. For instance, corn and sorghum starch is usually fermented less rapidly in the rumen than is starch from grains such as barley, oats, and wheat (10). Sugars such as those found in dried whey or molasses are fermented even more rapidly. Increasing the rate of carbohydrate fermentation in the rumen stimulates ruminal microbial protein synthesis, which can improve protein utilization by the animal and reduce the need for bypass proteins.

Processing of feed grains can also influence digestibility and utilization. For instance, high-moisture corn is usually more digestible than dry corn (11). Ground, cracked, or rolled grains are usually more readily utilized than are whole grains, and processing such as steam flaking may improve digestibility, especially for sorghum.

The interrelationships between carbohydrate and crude protein degradation in the rumen are important for optimal nutrient utilization and production by the cow. Ideally, one desires a balance between carbohydrate and crude protein degradation rates in the rumen that allows optimal microbial protein synthesis and volatile fatty acid production. Even with very high producing cows, it is still most advantageous to utilize the rumen to the fullest extent and to bypass the rumen with protein and energy sources only after ruminal production can no longer be increased.

PROTEIN

Requirements for protein increase even more dramatically at the onset of lactation than the increase in energy requirements because milk solids contain about 27% protein (1). Only small amounts of protein can be borrowed from blood, liver, and muscles for milk protein synthesis; thus, virtually all the required protein must be supplied by the diet. Protein requirement is really a requirement for amino acids by animal tissues. The amino acids are supplied by the digestion of microbial protein and feed protein that escapes microbial breakdown in the rumen.

In the typical dairy cow a sizable portion of the dietary crude protein is broken down or degraded by rumen microbial digestion to ammonia. For the ruminally degraded crude protein to be of any value to the animal, the ammonia must be converted to microbial protein, which in turn can be digested in the gastrointestinal tract (see Ch. 8). The relative proportion of ruminally degradable crude protein in the typical dairy cow diet is approximately 60%, but this proportion varies with feed intake, organic matter digestibility, feed type, protein level, and feeding system.

BYPASS PROTEINS AND AMINO ACIDS
Optimal protein utilization by the animal can best be achieved by supplying sufficient amounts of degradable crude protein and fermentable energy for maximum microbial protein synthesis and by supplying the remainder of the protein

needs with high-quality bypass (ruminally un-degradable) protein. The amount of microbial protein synthesized varies with the factors listed above but is likely limited to about 2 to 3 kg daily. The remainder of the required protein must be derived from bypass protein. High-producing cows (in general cows producing more than 5 kg of milk/100 kg of body weight) will likely benefit from diets formulated to contain greater than normal proportions of bypass protein.

All feed proteins are not degraded in the same extent in the rumen, as illustrated in Appendix Table 2. For instance, proteins in brewers grains, distillers grains, corn gluten meal, heated soybeans, and most animal by-product proteins are degraded less readily in the rumen than are most feed proteins. The protein in high-moisture corn is more degradable than that in dry corn, and the protein in grass or legume silage is more degradable than the protein in hay from the same forage. Crude protein from NPN sources such as urea is 100% degradable in the rumen.

It is an oversimplification to formulate protein in diets on the basis of ruminal degradability and undegradability only. In general, approximately 35% to 40% of dietary crude protein should be ruminally undegradable. However, digestibility of the bypass protein in the lower digestive tract as well as its amino acid composition are important factors, which, unfortunately, are often unknown. Thus, while many research studies showed increased production from feeding bypass proteins, many other studies showed no increase, and sometimes a decrease, in production (12).

There are several points to consider in the use of bypass proteins to assure that a positive response is obtained.

1. Don't bypass the rumen at the expense of rumen microbial production. If this occurs, the amount of microbial protein may be reduced to the same extent as the increased bypass protein with no net gain to the cow.
2. Some bypass proteins may be undigestible and therefore unusable in the lower digestive tract as well as in the rumen. Heat-damaged proteins and some by-product proteins may be in this category.
3. The quality (blend of amino acids) of the bypass protein should be at least as good as or better than the quality of rumen microbial protein. Several bypass proteins used in cattle feeds are very deficient in some dietary

essential amino acids. Even if the protein is digested, little is gained unless the supply of the amino acid(s) limiting production is increased. In some cases, treating proteins of known good quality, such as additional heat treatment of soybean meal or heated soybeans, may give a greater milk production response than substitution with a poorer-quality protein that is naturally less degradable in the rumen.

Supplementing the diet with ruminally protected amino acids can be another means to increase the amount of amino acids presented to the gastrointestinal tract for absorption and, ultimately, to increase milk production. However, to be successful with this approach, one must supplement with the most limiting amino acid(s). It is sometimes difficult to determine which amino acids are most limiting, because the amino acid composition of the portion of feed protein that escapes ruminal degradation and the proportion of microbial protein to escape protein are seldom known with any degree of certainty. With many diets, methionine, lysine, or one of several other dietary essential amino acids may be most limiting. Supplementing with a ruminally protected form of the limiting amino acid(s) may increase production (13); however, the response to the first limiting amino acid may be small if the second limiting amino acid soon becomes first limiting. One may obtain a greater production response by supplementing with two to four amino acids that are likely to be limiting than by supplementing with only one amino acid. However, the cost of such a supplement relative to feeding more of a quality feed protein may be a consideration. Technological advances may make the use of supplemental ruminally protected amino acids a feasible alternative in the future.

VITAMINS

VITAMINS A AND D Vitamin supplementation of diets usually has no direct effect on milk production, but there are situations in which supplementation of certain vitamins may be considered. Most nutritionists recommend providing supplemental vitamins A and D to ensure adequacy of the diet for the animal's health, even though no response in production is likely to occur. Supplemental vitamin A may be needed when cows are fed low-quality forages or low amounts of forage, primarily corn silage and a low-carotene concentrate mixture, or only feeds that have been stored for several months or longer.

Some reports suggest that β-carotene, the precursor to vitamin A, has a role in reproduction independent of other forms of vitamin A (1). However, conflicting research results indicate that additional studies are needed to clarify the physiological role of β-carotene before its supplementation can be recommended.

Supplemental vitamin D is unnecessary when animals consume sun-cured forage or are exposed to ultraviolet light or sunlight (1). In cows consuming adequate amounts of Ca, vitamin D supplementation permitted a positive Ca balance earlier in lactation. Continuous high doses of vitamin D (e.g., 70,000 IU of vitamin D/kg of concentrate) or massive doses (e.g., 20 million IU of vitamin D/day) starting 3 to 5 days before the expected calving date may help to reduce the incidence of parturient paresis. However, one should be cautious when using such procedures, because continuing the massive doses of vitamin D beyond 7 days can cause toxicosis.

VITAMIN E   Most feeds, even when stored for several months, contain sufficient amounts of vitamin E, although supplementation may be warranted under certain conditions. Vitamin E supplementation may be necessary if oxidized flavor of milk becomes a problem. This problem is most likely to occur when diets contain only stored feeds and/or when diets contain supplemental fats. If milk flavor problems occur, 400 to 1000 IU of additional vitamin E/head/day may be needed to correct the problem (1).

Recent research indicated that vitamin E and Se can help lower the incidence of mastitis (14). The vitamin E supplementation may be most helpful during the dry period, but supplementation during lactation may also be helpful. Selenium supplementation may also be helpful if diets contain less than 0.3 ppm of Se. The specific mechanism(s) in the mammary gland is not definitely known, although it is known that vitamin E and Se are involved in the protection of cells against oxidative damage caused by free radicals.

WATER-SOLUBLE (B) VITAMINS   The B vitamins are usually synthesized by rumen microorganisms in more than adequate amounts to meet the cow's needs. Thus supplementation with B vitamins or microbial products such as yeast to supply B vitamins is usually unnecessary and will not show any benefit. However, recent evidence indicates that supplementation with certain B vitamins may occasionally be beneficial.

Because niacin is involved intimately with energy metabolism, niacin supplementation may increase milk production by moderating effects of ketosis on energy-stressed high-producing cows. Niacin supplementation is most effective if initiated either before or immediately after calving and continued for 4 to 10 weeks. A level of 6 g/head/day effectively reduced the incidence of ketosis; feeding 12 g of niacin daily for several days to ketotic cows decreased blood and milk ketones and improved milk production. Several studies showed small positive responses in milk production and fat tests when feeding 6 g of niacin daily with all-natural proteins to early lactation cows (1). There were no benefits from such feeding in mid- to late-lactation cows, and cows fed NPN did not respond. A positive response in milk protein yield and content occurred in some research experiments, although most studies showed little or no response (9). NPN is one of the few nutritional additives that is likely to produce a positive response, assuming that a response in milk protein content does occur.

Choline supplementation will not affect milk production when diets contain adequate forage, but it may slightly increase milk fat percentages when diets are low in fiber (1). However, to improve animal performance, it may be necessary to provide the choline postruminally.

Vitamin B_{12} and/or Co supplementation will not likely be beneficial unless diets are deficient in Co (1). Cyanocobalamin, which contributes 10% to 20% of the vitamin B_{12} activity synthesized by rumen microorganisms, tends to be highest in the rumen of grazing animals and lowest in animals fed high-concentrate diets.

MINERALS

Supplemental amounts of Ca, P, and salt (NaCl) are needed in most rations, with supplemental amounts of some trace minerals possibly needed also, depending on regional soil deficiencies. These can often be supplied by 1% to 1.5% of a Ca–P supplement and 0.5% to 1% trace mineral salt in the concentrate mix, plus making these available for free-choice consumption. The source of Ca–P supplement can vary with the forages fed, which likely dictates the amounts of supplemental Ca and P needed. For instance, legume-based diets need primarily supplemental P, which can be supplied by a product such as dicalcium phosphate, monosodium phosphate, or monoammonium phosphate. In contrast, diets based on corn silage likely need large amounts of both Ca and P. The use of a trace-mineralized salt instead of merely NaCl often provides an inexpensive insurance that minerals needed in trace amounts are supplied.

CALCIUM AND PHOSPHORUS Requirements of Ca and P increase substantially with the onset of lactation because milk contains substantial amounts of both minerals. The inability of a cow to adjust rapidly to this increased Ca demand can result in milk fever. Dietary manipulations during the dry period to minimize the risk of milk fever were discussed previously. Diets for lactating cows should be formulated with the intent of meeting the cow's requirements for these two minerals; however, considerable resorption of Ca and P from bones can occur to make up for short-term deficits. In fact, many cows may be in a negative Ca and P balance during the first 3 to 5 months of lactation. Dietary Ca requirements may also increase slightly with high-fat diets because of increased fecal Ca losses.

The nutritional availability (true digestibility) of Ca may vary with different sources, so one may want to take this into account when formulating diets (1). Calcium from inorganic sources may be more available than that from organic sources. Of various organic sources, Ca from alfalfa, especially mature alfalfa, may be used less efficiently than that from other sources as a result of the relative indigestibility of Ca-containing crystals such as Ca oxalate. The nutritional availability for ruminants of various supplemental P sources varies somewhat (1), although these variations usually are not of great concern in formulating diets.

SODIUM AND CHLORINE Substantial amounts of Na and Cl are secreted in milk, thus making it essential to supply sufficient amounts in the diet. It is recommended that diets for lactating cows contain 0.18% Na and 0.25% Cl, which is about 80% higher than the concentrations recommended for dry cows (1). If most of the supplemental Na is supplied by NaCl, more than adequate amounts of Cl will be provided. However, if substantial amounts of the dietary Na come from other sources, such as $NaHCO_3$, it may be necessary to evaluate the Cl status of the diet.

OTHER MINERALS Forages generally contain more K than the estimated 0.8% of dietary dry matter required by lactating cows (1). However, stress, especially heat stress, increases the K requirement, possibly due to greater losses of K through sweat. While K deficiency was previously considered unlikely to occur, it is more likely to occur today because of several changes in rations in recent years that have tended to reduce the amount of K fed. Although most forages contain more K than required by cattle, a major forage source, corn silage, is low in K, some fiber sources such as cottonseed hulls and corn cobs are low in K, and many concentrates are deficient in K. Thus, feeding programs that include larger amounts of concentrates, more corn silage, and complete feeds using low-K fiber sources are likely to be bordering on K deficiency.

Iodine and Se are two trace minerals that are likely to be deficient in feeds grown in many areas of the world. Lactating cows require more I than do nonlactating cows because 10% or more of the intake is normally excreted in milk (1). Iodine is usually supplied as iodized salt to ensure adequate intake of 0.6 ppm I in diets; I compounds sometimes are used to prevent foot rot. However, excessive consumption of I may lead to dramatically increased I content of milk, because excretion of I via milk is one of the cow's means of getting rid of excess I. Supplementation of Se may be necessary to elevate dietary Se to 0.1 to 0.3 ppm.

BUFFERS Cows fed acidic diets, especially during the first few months of lactation, may benefit from added buffers in their diets. Times when buffers may help include when animals are being fed high corn silage diets or wet diets (more than 50% moisture), during heavy grain feeding (more than 2% of body weight), during heat stress, when the particle size of forages and grains is too small, when off-feed is a problem, such as during early lactation, or when milk fat tests are depressed. Under such conditions, adding buffers often improves feed intake, fiber digestibility, and microbial protein synthesis; increases the acetate to propionate ratio; and minimizes off-feed problems. $NaHCO_3$ at the rate of 0.75% to 1% of dietary dry matter or 1.25% of a 3 to 1 mixture of $NaHCO_3$ and MgO are effective buffers, as are similar mixes of similar buffering compounds.

WATER

An abundant supply of fresh water should always be available for cattle to ensure optimal production. Water is supplied by free drinking water, water in the feed consumed, and metabolic water produced by oxidation of organic nutrients. The amount of water a cow needs is influenced by milk production, ambient temperature, humidity, salt intake, dry-matter intake, and other factors (1). With large groups of cows, several watering stations may be needed, because cows may not drink sufficient amounts of water if they have to stand in line very long to wait for a drink. There should be one watering station for every 20 to 25 cows placed within 15 m of feeding areas.

OPTIMIZING FEED INTAKE DURING LACTATION

Water content of feed may be a factor to consider when attempting to achieve optimal intake. This is usually not important for diets containing primarily hay and concentrates, but it can become a limitation for diets containing mostly ensiled or fresh forages, coupled with other high-moisture feeds such as high-moisture corn, wet brewers grains, and liquid whey. Moisture's effect on dry-matter intake is less when present in the form of fresh forages than it is in the form of silage or other fermented feeds. Limited data indicate that total dry-matter intake decreases as the moisture content of the diet exceeds 50% from ensiled feeds (1). This response may be caused partially by chemicals in the feed rather than by moisture per se.

The number of times cattle are fed daily may influence total dry-matter intake. A minimum of four daily feed offerings, alternating between forages and concentrates, appears to be the most desirable method of feeding ration components individually (15). Frequent feeding of total mixed rations does not necessarily increase dry-matter intake; however, more frequent feeding may help to stabilize rumen fermentation, which may prevent drops in milk fat percentages and can minimize digestive upsets that may otherwise reduce feed intake. More frequent feeding of moist and ensiled feeds may be needed during warm conditions to maintain feed freshness and palatability.

High-producing cows should have access to their feeds for at least 18 to 20 h/day for maximum feed intake. When cows have access to their feeds, they will consume their daily intake in 12 to 22 meals. Feeding in groups, the most common practice in the United States, facilitates ready access to food (Fig. 15-6). Feeding more than 4.5 kg of concentrate mix/meal may cause acidosis. Many electronic grain feeders currently marketed are programmed to limit the amount of concentrate consumed by a cow within a short period of time, thus helping to maintain rumen stability under situations where concentrate mixes are fed separately from forages.

EXAMPLE RATIONS

Several example rations for lactating cows are listed in Table 15-4. These rations are formulated to approximately meet the nutritional needs of a 600-kg cow producing 40 kg of 3.5% fat-corrected milk daily. Such a cow requires a high protein and energy diet, and it is often difficult to meet these requirements and still maintain adequate fiber in the diet. It would be much easier to meet the nutritional needs of lower-producing cows. While numerous feed ingredients are available in various areas, several feed sources predominant in many areas were used in these examples to illustrate a couple of points.

Two somewhat extreme ration examples are ration 1, which contains only alfalfa as the forage, and ration 3, which contains only corn silage as the forage. Because corn silage contains more energy per kilogram than does alfalfa, a higher forage-to-concentrate ratio could be maintained and still provide slightly more energy than the alfalfa-based diet. However, because corn silage contains less crude protein and Ca than does alfalfa, corn-silage-based diets need much more protein and mineral supplementation.

Rations 6 and 7 were included as examples containing added fat from heated soybeans or cottonseed. Note that the protein content was raised slightly because the energy densities were also elevated slightly, and the Ca and Mg contents were raised.

FIGURE 15-6 Feeding a TMR with a mixer wagon.

TABLE 15-4

Example ration formulas for lactating cows

Ingredient	Ration (% of DM)						
	1	*2*	*3*	*4*	*5*	*6*	*7*
Alfalfa	50.0	55.0	—	27.4	50.0	25.0	45.0
Corn silage	—	—	60.0	27.4	—	25.0	—
High-moisture corn	41.2	—	17.2	32.9	—	29.3	35.3
Barley	—	38.0	—	—	42.0	—	—
Cottonseed meal	—	—	—	—	7.1	—	8.3
Cottonseed	—	—	—	—	—	—	10.0
Soybean meal	7.7	6.0	20.9	11.1	—	8.7	—
Heated soybeans	—	—	—	—	—	10.0	—
Dicalcium phosphate	0.3	0.5	0.5	0.7	0.4	0.4	0.3
Monosodium phosphate	0.3	—	—	—	—	—	—
Limestone	—	—	0.9	—	—	1.0	0.5
Trace mineral salt	0.5	0.5	0.5	0.5	0.5	0.5	0.5
MgO	—	—	—	—	—	0.1	0.1
Composition							
CP	17.0	17.0	17.0	17.0	17.0	18.0	18.0
Neℓ	1.67	1.65	1.71	1.69	1.63	1.73	1.70
ADF	17.5	20.0	19.4	18.3	20.8	17.6	20.2
Ca	0.80	0.86	0.64	0.64	0.83	0.90	0.91
P	0.42	0.42	0.43	0.44	0.42	0.41	0.42

FIGURE 15-7 Taken at the SDSU dairy research and training facility.

FIGURE 15-8 The SDSU Dairy Research and Training Facility. Curtained sidewalls are shown.

SUMMARY

Nutrient needs of dairy cows vary immensely between the dry period and peak lactation. Requirements for the former can often be met with forages alone, while the latter may require a considerable amount of high-energy feed sources, such as feed grains and added fat sources, and bypass proteins of good amino acid quality. A challenge is to formulate a diet that contains sufficient fiber to maintain normal rumen fermentation so as to prevent digestive upsets and to prevent milk fat depression. A successful feeding program meets a cow's nutritional needs for high production while minimizing body weight loss during early lactation, not causing digestive upsets, and maintaining health.

REFERENCES

1. Block, E. 1995. *J. Dairy Sci.* 77:1437.
2. Coppock, C. E. 1974. *J. Dairy Sci.* 51:926.
3. DePeters, E. J., and J. P. Cant. 1992. *J. Dairy Sci.* 75:2043.

4. Gibson, J. P. 1984. *Animal Prod.* 38:181.

5. Herrera-Saldana, R., and J. T. Huber. 1989. *J. Dairy Sci.* 72:1477.

6. Hogan, J. S., W. P. Weiss, and K. L. Smith. 1993. *J. Dairy Sci.* 76:2795.

7. Kennelly, J. J. 1994. Oilseeds and their effect on milk composition. In: *Proc. Canola Council of Canada Mini-Symposium,* 15th Western Nutr. Conf., p. 11, Winnipeg, Canada.

8. Moe, P. W., W. P. Flatt, and H. F. Tyrrell. 1971. *J. Dairy Sci.* 54:548.

9. NRC. 1984. *Nutrient requirements of beef cattle,* 6th rev. ed. Washington, DC: National Academy Press.

10. NRC. 1989. *Nutrient requirements of dairy cattle,* 6th rev. ed. Washington, DC: National Academy Press.

11. Palmquist, D. L., and T. C. Jenkins. 1980. *J. Dairy Sci.* 63:1.

12. Schingoethe, D. J. 1994. Dietary influence on protein level in milk and milk yield. In: *Proc. 15th Western Nutr. Conf.,* p. 1, Winnipeg, Canada.

13. Schingoethe, D. J., et al. 1988. *J. Dairy Sci.* 71:173.

14. Smith, J. F., and K. Becker. 1994. *Hoard's Dairyman* 139:780.

15. Smith, K. L., and D. A. Todhunter. 1982. In: *Proc. 21st Ann. Mtg.,* p. 97. Arlington, VA: Natl. Mastitis Council.

16

FEEDING DAIRY CALVES AND REPLACEMENT HEIFERS

Mark Aseltine

Dairying on a commercial scale in the United States makes it necessary to separate newborn calves from their dams as soon after birth as practical. Under present management systems, cows are milked in facilities dedicated to milking with no space for calves. Fresh cows require special nutrition and feeding facilities in order to take advantage of their inherited milking ability. Calves are therefore more efficiently housed in separate facilities.

The feeding management to be outlined in this chapter is from birth through weaning, puberty, breeding, and first freshening of heifers. At this time, male dairy calves (principally Holsteins) are being fed for slaughter for specialized markets on a greater scale than in the past. For this reason, a section of this chapter will be devoted to the principles of feeding Holstein steers for beef.

THE FIRST CRITICAL DAYS

Although birth might seem to be the logical starting point for a discussion of feeding dairy calves, in fact the health and vigor of calves at

birth depend on the nutrition of the cow during the last 60 days before freshening. During this period, approximately 70% of the birth weight of the calf is developed, and the cow must store nutrients for early lactation when milk production exceeds the cow's capacity for feed consumption.

The cow and the unborn calf share a vascular system. If the cow is on a lower than recommended plane of nutrition, the calf is apt to be born lacking vigor. Also, during this period, cows that are fed properly and immunized against locally prevalent pathogens, such as rotavirus, coronavirus, and infectious bovine rhinotracheitis, develop colostrum of good antibody quality. Cows not fed properly close to the end of gestation produce colostrum of poorer quality.

Because calves are born without a functioning immune system and cannot at first synthesize antibodies against disease, the importance of receiving colostrum at birth cannot be overemphasized. Colostrum not only supplies antibodies that a newborn calf lacks, but it is somewhat laxative to aid in starting the calf's digestive functions.

It is generally recommended that calves receive a minimum of 2 qts of colostrum in two feedings during the first 12 h of life.

Under commercial conditions, dairy calves rarely receive colostrum from their own dams. Fresh cows are milked out, with colostrum kept separate from milk for the market.

There is no apparent difference in the effectiveness of fresh colostrum versus frozen and thawed colostrum or fermented colostrum. Fermented colostrum is somewhat acidic, so it is usually recommended that a buffer (e.g., sodium bicarbonate) be mixed with fermented colostrum before feeding. Although in one investigation this did not change the rate of gain, it did improve colostrum intake (Fig. 16–1).

Average colostrum contains over four times as much protein (14.3%) as ordinary milk, but is lower in lactose than average whole milk.

Calves that do not receive colostrum during the first 12 hours of life cannot be expected to be thrifty or healthy or to grow at a normal rate.

THE FIRST 60 DAYS

Even though newborn calves are equipped with all the organs of a ruminant digestive system (reticulum, rumen, omasum, and abomasum) their digestive processes are similar to those of monogastric animals. The rumen is not populated with the microbes typical of a functioning rumen (bacteria, protozoa, and fungus) until the calf is close to 60 days of age. This makes it necessary to supply the calf with either whole milk or a milk replacer.

The calf must elevate its head to nurse when using a nipple. This activates the esophageal groove through which the milk flows directly to the omasum and abomasum and bypasses the rumen. Calves fed by lowering their heads to drink from a bucket do not acti-

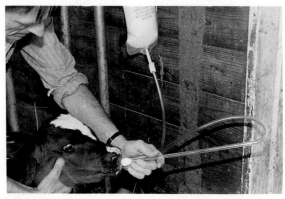

FIGURE 16–1 Administration of colostrum using an esophageal feeder. (Courtesy of E. P. Call.)

vate this esophageal groove; thus some, if not most, of the liquid will slop over into the rumen.

The more nearly calves are fed to approximate natural conditions, the more efficient their performance. Numerous feedings during the day is natural feeding, but not usually practical. The more frequently young calves can be fed, the more efficient their performance should be.

MILK REPLACER

When whole milk is not available, a milk replacer must be used. In fact, under most market conditions, reconstituted milk replacer is more reasonably priced than an equal amount of whole milk. Milk replacers are in a dry form and are generally recommended for use at the rate of 1 lb of dry milk replacer to 1 gal of water. Customarily, fluid milk replacers fed daily should not exceed 8% of a calf's body weight.

In some instances, milk from hospital pen cows can be used for raising calves. However, this milk can contain unsuitable levels of antibiotics or other medication and may even transmit disease. Certain products are available for use with whole milk, especially milk from cows in the hospital pen. These products contain scours inhibitors, as well as medication to reduce the possibility of disease transmitted by milk not fit for human consumption.

The selection of milk replacer is important for good results in raising dairy calves. High-quality milk replacers should be used for at least the first 3 weeks. Quality is determined primarily by the type and amount of protein and the type and amount of fat. Even though calves have no functioning rumen much before they are 60 days of age, milk replacers of relatively low quality can be used for healthy calves *after they are at least 3 weeks of age.*

In assessing milk replacer quality, milk source protein (dried whey, dried skim milk, casein) is superior to plant protein sources or to animal protein, such as fish protein concentrate. One possible disadvantage of fish protein concentrate is that it may contain polyunsaturated fat. Even at low levels (as little as 1.43% of the dry matter), polyunsaturated fat can interfere with vitamin E activity to produce degenerative muscle disorders such as white muscle disease. Feeding vitamin E supplement with fish protein can prevent this. There is also some evidence that fish protein feeds contain thiaminase, an enzyme that contributes to the degradation of thiamine (vitamin B_1). A lack of adequate thiamine can produce central nervous system disorders characterized by inco-ordination and, if untreated, even death. Most good-quality milk replacers contain added thiamine as thiamine mononitrate or thiamine hydrochloride.

Soybeans are the primary source of vegetable protein used in typical milk replacer. The main disadvantage of soybeans is that soy protein does not form a curd in the abomasum, as milk protein does. Curd formation is important in young calves, because it slows the time of passage through the alimentary canal and allows abomasum enzymes to digest proteins more completely.

Trypsin is an enzyme involved in protein digestion. A compound that can inhibit trypsin activity may be present in soybean protein products, but if proper heat treatment is used in processing soy protein, this can be deactivated. Another possible disadvantage of the use of soybean-source protein in milk replacers is that these products may contain allergenic material that can cause poor growth rates and persistent diarrhea. Soy protein supplements should be treated with alcohol to remove allergens or processed by precipitating protein with acid and then redissolving the protein with alkali. Products produced by these types of processing are soy flour and soy isolate. Improperly processed soybean products can be too high in undigestible carbohydrates. Young calves can utilize only limited amounts of starch or sugars other than lactose.

Examination of the ingredient section of a milk replacer label can help to identify whether soybean protein is the source of some of the protein. The amount used can be estimated from the fiber guarantee. Some soybean products contain as much as 7% fiber, but a total guarantee of ±0.5% fiber in the milk replacer is an indication that soybean protein is not a major portion of the milk replacer protein.

Milk replacer products now available are guaranteed to contain from 10% to 20% fat. There is apparently no advantage to the higher-fat products after calves are at least 3 weeks old and in healthy, thrifty condition. In extreme low temperatures the higher fat level (20%) is recommended. It is important to remember that milk replacer containing the higher level of fat should be formulated with more vitamin A than a product with only 10% fat. Fat from animal sources is preferred to vegetable oils for milk replacers.

The B vitamins play a critical role in calf health. Although dried milk and whey are good sources of the water-soluble (B) vitamins, it is a good idea to use a milk replacer that contains added B vitamins. The higher the content of nonmilk protein in the milk replacer, the greater the need for added B vitamins. Most milk replacers are formulated to approximate the nutrient content of cows' milk, so B vitamin deficiencies are seldom encountered.

The National Research Council (NRC) recommendations for nutrient density of milk replacer are shown in Table 16–1. This listing of recommendations was published in 1989. These recommendations should be used to evaluate milk replacer before purchase.

At 3 weeks of age, calves can be fed one of the available milk replacer products having lower protein and fat than listed. If a milk replacer is to be changed, the change should be made gradually, with the two products blended for a few feedings before the changeover is complete.

Many milk replacers contain antibiotics, with the tetracyclines being the most beneficial. Oxytetracycline and chlortetracycline can be added to milk replacer at the rate of 100 mg/lb for calves known to have received colostrum, and 250 mg/lb for calves that probably did not receive colostrum. Vitamin E is important to immune system function and has been recommended at levels as high as 135 units per day for calves during the time prior to weaning.

Diarrhea is a major concern in the care of calves prior to weaning. In an effort to decrease

TABLE 16–1

Recommended analysis for milk replacer

Nutrient	Recommended Density
Crude protein	22.00% (average)
Ether extract (fat)	10.00% (minimum)
Calcium	0.70%
Phosphorus	0.60%
Magnesium	0.07%
Potassium	0.65%
Sodium	0.10%
Sulfur	0.29%
Iron	100 ppm
Cobalt	0.10 ppm
Copper	10 ppm
Manganese	40 ppm
Zinc	40 ppm
Iodine	0.25 ppm
Selenium	0.30 ppm
Vitamin A	1730 IU/lb
Vitamin D	273 IU/lb
Vitamin E	18 IU/lb

These specifications should be considered as minimums. Many commercial milk replacers exceed the NRC recommendations on certain items.

fluid loss from excess diarrhea, some producers reduce milk replacer feeding. This is an erroneous practice that only makes the dehydration worse. Calves with diarrhea should be fed the usual amount of milk replacer plus solutions designed to replace minerals lost during diarrhea bouts.

Mineral excretion through the feces is normal, but in cases of diarrhea, Ca and Mg are excreted at 3.7 times the normal rate, Na and K are excreted at 11.3 times the normal amount, and P is excreted at 4.4 times the normal rate (8). This excess mineral loss also results in acidosis. Scouring calves need prompt replacement not only of fluids, but of minerals, and should be treated to correct acidosis.

Unless a poor-quality milk replacer is the cause of the scouring, a simple remedy for scouring can be made according to the following formulation and fed for at least two feedings along with the milk replacer in use.

One can of beef consommé

Two teaspoons of sodium bicarbonate

One package of pectin

One teaspoon light salt (KCl and NaCl)

Water sufficient to make 2 qt

This mixture should be given to the calf in addition to milk replacer until the manure is of normal consistency.

STARTER RATIONS

At about 1 week of age, calves should be offered a starter ration. Although they may not take much at first, it should be made available all the time. Because calves may refuse to consume the starter ration, it is important that it not be overfed. Unused portions left in the feeder can become oxidized and the fat rancid, and mold can grow in the dampness caused by saliva. This can make the feed unpalatable. Feeders should be cleaned after calves have had a reasonable length of time to consume feed. A calf hutch is shown in Figure 16–2. Along with the starter ration, good-quality alfalfa hay should be offered to calves in small amounts. Extremely leafy, fine-stemmed alfalfa may lead to scouring through excess protein intake.

Standard ingredients used in dairy feeds can be used for starter rations, but pellets or fine-ground feeds are not acceptable. Specifications for a starter feed are shown in Table 16–2.

FIGURE 16–2 Calf facilities used on a large calf-rearing ranch. (Courtesy of Steve Marks, Calico Ranches, Inc.)

TABLE 16–2

Starter ration specifications (as-fed basis)

16.0% Crude protein
2.5% Fat
0.6% Calcium
0.5% Phosphorus
0.7% Potassium
77 Mcal/cwt of NE_m[a]
48 Mcal/cwt of NE_g[b]

[a]Net energy for maintenance.
[b]Net energy for gain.

In addition to the listed specifications, starter feeds should contain trace minerals and vitamins at the same level as previously recommended for milk replacer and antibiotic at a rate designed to provide 50% the amount recommended for milk replacer.

Weaning should take place when calves are consuming approximately 1 lb of starter feed in addition to hay and milk replacer. There are two approaches to weaning calves from milk replacer. One is to abruptly remove the milk replacer so that calves have to utilize the starter ration. Another is to begin 2 weeks before weaning and gradually add increasing amounts of water to the milk replacer, until it is only water.

Before weaning and as soon as calves are eating regularly and are healthy, they should be moved from individual hutches to small group pens, with at least 100 ft² of space per calf and a maximum of 10 calves per pen. Weaned calves can be moved to larger group pens and offered growing rations. If possible, growing calves should be separated by size. If this is not practical, allow more space than recommended for calves in small group pens (100 ft²/calf).

HEIFERS: WEANING TO BREEDING

If heifers were properly introduced to solid feeds before weaning, the growing ration can be gradually changed so that they reach puberty at the age of 15 months. Table 16–3 is based on NRC recommendations for growing open large-breed heifers at the rate of 1.85 lb/day. Light heifers are expected to gain less than 1.85/day, while older, heavier heifers can be expected to gain 1.85/day, to achieve the ideal of 1.70-lb average daily gain needed to reach puberty and breeding size by 15 months. The goal of this program is to achieve maximum growth with minimum fat deposition. All recommendations are on an air-dry (89% to 90% dry matter) basis to facilitate ration formulation. The method for converting as-fed rations to air dry is discussed later in this section. Recommendations apply to Holstein and other large-breed heifers. Heifers of smaller breeds require proportionately smaller nutrient intakes.

Table 16–4 shows average nutrient values for typical feedstuffs. This information is to be used in balancing example rations. More specific values can be tolerated from the NRC publications or from local laboratory analyses of properly collected feed samples.

Formulating a ration for heifers or any other class of cattle involves setting up an example ration, making sure that the total feed intake (dry matter) does not exceed the probable capacity of the heifers being fed (approximately 2.7% of body weight), and selecting local feeds that should meet nutrient recommendations. A trial ration can be formulated, with nutrient intake calculated to determine which nutrients are limiting desired performance according to the recommendations.

Once the sample ration has been made, nutrient intake should be calculated using nutrient values from Table 16–3 or from one of the NRC publications. The following example can be used to test the probable success of the ration in meeting the production goal. A single feed ration has been used for this example to make the explanation of the method simple.

Fifteen pounds of alfalfa will be the total ration fed to a 500-lb heifer, with 1.85-lb daily gain as the goal. Table 16–4 provides nutrient analysis figures.

1. Determine crude protein (CP) content: 15 lb of alfalfa × 0.16 CP% = 2.40 lb CP.
2. Determine NE_m (Table 16–3): 5.03 Mcal NE_m/0.51 (Mcal NE_m per lb of alfalfa) = 9.86 lb of alfalfa needed for daily maintenance.
3. Determine alfalfa available for gain: 15.0 lb of alfalfa minus 9.86 lb needed for maintenance = 5.14 lb of alfalfa available for gain.
4. Determine NE_g: 5.14 × 0.23 (Mcal NE_g per lb of alfalfa) = 1.18 Mcal NE_g.
5. Determine Ca content: 15 × 0.012 (%Ca) = 0.18 lb Ca (81.72 g).
6. Determine P content: 15 × 0.0022 (%P) = 0.033 lb P (14.98 g).

Daily nutrient intake as calculated above can then be compared to a table of requirements (Table 16–3) in order to evaluate the ration. For the 100% alfalfa ration used as an example, the evaluation would be

The protein is unnecessarily high.

NE_g is lower than recommended.

Ca is unnecessarily high.

P is slightly low.

TABLE 16–3

Recommended daily nutrient intake

| | Average Body Weight (lb) | | | | |
	250	350	500	600	750
Crude protein, lb	1.23	1.43	1.70	1.81	2.24
Net energy					
NE_m, Mcal	2.98	3.84	5.03	5.76	6.81
NE_g, Mcal	1.81	2.08	2.48	2.72	3.02
Ca, g	19.00	22.00	23.00	26.00	27.00
P, g	11.00	13.00	16.00	19.00	21.00
Crude fiber, %	13.00	13.00	13.00	15.00	15.00

Source: NRC. *Nutrient Requirements of Dairy Cattle* 1989 (7)

TABLE 16–4

Typical nutrient analysis (as fed)

Type of Feed	Percent Crude Protein	Mcal/lb Ne_m	Mcal/lb NE_g	Percent Calcium	Percent Phosphorus
Alfalfa hay	16.00	0.51	0.23	1.20	0.22
Average grains[a]	9.00	0.83	0.53	0.03	0.29
Average grain hay[b]	8.50	0.50	0.23	0.35	0.25
Corn silage, 30% DM	2.20	0.23	0.13	0.10	0.05
Alfalfa haylage, 30% DM	6.40	0.21	0.11	0.51	0.12
Oil meals	41–44	0.77	0.51	0.16	0.70
Corn gluten meal	42.00	0.80	0.53	0.16	0.40
Limestone	—	—	—	38.00	—
Dical phosphate	—	—	—	22.00	—
Urea	281.00	—	—	—	—

[a]Oats, barley, grass.

[b]Corn, barley, oats, milo.

Note: Undegradable protein has been shown to be economically efficient in heifer rations only during the first month after weaning.

From this, it can be concluded that higher-protein feed can be replaced with higher-energy and phosphorus feed to arrive at a ration suitable to meet the set goal.

To adjust any single nutrient, the difference between the nutrient density of an available feed high in that nutrient and an available feed containing less of that nutrient can be determined using Table 16–4.

The difference between the NE_g of alfalfa (0.23 Mcal/lb) and the NE_g of corn (0.53 Mcal/lb) is 0.33 Mcal/lb. Therefore, for each pound of alfalfa hay replaced by corn, the ration would provide 0.33 Mcal of NE_g more than it previously provided.

The same method applies to rations formulated on the basis of percentage (or nutrient density in the case of energy). The two completed rations shown in Table 16–5 were formulated using the trial and error described above.

Feed concentration is shown in percentage for those enterprises utilizing a total mixed ration. Daily consumption as pounds of each ingredient can be calculated on the basis of body weight, with growing heifers expected to consume approximately 3% of their body weight on an air-dry basis. Air-dry feed is assumed to contain 89% dry matter. Refer to Table 16–5 for sample rations for open heifers, as well as calculated daily nutrient intake and performance.

Growing heifers use available nutrients in an irreversible order: (1) daily maintenance, (2) growth, and (3) ovulation and conception. This is why it is important to feed open heifers a ration that will allow normal growth with additional nutrients available for reproduction.

A good management practice is to **flush** heifers once they have reached the target

TABLE 16–5

Sample rations for open heifers and calculated daily nutrient intake and performance

Kind of Feed	Average Body Weight (lb) 350 (200–500)	Average Body Weight (lb) 600 (500–700)
Average grains	38.0%	15.3%
Average oil meal	4.7%	1.0%
Average grain hay	23.7%	22.5%
Alfalfa hay	33.3%	15.3%
Corn silage	0	45.8%
Mineral[a]	0.3%	0.1%
Estimated daily feed, lb	10.50	26.19
Crude protein, lb	1.32	1.82
NE_m, Mcal	3.84	5.76
NE_g, Mcal	1.60	2.96
Calcium, g	25.41	38.49
Phosphorus, g	14.38	20.42
Dry matter, %	89.29	61.84
Crude fiber, %	19.80	15.80
Estimated daily gain, lb	1.60 (7)	1.85 (7)

Source: NRC table of allowances (7).

[a]Contains 50% dicalcium phosphate, salt, trace minerals, and vitamins. Both protein and energy for gain are the limiting nutrients. Heifers gaining 300 lb (from 200 to 500 lb) at 1.6 lb/day and the same heifers gaining another 200 lb (from 500 to 700 lb) at 1.85 lb/day will average 1.70 lb gain/day for the entire growing period.

weight for breeding. Flushing involves increasing nutrient intake in order to assure that heifers reach the third level of use for ovulation and conception.

Flushing is accomplished by an increase in intake of all nutrients by increasing the nutrient density of the ration fed to heifers old enough and large enough for breeding. But there are other feeding management practices that can improve the possibility of early conception. Including sources of bypass protein in the ration is one. Protein from certain sources is unavailable for degradation by rumen microbes and bypasses the rumen to be hydrolyzed and subjected to the action of proteolytic enzymes in the abomasum and small intestine. Feeds known to contain a relatively high proportion of degradable protein may not be readily available to many heifer-raising enterprises, or it may not be practical to use them. It may only be practical and economical to utilize bypass protein sources in the rations for heifers during the important first breeding period and again shortly before first freshening, when both the rapidly developing fetus and the need to store nutrients to support early lactation increase protein demand.

In addition to changing heifer rations during first breeding by adding bypass protein, it might also be practical to supply at least part of added trace minerals in the form of proteinates. These are complexes of amino acids with trace minerals. An example is zinc methionine. Although efficacy data are not conclusive, including proteinates as at least part of the trace-mineral addition to heifer rations at breeding times may prevent subclinical deficiencies that could delay breeding. The most favorable results with proteinates are with extremely stressed cattle. However, because proteinated trace minerals are available, they will serve as one more ingredient that may improve breeding efficiency (9).

Another aid for improving heifer development is the use of ionophores (See Ch. 10). Two ionophores are now approved for use in rations for dairy replacement heifers. These are lasalocid and monensin. Both can be included in feed for heifers to improve rate of gain, which is an indicator of rate of growth.

Ionophores not only reduce waste caused by methane production, but they also reduce ruminal ammonia production, conserving intake protein. The net result of using an ionophore is that heifers use carbohydrate feeds more efficiently, saving more protein for growth. Growing heifers to reach breeding size by 15 months of age reduces the amount of milk that they must produce as cows to pay for their growth, compared to heifers fed for a slower rate of growth and sexual maturity.

In addition to having antibiotic action, ionophores are also coccidiostats when used as directed. Coccidiosis is almost universal and can cause general reduction in growth rate, which

could delay puberty. Chronic coccidiosis can also lead to reduced immune system function.

Results of numerous university studies show that ionophores used for developing heifers will return more than their cost in earlier puberty and lifelong production resulting from heavier body weight at first freshening.

GESTATION

When heifers are judged to be safe in calf, the feeding management can be changed. Nutrition of bred heifers can be considered in two phases: (1) from breeding to approximately 60 days prior to the expected calving date and (2) the last 60 days of gestation. During the first phase, rations should be designed for growth, with fat deposition avoided. Apparently, bred heifers of the dairy breeds fed high-energy, relatively low-protein rations tend to deposit fat in the udder, which limits future production capability (4).

Table 16–6 illustrates the differences between growing-heifer rations and rations for the same heifers after parturition. The second phase of bred-heifer nutrition involves a transition from one set of recommendations to the other.

In many areas, pasture or other feedstuffs low in dry matter are an important part of the feeding program for bred heifers. Table 16–4 does not include pasture. Pasture nutrient composition can vary from one day to the next. This variation is due to weather (temperature, moisture) or the species and maturity of the grasses. For this reason, attempting to balance a ration accurately with pasture as one of the feeds is not practical. However, using as-fed pasture analysis as shown in Table 16–7, example rations with or without pasture were formulated to demonstrate probable results in using pasture in a growing-heifer program.

Recommended daily nutrient allowances for bred heifers from breeding time up to 60

TABLE 16–6

Example gestation and lactation ration recommendations

Nutrient	Gestation	Lactation
Crude protein, %	12.00	17.00
Net energy, Mcal/lb	0.57	0.76
Fat, %	3.00	3.00
Calcium, %	0.39	0.77
Phosphorus, %	0.24	0.49
Potassium, %	0.65	1.00
Sulfur, %	0.16	0.25

Source: NRC, 1989 (7).

Cobalt (0.10 ppm), copper (10 ppm), manganese (40 ppm), and zinc (40 ppm) recommendations are the same for gestation and lactation.

TABLE 16–7

Average pasture analysis[a]

Dry matter	31.00%
Crude protein	3.90%
NE_m	0.19 Mcal/lb
NE_g	0.09 Mcal/lb
Calcium	0.15%
Phosphorus	0.10%

[a]Average of three common grasses in mid-bloom (neither early season nor late season and mature).

TABLE 16–8

Nutrient allowances for pregnant heifers

	Average Body Weight (lb)	
Nutrient	975	1150
Crude protein, lb	2.60	3.60
NE_m, Mcal	7.97	9.39
NE_g, Mcal (1.7 lb/day)	3.21	3.63
Calcium, g	29.00	29.00
Phosphorus, g	21.00	21.00

The optimum absorption of Ca and P is best when the ratio is two parts to one part (6). This is seldom practical, because of the limited availability of a wide range of feeds, but an attempt should be made to approximate this ration. This can be done through careful selection of mineral supplements.

days prior to calving and for first-calf heifers in late gestation in the last 60 days of pregnancy are shown in Table 16–8. The example assumes that the first group of heifers averages 975 lb, while the heifers in late-gestation average 1150 lb.

Consumption of high-moisture feeds, such as pasture and silage, can be estimated for dairy heifers by assuming that the heifers have the capacity to consume approximately 3.0% of their body weight in air-dry feed. To estimate consumption for a feed analyzing 31% dry matter, 89 (approximately DM of pasture) can be divided by 31 to arrive at the factor of 2.87. Dairy heifers whose only feed is 31% dry matter could be expected to consume 2.87 times 3% of their body weight. This can be considered as only a rough estimate, since physical capacity (body size and configuration) is also a factor in feed consumption. Refer to Table 16–9 to see example rations for heifers in early and late gestation and daily nutrient intake and performance calculations.

Rations fed to heifers in late gestation should contain a grain mix similar to that used when they enter lactation strings for these reasons:

1. To adjust the rumen population to increase microbes that ferment the specific feeds contained in the lactation ration
2. To increase nutrient intake for storage to support early lactation plus growth
3. To provide for the increased demand for nutrients by the rapidly developing fetus

The transition from an early gestation ration to a ration containing a grain mix designed to support lactation should be made gradually over a period of several feedings. Increases in the lactation ration should be based on manure consistency and acceptance of feed by the heifers being transitioned.

Although newborn calves contain about 70% moisture, approximately 18% of their weight is protein and 4.5% is in the form of mineral matter. During the last 60 days of gestation, a heifer needs to supply 75% of the calcium that her calf will have at birth (3). This is just one example of the need to formulate rations for heifers in late gestation with an adequate supply of all minerals, as well as protein, to assure that the developing calf and the still growing heifer are properly provided for. This is especially important for large-breed heifers, which continue to grow during the first and second lactations.

FEEDING DAIRY BREED STEERS FOR BEEF

It has been estimated that approximately 4 million Holstein steer calves are produced in the United States each year. A relatively small portion of these are slaughtered for veal. The rest are fed for the commercial beef market. A commercially fed Holstein steer will typically be started on a finishing regime at about 350 lb and be slaughtered at 1100 to 1200 lb.

Commercial cattle feeders are faced with an ongoing problem related to crossbreeding within beef cattle herds that supply feeder cattle. A group of cattle may have physical similarities (size, color, conformation, weight), but dissimilar genetic and nutritional backgrounds can result in variations in economically important factors such as feed efficiency and carcass quality.

Modern dairy animals are the end result of many generations selected for one factor—milk production. Also, dairy cattle, no matter their geographic origin, are raised from birth under a more or less standard set of conditions and feeding practices. Dairy cattle, especially

TABLE 16–9

Example rations for heifers in early and late gestation and daily nutrient intake and performance calculations (pound per day as fed)

| Type of Feed | Average Body Weight (lb) | | | |
| | 975 | | 1150 | |
	A	B	A	B
Pasture	46.00	0	0	0
Corn silage	0	35.00	25.00	0
Grain hay	7.50	0	10.00	15.00
Alfalfa hay	0	10.00	5.00	7.50
Barn grain[a]	0	0	12.00	12.00
Pasture supplement[b]	1.00	2.00	0	0
Crude protein, lb	2.64	2.70	3.97	4.25
NE_m, Mcal	7.97	7.97	9.39	9.39
NE_g, Mcal	2.70	3.63	7.22	6.47
Calcium, g	43.65	71.19	62.20	73.55
Phosphorus, g	31.55	22.20	37.91	40.18
Calculated gain/day, lb	1.40	1.85	1.70+[c]	1.70+[c]

[a]An average grain mix for lactating cows with these specifications: 16% crude protein; 80 Mcal/cwt of NE_m; 55 Mcal/cwt of NE_g; 0.10% calcium; 0.30% phosphorus, with added trace minerals and vitamins.

[b]50% Oil meal: 50 grain

[c]Actual gain per day includes the developing fetus.

cows during lactation, are confined most of the time and so are not selected as foragers. They do not differ greatly from one herd to the next in physical and physiological characteristics. Therefore, steer calves from this genetic background are also more uniform than steers of beef breeds. Variations across any group of dairy-breed feeder steers are smaller than variations across a group of beef-breed steers, whose environments, market, and feed situations have been much more diverse. Experienced dairy-steer feeders can design a ration program and predict performance more accurately than they can with an equal number of crossbred beef cattle.

Most beef cattle in the United States are from herds of less than 30 cows, with each small cattleman selecting replacements and buying bulls using their own unique criteria. For example, feed availability influences the type of cattle selected in most geographical regions. Cattle selected for ranches with low carrying capacity should have low mature weights. On these ranches, restricted feed supplies make large cattle very uneconomical. Cattle selected for adaptation to ranches with a temperate climate and relatively high rainfall leading to abundant feed can be of a larger mature weight.

On the average, Holsteins have been reported to gain at a faster rate with less feed required per pound of gain than beef steers. However, it should be remembered that dairy steers, specifically Holsteins, perform less effi-

ciently than beef-breed steers when conversion is measured in tissue gain per unit of energy consumed (5). Also, Holstein steers have a lower dressing percentage than average beef-breed steers. This is because a Holstein carcass tends to have a greater proportion of bone and less fat than typical beef-cattle carcasses.

The difference in feed efficiency means that rations for Holstein steers should be formulated differently than finishing rations for beef-breed steers. Since a greater proportion of a Holstein steer's nutrient intake goes into production of lean tissue and bone, Holstein rations should be formulated with a greater percentage of protein than steers of beef breeding.

Although there are almost as many programs for finishing Holstein steers as there are enterprises feeding Holsteins, three types of programs will be considered here.

1. Calves raised in hutches and small group pens, weaned along with replacement heifers, and then put on full feeding programs
2. Weaned calves going through an on-the-farm growing program before being put on a finishing regime
3. Weaned calves going to pasture before finishing

Table 16–10 shows the production goals listed in one review paper (2). These goals are achievable under most management systems.

TABLE 16–10

Performance goals for Holstein steers

| | Body Weight (lb) | | |
	400 (Weaned)	400–850	850–1100
Dry-matter intake, lb	7.5	16.0	18.3
Feed conversion	3.0:1.0	4.9:1.0	7.6:1.0
Average daily gain, lb	2.4	3.3	2.4

The table shows an increase in feed required per unit of gain by heavy cattle compared to lighter-weight cattle. One effort to overcome this characteristic has been an attempt to take advantage of compensatory gain. On these programs, steers in the middle weight group are purposely limited in their protein intake. Then, when the ration is increased in protein density, they seem to compensate with faster rates of gain than when fed recommended protein levels for the whole feeding period.

High roughage rations tend to increase rumen size and volume. This is desirable for heifers. But Holstein steers grown on pasture from weaning to about 750 lb before being put on a finishing program tend to consume more dry matter than they can convert to body-weight gain. For example, large Holstein steers from a pasture background consume more total feed in the drylot, but gain at the same rate or slower than similar steers fed in confinement from weaning to slaughter. Attempts have been made to limit-feed such cattle to improve feed conversion. Cattle fed free choice gain at a slightly higher rate, but also convert less efficiently than limit-fed cattle. Limit-fed programs usually feed 85% of what similar cattle are fed on a free-choice basis.

The theory of limit feeding seems to be valid. But Holstein cattle have relatively voracious appetites and tend to become distressed and nervous when feed is withheld. This apparently reduces the efficiency of limit-feeding programs when applied to Holstein

steers. When Holstein feeder steers come from a pasture background, overconsumption can be expected.

Table 16–11 shows a listing of nutrient restrictions for growing–finishing Holstein steers. No specific feedstuffs or body weights have been included. Feeds vary in availability depending on location, and all cattle are not started at a uniform body weight or nutritional background. For example, calves weaned and started directly on a concentrated feeding program probably would not require a starter ration; cattle grown in confinement before finishing would probably not need the transition ration. Cattle from a pasture background would require a complete three-step program. The restrictions listed are not definitive, but are meant to serve as a pattern for using available feeds and facilities.

Trace mineral and vitamin densities are not listed on the table of nutrient recommendations, since they are standard for all steps of a growing–finishing program. Because of the variation in available feeds, a mineral supplement providing macro (Ca, P, K) minerals and trace minerals cannot be standardized. If all available feeds are combined in a formulation that meets the protein and energy restrictions shown in Table 16–11, a mineral addition can then be formulated to bring the total density of minerals and vitamins to the recommended level. It is usually recommended that salt (NaCl) be included at 0.5% of the ration, so this would be part of an added mineral mix.

TABLE 16–11

Example nutrient restrictions (100% dry matter)

Nutrient	Start	Transition	Finish
Crude protein, %	17.00	13.50	12.00
NE_m, Mcal/lb	0.63	0.73	0.93
NE_g, Mcal/lb	0.93	0.55	0.66
Calcium, %	1.00	0.77	0.77
Phosphorus, %	0.33	0.33	0.33
Potassium, %	1.10	0.80	0.80
Crude fiber, maximum %	35.00	12.50	7.50

REFERENCES

1. Bethard, G. L., et al. 1994. *J. Dairy Sci.* 77: Supplement 1.

2. Chester-Jones, H. 1991. *Holstein beef symposium.* Pennsylvania NRAES 44.

3. Ellenberger, H. C., et al. 1950. *Vermont Agric. Exp. Sta. Bull.* 558.

4. Foldager and Sejrsen, 1982. *Proc. 12th World Congr. Diseases of Cattle,* p. 451.

5. Hooven, et al. 1971. *J. Animal Sci.* 34:1037.

6. Manston, R. 1967. *J. Agric. Sci. Camb.* 78:263.

7. NRC. *Nutrient Requirements of Dairy Cattle.* 1989. Nat. Research Council.

8. Radostits and Bell, 1970. *Canadian J. Animal Sci.* 50:405–452.

9. Spears, J. W., et al. 1991. *J. Amer. Vet. Assn.* 99:1731.

10. NRC. *United States–Canadian Tables of Feed Composition,* 3rd rev. ed. 1982. Nat. Research Council.

11. Van Nevel, C. J., et al. 1977. *Applied Environmental Microbiology* 34:251–257.

17

SHEEP NUTRITION

Rodney Kott

INTRODUCTION

Ultimately, the production of sheep is controlled by their economic efficiency in converting available feed resources into products of economic value, that is, meat, wool, and milk. In most situations, available feed resources dictate production levels. Throughout the world the common denominator in sheep production is pasture and forage. Productivity of the pasture, rangeland, or forage crop will largely dictate the maximum levels of productivity that the sheep producer can achieve. These vary from subsistence production levels on the African desert, to commercial milk sheep production systems in Mediterranean countries, to intensive lamb production systems where meat production is the primary goal in Great Britain, central Europe, and the United States. In all cases, end-product production objectives are closely related to the forage production capabilities of the land (Figs. 17-1 and 17-2).

Most of the world's sheep are located in arid to semiarid ecosystems, such as the arid rangelands of Australia, Africa, South America, and the southwestern United States. Extremely low rainfall in many of these regions results in limited feed resources that are marginally sufficient to support reproduction and lamb growth. In these situations, wool production is economically more important to the sheep enterprise. Because expected production outputs are low,

these production systems are extensive, external inputs to the system are minimized, and sheep numbers are generally large enough to offset low outputs per animal unit.

About 5% of all ewes fail to lamb and about 15% to 20% of all lambs born die between birth and weaning. Although there are many causes for production failure, faulty nutrition is certainly a major contributing factor. Supplying the nutrient needs of a sheep represents the single largest expense in the total cost of raising sheep. Because of this, a solid understanding of nutrition is necessary in order to minimize the annual cost of production, yet maintain optimum production levels. The National Research Council (NRC) in the United States (1) and the Agricultural Research Council (ARC) in Great Britain (2) have reviewed world research studies and established a fairly precise set of nutrient requirements for sheep for various stages of production and with different levels of productivity. These resources represent the most current understanding of the needs of sheep for specific nutrients such as energy, protein, minerals, and vitamins in order to meet clearly defined production objectives. The purpose of this chapter is not to restate information provided in these publications, but to develop an understanding of how these fundamental principles apply to the production systems in the various ecosystems in which sheep production may be a viable enterprise. The requirements presented

FIGURE 17-1 Ewes grazing western intermountain range.

FIGURE 17-2 Integrated crop–sheep production systems. Land not suited to crops or rotated to pasture is utilized by sheep along with crop residues and farm-grown feeds.

in these publications should be used as guidelines and not as rigid standards. In any flock, sheep are different sizes, have different levels of body condition, and are in different stages of production; it is not always possible to know at each feeding the exact nutrient composition of each feed. The most common misinterpretation of these recommendations is that each production system must provide for these nutrient levels and weight changes. This is not always possible under practical conditions. Deviations from this system are possible without greatly affecting sheep productivity; however, short-term deviations must be compensated for over the course of the entire production cycle if optimum production is to be maintained. Adequate quality and quantity of feed in conjunction with good management should achieve optimum production of a sheep flock.

NUTRIENT NEEDS

The nutrients of primary importance in sheep nutrition are water; energy as measured by to-

tal digestible nutrients (TDN), metabolizable energy (ME), or net energy (NE); protein, either crude or digestible protein; minerals, with salt, calcium, and phosphorus being the most critical; and vitamins, with vitamin A being of primary importance.

WATER

Water, although often overlooked, is one of the most important nutrients required for life. An adequate supply of clean, fresh water is essential to efficient sheep production. Inadequate water consumption reduces feed and forage intake and compromises performance. In fact, a deficiency of water will cause death much faster than a deficiency of any other nutrient. Water quality is probably more important to sheep than to other species of livestock. If water is stagnant or of poor quality or if water has an objectionable odor, temperature, or bacterial count, sheep will often not consume it in adequate amounts. Water containing salt or minerals in excess of 1.3% may not be palatable, thus affecting water intake and performance.

Daily water consumption by ewes varies from 0.72 gal during the cold winter months to 1.5 gal during the late winter months when temperatures begin rising to as high as 2.2 gal when sheep consume dry forage such as saltbush. A sheep's free water intake is greatly influenced by the amount and type of feed or vegetation consumed, protein intake, mineral intake, environmental temperature, water temperature, stage of production, and the amount of rain, dew, or snowfall (3).

DRY-MATTER INTAKE Water intake increases as dry-matter consumption increases. In general, water consumption is usually about two or three times the weight of dry-matter consumption. A practical rule of thumb is that a sheep consume 1 gal of water for every 4 lb of air-dry (90% dry matter) feed consumed.

PROTEIN INTAKE Protein intake that exceeds the animal's requirement increases water requirements. Additional water is needed for metabolizing and excreting the excess nitrogen that is not utilized for protein by the animal.

MINERAL INTAKE Excessive mineral intake significantly increases water consumption. The kidneys can only concentrate minerals to a specific percentage level and thus require water for secretion or excretion into urine. Salt is sometimes used to control consumption of concentrate supplements fed to sheep on range. When this is done, adequate water must be kept available at all times because the kidneys must eliminate this excess salt.

ENVIRONMENTAL TEMPERATURE One method by which body temperature is regulated is through the evaporation of water from the skin and lungs. Large increases in water needs can be expected when temperatures rise above 70°F. In contrast, water intake decreases during periods of extremely cold environmental temperatures.

WATER TEMPERATURE Water temperature affects water consumption during periods of extreme heat or cold. Livestock generally prefer water between 40° and 50°F. Consumption drops if temperatures deviate too much from this range. Water located in shade is more readily accepted on hot days. During winter months in northern climates, water heaters should be used to provide ice-free water.

STAGE OF PRODUCTION Water requirement for ewes increase dramatically during late gestation and lactation (Table 17-1). For optimum production, ewes should have an unlimited supply of good clean water where they are able to drink frequently during late gestation and lactation.

OTHER SOURCES OF WATER Sheep obtain water from moisture in feed, snow and dew, oxidation of other nutrients (metabolic water), as well as from drinking. Ewes and lambs on pastures where rain showers occur almost daily and there is a heavy dew each morning require very little additional water. Silage, succulent pasture, or range forage that is ex-

tremely high in moisture reduces water consumption. If ample amounts of soft, wet snow are present, sheep can usually meet their water needs without additional drinking water when grazing winter range during early and mid-pregnancy (4).

ENERGY

Insufficient energy probably limits performance of sheep more than any other nutritional deficiency (5). It may result from inadequate amounts of feed or from feeds of low quality. The energy requirements of a ewe vary greatly with her stage of production.

Adequate amounts of energy are extremely important during late gestation and lactation. Energy shortages are often complicated by protein or mineral deficiencies. An energy deficiency causes a reduction in body growth, reduced fertility, reduced wool quantity and quality, including breaks in the fiber, and, if the deficiency is severe enough, death. A sheep's energy needs can, in most instances, be supplied by feeding good-quality pasture, hay, or silage. Additional energy is generally needed immediately before and after lambing, in conditioning ewes and rams for breeding, and in finishing lambs. Grains such as barley, corn, wheat, oats, and milo are generally used to raise the energy level when supplementation is necessary. During lactation a ewe's metabolic energy requirements can at least partially be met by breaking down body fat reserves.

Special precautions, however, must be taken when feeding wheat grain. It requires a longer adaptation period than most other cereal grains. Lambs seem to be particularly susceptible to acute indigestion from overconsumption of wheat. Therefore, if used, it should be fed at a relatively low level (less than 50% of the grain portion of the diet) or increased gradually in the ration.

PROTEIN

In most situations the amount of protein supplied in the diet is more critical than protein quality (5). Ruminants have the ability to convert low-quality protein sources to high-quality proteins by bacterial action. Rumen microorganisms take the nitrogen portion of the proteins and build bacterial and protozoal protein, which is then digested in the intestines. Protein available for digestion in the small intestine

TABLE 17-1

Estimating water intake of sheep

Environmental Temperature (°F)	Pounds of Water Intake per Pound of Dry Matter Consumed for Ewes Carrying a Single Lamb[a]				
	Gestation, Month				
	1	2	3	4	5
Under 59°F	2.0	2.8	3.0	3.6	4.4
59°–68°F	2.5	3.5	3.8	4.5	5.5
Over 68°F	3.0	4.2	4.5	5.4	6.6

Source: Agricultural Research Council (2).

[a]Water intake of ewes with twins is about 20% greater in the third month, 25% greater in the fourth month, and 75% greater in the last month of gestation than for ewes with single lambs.

thus consists of microbial protein and feed protein that has escaped microbial breakdown in the rumen. Microbial protein synthesis is sufficient to supply the sheep's protein needs provided adequate precursors are available, except during lactation in high-milk-producing ewes and in very young lambs when rumen activity is limited.

When comprising the complete diet, green pastures provide adequate protein for most classes of sheep. When ranges are mature and bleached or have been dry for an extended period of time, and when grass hay or high-grain rations are fed, additional protein may be needed. High-protein feeds are often added to creep rations because they are usually extremely palatable and stimulate appetite and digestive activity. In some instances, it may be beneficial to feed proteins with a high bypass value.

Oil meals such as soybean meal or cottonseed meal contain 35% to 45% protein and are excellent sources of supplemental protein. Properly harvested legume hays such as alfalfa are often relatively high in protein content (up to 25% CP) and can be used effectively to supply supplemental protein to the sheep's diet. When protein supplementation is the primary objective, the cost per unit or pound of protein is the most important consideration.

In some instances sheep can utilize relatively inexpensive nonprotein nitrogen (NPN) sources such as urea to help meet their protein needs. Given adequate sources of energy and sulfur for optimal microbial growth and protein synthesis, rumen microbes can convert NPN to microbial protein. A number of rules of thumb should be followed when urea is used in sheep diets. Urea should not contribute more than one-third of the total nitrogen in the diet and should not be more than 1% of the total ration or 3% of the concentrate portion of the diet (1). Urea should not be used in the ration of young lambs or in creep rations. Urea cannot be utilized effectively when the rumen is not yet completely functioning. Also, urea is not usually recommended in range sheep rations, sheep feed low-energy feeds such as straw or poor-quality hay, or lambs that are on limited feed. In these situations, adequate energy may not be readily available for protein synthesis.

Protein supplements may be self-fed or hand fed. Hand-fed supplements are most conveniently fed in the form of pellets or cake. Protein blocks or liquid supplements can be self-fed. They are usually more expensive but tend to save labor. Intake of self-fed protein is sometimes limited by salt. Protein supplementation is often necessary during drought when water supply is also short. Increased salt intake requires increased water consumption.

MINERALS

Fifteen minerals have been demonstrated to be essential in sheep nutrition. They are Na, Cl, Ca, P, Mg, K, S, Co, Cu, I, Fe, Mn, Mo, Se, and Zn. Although relatively precise requirements have been published for the different minerals, it should be recognized that in practice the true dietary requirements vary greatly depending on the nature and amount of these and associated minerals in the diet. A number of mineral balances (e.g., Ca and P, Cu to Mo, Se and vitamin E) must be considered when establishing the actual requirements under specific conditions.

Most of these are met under normal grazing and feeding habits. In many situations, poor animal performance is attributed to a mineral deficiency when in fact it is due to something else. Under normal grazing situations, the minerals most likely to be deficient are salt (sodium chloride) and P.

Trace-mineralized salt is usually fed free choice to sheep. However, care should be taken to be sure that the trace mineral mixtures are specifically developed for sheep and do not contain the high levels of Cu commonly found in beef, dairy, swine, and poultry trace mineral mixes. Most trace-mineralized salt mixtures formulated for sheep provide 7 of the 15 essential minerals (Na, Cl, I, Co, Fe, Mn, and Zn). The minerals that are normally provided in sufficient amounts in natural feedstuffs include K, Mg, Fe, Cu, and Mn. It is important to note that trace mineral salt does not usually supply P.

SALT   Salt serves many functions in the body. When deprived of salt, sheep consume less feed and water. As a result, growth rate and milk production are reduced. As a general rule, sheep producers should provide supplemental salt to their sheep. Salt is generally fed to ewes at the level of 0.25 to 0.4 oz/head/day. It can be fed free choice or added to the feed mix at the rate of 0.5% of the entire ration or 1% of the concentrate portion of the ration. In alkaline areas, the water may contain enough salt to meet the requirements.

CALCIUM AND PHOSPHORUS   The need for Ca is highest in ewes during lactation. Most pastures, hays, and other forages contain adequate levels of Ca for sheep. Most sheep diets contain an excess of Ca rather than a deficiency, and thus Ca supplementation is seldom necessary. When pasture forage is low in Ca, it is almost always low in P and protein, and these nutrient deficiencies overshadow the effects of a Ca deficiency. However, grains are somewhat deficient in Ca and thus supplementation is often beneficial when sheep consume diets that consist primarily of grains or corn silage.

Mature pasture and range forage is often deficient in P. Grains, however, are relatively high in P content. A sheep's P requirement is highest during late gestation and lactation. If a sheep is fed a diet deficient in P, it can draw P stored in bone with minimal problems. However, extended periods of inadequate dietary P result in poor reproductive performance. Since in most situations a high percentage of a sheep's diet consists of roughage or pasture, P supplementation is often beneficial. Phosphorus is not very palatable and, therefore, must usually be mixed with an ingredient that sheep like to assure adequate consumption. The most desirable way to supply additional P, when needed, is by adding it directly to the feed mix. This, however, is not always practical or feasible. It is sometimes more convenient to supplement the sheep's diet with a high-P mineral mix.

The ratio between Ca and P must be considered when balancing sheep rations. Although ratios of 5 or 6 to 1 (Ca to P) seem satisfactory, a ratio of 2:1 is ideal for most sheep rations. Feedlot lambs or growing rams fed diets high in grain are prone to urinary calculi. In these situations the incidence of urinary calculi can be reduced by raising the Ca to P ratio to 3 or 4 to 1.

COPPER   There is a delicate balance between the Cu requirement and Cu toxicity in sheep. Sheep are more susceptible to Cu toxicity problems than other livestock species. The differential between Cu requirements and Cu toxicity is extremely narrow in sheep. Errors in feed mixing frequently result in death due to Cu toxicity.

Copper requirements of sheep depend on dietary and genetic factors, and therefore it is almost impossible to develop a set of well-defined requirements. In fact, it has been shown that dietary amounts of Cu that are adequate in one situation may be deficient in another and possibly toxic in a third (6). Concentration of Mo is a major dietary factor affecting the ewe's Cu requirement. Molybdenum forms an insoluble complex with Cu, which reduces its absorption, thus increasing the dietary levels needed to meet requirements. Also Merino breeds of sheep generally are less efficient in absorbing Cu from feedstuffs than British breeds of sheep.

Although it is impossible to give the exact requirements and toxic levels for Cu in the diet, several estimates have been made (Table 17-2). It should be stressed that these are just guidelines and may vary drastically from situation to situation. When selecting a trace mineral mix for sheep, it is generally recommended to choose one that contains no Cu. If sheep are grazing range forage containing normal amounts of Cu (8 to 12 mg Cu/kg) and not excessively low amounts of Mo (>1.0 mg/kg), they can usually be safely fed a trace mineral salt that contains relatively small amounts of Cu, given there is not other external sources of Cu (e.g., in protein supplement or drinking water).

Copper is found in adequate amounts in most of the United States. Deficient areas have been reported in Florida and in the coastal plains regions of the Southeast. Also, in several of the western states there are areas where an excess of Mo may induce a Co deficiency (7).

SELENIUM   In some areas of the United States the Se content of forage is below the dietary requirements of sheep. On the other hand, in some areas the Se concentration is so high that it causes Se toxicity. In sheep there is a very narrow range between the amount of Se that is required in the diet and that which is toxic. Diets containing less than 0.1 ppm Se are deficient, while those containing over 2 ppm are above the maximum tolerable level (1). White muscle disease in lambs results from a deficiency of Se and possibly vitamin E. A marginal deficiency in Se can result in reduced reproductive performance and increased lamb mortality. This deficiency can be prevented by giving injections of a commercial product containing both Se and vitamin E. Selenium and/or vitamin E can also be added to the entire ration, supplement, or salt–mineral mix of sheep. Probably the most practical and effective way of supplying Se to sheep is by feeding a salt–mineral mix containing Se. There are many excellent ones on the market. Do not try to mix your own. When supplementing Se (either by feeding or injection), producers should follow the manufacturers' or veterinarians' recommendations very closely. There may be some instances in sheep nutrition where "If a little is good, a lot is better." However, "a lot" of Se can be lethal.

TABLE 17-2

Recommended Cu allowance for sheep

Molybdenum Content in Diet (mg/kg)	Recommended Copper Allowance (mg/kg DM)		
	Growth	Pregnancy	Lactation
<1.0	8–10	9–11	7–8
>3.0	17–21	19–23	14–17

Source: NRC (1).

VITAMINS

Mature sheep require all the fat-soluble vitamins: A, D, E, and K. They usually do not require the B vitamins since these are synthesized in the rumen. Normally, the forage and feed supply all the vitamins in adequate amounts. Vitamin A can become deficient if sheep have been grazing on dry or winter pastures for an extended period of time. Sheep, however, store vitamin A for a considerable time, and if ewes have been on green forage or have had access to high-quality legume hay, vitamin A is usually not deficient. Vitamin D deficiencies may develop in confined sheep. Sheep raised outside usually have sufficient vitamin D because sunlight builds a store of this vitamin in the body.

BODY CONDITION SCORING

Body composition at a given point in the production cycle may influence both production response at that point and response to varying levels of nutrition. The most productive ewe in any flock of sheep is neither too thin nor too fat. Although measurements of body composition on live animals are estimates, producers must utilize the best system available. Body weight alone is inadequate because of apparent differences in mature body size among different breeds and individuals within a particular breed. The use of both body weight and condition scores can help producers make important feed management decisions. Body condition scoring is a simple but useful procedure that can help producers make management decisions regarding the quality and quantity of feed needed to optimize performance.

Condition scoring is a system of describing or classifying breeding animals by differences in relative body fatness (Table 17-3). It is a subjective scoring system but provides a fairly reliable assessment of body composition. Although scores of 1 to 10 can be used, the more accepted method of body condition scoring sheep throughout the United States is a system using scores between 1 and 5 (8). In this system the lower-scoring ewes are the least fat and the highest-scoring ewes are the fattest. A ewe in average body condition would have a score of 3. Usually, 90% of the ewes fall within the 2, 3, or 4 range, and 70% to 80% of the animals usually fall within a range of two condition scores.

A sheep producer will find that body condition scoring is relatively easy, and they will develop confidence in their ability relatively quickly. Condition scoring involves both visual and hands-on appraisal. Scoring is accomplished by using the hand to feel the fullness of muscling and fat cover over and around the vertebrae in the loin region. While the ewe is standing in a level and a relaxed position, the fingers and thumb are used to determine sharpness of the spine and transverse process (Fig. 17-3) behind the last rib and in front of the hip bone (loin area). In addition, it may be helpful to determine the extent of fat covering over the foreribs. After all factors have been evaluated, an overall condition score is assigned. As a general rule of thumb, mature ewes vary 6% to 7% in body weight for each half unit change in body condition score (9). For example, a ewe with condition score of 3 weighing 150 lb would weigh between 165 and 175 lb if she were in condition score 4. Examples of differing condition scores are shown in Figs. 17-4 and 17-5.

Body condition scoring is easy to learn and use. In general, young, developing, breeding animals (i.e., ewe lambs up to 2 years old) show less individual variation than older ewes. It may be impractical for large sheep producers to condition score all ewes; however, if a producer condition scored approximately 10% to 20% of the flock, this would be adequate to get an estimate of the condition of the entire flock. Regular condition scoring and action on the results will ensure healthier ewes and more pounds of lamb and wool marketed per year.

TABLE 17-3

Ewe body condition scores

Condition Score	Description
1	An extremely emaciated ewe, with no fat between skin and bone. Ewes in this body condition have no fat and very limited muscle energy reserves. They appear weak and unthrifty. Wool fleeces are often tender, frowsy, and lack luster.
2	Ewes in this body condition have only a slight amount of fatty tissue detectable between skin and bone. Spinous processes are relatively prominent. These ewes appear thrifty but have only minimal fat reserves.
3	Ewes in this body condition have average flesh but do not have excess fat reserves. This condition score includes ewes in average body condition.
4	This condition score includes ewes that are moderately fat. Moderate fat deposits give sheep a smooth external appearance.
5	Ewes that are extremely fat. Excess fat deposits can easily be seen in the brisket, flank, and tailhead regions. These ewes have excess fat reserves to the point that productivity may be impaired.

EWE NUTRITION

For optimum production, the sheep producer must realize that the nutritional status of the ewe may be critical at all stages of the production cycle. Lack of understanding on this point has caused much confusion among sheep producers. Optimum feeding systems in the sheep industry can vary from the intensive feeding of confined sheep, where they are entirely dependent on harvested feeds at one extreme, to the supplementation of flocks mainly dependent on range forage. An optimum feeding system consists of a planned nutritional regime that will result in an expected biological and economic response.

The nutritional status of the ewe at anytime during the year has an influence on productivity. Nutrition in the weeks just prior to and after breeding determines the number of lambs conceived. Nutrition during pregnancy determines the number of lambs born alive and lamb birth weights, which are directly related to subsequent lamb survivabil-

ity. Proper nutrition during lactation is critical for adequate milk production. After weaning, nutrition is important for replenishing body reserves, preparing the ewe for another production cycle.

Direct observation of a sheep's nutrient needs has provided a comprehensive framework for the formation of optimum feeding strategies. However, rarely do these strategies involve meeting the ewes exact nutrient requirements at each stage of her reproductive cycle. Instead, for economic, practical, and sound physiological reasons, they involve periods when the nutrient intake exceeds requirements and other periods when nutrient consumption is below the requirements. Body composition at a given point in the production cycle may influence both production response at that point and response to varying levels of nutrition. The goal is to achieve a balance in body composition over the reproductive cycle.

A ewe's nutritional needs are not static. They vary greatly with stage of production. The ewe's biological clock can be divided into several fairly well defined periods with differing nutritional needs (Table 17-4).

One of the best ways to determine how a ewe should be fed is by monitoring her changes in weight. Ideally, a ewe should lose 5% to 7% of her body weight during lactation, recover this during the postweaning period, and then gain weight during gestation (Fig. 17-6).

Nutritional requirements for a 150-lb ewe for dry matter, energy, and protein are shown in Figs. 17-7, 17-8, and 17-9, respectively. When utilizing these requirements, one must keep in mind the ultimate goals of the production unit. Any specific recommendations must also be

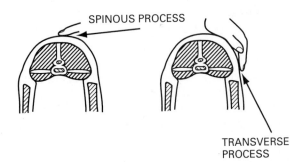

FIGURE 17-3 Feel the spine in the center of the sheep's back. Also feel the transverse process in the sheep's loin area (behind the last rib and in front of the hipbone).

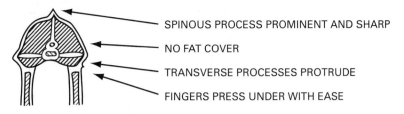

SPINOUS PROCESS PROMINENT AND SHARP

NO FAT COVER

TRANSVERSE PROCESSES PROTRUDE

FINGERS PRESS UNDER WITH EASE

Condition score 1. Ewe is thin and unthrifty. Skeletal features are prominent with no fat cover. The spinous processes are prominent and sharp. The transverse processes are also sharp and it is possible to feel between each process. Fingers can pass easily under the ends of the transverse processes.

SPINOUS PROCESS ROUNDED BUT SMOOTH

MUSCLE DEVELOPMENT FULL

NEED FINGER PRESSURE TO
FIND TRANSVERSE PROCESSES

Condition score 3. Sheep are thrifty with evidence of limited fat deposits in the forerib and over the top of the shoulder, backbone, and tailhead. Hip bone remains visible. The spinous and transverse processes are smooth and well covered and pressure is required to feel the ends. Loins muscles are full and have a moderate degree of fat covering.

SPINOUS PROCESS NOT DETECTABLE

TRANSVERSE PROCESS NOT DETECTABLE

Condition score 5. Ewe is extremely fat with excess detectable fat over the shoulder, backbone, rump, and forerib. The spinous process cannot be detected even with firm pressure. There is a depression between the layers of fat where the spinous processes would normally be felt, and the transverse processes cannot be felt. The loin muscles are full with a very thick fat cover.

FIGURE 17-4 Body condition scoring of ewes.

FIGURE 17-5 Ewes of similar frame size in condition score 2 (left) and condition score 4 (right).

adapted to local conditions, extensive versus intensive grazing programs, and the economics of recommendations versus expected production responses.

A sheep flock consists of ewes of different sizes, body conditions, and different levels of production and therefore varying nutritional needs. Although it is impossible to treat each ewe's needs separately, at times it is beneficial to divide the flock into groups of ewes having similar needs and to feed each group accordingly.

BREEDING AND FLUSHING

Flushing ewes is the practice of increasing the intake of ewes prior to and during mating. Its

TABLE 17-4

Ewes biological cycle

Period	Duration in Days
Breeding[a] (flushing)	34
Gestation	
Early and mid (first 15 weeks of gestation)	85
Late (last 6 weeks of gestation)	42
Parturition	1
Lactation[b]	
Early (first 6 to 8 weeks of lactation)	53
Late (last 4 to 6 weeks of lactation)	35
Postweaning[b]	115

[a]This period includes 14 days premating plus a 20-day breeding period.

[b]The duration of lactation and postweaning periods varies with differing management strategies. For example, intensified operations may wean lambs at 6 to 8 weeks of age while extensive range operations may wait till 6 to 8 months of age to wean lambs.

purpose is to increase the ovulation rate and, subsequently, the lambing rate. It can be accomplished by turning ewes onto a lush, high-quality pasture just prior to breeding. If such pasture is not available, the same result can be obtained by supplementing the ewes' regular diet with $\frac{1}{4}$ to $\frac{1}{2}$ lb of grain or pellets per head per day. Flushing usually begins about 2 weeks prior to joining with rams and continues for about 2 to 3 weeks into the breeding season, for a total flushing period of 4 to 6 weeks.

Much confusion exists on the relative importance of increased nutrition at breeding time on ovulation rate since the flushing effect seems to be confounded with ewe age, time of year, and most importantly ewe body condition or weight (1). Mature ewes appear to respond to flushing better than yearlings. Most data indicate that flushing is not economically beneficial in ewe lambs or yearlings that did not lamb the previous year. Also, flushing seems to be more beneficial in situations where producers are trying to breed early or late in the breeding season, rather than during the months when the ovulation rate is highest. Fat ewes will not respond to flushing.

The response to flushing can be divided into two components, the static effect of increased body weight or body condition, not specifically related to the breeding season, and the dynamic effect, which is specific to the breeding season. In most instances the flushing period is too short to drastically influence body weight or condition.

The effect of increased body weight on lambing rates has been well documented. Most studies have shown that, as a general rule, each 10-lb increase in body weight increases lambing rates by about 5% to 6%. The dynamic flushing effect, on the other hand, is distinguishable from the liveweight effect and is specific to the immediate premating and mating periods. Ovulation rate (10) appears to respond to short-term increased nutrition within a specific

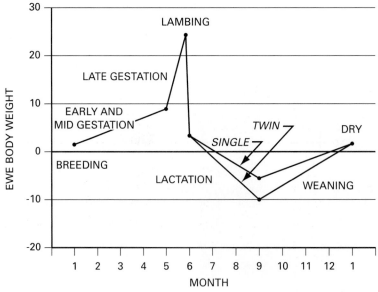

FIGURE 17-6 Weight changes expected in a 150-lb ewe throughout the different stages of her yearly production or biological cycle. [Adapted from NRC (1).]

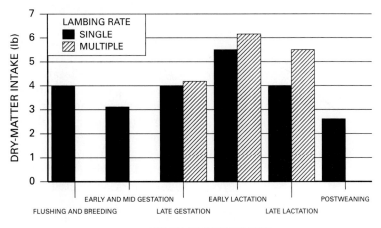

FIGURE 17-7 Estimated dry-matter intake for a 150-lb ewe at various stages of her biological cycle. [Adapted from NRC (1).]

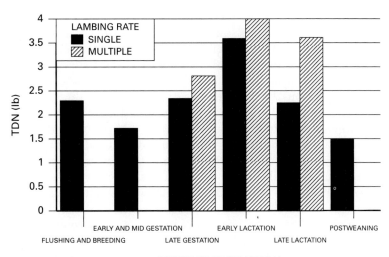

FIGURE 17-8 Energy requirements for a 150-lb ewe at various stages of her biological cycle. [Adapted from NRC (1).]

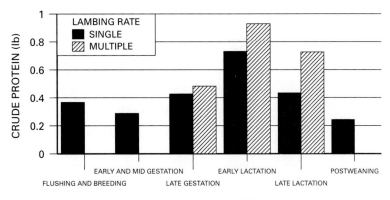

FIGURE 17-9 Protein requirements for a 150-lb ewe during various stages of her biological cycle. [Adapted from NRC (1).]

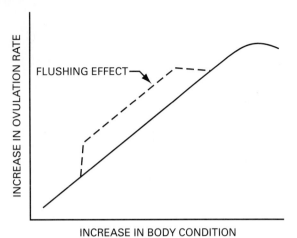

FIGURE 17-10 Effect of increased body condition on ovulation rate. [Adapted from Gunn (10).]

intermediate range of body condition (Fig. 17-10). Above and below this range, it is the body condition that is achieved that matters, and there is no additional positive or negative effect. It has been estimated (11) that for a 6-week flushing period (3 weeks prior to and 3 weeks during the breeding season) the dynamic flushing effect on lambing rate is about 6%-8% for flushed versus maintenance feeding or about 12% for flushed versus half-maintenance.

Although results vary greatly, most studies (12) suggest that flushing will improve lambing rates by 10% to 20%. In a 3-year study, Montana researchers (13) found that supplementing mature range ewes' diets with about 0.50 lb of a 20% to 30% protein pellet per head per day for about 20 days prior and 20 days after joining with the rams had a definite effect on ewe weights and lambing rates. In a similar study, Wyoming researchers (14) found that supplementation of range ewes with 0.25 and 0.75 lb of a high-energy supplement per head per day for 2 weeks prior and the first 3 weeks of breeding increased the lamb crop docked by 7.5% and 7.7%, respectively.

There seems to be different opinions on the relative contributions made by protein and energy to the flushing phenomenon. Energy intake (10) seems to be the single most important factor involved. Protein could be limiting at low levels, but provision in excess of maintenance produces no greater response (10). Research in the United States (15) suggests that there may be a benefit to increasing both energy and protein intake.

Another question frequently asked concerns the effect on lamb production of the length of the flushing period and the time of commencing the flushing treatments relative to the time of starting to breed. Researchers at the USDA Sheep Experiment Station in Dubois, Idaho (16), found that a short flushing period (17 days) immediately prior to breeding significantly increased lamb production. Extending this period an additional 17 days during breeding produced only a slight further increase in production. A further 17-day extension caused an apparent production decline from that observed by shorter flushing periods. The researchers suggested that a certain minimum time is required for a supplemental feed to produce a flushing effect, and the timing of such feeding must precede some definite and fairly early preovulatory phase of the ovarian cycle; and once flushing treatment has exerted its effect on the developing follicles, the effect persists until ovulation is completed for that cycle.

The practice (11) of intentionally bringing ewes down in condition in the postweaning period may not be necessary. Although the effect of flushing is usually greater in thin ewes than ewes in good condition (12% versus 5% or 6%), thin ewes, as a direct result of decreased body weight, usually have a 2% or 3% higher incidence of barrenness and a 4% to 5% lower lambing rate (a total of about a 7% decrease in lamb crop) than ewes in fair or good condition. In this case, the net increase in lamb crop is about 5% (12% − 7%). This increase is about the same as one would expect if the ewes were maintained in fair or good condition for the entire period and then flushed. From this discussion, one can conclude that it is not essential to deliberately keep ewes thin during the postweaning period in order to benefit from flushing.

Although it is not likely that all the benefits ascribed to flushing will be fully realized under all conditions, the general feeling persists that the practice results in more eggs being shed and therefore higher lambing rates, ewes coming in heat more promptly, and more certain and prompt conception, with lambs arriving in the early part of the lambing season.

GESTATION

Poor nutrition during pregnancy can lead to lamb deaths before, during, and soon after lambing due to numerous complex interactions. Many lamb deaths that occur shortly after birth can be attributed to nutritional factors during pregnancy, which influence placenta growth, fetal development, and ewe mammary gland development. Quite often cold weather is blamed for large lamb losses, when in fact the major

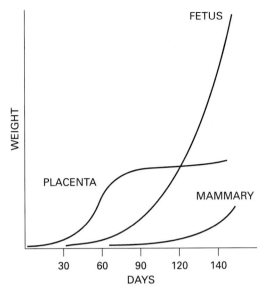

FIGURE 17-11 Placenta, fetal, and mammary growth during pregnancy.

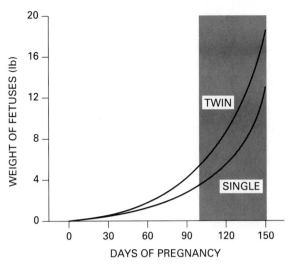

FIGURE 17-12 Growth of fetuses during pregnancy.

contributing factor was inadequate nutrition during pregnancy. Critical time periods for placental development, fetal growth, and mammary gland development are between day 30 and day 90, after day 90, and after day 120 of pregnancy, respectively (Fig. 17-11).

EARLY AND MID GESTATION   During pregnancy the ewe must be fed enough to meet her requirements for maintenance, fleece production, fetal and associated tissue development, and growth if the ewe is not fully mature. Since fetal growth is minimal during the first 15 weeks of pregnancy, the ewe's nutrient requirements during this time are only slightly higher than they are for maintenance. However, some important functions occur during this period and thus nutrition cannot be ignored. During this period the embryo becomes attached to the uterine wall (first 45 days of pregnancy). Extremes in nutrition (severe under- or overfeeding) during this period are detrimental to this process (17). Also, the majority of placental development occurs during this period (day 30 to 90 of pregnancy). Poor placental development results in lower fetal growth rates and reduced lamb survival rates. Good nutrition during late pregnancy is wasted if adequate placental development has not occurred.

LATE GESTATION   The last 6 weeks of gestation are the most critical period in ewe nutrition. During this period approximately 70% of fetal growth occurs (Fig. 17-12). Poor nutrition during late pregnancy causes lighter lambs at birth, uneven birth weights in twin and

triplet born lambs, reduced wool follicle development, and lower energy reserves in the newborn lamb. Lowered energy reserves in the newborn lamb result in increased perinatal lamb losses, especially in colder weather. Severe undernutrition leads to pregnancy toxemia and possibly ewe death.

Lamb birth weight (18) is a major factor affecting lamb mortality (Fig. 17-13). Birth weights vary from 3.5 to 20 lb. Although these differences are associated with breed, dam's age, and litter size, they are highly dependent on ewe nutrition and in particular energy intake during the last month of pregnancy. Inadequate energy intake during this period results in lowered birth weights, which in turn

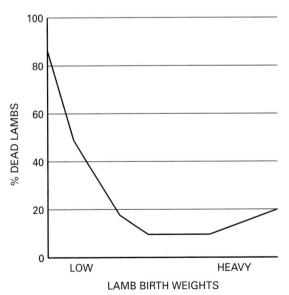

FIGURE 17-13 Birth weight and mortality from birth to weaning. [Adapted from Villette and Brelurut (18).]

is a major factor affecting lamb mortality. There may be as high as a 12% increase in lamb mortality for every 2-lb decrease in lamb birth weight. On the other hand, excessive levels of feeding may result in lambs with increased birth weights, leading to lambing difficulties.

The efficiency of converting dietary energy to energy gain in the fetus is only about 13% (9). This, coupled with the rapid increase in energy deposition in the fetus during the last weeks of pregnancy, results in large increases in the ewe's energy requirements. This comes at a time in the ewe's biological cycle when body fat reserves cannot safely be utilized to supply significant amounts of her energy needs.

Ewes in late pregnancy require 50% more feed if bearing a single lamb and about 75% more feed if bearing twin lambs than they do earlier in gestation. If the ewe is fed a high-roughage diet, she will usually not be able to consume enough to supply her requirements for energy. When on high-roughage diets, it is generally advisable to supplement the base ration with 0.5 to 1 lb of grain during the last 3 to 4 weeks of pregnancy. In situations where large numbers of multiple births are expected, it is often desirable to begin graining ewes as early as 6 weeks prior to lambing. All changes in feed should always be gradual.

During this period there is a limit to the extent to which body fat reserves can be utilized because excessive mobilization of fat results in pregnancy toxemia. Pregnancy toxemia (pregnancy disease, twin-lamb disease, or ketosis) is a result of improperly fed ewes in late pregnancy. Affected ewes are most often carrying multiple lambs.

The level of nutrition during this period has an added effect on the level of milk production of the ewe. The majority of mammary gland development occurs during the last 30 days of pregnancy and is extremely sensitive to underfeeding. Improved nutrition after birth cannot improve the milk-producing capability of the udder. In addition, ewe body condition at lambing may drastically affect her milk-producing capacity. As a general rule, producers provide enough nutrients during this period to build some fat reserves that can be utilized during early lactation.

PARTURITION

This is a very traumatic time for the ewe. Many hormone levels are abnormally high. Ewes require a little time to recover from the stress of lambing, and it normally takes several days for them to reach maximum milk production and for their lambs to consume large quantities of milk. It is generally recommended to provide them good-quality forage and plenty of fresh water and to start feeding grain or supplement about 12 to 24 hours after lambing.

LACTATION

After lambing, the feed allowance of the ewe should be increased according to her needs. A ewe usually reaches maximum milk production by 2 to 3 weeks after parturition. Milk production (19) generally declines fairly rapidly thereafter (Fig. 17-14). Assuming the ewe has the capacity to produce milk, she will produce at this level only if "challenged" by the lambs nursing her. Since single lambs normally are not able to consume all the ewe's milk, the ewe adjusts her milk production downward to the level the lamb is consuming. Ewes nursing multiple lambs produces 20% to 40% more milk than those nursing singles and thus have greater nutritional requirements. For maximum rate and efficiency of lamb gains, it is desirable to separate ewes with multiple lambs from ewes with singles and feed each according to their nutritional needs.

EARLY LACTATION In the first month after lambing the lamb's growth primarily depends on milk production. Milk is critical in the first 3 to 4 weeks of the lamb's life, and in this period the correlation between milk intake and liveweight gain is approximately 0.90 (20). Lambs receiving inadequate amounts of milk can compensate to some degree by increasing their consumption of feed. However, because of differences between the digestibility of milk and feed, dry-matter intake of feed must increase by about 3 to 5 units to compensate for each unit decrease in milk consumption (9).

A ewe suckling two lambs growing at 0.6 lb/day is as productive as a dairy cow yielding 65 lb of milk per day. To prevent loss of her body tissue, daily intakes of over 7.2 Mcal of ME energy (three times maintenance) are required for a 165-lb ewe (21). In practice, this is seldom achieved. As in high-producing dairy cows, it is difficult to feed a high-producing ewe enough feed to prevent body-weight loss during early lactation. Fortunately, early lactation is a period in which body fat can be safely used to meet some of the high-energy demands of lactation. During this period a loss in body condition score of 1.0, provided the ewe was in proper

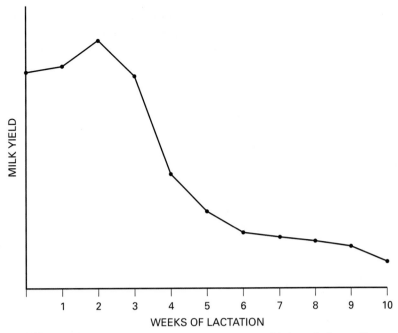

FIGURE 17-14 Ewe's milk production curve. [Adapted from Cowan et al. (19).]

body condition at lambing (3.5+), is quite acceptable. However, the ewe must have sufficient reserves of body fat to mobilize and use for milk production (22).

Fat can only be used efficiently for milk synthesis if the ewe is absorbing adequate amounts of amino acids from her diet (19). Thus,

protein intake (23) is critical during this period if maximum milk production is to be achieved in high-milk-producing ewes (Fig. 17-15). It appears that increased protein results in increased milk production, probably because the added protein increases the amount of energy available by increasing the rate of body fat loss.

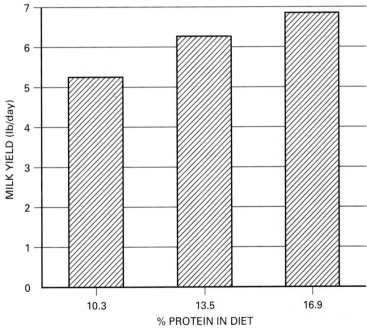

FIGURE 17-15 Effect of level of protein during lactation on milk production. [Adapted from Robinson et al. (23).]

NRC recommends feeding levels for ewes in early lactation based on a calculated ewe weight loss of 0.1/lb/day. Available forage must in many instances be supplemented to meet these requirements during the first 8 weeks of lactation. This is particularly the case in early lambing programs throughout the United States. In these cases, a feeding regime based on a planned weight loss of 0.25 lb/day may be more efficient. Approximately 12% less dietary energy is required for ewes losing 0.25 versus 0.1 lb/day (9).

LATE LACTATION (LAST 4 TO 6 WEEKS)
Although some ewes continue producing a good supply of milk throughout the nursing period, milk production in most ewes declines fairly rapidly after 2 to 3 weeks and is of minor importance after 8 to 10 weeks. Milk production during late lactation is relatively low, and nutrient needs are substantially lower than they were during early lactation. A ewe's requirements during this period can usually be met by good-quality pasture or range (Fig. 17-16).

POSTWEANING

This is a time of rest for the ewes. It is the period of time that the ewes' body condition can be adjusted so that they are in appropriate body condition at breeding. As a general rule, most ewes are overfed during this period. Mature ewes should be fed so that they gain back the weight lost during lactation. With a gestation period of 5 months and 2 to 4 months of lactation, the ewe has adequate time to recover (3 to 5 months) from weight loss that occurred during lactation before the next breeding season. It is desirable to utilize the poor-quality pasture or feed during this period, saving the better-quality forage for periods of the production cycle that are nutritionally more critical.

FIGURE 17-16 Ewe and lamb grazing lush range forage.

LAMB NUTRITION
NEWBORN LAMB

The lamb is born with a nonfunctional rumen that requires dietary nutrients to be provided mainly from milk or milk replacer. Proper growth and development for the first weeks of life are largely dependent on milk consumption. If lambs are provided access to dry feed immediately after birth, some degree of rumen functionality becomes apparent by 2 weeks of age (24). Continued access to dry feed produces an anaerobic microbial system in the rumen capable of digesting and synthesizing protein to the extent that milk is no longer essential for efficient production by 45 to 60 days of age. Table 17-5 shows how the rumen–reticulum to omasum–abomasum tissue weight changes with age (25).

Colostrum milk is a must for every newborn lamb. Colostrum provides immunoglobulin that cannot be supplied by any artificial or synthetic product. The antibodies obtained from the colostrum during the lamb's first few hours of life are its primary protection against infectious organisms until the lamb's own immune system is fully developed and operational (at about 3 weeks of age). In addition, colostrum has a laxative effect that appears to be essential to proper digestive functions. After 12 to 18 h the newborn lamb's ability to absorb these intact antibodies from its intestinal tract begins to diminish.

Cow colostrum can be used if no ewe colostrum is available. To be effective, the cow colostrum should be obtained from the cow as soon after caving (first milking), because the antibody content of colostrum drops off rapidly after 24 h postparturition. Excess ewe or cow colostrum can be frozen in small containers for future use. When thawing the colostrum, raise the temperature slowly (do not use microwave) and be careful not to heat above body tempera-

TABLE 17-5

Changes in tissue weight ratios of the rumen–reticulum (RR) to the omasum–abomasum (OA) at different ages of lambs

Age	RR:OA
Birth	1:2
30 Days of age	1.4:1
60 Days of age	2.6:1
Adult	2.7:1

Source: Church (25).

ture. In most cases, 4 to 6 oz of colostrum during the lambs first 18 h is adequate.

ARTIFICIAL REARING OF LAMBS

Raising lambs on milk replacers is usually a relatively time consuming and expensive process. Every attempt should be made to rear lambs on their mother or graft them onto a foster dam. However, in some situations there is no suitable alternative and the lamb must be raised on milk replacer. Feeding these lambs with a hand-held bottle is not a viable alternative. Lambs require small amounts of milk on a regular basis, and the sheep producer of today cannot afford the time required for a "bottle" operation to be a success. The technology and knowledge are available to successfully raise these orphan lambs (26). The key to doing this economically is to get them on dry feed quickly, allowing the milk feeding period to be as short as possible (Fig. 17-17).

SELECTING A MILK REPLACER Usually, the most economical milk replacer will be in the form of a powder to be mixed with water as needed. Selecting a suitable milk replacer is an essential step in assuring the success of an artificial rearing program. The nutrient content of a lamb milk replacer should be similar to that shown in Table 17-6. Milk replacers developed for calves usually contain less nutrient and energy concentrations than are satisfactory for lambs and should not usually be used. Good lamb milk replacers are expensive, but higher lamb survival and growth rates will compensate for the extra expense.

FIGURE 17-17 Lambs weaned at one day of age and reared on cold milk replacer continuously circulated to teat bar at the far end of the pen. (Courtesy of the director, ARI, Ottawa.)

TABLE 17-6

Composition of lamb milk replacers

Ingredient	% Range (Dry-matter Basis)
Crude fat	30–32
Crude protein	22–24
Crude fiber	0–1
Lactose	22–25
Ash	5–10

Source: Fredericksen (unpublished).

FEEDING THE MILK REPLACER To reduce labor and maximize milk intake, the lambs should be on a self-feeding system. The size and type of feeding system needed primarily depends on the number of lambs to be fed. The milk offered in the self-feeding device should be cold (33° to 40°F). There are three primary reasons for feeding lambs cold milk replacer.

1. Lambs drink less at each nursing, but drink more often. This imitates how lambs nurse when on the ewe and prevents excessive intake at any given nursing period.
2. Keeping milk cold reduces the problem of spoilage and reduces the frequency of equipment cleaning.
3. Lower milk temperatures lessen the separation of the milk replacer ingredients.

Lambs on a self-feeding system consume about 0.5 to 1 lb of milk replacer daily. Weight gains of about 0.5 lb plus per day are typical for healthy lambs on a high-quality replacer with free-choice solid feeds available at all times.

CREEP FEEDING

Creep feeding is the practice of providing nursing lambs the opportunity to eat supplemental feed. It is the first step in getting lambs to market as soon as possible. A creep feeding program is essential if lambs are going to be early weaned. Creep feeding ensures that the lamb is a functional ruminant that can perform well on complex rations at a relatively young age.

The advantages of a creep feeding program include the following:

Increased weight gains, especially in twin and triplet born lambs. Lambs usually will gain up to 0.25 lb/day more when they are creep fed.

Lambs use supplement feed more efficiently at this time than after weaning.

Lambs can be marketed at a younger age. Under normal circumstances lambs that are creep fed reach market weight 1 to 2 months earlier. Lambs can often be marketed at a time when prices are higher. Also, a lamb requires a certain amount of feed each day for body maintenance. Reducing the number of days until market can reduce the total maintenance feed requirements of the lamb. The risk of death, sickness, or injury to the lambs is lowered by reducing the number of days to market. Winter-born lambs will do better if they can be marketed before hot summer months.

Moving lambs to drylot at an early age enables producers to carry more ewes on available pasture.

Lambs will begin to nibble at a creep at about 1 week of age. However, up until about 3 to 4 weeks of age, the ewe's milk provides the majority of nutrients. Thus, daily consumption of creep feed is often less than average daily gain, and feed efficiency per unit of creep feed intake is extremely high. Although lambs will not consume significant amounts of feed until they are 3 to 4 weeks of age, the small amounts consumed are critical for establishing rumen function and the habit of eating. Lambs generally consume an average of 1.5 lb of creep ration per day (5) from 10 to 120 days of age (0.15 lb/day at 3 weeks of age, 1 to 1.5 lb/day at 40 to 50 days of age, and 3 lb/day at 120 days of age).

If creep feeding is to be successful, it is critical that lambs begin consuming creep as soon as possible. Creep rations do not have to be complex. Lambs often perform as well on simple ranch-formulated rations as they will on complex rations. However, palatability is a major consideration when formulating creep rations. Table 17-7 shows which of the common feeds lambs prefer to eat. Lambs seem to prefer soybean meal, sweet feeds, corn, and bran during the early stages of life. When formulating creep rations, every attempt should be made to include these feeds in the ration. Once lambs are eating the creep, ingredient composition can gradually be adjusted to more reflect the relative cost of the ingredients.

Creeps should be set up when the lambs are 7 to 10 days old. Locate the creep where the lambs will use it. It should be set up in an area where sheep generally spend the majority of their time. The creep area should be readily accessible to lambs and inaccessible to ewes. It should be placed in an area where ewe and lamb can easily see each other. At no time should the lambs feel that they are being penned away from their mothers. The creep area should be made as enticing to the lambs as possible. The normal tendency of the mother and lamb is to want to stay rather close to each other and, considering our goal of wanting to separate them and to encourage the lambs to spend time in the creep pen, we must realize that there is some conflict between our goal and the wishes of the ewe and her lambs. A sunny, warm, dry location may stimulate usage by lambs. The creep area should be well bedded and feeders should be kept clean. Water should be provided in the creep area or as close to it as possible. A light placed over the feeder has been shown to stimulate feed consumption. If the creep area is set up correctly, you should often find lambs laying or resting in it.

A trick to getting lambs on creep is to crowd ewes and lambs into a relatively small area around the creep pen at night. The lambs do not like to be squeezed by the ewes and take refuge in the creep pen. Some producers confine a ewe or two in the creep pen for short periods of time to help entice lambs to enter the pen. Be

TABLE 17-7

Voluntary intake of common feeds by lambs at 2-week intervals from birth to 10 weeks of age

Feed	Lamb Consumption (lb) by Two-week Periods				
	0–2	2–4	4–6	6–8	8–10
Oats	0.12	0.29	1.46	1.65	1.25
Corn[a]	0.11	0.30	2.77	7.83	8.02
Alfalfa hay	0.23	0.41	1.43	1.12	1.05
Alfalfa pellets	0	0.14	1.35	4.06	1.98
Bran	0.27	0.55	2.01	3.17	1.08
Soybean meal	0.81	1.70	6.94	11.01	10.63
Sweet feed	0.27	0.75	3.54	4.08	2.19

Source: Jordan Univ. of Minnesota (unpublished).

[a]Corn offered to lambs was whole; if corn is rolled or cracked (not ground), lambs usually consume it in much greater quantities.

sure only limited amounts of feed are available to the ewe.

The economics and merits of creep feeding lambs in Western range sheep flocks have long been debated by sheep producers. Most likely, inconclusive results have been caused by different conditions in feed supply and lamb and feed prices. Most range producers do not traditionally creep their lambs because their goal is to maximize the use of low-cost summer forage. Creep feeding is most profitable under range conditions when lamb prices are high relative to creep feed prices, when pastures begin to decline in quality and quantity, such as in a drought, in twin or triplet born lambs, or, at any other time when the lambs growth potential is not being met with milk and natural forage.

When summer forage quality and availability are excellent and lambs are going to be marketed at a relatively old age (5 to 6 months of age), most research suggests that there is minimal benefit to creep feeding range lambs. Lambs under this management scheme are not gaining near their potential, and by 5 to 6 months of age noncreep fed lambs seem to catch up.

However, under normal management (ewes and lambs turned to pasture together) creep feeding has proved beneficial, especially in twin and triplet born lambs and in drought conditions when forage quality and availability is low. In work conducted at Montana State University (27), creep-fed lambs grazing nonirrigated improved pasture under drought conditions weighed 12.76 lb (0.13 lb per day) more than noncreep-fed lambs at weaning at about 100 days of age. Similar trials in Minnesota (28) and Virginia (29) showed a 0.16- and 0.20-lb/day increase in average daily gain for creep-fed lambs, respectively. It is interesting that in the Montana study over 50% of the weight-gain advantage for creep-fed lambs occurred during the month of August when forage quality and quantity were dramatically reduced and ewes were in late lactation. Using the data from the Montana experiment (12.76 lb of gain advantage for creep- versus noncreep-fed lambs on 70.18 lb of creep) the dollar return over feed costs based on different feed cost and lamb price relationships can be calculated (Table 17-8). Given this level of performance, when lambs are selling for $70 cwt and creep feed is costing $140/ton, a return of $3.92 per lamb over feed costs would be generated. If lamb prices dropped to $60 cwt, return per lamb over feed costs would be reduced to $2.75. This table emphasizes the point that creep feeding is most profitable when lamb prices are high relative to creep feed prices.

Another management strategy for range sheep producers is to creep feed early-wean lambs and then move them directly to a feedlot. In this scenario, lambs are either never or only on pasture for a short period of time. This allows producers to market lambs earlier, taking advantage of higher prices. Traditionally, western range sheep operations lamb in late spring (March, April, and May) and market their lambs in late fall (September and October). Normally, lamb prices drop in the months of September and October (the months most range lambs are marketed). Creep feeding combined with a short feedlot phase allows range producers to market their lambs at similar weights but 1 to 2 months earlier.

The creep feeding–feedlot strategy may be an option for range sheep producers experiencing heavy predator losses. By not sending lambs to pasture, predator problems can usually be minimized. Lambs can be placed in an area where predator control programs can be concentrated. Also, by early weaning of lambs and placing them in a feedlot, producers only need summer pasture for the ewes. This allows producers to carry more ewes on available pasture. It is also an excellent tool to use to decrease grazing pressure in years when forage is limited.

TABLE 17-8

Dollar return over creep feed cost based on different feed and lamb price relationships[a]

	100	120	140	160	180	200
		(Return above Feed Cost per Head)				
50	2.87	2.17	1.47	0.77	0.06	−0.64
60	4.15	3.45	2.75	2.05	1.34	0.64
70	5.32	4.62	3.92	3.22	2.51	1.81
80	6.70	6.00	5.30	4.60	3.89	3.19
90	7.97	7.27	6.57	5.87	5.16	4.46
100	9.25	8.55	7.85	7.15	6.44	5.74

[a]Based on 12.76 lb of additional gain for creep-fed lambs on 70.18 lb of feed.

GROWING AND FINISHING OF LAMBS

Lamb feeding is a general term that includes numerous management and feeding strategies, all with the common objective of getting weaned lambs to sufficient weight and finish to produce acceptable slaughter lambs for the U.S. market. Under excellent forage conditions, lambs may reach slaughter weight and condition while nursing ewes. This is typical of high mountain ranges in the intermountain west. However, the majority of lambs produced in the United States are not at slaughter weight or slaughter grade when weaned. A few of these lambs are finished by producers on their own land. The majority are sold to commercial lamb feedlots or to farmer feeders where they are finished on high-concentrate rations. Many of these lambs are typically placed on high-quality pasture or crop aftermath such as alfalfa fields, brassica fodder crops, or beet tops prior to a short feedlot phase.

Most feeder lambs are available in the fall of the year when they are marketed from western and southwestern ranges. These lambs are usually about 5 to 6 months of age and weigh between 60 and 90 lb. Many producers are beginning to wean and market these lambs at an earlier age. This is especially true for fall- or winter-born lambs. Some producers are finding it profitable to wean their lambs at an early age, feed them theirselves, and market them as slaughter lambs. When gain prices are relatively low, this is an excellent way to market home-grown grains. Table 17-9 lists some estimated feed conversion ratios for selected weight ranges of early and late weaned lambs fed high-concentrate diets (5).

At slaughter the ideal market lamb should have between 0.1 and 0.2 in. of backfat.

TABLE 17-9

Approximate expected feed conversion ratios for selected weights of early and late weaned lambs fed high-concentrate diets

Weight (lb)	Weaning Date	
	Early (6 to 10 weeks of age)	Late (5 to 7 months of age)
40–60	3.50	
60–80	4.75	5.00
80–100	5.75	6.50
100–120	8.00	8.50
Over 120	9.00	10.00

Source: SID (5).

Because of differences in frame size among lambs, ideal slaughter weights vary between 95 and 145 lb. The feeder should plan to market the lambs when they have adequate finish regardless of their weight. Small-frame lambs should be marketed between 95 and 105 lb, medium-frame lambs between 105 and 120 lb, and large-frame lambs between 120 and 145 lb. Feeding beyond these weights for each frame size is too expensive because of the decline in feed efficiency and increase in carcass fat.

Many different feedstuffs can be used to provide all the nutrients required to economically finish lambs. It can be accomplished with rations ranging from all roughage to almost all concentrate. High-concentrate diets usually produce the fastest gain but may not be the most economical. Programs that use pastures for grazing or utilize various types of crop aftermath are often more economical, even though they produce slower rate of gains. Lightweight lambs (50–70 lb) can use more roughage, whereas heavier lambs (70–80 lb) require more concentrates and a shorter feeding period. Lightweight lambs are more desirable for pasture finishing, while heavier lambs perform best in the drylot. The most economic program depends on the relative costs of the feedstuffs, the type and weight of the lamb being finished, the management imposed, and the end product that the feeder wants to market.

HANDLING NEW LAMBS

The success of a feeder lamb finishing system depends on the first 2 to 3 weeks after arrival. Lambs have usually been transported many miles and will probably be scared, tired, hungry, and thirsty. Death loss can be extremely high during this period. Upon arrival, unload lambs into a dry, clean area and allow them to rest before processing. Provide shelter from wind, rain, and snow. Offer grass hay or mixed grass–legume hay. Clean, fresh, cool water must be available at all times to encourage feed consumption and maximize ration digestibility and nutrient metabolism. Getting lambs to drink water is the key to getting lambs on feed quickly. This is often a problem with western range lambs. Because western feeder lambs often have never watered from troughs, feeders sometimes try to attract lambs to water troughs by using running water systems where lambs can hear the water running. Feed consumption is closely related to water consumption. If lambs do not drink, they do not eat, and if they do not eat, they do not gain.

The method used in starting feeder lambs on feed varies considerably and is influenced by available labor, type of feed delivery system (feed bunks where lambs are fed daily to self-feeders), type of ration to be fed, and desired level of lamb performance. Starting diets usually have high concentrations of forage, with grain being added gradually to avoid digestive upsets (acidosis) and mortality from enterotoxemia. Adapt the lambs to their new ration slowly. If the lambs are to be finished in drylot, sort by weight and feed separately. Isolate sick and weak lambs daily. With good management, death losses should approximate 2% for the total finishing period, but can far exceed this if not properly handled.

Most feedlot operators try to shear all lambs sometime during the feeding period. Shearing prevents lambs from picking up mud and tags. Feed consumption and gains will also be increased, especially in warm weather. Lambs with number 1 pelts (0.5 to 1 in. of wool) usually sell at a premium to full-fleeced lambs. Usually, 40 to 60 days are required to produce a number 1 pelt after shearing. Based on expected gains, the feeder can predict the approximate time that the lambs will be ready for slaughter and schedule shearing so that they will have a number 1 pelt.

GROWING OR FINISHING LAMBS ON PASTURE

Feeders can economically fatten lambs or produce low-cost gains before putting lambs into the feedlot by utilizing pasture or field crop residues (Fig. 17-18). Lightweight lambs make the best use of pasture. They can be used to clean weed fields, fence rows, soybean stubble, and corn fields. Lush, cool-season grasses or alfalfa provide excellent fall pasture (and winter pasture in certain geographical areas). Rape and other brassicas can be used in the early fall. In certain areas of the northwest, sugar-beet tops or turnips provide excellent fall grazing for growing lambs.

Depending on the targeted marketing date, supplemental concentrates can be fed ad libitum throughout the finishing period or during the last 30 to 40 days. Lambs grazing cool-season grasses or alfalfa need only an energy source, such as whole cereal grains. Those cleaning weed fields, fence rows, soybean stubble, or corn fields usually benefit from a grain–protein supplement.

In many areas of the United States the most profitable pasture is winter wheat, oat, or

FIGURE 17-18 Finishing lambs with roughage rations. (Courtesy of M. E. Benson, Michigan State University, East Lansing, MI.)

rye. In some areas, however, the risk from nitrate poisoning precludes their use for grazing. Although lambs can be finished on small-grain pasture alone, gains can be increased significantly by supplementing with whole grain. The grain can be limit-fed throughout the finishing phase or self-fed during the last 30 to 40 days to produce optimally finished lambs. The choice of grain depends on relative dollar values. Supplemental protein is not required as long as lambs graze lush green pasture.

FINISHING LAMBS IN DRYLOT

Feedlots vary from large outside lots with self-feeders (Fig. 17-19) to complete confinement pens on slotted floors. Equipment used for feeding lambs varies from automated feeders, bunk fence-line feeders, where diets are fed twice daily, to self-feeders, to hay and grain feeders for hand feeding.

There is no best ration for feedlot lambs. There are many practical feeding programs. Rations are built around local feeds and price relationships. Depending on feed costs and desired

FIGURE 17-19 Finishing lambs with high-concentrate rations. (Courtesy of Harper Livestock Co., Eaton Co.)

gains, rations can consist of any combination of hay and grain, ranging from all hay to 20% hay and 80% grain. Although alfalfa hay is the most commonly used hay source, other forages, such as other types of hay, silage, or haylage, can be used. Corn, barley, milo, or wheat can be used for the grain portion of the diet. Depending on the types of forage and concentrates used, a protein source is sometimes added. Mineral and vitamin premixes plus some type of antibiotic are often added to the feed mix. Table 17-10 shows the estimated intake, average daily gains, and feed efficiencies of a 100-lb lamb fed varying concentrate-to-roughage rations. Lamb performance, both rate of gain and feed efficiency, tends to improve as the ratio of concentrate to roughage increases. Lambs fed free-choice hay plus grain generally consume a diet containing a 60% to 70% grain and 30% to 40% forage. High-concentrate diets such as those containing 70% to 80% grain are fairly typical for the final phases of the finishing period in most feedlots.

Hand-fed lambs are usually fed hay and supplemented with a grain–concentrate mix. Table 17-11 is a guide for hand feeding a diet of corn or barley, protein supplement, and hay (30). It is based on feeding alfalfa hay and a 34% to 36% protein supplement (at a constant 0.5 lb/lamb/day) and grain (at an increasing amount as the lamb gets larger). Protein levels are decreased and energy levels are increased in this diet regime as the lamb gets older and heavier. This program meets the basic requirements of the growing–finishing lamb.

Most feedlot rations, provided they do not contain silage or haylage, can be either bunk fed daily or self-fed. Both types of feeders require protection from rain or snow to maintain consistent intakes. If lambs are bunk fed, allow 12 in. of trough space per lamb. If self-fed, each linear foot can serve 6 to 10 lambs. Lambs can be fed in open fields. A windbreak or open shelter for winter feeding and a shade for summer feeding aid in producing maximum growth and feed utilization. The ideal drylot has some slope to the land to facilitate drainage. No more than 500 lambs should be fed per lot. Each lamb fed in this environment requires 15 to 20 ft^2 of lot space.

POTENTIAL NUTRITIONAL PROBLEMS

The incidence of nutritional problems with finishing lambs depends on the severity of stress, the innate ability of the lambs to withstand this

TABLE 17-10

Estimated production of a 100-pound lamb fed varying concentrate-to-roughage rations[a]

Grain[b] (%)	Hay[c] (%)	Intake (lb feed as fed)	Average Daily Gain (lb/day)	Feed Efficiency (lb feed/lb gain)
0	100	7.40	0.38–0.43	17.17–19.24
20	80	6.50	0.51–0.57	11.50–12.86
40	60	5.75	0.60–0.67	8.64–9.66
60	40	5.20	0.68–0.75	6.89–7.70
80	20	4.75	0.74–0.83	5.73–6.39

[a]Calculations are based on net energy equations reported in NRC, 1985 (1).
[b]Grain in example is barley.
[c]Hay in example is alfalfa.

TABLE 17-11

Feeding guide for hand feeding growing–finishing lambs, simplified ration

Lamb Weight (lb)	Daily Intake (lb/day)	Alfalfa Hay (lb/day)	Protein Supplement (lb/day)	Grain (lb/day)	Calculated Nutrient Composition			
					CP	TDN	Ca	P
50	2.0	0.4	0.5	1.1	18.4	73.4	0.85	0.46
70	2.5	0.5	0.5	1.5	16.7	75.4	0.73	0.43
80	3.0	0.6	0.5	1.9	15.6	76.1	0.65	0.40
90	3.5	0.7	0.5	2.3	14.8	76.6	0.60	0.39
100	4.0	0.8	0.5	2.7	14.2	77.0	0.55	0.37
105	4.5	0.9	0.5	3.1	13.8	77.3	0.52	0.36

Source: Thomas, 1990 (30).

stress, and the level of management imposed by the feeder.

Enterotoxemia type D is the most common nutritionally related problem encountered. Stress and sudden changes in rations precipitate this syndrome. Lambs to be early weaned should be vaccinated prior to weaning. Older feeder lambs that are transported to the finishing area should be vaccinated during the first 2 weeks after arrival, whether they are destined to be finished on pasture or in drylot. Mixing a buffer in the ration when putting lambs on feed when increasing the concentrate portion of the ration shortens the adjustment period by reducing the incidence of acidosis. Offering broad-spectrum antibiotics anytime concentrates are fed aids in the maintenance of healthy lambs and thereby allows them to better withstand any stress imposed during the finishing period.

Urinary calculi can be a common occurrence in rams or wethers in drylot. Maintenance of proper Ca to P ratios aids in prevention (31). The closer the ratio is to 1:1 the greater the probability of encountering urinary calculi. Most feeders formulate rations to contain 3 or 4 parts Ca to each part of P. Provision of a continual supply of fresh, clean, cool water with adequate waterer space aids in prevention. The addition of salt or trace mineral salt to the diet encourages maximum water consumption and thereby aids in the prevention of urinary calculi. Ammonium chloride or ammonium sulfate added to the complete diet at 0.5% helps to reduce the incidence of urinary calculi (32). It should be included even if the Ca:P ratio is optimum, salt intake is adequate, and consistent and high-quality water is available.

Feeding pelleted high-roughage rations may increase the incidence of rectal prolapse. Excessive ration dustiness may increase coughing, which can lead to rectal prolapse. Regardless of the ration fed, prolapse can still occur, since the tendency to prolapse may be genetically controlled. Lambs with short-docked tails are more prone to rectal prolapse than those with long docks.

FEED PROCESSING

With the exception of pelleting rations, processing feeds for feeder lambs does not markedly improve digestibility or lamb rate of gain. Pelleting bulky rations results in increased feed consumption and consequently increased rate of gain. If large amounts of hay are fed, pelleting increases consumption and performance (33). If hay is pelleted, it can be mixed efficiently with the whole-grain portion of the ration. Pelleting the entire ration for finishing lambs is usually not a viable option because the increased benefits are usually not large enough to offset the pelleting costs. Pelleting rations that contain over 50% concentrate usually do not result in a marked improvement in lamb performance. Whole barley, oats, corn, and wheat (34, 35, 36) fed to feedlot lambs usually result in decreased digestive disorders (particularly acidosis). Feed efficiencies from whole corn, barley, and oats are often higher than those obtained from the same ground grains. On the other hand, cracking, rolling, or flaking sorghum grains tends to increase efficiency of utilization.

The most critical factor to consider in determining whether to process grains for finishing lambs is the physical form of the complete ration. If protein, mineral, and vitamin supplementation is through pellets, they should be mixed with whole grains to prevent sifting in feeders and preferential consumption. Conversely, if the supplements are provided in meal form, the grains should be ground or cracked before mixing to encourage desired consumption of all nutrients.

Hay that is mixed with ration concentrates should be chopped or coarsely ground to facilitate mixing, encourage uniform consumption of ration ingredients, and reduce wastage when compared with feeding as long hay. Fine grinding of hay increases ration dustiness, reduces intake, and leads to poorer lamb performance.

NUTRITIONAL MANAGEMENT OF REPLACEMENT EWES

Sheep production economics are dominated by the overhead costs of maintaining the ewe. Postweaning nutritional needs of replacement ewe lambs vary with the lambs age at first breeding. Sheep producers have two management options with regard to breeding replacement ewes: they can breed ewe lambs to lamb at 1 year of age, or they can follow the more traditional management practice of breeding ewes as yearlings to lamb first as 2-year-olds.

Breeding ewe lambs is used to increase lifetime productivity of ewes, but it requires a higher level of management and feeding than if ewes are handled more traditionally. Breeding a ewe as a lamb can increase her lifetime lamb production by as much as 15% to 20%. However, breeding ewe lambs is not always economically advantageous. If ewes are to be successfully bred to lamb at 12 to 14 months of age, nutrition is critical. Sheep operations that have a high dollar input per ewe (intensive) are more likely to benefit from breeding ewe lambs than range sheep operations (extensive), where dollar input per ewe is normally lower.

BREEDING EWE LAMBS

Ewe lambs can be mated successfully without detrimental effects on subsequent reproductive performance provided they achieve a threshold body weight within the breeding season. However, ewe lambs that are to be bred to lamb at 1 year of age require special treatment if success is to be achieved. Many management factors affect successful breeding of ewe lambs. Because numerous factors influence conception rates among ewe lambs, it is possible for some sheep producers to get as high as 95% to 100% of their ewe lambs bred, while others only get 10% to 40% bred.

Age at puberty is influenced by both breed and nutrition as it affects the growth rate of the lamb. Good nutritional management is necessary for lambs to mature and develop sexually. It is important to develop a realistic and sound feeding program to ensure success and high fertility. There is no single correct management program for breeding ewe lambs. Management programs vary depending on the goals and objectives of the manager and farm or ranch resources (labor and feed availability).

In general, ewe lambs must weigh approximately 65% of their mature body weight at the start of the breeding season in order to ensure a high percentage of them breeding. However, for our more traditional breeds, such as Rambouillet, Targhee, Columbia, and Suffolk, a target weight of 70% of their mature body weight produces more satisfactory results. In contrast, breed and/or breed combinations that contain one-quarter or more Finn breeding (Polypay) can probably get by with a target weight of 60% to 65% of their mature body weight. For example, if a sheep producer raises Columbia sheep and ewes have a mature body weight of 165 lb, ewe lambs should attain an average weight of 115 lb at the start of the breeding season (165 lb × 70%). With good management, this should produce conception rates of 75% to 90%. If ewe lambs are one-quarter Finn and mature ewes weigh 145 lb, they would need to weigh between 87 and 94 lb at the start of breeding.

Range sheep producers often feel that it is not economically feasible for them to feed ewe lambs well enough to reach the desired target weight at the start of the breeding season and may accept a lower target weight. Producers making this management decision must be prepared to accept lower conception rates. Many times in range operations the biggest lambs are singles, and therefore selecting those ewe lambs that breed as lambs may encourage selecting for singles rather than multiple-birth ewe lambs. This would have a detrimental impact on overall flock prolificacy in the future.

Adequate nutrition is essential for lambs to reach the size and sexual development necessary for early breeding. Following weaning, the level of nutrition must remain high to assure adequate growth so that the lambs reach puberty at an early age. The type of feeding program required depends on the average weaning weight of the ewe lambs to be exposed and the anticipated breeding date. Most feeding programs should produce an average daily gain of 0.4 to 0.5 lb. This is not possible on good-quality forage alone; some grain must be fed. In general, if ewe lambs are fed only good-quality forage such as alfalfa hay, they will gain approximately 0.25 to 0.33 lb/day.

It is important to maintain ewe lambs separately from the mature ewe flock if possible. A 132-lb ewe lamb during the last 6 weeks of pregnancy requires 2.4 lb of TDN, while a 154-lb mature ewe only requires 2.3 lb of TDN. Therefore, pregnant ewe lambs require significantly more energy per unit of body weight than mature ewes. If ewe lambs and mature ewes are fed together, mature ewes are often overfed, while the ewe lambs are underfed. When ewe lambs and mature ewes are fed together, ewe lambs normally do not get their fair share at the feed bunk.

Immediately following breeding, ewe lambs should be fed the same ration that they were fed during breeding. Approximately 2 to 3 weeks following breeding, grain may be reduced if the ewes are becoming overconditioned (body condition score of 4 on a 1 to 5 scale). Excessive body condition during the developmental stages can result in reduced lifetime milk production. In most situations, grain may be reduced to 0.5 to 0.75 lb daily and 3 to 4 lb of a good-quality hay fed until lambing. Following lambing, feed the young ewes the same rations as mature ewes provided they were in good body condition at lambing. Give them additional feed if they are in poor condition.

Certain breeds mature earlier and conceive more readily than others (37). The Rambouillet is slow maturing and requires extremely good nutrition to breed successfully as a lamb. The Targhee and Columbia can be managed for successful early breeding with good nutrition. Breeds such as the Polypay or crosses with as little as 25% Finn breeding have very high conception rates. Breed selection is influenced by many factors other than early sexual maturity, and all breeds with good nutritional management can have acceptable conception rates.

BREEDING YEARLING EWES

Under range conditions, producers generally will not attempt to breed ewe lambs. Replacement ewes are generally bred first at 18 to 19 months of age to produce their first lambs when they are 2 years of age. With this management schedule, nutrition is not nearly as critical. Ewes have the opportunity to go through a green feed season before breeding.

Although nutritional needs of these ewes are not nearly as critical as if they were being bred to lamb at 1 year of age, they require good care and management. It is not unusual in range flocks for ewe lambs to weigh about the same at shearing time just before they are a year of age as they previously did at weaning time. Such ewes have a difficult time reaching mature size and also likely will not be at their peak ovulation rate by breeding time. Ewes that do not reach sexual maturity by breeding time, no matter at what age first breeding occurs, have reduced productivity during their first lambing as well as during their lifetime.

Generally, the nutritional needs of ewe lambs can fairly easily be met except when these ewe lambs are grazing on mature or weathered grasses during the winter. Some ewe lambs are moved to farm areas and grazed on crop aftermath or pastures or even fed forage diets in drylot. These ewe lambs usually grow out adequately and breed without problems by 18 to 19 months of age. Supplemental feeding is often necessary when ewe lambs are wintered on dryland pastures.

RANGE EWE NUTRITION

Rangelands provide the primary feed for many southwestern and western sheep enterprises. Rangelands in the intermountain areas of the west are usually only available to grazing during summer months and thus well suited to herded sheep operations. Conversely, other western and southwestern rangelands are usually grazed year round.

The goal of a range sheep enterprise is to maximize the use of range forage and minimize the use of harvested feed. In a range situation, a ewe's nutrient intake during the grazing season is extremely variable, and optimum production is often below genetic potential. Instead, for economical and practical reasons, a sheep's diet involves periods in which nutrient intakes are below requirements, followed by periods of relatively high nutrient intakes. Gain and loss of body fat is a natural and desired process in animals that are well adapted to range situations.

Supplemental feeding of grazing sheep is desirable only when performance from grazing alone is not adequate. The type and amount of supplement needed are extremely variable among ranges and years. Many of these decisions depend on weather and moisture conditions. Additional supplementation may be necessary at times of high nutrient demand (e.g., during lactation) or when winter weather is extremely harsh. During drought conditions, sheep may require supplementation year round.

Energy, protein, phosphorus, and vitamin A are the nutrients that may be limiting for grazing sheep. Often protein feeds such as cottonseed meal or soybean meal are more desirable supplements than energy-based supplements such as corn, wheat, or barley. Energy supplements fed at over 0.3% of body weight tend to decrease forage intake and digestibility and therefore may not increase the overall energy intake. Energy supplements are cost effective when supplemental objectives can be satisfied by feeding relatively small amounts of concentrates (less than 0.33 lb/head/day).

NUTRITION AND WOOL PRODUCTION

Although the wool-producing capacity of a sheep is determined by its genetic makeup, it can be influenced greatly by environmental factors. In one study in Australia (38), wool growth rates of sheep of similar genetic origin ranged from 1.6 to 20.2 g/day. These variations in wool production were largely a reflection in the nutritional status of the sheep.

An increase in energy intake, except at very low levels of protein content in the diet, usually has a direct positive effect on wool growth (39). Most research trials (38) suggest that an increase in digestible dry-matter intake will increase wool production. Allden (40) reported daily wool growth rates in South Australian

Merino wethers receiving a wide range of feed intakes from 3.1 to 16.8 g/day, the latter being close to the maximum genetic potential of 18.6 g/day (41). He found that these wethers produced 2.04 g of wool per every 100 g of digestible dry matter over the whole range of feed intakes. Other researchers (42) suggest that the efficiency of conversion of feed to wool has been slightly greater on low intakes than on high intakes. Williams and Winston (43) noted that when sheep lost 5% of body weight 1.07 g of wool was produced from 100 g of feed, and when body weight increased by 5%, only 0.78 g of wool was produced from 100 g of feed.

Rations containing protein levels of 100%, 80%, and 60% of NRC requirements were used to study the effect of protein level on wool production. The critical protein level appears to be near 80% of the recommended requirements or a diet containing 7% to 8% crude protein for a ewe in maintenance (44). Although feed intake can be partitioned into the supply of protein and energy-yielding nutrients, the separate effects of these components on wool growth are difficult to determine. While it is generally accepted that wool growth depends on energy rather than protein intake, this may be due to the normal degradation of protein in the rumen, since experiments in which the rumen is bypassed or the diet contains high concentrations of nonrumen degradable protein report significant responses in wool growth to protein levels (39). It is probable that the apparent response in wool growth to increased energy intake is a response to the increased supply of microbial amino acids and unfermented dietary protein reaching the duodenum.

Wool protein contains a high proportion of the high-sulphur amino acid cystine, and it has been shown that variation in the availability of the sulfur amino acids to the follicle can affect both fiber growth rate and fiber composition (45). Sulphur-containing amino acids play a major role in wool growth.

The optimum protein-to-energy ratio for wool growth is about 1.88 g of digested sulfur-containing amino acid per megacalorie of digestible energy. This ratio indicates that the supply of absorbed amino acids, particularly sulphur-containing amino acids, is the major component of feed intake that determines the rate of wool growth. This ratio also implies that an increase in protein absorption increases wool growth.

In the absence of unfermented dietary protein, the supply of microbial protein alone, which is about 182 g of digested protein per megacalorie of dietary energy, will meet the requirements for maintenance, slow growth, and early pregnancy, but will not meet the requirements for rapid growth, lactation, or wool growth (46). For maximum wool growth, therefore, a supply of unfermented dietary protein approximately equal to the nutrient supply from microbial protein must be made available to the sheep.

In general, nutritional limitations to wool growth mainly affect quantitative production (42). Both length and diameter are reduced in individual fibers to the extent that, in the whole fleece, a period of severe undernourishment is represented by a marked thinning down of all fibers at the corresponding part of the staple. Normal variations in nutrition can account for up to a 2- to 3-μm difference in fiber diameter, with reduced nutritional status producing the finer fibers. In severe cases, this thinning causes a break or weakness in the fleece, seriously reducing the manufacturing value of the fleece.

The physiological state of the animal has a significant effect on wool growth (46). The reproductive cycle in females can reduce wool growth significantly. During late pregnancy, depression of wool growth rate in the range of 20% to 40% has been reported. Where there are no restrictions to intake, voluntary feed intake during pregnancy is not reduced, except, perhaps, for a short period before parturition, so it is accepted that depression of wool growth is likely to be associated with partitioning of ingested nutrients. In lactating ewes, wool growth is also reduced by up to 30% or more despite the accepted increase in voluntary intake associated with lactation. It can be accepted that, as in pregnancy, the reduction in wool growth during lactation is influenced by changes in nutrient partitioning.

The nutrition of a ewe during pregnancy and lactation not only affects her immediate wool production, but can influence the lifetime wool production of her lambs (47). The amount of wool produced by a sheep is determined by the number, size, and type of wool follicles. Primary wool follicles are usually fully developed by the nineteenth week of gestation, whereas secondary follicles develop from that time until the lamb is born. The initiation of follicles and their subsequent development are closely related to fetal growth. Low levels of nutrient supply during prenatal life results in a restriction of body size and total number of follicles. Restriction on nutrient intake after birth reduces the capacity of individual follicles to produce fiber. Nutritional deprivation during early life has been shown to depress permanently the number of wool-producing follicles, thus influencing future productivity. Undernutrition in later life causes only a temporary reduction in the number of active follicles (47).

REFERENCES

1. NRC. 1985. *Nutrient requirements of sheep.* 6th ed. Washington, DC: National Academy Press.

2. ARC. 1980. *The nutrient requirements of farm livestock. No. 2: Ruminants.* 2nd ed. Slough, England: Commonwealth Agr. Bureaux.

3. Forbes, J. M. 1968. *Br. J. Nutr.* 22:33.

4. Butcher, J. E. 1970. *Natl. Wool Grower* 60:28.

5. SID. 1987. *Sheep production handbook.* American Sheep Industry Association, Production, Research and Education Council, Englewood, CO.

6. Wiener, G. 1979. *Livestock Prod. Sci.* 6:223.

7. Kubota, J. 1975. *J. Range Manage.* 28:252.

8. Russel, A. J. F., J. M. Doney, and R. G. Gunn. 1969. *J. Agric. Sci.* 72:451.

9. MLC. 1983. *Feeding the ewe.* Meat and Livestock Commission. Sheep Improvement Services, Queensway House, Bletchley, England.

10. Gunn, R. G. 1983. *The influence of nutrition on the reproductive performance of ewes.* In: *Sheep production,* W. Haresign, England: Butterworths.

11. Coop, I. E. 1966. *J. Agri Sci* 67:305.

12. Cole, H. H., and P. T. Cupps. 1977. *Reproduction in domestic animals.* 3rd ed. New York: Academic Press, Inc.

13. Hoversland, A. S., et al. 1956. *Proc. West. Sec. Amer. Soc. Animal Sci.* 7:1.

14. Kercher, C. J. 1980. *Forage and management systems for maximizing efficiency of western ewes.* Technical Report, Animal Sci. Division, University of Wyoming, Laramie, WY.

15. Torell, D. I., I. D. Hume, and W. C. Weir. 1972. *J. Animal Sci.* 34:479.

16. Hulet, C. V., et al. 1962. *J. Anim. Sci.* 21:505.

17. Robinson, J. J. 1983. Nutrition of the pregnant ewe. In: *Sheep production,* W. Haresign, England: Butterworths.

18. Villettee, Y., and A. Brelurut. 1980. *Bull. Techn. CRZV Theix. INRA* 40:5.

19. Cowan, R. T., et al. 1981. *Animal Prod.* 33:111.

20. Wallace, L. R. 1948. *J. Agric. Sci., Cambridge* 38:93.

21. Robinson, J. J. 1987. Nutrition of housed sheep. In: *New techniques in sheep production.* I. Fayez, M. Marai, and J. B. Owen, England: Butterworths.

22. Cowan, R. T., et al. 1980. *J. Agric. Sci., Cambridge* 95:497.

23. Robinson, J. J., et al. 1974. *J. Animal Prod.* 19:331.

24. Poe, S. E., et al. 1972. *J. Animal Sci.* 34:826.

25. Church, D. C., ed. 1975. *Digestive physiology and nutrition of ruminants. Vol. 1: Digestive physiology.* 2d ed. Corvallis, OR: O & B Books.

26. Frederiksen, K. R., R. M. Jordan, and C. E. Terrill. 1980. *Rearing lambs on milk replacer diets.* Farmers Bulletin No. 2270. USDA.

27. Thomas, V. M., and R. W. Kott. 1989. *Sheep research J.* 5:2, p. 1.

28. Jordan, R. M., and G. C. Marten. 1963. *J. Animal Sci.* 19:1307.

29. Carter, R. C., and J. S. Copenhaver. 1965. *Intensive lamb production methods.* Va. Agric. Exp. Sta. Livestock Res. Rpt.

30. Thomas, V. M. 1980. Feeding ewes and lambs. In: *A practical guide to sheep disease management* by N. Gates. Moscow, ID: News-Review Publishing Co.

31. Emerick, R. J., and L. B. Embry. 1963. *J. Animal Sci.* 22:510.

32. Crookshank, H. R. 1970. *J. Animal Sci.* 30:1002.

33. Thomas, V. M., and J. J. Dahmen. 1986. *SID Research Digest* 2:2, pp. 11–16.

34. Barns, B. J., and E. R. Ørskov. 1982. *World Animal Rev.* 42:38.

35. Ørskov, E. R., et al. 1974. *Brit. J. Nutr.* 32:59.

36. Ørskov, E. R. 1979. *Livestock Prod. Sci.* 6:335.

37. Hulet, C. V., and S. K. Ercanbrack. *Increasing market lamb production.* U.S. Sheep Station, Dubois, Idaho.

38. Black, J. L., and P. J. Reis. 1979. *Physiological and environmental limitations to wool growth.* Armidale, ID: Univ. of New England Publishing Unit.

39. Doney, J. M. 1983. Factors affecting the production and quality of wool. In: *Sheep Production.* W. Haresign, England: Butterworths.

40. Allden, W. G. 1968. *Aust. J. Agri. Sci.* 19:639.

41. Hogan, J. P., N. M. Elliott, and A. D. Hughes. 1978. Maximum wool growth rates from Australian Merino genotypes. In: *Physiological and environmental limitations to wool growth,* p. 43, eds. J. L. Black and P. J. Reis. Armidale, England: Univ. of New England Publishing Unit.

42. Allden, W. G. 1978. Feed intake, diet composition and wool growth. In: *Physiological and environmental limitations to wool growth,* p. 61, eds. J. L. Black and P. J. Reis. Armidale, England: Univ. of New England Publishing Unit.

43. Williams, A. J., and R. J. Winston. 1965. *Aust. J. Exp. Agri. Animal Husbandry.* 5:390.

44. Ferguson, K. A. 1959. *Nature* 184:907.

45. Reis, P. J. 1978. Effects of amino acids on the growth and properties of wool. In: *Physiological and environmental limitations to wool growth,* p. 223, eds. J. L. Black and P. J. Reis. Armidale, England: Univ. of New England Publishing Unit.

46. Corbett, J. L. 1979. Variation in wool growth with physiological state. In: *Physiological and environmental limitations to wool growth,* p. 79. J. L. Black and P. J. Reis, eds. Armidale, England: Univ. of New England Publishing Unit.

47. Lyne, A. G. 1964. *Aust. J. Agri. Research.* 15:788.

18

GOATS AND GOAT NUTRITION

J. E. Huston

INTRODUCTION

Goats are small ruminants that are distributed widely throughout the geographic and climatic regions of the world. Although the dry conditions, rough terrain, and sparsely distributed vegetation of the steppe regions are most suitable, goats adapt quickly and with proper management can be productive under almost any circumstance.

Goats in the United States include dairy (intensive), Angora (mostly extensive), and Spanish or "meat" goats (mostly extensive). The Angora, and to a lesser extent the meat goat, is found primarily in the Edwards Plateau region of Texas. Goats are valuable to that region as a source of income from the sale of fiber (mohair) and animals for slaughter; they are also important in stabilizing the vegetation by retarding encroachment of brush species. The dairy goat industry is expanding rapidly, but it remains very minor in comparison with the dairy cattle industry. Most large goat dairies are located near metropolitan areas. Many smaller herds (5 to 20 goats) are kept as backyard operations, producing milk that is sold to neighbors and friends.

Although goats are important worldwide because of their socioeconomic contribution (mainly in lesser developed countries), the most

unique thing about goats is their inquisitive feeding behavior. This alone is sufficient justification for dedicating this separate chapter to goats and goat nutrition.

FEEDING BEHAVIOR

Goats are unlike all other species of domestic livestock in choosing and consuming a diet. They discriminate between plant fractions or feed particles that appear identical, consuming one and leaving the other. They are insistent that their diet is fresh, clean, and previously untouched. They are versatile in diet selection.

FEEDING IN CONFINEMENT

Goats are very particular about what they eat even when fed a hay or mixed ration in a feed trough. Mobile lips and precise tongue movement (31) make it possible for the goat to sort through the offered feed, literally pushing the unchosen plant fragments and feed particles aside and taking only the most preferred (Fig. 18-1).

This particular behavior has two important effects on diet. First, diet composition differs from the feed offered. The values in Table 18-1 show the difference in the composition of hay offered and refused. In this case the diet of the goats was 12.5% and 16.5% greater in net

FIGURE 18-1 A dairy goat at a feed bunk containing stem refusals from an earlier offering of hay.

energy and protein, respectively, than the hay offered. Although the highly developed selectivity is usually advantageous to the goat, lower productivity or health problems could result from imbalances created by selective consumption. Pelleted rations offer less risk of such imbalances.

The second effect of goats' particular eating habits is that goats eat more if they have more from which to select (Fig. 18-2). At a lower feeding level, only a small amount is refused. As feeding rate increases, both refusals and net consumption also increase. It is important to not allow refusals to accumulate. Once the offering has been refused, little if any will be consumed at a later time without a serious effect on productivity.

DIET SELECTION AND COMPOSITION ON RANGE

The goat is active and inquisitive in its foraging behavior. Given an opportunity to be selective, the goat will graze (browse) from all plant types (trees, shrubs, herbaceous dicots, and grasses) and almost all species within a plant type. Oc-

casionally, a single plant will be identified and totally defoliated, while another of the same species will be completely avoided. However, both are exceptions to the general pattern of selecting a diversified diet from a relatively large area. Goats, known for their dexterity, often stand on high cliffs or climb onto low-hanging tree limbs. A common stance for browsing is on hind legs only with the front legs and head hidden in the lower branches of trees. The term **browse line** describes the appearance of a landscape where heavy use by goats has occurred (Fig. 18-3). Trees and shrubs take on a flattened-underline appearance 4 to 6 ft above the ground, the height that is within reach of the goat in a bipedal stance.

Although goats select all types of foliage, they are particularly attracted to trees and shrubs. This behavioral characteristic, unique among livestock species, has been exploited for control of undesirable brush or weed species in several countries. Depending on the circumstances, goats can remove undesired woody species from rangeland (35), reduce its presence and prevent encroachment (7, 25, 43), or just improve its value for browsing by other livestock (45). The meat goat appears to be a slightly more effective browser than the Angora (57). In some instances, income from brushy pastures is highest under goat browsing, in which case the shrubs should be maintained and used by goats, not overused to the point of their demise (52).

The nutritional value of a goat's selected diet is usually higher and more stable than that of the average available vegetation. Grasses and herbaceous legumes are generally considered higher in nutrients than shrubs and tree leaves and often comprise the major portion of the goat's diet. At other times the diet contains the most tender and palatable leaves from trees, leaf tips from shrubs, fallen mast (for example, oak acorns), flowers and/or

TABLE 18-1

Feeding value of alfalfa hay offered, refused, and eaten by dairy goats

	Offered	Refused	Ingested
Leaves, %	46.2	23.9	59.9
Stems, %	53.8	76.1	41.4
Nutrient value			
Energy, Mcal/kg DM	0.88	0.7	0.99
Digestible crude protein, %	12.1	8.5	14.1

Source: Computed from values reported by Morand-Fehr and Sauvant (39).

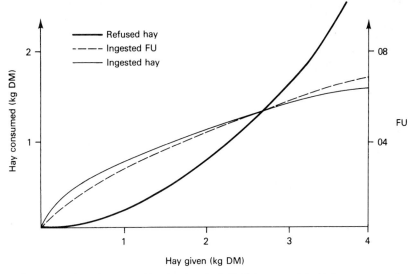

FIGURE 18-2 Influence of feeding level of alfalfa hay to adult Alpine goats on the relative amounts of hay refused and ingested and feed units (FU) ingested. [From Morand-Fehr (37).]

FIGURE 18-3 Meat goats foraging in the Edwards Plateau region of Texas. Note the browse line on the trees and associated shrubs. (Photo by Kenneth Bales.)

buds from xeric plants, grass seed heads, and so on. All these minor plant parts contain high concentrations of one or more required nutrients. If the range vegetation is comprised of a complex assemblage of plants, the goat's diet also will be complex. Moreover, the diet composition changes drastically, often from day to day, as the array of plants simultaneously emerge, mature, fruit, senesce, drop leaves, and so on (58). The net result is that the selected diet is higher than the average of the available vegetation and, compared with diets of less adaptable livestock species, is rather stable in nutrient composition. However, on pastures or ranges where there is little or no plant diversity, diets selected by goats, sheep, and cattle are similar.

PRODUCTIVITY OF GOATS

Goats are raised for milk, meat, and fiber (and hides). Although some overlap exists in roles of individual goat types (meat and milk, fiber and meat), goats have developed into specific types to supply primary products.

MEAT GOATS

Meat goats fit into a category that is highly variable in appearance. In some countries of the world, meat goat breeds have occasionally established characteristic appearances. The Boer goat of Africa serves as an example of an established meat goat breed. Until recently, meat goats in the United States have received little controlled selection; therefore, they are a myriad of color patterns, horn and waddle types, milking abilities, hair characteristics, and conformations. In 1993, three developments spurred a dramatic change in attitude toward meat goats: there was a realization that a large and increasing demand for goat meat existed in the United States (14); the Wool Act, which provided an incentive for mohair, began a 3-year phase out (29); and the Boer goat was introduced into the United States. Although meat from goats can come from Angoras or dairy goats, the generalized term meat goats refers to goats raised primarily, if not exclusively, for meat production. Currently, most of these goats are the nondescript Spanish goats, but a large impact from the Boer influence is expected.

Reproductive rate is relatively high in meat goats (26) and, except for a 2-month anestral period in March and April, these goats breed year round (51). Twinning is frequently observed, and litters of three or four offspring may occur. Although two sets of offspring within a 12-month period are theoretically possible, a reasonable expectation for reproductive rate is approximately 180 live kids per 100 females per year.

Does produce approximately 2 lb of milk per day at peak lactation (21 days postpartum) and gradually reduce milk production to about 0.7 lb/day at 120 days lactation (11). Higher milk production would be expected from meat goats having a substantial infusion of milk goat breeding.

Newborn kids weigh 5 to 7 lb and gain approximately 0.25 lb/day to 120 days of age. Thus kids weaned at 6 months of age should weigh 35 to 40 lb depending on sex of the kid (males are slightly heavier), number of kids in the litter, and environmental, primarily nutritional, conditions.

Slaughtered goats have a lower dressing percent (less than 50% yield) compared with sheep (55), partially because of a lower tendency to deposit fat (Fig. 18-4). Goat muscle is less ten-der (4) but is flavorful, especially when included in processed meat such as wieners, chili, and the like, irrespective of the age of the goat (32). Much of the meat goat product, especially that of young goats (cabrito), is used on a whole-carcass basis for outdoor, open-fire barbecues.

The Boer goat under South African conditions reportedly reproduces at a comparable rate (about 180%) with Spanish goats in the United States (6, 8). However, milk production is approximately double (averages 4 lb/day). In crossing with Spanish goats in the United States, the crossbred goats are slightly heavier at birth than the straight Spanish, grow approximately 30% faster under good nutrition (24), and produce a larger carcass at a given age (44, 60). Other quantitative and qualitative characteristics of goat carcasses are similar for the different breeds, except that Angora carcasses are slightly fatter (44). Although the Boer influence increases growth and production potential, increased productivity depends on a higher level of nutrition and may not be realized under low or marginal conditions (3, 36, 60).

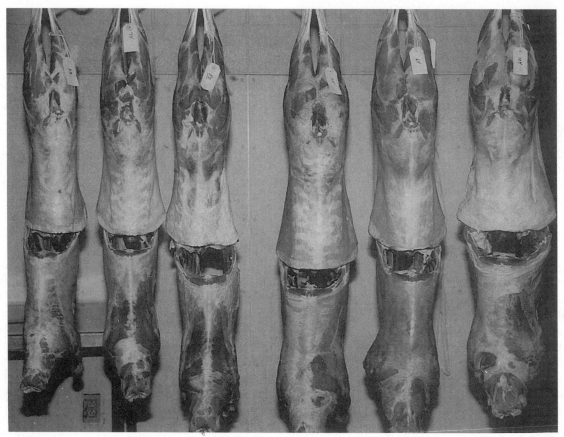

FIGURE 18-4 Carcasses of Spanish × Angora (left), Spanish (second, third, and fourth from left), and Spanish × Boer (two at right) goats at approximately the same age and following a similar development regime. Note the larger, more muscular appearance of the Spanish × Boer carcasses and the slight indication of increased fattening in the Spanish × Angora carcass. (Courtesy of D. F. Waldron.)

Hair on these meat goats is distinctly different from the mohair of the Angora (Fig. 18-3). Most meat goats produce two types of hair, a very coarse type (guard hair) that has no marketable value and a very fine fraction (cashmere) that is currently in great demand. Cashmere production is quite variable in meat goats and usually very low (less than 50 g/yr). However, there is presently much interest in developing a cashmere industry from selected individuals from within the meat goat population.

ANGORA GOATS

The Angora is the product of a long history of genetic selection for a single trait, mohair production. These goats are smaller than most other breeds, have higher nutrient requirements, especially for protein, and frequently suffer from less than optimal nourishment. They are more susceptible to parasitism and harsh environmental conditions (48).

The reproductive rate of Angora goats is low compared to that of the meat goat. Both the male and female have a restricted breeding season. Initiation of the estrous cycle in females and rutting in males usually corresponds with a cooling trend in late summer and continues until mid-winter or, for the female, until conception occurs. The moist active breeding period occurs between mid-September and mid-November. Although ovulation rate often exceeds 125% several adverse occurrences reduce weaning percentage to approximately one-half of ovulation (50). Major losses include failure to conceive, embryo death, abortion, perinatal losses, and predator losses. Many of these losses can be circumvented or reduced by good nutrition management.

Kids at birth are comparable in size to kids of meat goats (5 to 7 lb), but small, weak kids that die at birth are common in undernourished flocks. Kids weigh 25 to 45 lb at weaning (6 months) and 50 to 60 lb at first breeding (18 to 20 months). The period between weaning and first breeding is critical, and underdeveloped yearlings have reduced lifetime productivity (53).

Although the meat characteristics of Angora goats are comparable to those of other breeds (44), young Angoras are seldom slaughtered during periods of stable populations. Because of the low reproductive rates and high attrition rates in Angora flocks, essentially all female kids must be held for replacements. Male offspring that are determined to be unsuitable for breeding are castrated (6 to 18 months of age) and held as wether or "mutton" goats for mohair production. Therefore, only cull or aged Angoras go into the meat trade in any appreciable numbers.

Angoras are raised for mohair. This fiber has a diameter greater than cashmere and the fine wools but less than the coarse guard hairs of meat and milk goats and the wools of the long-wool breeds of sheep (Fig. 18-5). Average production of an Angora flock (including kids and yearlings) is about 7 lb (clean-fiber basis) per goat per year (56), but young adults (2 to 5 yr) exceed this level of production (Fig. 18-6). With advancing age, mohair production increases to a peak and then declines, fiber diameter increases but at a decreasing rate, and staple length remains relatively constant (30). Angora goats are shorn twice annually during late winter and late summer. Mohair is sold on the basis of fineness (fiber diameter), with the highest price paid for the finer grades. Three broad grades are used in the U.S. mohair trade. Kid hair (less than 30 μm) is the finest and demands the highest price, followed by young adult (30 to 34 μm) and adult (greater than 34 μm), respectively. In some areas of the world trade, objective measurements (fiber diameter, staple length, strength, etc.) are used in place of a grading system to assess value.

MILK GOATS

Goat dairies comprise a struggling yet promising industry in the United States (16). The goat, often described as the "poor man's cow," is extremely important in developing countries. In the United States, primary products of the dairy goat (whole milk, cheese, milk powder, yoghurt, etc.) are consumed in specialty markets in large metropolitan areas or by the owners of the goats and their friends and neighbors. Modern milking equipment is shown in Fig. 18-7. Between 1986 and 1993, an average of 792 dairy herds averaging 15 goats per herd participated in the Dairy Herd Improvement Association Program (DHIA) for goats (9). It is suggested that this is less than 1% of the estimated 1.5 million dairy goats in the United States (16). The state ranking in number of dairy goats is California (first by far), then Texas, Ohio, New York, Pennsylvania, Oregon, Washington, and Wisconsin.

The dairy goat breeds include the Alpine, LaMancha, Nubian, Oberhasli, Saanen, and Toggenburg. The most prevalent is the Nubian, which is a true dual-purpose breed and is often compared with the Jersey cow because of the

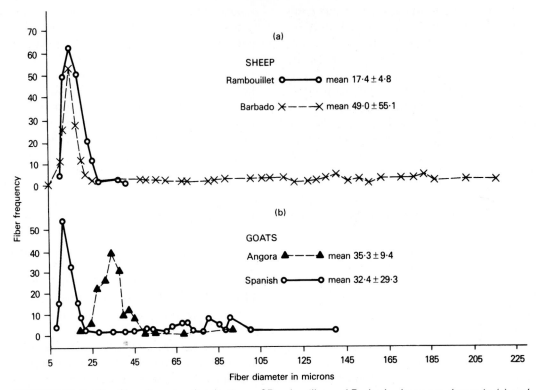

FIGURE 18-5 The fiber diameter distributions of Rambouillet and Barbado sheep are shown in (a) and Angora and meat goats in (b). [From Gallagher and Shelton (10).]

relatively high fat content in its milk (16). The Saanen, Alpine, and Toggenburg (Swiss breeds) are reportedly better milk producers (9). The LaMancha was developed in the United States and is easily recognized because it has only rudimentary external ears.

Similar to the Angora's being selected for mohair, milk goats are selected for milk yields with less emphasis placed on other characteristics. Reproductive rate in milk goats is higher than in Angoras but probably less than in meat goats (47). However, in milk goats reproduction is associated with freshening and thus with milk production. Seasonality of breeding is a significant constraint for dairies in the liquid milk market (2). Light modification to increase

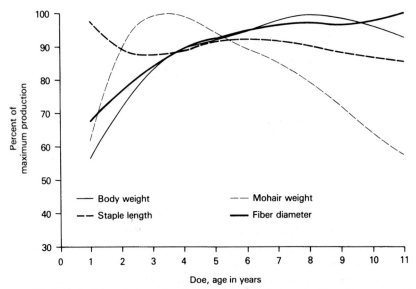

FIGURE 18-6 The relation of mohair weight, staple length, fiber diameter, and body weight to age in Angora females. [From Shelton (49).]

FIGURE 18-7 Milking dairy goats.

photoperiod in the spring months aids in breeding during this otherwise anestral period.

Birth weight and growth rate of kids vary among breeds, but both tend to be slightly higher than for meat goats. A kid weighing 8 to 9 lb at birth should gain 0.28 lb/day and reach 55 to 60 lb at 6 months of age (38). Meat characteristics are similar to those of meat goats except that carcasses are larger and higher yielding.

Milk yield is highly variable. Reported yields of 1300 to 1600 lb (9, 33) are far below those reported by individuals and what is considered potential production (15). Fat content of goat milk is reported at 4.2% (9). As in dairy cattle, milk yield and composition in goats are highly influenced by the level of nutrition in both prefreshening and lactation periods (33, 40).

NUTRITIVE REQUIREMENTS

Generally, goats must consume more dry matter (relative to body weight) or the dry matter consumed must contain higher concentrations of nutrients compared with the dietary requirements of other ruminant livestock. The reticulorumen of the goat is smaller according to body size, and retention time of feed particles tends to be shorter (Table 18-2). Whereas diet may be of similar digestibility (in vitro estimate), actual digestibility may be lower in goats (in vivo estimate) because of the shorter residence time in the reticulorumen. This allows a faster turnover of feed particles and an elevated level of consumption. The net result is a higher level of intake and lower digestibility, but a higher level of consumption of digested nutrients compared with other ruminant livestock (18). In humid tropic and desert environments, the goat

TABLE 18-2

Rumen retention time (RT), turnover rate (TR), and digestibility of diets selected by cattle, sheep, and goats from a common range in the Edwards Plateau region of Texas (values averaged over four seasons)

| Species | RT (h) | TR (%/h) | Digestibility (%) | |
			In vitro[a]	In vivo[b]
Cattle	36	5.2	55	51
Sheep	34	5.2	57	50
Goats	28	7.1	56	45

Source: Huston et al. (21).
[a]Calculated according to Van Soest et al. (59).
[b]Calculated according to Lippke et al. (27).

is reported to have greater ability to digest fiber, possibly because of its greater capacity to recycle nitrogen (34).

ENERGY REQUIREMENTS

Energy is required for body maintenance, growth, reproduction, lactation, fiber growth, and activity. These concurrent body functions and associated requirements influence the voluntary dry-matter intake dramatically (Fig. 18-8). Perhaps no other livestock species varies as widely in energy requirements as the goat. This variation in requirements is a result of the extremes of type, productivity, and activity level. The NRC published its first edition on the nutrient requirements of the goat and attempted to account for this recognized variation (42). Requirements tables from NRC are given in Appendix Table 21. Energy requirements are expressed according to the relationship of

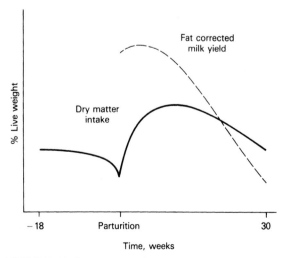

FIGURE 18-8 Variation of dry-matter intake in dairy goats yielding 750 kg of milk per lactation. [From Morand-Fehr and Sauvant (39).]

100 kcal of gross energy (GE) = 76 kcal digestible energy (DE) = 62 kcal metabolizable energy (ME) = 35 kcal net energy (NE). Also, 1 kg of digestible organic matter (DOM) = 1.05 kg total digestible nutrients (TDN) = 4.62 Mcal DE. The approach allows the user to pick and choose among size of goat, level of activity, physiological stage (pregnant or nonpregnant), rate of growth, level and composition of milk yield, and fleece production as they apply to a particular goat type.

Maintenance energy requirements vary with body size and are expressed according to the following relationship:

$$DE = 124 \, kcal/W^{0.75}$$

This expression should be considered to have some latitude of precision because the efficiency of use of DE is not constant for different feedstuffs. Also, the fractional power (0.75) is the accepted value for interspecies comparisons but differs from the average value of 0.64 reported by Brody for goats (5). Additional energy costs for activity are expressed as either 0.25, 0.50, or 0.75 times maintenance for goats in confinement, on relatively small pastures, or under extensive grazing on rough terrain, respectively. Whereas at minimal activity (stall-fed condition) the maintenance energy requirements of goats are comparable to those of other livestock species, they are somewhat elevated under the more extensive conditions. This reflects the goat's active foraging behavior. The energy requirement for growth in goats is estimated at 8.84 kcal DE/g of liveweight gain (42). Again, this estimate should be interpreted as an average of a relatively large range of values depending on type of goat, environment, and diet.

Successful reproduction in goats require adequate energy intake prior to and at breeding and elevated energy intake during the final 2 months of pregnancy. This 2-month period is especially critical in the Angora, which is subject to nutritional stress abortion (61). The NRC estimate of energy requirement for the last 2 months of pregnancy is 1.74 Mcal DE, irrespective of body size of the pregnant female (42). This level of energy intake is often difficult to achieve, especially in smaller goats. It is the author's opinion that this requirement for pregnancy is slightly overestimated for average reproducing goats. Small- and intermediate-sized goats carrying a single fetus require approximately half the estimated value. However,

the estimate is likely applicable for large goats and/or goats producing a high incidence of multiple births.

The energy requirements for lactation are expressed per kilogram of milk at 2.5% to 6% fat content. The values are based on an estimated 1.52 Mcal DE/kg of 4% fat-corrected milk (FCM), with an adjustment of 19.9 kcal for each 0.5% change in milk fat content (42).

The energy requirements for fiber production refer specifically to mohair produced by Angora goats (Fig. 18-9), although there appears to be an increasing interest in cashmere-producing goats. The low energy input into fiber relative to other body functions makes the direct measurement of energy costs difficult. The expressed values were derived from somewhat intuitive calculations (19). However, goats having high fiber production capability are recognizably benefited by elevated energy intake.

PROTEIN REQUIREMENTS

The protein requirements are expressed in Appendix Table 21 as total crude protein (TP) and digestible crude protein (DP) for different-sized goats and for different metabolic functions (42). The maintenance estimates reflect a calorie-to-protein ratio of 1 Mcal DE to 22 g DP. Although research data are not available on protein costs of activity in goats, estimates of requirements are increased correspondingly with increased energy costs. Requirements for growth equals 0.195 g DP/g liveweight gain. The pregnancy requirements are estimated at 57 g DP/day for all pregnant goats during the last 2 months of gestation. As suggested for energy, this level of protein for pregnancy applies to the large goat and/or a goat producing multiple births. Small goats and those producing single offspring re-

FIGURE 18-9 Mohair removal (shearing) from a typical adult Angora female at a Texas ranch.

quire approximately half the recommended protein for pregnancy. Digestible protein requirements for lactation were estimated at 51 g DP/kg of 4% FCM and were adjusted for fat content according to recommendations for dairy cattle (41). This value compares favorably with that determined recently by Lu et al. (28). Requirements for fiber growth were based on work by Huston et al. (19).

MINERAL REQUIREMENTS

The mineral requirements reported in NRC for goats were adapted from experimental data obtained using other domestic animals (42). It is almost certain that minor differences exist in some aspects of mineral metabolism between goats and other species, but, up to the present, non has been described that would justify changing the recommendations as presented in the NRC.

MACROMINERALS   Calcium and P are the two major minerals that should be given the most attention in diet formulation. The recommended daily requirements for these two minerals are included in Appendix Table 21. Both are important in bone development and maintenance and in milk synthesis. A deficiency of either or both in young animals results in poorly developed and malformed bones (rickets). In adult animals, especially those that are lactating, Ca and P are commonly mobilized from the bone to fulfill high-priority needs (such as milk synthesis). Normally, these bone deposits are repleted once dietary intake exceeds current needs. In a dietary deficiency, continued removal from the bone can result in a thinning and loss of strength (osteomalacia), resulting in brittleness of bones.

The Ca:P ratio should be maintained between 1.2 and 2.5. A lower ratio (below 1:1) exposes the animal to the development of urolithiasis (urinary calculi or kidney stones), especially in males. A high intake of Ca has not been associated with an acute hazardous effect, but it can interfere with absorption and cause deficiencies of P and certain divalent cations (Mg, Mn, and others).

High-producing dairy goats can suffer parturient paresis (milk fever) as a result of metabolic hypocalcemia (low blood Ca). Although this condition can be related to dietary Ca, it is confounded by action of the parathyroid hormone, which is involved in freeing Ca from the bone. A failure or delay of activity of the parathyroid hormone with the onset of lactation results in a drop in the blood Ca and the placid outward appearance of milk fever. The condition is less frequently observed in dairy goats than in dairy cows but, when observed, is treated similarly by intravenously administering a solution containing Ca and glucose.

Phosphorus is an extremely important mineral element for maintaining proper feed (forage) intake and intraruminal metabolism. In grazing cattle, low P is associated with a rough hair coat, unthriftiness, loss of weight, depraved appetite, and low reproductive rate. Providing supplemental P as various inorganic phosphates is a very effective and rewarding practice in reducing or eliminating these symptoms. Goats tend to be less affected by low P, but a proper intake of the element is still considered essential for satisfactory productivity. The goat's intensive selectivity undoubtedly results in a diet higher in P than would be apparent from gross forage analysis and higher than would be obtained by less discriminating grazers. Attention to P level (and Ca:P ratio) is more important in the formulation of complete diets using harvested or formulated feedstuffs.

Other macrominerals include Na, Cl, Mg, K, and S. Of these, only Na and Cl (salt) should be provided routinely free choice. A Mg deficiency can occur on lush pasture and cause grass tetany. Potassium is most likely deficient in animals grazing dormant vegetation, in which case intake and performance will be reduced. Sulfur can be low, primarily for Angora goats, which have a high requirement for S-containing amino acids. Most common forage and feedstuffs contain adequate Mg, K, and S.

MICROMINERALS   The microminerals of main concern include Fe, I, Cu, Mo, Zn, Co, Mn, and Se. None of these elements should be considered as routinely deficient. Potential deficiencies are regional, and inclusion of these elements in feed formulations should be on a regional basis. Because some microminerals have a narrow span between amounts producing deficiency and toxicity, specific circumstances should be considered before providing liberal addition of microelements to goat diets. Many question the sensitivity of goats to Cu toxicity. It is the author's opinion that goats are less susceptible to Cu toxicity than sheep, but levels greater than 15 ppm Cu in formulated goat diets should be avoided.

VITAMIN REQUIREMENTS

In monogastric animals (nonruminants), vitamins must be supplied in the diet. In the goat (ruminant), only a few are likely to be periodically limiting, since most are synthesized by ruminal microorganisms. The fat-soluble vitamins include vitamins A, D, E, and K. Of these, only vitamin A is likely to be deficient to the point of limiting productivity in goats. The B vitamins are all synthesized in adequate amounts by ruminal bacteria, with the possible exception of vitamin B_{12} in regions with a cobalt deficiency.

Vitamin A is not contained in plant tissue but is synthesized within animals from vitamin A precursors that are present in plants. The levels of these precursors are generally in proportion to plant pigments (green, yellow, etc.). Whereas actively growing plants are high in provitamin A activity, weathered vegetation contains very little. Once synthesized, vitamin A can be stored in the liver of animals and metabolized as needed. The seasonal, cyclic nature of forage growth and the selective grazing behavior displayed by goats almost preclude a vitamin A deficiency in free-ranging goats. A vitamin A deficiency can occur in pen-fed goats and in goats feeding in extended dry and/or cold conditions during which vitamin A stores are depleted and little or no green plant material is consumed. Requirements for vitamins A and D were estimated by NRC (42) and are included in Appendix Table 21.

Vitamin B_{12} is a large compound that includes cobalt in its central structure. Thus vitamin B_{12} is deficient in animals that are consuming diets low in cobalt. Again, this is a very regional problem and is not likely to occur under most circumstances.

WATER REQUIREMENTS

The quantity, quality, and placement of water are very important aspects of goat health and nutrition. Because water intake is related to feed intake and feed intake is correlated with productivity, the general recommendation is that goats be given free access to water to maximize water intake and thereby not limit feed (forage) intake. Goats are often more sensitive than other species to water quality and tend to refuse to drink water fouled by fecal and urine contamination. This is an interesting behavioral trait, because the goat may be the most incriminated among livestock species for actually fouling feed and water by direct defecation and urination. Designing the watering facilities to provide readily accessible water yet to protect it against contamination is crucial.

The quantity and frequency of water intake vary greatly among goat types, their locations, and associated diets. The small black Bedouin goat of the desert in Israel was reported to drink enormous amounts of water (up to 47% of dehydrated body weight) on an infrequent basis as a strategy to extend the grazing range to 2-days' distance from water (54). The goat is generally reported to consume less water than sheep and cattle on a metabolic size basis (12), but this may be influenced by diet. Huston et al. (22) found that goats consumed less water than sheep when both were fed wheat straw, but water intake was higher in goats when the diet was oat hay (higher protein and digestibility). In this circumstance, water intake was closely related to feed intake; that is, the relative level of feed intake of the goat to the sheep was lower when fed the low-quality forage and higher when fed the high-quality forage.

Lactation affects water consumption. Giger et al. (13) found that in a temperate climate the maintenance requirement for water by a lactating goat was 145.6 $g/kg^{0.75}$ and that an additional 1.43 kg of water was required for each 1 kg of milk produced.

Other factors such as water content of the vegetation, salt intake, environmental temperature, water temperature, and electrolyte concentration in the water affect water intake. However, it is still recommended that goats be allowed free access to clean, cool water that is as free of contaminants as possible.

FEEDING AND MANAGEMENT

Animals reach potential productivity only if provided adequate nutrients in proper balance in an acceptable setting. Such circumstances are rare in animal agriculture. More realistically, we search for the most economic return, which involves identifying the management strategy where the difference between value of products and cost of production is at a positive maximum. This section discusses the relative value of feedstuffs for goat diets and nutritional and management factors that affect productivity.

FEEDSTUFFS

Goats can utilize any feedstuff that is safe for ruminant consumption. Ingredients that appear in goat rations in the different countries of the world are those feedstuffs that are locally available and usually in low demand for human consumption. The most economic productivity results from maximizing the use of forages or roughage products combined with an adequate but minimal level of concentrates. Feedstuffs appearing in Table 18-3 are common ingredients used in ration formulation and serve as examples of feeds of varying nutrient content.

FORAGES Forages vary greatly in chemical composition and thus in nutrient content. Range vegetation is highly seasonal; that is, it is nutritious and palatable during the rapidly growing period (usually during spring–summer) and unpalatable and low in nutrients during the dormant period. In the Edwards Plateau region of Texas, range vegetation is comprised of five functional components (Table 18-4). Goats pick and choose from these plant groups, and if the range is sufficiently diverse, the nondairy goats can satisfy their nutrient requirements from the vegetation alone. Under a depleted circumstance as a result of drought, overgrazing, or the like, options for diet selection become limited and diet quality follows the pattern of the limited options. If only the production component is present, then diet quality fluctuates with the nutrient content of this component. Then, high productivity is achievable only with supplemental feeding of concentrates to bring the overall diet quality to a satisfactory level. Rangelands in other regions

of the world have assemblages of plants that differ from the one described for the Edwards Plateau region of Texas, but the principle of fluctuating quality and importance of diversity holds for all rangelands.

Forages in the temperate climates are likely grown on tame pastures (monocultures) that are either grazed directly at a relatively high stocking rate or harvested and fed as hay. These forages are characteristically higher in quality compared with range forages and typify forage fed to dairy goats. Alfalfa, various clovers, and ryegrass are examples of these forages.

ROUGHAGE PRODUCTS Roughage can be included in goat rations to reduce the risk of gastrointestinal disturbances and as a diluent to reduce the nutrient content of a mixture of ingredients to a desired level. There is no absolute requirement for roughage in goat rations, but as roughage is reduced below about 30% to 40% of the diet, greater attention to possible digestive upsets is needed. Products such as cottonseed hulls, peanut hulls, cereal straw, and corn cobs are low in nutrients but provide good roughage factors.

HIGH-ENERGY FEEDS Feedstuffs such as corn, barley, sorghum, other grains, and molasses that are high in soluble carbohydrates (starch and sugars) contain high concentrations of digestible energy. Grains require some kind of mechanical processing (grinding, cracking) for

TABLE 18-3

Levels of nutrients in some common feeds

	Energy		Protein		Minerals		
Feedstuff	TDN (%)	DE (Mcal/lb)	CP (%)	DP (%)	Ca (%)	P (%)	Carotene[a] (mg/lb)
Dry forage or roughage							
Bermuda grass hay	50	1.0	15	11	0.43	0.16	53
Annual ryegrass hay	53	1.1	9	5	0.53	0.29	113
Alfalfa hay	57	1.1	18	14	1.38	0.26	83
Wheat straw	39	0.8	4	0	0.16	0.04	1
Energy feeds							
Oats, grain	68	1.4	12	8	0.07	0.33	—
Sorghum, grain	78	1.6	10	7	0.03	0.28	—
Corn, grain	77	1.6	9	7	0.06	0.24	1
Molasses, cane	54	1.1	4	2	0.75	0.08	—
Protein feeds							
Fish meal	67	1.3	61	49	5.20	2.9	—
Soybean meal	79	1.6	45	42	0.30	0.6	—
Cottonseed meal	69	1.4	41	37	0.20	1.1	—
Sunflower meal	60	1.2	46	42	0.40	1.0	—

Source: NRC (42).

[a]Carotene contents are not reported for some feedstuffs. The expected levels in these feeds are usually negligible, but fish meal may be relatively high in Vitamin A.

TABLE 18-4

Nutritional components of range vegetation in the Edwards Plateau region of Texas

Component	Plant Type	Nutritional Description
Production	Perennial, warm-season grasses	Low to moderate in protein (3% to 13%) and digestibility (25% to 63%)
Quality	Perennial, warm-season forbs, legumes, and browse	Moderate to high in protein (10% to 30%) and digestibility (50% to 89%) when palatable
Level	Long-season or evergreen plants	Moderately low to moderately high in protein (6% to 12%) and digestibility (40% to 60%)
Bonus	All desirable annual plants	High in protein (15% to 30%) and digestibility (60% to 80%)
Toxic	All plant or plant parts that are poisonous or injurious to livestock	Low to high in all nutrients

Source: Huston et al. (20).

maximum utilization, but grain processing is less important for goats than for cattle. These high-energy feedstuffs are relatively low in protein content (usually less than 12% crude protein).

HIGH-PROTEIN FEEDS Protein concentrates are commonly by-products of an extraction process. Cottonseed meal, soybean meal, linseed meal, peanut meal, sunflower seed meal, and others are residues from the extraction of oil from the raw products. These by-products from plant sources are usually but not always more economical than animal by-product meals such as fish meal and meat meal. Either can be utilized effectively by goats, but because of the goat's particular nature, any ingredient possessing an objectionable odor may reduce consumption and should be kept at a low level in the diet. Protein concentrates are moderate in energy content.

NONPROTEIN NITROGEN (UREA) Any ammonia-releasing compound can be used to satisfy at least a part of the protein requirements in goats. Urea is the most commonly used nonprotein N product. It is recommended that no more than one-third of the protein in a goat diet be supplied by urea. Higher levels are possible but may not result in satisfactory performance and under some circumstances can become toxic.

FEEDING GOATS ACCORDING TO REQUIREMENTS

It is important that the goat's nutrient requirements and nutrients provided in the consumed diet be matched to within a narrow zone of safety. Underfeeding, either from amount of dry matter or the nutrient concentration, causes the animal to produce at a lower than expected level. Overfeeding is wasteful and can reduce

productivity also. Minor deviations should not be of concern because depletion and repletion of storage tissue are normal metabolic processes. A feeding plan should consider the nutrient requirements according to stage of production (dry, pregnant, lactating), level of feeding, and diet composition (or ration formulation).

Table 18-5 includes information derived from the larger and more detailed NRC requirements table (Appendix Table 21). The DE and CP requirements were calculated for dry and pregnant goats (non-Angora), for does in low (meat goats) and high (milk goats) lactation, for kids growing at three rates, and for lactating Angora does producing 8.8 lb of mohair per year. The necessary feed level and protein contents for diets of increasing energy density are listed in the six columns on the right-hand side of Table 18-5.

Nutrient requirements can be satisfied by either a large amount of a diet containing lower concentrations of nutrients or less of a diet containing higher concentrations of nutrients. A 60-lb dry doe (refer to Table 18-5) would receive equal protein and energy nutrition from either 1.7 lb of a 55% TDN (1.1 Mcal/lb), 7.7% CP feed, or 1.2 lb of a 75% TDN (1.5 Mcal/lb), 10.7% CP feed. Requirements increase as animals become larger and metabolic activity increases. A 100-lb doe requires more nutrients than does an 80-lb doe. A 20-lb kid requires more if gaining 150 g/day compared with only 50 g/day. However, a 60-lb doe that is lactating requires more nutrients than a 100-lb doe that is dry (3.24 versus 2.72 Mcal DE, respectively). Higher requirements can be satisfied by feeding more feed up to a level of maximum voluntary intake. Once maximum voluntary intake has been reached, additional intake of nutrients must be achieved by increasing the concentration of nutrients in the feed.

TABLE 18-5

Nutrient requirements and feeding levels for meat, milk, and angora goats (example)

Class and State	Body Weight (lb)	Daily Gain (g/day)	DE (Mcal/day)	CP (g/day)	Required Intake at Feed Energy Level (Mcal/lb)			Required CP at Feed[a] Energy Level (Mcal/lb)		
					1.1[b]	1.3[b]	1.5[b]	1.1[b]	1.3[b]	1.5[b]
					--------lb/d--------			----------%----------		
Does, dry and	60		1.85	59	1.7	1.4	1.2	7.7	9.2	10.7
early pregnant	70		2.08	67	1.9	1.6	1.4	7.7	9.2	10.7
	80		2.30	74	2.1	1.8	1.5	7.7	9.2	10.7
	90		2.51	80	2.3	1.9	1.7	7.7	9.2	10.7
	100		2.72	87	2.5	2.1	1.8	7.7	9.2	10.7
Does, late	60		2.83	90	2.6	2.2	1.9	7.7	9.2	10.7
pregnant	70		3.06	98	2.8	2.4	2.0	7.7	9.2	10.7
	80		3.28	105	3.0	2.5	2.2	7.7	9.2	10.7
	90		3.49	111	3.2	2.7	2.3	7.7	9.2	10.7
	100		3.70	118	3.4	2.8	2.5	7.7	9.2	10.7
Does, low	60		3.24	104	2.9	2.5	2.2	7.7	9.2	10.7
lactation	70		3.47	111	3.2	2.7	2.3	7.7	9.2	10.7
	80		3.69	118	3.4	2.8	2.5	7.7	9.2	10.7
	90		3.90	125	3.5	3.0	2.6	7.7	9.2	10.7
	100		4.10	131	3.7	3.2	2.7	7.7	9.2	10.7
Does, high	80		6.46	207	5.9	5.0	4.3	7.7	9.2	10.7
lactation	100		6.88	220	6.3	5.3	4.6	7.7	9.2	10.7
	120		7.28	233	6.6	5.6	4.9	7.7	9.2	10.7
	140		7.66	245	7.0	5.9	5.1	7.7	9.2	10.7
	160		8.03	257	7.3	6.2	5.4	7.7	9.2	10.7
Kids, growing	20	50	1.25	40	1.1	1.0	0.8	7.7	9.2	10.7
		100	1.69	54	1.5	1.3	1.1	7.7	9.2	10.7
		150	2.13	68	1.9	1.6	1.4	7.7	9.2	10.7
	30	50	1.54	49	1.4	1.2	1.0	7.7	9.2	10.7
		100	1.98	63	1.8	1.5	1.3	7.7	9.2	10.7
		150	2.42	77	2.2	1.9	1.6	7.7	9.2	10.7
	40	50	1.81	58	1.6	1.4	1.2	7.7	9.2	10.7
		100	2.25	72	2.0	1.7	1.5	7.7	9.2	10.7
		150	2.69	86	2.4	2.1	1.8	7.7	9.2	10.7
Angora does,	60		3.39	121	3.1	2.6	2.3	8.5	10.0	11.5
low lactation,	70		3.62	128	3.3	2.8	2.4	8.5	10.0	11.5
8.8 lb/yr mohair	80		3.84	135	3.5	3.0	2.6	8.5	10.0	11.5
	90		4.05	142	3.7	3.1	2.7	8.5	10.0	11.5
	100		4.25	148	3.9	3.3	2.8	8.5	10.0	11.5

Source: NRC (42).

[a]Protein content must be increased with increasing energy in diets in order to provide adequate protein at lower feeding levels.

[b]Feed energy levels 1.1, 1.3, and 1.5 Mcal/lb are approximately equivalent to 55%, 65% and 75% TDN, respectively.

Voluntary intake (self-feeding) is the target management strategy of many if not most goat production systems. Goats in range and pasture settings have free access to the standing forage (grasses, forbs, shrubs, mast) and consume to their voluntary limit. Whether the nutrient requirements are met depends on the nutrient composition of the forages selected. If one or more nutrients are too low, these can be provided supplementally. Goats fed in confinement depend on the skill of the manager in determining the proper formulation so that voluntary intake (free-choice feeding) results in proper nutrient intake. Unfortunately, voluntary intake is affected by many factors (diet digestibility, genotypic factors, environmental factors, etc.) and is very difficult to predict. For discussion purposes, the following rules of thumb for maximum voluntary intake are suggested (23).

Goat Class	Maximum Voluntary Intake (% of body wt)
Kids	4.5
Dry doe	2.8
Early pregnant doe	3.0
Late pregnant doe	2.7
Lactating doe, low	4.0
Lactating doe, high	5.0

The practical value of these numbers is in determining the appropriate minimum energy content of a ration for a goat of a particular size and in a particular productive state.

EXAMPLE 1

Dry doe weighing 80 lb.

$$\text{Maximum intake} = 80 \text{ lb} \times 2.8\%$$
$$= 2.2 \text{ lb/day}$$

Required nutrients/day: DE = 2.3 Mcal
CP = 74 g

Required intake at: 1.1 Mcal/lb = 2.1
1.3 Mcal/lb = 1.8
1.5 Mcal/lb = 1.5

Conclusion: Because 2.1 lb is less than the 2.2-lb maximum intake limit, a ration containing 1.1 Mcal/lb DE and 7.7% CP would adequately supply the required energy and protein for the dry doe.

EXAMPLE 2

Kid weighing 30 lb and gaining 100 g/day.

$$\text{Maximum intake} = 30 \text{ lb} \times 4.5\%$$
$$= 1.35 \text{ lb/day}$$

Required nutrients/day: DE = 1.98 Mcal
CP = 63 g

Required intake at: 1.1 Mcal/lb = 1.8
1.3 Mcal/lb = 1.5
1.5 Mcal/lb = 1.3

Conclusion: Because 1.3 is similar to the 1.35 lb/day maximum intake limit, a ration containing 1.5 Mcal/lb DE and 10.5% CP is adequate to achieve desired gain of 100 g/day.

EXAMPLE 3

Lactating Angora doe weighing 70 lb and producing 8.8 lb/yr mohair.

$$\text{Maximum intake} = 70 \text{ lb} \times 4.0\%$$
$$= 2.8 \text{ lb/day}$$

Required nutrients/day: DE = 3.62 Mcal
CP = 128 g

Required intake at: 1.1 Mcal/lb = 3.3
1.3 Mcal/lb = 2.8
1.5 Mcal/lb = 2.4

Conclusion: Because 2.8 lb equates to the required intake, a ration containing 1.3 Mcal/lb DE and 10% CP would provide the approximate required energy and protein for a lactating Angora.

For example 1, ingredients from Table 18-3 could be used to provide the indicated energy and protein requirements as follows:

A. Annual ryegrass, free choice (2.2 lb/day)

$$\text{DE} = 2.2 \text{ lb/day} \times 1.1 \text{ Mcal/lb}$$
$$= 2.4 \text{ Mcal/day}$$
$$\text{CP} = 2.2 \text{ lb/day} \times 9\%$$
$$= 90 \text{ g/day}$$
(Note: 1 lb = 454 g)

Both DE and CP exceed requirements.
B. Alfalfa hay, free choice (2.2 lb/day)

$$\text{DE} = 2.4 \text{ Mcal/day}$$
$$\text{CP} = 180 \text{ g/day}$$

Both DE and CP exceed requirements.
C. Bermuda grass hay–corn mixture (2.2 lb/day)

	Pounds	DE (Mcal)	CP (g)
Bermuda grass hay	2.0	2.0	136
Corn	0.2	0.3	8
Total	2.2	2.3	144

DE equals and CP exceeds requirements.
D. Wheat straw–corn–soybean meal mixture (2.2 lb/day)

	Pounds	DE (Mcal)	CP (g)
Wheat straw	1.50	1.2	27
Corn	0.59	0.9	24
Soybean meal	0.11	0.2	22
Total	2.20	2.3	73

Both DE and CP approximate requirements.

Both annual ryegrass and alfalfa hays contain adequate energy and excess protein concentrations for an 80-lb dry doe fed free choice. Bermuda grass hay contains excess protein but is slightly low in energy. Combining nutrients provided by 0.2 lb of corn with that in

2 lb of Bermuda grass hay satisfies energy and protein requirements. Wheat straw is low in both energy and protein, and provisions of both a high-energy and a high-protein ingredient is necessary to achieve the optimal balance. Rations that would satisfy the energy and protein requirements for examples 2 and 3 follow:

30-Pound kid gaining 100 g/day

	Pounds	DE (Mcal)	CP (g)
Alfalfa hay	0.35	0.38	29
Corn	1.00	1.60	41
Total	1.35	1.98	70

70-Pound Angora doe, lactating

	Pounds	DE (Mcal)	CP (g)
Alfalfa hay	0.28	0.31	23
Wheat straw	0.84	0.67	15
Corn	1.54	2.46	63
Soybean meal	0.14	0.22	29
Total	2.80	3.66	130

These examples consider only energy and protein required by the goat and supplied in the diet. Although these nutrients are the major considerations for selecting proper feedstuffs, minerals and vitamins must not be ignored. As suggested, Ca and P are the main mineral elements for consideration. The NRC table (Appendix Table 21) lists the actual requirements for the different classes of goats. In most cases, goats will not encounter a mineral imbalance if the diet contains at least one and one-half times as much Ca as P, and salt (NaCl) is offered free choice. In large, commercial operations it is suggested that a more precise formulation be exercised. Similarly, vitamin nutrition is probably adequate if green, leafy hays that are high in carotene (see Table 18-3) are included in the diet. If complete, mixed rations are fed, it is suggested that vitamin A be added at 2000 IU/lb.

SUPPLEMENTS FOR GOATS ON RANGELAND

Proper supplementation is achieved only if the range diet and supplemental feed combine to properly supply the required nutrients (Fig. 18-10). Usually, protein is the first limiting nutrient for animals (nonpregnant, nonlactating) producing at a low level. Thus a small amount

FIGURE 18-10 Adult Angora females in a 5-month fleece eating whole corn (*Zea mays*) off the ground. Note earthen pond in the background. (Photo by Kenneth Bales.)

(100 to 300 g/day) of a high-protein feed (32% to 45%) is usually adequate. High-quality forages (alfalfa) can occasionally be a good alternative, but rarely should a low-quality roughage be fed on rangeland. Animals in a high stage of production may need both supplementary protein and energy. The proper balance is difficult to estimate. The author prefers to feed a supplemental feed with a protein safety factor to assure that protein does not become limiting. Kids and yearlings on average to dormant rangeland should be fed up to 350 g/day of a 30% to 35% protein concentrate. Under the same circumstances, developing billies, breeding does, and does in late pregnancy should receive up to 500 g/day of a 25% protein concentrate. Dry does and does between breeding and the fourth month of gestation could be fed less of a higher-protein concentrate as indicated to prevent loss of weight.

FEEDING PLAN FOR DAIRY GOATS

Dairy goats require special consideration because of the large variation in nutrient requirements within the annual production cycle (40). Good nutrition management in dairy goats requires a keen understanding of forages and concentrates and their use in supplying (but not oversupplying) the needed nutrients.

The period of lowest requirements is during the dry, early-pregnant stage. Highest requirements occur between early and midlactation. Actually, energy requirements during peak lactation exceed the goat's capacity to take in energy. As a result, energy is mobilized from deposited body fat at a rapid rate. Because appetite is often depressed during late gestation and early lactation, fat reserves must be deposited during the low requirement periods (late lactation, early gestation).

A good feeding plan begins with a close scrutiny of body fatness during the latter stage of lactation (early pregnancy). If the doe is very thin, increase the level of nutrients (especially energy) in the diet. This can be accomplished either by feeding a high-quality hay, grazing a good pasture, feeding concentrates liberally, or supplying an appropriate combination of these. Once the doe has increased in condition to the desired level (well-conditioned but not overly fat), decrease the nutrient intake level by reversing the above strategy (decrease hay quality, lower concentrate feeding, etc.) so that weight is maintained or only slightly increasing.

The next decision point is at about 6 weeks before kidding. Again, the level of nutrition should be increased to adjust for the increasing requirements of fetal development and the tendency for the doe to have a reduced appetite. Good-quality hay should be fed along with concentrates, beginning at about $\frac{1}{3}$ lb/day and slowly increasing to 1 lb/day at kidding time. After kidding, concentrate feeding should continue, and after about 3 weeks it should reach 2 lb/day for average producers and 3 lb/day for high producers. Feeding of good-quality hay (or pasture) should continue. Concentrate feeding just before and just after parturition is important for the prevention of pregnancy toxemia and ketosis, respectively. After the peak lactation period (3 to 4 months), concentrate feeding can be decreased slowly, allowing the elevated forage portion of the diet to encourage continued milk yield.

The ingredients for the protein–grain ration should be determined based on the hay portion of the diet. Table 18-6 illustrates mixtures containing from 12% to 20% protein. Other feed ingredients having advantages of price and availability can be substituted. If grass hay is fed, the concentrate mixture should contain from 16% to 20%, because hay quality ranges from high to medium, respectively. Low-quality grass hay should not be fed. The 12% to 14% concentrates are adequate if excellent- to good-quality legume hay is fed.

FEEDING BREEDING MALES

Males should go into the breeding season in good, vigorous condition but not overly fat. Breeding males naturally reduce feeding time; thus, they are expected to lose weight. Although some weight loss is acceptable, a highly depleted state can result in low breeding efficiency. It is recommended that concentrate feeding be practiced (1 to 1.5 lb/day) for 3 to 4 weeks before the beginning of the breeding period until breeding has ended and body condition is recovered. Feeding between breeding seasons can be at a maintenance level.

UNIQUE MANAGEMENT NEEDS OF THE ANGORA

Early body development is important for good lifetime productivity of Angoras (49). The female kid should reach minimum weights of 35 lb (15.9 kg) at 6 months (weaning) and 60 lb (27.2 kg) at 18 months (first breeding). Optimal development (Fig. 18-11) varies according to the normal mature size of the flock, but goats that are underdeveloped at first breeding will be low producers during their productive lifetimes.

Reproductive rate in Angora females can be increased by flushing (increasing the feeding level or nutritional status about 3 weeks prior to breeding and continuing 3 weeks into the breeding season). This can be accomplished in several ways, including increasing the feeding rate, increasing the concentrate level of the feed, increasing the amount of concentrates fed on pasture or range, moving animals to a fresh pasture of higher quality, treating animals for internal parasites, and so on. Flushing will not give a consistent response. The most likely positive response occurs when the females are in fair to good (not poor to excellent) condition and have been slightly underfed. Goats that are in good to excellent condition and are well fed likely will not respond to increased concentrate feeding (17).

The Angora is very sensitive to rapid changes in environmental conditions. Abortion at about the fourth month of pregnancy and death from hypothermia following shearing are at least related to a nutrition and environment interaction. Abortion occurs when the glucose level drops and triggers an endocrine mechanism (61). Since the goat is so sensitive to rapid change, a change in diet, location, or weather, shearing, or drenching can cause the doe to change her eating pattern or to refuse to eat, resulting in a drop in blood glucose. The routine of Angora females should not be changed after the 80th day and until the 120th day of pregnancy, especially the feeding routine. Changes made after the 120th day should not increase the nutritional stress on the goats.

Death from hypothermia following shearing is as risky in the late summer as during winter. The shock comes as a result of the

TABLE 18-6

Sample grain mixtures for lactating goats

Ingredients	1	2	3	4	5
	Level of Protein in Finished Mix (%)				
	12	14	16	18	20
	(lb/ton)				
Rolled corn	1000	900	800	720	656
Crimped oats	451	421	300	240	200
Beet–citrus pulp	195	200	200	200	200
Dried brewers grain	—	—	150	200	200
40% protein supplement	170	300	—	516	—
Soybean meal	—	—	356	—	600
Molasses	150	150	150	100	100
Trace-mineralized salt	10	10	20	10	20
Dicalcium phosphate	—	—	10	10	20
Monosodium phosphate	20	15	10	—	—
Magnesium oxide	4	4	4	4	4

Source: Adapted from Adams et al. (1).

change (drop) in temperature (not just low temperature), wet conditions, and air velocity. A cold, blowing rain in August can kill freshly shorn goats at above 70°F. This is especially hazardous when goats that are freshly shorn and have been held off feed for several hours are returned to pasture without protection just before a cold, blowing rain. Freshly shorn goats should be either protected from the weather (shedded) and fed or turned into an area with natural protection (caves, cliffs, dense brush area) and with enough time to fill up with good-quality feed or forage.

SUMMARY

Goats are ruminant animals with an inquisitive and enterprising grazing–browsing behavior. They are uniquely attracted to trees and shrubs and are often used to reduce or prevent

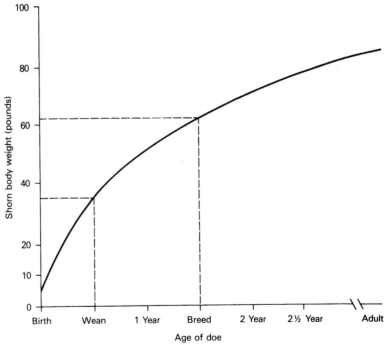

FIGURE 18-11 Optimal growth curve for Angora females. [From Huston et al. (19).]

encroachment of undesired, brushy plants. Goats are physically agile and able to reach areas that are too remote or otherwise unavailable for use by other livestock species.

Three general goat types include dairy, Angora, and Spanish (or meat) goats. Nutrient requirements for goats are determined as the sum of the requirements for the different physiological processes carried on simultaneously (maintenance, pregnancy, and lactation). The lactating dairy goat is the most productive and requires the highest levels of nutrients. The Angora is the most sensitive to dietary and environmental changes. The Spanish goat is more rugged and fertile than the Angora and tends to be a more aggressive and effective browser. The introduction of the Boer breed into the meat goat population increases meat production potential, but may also increase nutrient requirements and reduce the geographic areas of adaptation. Nutrient requirements and ex-

ample rations for different types and classes of goats were computed and discussed.

Certain periods of the annual cycle are critical, especially for the Angora, and require special managerial attention. Early life development (prior to first breeding) is important to establish an optimal adult body size and high lifetime productivity. The nutritional status at breeding should be good to assure proper ovulation and conception. Increased feeding at breeding (flushing) should be practiced when the goat has marginal body condition. The Angora goat is highly subject to nutritional stress abortion at about 90 to 100 days gestation. Changes in routine (feeding, watering, pasture location) should be avoided during this period. Perinatal losses of kids result if the does are underfed during late pregnancy. Dairy goats should be relatively fat at parturition to prepare for the mobilization of energy to support the high energy demands of early lactation.

REFERENCES

1. Adams, R. S., et al. 1984. In: G. F. W. Haenlein and D. L. Ace, eds. *Extension goat handbook: C-1.* Washington, DC: USDA Extension Service.

2. Ashbrook, P. F. 1982. *Proc. 3rd Int. Conf. Goat Prod. Dis.:* 153.

3. Blackburn, H. D. 1995. *J. Animal Sci.* 73:302.

4. Bowling, R. A., et al. 1976. *Tex. Agric. Exp. Sta.* PR-3408:48.

5. Brody, S. 1938. *Mo. Agric. Exp. Sta. Res. Bull.* 291.

6. Casey, N. H., and W. A. Van Niekerk. 1988. *Small Rum. Res.* 1:291.

7. Davis, G. D., L. E. Bartell, and C. W. Cook. 1975. *J. Range Manage.* 28:216.

8. Devendra, C., and G. B. McLeroy. 1982. *Goat and sheep production In the tropics: 271.* New York: Longman.

9. DHIA. 1995. *National cooperative dairy herd improvement program handbook.* Washington, DC: USDA Extension Service.

10. Gallagher, J. R., and M. Shelton. 1973. *Tex. Agric. Exp. Sta.* PR-3190:36.

11. Gathuka, Z. G., et al. 1982. *Tex. Agric. Exp. Sta.* CPR-4026:23.

12. Ghosh, T. K. 1987. *Proc. IV Int. Conf. on Goats* 2:1267, Brasilia, Brazil.

13. Giger, S., et al. 1981. In: P. Morand-Fehr, A. Bourbouze, and M. de Simiane, eds. Nutrition and systems of goat feeding, *Proc. Int. Symp.* (May 12–15): 254, Tours, France.

14. Glimp, H. A. 1995. *J. Animal Sci.* 73:291.

15. Haenlein, G. F. W. 1980. Nutrient requirements of dairy goats—past and present. *Int. Goat and Sheep Res.* 1:79.

16. Haenlein, G. F. W. 1996. *J. Animal Sci.* 74:——.

17. Hunt, L. J., et al. 1987. *SID Research Digest* (Summer): 18.

18. Huston, J. E. 1978. *J. Dairy Sci.* 61:988.

19. Huston, J. E., M. Shelton, and W. C. Ellis. 1971. *Tex. Agric. Exp. Sta. Bull.* 1105.

20. Huston, J. E., et al. 1981. *Tex. Agric. Exp. Sta. Bull.* B-1357:16.

21. Huston, J. E., et al. 1986. *J. Animal Sci.* 62:208.

22. Huston, J. E., B. S. Engdahl, and K. W. Bales. 1988. *Small Ruminant Res.* 1:81.

23. Huston, J. E. 1989. Unpublished data.

24. Huston, J. E., and D. F. Waldron. 1996. *Proc. Southeast. Reg. Meat Goat Prod. Symp.:* 12 (Feb. 21–24), Tallahassee, FL.

25. Lambert, M. G., D. A. Clark, and K. Betteridge. 1987. *Proc. IV Int. Conf. on Goats* 2:1307, Brasilia, Brazil.

26. Lawson, J. L., D. W. Forrest, and Maurice Shelton. 1983. *Tex. Agric. Exp. Sta.* CR-4171:3.

27. Lippke, H., W. C. Ellis, and B. F. Jacobs. 1986. *J. Dairy Sci.* 69:403.

28. Lu, C. D., T. Sahlu, and J. M. Fernandez. 1987. *Proc. IV Int. Conf. on Goats* 2:1229, Brasilia, Brazil.

29. Lupton, C. J. 1996. *J. Animal Sci.* 74:1164.

30. Lupton, C. J., et al. 1996. *J. Animal Sci.* 74:545.

31. Mahler, C. 1945. *East Afr. Agric. J.* 11:115.

32. Marshall, W. H., et al. 1976. *Tex. Agric. Exp. Sta.* PR-3410:51.

33. Martinez Parra, R. A., et al. 1981. *Proc. Int. Symp. Nutr. Syst. Goat Feed:* 369.

34. Masson, C., R. Rubino, and V. Fedele. 1991. In: P. Morand-Fehr, ed. *Goat nutrition:* 145. Wageningen: Pudoc.

35. McGee, A. C. 1957. *Tex. Agric. Exp. Sta.* MP-208:8.

36. McGowan, C. H., and A. McKenzie-Jakes. 1996. *Proc. Southeast. Reg. Meat Goat Prod. Symp.:* 12 (Feb. 21–24), Tallahassee, FL.

37. Morand-Fehr, P. 1981. In: C. Gall, ed., *Goat production:* 193. New York: Academic Press.

38. Morand-Fehr, P., et al. 1982. *Proc. 3rd Int. Conf. Goat Prod. Dis.:* 90.

39. Morand-Fehr, P., and D. Sauvant. 1985. In: D. C. Church, ed., *Livestock feeds and feeding,* 2nd ed: 372. Englewood Cliffs, NJ: Prentice Hall.

40. Morand-Fehr, P., and D. Sauvant. 1987. *Proc. 4th Int. Conf. Goats:* 1275.

41. NRC. 1978. *Nutrient requirements of dairy cattle.* 5th ed. Washington, DC: National Academy Press.

42. NRC. 1981. *Nutrient requirements of goats.* Washington, DC: National Academy Press.

43. Oates, A. V. 1956. *Rhodesia Agric. J.* 53:68.

44. Oman, J. S., et al. 1996. *Proc. Southeast. Reg. Meat Goat Prod. Symp.:* 15 (Feb. 21–24) Tallahassee, FL.

45. Provenza, F. D., et al. 1983. *Utah Science* (Winter): 91.

46. Rector, B. S. 1983. Diet selection and dietary forage intake by cattle, sheep and goats grazing in different combinations. Ph.D. dissertation: 173. Texas A&M Univ., College Station.

47. Riera, S. 1982. *Proc. 3rd Int. Conf. Goat Prod. Dis.:* 162.

48. Shelton, M. 1993. *Angora goat and mohair production.* San Angelo, TX: Anchor Publishing Co.

49. Shelton, M. 1961. *Tex. Agric. Exp. Sta.* MP-496.

50. Shelton, M., and J. R. Stewart. 1973. *Tex. Agric. Exp. Sta.* PR-3187:28.

51. Shelton, M., and D. Spiller. 1977. *Tex. Agric. Exp. Sta.* PR-3445:1.

52. Shelton, M., et al. 1982. *Tex. Agric. Exp. Sta.* CPR-4026:83.

53. Shelton, M., and J. Groff. 1984. *Tex. Agric. Exp. Sta.* B-1485.

54. Shkolnik, A., and N. Silanikove. 1981. In: P. Morand-Fehr, A. Bourbouze, and M. de Simiane, eds. Nutrition and systems of goat feeding. *Proc. Int. Symp.* (May 12–15): 236, Tours, France.

55. Smith, G. C., B. W. Berry, and Z. L. Carpenter. 1972. *Tex. Agric. Exp. Sta.* PR-3027:31.

56. TDA. 1985. *Texas livestock, dairy and poultry statistics.* Austin, TX: Tex. Dept. Agric.

57. Taylor, C. A., Jr. 1985. Multispecies grazing research overview. *Proc. Conf. on Multispecies Grazing:* 65. Morrilton, AR: Winrock Int.

58. Taylor, C. A., Jr. 1983. Foraging strategies of goats as influenced by season, vegetation and management. Ph.D. dissertation: 129. Texas A&M Univ., College Station.

59. Van Soest, P. J., R. H. Wine, and L. A. Moore. 1966. *Proc. 10th Int. Grassl. Congr.* 10:438.

60. Waldron, D. F., et al. 1995. *J. Animal Sci.* 73(Suppl. 1): 253.

61. Wentzel, D. 1987. *Proc. IV Int. Conf. on Goats* 2:571, Brasilia, Brazil.

19

FEEDING SWINE

Gary L. Cromwell

INTRODUCTION

Swine production represents an important segment of the food animal industry in the United States and throughout the world. Pork is an important source of dietary protein for humans and is the most widely consumed red meat in the world. Annual consumption of pork in the United States is approximately 65 to 70 lb per person. Today's modern pork is much leaner than it was in the past and is widely accepted by consumers.

Pigs are raised on about 250,000 farms in the United States. Most of the pigs are raised in the Midwest (in the corn belt), but certain areas in the Southeast (North Carolina) are expanding rapidly. Pigs on these farms are raised in a variety of ways, from outdoor lots to complete confinement. Hog farms vary in size from small to very large, with some of the largest farms having thousands of sows and producing hundreds of thousands of pigs annually. Some of the largest farms are owned by corporations, which also own their own feed mills and packing plants for slaughtering and processing. Many of these farms have separate site operations to control disease, practice early weaning, and utilize other intensive production practices.

Regardless of the size of the operation or the condition under which pigs are produced, nutrition and feeding management are very important aspects of swine production. Traditionally, feed represents 65% to 75% of the total costs of production. Therefore, it is extremely important that swine producers have a good understanding of the nutrient requirements of pigs during each phase of the life cycle, a knowledge of the feedstuffs that can be used in pig feeding, and an appreciation for the finer points of feeding management in order to raise pigs efficiently and economically.

The feeding program has a major impact on animal performance and on overall profitability of the swine herd. High-quality feeds are needed in order to attain an optimal rate and efficiency of growth from birth to market and a high level of reproductive performance in the breeding herd.

This chapter addresses the nutrient requirements of swine, identifies feedstuffs that can be used to meet these requirements, and describes feeding programs that optimize the growth and reproductive performance of pigs.

NUTRIENTS REQUIRED BY SWINE

Pigs have the ability to obtain nutrients from a wide variety of feedstuffs. Historically, wild pigs were omnivorous and consumed both vegetative and animal food sources. In the early days of our country, domestic pigs were allowed to forage on grass, roots, acorns, refuse, and whatever else was available for the taking.

Later, pigs were confined to dirt lots or pasture and fed ear corn and tankage or skim milk. Today, the majority of the pigs in the United States are raised in total confinement and fed highly fortified, grain-based diets in which the majority of the protein supplement consists of soybean meal.

Pigs are simple-stomached animals (see Ch. 2), so they must rely largely on feeds having readily digestible carbohydrates to meet their energy needs. The more complex carbohydrates (cellulose, hemicellulose) found in roughages and other fibrous feeds are broken down only by microbial fermentation. Pigs do not have a rumen; thus, the fibrous components of the diet are not utilized as efficiently as they are in ruminants. Also, simple-stomached animals like the pig depend on certain essential amino acids present in dietary protein from which to build their own body proteins. Unlike ruminants, pigs cannot synthesize the essential amino acids from poor-quality protein or from non-protein nitrogen (NPN) sources. Thus, the relative amounts of the essential amino acids in the protein are extremely important to pigs.

The nutrient requirements for growing pigs of various weights and the requirements of breeding swine are shown in Appendix Tables 22 through 28. These nutrient levels are established by the NRC (1) and are estimates of the minimum amount of a nutrient required by animals under average conditions. While the NRC standards serve as an excellent guideline, they do not include safety margins to allow for variability among animals or among feedstuffs. Therefore, recommended nutrient allowances made by nutritionists at universities and feed companies often are slightly higher than the NRC standards.

WATER

Water is so common that it is seldom thought of as a nutrient, yet it is one of the most important nutrient classes. From 80% (neonatal pig) to 55% (finishing pig) of the body is made up of water. Water is required in the body for hydration of cells and as a vehicle for moving nutrients into and wastes out of the body. Through evaporation (primarily in the lungs, because pigs do not sweat), water disperses surplus heat produced by the many metabolic processes. It also lubricates the joints, cushions the nerves, protects the developing fetuses, and is a major component of milk.

Swine of all ages should have free access to fresh, clean water at all times. Limiting wa-ter intake will result in reduced growth rate, reduced feed intake, poor efficiency of feed utilization in growing pigs, and reduced milk production in lactating sows. A severe limitation of drinking water can be fatal to pigs.

The requirement for water is influenced by many factors, including environmental temperature and humidity, water content of the feed, and weight of the pig. A useful guideline is that about 2 to 3 lb (or 1 to 1.5 qt) of water is required for every 1 lb of feed consumed. A lactating sow requires more water because of the high water content of the milk that she produces.

Water is often supplied from automatic drinking fountains (Fig. 19-1). In cold environments, water fountains need to be equipped with heaters to keep the water from freezing. In heated buildings or in the mild climates of the southern United States, nipple-type waterers are common (Fig. 19-2); they are less expensive and the water stays cleaner. When cup waterers are used, at least one cup should be provided for every 20 to 25 pigs. When nipple waterers are used, one nipple should be provided for every 15 pigs.

Quality of water is important. Water should be clean and relatively free of microbial contamination; otherwise, chlorination may be necessary. In some parts of the country, water is high in inorganic salts. Total dissolved solids (TDS) is a common measure of the mineral content of water. Ideally, water should have not

FIGURE 19-1 A freeze-proof bowl-type automatic waterer for pigs.

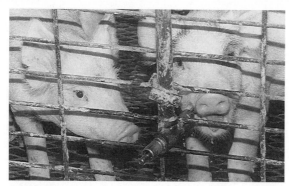

FIGURE 19-2 A nipple-type waterer for pigs.

more than 1,000 ppm of TDS. Higher TDS levels (2,000 to 3,000 ppm) can cause diarrhea or temporary water refusal. TDS levels above 5,000 ppm should be avoided, and TDS levels in excess of 7,000 ppm are unsuitable. Pigs will tolerate moderate levels of sulfates, but high levels (3,000 ppm or more) should be avoided.

ENERGY

Energy is not a nutrient, but it is an important end product of carbohydrate, fat, and protein metabolism in the body. The energy released from oxidation of these nutrients is used to maintain body temperature (approximately 102.5°F) or is stored in the form of high-energy compounds (e.g., ATP) for use in growth and other purposes. Energy is measured in kilocalories (kcal).

Energy is required for the basic body processes of pigs, such as breathing, heartbeat, and muscle movement. Energy is also necessary for protein synthesis, fat deposition, milk production, and other purposes. Accretion of lean tissue, which is largely protein and water, requires considerable energy (about 10.6 kcal/g of protein or 2.3 kcal/g of lean muscle). Accretion of fat tissue also requires energy expenditure (12.5 kcal/g of fat), but much of this energy is retained in the fat tissue. In young, rapidly growing pigs, energy may be more limiting than protein.

Nursing pigs derive most of their energy from fat and sugar (lactose) in milk. Most of the energy for growing pigs is derived from the metabolism of starch in cereal grains. Baby pigs are unable to utilize starch because of insufficient amylase and maltase (starch-digesting enzymes) in the small intestine. Protein in excess of the requirement can be used as an energy source, but it is too costly to be fed solely for energy. In older animals (sows and finishing pigs), a limited amount of energy can be derived from volatile fatty acids, products produced by bacterial fermentation of fiber in the large intestine.

Energy requirements of pigs are most commonly expressed as digestible energy (DE) or metabolizable energy (ME) and occasionally as net energy (NE). For all weight classes of pigs, the ME requirement is approximately 96% of the DE requirement. Energy requirements of pigs are influenced by their weight (which influences the maintenance requirement), by their genetic capacity for lean tissue growth or milk synthesis, and by the environmental temperature in which they are housed. The energy requirement for maintenance is directly related to metabolic body weight and is approximately 110 kcal of DE/kg body weight$^{0.75}$. (Metabolic weight can be determined on a hand calculator by multiplying pounds by 0.454; then cube the result and calculate the square root twice. For a 100-lb pig, the metabolic weight is $100 \times 0.454 = 45.4$ kg; $45.4 \times 45.4 \times 45.4 = 93,577$; $\sqrt{93,577} = 307$; $\sqrt{307} = 17.5$ kg). Thus, the maintenance requirement for a 100-lb pig is $110 \times 17.5 = 1,925$ kcal of DE per day. The total daily energy requirement for growing pigs is about 3.6 to 3.8 times the maintenance requirement in young pigs and 3.2 to 3.4 times the maintenance requirement in finishing pigs, when both are housed in a comfortable (thermoneutral) environment. Thus, the total daily energy requirement for a 100-lb pig is approximately $1,925 \times 3.6 = 6,930$ kcal of DE.

During the growth period, energy requirements increase as the pig increases in weight. For example, the ME requirement of a 200-lb pig (10,700 kcal/day) is greater than that of a 20-lb pig (2,100 kcal/day) because of its greater maintenance requirement and a faster rate of accretion of body tissues (fat and muscle). The ME requirement of a lactating sow (14,500 to 20,500 kcal/day) is much greater than that of a pregnant sow (5,800 to 6,500 kcal/day). In fact, high-producing lactating sows are capable of producing 18 to 24 lb of milk per day (or about 2.0 to 2.4 lb of milk per nursing pig), and they generally cannot consume enough energy to maintain energy balance; hence they lose weight during lactation.

Energy requirements are greater at low than at high environmental temperatures because of the need to produce heat for body warmth. For example, a 100-lb pig requires about 750 kcal more ME per day at 40°F compared with 60°F to achieve the same weight gain. When full fed, the pig simply consumes more feed in a cold environment to meet its increased energy needs. This is the reason why feed efficiency (i.e., the feed-to-gain ratio) of pigs is lower (or better) in summer than in winter.

Pigs that are full fed generally eat to meet their energy requirement, which means that they will consume more of a low-energy feed, such as a diet with oats or barley as the grain source, than they will consume of a moderate-energy diet, such as one based on corn. In contrast, if given a high-energy diet, such as a diet with added fat, their voluntary intake will decrease, because their energy requirement is met with less total feed. Therefore, feed conversion efficiency is influenced greatly by the energy density of the diet. Feeding high-energy diets results in a better feed-to-gain ratio; that is, less feed is required per unit of gain. Conversely, feeding low-energy diets increases the amount of feed required to produce a unit of body weight gain. Growth rate is influenced slightly by energy intake, but to a much lesser degree than is the feed-to-gain ratio.

Carcass quality also is influenced by energy intake of pigs. Reducing the energy intake by incorporating high levels of fibrous feedstuffs in the diet or by restricting feed intake will produce a leaner carcass. Adding fat to the diet tends to produce a slightly fatter carcass.

PROTEIN AND AMINO ACIDS

Body protein consists of 22 amino acids. About one-half of this protein is in the muscle tissues, with the remainder in the organs, digestive tract, blood, and hair. A small but important amount is found in the enzymes and hormones of the body. Protein makes up about 15% of the total body mass. Pig carcasses contain 45% to 55% muscle, approximately 22% of which is protein; thus, about 7% to 9% of the whole body (10% to 12% of the carcass) is in the form of edible protein.

For protein synthesis (or growth) to occur at a rapid rate in pigs, the diet must supply sufficient amounts of 10 of the 22 amino acids. These 10 amino acids are termed as dietary essential, because they cannot be synthesized by pigs. The other 12 can be synthesized as long as sufficient N is present.

Lysine is generally the first limiting amino acid in pig diets for two reasons: the lysine requirement is high because of the relatively high concentration of lysine in muscle (about 7%), and the lysine content of most feedstuffs is low. The requirements for the other amino acids are often expressed in relationship to lysine, with the ratio of each amino acid to lysine approximating the ratios in the whole body protein. Table 19-1 shows the ideal amino acid ratios for three weight classes of pigs. An ideal protein would be one that has all the amino acids in the exact pattern required by the pig. This, of course, does not happen in practice.

A good-quality protein source is one that provides the 10 essential amino acids in the amounts and proportions necessary for the particular need of the pig (growth, reproduction, lactation). Unfortunately, the protein in cereal grains is of very poor quality (see Ch. 7). Table 19-2 shows that 6 of the 10 amino acids required by a 50-lb growing pig are deficient in corn. Failure to supplement corn with sufficient amounts of a high-quality protein source, such as soybean meal, results in poor growth, inefficient feed utilization, an increase in carcass fatness, and general unthriftiness (Table 19-3).

Traditionally, swine diets have been formulated on the basis of crude protein. Protein levels are established for the various weight classes of pigs so that the most limiting amino acid (lysine) is present in adequate amounts. This system works well when corn and soybean meal are the major sources of energy and protein. But when other ingredients are used, diets should be formulated on a lysine basis. Table 19-2 illustrates that supplementing corn with soybean meal (79.0% corn, 19.5% soybean meal) to provide a 15.3% protein level meets the lysine and other amino acid requirements of a 50-lb pig.

Amino acid requirements are influenced by age and weight of the pig. The daily requirements are a function of the lean tissue (protein)

TABLE 19-1

Ideal pattern of essential amino acids for pigs of three weight categories

Body Weight (lb):	10–45	45–110	110–220
Amino Acid		% of Lysine	
Lysine	100	100	100
Arginine	42	36	30
Histidine	32	32	32
Isoleucine	60	60	60
Leucine	100	100	100
Methionine + cystine	60	65	70
Phenylalanine + tyrosine	95	95	95
Threonine	65	67	70
Tryptophan	18	19	20
Valine	68	68	68

Adapted from Baker and Chung (2).

TABLE 19-2

Essential amino acids likely to be deficient in corn and corn plus soybean meal (44% protein) for the growing pig

Item	Corn	Corn + Soybean Meal (79.0% corn + 19.5% SBM)	Requirement[a] (50-lb Pig)
Protein, %	8.5	15.3	15.5[b]
Amino acids, %			
Arginine	0.40	0.94	0.25
Histidine	0.24	0.42	0.22
Isoleucine	0.28[c]	0.61	0.46
Leucine	0.99	1.45	0.60
Lysine	0.25[c]	0.75	0.75
Methionine + cystine	0.37[c]	0.57	0.41
Phenylalanine + tyrosine	0.72	1.29	0.66
Threonine	0.29[c]	0.57	0.48
Tryptophan	0.06[c]	0.16	0.12
Valine	0.39[c]	0.71	0.48

[a]From NRC (1).

[b]The pig does not have a protein requirement as such, but a corn–soybean diet with this level of dietary protein will meet the requirement for the most limiting amino acid, lysine.

[c]Deficient.

accretion pattern of the pig. The accretion of protein increases with body weight, tends to maximize at about 90 to 110 lb of body weight, then decreases with increasing body weight (Fig. 19-3). Since dietary protein is used mostly for muscle growth, the daily requirements follow roughly the same pattern. The rate of lean growth (i.e., the rate of body protein accretion), which is influenced by genetics, health, gender, and environment, also affects the protein and amino acid requirements of pigs. For example, a pig gaining at a rate of 0.90 lb of carcass lean per day requires about 25% more dietary protein than one gaining at a rate of 0.70 lb of carcass lean per day.

When expressed on a daily basis, the protein and amino acid requirements increase with increasing body weight (at least up to 150-lb body weight), but when expressed as a percent of the diet, the requirements decrease with increasing pig weight. For example, a 150-lb pig

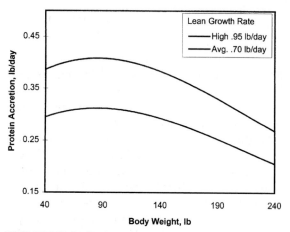

FIGURE 19-3 Daily protein accretion rate of pigs with average lean growth rates of 0.95 and 0.70 lb/day from 40 to 240 lb. The maximum lean accretion occurs between 90 and 110 lb of body weight.

requires more grams of protein per day than a 20-lb pig (375 versus 130 g/day), but when the requirements are expressed on a percentage basis, the requirement of a 20-lb pig is higher than for a 150-lb pig (20% versus 14%). This is because a 150-lb pig consumes about four times as much feed as a 20-lb pig (6.0 versus 1.5 lb).

MINERALS

Minerals have many very important functions in the body. Compared with energy and protein, minerals are required in very small amounts. Thirteen minerals are required in the diet by pigs (Table 19-4). The minerals most likely to be deficient in grain–soybean meal diets are calcium (Ca), phosphorus (P), sodium (Na), chlo-

TABLE 19-3

Effects of feeding a protein-deficient diet on performance of growing pigs[a]

Item	Corn,[b] 8% Protein	Corn–Soybean Meal,[b] 16% Protein
Daily gain, lb	0.37	1.85
Daily feed, lb	2.50	4.83
Feed-to-gain ratio	6.76	2.61

[a]Summary of three University of Kentucky experiments, 12 pigs per treatment, 47 to 80 lb.

[b]Both diets supplemented with minerals and vitamins.

TABLE 19-4

Minerals likely to be deficient in corn and corn plus soybean meal (44% protein) for the growing pig

Item	Corn	Corn + Soybean Meal (79.0% corn + 19.5% SBM)	Requirement[a] (50-lb pig)
Major minerals, %			
Calcium	0.01[b]	0.06[b]	0.60
Phosphorus	0.28[b]	0.34[b]	0.50
Phosphorus, available	0.03[b]	0.05[b]	0.23
Sodium	0.01[b]	0.02[b]	0.10
Chlorine	0.05[b]	0.05[b]	0.08
Magnesium	0.11	0.14	0.04
Potassium	0.33	0.66	0.23
Sulfur	0.11	0.17	—[c]
Trace minerals, ppm			
Copper	3.5[b]	7.2	4.0
Iron	33.0[b]	53.0[b]	60.0
Manganese	5.7	10.3	2.0
Zinc	9.0[b]	25.0[b]	60.0
Iodine	0.03[b]	0.06[b]	0.14
Selenium[d]	0.06[b]	0.09[b]	0.15

[a]From NRC (1).

[b]Deficient.

[c]The requirement is unknown but is met by the sulfur in methionine and cystine.

[d]Level in corn is variable, depending on area where grown.

rine (Cl), and the trace minerals iron (Fe), zinc (Zn), iodine (I), and selenium (Se).

The two minerals required in the greatest amounts by pigs are Ca and P. Adequate levels of both Ca and P must be included in the diet for strong skeletal structure. If either Ca or P is deficient, poor bone mineralization will occur, which can lead to deformation of the legs (rickets), lameness, or bone fractures. Sound feet and legs are especially important in pigs reared on concrete or slatted floors in confinement facilities. A deficiency of P will also result in slow and inefficient growth. Calcium deficiency seldom depresses growth, unless the deficiency is severe.

Calcium and P need to be kept in proper balance in the diet. An excess of Ca can cause problems, especially if the P level is marginal. The most favorable ratio of Ca:P is between 1:1 and 1.25:1. The maximum ratio of Ca:P is 1.5:1.

Feed grains are extremely deficient in both Ca and P (Table 19-4). In addition, most (60% to 80%) of the P in grains and oilseed meals is in the form of phytate, an organic complex that is not well utilized by pigs. The total P supplied by corn and soybean meal amounts to about 0.32% of the diet, but only about 0.05% is available to the pig. The requirement for total P takes into account that a large portion of the P in the natural feedstuffs is unavailable.

The amount of P in feedstuffs that is actually available to the pig is quite variable (Table 19-5). For many years, nutritionists used a guideline that one-third of the P in feedstuffs of plant origin was available to the pig. But now it is known that this rule of thumb is not valid (3). For example, in cereal grains the bioavailability of P in corn is low (12% to 14%), it is intermediate in grain sorghum, oats, and barley (20% to 30%) and it is high in wheat, triticale, and high-moisture grain (46% to 53%). The P in sunflower meal, peanut meal, and cottonseed meal is poorly available (1% to 12%), in soybean meal it is intermediate in bioavailability (23% to 31%), and in most animal protein sources the P is more highly available (66% to 96%). The P in inorganic phosphate supplements is considerably more bioavailable than that in natural feedstuffs. Formulating swine diets on an available P basis is recommended when ingredients are used that have P availability values that are considerably different from those of corn and soybean meal.

Phytase is now available for addition to pig diets (see the section on feed additives). This enzyme degrades phytate and increases the bioavailability of P in corn, soybean meal, and other feedstuffs of plant origin. Use of phytase in the diet allows for the feeding of lower P diets, which reduces the amount of P excreted by pigs. A reduction in P in pig manure has important implications, because excess P from animal wastes is considered a potential environmental pollutant.

TABLE 19-5

Biological availability of the phosphorus in feedstuffs for pigs

Ingredient	Bioavailability of P (%)[a]	Ingredient	Bioavailability of P (%)[a]
Grains		High-protein meals, plant origin	
Corn, pelleted	12	Cottonseed meal	1
Corn	14	Sunflower meal	3
Grain sorghum	20	Palm kernel cake	11
Oats	22	Peanut meal	12
Barley	30	Canola meal	16
Grain sorghum, high moisture	43	Soybean meal, dehulled	23
Triticale	46	Soybean meal, 44%	31
Wheat	50	High-protein meals, animal origin	
Corn, high moisture	53	Meat and bone meal	66
Grain by-products		Dried skim milk	91
Oat groats	14	Blood meal	92
Hominy feed	14	Fish meal	93
Corn gluten meal	25	Dried whey	96
Rice bran	25	Inorganic phosphates	
Wheat bran	29	Steamed bone meal	82
Brewers grains	33	Defluorinated rock phosphate	87
Wheat middlings	41	Monocalcium phosphate	100
Corn gluten feed	59	Dicalcium phosphate	100
Distillers grains + solubles	76		
Soybean hulls	78	Phytase addition	
Miscellaneous		Corn–soybean meal	15
Alfalfa meal	>100	Corn–soybean meal + phytase	45

Source: Cromwell and Coffey (3).

[a]Relative to the availability of P in monosodium or monocalcium phosphate, which is given a value of 100. All values are based on slope-ratio assays of bone ash and/or bone-breaking strength in pigs.

Sodium and Cl are also required for normal growth and body functions. Most feeds are low in Na and marginal in Cl. The requirement for both minerals can be met by supplementing the diet with 0.25% to 0.50% salt. Failure to supply adequate salt to pigs causes depressed feed intake and poor growth rate. Pigs are able to cope with high levels (5% to 7%) of salt in the diet as long as they have free access to fresh water.

Potassium (K), magnesium (Mg), and sulfur (S) are required by pigs, but they do not need to be supplemented because the natural feedstuffs contain ample amounts of these minerals (Table 19-4).

Zinc is required for normal growth and healthy skin. A deficiency of Zn causes slow growth and a scabby appearance of the skin, a condition called parakeratosis. This condition is similar to the appearance of a pig heavily infested with mange. High levels of dietary Ca tend to interfere with Zn absorption, so if high levels of Ca are fed, Zn levels should also be increased. Also, high levels of phytate tend to bind Zn and make it unavailable. Thus, the Zn requirement is much higher in a grain–soybean meal diet (which is high in phytate) than in a diet in which a large portion of the protein is of

animal origin. As a guide, a grain–soybean meal diet should contain 100 ppm of supplemental Zn for each 1% Ca in the diet. Pharmacological levels of Zn (3,000 ppm, as Zn oxide) have recently been shown to stimulate growth and feed intake in early-weaned pigs. This will be discussed further in the section on feed additives.

Iron is needed for hemoglobin synthesis in pigs. A deficiency of Fe results in depressed hemoglobin levels in the blood, a condition called anemia. Baby pigs develop anemia in the first week of life if not given supplemental Fe. In white pigs, anemia shows up as a pale color around the eyes, nose, and ears, tissues that otherwise are pink. In advanced stages of anemia, "thumping" or spasmodic breathing may occur. Growth rate is seldom depressed; in fact, the largest pig in the litter is often the first one to show symptoms of anemia because of its fast growth rate and the rapid dilution of its body Fe supply.

There are several reasons why the baby pig is very susceptible to anemia. The newborn pig grows exceptionally fast, quadrupling its weight in only 2 to 3 weeks. During this time, its only food is milk, which is extremely deficient in Fe. The pig has relatively low Fe reserves at birth and has little opportunity to obtain any Fe from its environment when it is raised in confinement. Pigs raised in dirt lots generally obtain sufficient Fe from the soil to meet their needs.

Anemia can be prevented by giving an intramuscular injection of 100 to 200 mg of Fe (as Fe-dextran, Fe-dextrin or gleptoferrin) before 3 days of age (Fig. 19-4) or by oral administration.

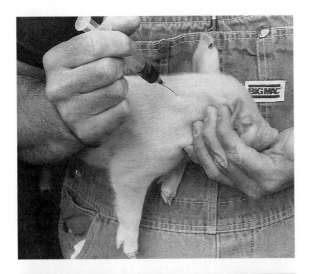

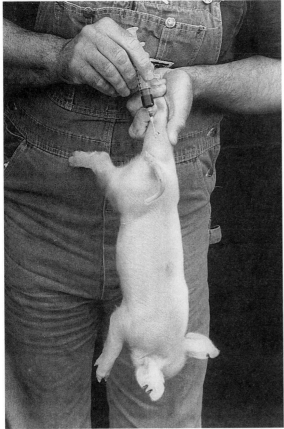

FIGURE 19-4 Anemia can be prevented by giving baby pigs an iron solution injection in the neck or ham.

Iron deficiency is seldom encountered in older pigs because most feedstuffs contain some Fe, and the P supplements (such as dicalcium phosphate and defluorinated rock phosphate) contain relatively high levels of Fe.

Copper also is needed for hemoglobin synthesis, but the Cu requirement is much less than the Fe requirement. Copper deficiency is seldom seen in pigs. This mineral, like Zn, stimulates growth and feed intake when pharmacological levels (100 to 250 ppm) are fed, with the response being much like that obtained when antibiotics are fed. This will be discussed later in the section on feed additives.

Iodine is deficient in feedstuffs grown in certain areas of the country. In addition, many feedstuffs (soybeans, rapeseeds, linseed, etc.) contain goitrogens, compounds that interfere with I absorption or the incorporation of I into thyroxine. Failure to supplement a grain–soybean meal diet with I will cause a four- to fivefold enlargement of the thyroid, a condition called goiter. Iodine can be provided conveniently by adding the recommended level of iodized salt to the diet.

Selenium deficiency is frequently a problem in swine. Crops grown in the Pacific Northwest, the Northeast, and the states around the Great Lakes are very low in Se, whereas crops grown in the South Dakota, Nebraska, and certain areas in the western United States are high in Se. A recent survey of corn and soybean meal samples obtained from 15 states in the Midwest (4) showed a 15-fold range in the Se content of corn (0.02 to 0.30 ppm) and a 12-fold range in the Se content of soybean meal (0.08 to 0.95 ppm).

Failure to supplement diets containing low-Se grain and low-Se protein sources with Se can result in liver necrosis, heart lesions, pale muscles, and sudden death in pigs, a condition called hepatosa dietetica. This trace mineral is regulated by the FDA, and the maximum level of supplementation is 0.3 ppm in swine feeds.

Chromium (Cr) is an important regulator of glucose metabolism and is probably required by pigs at very low dietary levels. Several recent research studies have shown improvements in carcass leanness when Cr (as Cr picolinate) was added to the diet of growing–finishing pigs (5). There is also evidence that feeding Cr picolinate during the developmental stage may improve reproductive performance in sows (6); however, more research is needed to confirm this finding.

One additional mineral, fluorine (F), is not considered as essential, but care should be taken to prevent excessive levels in the diet. Excessive F intake is generally associated with the feeding of unprocessed, high-F rock phosphate as a P source. Excessive F can cause mottling and erosion of teeth and abnormal bone formation. Pigs can tolerate up to 225 ppm F when rock phosphate is the source and up to 150 ppm F when the F is in a more soluble form, such as sodium fluoride.

VITAMINS

Vitamins serve many important and varied functions in the body. Fourteen vitamins are required by pigs, all in very small amounts. Cereal grains and plant protein supplements are very poor sources of many of the vitamins (Table 19-6). Those most likely to be deficient in grain–soybean meal diets are vitamins A and D and, of the B-complex vitamins, riboflavin, pantothenic acid, niacin, and vitamin B_{12}.

Vitamin A occurs as its precursor, carotene, in yellow corn and green plants. However, it is risky to rely on the carotene in corn to meet the vitamin A requirements of pigs. Some of the carotenes in corn are not converted efficiently to vitamin A. Also, heat damage or the use of organic acids to preserve high-moisture corn destroys much of the carotene. From a practical standpoint, vitamin A should be added to

pig feed (as vitamin A palmitate) at levels of at least two to three times the requirement; first, because it is relatively cheap, and second, because vitamin A is relatively unstable in mixed feeds.

Vitamin D is essentially absent from most feedstuffs that are commonly fed to pigs. Swine can synthesize vitamin D when exposed to sunlight; however, because many pigs are raised in confinement and away from direct sunlight, vitamin D should be supplemented. Although pigs can utilize either the form found in plants (D_2) or the form found in animal products (D_3), there is evidence that pigs, like chicks, utilize vitamin D_3 better than vitamin D_2.

Vitamin E may be adequate in grain and plant protein sources, but is likely to be marginal or deficient in some instances. Therefore, it is recommended that pig diets be fortified with vitamin E. This vitamin serves as a natural antioxidant in feeds and, when deficient, will cause symptoms that are similar to those of Se deficiency (liver necrosis, pale muscles, etc.). High levels of vitamin E fortification of sow diets have recently proved beneficial to reproductive performance (7).

Vitamin K is present in some feeds and is synthesized by microorganisms in the hind gut of the pig. Generally, microbial synthesis is assumed to provide adequate levels of vitamin K for pigs, provided that the pigs have access to

TABLE 19-6

Vitamins likely to be deficient in corn and corn plus soybean meal (44% protein) for the growing pig

Vitamin	Corn	Corn + Soybean Meal (79.0% corn + 19.5% SBM)	Requirement[a] (50-lb pig)
Fat-soluble vitamins			
Vitamin A, IU/lb	200[b]	158[b]	590
Vitamin D, IU/lb	—[b]	—[b]	68
Vitamin E, IU/lb	10	7.7	5
Vitamin K, mg/lb	—	—	0.2[c]
Water-soluble vitamins, mg/lb			
Vitamin C	—	—	?[d]
B-complex			
Riboflavin	0.5[b]	0.64[b]	1.1
Pantothenic acid	2.3[b]	3.2[b]	4.6
Niacin, available	—[b]	2.5[b]	4.5
Vitamin B_{12}	—[b]	—[b]	0.005
Choline	227	404	136
Pyridoxine	2.8	2.7	0.5
Thiamin	1.7	1.9	0.5
Folacin	0.14	0.16	0.14
Biotin	0.03	0.05	0.02

[a]From NRC (1).
[b]Deficient.
[c]Requirement generally met by microbial synthesis.
[d]Requirement unknown. Sufficient amounts are present in natural ingredients and/or are synthesized in the body to meet the requirements.

their feces. However, deficiencies have been encountered when pigs are fed moldy grain or other feeds with mold damage or when fed diets with excessive levels of Ca. Symptoms include prolonged blood clotting time, resulting in excessive bleeding and bruising. Death often results from internal hemorrhage. The deficiency can be corrected by injection of vitamin K or by including it in the feed. For insurance purposes, vitamin supplements should contain a small amount of vitamin K.

Cereal grains and plant protein supplements are very poor sources of riboflavin, niacin, and pantothenic acid, and they contain no vitamin B_{12}. In addition, the niacin in cereal grains is chemically bound and totally unavailable to the pig. A deficiency of one or more of these vitamins can cause retarded growth, and if the deficiency is severe, chronic diarrhea can occur. Vitamin supplements should always contain these four B vitamins.

Two other B vitamins, biotin and folacin, generally do not need to be added to growing pig diets. However, biotin may become a problem when wheat or barley is fed because the biotin in these two cereal grains is much less available than the biotin in corn. Also, there is clear evidence that reproductive performance in sows is improved when biotin and folacin are supplemented (8, 9). Thus, most vitamin premixes generally contain both of these vitamins. Biotin deficiency causes cracking of the hooves. The importance of feet and leg soundness in pigs and breeding animals housed in total confinement is another impetus for biotin supplementation.

The choline requirement of growing pigs is generally met by natural ingredients, but choline supplementation has been shown to improve litter size in pregnant sows (10). Under experimental conditions, choline deficiency in sows has been shown to cause a spraddle-legged condition in newborn pigs. Choline supplementation of a corn–soybean meal diet for sows has not been effective, however, in preventing the spraddle-legged pig condition occasionally seen in practice.

Thiamin (vitamin B_1) and pyridoxine (vitamin B_6) are adequately supplied to the growing pig by the natural ingredients, and supplementation is not necessary. Ascorbic acid (vitamin C) is synthesized in adequate amounts by pigs, and most research has not shown a benefit from the addition of this vitamin. An exception might be under high-stress conditions, such as immediately postweaning, in which case vitamin C supplementation occasionally has been shown to improve performance of early-weaned pigs.

FEEDSTUFFS USED IN SWINE DIETS

This section will describe the more common feed ingredients used in swine diets in the United States. Additional details on these and other feedstuffs not covered in this section are provided in Chs. 4 to 10 and in other publications (11–14).

ENERGY SOURCES

CEREAL GRAINS Corn accounts for over 85% of the grain fed to swine in the United States. It is an excellent feed for swine of all ages because it is high in digestible starch, is low in fiber, is very palatable, and can be fed in a variety of ways. For these reasons, corn is the standard to which other grains are compared (Table 19-7). In spite of its many virtues, corn is low in protein (8.5%), is extremely deficient in

TABLE 19-7

Relative value and maximum replacement for corn of energy feedstuffs for swine

Feedstuff	Relative Value[a]	Maximum Replacement[b]
Corn	100	100
Fat, feed grade	230	5–10
Rice polishings	100	100
Oat groats	100	100[c]
Bakery waste, dehydrated	100	50–80
Wheat	98	100
Hominy feed	97	100
Grain sorghum (milo)	97	100
Rice, polished, broken	96	100
Triticale	89	75
Millet, proso	89	100
Barley	89	100
Rye	88	25
Wheat middlings	87	30
Wheat shorts	84	30
Rice bran	84	10
Oats	80	25
Beet pulp	79	10
Wheat bran	63	10
Molasses, cane	59	5
Alfalfa meal	50	5
Potatoes, cooked	25	25

[a]Based on ME value.

[b]Maximum replacement for corn in diets for growing–finishing swine. Two to three times these levels of the fibrous feedstuffs (oats, wheat bran, rice bran, alfalfa meal, beet pulp) can be used in sow diets.

[c]Generally used only in starter diets. Cost prohibits the inclusion of more than 10% to 20%.

lysine (0.25%) and tryptophan (0.06%), and is deficient in many of the essential minerals and vitamins as shown in Tables 19-2, 19-4, and 19-6. However, when these deficiencies are corrected with appropriate amounts of protein, mineral, and vitamin supplementation, corn makes an excellent feed for pigs.

Several genetic mutants of corn have been identified that are higher in lysine, tryptophan, and other amino acids as compared with normal hybrid corn (see Ch. 7). The lysine in these mutant cultivars varies from 0.30% to 0.40%, compared with about 0.25% lysine in normal corn. Because of the higher lysine content, less protein supplement is required to meet the amino acid requirements. Unfortunately, most high-lysine varieties of corn are lower yielding and more susceptible to harvesting damage because of the softer, more floury kernel. Cultivars of corn selected for high oil content have been developed and are now commercially available. These corns are higher in energy because of their higher oil content (6 to 8 percent) as compared with normal corn (3% to 4%).

Grain sorghum (milo) is grown in the southern and southwestern states and is the second most common cereal grain fed to pigs. The energy value of grain sorghum for swine is similar to corn, and it contains about the same or slightly lower levels of protein and lysine as corn. The overall feeding value of grain sorghum is nearly equivalent to corn for growing–finishing pigs (Table 19-8). Some grain sorghums have been developed that are resistant to bird damage, but the feeding value of most bird-resistant varieties is considerably lower than conventional grain sorghum because of their high tannin content. The bird-resistant varieties are seldom grown in the major grain sorghum producing areas of the United States.

Barley is commonly fed to pigs in the northern states and in Canada. Barley is higher in fiber and lower in energy as compared with corn; therefore, it has a lower feeding value (Table 19-8). Also, barley is quite variable in its composition and feeding value. Some of this variability is attributed to the amount of beta-glucans present in the barley. Beta-glucans are complex carbohydrates (polymers of glucose) that cause gels to form in the gastrointestinal tract that interfere with nutrient digestibility.

Barley and oats are generally considered to have about 85% to 90% and 75% to 80%, respectively, of the feeding value of corn (Table

TABLE 19-8

Comparison of corn, wheat, grain sorghum (milo), and barley as energy sources for growing–finishing pigs

Comparisons	Corn	Other Grains
Corn versus grain sorghum (milo)[a]		
Daily gain, lb	1.60	1.57
Feed:gain ratio	3.33	3.44
Corn versus barley[b]		
Daily gain, lb	1.76	1.69
Feed:gain ratio	3.13	3.45
Corn versus wheat[c]		
Daily gain, lb	1.67	1.68
Feed:gain ratio	3.19	3.17

[a]Ten experiments, 508 pigs, 51 to 221 lb.
[b]Four experiments, 280 pigs, 42 to 225 lb.
[c]Fifteen experiments, 984 pigs, 47 to 215 lb.

19-7). However, both are slightly higher in protein and lysine than corn. Barley can replace a part or all of the corn in the diet, but oats should not replace more than 25% of the corn for growing–finishing pigs. Higher levels can be used in sow diets.

Wheat is an excellent feed for pigs (Table 19-8), but generally it is too expensive to be used as a feed ingredient. However, at certain times, wheat compares favorably with corn on a cost basis. Wheat is similar to corn on an energy basis and is higher in protein, lysine, and available P than corn. On an energy basis, wheat has slightly less value than corn (98%), but when the additional lysine in wheat is taken into account, wheat has about 105% of the value of corn. Furthermore, on a bushel basis, wheat has about 112% of the value of corn, because of its greater weight per bushel (60 versus 56 lb/bu). Wheat that is damaged moderately by disease, insects, or frost or wheat that is contaminated with wild garlic bulblets cannot be used for milling, but it makes excellent feed for swine. Wheat can replace a part or all of the corn in the diet.

Rye is quite high in energy, but it is not as palatable as the other cereal grains. Also, rye contains high levels of pectin, a complex carbohydrate, which reduces its feeding value. In addition, rye is susceptible to ergot (a fungus) infection, which can cause abortion in sows and markedly reduce performance in growing pigs.

Triticale has a feeding value that is better than rye, but somewhat less than wheat. Triticale possesses a trypsin inhibitor, which limits its feeding value if fed as the only source of grain to pigs. Certain cultivars of triticale have been developed for the southern United States and have been shown to give excellent performance when fed to pigs.

It is very important to remember that, when cereal grains other than corn are used, either the grain should be substituted for corn on a pound-for-pound basis, or the diet must be formulated on the basis of the lysine content of the grain, not on the basis of its protein content. Because most grains are higher in protein than corn, formulating the diet on a protein basis will result in a deficiency of lysine. An example of the difference in pig performance when diets are formulated on a protein basis instead of a lysine basis is shown in Table 19-9.

DAMAGED GRAIN Cereal grains can be damaged by a number of different causes, one of which is a lack of rainfall during the growing season. Drought-damaged grain is lighter than normal in test weight, but its feeding value is generally equal to normal grain when compared on a weight basis.

Grains can also be damaged by molds (see Ch. 4). Mold growth can occur in the field prior to harvest or after the grain is in storage. Grains are particularly susceptible to mold damage if they are not properly dried or if stored under warm (80° to 100°F), humid conditions. Broken kernels or insect damage also makes grain more vulnerable to mold infestation. Besides cereal grains, peanuts and cottonseeds (thus, peanut meal and cottonseed meal) are also quite susceptible to certain molds. Molds themselves are not harmful, but they produce toxins (mycotoxins) that can be harmful to animals.

One of the major mycotoxins of concern is aflatoxin, produced by the mold *Aspergillus flavus*. Several aflatoxins have been identified, with four of the more common ones being B_1, B_2, G_1, and G_2. Corn may be infected with this fungus, yet not be visibly moldy. The presence of the fungus can be detected by placing corn under ultraviolet light (black light), in which case a bright, greenish-yellow fluorescence will occur if the mold is there. The detection of the mold by the screening procedure does not necessarily mean that the grain is contaminated with aflatoxin, and a further chemical test must be conducted to determine whether aflatoxin is present and at what levels.

Aflatoxin is a potent carcinogen, and the FDA has placed certain restrictions on the interstate movement of grain containing this mycotoxin. At low levels (20 to 50 ppb), aflatoxins do not cause any problems in growing pigs, but high levels cause reduced gain, lower resistance to disease, prolonged blood clotting time, and liver damage. Grain with up to 200 ppb aflatoxin can be fed to finishing pigs, but it should not be fed to younger pigs or breeding animals.

Certain zeolites or clays, such as hydrated Na–Ca aluminosilicate (trade name, NovaSil®) and Na bentonite protect against aflatoxin toxicity when added to aflatoxin-contaminated diets (16). These compounds ordinarily are added to feed ingredients for improved flowability or to feeds as pelleting agents. Apparently, the zeolites bind the aflatoxins and prevent their absorption.

The trichothecenes, produced by *Fusarium* fungi, are potent mycotoxins. Deoxynivalenol (commonly called vomitoxin or feed refusal factor), T-2 toxin, and diacetoxyscirpenol are examples of toxins in this class. Grains containing these toxins cause reduced feed intake, reduced growth, impaired immune function, nervous disorders, and vomiting. As

TABLE 19-9

Effects of method of diet formulation on performance of finishing pigs fed corn and wheat-based diets

Method of Diet Formulation:	Corn	Wheat (weight basis)[a]	Wheat (lysine basis)[a]	Wheat (protein basis)[a]
Grain in diet, %	83.0	83.0	86.0	90.6
Soybean meal in diet, %	14.5	14.5	11.6	6.9
Protein in diet, %	14.0	16.8	15.7	14.0
Lysine in diet, %	0.66	0.74	0.66	0.53
Daily gain, lb	1.65	1.69	1.67	1.38
Feed:gain ratio	3.17	3.18	3.26	3.63

Source: Cromwell et al. (15). Two trials, 36 pigs per treatment, 90 to 217 lb.

[a]Method of diet formulation.

little as 5 ppm vomitoxin has been found to reduce feed intake by 50% and levels of 200 ppm or higher may result in nearly total feed refusal. The tolerance level for vomitoxin is about 1 ppm (although this level will reduce feed intake by 5% to 10%), whereas the tolerances for T-2 toxin and diacetoxyscirpenol are not known.

Zearalenone, one of the most frequently encountered mycotoxins, is produced by *Fusarium roseum* and *Gibberella zeae*. This compound produces estrogenic effects, such as reddening and swelling of the vulva and nipples, irregular estrual cycles, and pseudopregnancy in gilts and sows. High levels can cause vaginal and rectal prolapses. The tolerance for zearalenone in contaminated feed is 20 ppm for growing swine and considerably less for breeding animals. From day 7 to 10 after breeding (i.e., at about the time of implantation of the embryos into the uterine wall) is an especially critical time; if bred animals consume zearalenone at that time, they likely will lose their litters and not recycle for a considerable length of time.

Ochratoxin and citrinin are less common mycotoxins that may be present in mold-infected barley and other small grains. Ergot is an alkaloid like mycotoxin produced by fungi that may affect rye or wheat. Ergot is quite toxic and ergot-contaminated grain should not be fed to pigs.

If mycotoxins are suspected in grain or other feeds, they should be subjected to a laboratory test to determine which ones are present and at what levels. If levels of mycotoxins are within the tolerance levels, the grains may be fed to finishing pigs, but it is best not to feed mycotoxin-contaminated feeds to young pigs. It is advisable not to feed mycotoxin-suspect feed to replacement gilts, and it should never be fed to pregnant animals. Blending of contaminated grain with sound grain to bring the grain within the tolerance range is an option when the grain is fed to hogs on the farm.

GRAIN BY-PRODUCTS Wheat shorts, bran, and middlings are higher in fiber and lower in energy than wheat, but often they are an economical ingredient to use in pig feeds. Feed manufacturers like to use small amounts (5% to 10%) of wheat middlings when feeds are pelleted because they have good binding properties and give a hard, cohesive pellet. Wheat bran is often used in sow farrowing feed because of its laxative properties. Rice bran is another fibrous feed that can be used in sow feeds. Rolled oats are very palatable and make an excellent addition to starter feeds for pigs, but their high cost limits their usage in other pig feeds.

PURIFIED SUGARS Sucrose, lactose, and other pure sugar sources are good energy sources for young pigs. Lactose is very well utilized by early-weaned pigs and is most commonly used in prestarter diets. Dried whey, a by-product of the cheese industry, is high in lactose (68%) and is widely used in prestarter and starter diets. Sucrose is often used in young pig diets for improved palatability. Crude forms of sugar, such as cane or beet molasses, are used in some parts of the world, but not to much extent in the United States.

FAT Feed-grade fat can be used as an energy source in swine diets. As a rule of thumb, for fat to be economical as a feed ingredient, it must be no more than 2.5 to 3.0 times the cost of corn. Table 19-10 shows that when fat is added at a level of 5%, growth rate is improved slightly, feed intake is reduced, and feed efficiency is improved. However, when fed fat, pigs tend to consume more calories, so backfat is often increased slightly. With 5% added fat, a 0.1-in. increase in backfat should be expected. Because feed intake is reduced when fat is added, it is important to increase the level of protein so that the daily intake of protein is maintained. As a general rule, the protein level should be increased by 0.2 of a percentage unit for every 1% addition of dietary fat.

Fat is utilized more efficiently in a hot environment than in a moderate or cold environment. This is so because the amount of heat produced when fat is metabolized is less than when carbohydrate or protein is metabolized. Therefore, fat is most beneficial when added to

TABLE 19-10

Supplemental fat for growing–finishing swine fed a corn–soybean meal diet[a]

Item	Added Fat (%)	
	0	5
Daily gain, lb	1.68	1.77
Daily feed, lb	5.44	5.22
Feed:gain ratio	3.24	2.95
Carcass backfat, in.	1.20	1.31
Ham–loin, % of carcass	43.4	42.3

[a]Five experiments, 88 pigs per treatment, 57 to 208 lb. University of Kentucky and University of Nebraska.

diets in the summer or when used in the hotter regions of the country.

The maximum amount of fat that can be added to a diet is about 6% to 7%; higher levels cause the feed to bridge up in self-feeders. Small amounts of fat (1% to 2%) are commonly added to commercial hog feeds to improve the physical characteristics of the feed, to facilitate pelleting, to control dust in feed mills, and to reduce segregation of fine particles in mixed feed.

The inclusion of 3% to 5% fat in swine feeds has another major benefit in that it reduces feed dust in confinement hog buildings. Research has shown that this reduced dustiness and improved air quality in hog buildings results in fewer respiratory problems in confinement-reared pigs and in people who work in hog buildings.

There is some interest in feeding fat to sows during late gestation and early lactation, in that it results in improved survival of newborn pigs, especially if the pigs are small at birth (17).

Tallows, white or yellow greases, hydrolyzed animal–vegetable fat (a by-product of the soap industry), or commercial blends of these fats are the types of fat most widely used in swine feeds. Liquid fat sources (e.g., soybean oil) are also used some, but they are more expensive than solid or semisolid fat sources. Most of the types of fat used in farm-mixed feeds require heating prior to mixing. Some suppliers of fat provide heated storage tanks for this purpose.

MISCELLANEOUS ENERGY SOURCES
Bakery wastes, processed dried food wastes (macaroni, breakfast cereal fines, etc.), and similar products can be used as energy sources for pigs. Certain roots and tubers (potatoes, Jerusalem artichokes, cassava) are other examples of energy sources that can be used in swine feeds. Potatoes need to be cooked in order to destroy enzyme inhibitors and to maximize their feeding value.

PROTEIN SOURCES

Feedstuffs that are high in protein (>20%) are added to swine diets to correct the amino acid deficiencies of the cereal grains and other energy sources. The relative value of various protein supplements is related to the overall protein content, as well as to the proportions of the essential amino acids in the protein. One can roughly assess the relative value of protein supplements based on their lysine content (Table 19-11), because lysine generally is the first limiting amino acid in pig diets.

PLANT PROTEINS   Soybean meal is the major protein source used in pig diets, accounting for over 85% of all protein supplements fed to pigs. Soybean meal is unsurpassed by any other plant protein in terms of its biological value, and therefore it is the standard to which other protein sources are compared (Table 19-11). It has an excellent balance of amino acids and is especially high in lysine, tryptophan, and threonine, the amino acids that are most deficient in cereal grains. Soybean meal is very palatable and is readily available throughout the major swine-producing areas in the United States. It is available in two forms, 44% protein meal and dehulled meal, which ranges in protein from 48% to 50%. The dehulled meal is lower in fiber, 3.4% versus 7.3%, so it is more desirable for young pigs. Dehulled soybean meal contains about 3.10% lysine, and the 44% protein meal contains about 2.85% lysine. When soybean meal is blended with corn to meet the recommended protein level for the various weight classes of swine, all the amino acid requirements are met (Table 19-2).

Soybean meal possesses certain factors that are undesirable, especially when fed to very young pigs. Certain oligosaccharides (carbohydrates) in soybean meal are indigestible and can result in excessive fermentation in the hind gut. When diets containing high levels of soybean meal are consumed by young pigs, the gut lining develops a hypersensitivity to antigenic factors in the soybean meal, resulting in abnormal morphology of the gut. For this reason, levels of soybean meal (in excess of 20% to 25%) are generally not recommended in starter diets. Soy protein concentrate is a product that has undergone processing to remove or reduce these factors. This product, although rather expensive, can be used in pig starters to replace some of the soybean meal.

Whole soybeans contain about 35% to 37% protein and 18% to 19% oil. Whole soybeans can be fed to pigs, but they must first be heated to a temperature of above 250°F to destroy the trypsin inhibitor and other enzyme inhibitors (see Ch. 8). Young pigs do not digest or utilize raw soybeans very well because of these inhibitors, although older animals appear to tolerate raw soybeans better than younger animals. During gestation, sows can be fed raw soybeans as the only source of supplemental protein with no detrimental effects.

TABLE 19-11

Relative value of protein sources for swine

Ingredient	Protein (%)	Lysine (%)	Relative Value[a]	Maximum Inclusion Rate (%)[b]
Plant sources				
Soybean meal, dehulled	48.5	3.10	100	20
Soybean meal	44.0	2.80	100	20
Peas, cull	23.8	1.52	100	80
Soybean, full fat, cooked	37.0	2.40	100	25
Beans, cull	26.7	1.68	99	50
Canola meal	38.0	2.27	94	10
Alfalfa meal	17.5	0.85	76	5
Safflower meal	29.0	1.30	71	8
Cottonseed meal, low gossypol	41.7	1.70	64	8
Sunflower meal	45.5	1.68	58	8
Linseed meal	33.0	1.20	57	5
Brewers dried grains	27.3	0.88	51	15
Copra meal	20.0	0.64	50	5
Peanut meal	49.0	1.45	47	10
Sesame meal	45.0	1.26	44	5
Corn gluten feed	23.3	0.64	43	10
Distillers dried grains + solubles	27.0	0.70	41	20
Corn gluten meal	61.2	1.03	26	5
Animal sources				
Blood cells, spray dried[c]	92.0	9.00	153	3[d]
Plasma, spray dried[c]	78.0	6.90	138	6[d]
Blood meal, spray dried[c]	86.0	7.44	136	3
Fish meal, menhaden	61.2	4.74	122	5
Skim milk, dried	33.3	2.54	120	20[d]
Whey, dried	13.3	0.94	111	35[d]
Meat and bone meal[c]	50.9	2.89	89	5
Meat meal[c]	55.6	3.09	87	5
Feather meal	84.9	1.67	31	2
Synthetic sources				
Lysine·HCl	—	78.0	—	0.20[e]

[a]Based on the lysine content as a percent of the protein.

[b]Maximum inclusion in grower–finisher diets without depressing performance.

[c]Although high in lysine, these sources are relatively low in tryptophan, isoleucine, and/or methionine.

[d]Applies to starter diets.

[e]This level supplies 0.15% lysine.

Cooking or roasting of whole soybeans and toasting of soybean meal must be controlled carefully, because overheating ties up some of the lysine and reduces the biological value of the meal. When properly roasted, ground, full-fat soybeans make an excellent protein supplement for pigs. In addition, feeding whole soybeans allows one to take advantage of the benefits of fat (discussed previously) supplied by the whole bean.

Cottonseed meal is considerably lower in lysine than soybean meal (Table 19-11) and is higher in fiber. Also, cottonseed meal contains gossypol, a compound that is toxic to pigs. Only low-gossypol cottonseed meal (<0.04% free gossypol) should be used in pig feeds. To protect against gossypol toxicity, Fe sulfate should also be added to provide the same level of Fe as the free-gossypol level. The Fe binds with gossypol and prevents its absorption.

Rapeseed meal is slightly lower in lysine and higher in fiber than soybean meal, but it serves as an excellent supplement when substituted for up to half of the soybean meal, provided the newer varieties, called Canola, are used. Canola includes those cultivars that have been specifically bred to be low in antinutritional factors (see Ch. 8).

Peanut meal is relatively high in protein but is much lower in lysine than soybean meal. Its usage in the United States is mostly limited to the peanut-producing states in the Southeast. Peanut meal is quite susceptible to the fungus *Aspergillus flavus*, which is capable of producing alfatoxin.

Other plant proteins that have limited usage in pig feeds in the United States include sunflower meal, sesame meal, linseed meal, safflower meal, and copra (coconut) meal. Most of these have only average protein quality, in that the lysine is quite low compared with soybean meal.

Cull peas and beans are often an economical feed ingredient in the Pacific Northwest, and they can serve as a source of both energy and protein. Certain types of beans (kidney, navy, pinto) and peas (cowpeas, pigeon peas) require heat treatment to destroy enzyme inhibitors, while others (field beans, horsebeans, field peas) do not require heating. When the majority of the protein is supplied by these products, methionine supplementation may be necessary.

Protein sources derived from grain, such as corn gluten meal, distillers dried grains, and brewers dried grains, can be used to supply a portion of the protein supplement, but the biological value of the protein is rather poor. The amino acid pattern is similar to the grain from which they are produced.

ANIMAL PROTEINS Animal proteins are good sources of lysine and other amino acids. Also, the animal protein sources are much higher in Ca, P, and B vitamins (particularly vitamin B_{12}) than the plant proteins. However, they are more variable in nutrient content and are subjected to high drying temperatures during processing for dehydration and sterilization. If drying time and temperature are not controlled carefully, the bioavailability of lysine, tryptophan, and other amino acids can be reduced markedly.

By-products of the meat packing industry, such as meat meal, meat and bone meal, and tankage, are high in lysine but relatively low in tryptophan. In addition, the biological availability of tryptophan in these products is quite low. Also, these products tend to be somewhat unpalatable, so diets for growing pigs should not contain more than 4% to 5% meat by-products or performance may be reduced. Higher levels can be used in growing pig diets if supplemental tryptophan is provided. Sow feeds may contain higher levels, up to one-third of the protein supplement.

Dried blood meal is very high in protein (85% to 90%) and in lysine (7% to 8%). Much of the lysine was destroyed by the older drying methods, but the newer drying processes (ring drying or flash drying) result in a product with a high level of available lysine. Blood meal tends to be low in the amino acid isoleucine, and this protein source is not very palatable, so it should not be added at levels exceeding 3% of the diet.

Two relatively new products are made by separating blood from hog slaughter plants into plasma and blood cells and carefully drying the two products. Spray-dried porcine plasma is an excellent protein source for early-weaned pigs. Aside from its amino acid content, the globular proteins (including immunoglobulins) in dried plasma stimulate growth and feed intake during the critical postweaning stage. This product, although quite expensive, is now commonly used in pig starters for the first 7 to 10 days following weaning. Dried blood cells, when used, are included in starter diets.

Fish meal is an excellent protein supplement for pigs; however, the high cost of fish meal in the United States limits its use. Fish meals are quite variable, depending on the type of fish used and the type of processing methods. Some marine meals are made from residues, such as shrimp meal, and these are less nutritious. Certain long-chained fatty acids in fish oil can cause a fishy flavor in pork, so the amount of fish meal in pig feed should not exceed 6% to 7%. Inclusion of fish meal or fish solubles in starter diets has recently been shown to improve performance of early-weaned pigs.

Dried milk products are excellent protein supplements. They are very palatable, highly digestible, and have an excellent balance of amino acids; however, they are very expensive ingredients (especially dried skim milk). Dried whey, an excellent source of both carbohydrate (lactose) and protein, is commonly included in starter diets for young pigs, but it is too expensive to be used in feeds for older pigs.

Liquid skim milk, buttermilk, or whey is sometimes available at low cost around processing plants. Although these products are excellent sources of protein and other nutrients, they are so high in water that pigs are not able to consume enough to meet their dry-matter requirements. However, if transportation costs are not too great and if proper feeding equipment is available, these products make excellent protein supplements.

OTHER PROTEIN SOURCES Bacteria growth on petroleum and other wastes have been shown to be a good protein supplement for pigs. However, the cost of producing single-cell protein is currently too expensive to justify its use as a feedstuff.

SYNTHETIC AMINO ACIDS Pigs can utilize synthetic amino acids to meet a portion of their dietary requirements. Until recently, lysine and methionine were the only amino acids that were economically feasible to use in feeds. Generally, methionine is the fourth limiting amino acid in most pig diets, so there is little need for supplemental methionine in practical swine diets. An exception is in starter diets that contain dried plasma (which is low in methionine). New biotechnology procedures have been developed for the production of tryptophan and threonine, so they may be used more extensively in the future.

Lysine is commercially available as lysine hydrochloride. When lysine·HCl (which contains 78% lysine) is used as a supplement to provide 0.15% added lysine, the protein level of a grain–soybean diet can be reduced by 2 percentage units. Put another way, 96 lb of corn and 4 lb of lysine·HCl can replace 100 lb of dehulled soybean meal in a ton of swine feed. With this amount of protein reduction and lysine supplementation, there is no reduction in growth performance. Greater reductions in protein, however, reduce pig performance, even if additional amounts of lysine are added. This is due to the fact that other amino acids become limiting when the protein level is lowered. A 4-percentage-point reduction in dietary protein with added lysine, tryptophan, threonine, and methionine results in optimal growth, but presently the cost of tryptophan and threonine is prohibitive for least-cost performance.

There is renewed interest in amino acid supplementation of low-protein diets because this is an effective means of reducing N excretion by pigs. Some farms do not have sufficient land area to handle all their wastes, and a reduction in manure N may be of significant benefit to these farms. Under these conditions, there may be incentive to use amino acids, even if they result in a more costly diet. Low-protein, amino acid supplemented diets also reduce ammonia emissions from swine manure, resulting in less odor, another potential environmental problem.

NPN sources, such as urea, are of no practical nutritional value to pigs.

MINERAL SOURCES

Most swine diets need to be supplemented with minerals, especially if the diet consists mainly of plant proteins. Mono- and dicalcium phosphate, defluorinated rock and steamed bone meal are good sources of highly available Ca and P. High-F rock phosphate, soft rock phosphate, and fertilizer-grade phosphates should be avoided because the P is less available and the high F level can be toxic. Ground limestone (calcium carbonate) is commonly used as an additional source of Ca. Oyster shell flour, aragonite, marble dust, and gypsum also are good sources of Ca, but dolomitic limestone, which is high in Mg, should be avoided because the Ca is poorly available.

The inclusion of ordinary salt in the diet is the most practical way of meeting the dietary requirements for Na and Cl. If iodized salt is added to meet the Na requirement of pigs, the I needs of pigs will also be met.

Generally, a trace mineral mix is added as a premix to swine diets to supply the necessary trace minerals. An example of a trace mineral mix is shown in Table 19-12. This mix can be added at variable levels (2 lb/ton in starter and sow feeds; 1.5 lb/ton in grower feeds; 1 lb/ton in finisher feeds) to meet the trace mineral requirements. Another means of providing trace minerals is in the form of trace-mineralized salt.

Pigs that are self-fed shelled corn and a protein supplement can also be self-fed a mineral mixture consisting of 50% dicalcium phosphate or defluorinated rock phosphate, 30% ground limestone, and 20% trace-mineralized salt. However, if appropriate levels of minerals are provided in the diet, the free-choice feeding of additional minerals is not necessary.

VITAMIN SOURCES

Alfalfa meal, fermentation by-products, and animal protein sources are good sources of many of the vitamins. However, synthetic vitamins

TABLE 19-12

Trace mineral premix[a]

Mineral Element[b]	Concentration in Premix (%)	One Pound of Premix Supplies to Diet (ppm)
Copper	1.0	5.0
Iodine[b]	0.10	0.50
Iron	15.0	75.0
Manganese	2.0	10.0
Selenium[c]	0.03	0.15
Zinc	15.0	75.0

[a]Use 2 lb/ton in prestarter, starter, gestation, and lactation diets, 1.5 lb/ton in grower diets, and 1 lb/ton in finisher diets. These amounts should be mixed in 10 to 20 lb of corn prior to blending in a ton of feed.

[b]Iodine not needed if iodized salt is fed.

[c]Selenium not needed in high-selenium areas.

are relatively inexpensive, so they are generally added to swine diets. An example of a vitamin premix is shown in Table 19-13. This supplement can be added at various levels (1 to 2 lb/ton) to meet the requirements during the different stages of growth and reproduction.

Choline is commonly added to sow feed as a 50% or 60% choline chloride supplement (43% or 51% choline). Choline should not be combined with the other vitamins in a concentrated premix because it tends to promote the destruction of other vitamins, especially vitamin A.

FEED ADDITIVES

Feed additives used in swine diets include antimicrobial agents (antibiotics and chemotherapeutics), anthelmintics (dewormers), and a number of others. Antimicrobial agents are added to pig feeds to stimulate growth and efficiency of feed utilization and reduce mortality and morbidity. They are commonly added to feed at low (subtherapeutic) levels for growth promotion, at moderate-to-high (prophylaxis) levels for the prevention of disease in exposed animals, and at high (therapeutic) levels for the treatment of disease. Anthelmintics are used to prevent or remove internal parasites (e.g., roundworms and lungworms), and one (Ivermectin) also acts as a systemic to control external parasites (lice, mange). These two groups of compounds are considered drugs, and their use is regulated by the FDA (see Ch. 10).

Only certain types, levels, and combinations of antimicrobial agents and anthelmintics are approved for swine (18). Some require withdrawal from the feed prior to slaughter in order to prevent carcass residues (see Ch. 10). It is im-

portant that swine producers who mix their own feed be familiar with proper usage of antimicrobial agents and that they mix feeds properly to prevent carryover of drugs from one feed to the next. This is especially important when feeds containing sulfamethazine are mixed. The sulfa drugs are electrostatic and tend to accumulate in feed dust. Failure to clean mixers and feed-conveying equipment can result in cross contamination of feed with drugs, and very small amounts of certain drugs (such as sulfamethazine) cause harmful residues in pork carcasses (19).

The greatest benefit from antimicrobials is in the very young pig. Young pigs are more susceptible to stress and disease organisms than are older pigs. The passive immunity that the baby pig acquires from the sow's colostrum nearly disappears by the time it reaches 3 weeks of age (Fig. 19-5). Since the baby pig does not synthesize antibodies very well, its immunity remains low following weaning. Eventually, following exposure to antigens from the sow and the environment, they produce specific immunoglobins that give them protection against disease-causing organisms. Table 19-14 shows the relative response to antimicrobials in pigs during the starter phase, the grower phase, and for the entire growing–finishing stage.

The overall health of the pigs and the cleanliness of the environment also influence the pig's response to antimicrobial agents. In general, the higher the disease level is, the less sanitary the environment and the poorer the management, the greater is the response from antimicrobial usage. However, antimicrobials

TABLE 19-13

Composition of vitamin premix[a]

Vitamin	Amount in 1 Pound of Premix	One Pound of Premix per Ton Supplies These Levels per Pound of Diet
Vitamin A	3,000,000 IU	1500 IU
Vitamin D_2 or D_3	200,000 IU	100 IU
Vitamin E	10,000 IU	5 IU
Vitamin K	2 g	1 mg
Riboflavin	4 g	2 mg
Pantothenic acid	12 g	6 mg
Niacin	16 g	8 mg
Vitamin B_{12}	12 mg	6 µg
Biotin	100 mg	50 µg
Folacin	500 mg	250 µg
Carrier (to 1 lb total)	+	

[a]Use 2 lb/ton in prestarter, starter, gestation, and lactation diets, 1.5 lb/ton in grower diets, and 1 lb/ton in finisher diets. These amounts should be mixed in 10 to 20 lb of corn prior to blending in a ton of feed.

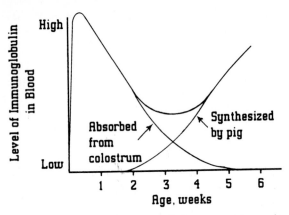

FIGURE 19-5 Development of disease resistance in young pigs.

should never be considered as a substitute for good management.

Antimicrobials have been shown to be beneficial for sows and gilts at breeding time in that they tend to improve conception rate and litter size (Table 19-15). A high level (0.5 to 1 g/day) of an absorbable antibiotic (such as one of the tetracyclines) is recommended for this purpose.

Antimicrobials are of little value during gestation, but they are generally considered to be beneficial just before and after farrowing because this is a high-stress period. Antibiotics have been shown to reduce the incidence of agalactia (lack of milk) and uterine infections that occur occasionally in sows shortly after farrowing. The data in Table 19-16 suggest that pig survival and pig weaning weights are improved slightly when antimicrobials are included in the lactation diet.

Several reviews on antibiotics in animal feeds have been written by CAST (20), Hays (21), Zimmerman (22), and Cromwell (23).

Two minerals, Cu and Zn, when fed at pharmacological levels in the diet, act in a simi-

TABLE 19-15

Effects of antibiotics in the feed at breeding on reproductive performance

Item	Control	Antimicrobial[a]
Farrowing rate, %[b]	75.4	82.1
Live pigs per litter	10.0	10.4

Source: Cromwell (23). Data from nine experiments involving 1,931 sows.

[a]In most cases, 0.5 to 1.0 g of antibiotics per sow per day.
[b]Number of sows farrowed divided by the number bred.

TABLE 19-16

Antimicrobial agents in the prefarrowing and lactation diet for sows

Item	Control	Antimicrobial[a]
Live pigs born per litter	10.3	10.6
Pigs weaned per litter	8.2	8.6
Survival, %	84.9	87.1
Average pig weaning weight, lb	10.23	10.34

Source: Cromwell (23). Summary of 11 experiments with 2,105 litters.

[a]Tetracyclines, ASP-250, tylosin, bacitracin, or copper sulfate fed from 3 to 5 days prepartum through 7 to 21 days of lactation.

lar manner as antimicrobials with respect to stimulation of growth in young pigs. The responses to high levels of copper sulfate (125 to 250 ppm of Cu, or 1 to 2 lb of copper sulfate per ton) of both weaning, growing, and growing–finishing pigs are shown in Table 19-17. Like with other antimicrobials, the greatest response to Cu is in the young pig. Zinc oxide, when added to diets to supply 3,000 ppm Zn (or 8 lb of zinc oxide/ton) for young pigs, has a similar effect as high levels of Cu (Table 19-18). Also, there is some evidence from research abroad that these levels of zinc oxide also help

TABLE 19-14

Effects of age and weight of pigs on response to antibiotic feeding[a]

Growth Stage	Control	Antibiotic	Improvement (%)
Starter phase (16–55 lb)			
Daily gain, lb	0.86	0.99	16.4
Feed:gain ratio	2.28	2.13	6.9
Grower phase (37–108 lb)			
Daily gain, lb	1.30	1.45	10.6
Feed:gain ratio	2.91	2.78	4.5
Grower–finisher phase (53–197 lb)			
Daily gain, lb	1.52	1.59	4.2
Feed:gain ratio	3.30	3.27	2.2

Source: Hays (21) and Zimmerman (22).

[a]Data from 1194 experiments involving 32,555 pigs.

TABLE 19-17

Effect of copper sulfate on performance of weanling and growing–finishing pigs

| Growth stage | Copper (ppm)[a] | | Improvement (%) |
	0	250	
Starting period (15–31 lb)			
Daily gain, lb	0.51	0.62	24.0
Feed:gain ratio	2.04	1.86	9.7
Growing period (40–123 lb)			
Daily gain, lb	1.47	1.57	6.9
Feed:gain ratio	2.80	2.70	3.6
Growing–finishing period (40–205 lb)			
Daily gain, lb	1.57	1.62	3.1
Feed:gain ratio	3.18	3.10	2.5

Source: Cromwell (23). Summary of 12 starter experiments involving 462 pigs and 18 growing–finishing experiments involving 672 pigs.

[a]Does not include the copper supplied by the trace mineral mix.

to control diarrhea in young pigs. A recent study (24) shows that combinations of high Cu and high Zn are no more effective than when either is fed alone (Table 19-18). High Cu makes the manure very black in color, but Zn does not.

How these minerals work to improve performance in young pigs is not known. Copper and antimicrobials seem to work differently, because a combination of Cu and antimicrobials stimulates growth to a greater degree than when either Cu or antimicrobials are added singly (Table 19-19). The feeding of high-Cu diets to sows has been shown to increase birth weights and weaning weights in their pigs (25).

Copper can be toxic if high levels (greater than 250 ppm Cu) are fed for extended periods of time; so if Cu sulfate is used as a supplement, it should never exceed 2 lb/ton. A potential environmental problem associated with the feeding of high-Cu and high-Zn diets is the excessive excretion of Cu and Zn in the manure.

Several other feed additives are sometimes included in swine feeds for various purposes. They include microbial supplements, enzymes, organic acids, oligosaccharides, flavors, odor-control agents, antioxidants, pellet binders, flow agents, and carcass modifiers.

Microbial supplements, commonly called probiotics, include *Lactobacillus acidophilus*, *Streptococcus faecium*, *Saccharomyces cerevisiae* (yeast), and similar microbial products. These products are added in an effort to promote the colonization in the gut of desirable microorganisms and reduce the numbers of potentially undesirable ones (such as pathogenic *Escherichia coli*). In some instances, these products have been reported to benefit young pigs under high-stress conditions; however, most controlled experiments at research stations have failed to show any consistent beneficial response from their inclusion.

Mixtures of enzymes are available as feed additives, but most research shows that they are not generally very effective for the types of feeds commonly fed in the United States. In some instances, "cocktails" of cellulases, hemicellulases, and proteases have been reported to improve digestibility of high-fiber diets, the types of diets more commonly fed in other countries. In some instances, β-glucanase has been shown to improve the utilization of barley when

TABLE 19-18

Effect of pharmacological levels of zinc and copper on performance of weanling pigs from 14 to 38 pounds

Zinc oxide (3000 ppm Zn):	−	+	−	+
Copper sulfate (250 ppm Cu):	−	−	+	+
Daily gain, lb	0.78	0.90	0.88	0.89
Daily feed, lb	1.35	1.48	1.43	1.46
Feed:gain ratio	1.74	1.66	1.65	1.65

Source: Hill et al. (24). Summary of 10 experiments involving 1,156 pigs from 22 to 50 days of age. Chlortetracycline (200 g/ton) was included in all diets.

TABLE 19-19

Effect of copper and antibiotics on performance of weanling pigs from 15 to 40 pounds

Antibiotics:	−	+	−	+
Copper sulfate (250 ppm Cu):	−	−	+	+
Daily gain, lb	0.57	0.66	0.68	0.75
Daily feed, lb	1.19	1.28	1.30	1.40
Feed:gain ratio	2.10	1.95	1.91	1.84

Source: Cromwell (23). Summary of 14 experiments involving 1,700 pigs conducted at six experiment stations.

the barley is high in β-glucans and low in naturally occurring β-glucanase.

An enzyme that is very efficacious when added to pig feeds is phytase. This enzyme (produced by *Aspergillus niger*) degrades phytate P in cereal grains and oilseed meals, making the P more highly available to pigs (Table 19-5) and reducing the amount of P excreted in the manure (Fig. 19-6). Because of the concern of the potential polluting effects of excess P in animal manure, increased use of phytase in feeds will likely occur in the near future. The FDA recently approved phytase (produced by genetically engineered microorganisms) for use in the United States. This product is widely used in several European countries where application of manure to cropland is regulated based on the P content of the manure. If feeds are pelleted, phytase must be sprayed onto the pellets after the feed is pelleted, because the high heat generated during pelleting destroys the activity of the enzyme.

Two organic acids, citric acid and fumaric acid, have been shown to benefit early-weaned pigs when included in starter diets at levels of 1% to 3% (Table 19-20). The mode of action of these acids is not known, but it is thought that they may lower the pH of the stomach of the weaning pig, thereby increasing the activity of the digestive enzyme pepsin. The lower pH may also help to reduce the proliferation of undesirable bacteria in the stomach and small intestine.

Flavors are sometimes added to feeds to improve palatability or to mask off-flavors or off-odors. When young pigs are given a choice, they generally select a diet with added flavors. But when not given a choice, there generally is no benefit in pig performance from feeding a diet with added flavors. Some argue that flavors are often added simply to make the product smell good to the person purchasing the feed.

Certain feed additives (e.g., sarsaponin, an extract from the yucca plant) are claimed to reduce odor in swine manure. Some do appear to reduce ammonia emission from manure, apparently due to an inhibition of urease activity in the waste. Some of the zeolites also seem to reduce odor and N volatilization.

Certain zeolite clays (bentonite) are sometimes added to commercial feeds prior to pelleting to improve pellet quality. Other zeolites are sometimes added to certain feed ingredients to aid in flowability. Both of these compounds reduce alfatoxin toxicity (see previous section on damaged grains), but they are not approved by the FDA for this purpose.

Antioxidants (e.g., ethoxyquin, butylated hydroxytoluene) are generally added to high-fat feeds to reduce the chances for oxidative damage. They also protect vitamins from oxidation.

Finally, certain β-adrenergic agonists (e.g., clenbuterol, cimaterol, ractopamine) increase carcass leanness when included in the diet; however, they are not currently approved by the FDA. Under some conditions, betaine and carnitine have been found to improve carcass leanness. The trace mineral Cr also improves

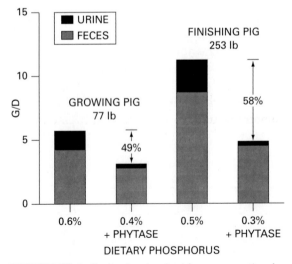

FIGURE 19-6 Reduction in phosphorus excretion by pigs fed low-phosphorus diets supplemented with phytase. [Carter et al. (26).]

TABLE 19-20

Efficiency of acidifying agents in pig starter diets

Item	Basal	+ Acid	Improvement, %
Studies with citric acid[a]			
Daily gain, lb	0.65	0.69	5.2
Feed:gain ratio	1.88	1.75	7.2
Studies with fumaric acid[b]			
Daily gain, lb	0.68	0.71	4.1
Feed:gain ratio	1.86	1.73	6.9

[a]Summary of five trials, 311 pigs, 19 to 35 lb, 24 days on test.
[b]Summary of five trials, 386 pigs, 20 to 38 lb, 27 days on test.

carcass leanness when added to the diet as chromium picolinate.

PROCESSING FEEDS

GRINDING

Feed grains must be ground before they are mixed into diets. Corn and sorghum should be ground medium to fine, with an average particle size of 575 to 600 μm. Wheat should be ground slightly coarser (675 to 700 μm). A number of commercial laboratories will determine particle size for a modest fee. Grinding is accomplished with a hammer mill, using a $\frac{3}{16}$-, $\frac{1}{4}$-, or $\frac{3}{8}$-in. screen. Research has shown that finely ground corn is utilized more efficiently than coarsely ground corn (Table 19-21). However, exceptionally fine grinding requires more electrical energy because of increased grinding time, it may cause feed to be dusty, and it may cause bridging of the feed in feeders. Also, exceptionally fine grinding has been associated with an increased incidence of stomach ulcers in finishing pigs. Alternatively, if grain is ground too coarse, pigs will sort out the larger particles and wastage becomes a problem. If grains are processed in a roller mill, the rollers need to be set relatively close.

PELLETING

Pigs that are fed pelleted diets tend to gain slightly faster and utilize their feed more efficiently than pigs fed diets in meal form (Table 19-22). Pelleting especially improves feed utilization in diets containing barley, oats, or other fibrous feeds. Pelleting also reduces dustiness and helps to prevent segregation and wastage of feed. It is very important that pellets be hard and not crumbly. A soft pellet breaks up when handled mechanically, and variable particle size makes it difficult, if not impossible, to adjust feeders properly to prevent feed wastage.

COOKING AND ROASTING

Grains are not improved sufficiently by cooking or roasting to offset the cost. Micronizing is a processing method that may improve the utilization of grain sorghum, but it is also an expensive process. Soybeans and certain types of peas and beans are improved greatly by roasting or other forms of heating. Potatoes must be cooked before feeding them to pigs. Where garbage feeding is permitted, laws require that it be cooked to control transmission of trichinosis and other diseases.

LIQUID FEEDING

Several mechanical feeders are being marketed that allow feed and water to be mixed in a gruel. Liquid feeding may help to prevent feed wastage, but there is very little research evidence to indicate that performance of pigs is improved by liquid feeding. Therefore, liquid feeding should be evaluated on its merits as a means of dispensing feed to pigs. Wet feeding is sometimes used to increase feed intake in lactating sows when the environmental temperature is high.

TABLE 19-21

Effects of fineness of grind on performance of growing–finishing pigs

	Coarse (>1,000 μm)	Medium (700–900 μm)	Fine (<600 μm)
Daily gain, lb	1.66	1.75	1.71
Feed:gain ratio	3.57	3.51	3.36

Source: Wondra et al. (27). Five experiments involving 740 pigs, 76–198 lb.

TABLE 19-22

Effect of steam pelleting of corn– or grain sorghum–soybean meal diets on performance of growing–finishing pigs

Item	Meal	Pellet	Improvement (%)
Daily gain, lb	1.64	1.71	4.3
Feed:gain ratio	3.35	3.19	4.8

Source: Wondra et al. (27) and unpublished data. Summary of 33 experiments involving 3,505 pigs, 98 to 221 lb.

HIGH-MOISTURE GRAINS

High-moisture shelled corn and other grains are essentially equal to dry grain on a dry-matter basis. High-moisture grain can be fed free choice with a complete supplement, or the grain can be rolled and mixed with supplement. However, it must be prepared daily or it will heat, spoil, and cake up in feeders. Propionic acid-treated, high-moisture grain can be kept for longer periods of time without spoilage. Less supplemental P is needed when high-moisture grain is fed because the P is three to four times more available in high-moisture versus dry grain.

FEEDING MANAGEMENT

PREBREEDING

Gilts that are kept for breeding purposes should be selected at market weight and then removed from the self-feeder and hand-fed 6 to 7 lb of feed per day. They should be kept in a good gaining condition, but not allowed to become overfat.

Gilts are normally bred at 7 to 8 months of age. Earlier breeding is possible, but it will result in smaller litter size. If gilts are thin at time of breeding, they may be flushed by full-feeding them for 1 to 2 weeks prior to breeding. Flushing increases the number of eggs ovulated and may result in larger litters. If gilts are flushed, they should be returned to a moderate feeding level (4 to 6 lb/day) immediately after breeding. A high level of an absorbable antibiotic should be included in the breeding diet (e.g., 200 to 300 g/ton of chlortetracycline) to maximize conception rate and reduce embryonic mortality (Table 19-15).

GESTATION

The nutrient requirements of the pregnant sow are low, particularly during the first two-thirds of gestation. The majority of the fetal development occurs during the last month of gestation

(Fig. 19-7). Sows should be fed at a level to accommodate the growth of the fetuses and the placenta, approximately 44 lb (Table 19-23), along with an increase of 30 to 35 lb of their body weight. Gilts should be fed to gain more of their own body weight (60 to 65 lb) because they have not yet reached their mature size. Thus, the targeted pregnancy weight gain should be 75 lb for sows and 100 lb for gilts.

Both overfeeding and underfeeding should be avoided. Overfeeding is costly and wasteful, and overfat sows may have smaller litters and more farrowing difficulty and are more likely to crush their pigs by overlay. If sows are underfed, pigs are smaller and weaker at birth, and the sow may not have sufficient body stores to produce sufficient milk for a large litter. In addition, she may be too thin at weaning and is less likely to recycle.

Approximately 5,800 to 6,500 kcal/day of ME or about 4 to 4.5 lb/day of corn–soybean meal diet (containing 1,450 kcal ME/lb) in a moderate climate will generally result in the right amount of gain in gilts and sows. If lower-energy feeds such as oats or barley are fed, more feed is necessary. The feeding level should be increased to 5 to 6 lb/day for sows that are housed outside in a cold environment. Conversely, less feed (3 to 3.5 lb/day) may be all that is necessary to produce adequate weight gains during gestation for animals in warmer climates or for sows housed in individual crates in confinement gestation houses (Fig. 19-8).

If sows are not penned individually, the use of feeding stalls is recommended in order to prevent boss sows from taking more than their share of feed (Fig. 19-9). Sows should be penned separately from gilts when animals are group penned. Once a day feeding is recommended.

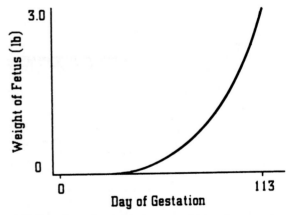

FIGURE 19-7 Fetal development during the gestation period.

TABLE 19-23

Components of gestation weight gain in gilts and sows

Component	Gilt (lb)		Sow (lb)	
Pigs and placenta				
Pigs, at 3 lb	(10 pigs)	30	(12 pigs)	36
Placental membranes		5		5
Placental fluids		3		3
		38		44
Sow				
Uterus		7		8
Udder and blood		8		9
Body muscle and fat		47		14
		62		31
Total gestation weight gain		100		75

Source: Adapted from NRC (1) and Whittemore and Elsley (28).

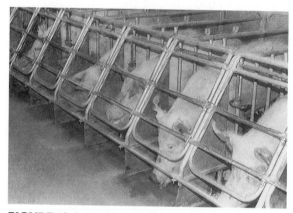

FIGURE 19-8 Sows in confinement gestation stalls.

A protein level of 14% and Ca and P levels of 0.80% and 0.65%, respectively, are recommended when a 4-lb/day feeding level is used. If sows are fed less than 4 lb/day, then the nutrient percentages should be adjusted upward so that sows receive a minimum of 250 g of protein, 14.5 g of Ca, and 11.5 g of P per day (Table 19-24). Sows may either be fed a complete-mixed diet or they can be fed 3 to 4 lb of grain and 1 lb of a complete protein supplement daily.

Examples of diets are shown in Table 19-25. A grain–soybean meal diet fortified with minerals and vitamins is adequate for sows; however, some animal protein, alfalfa meal, or by-product feedstuffs may also be included in the diet.

Sows are able to utilize roughages better than young pigs. Corn silage, haylage, alfalfa hay, and good-quality grass–legume pasture can be used to provide a portion of the nutrients of the sow (Fig. 19-10). Generally, 8 to 12 sows/acre is the recommended stocking rate. The feeding of 12 to 15 lb of corn silage plus 1 lb of complete supplement per day or self-feeding high-quality alfalfa hay to sows during gestation has resulted in good reproductive performance.

Attempts to devise a system for self-feeding of sows have met with limited success. Allowing sows access to a self-feeder for 2 to 4 h every third day will result in weight gains and reproductive performance similar to hand-fed sows, but a system like this needs to be managed very carefully. Self-feeding a bulky diet does not seem to work unless extremely high levels of a

FIGURE 19-9 Individual sow feeding stalls.

TABLE 19-24

Percentages of protein, calcium, and phosphorus at three feeding levels for sows during gestation

Daily Feeding Level (lb)	Dietary Levels Needed (%)[a]		
	Protein	Ca	P
3	17	1.10	0.85
4	14	0.80	0.65
5	12	0.65	0.55

[a]These levels provide a minimum daily amount of 250 g (0.55 lb) of protein, 14.5 g of Ca, and 11.5 g of P.

TABLE 19-25

Gestation diets for pregnant sows and gilts[a]

Ingredient	1	2	3	4	5
Corn or milo, ground	1,617	—	1,147	1,584	1,498
Wheat or barley, ground	—	1,693	—	—	—
Oats, ground	—	—	500	—	—
Soybean meal (44%)[b]	314	242	285	176	194
Meat and bone meal	—	—	—	110	—
Alfalfa meal	—	—	—	100	—
Wheat middlings	—	—	—	—	200
Fish meal	—	—	—	—	50
Dicalcium phosphate	37	29	36	14	26
Limestone, ground	16	20	16	—	16
Salt	10	10	10	10	10
Vitamin mix[c]	2	2	2	2	2
Choline mix[c]	2	2	2	2	2
Trace mineral mix[c]	2	2	2	2	2
Total	2,000	2,000	2,000	2,000	2,000

[a]Provides 1400 to 1485 kcal/lb of ME, 14% to 15% protein, 0.65% lysine, 0.80% Ca, and 0.65% P. Feed at a level of 4 lb/day.
[b]If dehulled soybean meal is used, reduce level by 10% and replace difference with grain.
[c]See Tables 19-12 and 19-13.

FIGURE 19-10 Pasture provides additional nutrients for sows.

low-energy ingredient (e.g., ground hay or sawdust) are included in the feed.

There has been recent interest in the use of high levels of supplemental fat in the sows' feed during late gestation. Research has shown that birth weights of pigs are improved slightly and pig survival is increased when sows are fed fat. Some of the benefit is attributed to a higher fat level in the colostrum and milk and a slight increase in milk production as a result of feeding fat to sows (30). A relatively new compound, 1-3 butanediol, is available for use in sow feed during late gestation. This compound, like fat, improves survival in young pigs.

FARROWING

The farrowing diet should be laxative in nature to prevent constipation, a problem often encountered when sows are moved into the far-

rowing crate. Wheat bran or dried beet pulp can be added to the diet at levels of 10% to 15% to provide a laxative effect. As an alternative, magnesium sulfate (Epsom salts) or KCl, added at levels of 0.75% to 1.0% (15 to 20 lb/ton) will also serve to soften the feces. An advantage of these chemical laxatives is that they do not reduce the energy content of the diet. Examples of farrowing diets are given in Table 19-26.

A high level of an absorbable antibiotic is recommended in the prefarrowing and postfarrowing diet. Antibiotics tend to reduce the incidence of MMA (mastitis, metritis, agalactia) that sometimes occurs shortly after a sow farrows. In addition, small amounts of the antibiotic may be passed on to the pigs through the milk.

LACTATION

The nutrient requirements of the sow during lactation are three to four times higher than the requirements during gestation because of the high demands for milk production. While milk production is affected by the genetic capability of the sow and the diet consumed, the major factor that influences milk production of a lactating sow is the number of pigs that are nursing the sow. With an adequate diet, modern-type sows produce about 2.0 to 2.4 lb of milk per nursing pig. In other words, a sow with 10 pigs produces 20 to 24 lb of milk per day, and she reaches that level of production by the second week of lactation (Fig. 19-11). Also, sow's milk

TABLE 19-26

Farrowing and lactation diets for sows and gilts[a][b]

Ingredient	1	2	3	4	5	6
Corn or milo, ground	1,533	—	1,580	1,343	1,535	1,615
Wheat or barley, ground	—	1,299	—	—	—	—
Oats, ground	—	400	—	—	—	—
Soybean meal (44%)[c]	300	236	186	290	320	316
Meat and bone meal	—	—	100	—	—	—
Wheat bran	100	—	—	—	—	—
Alfalfa meal	—	—	100	—	—	—
Wheat middlings	—	—	—	200	—	—
Fat	—	—	—	100	—	—
Potassium chloride or magnesium sulfate	—	—	—	—	15	—
Dicalcium phosphate	33	30	16	32	38	37
Limestone, ground	18	19	2	19	16	16
Salt	10	10	10	10	10	10
Vitamin mix[d]	2	2	2	2	2	2
Choline mix[c]	2	2	2	2	2	2
Trace mineral mix[d]	2	2	2	2	2	2
Antibiotics[e]	+	+	+	+	+	+
Total	2,000	2,000	2,000	2,000	2,000	2,000

[a]Provides 1400 to 1560 kcal/lb of ME, 14% to 15% protein, 0.65% lysine, 0.80% Ca, and 0.65% P. Feed 4 to 5 lb/day prior to farrowing and full-feed after farrowing.

[b]For sows nursing large litters (nine or more pigs), substitute 100 to 150 lb of additional soybean meal for the grain. This will increase the protein level by 2% to 3%.

[c]If dehulled soybean meal is used, reduce level by 10% and replace difference with grain.

[d]See Tables 19-12 and 19-13.

[e]Broad-spectrum absorbable antibiotics are recommended. Add to supply 50 to 150 g/ton.

is much richer in fat, protein, and total dry matter than cow's milk (Table 19-27).

Although the daily requirements for all nutrients are higher during lactation than during gestation, they are not all that much different as during gestation when the requirement is expressed as a percentage of the diet. In general, the requirements of the lactating sow can be met by full feeding the gestation diet during lactation. However, highly prolific sows nursing large litters (10 or more pigs) require 2 to 4 percentage units more protein during lactation. Recent research indicates that feeding 16% to 18% protein maximizes milk production and pig gain and minimizes sow body weight loss during lactation (31, 32). If less protein is fed, sows simply mobilize protein from their body reserves to produce milk protein. Excessive mobilization of protein from body sources, however, can be detrimental if lactation is prolonged. Generally, sows consume 12 to 16 lb of feed daily during lactation.

Failure to give high-producing sows all they will consume can result in excess losses of body weight and condition, especially if they are thin at farrowing. Feeding a diet too low in protein will do the same. Conversion of body fat to

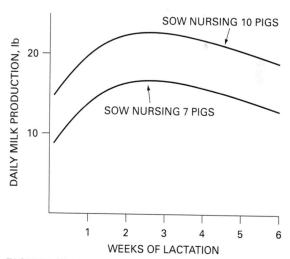

FIGURE 19-11 Lactation curves of sows nursing 7 or 10 pigs. [Toner et al. (29).]

milk energy is very efficient in sows, but drawing from body protein reserves to meet milk protein needs is not as efficient. Also, if feed is restricted during lactation, bone mineral reserves can be depleted, especially if the diet is marginal in Ca and P. Bone fractures or paralysis can occur in mineral-depleted sows toward the end of lactation or during the rebreeding

TABLE 19-27

Composition of sow's milk and cow's milk

Item	Sow Milk (%)	Cow Milk (%)
Fat	8.0	3.5
Protein	5.8	3.3
Lactose	5.4	5.0
Ash	0.8	0.7
Total solids	20.0	12.5

period, particularly if animals are penned together and fighting occurs. Sows should not be allowed to get excessively thin during the lactation period; otherwise, they may have extended delay to estrus following weaning.

Some producers gradually increase the sow's feed after farrowing until they are on full feed within 5 to 7 days. Others begin full feeding immediately after farrowing. Either practice is acceptable. Feed intake of lactating sows often will be reduced if the farrowing house is hot. Supplemental fat, either incorporated in the diet at 5% to 8% or top dressed on the feed will help to offset the reduced energy intake. Drip cooling (a system that drips water on the sow's neck and shoulders) also is beneficial in this regard.

Protein level of 14% (for litters of eight or less) to 16% (for litters of nine or more) are recommended for the lactating sow. Calcium and P are critical and should be at least 0.80% and 0.65%, respectively. Examples of lactation diets are given in Table 19-26.

NURSING PIGS

The baby pig receives all its nutrients from the sow's milk during the first 2 weeks of life (Fig. 19-12). The first milk secreted by the sow (colostrum) is exceptionally rich in immunoglobulins (antibodies). The newborn pig is capable of absorbing these antibodies intact for the first 12 to 24 h of life (Fig. 19-5), which gives it passive immunity. Newborn pigs depend entirely on colostrum for their immunity, and those that do not receive colostrum stand little chance of survival. During the first 24 to 48 h postpartum, the colostrum gradually changes to a composition that is typical of milk (Table 19-27).

The requirements for protein, minerals, and other nutrients are quite high at birth, but these requirements are adequately met by milk alone, because milk is quite high in protein, energy, Ca, P and other nutrients on a dry-matter basis (Table 19-28).

FIGURE 19-12 Baby pigs receive all their nutrients from sow milk.

Iron is the only nutrient required by baby pigs that is deficient in milk. The pig is born with about 50 mg of Fe in its body. They require about 6 to 8 mg of Fe per day for hemoglobin synthesis, and milk supplies only about 1 mg/day. So if not given supplemental Fe, the pigs will become anemic within 3 to 4 days. The simplest method of supplying supplemental Fe is to give an intramuscular injection of 100 to 200 mg of Fe (as Fe-dextran, Fe-dextrin, or gleptoferrin) before the pig is 3 days of age. The injection can be given in the ham or in the neck muscles (Fig. 19-4). This will generally meet the pig's need until it reaches 2 to 3 weeks of age and begins to eat dry feed. Sprinkling some Fe-rich feed on the floor of the pen or swabbing the sow's udder with an Fe-rich solution is effective, but it must be done on a daily basis. Feeding a high level of Fe to the sow also is effective, not because it increases the Fe content of her milk, but because the pen becomes contaminated with Fe-rich feces from the sow.

If a sow does not milk properly, the pigs should be transferred to another sow. The transfer should be made within 24 h after the

TABLE 19-28

Comparison of sow colostrum and sow milk on a wet and dry basis

Item	Colostrum		Milk	
	Wet Basis	Dry Basis	Wet Basis	Dry Basis
Total solids, %	25	100	20.0	100
Protein, %	15	60	5.8	29
Fat, %	5	20	8.0	40
Lactose, %	4	16	5.4	27
Calcium, %	0.05	0.20	0.22	1.1
Phosphorus, %	0.10	0.40	0.16	0.8

foster dam has farrowed. Transferring a few pigs from an excessively large litter to a sow with a small litter is a recommended procedure in that it will result in more uniform weaning weights of the pigs. Pigs can be bottle fed or pan fed cow's milk, but these procedures are very laborious and are successful only if excellent sanitation is used and only if pigs first receive colostrum. Sow's milk contains an appreciably higher nutrient concentration than cow's milk, so if cow's milk is used, condensed milk or dry milk replacer should be added to increase the dry-matter content to 20% (Table 19-28). Commercial equipment is available for mechanically feeding artificial milk diets to baby pigs.

Pigs begin to nibble on dry feed at about 2 weeks of age. If pigs are weaned at 2 or 3 weeks of age, creep feeding is probably not necessary. But if pigs are weaned at an older age, creep feeding is advised. The creep feed should be palatable, placed in a readily accessible place for the pigs, and kept fresh. A commercial pelleted starter is recommended, but a ground and mixed diet is acceptable. The creep feed does not necessarily need to be excessively high in protein, because the sow supplies ample protein in her milk. However, some producers feel that supplying a creep feed of the same makeup as the diet that pigs will receive following weaning helps to accustom the pig to its post-weaning diet.

Pigs can be weaned from a few days of age up to 6 to 8 weeks of age. Later weaning is possible, but milk production falls off and sows can become excessively thin with long lactation periods. The younger a pig is weaned, the more critical are the environmental needs and the better the feeding program must be. Early weaning is becoming more popular because it allows the sow to get back into the business of producing pigs sooner and increases the potential for increasing the number of pigs produced per sow per year. Most medium to large producers wean at 3 to 4 weeks of age. Many of the very large producers are now weaning at 1½ to 2 weeks of age and moving the pigs to an off-site nursery. The system is called segregated early weaning (SEW). The advantage of SEW is that pigs are separated from the sow before their immune system becomes activated as a result of exposure to antigens from the sow. SEW pigs having low-antigen challenge backgrounds gain weight extremely rapid as long as they are kept in clean, off-site nursery facilities that are managed on an all-in, all-out basis. This improvement in growth is carried on through market weight if the pigs are moved into clean growing–finishing buildings that, like nursery buildings, are handled as all-in, all-out facilities.

WEANLING PIGS

The starter phase of production is the period from weaning until about 40 to 50 lb. Pigs weaned at 1½ to 2 weeks of age (SEW) weigh less than 10 lb, whereas those weaned at 3 to 4 weeks of age weigh 12 to 16 lb at weaning. Pigs need a warm and exceptionally dry and draft-free environment during this stage. The younger the pig at weaning, the warmer the environment must be. Temperatures of 90° to 95°F during the first week after weaning, with a 5°F/week reduction in temperature for 2 to 3 weeks are recommended for very young pigs. Rearing pigs on raised decks with totally slotted floors (Fig. 19-13) is recommended for this stage of production because the pens can be kept warmer and the pigs' dunging habits are not yet established.

Feeding programs for weanling pigs vary depending on weaning age. A single starter (1.0% lysine) is adequate for pigs weaned at 6 weeks or older. Pigs weaned at 3 to 4 weeks are generally fed two starters, the first one (1.2% lysine) to 25 lb, then a second one (1.0% lysine)

FIGURE 19-13 Early-weaned pigs in raised, decked pens (top) and in tiered decks (bottom).

to 40 to 50 lb. For pigs weaned at 2 to 3 weeks, a three-phase starter program is common. A phase 1 starter (1.4% lysine) is generally fed for 7 to 10 days until the pigs reach about 15 lb. Then a phase 2 starter (1.2% lysine) is fed for 2 weeks (to 25 lb), after which a phase 3 starter (1.0% lysine) is fed to 40 to 50 lb. In SEW programs, where pigs are weaned at 1½ to 2 weeks, they are generally started on a nutrient-dense prestarter diet (1.6% lysine) for the first week postweaning and then switched to the three-phase starter regimen.

Several unique ingredients are used in prestarter and starter diets. Dried whey is commonly included at high levels (25% to 35%) in prestarter and phase 1 starter diets and at lower levels (10% to 20%) in phases 2 and 3 starters. Lactose and dried skim milk are sometimes used in prestarters. Dried plasma protein is almost always included in prestarter and phase 1 starters, and dried blood meal, dried blood cells, or fish meal are often included in phase 2 starters. Amino acids (in particular lysine) and/or soy protein concentrate are often used in order to keep the amount of soybean meal from becoming excessive (due to its antigenicity). Commercial starters often contain some sucrose and/or flavoring agents to encourage pigs to eat. Whether they are actually of any benefit is questionable. Prestarter and starter diets are always highly fortified with antimicrobials, and pharmacological levels of copper sulfate or zinc oxide are commonly included in one or more phases. Compositions of various prestarter and starter diets are shown in Table 19-29.

Starter diets may be purchased or mixed. Commercial starters are generally pelleted to give a small diameter pellet. Small, nursery-type feeders (Fig. 19-14) and nipple waterers work well in nursery pens for young pigs. Care should be taken to keep the feeders clean and to prevent the feed from becoming stale. Diarrhea, if it occurs, should be treated quickly. The use of in-line water medicators are very useful in this regard.

GROWING–FINISHING PIGS

From 50 lb to market weight, pigs are generally full-fed complete mixed diets in self-feeders (Fig. 19-15). There should be a feeder space for every four to five pigs. Grain–soybean meal based diets that are adequately fortified with minerals and vitamins are common for this stage of production. Overcrowding of growing–finishing pigs is not uncommon as producers attempt to maximize their building space and often take pigs to heavier slaughter weights. Overcrowding, however, has pronounced negative effects on pig performance and should be avoided unless absolutely necessary.

In general, the recommended protein level for corn– or sorghum–soybean meal diets is 16% for average pigs for the growing phase (to 100 to 125 lb) and 13% to 14% for the finishing phase (thereafter to 240 lb). Recommended lysine levels for these two stages are 0.80% and

TABLE 19-29

Prestarter and starter diets for young pigs

Ingredient	Prestarter[a,b]		Phase 1 Starter[a,c]			Phase 2 Starter[a,d]			Phase 3 Starter[a,e]		
	1	2	1	2	3	1	2	3	1	2	3
Corn or milo, ground	521	701	737	975	790	1071	1140	998	1282	1309	1252
Oats groats, rolled	—	—	—	—	—	—	—	200	—	—	200
Soybean meal (48.5%)	356	384	276	196	350	312	430	400	280	404	400
Soy protein concentrate	—	150	—	100	—	—	—	—	—	—	—
Fish meal	50	—	50	—	—	50	40	—	—	—	40
Dried whey	400	500	400	500	600	400	300	200	300	200	—
Dried skim milk	—	150	—	—	100	—	—	—	—	—	—
Dried plasma protein	150	—	120	120	—	—	—	—	—	—	—
Dried blood cells	—	—	—	—	—	60	—	—	40	—	—
Dried blood meal	—	—	—	—	50	—	40	40	—	30	—
Lactose	400	—	300	—	—	—	—	—	—	—	—
Sucrose	—	—	—	—	60	60	—	100	—	—	50
Fat	60	60	60	60	60	60	—	—	40	—	—
Lysine·HCl (78% lysine)	4	—	4	—	4	—	—	4	2	—	2
Methionine	2	—	2	—	2	—	—	2	—	—	—
Dicalcium phosphate	23	23	21	20	22	25	26	24	25	24	23
Limestone, ground	14	12	10	15	8	8	10	18	17	19	19
Salt, iodized	8	8	8	8	8	8	8	8	8	8	8
Vitamin mix	2	2	2	2	2	2	2	2	2	2	2
Trace mineral mix	2	2	2	2	2	2	2	2	2	2	2
Copper sulfate	—	—	—	—	—	—	—	—	—	—	—
Zinc oxide	8	8	8	—	—	—	—	—	—	—	—
Antibiotics	+	+	+	+	+	+	+	+	+	+	+
Total	2,000	2,000	2,000	2,000	2,000	2,000	2,000	2,000	2,000	2,000	2,000

[a]The prestarter and three-phase starters contain, respectively, 1.60%, 1.40%, 1.20%, 1.00% lysine; 21% to 23%, 19% to 20%, 19% to 20%, 16% to 18% protein; 0.90%, 0.80%, 0.80%, 0.80% Ca; 0.75%, 0.70%, 0.70%, 0.70% P, and 1480 to 1570 Mcal/lb of ME.

[b]For pigs weaned at 1½ to 2 weeks (5 to 10 lb), then feed phase 1 (10 to 15 lb), phase 2 (15 to 25 lb), and phase 3 (25 to 40 lb) starters.

[c]For pigs weaned at 2 to 3 weeks (10 to 15 lb), then feed phase 2 (15 to 25 lb) and phase 3 (25 to 40 lb) starters.

[d]For pigs weaned at 4 to 5 weeks (15 to 25 lb), then feed phase 3 (25 to 40 lb) starter.

[e]For pigs weaned at 6 to 8 weeks (25 to 40 lb).

[f]See Tables 19-12 and 19-13.

[g]Antibiotics added to provide 100 to 250 g/ton.

FIGURE 19-14 Nursery-type self-feeder.

0.65%, respectively. Some producers partition the growing–finishing period into three phases (50 to 100 lb, 100 to 170 lb, and 170 to 240 lb) or even more phases to more closely match the nutrient requirements with the growth of the pigs. When this is done, higher levels of protein are fed earlier in the growth phase and lower levels are fed during the later stages. Other grains or protein supplements can be used when economically feasible. Examples of grower and fin-

FIGURE 19-15 Two types of self-feeders for growing–finishing pigs.

isher diets are shown in Tables 19-30 and 19-31, respectively.

Until recently, feeding programs for growing–finishing pigs were based simply on the body weight of the pig, without regard to gender (sex), lean growth potential, or health status. Today, nutritionists realize that these factors need to be considered when feeding growing–finishing pigs. From about 100 lb to market weight, barrows grow faster and eat more feed than gilts, but gilts are more efficient (lower feed-to-gain ratio) and produce leaner carcasses than barrows (Table 19-32); thus, their protein requirements are higher (33). Protein requirements of pigs are closely associated with their lean growth curve (Fig. 19-3), which, in turn, is influenced by genetics and stage of growth. Pigs with a high lean growth rate have higher protein requirements than those with low or average lean growth rates.

Swine producers must know what the genetic potential of their pigs is in order to take advantage of diet adjustments for pigs that are generally superior. Lean growth rate can be determined by keeping track of the days in the growing–finishing period and by obtaining carcass data from a packing plant. For example, assume that it takes 100 days for a group of 40-lb pigs to reach 240 lb (i.e., 200 lb of gain). Assume that the carcass kill sheets indicate that the average carcass yield (hot carcass weight per live weight) was 74% and the carcasses averaged 54% lean. The lean gain then is final pound of carcass lean minus initial pound of carcass lean divided by the number of days. Initial carcass lean is about 40% of live weight. Thus, lean gain would be

(240 lb of final weight × 74% carcass yield × 54% lean in the carcass = 96 lb of lean) − (40 lb × 40% lean = 16 lb of lean)/100 days; or (96 − 16)/100; or 0.80 lb of lean/day.

Pigs with a lean growth rate of 0.80 lb/day or higher are considered high-lean-growth pigs. A lean growth rate of 0.65 to 0.75 is considered as average, and less than 0.65 is considered as low lean growth rate. High-lean-growth pigs should be fed diets that are at least 2 percentage points higher in protein (or 0.15 percentage points higher in lysine) than pigs with a moderate lean growth rate. Exceptionally lean, fast-growing pigs may need even higher levels of protein to maximize their lean growth potential.

Because of their greater leanness and reduced feed intake, gilts require a higher dietary

TABLE 19-30

Diets for growing pigs (50 to 125 pounds)[a,b]

Ingredient	1	2	3	4	5	6	7	8	9
Corn or milo, ground	1562	1522	—	1243	1497	1631	1388	1639	1575
Wheat or barley, ground	—	—	1597	—	—	—	—	—	—
Oats, ground	—	—	—	300	—	—	—	—	—
Soybean meal (48.5%)	390	—	—	—	—	—	—	—	—
Soybean meal (44%)	—	430	360	410	356	247	415	308	312
Meat and bone meal	—	—	—	—	60	—	—	—	—
Wheat middlings	—	—	—	—	60	—	—	—	—
Fish meal	—	—	—	—	—	40	—	—	—
Blood meal	—	—	—	—	—	40	—	—	—
Distillers grains/solubles	—	—	—	—	—	—	100	—	—
Fat	—	—	—	—	—	—	50	—	60
Lysine HCl (78% lysine)	—	—	—	—	—	—	—	4	4
Dicalcium phosphate	25	24	16	23	9	22	23	27	27
Limestone, ground	15	16	19	16	10	12	16	14	14
Salt	5	5	5	5	5	5	5	5	5
Vitamin mix[c]	1.5	1.5	1.5	1.5	1.5	1.5	1.5	1.5	1.5
Trace mineral mix[c]	1.5	1.5	1.5	1.5	1.5	1.5	1.5	1.5	1.5
Antibiotics[d]	+	+	+	+	+	+	+	+	+
Total	2,000	2,000	2,000	2,000	2,000	2,000	2,000	2,000	2,000

[a]Provides 1455 to 1555 kcal/lb of ME, 16% to 17% protein (14%, diets 8 and 9), 0.8% lysine, 0.65% Ca, and 0.55% P.

[b]For high lean growth pigs (0.80 lb/day or greater), substitute an additional 100 lb of soybean meal for the grain.

[c]See Tables 19-12 and 19-13.

[d]Add to provide 50 to 100 g/ton. Copper sulfate may also be included to provide 100 to 250 ppm copper.

protein level than barrows. Often, swine producers pen the sexes separately and feed the gilts diets that are 2 to 3 percentage points higher in protein (0.15 to 0.25 percentage points higher in lysine) than the barrows. This procedure is commonly referred to as *split-sex feeding*. Table 19-33 shows adjustments in dietary protein and lysine for split-sex feeding and for pigs of average and high lean growth potentials.

Diets for growing–finishing pigs can be purchased as mixed feed from a feed manufac-

TABLE 19-31

Diets for finishing pigs (125 to 240 pounds)[ab]

Ingredient	1	2	3	4	5	6	7	8	9
Corn or milo, ground	1671	1644	—	1460	1617	1578	1547	1759	1691
Wheat or barley, ground	—	—	1722	—	—	—	—	—	—
Oats, ground	—	—	—	200	—	—	—	—	—
Soybean meal (48.5%)	285	—	—	—	—	—	—	—	—
Soybean meal (44%)	—	312	239	297	240	318	—	192	200
Soybeans, full fat, roasted	—	—	—	—	—	—	410	—	—
Meat and bone meal	—	—	—	—	60	—	—	—	—
Wheat middlings	—	—	—	—	60	—	—	—	—
Fat	—	—	—	—	—	60	—	—	60
Lysine HCl (78% lysine)	—	—	—	—	—	—	—	4	4
Dicalcium phosphate	22	21	12	20	6	22	20	24	24
Limestone, ground	15	16	20	16	10	15	16	14	14
Salt	5	5	5	5	5	5	5	5	5
Vitamin mix[c]	1	1	1	1	1	1	1	1	1
Trace mineral mix[c]	1	1	1	1	1	1	1	1	1
Antibiotics[d]	+	+	+	+	+	+	+	+	+
Total	2,000	2,000	2,000	2,000	2,000	2,000	2,000	2,000	2,000

[a]Provides 1470 to 1560 kcal/lb of ME, 14% to 15% protein (12%, diets 8 and 9), 0.65% lysine, 0.6% Ca, and 0.5% P.

[b]For high lean growth pigs (0.80 lb/day or greater), substitute an additional 100 lb of soybean meal for the grain.

[c]See Tables 19-12 and 19-13.

[d]Add to provide 20 to 50 g/ton.

TABLE 19-32

Comparison of finishing barrows and gilts

	Barrows	Gilts
Daily gain, lb	1.88	1.74
Daily feed, lb	6.38	5.65
Feed:gain ratio	3.39	3.25
Loin eye area, in.^{2}	5.01	5.41
Backfat, 10th rib, in.	1.17	0.98
Carcass muscle, %	52.8	55.1

Source: Adapted from Cromwell et al. (33). Based on 12 experiments involving 2318 pigs from 103 to 227 lb.

turer or diets can be prepared on the farm. Stationary screw mixers, volumetric proportional mixers, and tractor-driven, portable grinder–mixers are popular means of on-farm mixing (see Ch. 11).

A common practice among larger swine producers who mix their own feed is to purchase soybean meal in bulk and mix it on the farm with ground corn and a base mix. A base mix is a mineral–vitamin pack added to the grain–soybean meal mixture to make the diet nutritionally complete. Special base mixes that contain milk-based ingredients are available for mixing of starter diets. Examples of base mixes are shown in Table 19-34.

Another system commonly used by smaller producers is to purchase a commercial supplement, which is then mixed with grain on the farm. These types of supplements contain various protein, mineral, and vitamin sources and are formulated to supply all the pig's nutritive requirements when blended with grain in the proper proportion. Examples of a complete supplement are shown in Table 19-35. One part of a complete supplement containing 40% protein is mixed with 3.2 parts of corn to provide a 16% protein level and mixed with 4.7 parts of corn to provide a 14% protein level.

Still another feeding system is to give pigs free-choice access to shelled corn and a commercial supplement. This system works better for corn than for other grains, but it requires close management to ensure that pigs consume from $\frac{3}{4}$ to 1 lb of protein supplement daily. This system works better for finishing pigs (125 to 235 lb) than for younger pigs.

Mechanical feeding systems are commercially available that dispense feed on the floor or in troughs at prescribed time intervals (Fig. 19-16). These systems generally necessitate a restricted feeding regimen in order to prevent wastage of feed. Pigs that are restricted to 85% to 90% of full-feed gain slower and require more days to reach market weight, but have leaner carcasses. The effects of feed restriction on efficiency of feed conversion are sometimes (but not always) improved.

Regardless of the feeding system used, it is very important that self-feeders be adjusted properly to prevent feed wastage. This is especially important when pigs are kept on slotted floors because considerable amounts of wasted feed can be lost through the slats and go undetected.

BOARS

The requirements of breeding boars are not well known but are assumed to be similar to sows. Boars should be hand fed 4 to 6 lb daily of the sow's gestation feed. Boars should be kept in fairly trim condition, not too thin and not too fat. During heavy use, breeding boars may require a bit of extra feed to remain in good condition.

Purebred breeders who raise boars should feed diets that are about 2% to 3% higher in protein (i.e., 0.15% to 0.25% higher in lysine) and 0.1% to 0.2% higher in Ca and P during the development period than the levels used in conventional growing–finishing diets. Boars grow faster, consume less feed per unit of gain, and are leaner than barrows and gilts, so their nutritional requirements are higher. Boars are raised for meat in some parts of the world (e.g., Australia and the United Kingdom), but they are generally slaughtered at lighter weights before a problem of taint (off odor) in the meat develops. Also, consumers in those areas tolerate the occasional taint problem more than do consumers in the United States. Packing plants in this country do not slaughter boars for meat.

FEED REQUIREMENTS

Records indicate that it takes between 350 and 375 lb of feed for each 100 lb of pork marketed in a well-managed, farrow-to-finish swine operation. This value may be as low as 325 or even less in highly productive, well-managed herds. This includes the feed consumed by the breeding herd and that consumed by the pigs from birth to market.

The approximate distribution of feed required to produce a 240-lb market hog is

TABLE 19-33

Adjustments in dietary protein and lysine levels in three feeding programs for pigs of differing lean growth rate and gender

Program	(Protein, %)[a]						(Lysine, %)					
	Average Lean Gain			High Lean Gain			Average Lean Gain			High Lean Gain		
	Mixed	Barrows	Gilts	Mixed	Barrows	Gilts	Mixed	Barrows	Gilts	Mixed	Barrows	Gilts
Two phase (lb)												
40–125	16	16	16	18	18	18	0.80	0.80	0.80	0.95	0.95	0.95
125–240	14	13	15	16	15	17	0.65	0.57	0.72	0.80	0.72	0.87
Three phase (lb)												
40–100	16	16	16	18	18	18	0.80	0.80	0.80	0.95	0.95	0.95
100–170	14	13	15	16	15	17	0.65	0.57	0.72	0.80	0.72	0.87
170–240	13	12	14	15	14	16	0.57	0.50	0.65	0.72	0.65	0.80
Four phase (lb)												
40–80	17	17	17	19	19	19	0.87	0.87	0.87	1.00	1.00	1.00
80–130	15	14	16	17	16	18	0.72	0.65	0.80	0.87	0.80	0.95
130–180	14	13	15	16	15	17	0.65	0.57	0.72	0.80	0.72	0.87
180–240	12	12	14	14	14	16	0.50	0.50	0.65	0.65	0.65	0.80

[a]Protein level applies to a corn–soybean meal diet.

TABLE 19-34

Composition of base mixes for weanling, growing, and finishing pigs and gestating and lactating sows[a,b,c,d]

Ingredients	1	2	3
Dried whey	—	—	243
Defluorinated rock phosphate	32	—	—
Dicalcium phosphate	—	26	33
Ground limestone	10	16	12
Salt	5	5	8
Vitamin mix	1.5	1.5	2
Trace mineral mix	1.5	1.5	2
Total	50 lb	50 lb	300 lb

[a]Mix 50 lb of base mix 1 or 2 with 1950 lb of grain and soybean meal for growing–finishing pigs. The diet will contain 0.65% to 0.75% Ca, 0.55% to 0.65% P, and 0.25% salt.

[b]Mix 75 lb of base mix 1 or 2 with 1925 lb of grain and soybean meal for gestating and lactating sows. The diet will contain 0.95% to 1.05% Ca, 0.65% to 0.75% P, and 0.38% salt.

[c]Mix 300 lb of base mix 3 with 1700 lb of grain and soybean meal for weanling pigs. The diet will contain 0.80% to 0.95% Ca, 0.70% to 0.75% P, and 0.40% salt.

[d]An antibiotic fortification pack should be added in diets for weanlings and in other diets where appropriate.

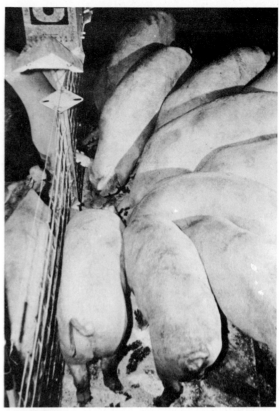

FIGURE 19-16 Floor feeding system for limit-fed pigs.

shown in Table 19-36. About 15% to 20% of the feed is consumed by the breeding herd that produces the pig and about 80% to 85% is consumed by the pig itself. Analysis of farm records shows that those farms having the best herd feed conversion are the ones that are the most profitable.

A number of factors can influence feed conversion in a swine herd. Genetically lean pigs are more efficient in converting feed to weight gain than fat hogs, because lean tissue is more energetically efficient to produce than fat tissue. Marketing hogs at heavy weights has

TABLE 19-35

Complete supplements for growing–finishing pigs and sows

Ingredient	Growing–Finishing		Sows[b]	
	1	2	1	2
Soybean meal (44%)	1284	—	757	1202
Soybean meal (48.5%)	—	1341	—	—
Meat and bone meal	400	200	600	300
Alfalfa meal	—	160	500	—
Wheat middlings	200	120	—	250
Dicalcium phosphate	36	85	63	115
Limestone, ground	42	56	—	53
Salt	30	30	50	50
Vitamin mix[c]	4	4	10	10
Choline mix[c]	—	—	10	10
Trace mineral mix[c]	4	4	10	10
Antibiotics[d]	+	+	—	—
Total	2000	2000	2000	2000

[a]Contains 40% protein, 3.3% Ca, and 1.75% P. Mix 1 part supplement with 3.2 parts grain for growing pigs (50 to 125 lb) and 1 part supplement with 4.7 parts grain for finishing pigs (125 to 240 lb).

[b]Contains 36% protein, 4.0% Ca, and 2.25% P. Mix 1 part supplement with 4 parts grain.

[c]See Tables 19-12 and 19-13.

[d]Add to provide 100 to 300 g/ton.

TABLE 19-36

Total feed required to produce a 240-pound market hog[a]

Stage	Lb
Boar feed	8
Sow gestation feed	75
Sow lactation feed	45
Starter feed (15–40 lb)	50
Grower feed (40–125 lb)	240
Finisher feed (125–240 lb)	400
Total	818[b]

[a]Based on 2.25 litters/sow/yr and 9.0 pigs marketed per litter (20 pigs/sow/year).

[b]Herd feed conversion 818/240 = 3.40. Grow–finish feed conversion = 640/200 = 3.20.

a negative effect on feed conversion; conversely, marketing hogs at lighter weights improves overall herd feed conversion. As hogs increase in weight, a greater proportion of the gain is fat and a lesser proportion is lean. Other factors that reduce efficiency of feed conversion are cold or excessively hot environmental tempera- tures, poor housing, failure to control feed wastage, overcrowding, feeding low-quality diets, and failure to maintain parasite and dis- ease control in the herd. Increasing the pro- ductivity of the sow herd, such as increasing the number of pigs marketed per sow per year, and effectively culling open, nonproductive sows also improves overall herd feed conver- sion efficiency.

SUMMARY

A sound nutrition program during all phases of the life cycle is essential for raising swine prof- itably. The current trends in the swine industry such as early weaning, confinement rearing, and intensive production systems place added importance on the quality of the feeding pro- gram. Raising pigs will continue to be a prof- itable undertaking for those producers who use top-level management in their operation. Effi- cient swine production will ensure a readily available supply of pork as a palatable and nu- tritious food for humans.

REFERENCES

1. NRC. 1988. *Nutrient requirements of swine.* 8th ed. Washington, DC: National Academy Press.

2. Baker, D. H., and T. K. Chung. 1992. *Ideal pro- tein for swine and poultry.* Biokyowa Tech. Rev. − 4. BioKyowa, Chesterfield, MO.

3. Cromwell, G. L., and R. D. Coffey. 1993. An as- sessment of the bioavailability of phosphorus in feed ingredients for nonruminants. *Proc. Maryland Nutr. Conf.,* Baltimore, MD, pp. 146–158.

4. North Central Region-42 Committee on Swine Nutrition. 1993. *J. Animal Sci.* 71 (Suppl. 1):67 (abstr.).

5. Page, T. G., et al. 1993. *J. Animal Sci.* 71:656.

6. Lindemann, M. D., et al. 1995. *J. Animal Sci.* 73:457.

7. Mahan, D. C. 1995. *J. Animal Sci.* 72:2870.

8. Tribble, L. R., J. D. Hancock, and D. E. Orr, Jr. 1984. *J. Animal Sci.* 59(Suppl. 1):245 (abstr.).

9. Lindemann, M. D., and E. T. Kornegay. 1989. *J. Animal Sci.* 67:459.

10. North Central Region-42 Committee on Swine Nutrition. 1976. *J. Animal Sci.* 42:1211.

11. National Pork Producers Council. 1996. *Pork in- dustry handbook.* Cooperative Ext. Serv., Purdue Univ., W. Lafayette, IN.

12. Pond, W. G., and J. H. Maner. 1974. *Swine pro- duction in temperate and tropical environments.* San Francisco, CA: W. H. Freeman.

13. Miller, E. R., D. E. Ullrey, and A. J. Lewis. 1991. *Swine nutrition.* Boston: Butterworth Heine- mann Publ.

14. Thacker, P. A., and R. N. Kirkwood. 1990. *Non- traditional feed sources for use in swine production.* Boston: Butterworths Publ.

15. Cromwell, G. L., T. S. Stahly, and H. J. Monegue. 1984. *J. Animal Sci.* 59(Suppl. 1):103 (abstr.).

16. Schell, T. C., et al. 1993. *J. Animal Sci.* 71:1226.

17. Pettigrew, J. E. 1981. *J. Animal Sci.* 53:107.

18. *Feed Additive Compendium.* 1997. Minneapolis, MN: Miller Publ.

19. Cromwell, G. L., et al. 1981. *J. Animal Sci.* 53(Suppl. 1):95 (abstr.).

20. Council for Agricultural Science and Technology. 1981. *Antibiotics in animal feeds.* Rpt. No. 88. Iowa State Univ., Ames, IA.

21. Hays, V. W. 1981. *The Hays report.* Long Beach, CA: Rachelle Laboratories.

22. Zimmerman, D. R. 1986. *J. Animal Sci.* 62 (Suppl. 3):6 (abstr.).

23. Cromwell, G. L. 1991. Antimicrobial agents. In: *Swine nutrition* (E. R. Miller, D. E. Ullrey, and A. J. Lewis, eds). Stoneham, MA: Butterworth-Heine- mann Publ.

24. Hill, G. M., et al. 1996. *J. Animal Sci.* 74 (Suppl. 1):181.

25. Cromwell, G. L., H. J. Monegue, and T. S. Stahly. 1993. *J. Animal Sci.* 71:2996.

26. Carter, S. D., et al. 1996. *J. Animal Sci.* 74(Suppl. 1):59 (abstr.).

27. Wondra, K. J., et al. 1996. *Feedstuffs* (January 29) 68:5:13.

28. Whittemore, C. T., and F. W. H. Elsley. 1977. *Practical pig production.* Ipswich, Suffolk, England: Farming Press Ltd.

29. Toner, M. S. et al. 1990. *J. Animal Sci.* 74:167.

30. Stahly, T. S., G. L. Cromwell, and H. J. Monegue. 1985. *J. Animal Sci.* 61:1485.

31. Stahly, T. S., G. L. Cromwell, and H. J. Monegue. 1992. *J. Animal Sci.* 70 (Suppl. 1):238 (abstr.).

32. Monegue, H. J., et al. 1993. *J. Animal Sci.* 71 (Suppl. 1):67 (abstr.).

33. Cromwell, G. L., et al. 1993. *J. Animal Sci.* 71:1510.

20

FEEDING POULTRY

Danny M. Hooge

INTRODUCTION

The term poultry has come to mean any of the domesticated and commercialized types of birds (avian species) used for production of eggs and/or meat for human food, and in certain cases feathers for fly-tying and ornaments, skin for leather goods, and/or blood plasma for medical purposes. The classification now includes mainly chickens, turkeys, pigeons, guinea fowl, peafowl, ducks, geese, upland game birds (quail, pheasants, chukars, and partridges) and ratites, primarily ostriches, emu, and rheas.

Nutrition and feeding must be considered together when discussing the general nutritive care of poultry. Dietary nutrients must be provided in adequate supply with sufficient digestibility and availability to the birds to support normal growth, feathering, disease resistance and immunity, stress resistance (temperature, crowded conditions, etc.), egg production, reproduction, and hatchability. A *total* content of some nutrient in the feed may not express the *available* content of that nutrient to the bird, for example, nonphytate and phytic acid-bound phosphorus or natural and carbohydrate-bound lysine (such as from the browning reaction or overcooking). A nutrient may be absorbable from the intestine but not metabolizable by the bird, as in the case of bound lysine. The effects of aflatoxin on methionine and lysine requirements of broiler chicks are shown in Figure 20-1.

Quality of ingredients as well as quantity of nutrients present must be evaluated. Certain substances such as trypsin inhibitor in under-processed soybean meal; thiaminase in certain types of fish meal, especially when not adequately processed; and peroxides in rancid fats can cause indigestibility, nutrient destruction, or absorption problems (see Chs. 7 and 8). Variability among plant and animal (including fish) by-product ingredients can be minimized by sampling and analysis, working closely with suppliers to obtain consistency, and eliminating ingredient sources that do not meet established criteria.

The commercial poultry industry has been an innovator and applicator of advancing technology and knowledge in the areas of genetics, veterinary medicine, and nutrition to keep meat and egg prices relatively constant for decades. Because feed expense is the largest single cost item in poultry production, amounting to 60% to 75% of the cost of getting poultry from hatching egg to processing plant, much emphasis has been placed on least-cost feed formulation and on obtaining the lowest feed expenditure per unit of salable meat and egg products. As a result, refinement of nutrient requirements, genetic progress, advances in disease control, and development of new types of equipment and housing have occurred (see Figures 20-2 through 20-5). These efforts have led to steady improvements in growth rate, feed conversion, and livability under intensive commercial conditions. The nutrition

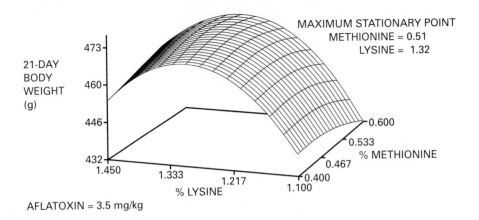

21-DAY BODY WEIGHT (g)

MAXIMUM STATIONARY POINT
METHIONINE = 0.51
LYSINE = 1.32

473
460
446
432

1.450 1.333 1.217 1.100
% LYSINE

0.600
0.533
0.467
0.400
% METHIONINE

AFLATOXIN = 3.5 mg/kg

FEED:GAIN RATIO (0–21 days)

MAXIMUM STATIONARY POINT
METHIONINE = 0.53
LYSINE = 1.34

1.757
1.686
1.615
1.544

1.450 1.333 1.217 1.100
% LYSINE

0.600
0.533
0.467
0.400
% METHIONINE

AFLATOXIN = 3.5 mg/kg

FIGURE 20-1 Increased dietary methionine and lysine levels above NRC 1994 levels (0.50% Met and 1.10% Lys) are required during aflatoxicosis for optimal body weight and feed-to-gain ratio of young broiler chicks (30).

FIGURE 20-2 Broiler houses typical of large operations in the United States.

FIGURE 20-3 Broilers under intensive confinement conditions.

FIGURE 20-4 Turkey poults, such as these grown in environmentally controlled buildings, are slightly more difficult to start than broiler chicks.

FIGURE 20-5 Market turkeys can be grown outdoors in the milder seasons of the year, although the majority are raised inside in the United States.

and feeding of turkeys and ducks have improved markedly as these industries have followed the broiler chicken industry in expansion.

The purpose of this chapter is to present updated practical information about nutrition, feed formulation, and feeding of current strains of poultry for optimal live performance, egg production, fertility and hatchability, processing yields, and profitability. Emphasis is on nutrition of chickens, turkeys, and ratites.

PRACTICAL NUTRIENT REQUIREMENTS

An important reference work for poultry nutrient recommendations is the NRC publica-

tion *Nutrient Requirements of Poultry* (1). It provides a basis for setting the minimum nutrient levels in poultry feeds. Many of the requirements have been developed in university facilities with low levels of disease and environmental stresses, so vitamin levels must be increased substantially under commercial poultry conditions, keeping in mind that some of these (vitamins A and D in particular) can be toxic when given in great excess. Although a single level of nutrients is recommended for all stages of egg production in egg-type strains of chickens, the commercial egg industry uses *phase feed* changes, attempting to vary the level of nutrient intake with age of bird, season, egg production, and egg size to lower feed expenses. The booklet is excellent for its many references citing the sources on which each nutrient recommendation was based.

Because of the ease of global communications and travel today, estimates of poultry nutrient requirements are shared around the world. From North America, the journals *Poultry Science* and *Journal of Applied Poultry Research* provide academic and industry research results. *British Poultry Science, Japanese Poultry Science* (with English summaries), and *World's Poultry Science Journal* are three other referred publications with information for the poultry professional. Rhone-Poulenc, Inc., of Commentry, France, with a corporate office in Atlanta, Georgia, periodically updates their booklet *Animal Feeding,* giving current poultry nutrient requirements. Miller Publishing Company, Minnetonka, Minnesota, updates poultry nutrient recommendations by species each year in the *Feedstuffs Reference Issue.* Broiler chicken amino acid requirements are published periodically by the University of Maryland (Owen P. Thomas) in the proceedings of the Maryland Nutrition Conference. Poultry nutritionists make presentations each year at many conferences around the globe, and printed proceedings are available from most of these.

Commercial strains within each type of poultry may vary significantly in the nutrient levels required for growth or egg production. Genetic progress is being made annually in the performance of chicken, turkeys, and guinea fowl and in other classes of poultry where selection pressure is being applied. Genetic differences, including growth rate and pattern of growth, rate of egg production, egg size, and shell quality, can have marked effects on the nutrient needs of poultry at a given age. A good illustration of this is the use of lower-energy, higher-protein turkey feeds in Europe with the British United turkey strain and higher-energy feeds in North America with the Nicholas turkey strain, an earlier maturing market turkey. It is important to know as much as possible about the strain of poultry being fed in order to more satisfactorily meet nutrient needs. For this reason, nearly every primary breeding company provides nutrition and feeding guidelines for their stock.

VITAMIN AND TRACE MINERAL PREMIXES

Although supplemental vitamins and trace minerals contribute a relatively small amount to the cost of complete feed (about $2.50 to $7.50 per ton), these approximately 20 essential nutrients play major roles in the metabolic functions of poultry. Because practical feed ingredients vary in their vitamin and trace mineral contents from year to year and the bioavailabilities of all vitamins and trace minerals are not well established for many feedstuffs, poultry nutritionists supplement feeds with custom vitamin and trace mineral premixes to assure adequacy.

Results of a survey of vitamin levels used commercially for broiler and breeder chickens, market and breeder turkeys, and laying hens are given in Tables 20-1 and 20-2. These averages can be used as a basis for formulating custom premixes. The BASF Corporation survey by Ward (2) did not include ascorbic acid (vitamin C) or choline, so suggested practical levels for these vitamins and for trace minerals are given in Table 20-3. Practical vitamin and trace mineral supplementation levels for ratite diets are presented in Table 20-4. Premixes with low inclusion rates per ton are best.

In cases where mash (i.e., ground and mixed) feeds are used for chickens in egg production, some of the extra B vitamins added to feed intended for steam pelleting and crumbling are not needed. There are usually sufficient quantities of thiamine, pyridoxine, and folic acid in practical feed ingredients to supply laying hens via mash feeds. The use of expanders in mills for steam-conditioning feeds has been growing in popularity because of reduction or elimination of *Salmonella* and other potentially harmful species of bacteria. Whenever steam and high temperatures are employed for pelleting, expanding, or extrusion (usually not done to poultry feeds due to its high cost), there may be some destruction of vitamins in feed. Vitamin A content is a good indicator of any reduction in vitamin activity because it is one of the first to be oxidized and show a loss of potency.

Trace minerals complexed with amino acids or proteins (peptides) have increased in commercial use in recent years because bioavailability is often better than from inorganic sources. In this category, zinc methionine (Zinpro[R], Zinpro Corporation, Edina, Minnesota) is one of the most widely utilized and extensively researched. It has been shown to improve skin

TABLE 20-1

Average vitamin supplementation levels for broiler and broiler breeder chickens based on an industry survey sponsored by BASF Corporation

Per Ton of Complete Feed	Broiler				Broiler Breeder
	Starter	Grower	Finisher	Withdrawal	
Vitamin A, MIU	8.02	7.36	6.31	4.51	10.03
Vitamin D3, MIU	2.55	2.33	2.00	1.45	2.78
Vitamin E, KIU	16.29	14.30	12.41	8.86	24.17
Niacin supplement, g	41.51	39.50	34.27	22.30	31.56
Pantothenic acid, g	10.92	9.90	8.53	6.81	12.72
Riboflavin, g	6.44	5.84	5.08	3.85	7.42
Menadione, g	1.68	1.48	1.88	0.97	1.62
Thiamine, g	1.61	1.27	1.06	0.89	1.85
Pyridoxine, g	2.38	2.04	1.79	1.48	3.17
Folic acid, mg	780	682	578	405	899
Biotin, mg	72.3	63.5	55.3	46.8	147.8
Vitamin B_{12}, mg	12.5	11.3	9.74	6.64	13.1

Source: Ward (2). Used by permission of the *Journal of Applied Poultry Research.*

The withdrawal feed usually has coccidiostat medication withdrawn, because it takes about 7 days for a coccidiosis problem to develop, and the withdrawal period is usually this length or shorter. Only those companies adding vitamins to the withdrawal feed were included (some used no vitamins). MIU = million international units; KIU = thousand international units. Niacin or niacinamide could be used. Menadione is vitamin K3.

TABLE 20-2

Average vitamin supplementation levels for market and breeder turkeys and chicken laying hens based on an industry survey sponsored by BASF Corporation

Per ton of Complete Feed	Turkey				Laying Hens
	Starter	Grower	Finisher	Breeder	
Vitamin A, MIU	11.62	8.65	6.13	12.06	7.38
Vitamin D3, MIU	4.61	3.46	2.44	4.95	2.44
Vitamin E, KIU	34.70	24.50	14.39	48.11	7.52
Niacin supplement, g	84.70	59.10	43.54	78.70	24.70
Pantothenic acid, g	19.20	14.21	10.02	21.74	7.10
Riboflavin, g	8.94	6.68	4.62	11.44	4.60
Menadione, g	2.40	1.71	1.39	2.55	1.00
Thiamine, g	2.63	1.62	1.11	3.23	1.23
Pyridoxine, g	3.94	2.59	1.82	5.98	1.90
Folic acid, mg	1722	1119	762	2344	306
Biotin, mg	174.6	101.2	73.1	292.4	64.5
Vitamin B_{12}, mg	19.4	12.8	9.70	30.4	7.7

Source: Ward (2). Used by permission of the *Journal of Applied Poultry Research*.

Only those companies adding thiamine, pyridoxine, folic acid, and biotin were averaged for laying hens (some companies do not add these vitamins). MIU = million international units; KIU = thousand international units. High-producing older chicken hens may require higher vitamin D3 levels (about 2.75 to 3 million IU per ton) for shell quality. Niacin or niacinamide can be used. Menadione is vitamin K3.

and feathering, immune functions, growth, feed conversion, and other performance factors. Selenium methionine (Selplex[R], Alltech, Inc., Nicholasville, Kentucky), a yeast-derived product also containing selenium cystine (cysteine) and some other amino acids, has been shown to improve the feathering on broiler chickens.

Some products are known as *chelates* when there is a definite complex of compounds around the element and *proteinates* when some complexation also occurs, but this is sometimes difficult to quantify. Determination of extent of

the complexing or chelation has become easier with the more sophisticated laboratory equipment available today.

Typically, when organic trace mineral products are used they supply a portion, such as 20% to 40% of the element, and the remainder is provided from an inorganic source. The organic trace mineral supplements are often priced proportionally higher than their demonstrated increase in bioavailability, and manufacturers sometimes cite more or less intangible benefits that are hard to pinpoint

TABLE 20-3

Supplemental ascorbic acid, choline, and trace mineral levels for various poultry feeds[a]

Per Ton of Complete Feed (relative)	Meat: Eggs:	Starter Breeder and Prelay (100%)	Grower (85%)	Finisher Layer (75%)	Maintenance Postmolt (60%)
Additional vitamins					
Ascorbic acid, g		90	76.5	67.5	54
Choline, g		375	318.8	281.3	225
Trace minerals					
Zinc, g		110	93.5	82.5	66
Manganese, g		100	85	75	60
Iron, g		35	29.7	26.3	21
Copper, g		7.5	6.4	5.6	4.5
Iodine, g		2.25	1.91	1.69	1.35
Selenium, mg		272.2	231.4	204.2	163.3

[a]Ascorbic acid and choline were not included in the BASF Corporation surveys (Tables 20-1 and 20-2). Starting turkey poults need about 500 g of choline added per ton of feed, and growing turkeys need slightly more choline (about 375 g added per ton) than chickens. An added choline level of about 250 g per ton of feed is typical for laying hens. For economy, some egg companies use trace mineral levels as indicated in the 60% column for laying hens. Ethoxyquin is recommended for combined vitamin and trace mineral premixes at 2.25 g/lb of premix to protect vitamins (and at 50 g/ton complete feed when significant levels of fat containing no antioxidant are included in the diet). Extra copper (125 to 185 ppm) is commonly added to broiler chicken and market turkey feeds in the form of copper sulfate (1 to 1.5 lbs/ton) as a growth promoter and "pass through" fungicide for litter (see also Figure 20-6).

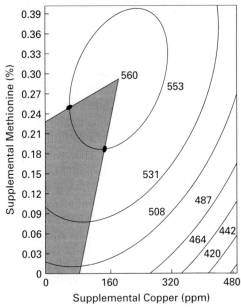

FIGURE 20-6 Response surface (contour plots) of effects of supplemental methionine and copper, from copper sulfate, on 21-day broiler chick weights. Shaded area is zone of substitution, where 0.25% Met and 60 mg Cu/kg equals response of 18% Met and 130 mg Cu/kg, for example. [Wang (31).]

(increased vigor, healthier, able to resist stressful conditions such as poor management, etc.). In addition to growth, disease resistance, and processing yield and grade improvements, improved reproductive performance in poultry has been indicated with certain organic trace minerals or blends.

There is some interest and experimentation currently on the "ultratrace minerals," which may eventually be determined to be required in very small amounts, such as parts per billion (ppb). Trivalent chromium (CR^{3+}) at 200 ppb has shown efficacy in several studies for reducing carcass fat and improving lean gain. It is known to be a component (cofactor) in the glucose tolerance factor needed to potentiate or assist insulin in moving glucose from circulation into peripheral tissues. Trace minerals that someday may be demonstrated to be required include arsenic, boron, nickel, silicon, and others.

METABOLIZABLE ENERGY

The energy-yielding nutrients are carbohydrates, fats, and proteins. The energy derived

TABLE 20-4

Suggested vitamin and trace mineral supplementation levels for ratite feeds[a]

Per Ton of Complete Feed (relative)	Starter and Breeder (100%)	Grower (85%)	Holding or Maintenance (60%)
Vitamins			
Vitamin A, MIU	9.50	8.08	5.70
Vitamin D3, MIU	2.55	2.17	1.53
Vitamin E, KIU	40	34	24
Niacin suppl., g	40	34	24
Panto. acid, g	12	10.2	10.2
Riboflavin, g	5.6	4.76	4.76
Menadione, g	0.65	0.55	0.39
Thiamine, g	1.67	1.42	1.00
Pyridoxine, g	2.67	2.27	1.60
Folic acid, mg	800	680	480
Biotin, mg	100	85	60
Vit. B_{12}, mg	11.3	9.6	6.78
Ascorbic acid, g	113.4	96.4	68
Choline, g	417	354	250
Trace minerals			
Zinc, g	110	93.5	66
Manganese, g	100	85	60
Iron, g	35	29.8	21
Copper, g	7.5	6.38	4.5
Iodine, g	2.25	1.91	1.35
Cobalt, g	0.90	0.77	0.54
Selenium, mg	272.2	231.4	163.3

[a]MIU = million international units; KIU = thousand international units. Ratites need approximately the same protein, energy, and macromineral levels as replacement pullet chickens and breeding chickens. Overfortification of protein and energy can lead to leg and joint problems, similar to turkeys and large and giant breeds of dogs.

from chemical bonds through metabolic activities mainly involves carbon and nitrogen metabolism. There is a comfort zone of about 68°F (20°C) to 82°F (27.8°C), at which there is an optimizing of efficiency of metabolic activity without panting, cold distress, and the like. However, very young birds require a warmer environment until their systems are able to regulate and maintain a desired body temperature at about 10 days of age. The optimal environmental temperature therefore decreases from hatch to adult ages; consequently, younger birds are more tolerant of heat stress than older birds. Broilers over 4 weeks of age and turkeys over about 10 weeks of age are most susceptible to heat stress, causing loss in weight gain and eventually mortality.

Feed intake for a poultry flock varies from day to day, even in adult birds at mature weight, due to environmental fluctuations. Feed consumption increases in cold weather and decreases in hot weather. A flock under a given set of circumstances has a *set point* associated with the Na:Ca ratio in the brain hypothalamus and cerebral fluid, to which they will consume energy. This fact is useful in research for determining the comparative energy contents of different feeds or feed ingredients. Commercially, when accurate energy levels are known for feedstuffs, ingredient substitutions can be made as costs and supplies dictate without altering feed intakes or efficiencies of meat and egg production.

Apparent metabolizable energy (AME or ME) is the main calorie system used for formulating poultry feeds. The AME is the gross energy of the feedstuff or feed minus the energy lost as feces, urine, and combustible gases when the product is eaten. The feces and urine, voided together in birds, are termed **excreta.** The gaseous losses in poultry are small and invariably ignored. The units of energy measurement in the United States are kilocalories per pound (kcal/lb), but many other countries use kilocalories per kilogram (kcal/kg) or kilojoules per kilogram (kJ/kg). One kcal equals 4.184 kJ. The calorie system seems to have been developed to measure energy involved with temperature, and the joule system apparently originated as a scale to measure energy required to accomplish work involving force.

A true metabolizable energy or TME procedure was published in 1976 by Sibbald (3) with Agriculture Canada. It involves fasting (24 to 48 h), forced feeding of a precise quantity of material, and collection of excreta from adult roosters. Negative control birds are given the initial fast and are continued in a fast until the experiment ends for collection of fecal metabolic energy (FMe) and urinary endogenous energy (UEe). The two together are called **endogenous energy loss** (EEL).

In the TME system, only the energy of undigested and unmetabolized residues emanating directly from the feedstuff or feed are subtracted from the energy intake, not total fecal and urinary energy as in the AME system. The least variable and most repeatable TME measurements can be obtained by using each bird as its own negative control, but for time and economic reasons, separate groups of roosters are used for EEL and TME determinations.

Farrell et al. (4) have pointed out that the EEL observed with the TME method indicates more catabolism of body stores in fasted birds, especially at low feed intakes. Interestingly, they concluded from research with a continuous feeding method that no EEL occurred. Therefore, in continuously fed birds, AME and TME must theoretically be the same. The AME values determined with young poultry (e.g., broiler chicks) are usually smaller than with adult roosters using the same feed ingredient or diet. Lower feed intakes cause lower energy estimates by both AME and TME methods, and TME values are generally larger than AME figures.

McNab and Fisher (5), on the other hand, strongly recommended measurement of EEL in AME and TME assays, along with a 48-h fast and 50-g intake for the TME procedure. The EEL can be derived in three possible ways: with starved birds; with birds given an energy source, such as 25 to 50 g glucose in an aqueous solution, to minimize protein catabolism and weight loss; or by extrapolation to zero intake of a line relating energy intake to energy excretion. Use of the glucose solution is preferable according to these researchers.

A second correction, for retained nitrogen (called the correction to *zero nitrogen balance*), may also be done using excreta (i.e., fecal and urinary) nitrogen from unfed control birds and is written AMEn, MEn, or TMEn (see related Figure 20-7). The need for this correction arises because catabolism of any retained nitrogen compounds releases energy, and the nitrogen is ultimately excreted in the urine. Nitrogen excretion in adult Leghorn roosters is about 0.75 g/day. The two nitrogen-correction factors in common usage are by Titus (6), 8.73 kcal/g of urinary nitrogen, and Hill and Anderson (7),

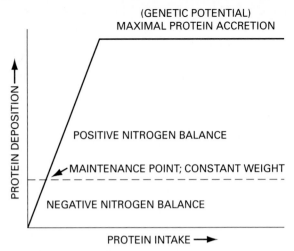

FIGURE 20-7 Nitrogen balance and protein utilization in poultry. [Adapted from Firman (25).]

8.22 kcal/g of nitrogen as uric acid. Nitrogen-corrected values are generally considered to be more accurate than uncorrected figures. Mc-Nab and Fisher (5) have indicated that "AME values are corrected to zero nitrogen balance to make them more typical of the foods rather than of the conditions under which they were fed." The AMEn values tended to be higher than AME figures, whereas TMEn estimates were lower than TME results in work reported by Wolynetz and Sibbald (8).

The TME method, whether Sibbald's or a modified version of his procedure, is quick (a few days), relatively inexpensive, and repeatable. It is valuable for intracompany evaluations of feed ingredients and comparing values between ingredients. Typically, when doing either conventional AME or TME evaluations, corn is used as a control for grains or other energy sources and soybean meal is used as a control for protein supplements. In many cases, AME and TME values obtained for the same ingredient are close. Fat and fiber contents may have impacts on these tests because fat slows and fiber speeds up the passage of digesta through the intestinal tract. Fat has been shown to increase the AME level of animal by-product ingredients, for example.

The recommended energy levels for poultry feeds are presented in the following section in conjunction with protein requirements. It is a generally accepted practice to raise energy contents of feeds as the protein levels are increased because there appears to be an optimal ratio of calories to protein in the diet. Conversely, as protein levels decrease, energy may also be increased except in special cases such as turkey holding (maintenance) diets, where high energy is used with relatively low protein (10% to 12%) in order to minimize feed and protein intakes and keep fleshing optimal for reproduction.

FORMULATING FOR PROTEINS AND AMINO ACIDS

Dietary proteins mainly from plant sources such as oilseed meals have been used to provide commercial poultry with a ready source of essential and dispensable amino acids for development of feathers, muscle tissue for meat, and egg proteins. Traditional formulation based on crude protein levels has given way to a more sophisticated approach involving supplementation with commercially produced and affordable (i.e., feed grade) amino acids, including DL-methionine, L-lysine HCl, and L-threonine. L-tryptophan is also available for use in feeds, but is considerably more expensive than the other three amino acids. These amino acids become more economically advantageous whenever the main protein supplements such as soybean meal and meat and bone meal are scarce and prices are high.

A method for determining *true digestibility* of amino acids in feedstuffs has been developed in conjunction with the TME procedure. True amino acid digestibility has been determined with adult male White Leghorn or broiler breeder roosters, either cecectomized or with intact cecas. It involves force feeding about 25 to 75 g of material per bird by funnel into the crop or training birds to consume a given amount of material within a short time period, for example, 1 h, and collecting the excreta containing fecal (intestinal and cecal) and urinary wastes. Some of the birds used as negative controls are fasted and excreta collected so that an amino acid correction may be done. Amino acid assays of the feed ingredient and the excreta samples allow the calculation of digestibility for each amino acid evaluated.

Although a few companies and universities have been doing true amino acid digestibility work for many years, commercial use on more than a limited scale has only begun recently. This has been possible largely through the efforts of Heartland Lysine, Inc. (8430 W. Bryn Mawr Avenue, Suite 650, Chicago, Illinois 60631), an amino acid supplier, which gathered data from several sources and publishes annual update charts of digestibilities of essential amino acids for poultry (see Table 20-5) and swine. Cystine values reported in the table are

TABLE 20-5

True digestibility of essential amino acids in poultry feed ingredients[a]

Feedstuff as Fed	Crude Protein (CP), %	True Digestibility (%)										
		Arg	Cys	His	Ile	Leu	Lys	Met	Phe	Thr	Trp	Val
Alfalfa meal	16.19	80.3	42.4	74.1	76.6	78.9	61.1	74.0	77.7	70.6	n.a.	75.3
Algae	33.64	87.1	83.2	82.4	81.2	82.9	82.5	83.4	83.4	78.0	n.a.	81.1
Bakery meal	10.06	80.5	75.3	77.9	81.7	84.2	59.8	83.1	85.7	69.4	84.0	76.7
Barley, <10% CP	8.74	78.4	75.5	81.0	72.2	78.2	70.2	75.5	76.3	67.2	n.a.	72.0
Barley, >10% CP	11.94	84.5	82.0	86.4	82.1	85.7	79.2	78.8	86.8	77.0	n.a.	81.4
Blood meal												
Flash dried	87.75	89.4	80.5	84.2	82.5	91.1	88.5	91.6	91.9	89.0	88.8	89.8
Vat dried	80.87	66.5	36.6	58.2	56.2	57.6	58.1	67.4	60.5	59.3	66.8	55.7
Bone meal	39.80	70.5	48.7	60.5	76.2	77.9	69.2	79.8	77.5	75.4	n.a.	75.6
Brewers grains	22.70	63.2	75.0	73.0	80.1	81.6	61.9	77.6	79.8	72.5	n.a.	78.1
Canola meal	36.07	88.6	73.1	85.7	82.9	86.4	78.4	88.4	86.3	77.5	82.3	81.0
Canola seed	21.67	91.0	72.5	82.0	86.0	89.0	88.5	83.5	91.0	79.0	n.a.	86.5
Casein	74.58	95.7	87.5	78.7	94.2	94.6	92.8	93.8	95.1	92.8	91.1	93.7
Coconut meal	21.77	83.7	55.2	72.0	76.6	79.7	50.3	83.6	83.7	63.0	51.1	78.9
Cookie meal	12.11	89.1	86.3	87.1	88.6	91.1	74.9	89.1	88.8	82.0	87.6	88.5
Corn yellow	8.52	90.2	82.9	88.1	88.1	93.7	80.8	91.4	90.8	82.6	78.5	87.5
Corn, high lysine	9.69	88.8	84.0	90.0	76.1	84.0	71.7	83.1	82.5	77.1	n.a.	82.8
Corn, high protein	11.37	89.9	84.5	91.8	83.4	91.7	73.9	89.6	89.8	81.4	n.a.	85.8
Corn germ meal	21.43	90.3	57.8	83.7	86.2	87.2	83.0	85.6	88.0	78.6	n.a.	85.6
Corn gluten fd	21.88	88.1	63.6	82.8	82.0	89.0	70.6	84.0	86.0	74.9	n.a.	81.7
Corn gluten ml	62.43	94.5	86.6	91.8	94.3	96.9	88.1	95.8	96.0	91.3	84.8	94.2
Cottonseed ml	40.74	86.1	67.2	65.5	72.4	75.3	61.2	74.2	84.8	66.5	84.1	74.9
DL-methionine	58.10	—	—	—	—	—	—	100.0	—	—	—	—
Distillers dried grains w/solubles	26.11	68.4	76.7	74.0	80.1	88.9	57.3	84.7	86.1	70.6	n.a.	79.3
Feather meal	84.11	83.3	62.0	70.6	85.8	82.2	65.2	74.4	84.3	73.6	82.5	81.9
Field beans	22.38	89.7	62.8	79.3	85.0	86.3	86.9	71.0	86.8	78.4	n.a.	81.1
Fish meal analog												
High quality	62.75	89.0	79.2	80.6	87.9	91.0	89.4	94.6	91.7	86.7	91.6	91.2
Low quality	63.64	88.4	75.3	74.2	88.1	88.0	84.7	92.6	89.0	82.6	82.4	88.0
Fish meal	62.96	92.5	77.7	89.0	93.4	93.7	90.2	92.8	91.6	91.2	88.3	92.4
Flax meal	39.00	95.0	73.0	87.0	86.0	87.0	87.0	82.0	89.0	n.a.	n.a.	83.0
Flax seed	25.00	91.0	80.0	90.0	86.0	90.0	89.0	79.0	88.0	n.a.	n.a.	81.0
Hair hydrolyzed	66.07	62.5	n.a.	51.9	64.0	76.1	61.2	73.2	70.1	33.3	n.a.	54.6
L-lysine HCl	94.40	—	—	—	—	—	100.0	—	—	—	—	—
Liver meal	73.80	76.3	n.a.	73.1	73.8	73.8	69.5	75.0	73.3	71.3	n.a.	75.7
Lupin seed meal	35.71	92.9	87.9	88.6	90.6	91.7	88.7	83.5	91.4	87.2	n.a.	88.0

TABLE 20-5 (cont.)

Feedstuff as Fed	Crude Protein (CP), %	True Digestibility (%)										
		Arg	Cys	His	Ile	Leu	Lys	Met	Phe	Thr	Trp	Val
Meat and bone meal												
High quality	50.57	88.9	67.8	83.8	87.3	88.5	86.0	89.3	88.0	84.6	82.2	86.7
Low quality	51.64	80.1	57.6	75.7	79.2	79.8	74.5	80.9	79.0	74.2	74.5	77.0
Oat groats	16.14	91.9	83.9	90.6	88.0	88.2	79.8	89.6	91.6	83.2	n.a.	87.9
Oats	10.78	93.2	83.9	91.6	88.1	90.4	86.7	85.9	92.4	83.0	n.a.	86.5
Palm kernel meal	13.21	88.6	66.6	80.3	81.0	85.0	58.9	83.7	85.3	69.2	n.a.	80.1
Peanut meal	44.81	89.0	79.2	85.0	88.2	90.1	76.6	86.3	91.9	83.2	75.6	87.7
Poultry by-product												
High quality	56.76	89.6	72.1	82.6	88.3	87.6	83.2	89.2	87.5	83.5	80.1	86.6
Low quality	54.13	86.7	64.9	75.7	83.6	82.4	76.8	82.7	81.6	77.3	77.1	81.9
Rapeseed meal	38.75	85.4	74.5	89.9	83.7	85.7	79.1	93.4	85.4	78.9	80.4	82.8
Rice (rough)	7.29	92.4	85.3	91.3	84.0	87.7	79.7	89.6	85.2	81.7	90.8	87.2
Rice bran												
Defatted	13.82	85.5	66.6	80.3	73.5	73.5	73.0	76.8	74.1	68.2	79.0	74.5
Parboiled	17.37	90.9	88.4	88.4	85.5	86.7	84.4	83.8	87.1	79.0	n.a.	87.0
Safflower meal	24.39	91.7	69.4	86.3	85.1	87.4	83.4	88.4	85.5	79.6	86.4	73.8
Sesame seed meal	65.10	77.6	81.7	71.0	72.1	73.1	58.4	76.6	74.9	60.7	n.a.	73.0
Shrimp meal	64.50	93.5	78.6	91.0	95.1	95.2	90.0	96.2	n.a.	91.1	n.a.	93.5
Single-cell protein	68.18	88.6	66.9	88.7	87.8	88.0	86.6	88.8	80.9	84.5	n.a.	85.8
Sorghum (milo)	9.79	79.0	75.0	81.2	84.5	88.2	72.9	84.7	86.1	76.0	79.1	82.6
Soybeans, cooked	36.58	92.9	88.4	92.9	94.1	94.4	91.5	91.9	93.5	89.7	n.a.	92.4
Soybeans, raw	38.16	74.4	61.3	72.7	73.0	70.2	72.7	67.8	70.5	67.5	51.6	71.1
Soy meal, 44% CP	42.91	86.3	73.6	84.3	83.9	84.5	83.4	82.6	85.6	80.9	70.5	82.8
Soy meal, 48% CP	48.96	93.0	83.8	89.5	92.2	92.2	90.6	91.9	92.3	88.4	87.8	90.9
Sunflower meal	31.38	94.1	79.7	87.3	90.1	90.0	80.7	92.3	91.9	84.8	n.a.	85.9
Triticale	16.28	85.8	n.a.	91.6	91.8	94.6	92.3	95.7	95.0	89.5	n.a.	88.7
Wheat	15.77	87.4	87.7	90.6	88.1	90.2	81.8	87.0	91.6	82.3	n.a.	86.1
Wheat bran	15.60	83.7	75.7	82.0	78.4	79.6	74.7	79.3	83.2	74.8	82.9	76.9
Wheat middlings	17.03	86.5	83.3	86.8	79.6	83.1	79.0	79.5	84.3	77.6	83.9	79.3
Wheat screen	13.64	89.4	74.3	85.5	84.4	87.1	79.3	82.4	87.3	79.8	n.a.	83.6
Wheat shorts	17.39	86.8	70.5	84.1	82.5	84.4	81.1	79.6	85.6	79.0	84.6	81.8
Yucca	1.55	94.7	n.a.	88.3	n.a.	73.4	76.1	n.a.	n.a.	54.6	n.a.	n.a.

aUsed by permission of Heartland Lysine, Inc., Chicago (9). Arg = Arginine; cys = Cystine; his = histidine; Ile = isoleucine; and leu = leucine; Lys = lysine; Met = methionine; Phe = phenylaline; Thr = theorine; Trp = tryptophan; Val = valines. Combination of data from conventional and cecectomized precision-fed rooster assays. n.a. = information not available. Safflower meal values from a private source.

actually half-cystine or cysteine, but it has been standard in poultry nutrition for decades to use the term cystine.

Digestible amino acid values for all ingredients used in a least-cost formulation are needed. Occasionally, in using such information a nutritionist is missing data from one or more by-product ingredients that have not been studied, and he or she must obtain such data from private sources or make an estimate. For example, if cull peas with about 23% crude protein are being used as a protein source in the Pacific Northwest, it is not listed in the Heartland Lysine (9) true digestibility chart, and a total true digestibility for each amino acid in a formula cannot be calculated. Only a partial total can be obtained, and this is useless. The essential amino acid digestibilities in cull peas then have to be obtained from a private source, or the ingredient must be submitted to a poultry laboratory set up to conduct such procedures, or, as a last resort, the digestibilities can be estimated from other products (e.g., field beans).

In advancing from traditional crude protein or crude protein and amino acid to true amino acid digestibility formulation, a good idea is to use formulas that have given excellent performance in the past as a basis and to obtain both total and true digestible amino acid values on future printouts for a while. You will notice over time what true digestible amino acid minimums appears acceptable and you may begin to use them rather than total amino acid minimums to formulate feeds.

The use of true digestible amino acids allows more consistency in providing minimum levels of amino acids to poultry. This theoretically results in less variable (i.e., smoother) growth and egg production curves, which are sometimes depicted graphically with jagged sawtooth edges, by reducing nutrient variations. It should also improve feed conversion and reduce feed expense for producing meat and eggs by giving poultry the same well-balanced protein from one week's feed formula to the next.

IDEAL PROTEIN PROFILE FOR BROILERS

David H. Baker (10), at the University of Illinois, in 1995 presented an ideal protein profile for commercial broiler chickens (see Table 20-6). Because solid research data were not available on requirements for most amino acids, particularly during the growth period beyond 21 days of age, it made sense to use the large body of research on amino acid needs relative to lysine to develop an ideal amino acid balance or profile, rather than set individual amino acid requirements. It is known that female broilers have lower amino acid requirements at similar weights to males, because females have higher body fat and lower protein contents. Broiler chickens appear to need higher levels of some amino acids (cystine, threonine, tryptophan, valine, arginine, and isoleucine) relative to lysine during the growth period from 21 to 42 days of age, so separate ideal amino acid ratios were developed for starting and growing (early and late growth).

Lysine was chosen as a reference amino acid because it is second limiting in broiler diets, it is economically feasible to add as a dietary supplement, its analysis is direct and straightforward, it has only one function in the body—protein accretion, and a good deal of information is known about lysine requirements of poultry under a wide variety of dietary, environmental, and body compositional circumstances.

To illustrate the ideal protein ratio, Baker stated that if the ideal profile is 67% digestible threonine to 100% digestible lysine from 0 to 21 days of age, this ratio would be valid for both males and females, diets high or low in energy, diets high or low in protein, strains with high or low lean growth potential, and presumably under all environmental conditions as well. Therefore, the ideal ratio can be factored up or down, as the case may be, to meet the bird's requirement under a given circumstance.

Because supplemental methionine is only 81.2% efficient in providing cystine, the methionine and cystine deficits must be considered separately in feed formulation. For example, a 0.0812 shortage of methionine can be met with 0.0812% methionine, but a a deficiency of 0.0812% cystine would need to be met with 0.10% methionine. In the case of both deficiencies, a supplemental methionine level of 0.1812% would be needed instead of 0.1624% as conventionally figured. An adjustment for DL-methionine 99% or liquid methionine hydroxy analog 88% purity would need to be factored in as well.

Owen P. Thomas (11) at the University of Maryland indicated in a 1995 conference presentation that male broiler chickens given 1450 kcal ME/lb of feed (assuming a 70°F environment) need minimum dietary glycine plus serine levels of 1.45%, 1.16%, and 1.07% during the starter (0 to 21 days), grower (21 to 35 days), and finisher (35 to 49 days) periods, respectively. Females require levels of at least 1.36%, 1.07%, and 0.96%, respectively, under similar

TABLE 20-6

Dietary ideal amino acid ratios and digestible amino acid requirements of male and female broiler chickens from 0 to 21 days and 21 to 42 days of age

Amino Acid	Starter Period (0–21 days)			Grower Period (21 to 42 days)		
	Ideal Ratio	Requirement		Ideal Ratio	Requirement	
		Male	Female		Male	Female
		(% of diet)			(% of diet)	
Lysine	100	1.12	1.02	100	0.89	0.84
Methionine + cystine	72	0.81	0.74	75	0.67	0.63
Methionine	36	0.405	0.37	36	0.32	0.30
Cystine	36	0.405	0.37	39	0.35	0.33
Arginine	105	1.18	1.07	108	0.96	0.91
Valine	77	0.86	0.79	80	0.71	0.67
Threonine	67	0.75	0.68	70	0.62	0.59
Tryptophan	16	0.18	0.16	17	0.15	0.14
Isoleucine	67	0.75	0.68	69	0.61	0.58
Histidine	32	0.36	0.33	32	0.28	0.27
Phenylalanine + Tyrosine	105	1.18	1.07	105	0.93	0.88
Leucine	109	1.22	1.11	109	0.97	0.92

Source: Baker (10). There is some research evidence from Degussa Corporation to indicate that the methionine to cystine ratio may be as high as 62% to 38%, so this topic is somewhat controversial at present. Additional methionine supplementation to commercial broiler chicken feeds is known to improve meat yields.

conditions. In his estimates, Thomas assumed a chemical molar ratio of 50%:50% glycine to serine in the diet.

He also suggested that, although the optimum time for changing from the first to second broiler feed is about 14 days of age, chicks this young are unable to consume regular-sized pellets. Consequently, feed changes are made at 18 to 21 days in actual practice. With the fast-growing birds that are now available, it would appear that the birds should be switched to the last phase at about 35 days of age.

The effects of protein and amino acid levels on abdominal fat, breast yield, and nitrogen excretion are receiving more attention. Male broilers are affected more than females by reduced amino acid levels. Any time that crude protein levels can be reduced while meeting amino acid requirements, nitrogen content in the manure and litter will decrease. The amino acid requirements for maximum feed efficiency and breast yield have often been reported in the literature to be higher than those needed for maximum weight gain.

PROTEIN FOR BROILER BREEDERS AND REPLACEMENTS

Broiler breeder replacements do well when fed 18% to 20% protein starter feed for at least 3 weeks, preferably for 4 to 6 weeks (1270 to 1365 kcal ME/lb or 2800 to 3009 kcal ME/kg). The same essential amino acid profiles recommended for broilers are applicable for broiler breeder replacements. In most breeds, the males grow slower than the females for the first few days. The male body weight at 4 weeks of age is critical, as males appear to function best when they have rapid early growth and high 4-week body weights. Full feeding to this point is important. See Table 20-7 for a schedule developed by Brake (12), which is useful in feeding broiler breeder replacement males and females.

Ideally, it is desirable for males to be reared separately because, when cockerels and pullets are brooded and fed together, males are often observed to weigh less than their female penmates at 4 weeks of age. This is inevitably associated with poor fertility later. Separate rearing means solid-partition separation and separate feeding systems. Field experience shows that for most breeds the minimum 4-week body weight to obtain reasonable fertility is 1.3 lb or 590 g, and the optimal weight may vary by breed but should be around 1.5 lb or 680 g. Males have slightly higher protein requirements than females during rearing.

Cockerel and pullet grower–developer feeds should contain 14.5% to 15.5% crude protein (1270 to 1315 kcal ME/lb or 2800 to 2900 kcal ME/kg). Some managers find it difficult to provide completely separate rearing facilities for males. In these cases, it is recommended that males be grown separately to at least 6 weeks of age before mixing with females. It is important to reach target 20-week body weights. Skip-a-day, 4 days on/3 days off, and

TABLE 20-7

Broiler breeders replacement male and female feeding programs[a]

Age (weeks)	Male Feed Preferred Type	Male Feed Daily Amount (g)	Female Feed Preferred Type	Female Feed Daily Amount (g)
1	S	Ad libitum	S or G	Ad libitum
2	S	Ad libitum	S or G	27
3	S	Ad libitum	S or G	29
4	S	Ad libitum	G	32
5	G	60	G	35
6	G	65	G	38
7	G	69	G	41
8	G	73	G	44
9	G	77	G	47
10	G	80	G	51
11	G	83	G	55
12	G	86	G	59
13	G	89	G	63
14	G	92	G	67
15	G	95	G	72
16	G	98	G	77
17	G	101	G	83
18	G	104	G	89
19	G	107	P	95
20	G	110	P	102
21	M or G	115	P	109
22	M or G	120	P	116
23	M or G	120	P	123
24	M or G	120	P	130
25	M	120	P or B	137
26	M	120	B	Increase feed according to production

Source: Brake (12).

[a]Feed types: S = starter, G = grower; M = male; P = prebreeder; B = breeder. For females, feed distribution so that each bird receives her appropriate amount is a major problem. Every day limited feeding after 4 weeks of age is more theoretical (i.e., an ideal) at this time than practical. Commercially, skip-a-day or various off-feed programs (e.g., 4:3, 5:2, and 6:1 days per week on feed:off feed) are used to restrict weights to target levels. The three or two off-feed days per week are not consecutive. It should be noted that a 1-day skip really amounts to about 48 h off feed if feeding is once a day at, for example, 8:00 A.M. More research on feeding breeder pullets is definitely needed because of its large impact on the number of hatching eggs and chicks produced later.

5 days on/2 days off are some of the feed restriction programs used. For males, the objectives are to maintain large frames and to redistribute breast muscle fleshing to develop overall well-fleshed, lean, muscular roosters that stand erect and possess sharply angled breasts with a prominent keel when handled. In females, extremely poor breast fleshing cannot be compensated for by a higher-protein feed from 18 to 36 weeks of age. A minimum 14% protein grower–developer feed is required for satisfactory fleshing and subsequent egg production during lay.

A prebreeder (prelay) hen diet containing 16% to 18% crude protein, with about 1335 kcal ME/lb (2943 kcal ME/kg), is necessary from the time of photostimulation (18 or 19 weeks) to sexual maturity (24 to 25 weeks) when egg production begins. Breeder hens should be given 17% protein feed initially, decreasing to 16%, then 15% protein on a three-phase program (1300 kcal ME/lb or 2866 kcal ME/kg).

Broiler breeder males should be fed in separate feeders (female-only grills are available) a diet containing 12% to 14% crude protein with about 1280 kcal ME/lb (2822 kcal ME/kg) (see Figure 20-8). They should receive 0.9% rather than 3.2% Ca in complete broiler breeder hen feeds. A grill opening of $1\frac{11}{16}$ in. (4.29 cm), center to center, will prevent the males from eating hen feed. Female broiler breeder hens should reach full feeding at 75% egg production.

PROTEIN NEEDS OF TURKEYS

Crude protein minimums can be reduced by about 2% (actual) when utilizing digestible

FIGURE 20-8 Separate feeding of about 12% protein diets to broiler breeder males and 16% protein diets to females improves fertility. Note male-exclusion grills on hen feeders at bottom of picture (28).

oxygen be delivered by the heart and lungs, which are somewhat constrained by their predetermined thoracic frame. Thus, due to the high rate of metabolic activities a considerable demand is always placed on the cardiopulmonary system of turkeys. Cool temperatures sufficient to dissipate excess body heat and adequate oxygen help to assure optimal growth of turkeys.

Weight management of male and female turkey breeder candidates is a primary concern. Skeletal growth of female and male turkeys is nearing completion by about 17 and 24 weeks of age, respectively, although long bones continue to grow in thickness to offer support for the muscles, which increase in weight with age. Feed restriction after selection limits muscle mass, reducing the weight stress put on the skeletal frame and minimizing deposition of abdominal, mesenteric, and gizzard fat. Selection is usually done at around 28 to 30 lb for males and 14 to 15 lb for females.

Under a quantity-restricted program where approximately 0.8 to 1.2 lb of feed is fed per tom per day, it is critical that the toms never lose weight. Some individual birds may lose weight, but the flock in total should not lose weight. The toms should continue to gain weight slightly

amino acid formulation versus traditional total amino acid formulation (Table 20-8). Market turkeys require higher protein and amino acid levels for starting and growing than do chickens, although protein needs during the finisher phases of each are comparable. Breast meat yield is a major consideration with birds marketed as whole body or further processed. Muscle growth, especially the breast mass with its limited vascular network, requires that ample

TABLE 20-8

True digestible amino acid recommendations for turkey growth, maintenance, and breeder feeds[a]

| | Market Turkeys | | | | | | Breeder |
Age (weeks):	0–4	4–8	8–12	12–16	16+	Hold	Hens
Metabolizable energy (ME):							
kcal/kg	2900	3000	3100	3200	3300	2750	3000
kcal/lb	1315	1361	1406	1452	1497	1247	1361
Protein, %[b]	26.0	24.0	21.5	18.5	15.0	12.0	13.5
Methionine + cystine, %	0.97	0.88	0.79	0.70	0.62	0.44	0.53
Methionine, %	0.56	0.51	0.46	0.41	0.36	0.23	0.29
Cystine, %	0.41	0.37	0.33	0.30	0.26	0.21	0.24
Lysine, %	1.65	1.49	1.29	1.05	0.90	0.57	0.65
Threonine, %	0.88	0.79	0.70	0.61	0.52	0.35	0.40
Arginine, %	1.49	1.34	1.19	1.03	0.90	0.60	0.65
Tryptophan, %[b]	0.25	0.225	0.20	0.175	0.15	0.09	0.12
Histidine, %	0.54	0.49	0.43	0.38	0.32	0.24	0.28
Isoleucine, %	0.99	0.89	0.79	0.68	0.58	0.41	0.46
Leucine, %	1.74	1.56	1.39	1.21	1.05	0.71	0.90
Phenyl-alanine, %	0.93	0.84	0.74	0.65	0.56	0.37	0.53
Valine, %	1.07	0.98	0.86	0.75	0.64	0.48	0.54

[a]Used by permission of Heartland Lysine, Inc., Chicago, Illinois. Assumed 70°F temperature during growing and finishing periods. For each 10°F increase in average daily in-house temperature, increase amino acid levels by 3% of value. Commercially, energy levels range from about 1275 kcal ME/lb (2811 kcal ME/kg) in starter to about 1550 kcal ME/lb (3417 kcal ME/kg) in finisher. Holding and breeder diets have about 1330 kcal ME/lb (2932 kcal ME/kg) and 1350 kcal ME/lb (2976 kcal ME/kg), respectively.

[b]Protein and tryptophan are total, not digestible, values.

[c]Firman (25) proposed an ideal turkey dietary digestible amino acid profile as follows: lys, 100%; met + cys, 66%; thr, 55%; val, 76%; arg 105%; his, 36%; ile, 69%; phe + tyr, 105%; trp, 16%; and leu, 124%.

throughout their reproductive life. Based on industry-wide observations, a tom restriction feed of between 10% and 14% protein is acceptable for most climates. In some instances, such as extreme cold and/or unheated tom barns, males can be switched to a 16% to 17% protein breeder feed to avoid weight loss for short periods of time. A nutrient-restriction program using 8% to 10% protein diets is risky because this may not suffice for maintenance, optimal health, and semen production.

After selecting females, they should be placed on about a 14% protein holding feed for about 2 weeks. Higher or lower protein levels may be used if the flock is behind or ahead on target weights. If candidates continue to gain weight too rapidly, place them on a 12% protein holding feed. Hen lines have ability to gain weight well beyond what is needed for the onset of egg production. Feeding of whole oats with 8% to 10% crude protein as early as 14 weeks of age, but typically about 18 weeks of age, can be done on a regular basis to obtain optimal body weights. Free-choice feeding of grit and the use of a water-soluble vitamin pack can accompany the feeding of whole oats. Restricted birds may also eat litter and residual feathers. Another approach is to mix the standard vitamin and trace minerals with whole oats to be used as the feed source.

Optimum egg production, fertility, semen production, and hatchability are closely correlated with mature body size and condition of breeder turkeys. Managers should always strive to have birds at target weight throughout rearing, since it is difficult to correct underweight problems in the last few weeks of growing.

For breeder hens, a prebreeder diet with 16% crude protein and properly balanced essential amino acids should be used during late rearing. This protein level is higher than in conventional holding diets and prepares candidates for impending lay. Indeed, the greatest benefit may be seen as increasing the growth rate of underweight flocks after 26 weeks of age. A series of breeder hen diets with 17%, 16%, and 14% protein (1350 kcal ME/lb or 2976 kcal ME/kg) should be initiated at the time of photostimulation for egg production.

For breeder toms, which usually enter stud pens around 27 to 30 weeks of age at about 45 lb live weight, a daily intake of a 12% protein feed is recommended. Young toms may be restricted fed this diet, permitting 0.75 to 1 lb of growth per week to 40 weeks of age. Toms must not lose weight from one week to the next. Be-

yond 40 weeks of age, growth rate may be slowed to between 0.5 and 0.75 lb per week by restricted feeding or use of a 10% protein tom feed on a free-choice basis. Toms feeds should contain about 1330 kcal ME/lb (2932 kcal ME/kg) (13, 14).

OTHER NUTRIENTS

CALCIUM AND PHOSPHORUS

Although surface response studies have shown that broiler chickens need about 1.3% Ca and 0.90% P in starter feed for maximum growth, commercial feeds typically contain about 1% Ca and 0.8% total P. These values decrease with age of the birds, as recommended in the NRC *Nutrient Requirements of Poultry* (1), due to the skeleton being more nearly completely formed with each phase of growth.

Turkeys have higher starting Ca and P requirements than chickens, with about 1.4% Ca and 0.85% total P being typical for rapidly developing turkeys today. Phosphate is more poorly utilized by turkeys than by chickens.

Caged layers, broiler breeder hens, and turkey breeder hens need increased Ca levels with the onset of egg production, termed *sexual maturity*. It is customary to increase dietary Ca from about 2.5% in prelay feeds to about 3.75% at peak production and around 4.00% later in lay. See related figures 20-9 and 20-10. It is typical in the United States to molt hens by feed withdrawal to rest them for a few weeks. Egg production reaches peak (about 85% of original peak) more rapidly in molted hens than

FIGURE 20-9 Caged laying hens should be fed about every 1.5 h when lights are on (five to eight times per day) with automatic feeders.

FIGURE 20-10 To minimize egg breakage in mechanical collection systems, layer diets must have adequate Ca (3.5% to 4.25%) and available P (0.50% to 0.25%) for their age and production.

in pullets, so the nutritionist has to watch Ca levels more closely. Free choice-oyster shell or limestone is often used for breeders.

In feed formulation, total P and available P are used. The available P can be determined as nonphytate P in the lab. Phosphorus is important in shell formation, because high blood P levels seem to be the signal to cause cessation of egg shell calcification. Therefore, minimizing dietary P for layers is important, and levels of about 0.55% to 0.35% available P are used. On the other hand, having adequate P during heat stress is critical because mortality rates of about 10% in some cases have been reported with P deficiency. A level of around 0.25% available P appears to be the critical lower limit in practice, but feed intake is another important consideration.

MACROMINERAL ELEMENTS: CATIONS AND ANIONS

Protein (nitrogen) and energy (carbon) metabolism and acid–base regulation are interrelated processes that can influence the performance of poultry (15). Feed consumption, water intake, and water balance are directly affected by these functions. Growth itself is an endogenous acid-producing process. Heat stress can bring on respiratory alkalosis in blood and metabolic acidosis in soft tissues simultaneously. Normal acid–base balance, which is maintained by natural buffering systems, can also be disturbed in poultry during disease, nutrient deficiencies (e.g., K or vitamin D), fasting, live haul,

and other conditions. Any time acid–base balance in the bird deviates substantially from the normal conditions toward alkalosis or acidosis and most of the metabolic pathways (i.e., enzyme systems) do not have their required optimal conditions to function efficiently, the overall result is depressed growth.

Cations are positively charged and anions are negatively charged elements. The phrase cation–anion balance includes all macromineral elements. Electrolytes are all compounds capable of dissolving and dissociating into positively and negatively charged ions in a suitable medium and having the capacity to conduct an electrical current. The term, as it is now used in animal nutrition, primarily refers to sodium (Na), potassium (K), and chloride (Cl). Dietary electrolyte balance (DEB) expressed as Na + K − Cl in milliequivalents (mEq)/kg of diet (or mEq/100 g, convenient for caged layers consuming about this amount per day) has a direct effect on acid–base balance within birds, with increased Na or K causing pH of body fluids to move toward alkalosis and increased C1 promoting acidosis.

Convenient factors for use in calculating DEB in poultry feeds are (percent of electrolyte mineral in diet times): 434.98 for Na, 255.74 for K, and 282.06 for Cl. It is customary to obtain the DEB on poultry formula printouts, but setting minimums for DEB is not a common practice because the requirements are so vague at this time and further research is needed. It has been found that studies exploring one variable at a time (e.g., Na, K, or Cl) tend to give lower estimates of electrolyte requirements than experiments in which interactions are considered (e.g., Na × Cl, or K × Cl, or Na × K × Cl). The use of response surface methodology in electrolyte trials is excellent for studying interactions.

The absorption coefficients for dietary cations and anions for human infants, which should be relatively similar to poultry, have been published by the [European] Committee on Nutrition of the Preterm Infant (16). These are presented in Table 20-9. Fixed mineral salts such as NaCl and KCl contain the primary elements, while salts containing a cation or anion such as ammonium of bicarbonate that can be further broken down are termed *metabolizable*. Because of this and differences in cation and anion absorption coefficients, a macromineral salt that is chemically neutral (e.g., KCl) may cause an acidifying or alkalinizing effect on body fluids. An alkalinizing effect during respi-

TABLE 20-9

Absorption coefficients for dietary mineral electrolytes for human preterm infants as an approximation of bioavailabilities for poultry

Mineral Electrolyte	Symbol	Valence	Absorption Coefficients	
			Range	Center
Sodium	Na	+1	0.80–0.90	0.85
Potassium	K	−1	0.72–0.80	0.76
Calcium	Ca	+2	0.40–0.60	0.50
Magnesium	Mg	+2	0.45	0.45
Phosphorus	P	−1.8[a]	0.90	0.90
Chloride	Cl	−1	0.99	0.99

Source: Committee on Nutrition of the Preterm Infant (16).

[b]Phosphorus is in the form of $H_2PO_4^{-1}$ (about 20%) and HPO_4^{-2} (about 80%).

ratory alkalosis can further raise blood pH, creating problems, while an acidifying effect during disease or other stress may further depress performance. Generally, supplements that are alkalogenic have been most beneficial for improving poultry performance, except in the case of severe heat stress involving panting.

Dietary Na, K, and Cl have shown independent effects for increasing water intake, fecal moisture, and litter moisture of poultry when these electrolytes have been included at levels above the requirements. Younger chickens require higher levels of each of these macrominerals than do older growing chickens to cause increased water consumption. For example, 2-week-old chicks require over 1.15% K before they exhibit increased water intake, but broiler chickens near market age show increased water intake, with over about 0.85% K reportedly in commercial flocks (i.e., wet litter problems).

Replacing some of the salt (NaCl) in the diet with sodium bicarbonate (−1 part salt: +2 parts Na bicarbonate) lowers the dietary Cl content while keeping Na nearly constant and reduces water intake by poultry. This is valuable for improving litter condition, which has been directly related to disease status and processing grade out. In addition, an ionophore coccidiostate potentiating effect for extra sodium (particularly from sodium bicarbonate) has been well documented, improving broiler chicken and market turkey body weights, feed efficiencies, livabilities, and processing yields and reducing coccidial lesion scores.

About 4 to 6 lb/ton (0.2 to 0.3%) of sodium bicarbonate is recommended in growing chicken and turkey diets, while 2 to 4 lb/ton (0.1% to 0.2%) reportedly is beneficial for laying and breeding hens and breeder turkeys. With laying hens, firmer droppings have been no-

ticed, a big plus in stacked deck operations with manure cleanout belts under the cages. In some cases, thin or rough shells have been improved with better dietary electrolyte balance. At 8 to 10 lb/ton (0.4% to 0.5%) during heat stress (panting), improvement in livability of broiler chickens between 4 and 7 weeks of age has been observed with sodium bicarbonate. Heat stress usually has the most detrimental effect on growing turkeys after around 10 weeks of age.

The K requirements of poultry have been less intensively researched than the Na needs, but it has become apparent with the use of more animal and fish by-products and less soybean meal in commercial feeds that a potential need for K supplementation can develop. Because soybean meal usually provides a large portion of the K and protein in poultry feeds, as protein level declines in phase feeds, the K is reduced also. Potassium supplements, except occasionally potassium or magnesium sulfate to provide sulfate for a slight methionine–sparing effect, have not been used. This should change in the future as K needs and interactions with Na and Cl are more clearly defined. From an acid–base balance standpoint, once the Cl requirement has been met, it is desirable to use an alkalogenic rather than an acidogenic K supplement. The least expensive alkalogenic K supplement is potassium carbonate.

Some other electrolyte treatments that have been effective for broiler chickens during hyperthermia include the use of 1% ammonium chloride and 0.50% sodium bicarbonate in feed (17) or 0.20% ammonium chloride and 0.30% potassium chloride in drinking water (18). These should primarily be used during the time that broilers are panting (e.g., 12 noon to 8 P.M.). These combinations are acidogenic and lower the pH of blood by about 0.1 unit, but this is desirable during heat stress with respiratory alkalosis when blood pH has been raised 0.1 unit due to panting. These products can be detrimental when administered under thermoneutral conditions. These electrolyte blends have proved successful in research studies for improving gain and minimizing mortality of broiler chickens after 4 weeks of age during severe heat stress.

Research by McCormick (19) has indicated that supplemental P (0.55% available P versus 0.35% available P) along with reduced dietary Ca helped broiler chickens to better withstand heat stress. Survival time was increased from 165 min with 1% Ca and 0.35% available P to 210 min with 0.55% available P and to 275 min with 0.30% Ca and 0.55% available P. The 0.30% Ca level is quite low for commercial broiler chicken feeds. It would prove beneficial during hot weather to assure that dietary P levels in feeds are not deficient and tend to be high if possible.

ENZYME AND DIRECT-FED MICROBIAL SUPPLEMENTS

Years of continuing research in the areas of enzymes to enhance the digestibility of feeds and direct-fed microbials (formerly called probiotics) to maintain gut health with constant feed changes and disease challenges have brought forth products that are beginning to be cost effective and efficacious. Multienzyme blends are now commercially available from several companies to suit special needs, such as wheat- or barley-based diets. Phytase enzyme products are useful in improving P utilization and digestibility of other nutrients, enhancing overall performance, and allowing reductions in dietary Ca and P. Direct-fed microbials able to withstand higher temperature and moisture conditions of expanders and steam pellet mills are now on the market.

FEED INGREDIENTS

With low feed costs, it is common to use corn- and soybean-meal-based diets primarily, but with high feed costs the use of many replacement ingredients occurs. See example feed formula in Table 20-10. In some areas of the country, such as the West Coast and Hawaii, which are far away from the Cornbelt States, utilizing alternative ingredients is commonplace. Formula costs can be lowered by offering more ingredients to the computer for least-cost formulation, but this presents problems for the mill. There is a need for more storage bins and just-in-time delivery schedules. With more pressure being put on feed ingredients by the rising human population worldwide, increasing animal production, political trade barriers to global trade, labor disputes disrupting transportation, earthquakes that are increasing in frequency and severity, and other factors, it is imperative that feed mill managers make strategic plans for continued supplies of quality feed ingredients.

Reference sources for information on the nutrient profiles of a wide variety of feed ingredients in North America include the NRC *Nutrient Requirements of Poultry* (1) and *United States–Canadian Tables of Feed Composition* (20), *Special Report: Energy Values of Alternative Feed Ingredients* by Dale and Fuller (21) at the University of Georgia, *IFI Tables of Feed Composition* (22), *Raw Material Compendium* by Novus International, Inc. (23), and *Feedstuffs Annual Reference Issue* (24).

SUMMARY

Classes of poultry today include avian species used for the production of meat, eggs, and specialty products such as feathers, skins, and blood plasma. For optimal nutrition, nutrient requirements and nutrient contents of diets must be accurately known for the particular type, breed, and age of poultry, as well as season, stage of production, and desired production outcome. The energy-yielding nutrients (carbohydrates, fats, and proteins) are the greatest expense in poultry feeds. The primary energy systems used in poultry feed formulation are ME (AME) and TME. True digestibility of dietary amino acids and protein is now used in feed formulation to obtain consistent poultry performance regardless of frequent formula changes.

Macrominerals, trace minerals, and vitamins must be included at adequate, but not deficient or toxic levels, to support metabolic activities. Dietary electrolyte balance (Na, K, and Cl) has become increasingly more important in recent years as genetic gains, higher stocking densities, and environmental extremes have put stresses on poultry. Deviations in pH of body fluids away from normal, either toward acidosis or alkalosis, affect metabolic enzyme systems and depress performance.

Special-purpose feed additives, such as antibiotics (see Figure 20-11), enzymes, antioxidants (see Figure 20-12), mold inhibitors

TABLE 20-10

Examples of broiler starter, turkey starter, and caged layer peak egg production feeds[a]

	Broiler Starter (%)	Turkey Starter (%)	Layer Peak (%)
Ingredient			
Corn, yellow	56.45	47.75	60.50
Soybean meal (47.5% CP)	27.33	38.83	21.50
Meat and bone meal (50% CP)	7.00	—	5.09
Meat meal (56% CP)	—	9.50	—
Bakery by-product	6.00	—	—
Animal–vegetable fat	1.82	0.31	3.00
Limestone (or oyster shell)	0.49	0.81	8.66
Dicalcium phosphate	0.13	1.54	0.49
Salt	0.10	0.09	0.20
Sodium bicarbonate	0.20	0.20	0.20
Copper sulfate	0.05	0.05	—
Vitamin–mineral premix	0.25	0.25	0.25
DL-methionine (99%)	0.17	0.24	0.11
L-lysine HCl (78.4% lysine)	—	0.23	—
Bacitracin-MD (50 g/lb)	0.05	0.05	—
Coban (monensin) 30 g/lb	—	0.10	—
Nicarbazin (25%)	0.05	—	—
Liquid mold inhibitor	0.05	0.05	—
Calculated analysis			
Protein, % (N × 6.25)	22.50	28.00	18.00
ME, kcal/lb	1425	1280	1320
Lysine, %	1.21	1.80	0.94
Methionine + cystine, %	0.92	1.10	0.71
Ca, %	0.95	1.45	3.80
Available P, %	0.48	0.83	0.45
Na, %	0.20	0.19	0.18
K, %	0.83	0.94	0.68
Cl, %	0.25	0.24	0.19

[a]CP = crude protein; ME = metabolizable energy; N = nitrogen.

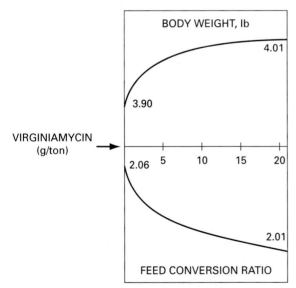

FIGURE 20-11 Dose–response curves for virginiamycin in broiler feed. [Miller and Free (26).]

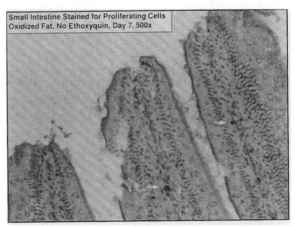

FIGURE 20-12 Destructive effects of rancid fat on small intestinal villi of chicks. (This is prevented by proper addition of an antioxidant to the feed.) [Dibner et al. (27).]

(see Figure 20-13 through 20-15), copper sulfate, and sodium bicarbonate, which have given proven performance benefits, can be used to improve profits. Experience has shown that, as feed expenses and poultry product prices rise, feed additives have increasing benefit-to-cost ratios.

Some anticipated areas of poultry nutrition research over the next few years include feed-processing technologies to improve nutrient digestibilities, more precisely defining nutrient requirements, feeding ratites, emphasis on efficiency rather than maximizing production, and greater use of alternative feedstuffs. As the human and domestic animal populations continue to increase and compete for grains and oilseed crops, feed ingredient price and supply problems mandate more emphasis on purchasing, forward contracts, storage, and other strategies (including marketing of less meat and fewer eggs) to remain profitable.

FIGURE 20-14 Dual feed bins allow feed changes to be made at proper ages and provide extra storage during inclement weather. Having two bins makes cleanout more convenient as well.

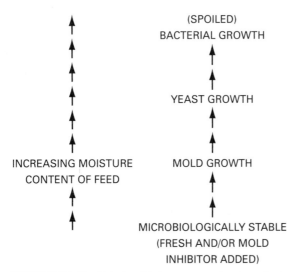

FIGURE 20-13 Relationship between moisture and microbial degradation of feed. [Nelson (29).]

FIGURE 20-15 Cleaning out the boot and feed bin and disinfecting with dilute bleach between flocks helps to minimize mycotoxin problems from moldy feed.

REFERENCES

1. National Research Council. 1994. *Nutrient requirements of poultry,* 9th rev. ed. Washington, DC: National Academy Press, 148 pp.

2. Ward, N. E. 1993. *J. Appl. Poultry Res.* 2:286.

3. Sibbald, I. R. 1976. *Poultry Sci.* 55:303.

4. Farrell, D. J. et al. 1989. *Recent advances in animal nutrition in Australia,* April 16–19. Univ. of New England, Armidale, NSW, Australia, D. J. Farrell, ed, pp. 269–284.

5. McNab, J. M., and C. Fisher. 1984. An assay for true and apparent metabolisable energy. *Proc. Abstr. XVII World's Poultry Congress Exhibition,* Helsinki, Finland, Aug. 8–12, Nutr. Sect., pp. 374–375.

6. Titus, H. W. 1956. Energy values of feedingstuffs for poultry. *Proc. Semi-Annual Meeting Nutr. Council,* American Feed Manuf. Assoc., pp. 10–14.

7. Hill, F. W., and D. L. Anderson. 1958. *J. Nutr.* 64:587.

8. Wolynetz, M. S., and I. R. Sibbald. 1984. *Poultry Sci.* 63:1386.

9. Heartland Lysine, Inc. 1995. True digestibility of essential amino acids for poultry—1995. Folding chart, 1 p.

10. Baker, D. H. 1995. Ideal protein for broiler chicks. *Multi-State Poultry Feeding and Nutrition Conference,* Indianapolis, IN, 13 p.

11. Thomas, O. P. 1995. Broiler nutrition. *Multi-State Poultry Feeding and Nutrition Conference,* Indianapolis, IN. May 16–18, 1995, 5 p.

12. Brake, J. T. 1993. Nutrition and feed management programs for optimum broiler breeder performance. *Proc. Nutr. Tech. Symp.,* Novus International, Inc., Springdale, AR (Nov. 16) and Atlanta, GA (Nov. 18), pp. 56–72.

13. *Nicholas turkey breeding farms management guidelines.* 1992. Sonoma, CA.

14. *British United Turkeys of America (BUTA) management guide.* 1994. Nutrition, sections 6.1 through 6.4.

15. Hooge, D. M. 1995. Dietary electrolytes influence metabolic processes of poultry. *Feedstuffs* 67(50):14–15, 17–19.

16. Committee on Nutrition of the Preterm Infant. 1987. European Society of Paediatric Gastroenterology and Nutrition, Oxford, Blackwell, England. In: *J. Nutr.* 119 (12S, Supplement A):1789–1798, December, 1989.

17. Teeter, R. G. et al. 1985. *Poultry Sci.* 64:1060.

18. Teeter, R. G., and M. O. Smith. 1986. *Poultry Sci.* 65:1777.

19. McCormick, C. C. 1981. Nutritional status and tolerance of the chickens to high temperature. *Proc. Cornell Nutr. Conf.,* pp. 81–88.

20. National Research Council. 1982. *United States–Canadian tables of feed composition.* 3rd rev. ed. Washington, DC: National Academy Press, 148 p.

21. Dale, N., and H. L. Fuller. 1987. *Special report: Energy values of alternative feed ingredients* (Project 139). University of Georgia, 30 p.

22. Fonnesbeck, P. V., and H. Lloyd 1984. *IFI tables of feed composition.* Utah State University, 607 p.

23. de Mol, J. 1994. Raw material compendium. 2nd ed. Novus International, Inc., 541 p.

24. *Feedstuffs annual reference issue.* 1995. Vol. 67, No. 30, 264 p. Minnetonka, MN: Miller Publishing Co.

25. Firman, J. D. 1995. Reduced protein diets for turkeys. *Elanco Turkey Tech. Symp.,* Indianapolis, IN, pp. 1–13.

26. Miller, C. R., and S. M. Free. 1983. Growth promotion characteristics of virginiamycin in broilers. *Virginiamycin Tech. Seminar,* Atlanta, GA, Jan. 26, 8 p.

27. Dibner, J. J. et al. 1994. Feeding of oxidized fats to broilers: Poor performance is associated with increased enterocyte turnover and hepatocyte proliferation. *Novus Tech. Symp.,* Vancouver, B.C. pp. 14–22.

28. Anonymous. 1985. Separate tables for breeder hens and males. *Poultry Digest* 44(253):384–388.

29. Nelson, C. E. 1990. The implications of mold growth in poultry feeds. Kemin Industries, Des Moines, Iowa. 2 p.

30. Anonymous, undated. Poultry research report 8. Alfatoxicosis, lysine and methionine levels. Heartland Lysine, Inc., 4 p.

31. Wang, J.-S., et al. 1987. *Poultry Sci.* 66:1500.

21

FEEDING HORSES

Laurie Lawrence

Improper feeding and watering will doubtless account for over one-half of the digestive disorders met with in the horse, and hence the reader cannot fail to see how very important it is to have some proper ideas concerning these subjects. (Michener, 13)

THE DIGESTIVE SYSTEM

Digestion in the horse begins in the mouth, and the condition of a horse's teeth may affect its ability to obtain and digest food. For horses on pasture, the incisors are very important for nipping or tearing plant parts. The molars are used for grinding feed and increasing the surface area of the feed that will be exposed to enzymatic and microbial digestion in the subsequent portions of the digestive tract. The ability of the horse to grind its feed is particularly important when whole grains with hard seed coats (such as corn or milo) are fed. It is not uncommon for horses to develop sharp edges (points) on their molars that can irritate the tongue or cheeks. The points on molars can be rasped off by a veterinarian. This process is referred to as *floating* the teeth and often improves a horse's ability to grind its feed properly.

Very little actual digestion occurs in the stomach, but a few items about the stomach are worth noting. First, the stomach of the horse is small relative to its body size. Horses evolved as grazing animals, eating many small meals during a day, rather than one or two large meals.

Second, horses cannot vomit, and therefore the stomach is susceptible to distention and rupture under certain conditions.

Enzymatic digestion of starch, protein, and fat occurs in the small intestine. However, if an excessive amount of a cereal grain or other high-starch feed is consumed, the capacity of the small intestine to digest and absorb the starch may be exceeded, and some of the starch passes into the large intestine. Most digestion in the large intestine occurs in the cecum and the large colon. These structures contain microbial populations similar to those found in the rumen of sheep and cattle. If a large amount of starch reaches the large intestine, the microbial population may be negatively affected, and digestive problems such as colic, diarrhea, or laminitis may result.

The primary function of the microbial population in the large intestine is to digest the fiber found in common horse feeds such as pasture and hay. Although microbial digestion of fiber in the horse is not as efficient as in ruminants, it still plays an extremely important role in the nutrition of the horse. Fiber is fermented by the microbial population to volatile fatty acids (VFA), primarily acetate and propionate, which are then absorbed through the large intestinal wall. Acetate can be utilized for energy immediately or can be used for fat synthesis. Propionate may also be used for energy, but can also be used to synthesize glucose. Because horses are capable of utilizing fibrous feeds for energy, many horses at maintenance can meet their energy requirements from hay or pasture alone.

NUTRIENT REQUIREMENTS

WATER

The nutrient most crucial for survival is water. Most horses consume about 2.5 to 3.5 kg of water for every kilogram of feed dry matter. Water consumption increases if horses are exercising or lactating or if the environmental temperature is high. Table 21-1 provides an estimate of daily water requirements for different types of horses. Water consumption may be increased if diets contain large amounts of dry roughages, high salt levels, or excessive protein levels.

Clean, fresh water can be provided in buckets, troughs, ponds, or streams, (Figure 21-1). A stagnant pond is not a suitable water source, nor is snow an adequate water source. Most horses

TABLE 21-1

Expected levels of daily water consumption

Maintenance, 500 kg, thermoneutral environment	6–8 gal
Maintenance, 500 kg, warm environment[a]	8–15 gal
Lactating mare, 500 kg	10–15 gal
Working horse, 500 kg, moderate work	10–12 gal
Working horse, 500 kg, moderate work, warm environment[a]	12–18 gal
Weanling, 300 kg, thermoneutral environment	6–8 gal

[a]Ambient temperature of 30° to 35°C; water requirement is also affected by ambient humidity and availability of shade.

FIGURE 21-1 Automatic waterers are a convenient means of providing fresh, clean water to horses in stalls or pastures. Waterers should be checked at least once a day to make certain that they are clean and functioning properly.

should have water available at all times. The only exception would be horses that have just completed a hard workout. After working, horses should be "watered out" gradually, usually over a period of about an hour. During this time the horse should be allowed to drink a small amount every 5 to 10 min. When horses are worked for long periods of time (several hours), it is advisable to allow them to stop and drink every 30 to 45 min to prevent dehydration.

ENERGY

The daily energy requirements for horses are expressed in megacalories (Mcal) of digestible energy (DE) needed per day (see Appendix Tables 45 and 47). The DE requirements listed in Appendix Table 45 for horses at maintenance are scaled to body weight, but, in practice, additional adjustments must be made to account for differences between individuals. The term *easy keeper* is often used to describe a horse that can maintain body weight on less than average amounts of dietary energy, whereas a *hard keeper* requires more than average amounts of energy to maintain body weight. In cold climates, winter weather can have a large effect on energy requirements for maintenance, particularly if horses do not have shelter. Figure 21-2 illustrates the anticipated effect of cold stress on the DE requirements of horses. Horses can acclimate to fairly low temperatures, but energy requirements increase once temperatures drop below the lower critical temperature (LCT). Cold stress can be created by an environmental temperature below the LCT or by a combination of cold, wind, and precipitation.

Horses obtain energy from the digestion of a variety of feed components. At maintenance, horses may derive the majority of their daily energy requirement from the digestion of fiber in the large intestine. Fiber digestion alone is not usually sufficient to meet the energy requirements of lactating, growing, or exercising horses, and the addition of cereal grains to a horse's diet may be necessary. The primary energy source in cereal grains is starch. Most starch is broken down in the small intestine and absorbed as glucose. Replacing a portion of the horse's daily allotment of hay with cereal grains results in a more energy dense ration (more DE per kilogram of feed). Another way to increase dietary energy density is to add fat to the ration. Because fat contains about 2.25 times the energy of carbohydrate, the addition of high-fat ingredients to horse diets has become popular in recent years.

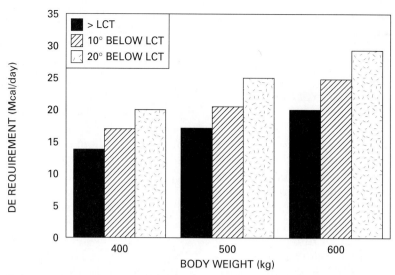

FIGURE 21-2 Effect of cold stress on the digestible energy (DE) requirements of mature horses. The solid bars indicate the DE requirement of horses in a thermoneutral environment. The striped and dotted bars represent the DE requirement at 10° and 20°C below the lower critical temperature (LCT), respectively. In cold-adapted horses, the LCT may be as low as −10°C. Cold stress may include effects of wind and precipitation in addition to environmental temperature.

Although it is possible to meet a horse's energy requirement with a diet of only cereal grains, this type of diet should never be fed. Diets containing large amounts of starch and low amounts of fiber are associated with an increased incidence of colic and laminitis. The minimum acceptable level of fiber in horse diets has not been established, but may be about 12% to 15%.

Energy deficiency in mature, nongestating horses results in a loss of weight and body condition. In gestating mares, an energy deficiency may cause weight loss or lower weight gains than expected. In growing horses, an energy deficiency may result in weight loss (severe restriction of energy) or a reduced growth rate. When horses cannot be weighed regularly, a subjective condition scoring system can be used to monitor body condition. One of the most common systems rates horses on body fatness using a 1 (very thin) to 9 (very fat) scale (3). A summary of the system is given in Table 21-2. Most horses should be maintained at a condition score of at least 4 and not exceeding 7 (Fig. 21-3). A change in condition score indicates fat gain or loss and should be a signal to evaluate the energy content of the diet.

PROTEIN

The protein requirements of horses are usually expressed as grams of crude protein (CP) required per day (Appendix Tables 45 and 47). Mature horses at maintenance have relatively low protein requirements, and protein deficiency is rare if energy requirements are met. A mature 500-kg horse requires about 660 g/CP per day, which can be supplied by just 8 kg of a hay containing 8.25% CP. Young horses, lactating mares, and mares in late gestation require diets with higher protein quantity and quality. Protein quality (the amount and balance of essential amino acids) is an important consideration in growing horses. Lysine appears to be the first limiting amino acid in diets for growing horses, and threonine has been suggested as the second limiting amino acid. If young horses are fed low-lysine diets, growth will be reduced. It is likely that horses can absorb some nitrogen, probably as ammonia, from the large intestine, which can be used for the synthesis of nonessential amino acids. The microbial population of the large intestine can synthesize some essential amino acids, but the ability of the horse to absorb these amino acids is limited. Therefore, the essential amino acids must be obtained from the diet.

In broodmares, dietary protein is used for the deposition of fetal tissues and for milk production. Dietary protein requirements for broodmares have been estimated by adding the mare's maintenance requirement to the amount of protein needed to meet the needs of

TABLE 21-2

Body Condition Scoring System

Score	Description
1	No fat covering on ribs, spine, pelvis, etc. All boney structures in trunk extremely apparent. Bones in neck may be visible.
2	Ribs, shoulder, spine, and pelvis structures are prominent. Neck is very thin.
3	Ribs, shoulder, spine, and pelvis are clearly discernable but some fat cover can be felt. Neck is thin.
4	Ribs are slightly visible. Some fat cover can be felt on ribs, spine, tailhead. Neck is not thin.
5	Some fat covering over ribs, spine, shoulder, and pelvis. Ribs not visible; back (loin) may be flat. Body parts are distinct but blend together smoothly.
6	More fat accumulation on ribs and at tailhead. Back (loin) may be flat or have a slight crease. Neck starting to fill out.
7	Fat accumulating on ribs and tailhead feels soft. Definition between body parts (neck to shoulder; shoulder to ribs) is decreased. Neck is filled in over crest.
8	Ribs are difficult to feel. Back (loin) has noticable crease. Neck is thick. Tailhead is very fat and soft.
9	Soft, thick fat accumulated on ribs, shoulder, and tailhead. Neck is very thick and withers have lost definition. Fat accumulated between thighs, in flank area, and behind shoulder.

Source: Adapted from Henneke et al. (3).

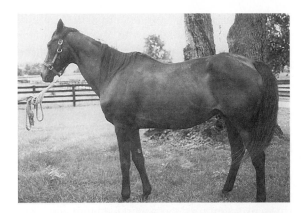
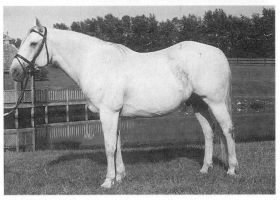

FIGURE 21-3 Horses exhibiting different levels of body condition. The mare at the top would receive a condition score of 4. Although her ribs are not readily visible, notice the prominent bone structure in the shoulder, withers, spine, and hip areas. Her neck is also somewhat thin. The gelding at the bottom would receive a condition score of 7. For the details of the condition scoring system, see Table 21-2.

fetal growth or milk secretion. For example, if mare's milk contains about 2.0% protein and a 500-kg mare produces 15 kg of milk, about 300 g of protein will be secreted daily in the milk. The conversion of dietary protein to milk protein is not very efficient, so approximately 1430 g of dietary CP is needed by a 500-kg mare for maintenance and milk production.

MINERALS

The minerals required by horses are listed in Appendix Tables 45 through 48. Calcium and phosphorus are of special importance in horses because of their role in the formation and maintenance of normal bone. The development of quality bone is more important in horses than other livestock species because horses are used for athletic purposes, which may put more stress on bones. Equine bone is about 35% Ca and 16% P. Deficiencies or imbalances in di-

etary Ca and P can result in a variety of bone disorders. High levels of P can impair the absorption of Ca, and it is recommended that the concentration of P in the diet not exceed the concentration of Ca. All horses require some Ca and P in their diets, but the requirements (as a percent of diet) are highest in growing horses and lactating mares. The Ca and P recommendations in Appendix Tables 45 and 47 take into account the availability of Ca and P in the horse feeds commonly used in the United States. The availability of Ca in most horse feeds is about 50% to 70%, whereas the availability of P is about 35% to 45%. When unusual feeds are used, differences in Ca or P availability must be considered. For example, if the availability of Ca in a particular feed is only 25% (instead of 50%), the total amount of Ca in the diet must be increased to ensure that adequate Ca is available to the horse.

Sodium, K, and Cl, function as electrolytes and are essential nutrients for all classes of horses. Electrolytes play an important role in maintaining fluid balance and are necessary for nerve and muscle function. Most nonworking horses obtain enough Na and Cl to meet their needs if they have access to a salt block or to loose salt. Potassium needs are usually met by the naturally occurring K found in hay and pasture. Working horses, lactating mares, and horses that are exposed to high environmental temperatures will have elevated requirements for these nutrients and may require some supplementation. When horses are deficient in these nutrients, a decrease in feed and water intake may be observed. In addition, the horses may exhibit unusual oral behavior, such as licking of stall surfaces. In working horses, electrolyte depletion can impair muscle function and fluid balance.

Horses require all the minerals needed by other livestock. Iodine is necessary for formation of the thyroid hormones. Instances of both I deficiency and I toxicity have been reported in horses. The I content of common horse feeds can vary considerably and can be affected by soil I content. Salt blocks often contain a small amount of I. Some nontraditional equine feed supplements, such as kelp, may be very high in I and should be used cautiously.

Iron is used in the formation of hemoglobin, which is involved in oxygen transport in the body. The dietary Fe requirement is estimated at 40 to 50 mg/kg DM (14), which is almost always met by typical feed ingredients. For example, alfalfa hay is commonly high in Fe (>200 mg/kg DM). The availability of Fe in grains and forages may be low, but Fe deficiencies are rarely reported. Providing horses with additional Fe supplements has not been proved to be of benefit.

Copper is involved in cartilage formation and development and thus is of particular concern in growing horses. The level of Cu recommended by the NRC in 1989 was 10 mg/kg DM (14). The level and availability of Cu is very low in many forages, and it is common practice to formulate grain mixes to contain 20 to 30 mg Cu/kg DM. Forages may also be low in Zn. Zinc is required for several biochemical processes in the body, and Zn deficiency results in reduced growth in young horses. The current recommendation for Zn in the total diet dry matter is 40 ppm (14), but grain mixes are often supplemented to a level of 80 to 100 mg of Zn/kg DM.

Many regions of North America have low levels of selenium (Sw) in the soil and produce feeds that are also low in Se. Selenium deficiency can result in myopathy and may be exacerbated by low vitamin E status. Selenium supplementation is often necessary but should be done carefully, because high Se intakes (2 mg Se/kg DM) can be toxic (14). Signs of chronic Se toxicity include rough hair coats, hair loss, and cracked hooves. The recommended minimum level of Se in the total diet is 0.1 ppm.

VITAMINS

Of the fat-soluble vitamins (A, D, E, and K), A and E are of the most practical significance in horse diets. Although vitamin D is necessary for bone development, vitamin D supplementation of horses kept outside is not usually necessary. Horses can synthesize the active form of vitamin D as long as they are exposed to sunlight. A dietary vitamin D level of 300 to 800 IU/kg diet DM is recommended for horses not exposed to sunlight (14). A dietary requirement for vitamin K has not been established in the horse. The microbial population in the intestine can synthesize compounds with vitamin K activity, and the horse can obtain substances with vitamin K activity from hay and pasture.

Vitamin A is necessary for vision, bone development, and maintenance of skin and other tissues. When horses receive diets consisting of natural ingredients, vitamin A is derived from the breakdown of β-carotene in the intestinal wall. Feed composition tables often list the β-carotene content of a feed rather than the vitamin A activity. Approximately 400 units of vitamin A activity is obtained from the conversion of 1 mg of β-carotene by the horse.

One of the richest sources of β-carotene in horse diets is green pasture. Several studies have documented that plasma vitamin A concentrations are highest when horses are consuming pasture during the late spring and summer and lowest when they are consuming hay and grain in the winter. Once grasses or legumes are cut to make hay, the vitamin A activity decreases rapidly. As long as horses receive pasture for several weeks a year, adequate vitamin A status is probably maintained. However, if horses are kept in confinement and fed only hay and cereal grains, vitamin A supplementation is necessary. Vitamin A can be added to horse diets in several forms, including retinyl acetate and retinyl palmitate. The recommended minimum level of vitamin A in horse diets is 3000 IU/kg DM for broodmares and 2000 IU/kg DM for all other classes (14). Vitamin A levels as high as 7000 IU/kg DM have been suggested as optimal for growing horses (6). Signs of vitamin A deficiency include reduced growth, decreased reproductive efficiency, night blindness, and hyperkeratinization. Vitamin A toxicity can occur with high levels of supplementation.

Although β-carotene is known primarily for its function as a vitamin A precursor, there is now some indication that it has its own role in animal physiology. Studies with several species have suggested a link between β-carotene supplementation (usually by injection) and reproductive performance in females. There are no reports linking dietary β-carotene supplementation to enhanced reproductive performance in mares receiving adequate levels of dietary vitamin A.

Vitamin E is the other fat-soluble vitamin of interest in horse nutrition. Vitamin E acts as an antioxidant and helps to maintain the integrity of cell membranes. Vitamin E may also be necessary for optimal immune function. Deficiencies of vitamin E and Se have been linked to white muscle disease in foals. The minimum recommended level of vitamin E is 50 mg/kg DM for horses at maintenance and 80 mg/kg DM for all other classes of horses (14). Horses affected by equine degenerative myeloencephalopathy, a specific neuromuscular disease, may require very high levels of dietary vitamin E, and a supplementation rate of 6000 IU/day has been suggested for such horses. Vitamin E activity in feeds is derived from several tocopherols and tocotrienols, but α-tocopherol is the most biologically active form. The vitamin E activity of forages is highest when plants are in an early stage of maturity and pasture may contain 400 IU vitamin E/kg DM. Once a plant is harvested for hay, vitamin E activity decreases. Vitamin E content of hay can vary from 10 to 200 IU/kg DM. The vitamin E content of most cereal grains is very low (<30 IU/kg DM). If horses are consuming stored hay and an unsupplemented grain ration, it is likely that their diet is inadequate in vitamin E. Vitamin E can be added to diets in several forms, including DL-α-tocopheryl acetate, D-α-tocopheryl acetate, D-α-tocopheryl succinate, and D-α-tocopheryl nicotinate. The relative bioavailability of these compounds in the horse is not known. Generally, 1 mg of DL-α-tocopheryl acetate is believed to be equivalent to 1 unit (IU) of vitamin E activity.

Little information is available concerning the horse's dietary requirement for the water-soluble vitamins. A dietary requirement for vitamin C has not been determined. The microbial population in the large intestine appears to be capable of synthesizing several B vitamins, and dietary requirements are listed for only riboflavin and thiamin (14). Many of the B vitamins are involved as cofactors in energy-producing reactions in the horse, and thus the need for these vitamins may increase as energy requirements increase. When fed at a rate of 10 to 30 mg/day for several months, biotin may improve hoof quality in some horses.

FEEDS

FORAGES

Forages are the foundation of horse feeding programs and can provide many of the essential nutrients required by horses. The fiber in forages assists the horse in maintaining gastrointestinal health. Horses should receive 1.0 to 2.5 kg of good-quality hay (or pasture equivalent) per 100 kg of body weight each day.

Pasture should be utilized whenever possible because it reduces the labor associated with feeding and can provide a high-quality feed source. During most parts of the growing season, 2.5 to 3.0 kg of fresh pasture are equivalent to about 1 kg of good hay. In the early part of the growing season, the fiber content in pasture may be low and horses may require some additional fiber from hay. When given access to average- or good-quality pasture, horses consume about 2 to 3 kg of fresh pasture per hour. However, intake may be much higher when horses are first let into a pasture or if the pasture is very lush. To prevent overeating, horses should be introduced to lush, rapidly growing pastures gradually.

The appropriate grazing intensity (acres per horse) depends on the growing conditions, intensity of pasture management, and plant species found in the pasture. During peak growing periods, well-managed pastures may support 1 horse per acre, but a more typical range would be 1 horse for every 2 to 4 acres. Lactating mares require a greater allotment than most other classes of horses. Overgrazing should be avoided because it damages the vigor of the desirable pasture plants and enhances the growth of many weeds.

The productivity (kilograms of consumable nutrients per acre) of horse pastures can be maximized by weed control and appropriate fertilization. In addition, many pastures benefit from regular mowing. Mowing is an effective weed control procedure, but also helps to maintain plants in a state of high nutrient content. As plants develop seed heads and reach maturity, nutrient availability and palatability decline (Fig. 21-4). By keeping pastures clipped, the onset of maturity (seed head formation) is delayed and pasture productivity is increased.

The plant species best suited for use in horse pastures vary according to climate and soil conditions. In the southern United States, perennial warm-season grasses such as Bermuda grass and Bahia grass are popular for horse pastures. In the remainder of the United States, horse pastures often contain perennial cool-season grasses such as orchard grass, Kentucky bluegrass, bromegrass, and tall fescue. Bromegrass, tall fescue, and orchard grass are bunch grasses, whereas Kentucky bluegrass is a sod-forming grass. Orchard grass and bromegrass are more heat tolerant than tall fescue or Kentucky bluegrass, but they are not as resistant to overgrazing. Tall fescue is a hardy and productive forage that is often infected with an endophytic fungus, *Acremonium coenophialum*. Pregnant mares grazing endophyte-infected tall fescue may experience prolonged gestation, stillborn foals, dystocia, and reduced milk production. Problems associated with endophyte-infected tall fescue can be minimized if mares are removed from the pastures during the last 90 days of gestation. Endophyte-infected tall fescue pastures may be used by other classes of horses and are often more productive and hardy than pastures with other plant species. It is common to interseed

some type of legume, usually clover, into grass pastures (Fig. 21-5). Clovers are productive in warm weather, when the growth of cool-season grasses is reduced. Red clover is sometimes affected by a mold that causes horses to slobber.

Annual grasses may also be used for horse pasture. In the south, ryegrass, which is a winter annual, is often used during the cool months when the warm-season perennial grasses are dormant. Annual grasses can also be used for temporary pastures. Summer annuals that are used for temporary pastures include sorghum, Sudan grass, Johnson grass, and some millets. When drought-stressed, frosted, or fertilized with high levels of nitrogen, these annuals may concentrate nitrate, which can cause prussic acid poisoning. Sorghum and Sudan grass have also been reported to occasionally result in kidney and/or urinary tract disorders. Because of the potential for problems to occur with many of the summer annuals, their use is sometimes avoided.

When pasture is not available, horses must be fed hay or another roughage source. Horses can be fed different types of hay, but all hay should be free from dust and mold. Heaves, or chronic obstructive pulmonary disease, can result if horses develop an allergic response to dusty or moldy hay. Horses with heaves are usually exercise intolerant and may be of limited use for any type of strenuous work. Moldy hay is produced by baling the hay at a high moisture level. Mold formation in hay can be

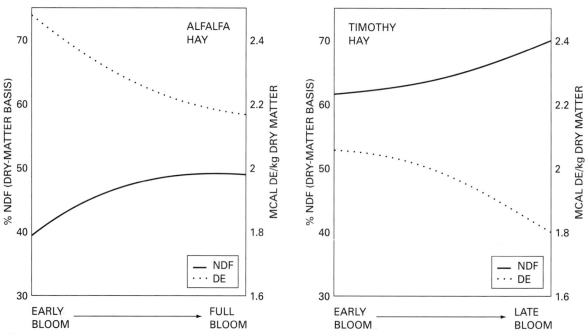

FIGURE 21-4 Effect of plant maturity on neutral detergent fiber (NDF) and digestible energy (DE) content of alfalfa and timothy hay. As NDF content increases with plant maturity, DE content decreases. [NRC (14).]

FIGURE 21-5 Pasture can be an excellent source of many nutrients. This pasture contains a mixture of cool-season perennial grasses and clover.

prevented by spraying the hay with a chemical preservative at the time of baling. Treating hay with a preservative consisting of a combination of propionic and acetic acid did not affect feed intake or weight gain in yearling horses (7).

Alfalfa is a highly palatable and nutritious hay for horses. It is the most common legume hay and is high in crude protein and Ca. If alfalfa is cut in an early stage of maturity (early bloom), it contains more energy than most other types of hay. Alfalfa hay can be used for all classes of horses, but to make the best use of alfalfa hay, horse owners and managers must be able to match the nutrient content of the hay to the nutrient requirements of the particular horse (or horses) being fed (Table 21-3).

Alfalfa hay provides excess protein and Ca for mature horses at maintenance, mares in early gestation, and mature horses in light work. Alfalfa hay cut at a later stage of maturity (mid to full bloom) is most appropriate for horses in these groups. If early-bloom alfalfa hay is fed free choice to the above classes of horses, it is probable that energy requirements will be exceeded and the horses may become too fat. Early bloom alfalfa hay is a more appropriate feed source for horses with high nutrient requirements and may be the hay of choice for growing horses, performance horses, and broodmares in late gestation or early lactation. When alfalfa is fed to growing horses, the total ration must be carefully balanced for Ca and P. Alfalfa hay that is contaminated with blister beetles should not be fed to horses. Blister beetles contain a chemical (cantharidin) that is toxic to horses. Alfalfa harvested before or after blooming, harvested in early or late summer, or harvested without crimping at the time of harvest is less likely to contain blister beetles (11). Other legume hays that may be used for horses include red clover, lespedeza, and bird's-foot trefoil.

Several types of grasses are used to make horse hay. Grass hays are usually much lower in protein and Ca than alfalfa hay, but may have a more balanced Ca:P ratio. If grass hays are harvested in early maturity, palatability and digestibility are comparable to mid- or full-bloom alfalfa hay. However, if grass hays are harvested in late maturity, fiber levels will be increased and digestibility will be much lower (Fig. 21-4). Because late-maturity grass hays are low in palatability, horses waste more of these hays.

Timothy hay has been a traditional grass hay for horses and is recognizable by its distinctive, cylindrical seed head. A large seed head indicates that the hay was cut in late maturity and will have a relatively low feeding value. Orchard grass, bromegrass and Bermuda grass are also used for hay. As with timothy, the feeding value of these hays decreases greatly with

TABLE 21-3

Matching nutritional value of alfalfa hay to nutritional need in horses

Type of Hay	Description[a,b]	Type of Horse
Prebloom alfalfa	Very leafy, very fine stems; very high nutrient density and palatability	Weanlings, some yearlings, some performance horses, poor eaters
Early-bloom alfalfa	Leafy, relatively fine stems, high nutrient density and palatability	Weanlings, yearlings, lactating mares, some performance horses
Mid-bloom alfalfa	Somewhat stemmy, stems not fine; moderate nutrient density and palatability	Some weanlings, yearlings, lactating and gestating mares, some performance horses, horses at maintenance
Late to full-bloom alfalfa	Very stemmy, stems are thick; less palatable, especially to young horses	Early gestating mares, mature horses in light work, horses at maintenance

[a]All hays for horses should be free of weeds, dust, and mold.
[b]When alfalfa hay is fed, the calcium and protein requirements of many horses will be exceeded.

stage of maturity. Mature grass hays may contain only 7% to 8% CP and 1.8 Mcal DE/kg DM. Bahia grass is low growing and not used as a hay plant. Similarly, Kentucky bluegrass is not used as a hay plant. Tall fescue can be used for hay, but if it is infected with endophyte, it will have the same effects on mares as tall fescue pasture. Tall fescue hay is often less palatable to horses than timothy, orchard grass, or bromegrass hay.

Good-quality grass hay can be fed to all classes of horses, but must be supplemented with energy, protein, and minerals when fed to growing horses, broodmares, and working horses. The level of supplementation depends on the stage of maturity of the plant at the time of harvest and the nutrient requirements of the horses. For example, if a late-maturity timothy hay is being fed, horses in race training and lactating mares will require a much higher level of supplementation than horses in light work or mares in early gestation. The composition of the concentrate feed or supplement used with grass hays will almost always differ from the composition of those used with alfalfa hays. When alfalfa hay is fed, the need for protein and Ca supplementation is usually reduced.

The exact amount of hay that should be fed to various types of horses varies with hay quality, plant species, and individual horse. Table 21-4 provides some hay feeding guidelines for horses of various ages and physiological classes. Horses are usually fed long hay rather than chopped hay. Long hay is usually fed from 25- to 35-kg rectangular bales. Large round bales can be used for feeding horses if they are baled and stored under dry conditions.

Round bales are useful for feeding large groups of horses on winter pasture. Chopped hay (or chaff) is fed in small amounts in the United Kingdom, but is not common in the United States. Hay cubes, usually made from alfalfa, are available for horses in the United States and can be fed on a weight-equivalent basis to long hay. Cubes may be more expensive than long hay, but may be more convenient to handle and store. Haylage is occasionally fed to horses, but its use is not common.

ENERGY FEEDS

When a horse cannot meet its energy need from the consumption of forage alone, an energy feed must be added to the diet. The most common energy feeds are cereal grains and grain by-products. Table 21-5 compares the nutrient content of several common energy feeds. Of the cereal grains and grain by-products, corn contains the highest amount of energy. Corn is low in fiber and high in starch. Corn is also relatively low in protein and low in the essential amino acid lysine. Under some growing conditions, corn may be contaminated with a mold, *Fusarium monilaforme*, that produces a mycotoxin, fumonisin B_1. Fumonisin B_1 can cause leukoencephalomalacia, a deterioration of the brain, and eventual death. Corn containing fumonisin B_1 should not be fed to horses.

As the traditional feed for horses, oat grain is slightly lower in energy than corn, but is somewhat higher in protein quantity and quality. Some horse owners believe that oats cause fewer digestive problems than corn. The fiber content of oats is higher than for corn, which may contribute to its reputation as a

TABLE 21-4

Approximate daily hay consumption of horses with average mature weight of 500 kilograms

Type of Horse	Hay Intake[a] (kg/day)	Comments
Maintenance, very light work early gestation	9.5–12	More hay may be needed in cold temperatures; if alfalfa hay is fed, concentrate is usually not necessary
Late gestation, light work	7.5–11	Most horses require a small amount of concentrate (1.5 to 3 kg) in addition to hay
Lactation	9.5–13	Most lactating mares require at least 3 to 4 kg of concentrate grain in addition to hay
Yearling	7–11.5	Amount of concentrate varies (2 to 6 kg) depending on age, type of hay, and situation (sale preparation, breaking, etc.)
Weanling	3.5–7	Amount of concentrate varies (2 to 4 kg) depending on age of horse and type of hay
Performance horse	7–12	Amount of hay and grain varies depending on level of work; most will receive 3 to 6 kg of grain per day

[a]Assumes that horses have no access to pasture. When pasture is available, the amount of hay needed will be reduced.

TABLE 21-5

Composition of cereal grains and grain by-products fed to horses (100% dry-matter basis)

Feed	DE (Mcal/kg)	%CP	%Ca	%P	Cu (ppm)	%Zn (ppm)	%CF
Barley	3.68	13.2	0.05	0.38	9.2	19	5.6
Corn	3.84	10.4	0.05	0.31	4.2	22	2.5
Oats	3.20	13.3	0.09	0.38	6.0	35	10.7
Sorghum	3.56	12.7	0.04	0.36	6.0	30	2.8
Wheat	3.86	14.6	0.05	0.42	5.5	37	2.8
Wheat middlings	3.47	17.3	0.11	1.13	20.6	109	9.1
Wheat bran	3.3	17.4	0.14	1.27	14.2	110	11.3
Vegetable Oil	4.09	—	—	—	—	—	—

Source: NRC (14).

"safe" feed for horses. One of the factors that affects the energy content of oats is the bushel weight. Heavy oats (>32 lb/bu) contain less fiber and more digestible energy than regular oats. The nutrient composition of barley is similar to the composition of oats. Barley is most popular as a horse feed in areas where it is grown and thus available at an economical price. Wheat and rye are not used to a large extent in feeding horses, but milo is sometimes used. Milo (sorghum) is a small grain with a hard seed coat that should be processed prior to feeding to horses.

A number of grain by-products can be used as horse feeds. Wheat middlings (mids) are used in pelleted horse feeds and may be an economical source of energy and protein. Wheat bran can be considered an energy feed, but it is very high in fiber. The main use of wheat bran is as an ingredient in hot mashes. A hot mash, made with wheat bran, hot water, and other feed ingredients is highly palatable and may be prepared for horses with poor appetites. Some corn by-products can be used for horses, such as corn gluten feed, but corn screenings should not be fed to horses. Corn screenings tend to have higher levels of fumonisin B_1 than whole corn grain and thus carry a larger risk of causing leukoencephalomalacia.

All the cereal grains and by-products listed in Table 21-5 are low in Ca and have an unbalanced Ca:P ratio. Wheat bran is particularly high in P. If cereal grains or their by-products are fed in large amounts to horses without Ca supplementation, bone problems may develop, particularly in young horses. Years ago, a problem called "big head" disease was relatively common. The affected horses had been fed by-products of the grain milling industry, and thus the problem was also referred to as *miller's disease*. The problem was eventually traced to the high P and low Ca levels in the diet. When horses receive Ca-deficient diets, Ca is mobilized from bone to maintain blood Ca levels. After receiving a Ca-deficient diet for a long period of time, Ca was mobilized from the facial bones in the horses and these areas filled with connective tissue, resulting in a swollen appearance. Frequently, horses with big head also had lameness problems. Today, big head disease is relatively uncommon in the U.S. horse population, but instances of Ca deficiency still occur. The most common problems arise when young horses are fed a diet of grass hay and cereal grains without any additional Ca. Signs of a Ca deficiency include enlarged joints, lameness, and angular limb deformities (crooked legs). Calcium deficiency is not common when horses receive diets containing alfalfa hay.

Animal fat and vegetable fat are also fed for energy. Fat is well digested by horses, but may be less palatable than cereal grains, particularly at high levels of inclusion in the diet. Total fat intake is usually limited to about 5% to 10% of the total diet. Corn oil and soybean oil tend to be more palatable to horses than animal fat.

PROTEIN SUPPLEMENTS

A number of protein supplements may be used in horse rations. Soybean meal has excellent protein quality and is readily available. Cottonseed meal and linseed meal can also be used in horse rations, but both are slightly lower in protein quantity and quality than soybean meal. Most cottonseed meal contains gossypol, which has not been reported to be toxic in horses but negatively affects other species. A maximum level of 0.03% gossypol in diets for young horses has been proposed (11). Milk products are excellent protein sources, but are usually more expensive than soybean meal. Milk protein sources are rarely used in rations for mature horses, but are often included in foal feeds.

MINERAL SUPPLEMENTS

Forages are good sources of K. Na and Cl must be added to most rations, usually with a salt block or loose salt lick. Most horses must consume about 28 g (1 oz) of salt each day to meet their maintenance requirement. If voluntary salt consumption is less than 28 g/day, then it may be necessary to add salt to the grain. The salt block or lick may contain plain salt, iodized salt, or trace-mineralized salt. The trace minerals found in salt blocks usually include Fe, Cu, Zn, Mn, and I. Some salt blocks contain Co and Se. Selenium-containing salt blocks should be used cautiously in areas with high soil Se or if the horses are receiving other sources of supplemental Se.

Supplements containing specific trace minerals are also used in horse rations. Trace minerals may be provided from inorganic sources or as chelates. Common inorganic forms of some trace minerals include zinc chloride, zinc sulfate, sodium selenite, and copper sulfate.

When diets are low in Ca and/or P, dicalcium phosphate or bone meal may be added to the diet. Limestone is used as a Ca supplement, and monosodium phosphate is used to supplement P. Dolomitic limestone also contributes Mg to a ration.

OTHER FEED INGREDIENTS

A number of nontraditional feeds can be used for horses. For example, horses can be fed rice bran, rice hulls, citrus pulp, or soybean hulls. The availability and cost of these feed ingredients often influences their use. Some may have nutritional or palatability characteristics that limit the amount that can be fed to horses. Beet pulp is a feed ingredient that has gained popularity in recent years. It has a relatively high fiber content, but the fiber is well digested by the horse and, as a result, beet pulp has a fairly high energy value. Because of its high fiber content, beet pulp is often used to replace hay in rations for horses with heaves. Rice hulls and soyhulls have also been used as fiber sources in horse feeds, but their feeding values are lower than for mature grass hay.

MANUFACTURED FEEDS

A variety of manufactured feeds is available for feeding horses. These feeds can be grouped into three general categories: fortified grain mixes, supplements, and complete feeds. Fortified grain mixes are used widely throughout the horse industry. These mixes are formulated to meet the nutrient needs of horses when fed with forage alone. The ingredients in fortified grain mixes usually include a combination of grain or grain by-products, a protein supplement, Ca and P sources, trace minerals, vitamins, and salt. Some fortified grain mixes contain vegetable oil. Molasses is often added to grain mixes to improve palatability and to keep all feed ingredients uniformly mixed. The amount of each ingredient included in a specific feed is dictated by the nutrient requirements of the horse for which the feed is intended. Most manufacturers offer at least three separate formulations. One product is formulated for horses at maintenance or light work and usually contains about 10% to 12% crude protein. A second product is formulated for performance horses, broodmares, and yearlings and contains 13% to 14% crude protein. The third product is formulated for lactating mares, weanlings, and yearlings and contains 15% to 16% crude protein and the highest level of vitamin and mineral fortification. Many manufacturers also offer specialty formulations to meet more specific needs or to be fed under specific conditions.

In addition to being available in several formulations, fortified grain mixes are available in various physical forms. The most common physical form is a coarse mix, sometimes referred to as sweet feed or textured feed. Coarse mixes are composed of whole or partially processed grains (cracked corn, rolled oats, etc.) and other ingredients (soybean meal, dicalcium phosphate, salt) mixed together with molasses. Frequently, a coarse mix includes a pellet or crumble that contains a protein supplement, vitamins, and minerals. One disadvantage of a coarse mix is that some horses sort the ingredients. Ingredient sorting can be decreased by feeding a pelleted feed. Ground by-product feeds are not easily incorporated into coarse mix feeds but can be used in pelleted feeds. In pelleted feeds, the ground, mixed ingredients are pressed through a die, usually with the assistance of steam heat (see Ch. 11). The steam heat used in the pelleting process can destroy some vitamins, so pelleted feeds often contain added vitamin fortification to compensate for activity lost during the pelleting process. Pellets should have a firm consistency that resists crumbling, but should not be too hard. Pellet diameter may vary from $\frac{1}{4}$ to $\frac{1}{2}$ in. Larger pellets (sometimes referred to as cubes) are easier for horses to pick up off the ground and thus may reduce waste.

Fortified grain mixes may also be extruded. In the extrusion process, the ground and mixed ingredients are processed under pressure. The resulting product tends to have an expanded appearance and to be less dense (lower weight per volume) than other physical forms. Extrusion may increase the digestibility of a fortified grain mix to a slight extent.

The use of fortified grain mixes has increased in the horse industry, but some managers still prefer to hand-mix feed ingredients and add one or more supplements to a diet of hay and plain cereal grains. Supplementation of any nutrient beyond the requirement will not benefit the horse, and excessive supplementation may create toxicity problems. Therefore, supplements should be used to address specific deficiencies. When horses are fed diets consisting of pasture or hay and plain cereal grains, the potential for mineral and/or vitamin deficiencies exists (Table 21-6). These deficiencies may be overcome by using multiple single-nutrient supplements or a multinutrient supplement specifically formulated to compensate for expected deficiencies in typical hay–grain or pasture–grain diets. For example, when pastures are lush enough to meet all a horse's energy requirement, a small amount of supplement may be fed every day to meet the needs for minerals, protein, and vitamins.

A *complete* manufactured feed contains a roughage source and is formulated to be fed without any forage. The fiber level in a complete feed is usually at least 12% to 15% and may be above 20%. Dehydrated alfalfa meal and beet pulp are common fiber sources in complete feeds. Complete feeds containing alfalfa meal are often pelleted, but those containing beet pulp may be manufactured in a textured form. Complete feeds are useful in situations where good-quality hay or pasture is not available. They are also useful for horses that have difficulty utilizing common forage sources, such as older horses with poor teeth that cannot graze well or horses with respiratory allergies to hay.

FEEDING GUIDELINES

BROODMARES

Most mares do not enter the breeding herd until they are at least 3 or 4 years old, and successful performance horses may not be bred until they are more than 8 years old. To maximize production efficiency, it is desirable for mares to produce a foal every year. Mares do not

TABLE 21-6

Possible deficiencies in a diet containing 6-kg midbloom bromegrass hay and 4-kg oat grain when fed to a 500-kg horse in moderate work

Nutrient (per day)	Requirement	Amount in Diet[a]	Comment
DE, Mcal	24.6	25.6	Adequate
CP, g	984	1300	Adequate
Ca, g	30	21	Deficient
P, g	21	32	Unbalanced Ca:P ratio
Mg, g	11.3	12	Adequate
Na, g	30	3	Requirement would be met with 70 g NaCl
K, g	37.4	135	Adequate
S, g	15	16	Adequate
Fe, mg	400	>800	Adequate
Mn, mg	400	400	Just adequate
Cu, mg	100	175	
Zn, mg	400	340	Marginal
I, mg	1	>0.52[b]	Adequate
Co, mg	1	3.5	Adequate
Se, mg	1	1	Just adequate
Vitamin A, IU	22,000	>60,000	Adequate
Vitamin E, IU	800	200	Marginal or deficient
Thiamin, mg	50	26	Synthesis in hind gut may result in adequacy
Riboflavin, mg	20	7	Synthesis in hind gut may result in adequacy

Source: NRC (14).

[a]Calculated from average values as reported by NRC (14).

[b]Accurate values for I content in bromegrass hay are not available.

commonly exhibit lactational anestrus, so most mares can be rebred within a few weeks of foaling, just as they reach peak lactation. Proper nutrition is important for maintaining high reproductive efficiency in broodmares. If mares do not consume adequate nutrients in their diets during gestation and lactation, they will use body stores to meet the nutrient demands of fetal growth and milk production. Loss of weight or body condition is an indication that the nutrient content of the diet is not meeting the mare's requirement. Contrary to common belief, mares that are very fat have not been shown to have lower reproductive efficiency than mares in thinner body condition. In fact, the opposite is true; mares with body condition scores (Table 21-2) below 5 have lower reproductive efficiency than mares carrying more body fat (4). In addition, mares that do not receive adequate Ca during lactation may mobilize Ca from their bones to meet milk demands. Many mares remain in the breeding herd until they reach their late teens or early twenties and may have as many as 12 foals. The cumulative effects of inadequate Ca nutrition over several gestation and lactation cycles could be significant to the long-term soundness of mares.

Gestation and lactation often overlap in the broodmare, and mares are often rebred for the next foal within a few weeks of parturition. The gestation period in the mare averages about 340 days. During the first two-thirds of gestation, the amount of fetal tissue accumulated is very small, and the mare's nutrient requirements are not substantially increased above maintenance by fetal growth. If a mare has lost weight during lactation and enters the second trimester of gestation in suboptimal body condition, her nutrient intake should be increased by 10% to 20% above maintenance. Pregnant mares should be maintained at a condition score of at least 5 throughout gestation. In areas with severe winters, it may be desirable to have mares at higher condition scores at the beginning of the winter.

Changes in body weight do not appear to coincide precisely with the fetal growth curve (Fig. 21-6). The majority of fetal growth occurs in the last third of gestation, but mares may experience the greatest gains in the second third of gestation. It is not known if these gains are associated with placental tissues and fluids or if they are actual increases in body flesh, but this pattern of weight gain appears normal (10). Over the course of a 340-day gestation, a 500-kg mare is expected to gain about 50 to 70

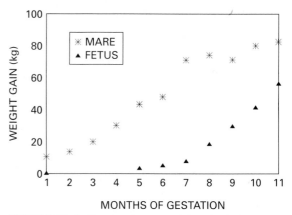

FIGURE 21-6 Fetal and mare weight gain during gestation. The weight gains observed in mares (10) do not exactly correspond to the weight gains observed in the fetus (12). Some of the difference in weights can be attributed to placental tissues and fluids.

kg. At birth, the foal will weigh about 40 to 55 kg. The remainder of the weight gain is attributed to placental tissues and fluids. Regular weighing of mares during gestation assists managers in evaluating the adequacy of the diet. It is also helpful to condition score mares regularly during the gestation period. Gestating mares may lose condition even though they are gaining weight if they are converting energy and protein in their body tissues to fetal tissue. Loss in body condition or weight suggests that the ration is not meeting the need for one or more nutrients (especially energy).

The nutrient requirements of gestating mares are shown in Appendix Tables 45 and 46. Nutrient requirements are highest during the last two months of gestation. It may be possible to meet the mare's requirements for late gestation by increasing the amount of feed that the mare is receiving. However, many mares consume less feed in late gestation than at other times, and thus it is often necessary to alter the composition of the ration to meet requirements. Typically, these alterations include a reduction in amount of forage and an increase in concentrate (Table 21-4).

GROWING HORSES

Foals begin to nibble on solid food within a few days of birth and will consume significant amounts of hay, pasture, or grain by 2 months of age. Figure 21-7 illustrates growth rates in Thoroughbred and Quarter Horse foals from birth to 4 months of age. Average daily gain ranges from 1.2 to 1.6 kg/day in the first month and then gradually declines to about 1.0 kg (2.2 lb)/day at 4 months of age. If foals are not grow-

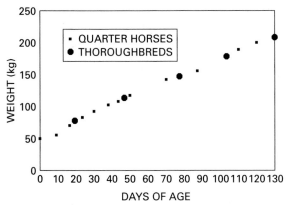

FIGURE 21-7 Representative growth curves for foals with an expected mature weight of 500 to 600 kg. The curves illustrate the growth of Quarter Horse foals (8) and Thoroughbred foals (17) prior to weaning, through 120 days of age.

FIGURE 21-9 This 3-month-old foal is almost ready to be weaned. Most foals are weaned between 4 and 6 months of age. The mare in this picture is wearing a fly mask.

ing at an acceptable rate, it may be advisable to provide a creep feed at about 2 months of age. Foals should be fed a diet formulated to meet their needs, rather than a diet formulated for mares. Initially, foals can be fed about 0.5 kg of feed per month of age (15). An acceptable creep feed for nursing foals contains 16% crude protein, 0.9% Ca and 0.6% P, and a high-quality protein source should be used (11). Foals should not be given unlimited access to creep feed. If several foals are fed together, they should be closely observed to ensure that each has equal access to the feed. Creep feeders can be constructed from a variety of designs, but the key feature is an entryway that allows the foals to enter but excludes the mares (Fig. 21-8). An alternative to creep feeding is bringing each mare and foal into a stall every day and restraining the mare while the foal consumes its own feed.

Most foals are weaned at 4 to 6 months of age (Fig. 21-9). By 4 months of age, the con-

tribution of milk to the total nutrient requirements of the foal starts to decline, and supplemental feed is necessary even if the foal is not weaned. By 6 months of age, milk provides less than 50% of the foal's daily nutrient intake (Fig. 21-10). If the mare has been rebred, she will be entering the middle of her gestation period and weaning will enable her to redirect nutrients to the developing fetus and to the replenishment of her body stores. Weaning can be stressful for many foals, and a period of markedly reduced gain is expected. The depression in rate of gain may be as brief as a week if foals are accustomed to eating significant amounts of pasture, hay, or grain prior to weaning (Fig. 21-11).

After weaning, foals should be fed a diet that meets or exceeds the nutrient requirements specified in Appendix Tables 47 and 48. Rate of gain may be affected by energy intake, but if a rapid rate of gain is desired, the intake of other

FIGURE 21-8 One type of creep feeder that allows the foal access to feed while excluding the mare.

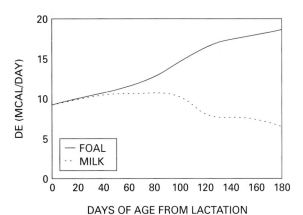

FIGURE 21-10 Comparison of the digestible energy (DE) needs of a growing horse and the digestible energy contributed by the mare's milk. Foals should be allowed access to supplemental feed by 2 months of age.

FIGURE 21-11 A depression in growth rate at weaning may be minimized if foals are accustomed to eating adequate amounts of forage and concentrate prior to weaning. This foal is being accustomed to eating concentrate from a fence feeder where he will be fed with other foals after weaning.

nutrients must be increased also. Appendix Tables 47 and 48 list nutrient recommendations for young horses growing at two rates of gain, which can be classified as moderate or rapid growth. Many horses intended for show or sale as weanlings or yearlings are fed for rapid growth. Rate of gain during the first 2 years of life will not affect final mature size, but may affect the age at which the animal reaches maturity. In addition, very rapid rates of growth have been associated with an increased incidence of bone and joint problems. When rapid growth is desired, the potential for problems can be minimized by feeding diets that have adequate levels of all required nutrients, not just energy.

Bone and joint problems are often referred to as developmental orthopedic disease (DOD) and include such entities as osteochondrosis, physitis, cervical vertebral malformation, and angular limb deformities. Osteochondrosis and physitis are probably the two most prevalent problems. Osteochondrosis occurs when growing cartilage in the young animal fails to mineralize normally. Osteochondrosis most commonly affects the articular region of long bones and may result in significant lameness. Physitis occurs when a growth plate becomes inflamed. Physitis is most common in the growth plates of the long bones in the forelegs. Young horses with physitis may have enlarged joints (particularly fetlocks and knees) and may have pain in these joints. Nutritional factors that may contribute to DOD include low dietary Ca, low or high dietary P, an unbalanced Ca:P ratio, low dietary Cu, very high dietary Zn, and excessive dietary energy. Providing young horses with balanced diets minimizes the potential for osteochondro-

sis and physitis, but will not necessarily prevent all cases. Other factors such as environment and genetics may also be important.

PERFORMANCE HORSES

The feeding management and dietary requirements of performance horses varies with the type and level of activity. The spectrum of performance ranges from the racing Quarter Horse that will compete at maximal speed for 400 yards or less to the Arabian endurance horse that will compete over distances of 50 to 100 miles. Feeding programs for performance horses must also take into account such factors as environment, stage of training, and age. The recommendations listed in Appendix Tables 45 and 46 are based on level of work performed (light, moderate, or intense), with few allowances for many of the factors listed above. Since the publication of these recommendations in 1989, considerable research on the nutrition of performance horses has been conducted, and it is now possible to make more specific recommendations for horses performing different activities.

The amount of energy required by performance horses is affected by two primary factors: the intensity of the daily exercise and the duration of the daily exercise. The DE required for exercise (above maintenance) can be more precisely calculated if the speed of travel is known by using the following equation:

$$DE_{(kcal/kg/min)} = \frac{e^{(3.02 + .0065x)} - 13.92}{0.57}$$

where kg is the weight of the horse (with rider) and x is the speed of travel in meters per minute (16). Although this equation is useful in some situations, it does have some pitfalls. This equation was derived using slow to moderate speeds on a flat surface and may underestimate the energy required by horses that are worked at racing or near-racing speeds, going up and down slopes, working in mud or sand, pulling weights, or jumping. In addition, most performance horses do not receive the same amount of work every day, and it is not convenient to recalculate energy on a daily basis. Additional allowances may be necessary for increased stall activity, effects of frequent travel, and thermoregulation.

The recommendations for DE intake listed in Appendix Table 45 are based on general categories of work effort and estimate that increases in DE of 25%, 50%, and 100% above maintenance are necessary for horses involved in light, moderate, and intense exercise, respectively. Light work is described as pleasure riding, whereas intense work includes race training and polo. In practice, most performance horses are fed to maintain body weight at a body condition appropriate to their type of event and level of competition. Table 21-7 lists the results of several studies that evaluate body condition in horses engaged in various events. All of the investigators used the body condition scoring system presented in Table 21-2. The values in Table 21-7 do not necessarily represent optimum body condition scores, but instead represent the common range of scores found in horses in the listed activities.

Horses performing moderate or intense work cannot consume sufficient forage to meet their energy needs. Total dry-matter consumption of performance horses usually ranges from 2.0% to 2.5% of body weight. The lightly worked horse requires a minimal amount of concentrate, but as the amount of daily work increases, the amount of concentrate necessary in the diet also increases. At high levels of activity, the daily DE requirement may be very high, and concentrate will comprise a large portion of the diet. To prevent the potential for starch overflow to the large intestine, concentrate intake should be limited to less than 1.25 kg/100 kg of body weight or 50% of the total dry matter. When horses are unable to maintain their body weight on a diet of 50% roughage and 50% concentrate, several dietary modifications are possible. First, roughage quality should be evaluated and upgraded if possible. Early-maturity hay is higher in energy density than late-maturity hay and is also much higher in palatability. If hay quality and intake have been maximized, then the addition of fat to the diet should be considered. The addition of fat to a concentrate feed markedly increases energy density without increasing starch intake. Many commer-

cially manufactured horse feeds now contain 4% to 8% added fat in a form that is palatable to the horse. Fat can also be added directly to the diet by top dressing a vegetable oil on the concentrate portion of the diet. Horses accept the vegetable oil most readily if it is introduced to the diet gradually. If a change in hay quality and/or addition of fat to the diet does not improve the horse's energy balance, then larger amounts of concentrate can be fed. When horses are fed more than 6 kg of concentrate per day, the concentrate should be fed in at least three meals so that no single meal exceeds 3 kg. Diets for endurance horses depend more heavily on energy from forage than diets for racehorses. High-forage diets encourage water consumption and may assist horses in maintaining hydration status duing long-distance exercise when high sweat losses are common. In addition, fermentation of forage in the large intestine provides a continual supply of VFA that can be used for energy during long-term exercise. Feeding programs for Thoroughbred and Standardbred racehorses often restrict forage intake the night or morning before a race to decrease bulk in the gastrointestinal tract.

It has been recommended that working horses should receive more dietary protein than sedentary horses (Appendix Table 45). Protein is lost in sweat, and limited data suggest that a small amount of protein may be broken down during exercise. In addition, horses in training may retain slightly more nitrogen than horses at maintenance. Thus, regular exercise probably increases the dietary protein requirement, but the magnitude of the increase is relatively small, and the values in Appendix Table 45 may slightly overestimate the protein needs of working horses. In situations where the increased energy demand is met by increased feed intake, the protein requirement is almost always satisfied. If feed intake does not increase and energy needs are met by adjusting the roughage-to-concentrate ratio, the concentration of crude protein in the ration must be increased slightly above maintenance to meet the protein requirement.

Exercise markedly affects the amount of electrolytes required by horses. Horse sweat is very high in Na, K, and Cl, and any exercise that results in significant sweat losses affects the daily need for these minerals. Electrolytes cannot be stored in the body, so the requirement must be met on a daily basis. The exact composition of horse sweat appears to vary with type of exercise, individual, site of sampling, and

TABLE 21-7

Body condition of horses competing in various activities

Type of Activity	Mean Condition Score	Range of Condition Scores
Thoroughbred racing[a]	5.0	5–6
Standardbred racing[b]	5.7	5–7
Endurance[c]	4.67	3.6–5.2
Three-day Event[d]	4.9	3.5–6.0

[a]Gallagher et al. (1).
[b]Gallagher et al. (2).
[c]Lawrence et al. (9, 10).
[d]Taylor, L. (personal communication).

TABLE 21-8

Estimated electrolyte losses in horses performing various types of exercise[a]

Type of Activity	Na g	K g	Cl g
Thoroughbred racehorse (daily work)	14	5	28
Standardbred racehorse (race)	35	12	71
Three-day eventing horse (second day)	69	23	142
Endurance horse (50 to 60 mile race)	83	28	170

[a]Using an approximate sweat composition of Na, 150 mmol/l; K, 30 mmol/l; Cl, 200 mmol/l; and water losses of 4, 10, 20, and 24 kg for the Thoroughbred, Standardbred, three-day, and endurance horses, respectively.

other factors. Although it is difficult to precisely determine the requirement for these minerals, Table 21-8 provides some comparative estimates of Na, K, and Cl losses of horses performing different amounts of work. These amounts were calculated using estimated sweat losses during exercise and a sweat composition of 150, 30, and 200 mmol/l for Na, K, and Cl, respectively. The largest losses are usually sustained by endurance horses competing in long-distance races, particularly in hot weather. The preferred method of replacing electrolytes in these horses is to give small amounts of a balanced supplement at regular intervals during a long ride. This goal is usually fairly easy to accomplish, because most endurance competitions include mandatory stops at regular intervals when the horses are watered and examined by a veterinarian.

Little is known about the effect of exercise on the requirements for other minerals. Calcium and P intake are most critical for young horses just entering training because many may still be growing. In addition, regular exercise may induce a bone remodeling response. A reasonable guideline for mineral requirements in exercising horses is to increase the mineral intake (in grams per day) in proportion to the energy intake. For example, if a horse requires 30% more DE to meet its energy need for exercise, then the amount of Ca should also be increased by 30%. Exercise may increase the requirement for other minerals, but this area has not been well studied in the horse. Attempts to increase erythrocyte numbers or blood hemoglobin levels with Fe supplementation in controlled studies have not generally been successful.

Performance horses may require higher dietary concentrations of some vitamins than sedentary horses. Many B vitamins are synthesized by the microbial population in the large intestine, but it may be prudent to ensure that at least 50% of the expected requirement is provided by the diet. Table 21-9 lists the estimated requirements for several B vitamins of importance in performance horses. Thiamin, niacin, riboflavin, pantothenic acid, choline, and biotin function as cofactors in energy metabolism, whereas folic acid and B_{12} are important for erythrocyte formation.

Vitamin E protects cell membranes from oxidative damage and may affect the susceptibility of muscle to damage from substances produced during exercise. However, a direct link between exercise-induced muscle damage and vitamin E intake has not been demonstrated in the horse. It is recommended that working horses receive 80 mg of vitamin E/kg of diet dry matter (14).

TABLE 21-9

Estimated requirements of horses for selected B vitamins

Vitamin	Estimated Requirement	Comment
Thiamin	36 mg	Hard working horses may require up to 60 mg/day
Riboflavin	60 mg	
Niacin	300 mg	
Pantothenic acid	150 mg	
Folic acid	10 mg	
Biotin	2 mg	Supplementation rates of 10 to 30 mg/day have been reported to improve hoof quality in horses with poor hooves
Choline	10,000 mg	
B_{12}	200 μg	

Source: Ott (15).

COMMON FEED-RELATED PROBLEMS IN HORSES

Some, but not all, cases of colic may be caused by dietary factors. Colic is characterized by severe abdominal pain in the horse. Horses with colic may kick at their abdomens, roll, or repeatedly attempt to urinate. Sweating and general signs of anxiety and discomfort are also common with colic. Some cases of colic are mild and resolve quickly, while other cases require surgery or may result in death. Horses with 24-hr access to pasture may have a decreased incidence of colic, whereas horses receiving high levels of concentrate intake may have an increased incidence of colic (18). Other dietary factors that may increase the incidence of colic include sudden changes in diet and lack of water availability. Many factors other than diet can contribute to colic, but to minimize the potential for problems, horses should be provided adequate water and forage, and any diet changes should be made slowly.

Laminitis, or founder, may be caused by overconsumption of concentrate or lush-growing pasture. Horses with acute laminitis exhibit pain and heat in the hooves and a reluctance to move. Laminitis often results in permanent lameness and may cause death in some cases. To reduce the potential for laminitis, horses should be accustomed to lush pastures gradually, and energy needs should be met with roughage instead of concentrate whenever possible. If a horse requires large amounts of concentrate to meet its energy needs, it should be adjusted to the diet gradually and should not be given more than 3 kg of concentrate at any meal. Once a horse has foundered, it may be more susceptible to subsequent episodes with less provocation.

Horses are susceptible to plant poisoning from a variety of trees, weeds, and ornamental shrubs (5). Some plants, such as thistles, nettles, and burrs, are mechanically injurious to horses and may cause damage to the nose and mouth. St. John's wort and buckwheat may produce photosensitization and dermatitis. Photosensitization is most commonly observed on unpigmented areas, such as white markings on the face or legs. Mountain laurel, azalea, jimson weed, oak, field blindweed, buttercups, and a number of other plants may cause colic or diarrhea. Some of these plants may be fatal if sufficient quantities are consumed. Other plants that may be fatal include serviceberry, elderberry, foxglove, oleander, and yew. Horses may become poisoned when trimmings from trees or ornamental plants are placed in the pasture. In addition, if pasture quality is poor, some horses may consume leaves from trees or shrubs in a fenceline or weeds in the pasture that they would normally ignore. Mixed-grain feeds can occasionally result in a poisoning situation when a mixing error is made, but the most common problems arise when grain feeds mixed for other species are fed to horses. Cattle feeds containing ionophores should not be fed to horses. Similarly, feeds mixed for other species that contain antibiotics should not be fed to horses.

FEEDING MANAGEMENT

Many pleasure and performance horses are maintained in situations where they are housed in box stalls and have little access to pasture. This housing arrangement makes it easy to provide each horse with a ration that is specific to its needs, but it may have disadvantages also. When horses graze in a natural state, eating and noneating periods are interspersed, and eating periods are rarely separated by more than 2 or 3 h. In addition, grazing horses spend about 50% to 60% of their time eating. When horses are kept in stables, they are usually fed only two or three times per day, which results in long periods between meals. In addition, horses consuming a typical hay and concentrate diet may spend less than 6 h a day in eating activities, compared to the 12 to 14 h that might be spent in a pasture environment. To allow stabled horses to have a more natural feeding environment, hay availability should be maximized. When possible, hay should be offered at least 1 h before the concentrate to encourage the horse to consume the hay first. No more than 3 kg of concentrate should be fed at one time to a mature horse (500 kg). If it is necessary for a horse to receive more than 6 kg of concentrate per day, the concentrate should be divided into at least three meals.

When mature horses are maintained on good-quality pasture, they should have access to a salt block and a source of clean fresh water. In many cases, additional supplementation is not necessary. However, if the pasture is not sufficient to meet nutrient needs, horses should receive supplemental feed. The most desirable method of providing concentrate to several horses kept together on pasture is to bring the horses into stalls for individual feeding once or twice a day. When this practice is not possible, several horses may be fed concentrate together in the pasture if they have similar nutrient needs. Feeding management is always simplified if horses with similar nutrient requirements are grouped together. Grouping horses by nutrient need minimizes the potential for underfeeding some horses and overfeeding others.

In a pasture environment, horses may be fed concentrate from individual fence feeders, individual ground feeders, or feed troughs. If fence feeders or ground feeders are used, they should be spaced far enough apart to limit fighting among the horses. Feeders may be placed 4 to 6 m apart for mature horses and 3 to 5 m apart for young horses. If a long trough is used for feeding concentrate to a group of horses, close observation is necessary to ensure that one horse does not dominate the area and consume all the concentrate. Trough feeders are more effective for young horses than for mature horses that have more defined social hierarchies. Concentrate may be fed directly on the ground (without a feeder), but the amount of feed wastage will be greater and there is increased potential for transmission of parasites.

If horses are kept in a dry lot or if pasture quantity is not sufficient, they will need to be fed hay. When hay is fed in a group situation, waste will probably be increased and feeding allowances should be increased by at least 10% to 20% beyond the amount that would be required if horses were individually fed. Hay should be placed in feeders that provide enough room for all horses in the pasture to eat at once. Hay feeders designed for horses are usually safer than feeders designed for cattle or other livestock.

SUMMARY

Forage is the primary component of most horse diets and many horses require only good quality forage, salt and water to meet their nutrient needs. Pasture and hay are the most common forage sources; haylage and silage are not commonly fed to horses. Broodmares, growing horses and working horses will require other feeds in addition to forage to meet their nutrient requirements. Commercially manufactured concentrate feeds that have been fortified to meet the nutrient requirements of various classes of horses are commonly used in the horse industry. These feeds are formulated to be fed with forage. Whenever the daily intake of a fortified concentrate feed or a plain cereal grain exceeds 3 kg, the feed should be divided into at least two meals. Excessive intake of high starch feeds such as fortified concentrates, oats, corn or barley can lead to serious digestive and metabolic problems in horses such as diarrhea, colic and founder (laminitis). To minimize digestive problems, forage should comprise at least 50% of the diet dry matter.

REFERENCES

1. Gallagher, K., et al. 1992. *J. Equine Vet. Sci.* 12: 43.

2. Gallagher, K., et al. 1992. *J. Equine Vet Sci.* 12: 382.

3. Henneke, D. R., et al. 1983. *Equine Vet. J.* 15: 371.

4. Henneke, D. R., G. D. Potter, and J. L. Kreider. 1984. *Theriogenology* 21: 297.

5. Knight, A. P. 1995. In: *Equine clinical nutrition.* Media, PA: Williams and Wilkens.

6. Kronfeld, D. S., T. N. Meacham, and S. Donoghue. 1990 *Vet. Clinics North America: Equine Practice.* 6: 451.

7. Lawrence, L. M., et al. 1987. *Can. J. Animal Sci.* 67: 217.

8. Lawrence, L. M., et al. 1991. *Equine Pract.* 13: 19.

9. Lawrence, L. M., et al. 1992. *J. Equine Vet. Sci.* 12: 320.

10. Lawrence, L. M., et al. 1992. *J. Equine Vet. Sci.* 12: 355.

11. Lewis, L. D. 1995. *Equine clinical nutrition.* Media, PA: Williams and Wilkens.

12. Meyer, H., and L. Ahlswede. 1978. *Animal Res. Dev.* 8: 86.

13. Michener, C. B. 1907. In: *Diseases of the horse.* Washington, DC: Government Printing Office.

14. NRC, 1989. *Nutrient requirements of horses.* Washington, DC: National Academy Press.

15. Ott, E. A. 1991. In: *Feeds and feeding.* Englewood Cliffs, NJ: Prentice Hall.

16. Pagan, J. D., and H. F. Hintz. 1986. *J. Animal Sci.* 63: 822.

17. Ruff, S. J., et al. 1993. *J. Equine Vet. Sci.* 13: 596.

18. White, N. A., et al. 1993. *Proc. Amer. Assoc. Equine Pract.* 39: 97.

22

FEEDING AND NUTRITION OF THE DOG AND CAT

Diane A. Hirakawa

INTRODUCTION

Dogs and cats have lived as companions to humans for thousands of years. Both domestic dogs and cats are members of the order *Carnivora,* which are flesh-eating mammals. Living families of the present-day order *Carnivora* are *Canidae* and *Felidae,* commonly referred to as the dog and cat families.

Carnivora possess anatomical features that have supported their carnivorous feeding behavior through evolution (Fig. 22-1). The canine teeth allow these animals to successfully catch and consume prey. The carnassids, flat molars, facilitate the reduction in food particle size to ease the swallowing of prey.

Archeological records indicate that the special relationship between humans and dogs is at least 12,000 years old. In what is now known as northern Israel, the remains of a wolf or dog pup were discovered with the 12,000-year-old burial remains of a human child.

Archeological evidence suggests that the first domesticated canid appeared before the agricultural phase, making the dog a likely candidate as the first animal domesticated by humans. It is believed that the wild canid ran in packs for hunting and for protection. First brought into close contact with humans by hunting the same prey that humans hunted, the wild canid probably learned to scavenge scraps left by human hunting parties. More than likely, it was the young offspring of these scavengers who were adopted and hand raised by humans, which led to taming and eventually domestication.

The domesticated dog continued to serve humans on the hunt and in protective capacities even when agricultural development progressed. It is likely that selective breeding of dogs was routinely practiced at this time.

The tremendous breed variation seen today is a result of 12,000 years of planned matings. Today's dog may weigh as much as 100 kg or as little as 1 kg. Hair length ranges from that of the afghan's very long fibers to that of the Mexican hairless's coat, and nose shapes range from the collie's to the pug's. Some breeds, such as the retriever, have retained their innate desire for the hunt. Other breeds, such as the toys, are better suited for pampering.

A younger relationship exists between humans and cats. Findings in Egypt dated to approximately 3000 years ago suggest that cats were kept in captivity. However, unlike the dog, few anatomical changes have occurred in the cat during its domestication; therefore, the precise date of a clear association with humans is

FIGURE 22-1 Evolution of the domestic dog. (Courtesy of the Iams Co., Dayton, OH.)

unclear. By 1600 B.C. cats were surely domesticated and considered as sacred beings by the Egyptians. Whether the first domesticated cat was revered as a sacred animal, kept as a beloved companion, or kept for its keen hunting capabilities as a method of rodent control remains to be elucidated.

Although both the dog and cat are classified in the order *Carnivora,* the divergence of the order occurred early in the evolutionary pathway; therefore, considerable distinction between the domestic dog and cat is present today. It is the evolution of the domestication process for the dog and cat that enhances our current understanding of their nutritional requirements. The domestic cat (*Felis silvestrus catus*) is a strict carnivore whose nutritional requirements have remained quite specialized through domestication. On the other hand, the domestic dog (*Canis familiaris*) is often considered an omnivore. Considering the phenotypic diversity among dog breeds, it is evident that the dog has acquired anatomical and physiological differences that directly affect nutritional requirements.

ECONOMIC AND SOCIAL SIGNIFICANCE OF DOGS AND CATS

In 1994, 38.2% of U.S. households owned at least one dog and 31.3 percent owned at least one cat. This is equivalent to a total of 77.8 million dogs and 63.8 million cats. The cat population has surpassed that of dogs for several years. This increase is most probably due to the changing lifestyle of today's fast-paced family. Owning a cat yields animal companionship; however, the independent behavior of the domestic cat allows it to adjust to changing life-styles.

Recent research efforts have focused on the importance of the human–animal bond. In addition to filling traditional roles as hunting companions, guard dogs, guides for the sight- and hearing-impaired, service dogs for the physically disabled, or scent dogs for narcotics, explosives, and missing people, the companionship of dogs and cats has been shown to lower blood pressure, have a positive effect on recovery from illness, and enhance the psychological health of humans. Companion animals are currently being used, for example, in therapy programs for elderly individuals in long-term-care facilities, emotionally disturbed children, and criminally insane prisoners.

In 1994, retail sales of pet foods was $8.8 billion, showing a 4% increase from 1993 (1). This increase was due primarily to an increase in pet ownership and to increased sales of premium pet foods. Because domestic animal research is focused on efficient use of foodstuffs, it seems reasonable that nutritional research with companion animals should be viewed with as much interest as that existing in nutritional research with food-producing animals.

Establishment of the economic and social importance of companion animals in our environment strongly justifies future experimentation concerning nutritional requirements and factors affecting these requirements, although this is not an easy task. There are no economic benefits to increased growth rates or improved reproductive efficiency, per se, in companion animals. Thus, it is difficult to define response criteria to assess nutrient requirements in this class of animals. The nutritional efficacy of a companion animal diet is based on maintenance of optimal health, well-being, and longevity, an unusual response parameter for a nutritionist, except when dealing with other pets or animals used for exhibition, but an area that warrants investigation.

NUTRIENT REQUIREMENTS OF DOGS AND CATS

The nutrient requirements for today's dog and cat can be supplied in a variety of ways through the use of commercial diets available in most grocery and pet supply stores. There are hundreds of brands of commercial pet food to choose from. They vary from 10% to 78% moisture, 12% to 35% protein, 5% to 30% fat, and 1% to 10% fiber. How does a pet owner provide complete and balanced nutrition and understand the options without becoming a specialist in companion animal nutrition?

The dietary requirements for dogs and cats were originally established by the National Research Council (NRC). These recommendations were revised and updated in 1985 and 1986 for the dog and cat, respectively (2,3). The NRC recommendations for each individual nutrient are based on minimum dietary requirements that were founded primarily on research employing the growing puppy or kitten. The experimental diets were often purified in composition, containing crystalline amino acids, vitamins, and minerals, thereby providing nutrients with high availability. While these studies provide valuable information about the minimum requirements of dogs and cats for available nutrients, they cannot be used as guidelines for the formulation of commercial pet foods without the inclusion of safety margins. Safety margins are needed to account for the decreased availability of nutrients in practical ingredients, variability of nutrient availability among ingredients, and loss of nutrients due to processing.

The Association of American Feed Control Officials (AAFCO) is an advisory agency composed of representatives from each state. Within the pet food industry, AAFCO functions to ensure that pet foods marketed nationally are uniformly labeled and nutritionally adequate. In the early 1990s, AAFCO established pet food committees consisting of canine and feline nutritionists from universities and the pet food industry (4). These committees developed two sets of standard nutrient profiles: one for dogs and one for cats. The profiles are based on ingredients that are commonly included in commercial foods. Minimum nutrient levels to be included in the pet food are provided for two categories: growth and reproduction and adult maintenance. Maximum levels are suggested for nutrients that have been shown to have the potential for toxicity or when overuse is a concern. These nutrient profiles have replaced the NRC recommendations as the standard to be used by pet food manufacturers. Like the NRC recommendations, the AAFCO profiles provide nutrient requirement estimates for dogs and cats. However, the primary difference between the two is that the AAFCO profiles provide suggested ranges of nutrients to be included in foods, rather than minimum nutrient requirements (Tables 22-1 and 22-2).

Nutrient requirements cannot be defined simply as being at a single level; requirements should be given as a range that provides sound nutrition, thereby avoiding states of nutrient deficiencies or toxicities. Optimal nutrition for dogs and cats often requires nutrients above minimum requirements. These depend on other nutrients in the diet, the individual animal being fed, and the desired response criteria as measured and seen by the pet owner.

The suggested wide range of nutrient requirements is not surprising when one considers breed diversity of the canine species. The mature adult weight of the canine can vary 100-fold, from the chihuahua, weighing as little as 1 kg, to the great dane weighing more than 100 kg. In addition, bone length and density, hair type and length, desired skin condition, muscle tone, and other physiological differences can contribute to difficulty in the assessment of nutrient requirements. Environmental factors and physiological phases such as growth, gestation, lactation, and physical stress must be considered as well. A paucity of information exists on definitive nutrient requirements related to breed, age, and sex. Lack of uniformity in the canine species makes it difficult, but by no means impossible, to establish such requirements.

Nutrient requirement estimates for dogs and cats have been established for energy, protein, fat, vitamins, minerals, and, of course, water. Rather than address the six nutrient classes and respective minimum requirements within each class, tables of reference are provided for dogs (Table 22-1) and cats (Table 22-2). The nutritional information emphasized here is intended to enhance the reader's general knowledge of canine and feline nutrition and to impart information about specific nutritional idiosyncrasies of feline diets.

WATER

Water, although often overlooked, is of utmost importance for life. It is known that an animal can survive approximately 10 times longer without food than without water. The

TABLE 22-1

AAFCO nutrient profiles: dog foods[a]

Nutrient	Units per Day (DMB)[b]	Growth and Reproduction (minimum)	Adult Maintenance (minimum)	Maximum
Protein	%	22.0	18.0	
Arginine	%	0.62	0.51	
Histidine	%	0.22	0.18	
Isoleucine	%	0.45	0.37	
Leucine	%	0.72	0.59	
Lysine	%	0.77	0.63	
Methionine–cystine	%	0.53	0.43	
Phenylalanine–tyrosine	%	0.89	0.73	
Threonine	%	0.58	0.48	
Tryptophan	%	0.20	0.16	
Valine	%	0.48	0.39	
Fat	%	8.0	5.0	
Linoleic acid	%	1.0	1.0	
Minerals				
Calcium	%	1.0	0.6	2.5
Phosphorus	%	0.8	0.5	1.6
Ca:P ratio	%	1:1	1:1	2:1
Potassium	%	0.6	0.6	
Sodium	%	0.3	0.06	
Chloride	%	0.45	0.09	
Magnesium	mg/kg	0.04	0.04	0.3
Iron	mg/kg	80	80	3000
Copper	mg/kg	7.3	7.3	250
Manganese	mg/kg	5.0	5.0	
Zinc	mg/kg	120	120	1000
Iodine	mg/kg	1.5	1.5	50
Selenium	mg/kg	0.11	0.11	2
Vitamins				
Vitamin A	IU/kg	5000	5000	50,000
Vitamin D	IU/kg	500	500	5000
Vitamin E	IU/kg	50	50	1000
Thiamin	mg/kg	1.0	1.0	
Riboflavin	mg/kg	2.2	2.2	
Pantothenic acid	mg/kg	10	10	
Niacin	mg/kg	11.4	11.4	
Pyridoxine	mg/kg	1.0	1.0	
Folic acid	mg/kg	0.18	0.18	
Vitamin B$_{12}$	mg/kg	0.022	0.02	
Choline	mg/kg	1200	1200	

Source: Reprinted with permission from the 1994 AAFCO Official Publication. Copyright 1994 by the Association of American Feed Control Officials.

[a]Presumes an energy density of 3.5 kcal ME/g DM. Rations greater than 4.0 kcal/g should be corrected for energy density.

[b]Dry-matter basis.

amount of body water is inversely related to body fat or positively correlated to lean body mass. Thus, it is not surprising that both growing puppies and kittens and racing sled dogs have a high lean body composition and also a high body water content. Therefore, hydration or dehydration is a primary concern in these animals.

The water requirement of an animal is provided by the moisture content of food, water of metabolism (primarily oxidation), and the consumption of drinking water or other fluids. Water content of commercial diets can range from 10% to 78%; thus, the consumption of water will vary accordingly. Needed water intake is affected by the amount of water lost in the process of thermoregulation (primarily excretion via the lungs) and excretion of metabolic waste products through the urine. The quality of drinking water may be as important as the nutrient content of the diet. For example, hard and soft water may contain high levels of specific minerals that may adversely affect the mineral balance of the diet.

TABLE 22-2
AAFCO nutrient profiles: cat foods[a]

Nutrient	Units per Day (DMB)[b]	Growth and Reproduction (minimum)	Adult Maintenance (minimum)	Maximum
Protein	%	30.0	26.0	
Arginine	%	1.25	1.04	
Histidine	%	0.31	0.31	
Isoleucine	%	0.52	0.52	
Leucine	%	1.25	1.25	
Lysine	%	1.20	0.83	
Methionine–cystine	%	1.10	1.10	
Methionine	%	0.62	0.62	1.5
Phenylalanine–tyrosine	%	0.88	0.88	
Phenylalanine	%	0.42	0.42	
Taurine (extruded)	%	0.10	0.10	
Taurine (canned)	%	0.20	0.20	
Threonine	%	0.73	0.73	
Tryptophan	%	0.25	0.16	
Valine	%	0.62	0.62	
Fat	%	9.0	9.0	
Linoleic acid	%	0.5	0.5	
Arachidonic acid	%	0.02	0.02	
Minerals				
Calcium	%	1.0	0.6	
Phosphorus	%	0.8	0.5	
Potassium	%	0.6	0.6	
Sodium	%	0.2	0.2	
Chloride	%	0.3	0.3	
Magnesium	%	0.08	0.04	
Iron	mg/kg	80	80	
Copper	mg/kg	5	5	
Iodine	mg/kg	0.35	0.35	
Zinc	mg/kg	75	75	2,000
Manganese	mg/kg	7.5	7.5	
Selenium	mg/kg	0.1	0.1	
Vitamins				
Vitamin A	IU/kg	9000	5000	750,000
Vitamin D	IU/kg	750	500	10,000
Vitamin E	IU/kg	30	30	
Vitamin K	mg/kg	0.1	0.1	
Thiamin	mg/kg	5.0	5.0	
Riboflavin	mg/kg	4.0	4.0	
Pyridoxine	mg/kg	4.0	4.0	
Niacin	mg/kg	60	60	
Pantothenic acid	mg/kg	5.0	5.0	
Folic acid	mg/kg	0.8	0.8	
Biotin	mg/kg	0.07	0.07	
Vitamin B_{12}	mg/kg	0.02	0.02	
Choline	mg/kg	2400	2400	

Source: Reprinted with permission from the 1994 AAFCO Official Publication. Copyright 1994 by the Association of American Feed Control Officials.

[a]Presumes an energy density of 4.0 J/kcal/g ME, based on the modified Atwater values of 3.5, 8.5, and 3.5 kcal/g for protein, fat, and carbohydrate (nitrogen-free extract, NFE), respectively. Rations greater than 4.5 kcal/g should be corrected for energy density; rations less than 4.0 kcal/g should not be corrected for energy.

[b]Dry-matter basis.

ENERGY

Following water, dietary energy is the nutrient required in largest amounts. Energy requirements are influenced by the animal's metabolic efficiency, environmental factors (temperature, humidity), physical exercise and activity level, the animal's age (Fig. 22-2), and the stage of production (Fig. 22-3). In addition, the energy requirement of a dog and cat (as with any other warm-blooded animal) per unit of body weight decreases as the size of the animal increases.

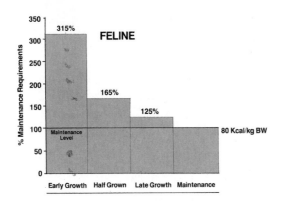

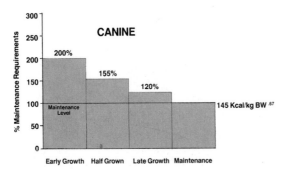

FIGURE 22-2 Feline and canine metabolizable energy requirements for growth.

Traditionally, commercial pet food analysis has been presented so that the percentage composition of major nutrients in the food were listed. While this accurately shows the chemical breakdown of the diet, it does not fully explain the balance of the various nutrients to a common factor other than "as fed" or dry matter.

Almost 20 years ago the concept of nutrient digestibility was introduced to the pet food market. Pet food digestibility must be considered when judging pet food quality. The higher the digestibility is the better the nutritional quality of the diet (see Chs. 2 and 3). This principle holds true for all nutrients. Some pet food manufacturers have begun to provide nutritional information based on metabolizable energy (ME). The concept has been in existence for decades among nutritionists, but the need for educating consumers about ME is recent (see Ch.3).

Animals offered a balanced diet tend to eat primarily to satisfy their energy need; thus, the logical comparison involves evaluating a diet on the basis of units of a nutrient per unit of energy. Separate diets can then be compared more accurately, and actual nutrient intake can be determined knowing the caloric content.

A study employing growing puppies fed four diets containing different levels of gross energy (GE) illustrated the phenomenon that puppies consume feed to meet their ME needs. Interestingly, on one commercial diet, puppies consumed 23% more GE to rectify the inefficient digestibility of the fibrous feedstuffs (5).

CARBOHYDRATES

Grain starches provide an important and economical source of dietary energy in most pet foods. Whether the carbohydrate source is supplied by corn, rice, wheat, or oats is less important than the method of processing. Utilization of starch in cereal grains is enhanced significantly when grains are finely ground and properly heat treated (see Ch. 11). Mild heat treatments allow the starch molecule to swell and enhance its digestion. Improper heat treatment leaves raw starches that may ferment in the intestinal tract, thereby creating flatulence and by-products of potential detriment to the efficiency of the digestive process.

The liver of most omnivorous animals, including the domestic dog, has two enzymes that metabolize glucose. One of these enzymes, hexokinase, functions when low levels of glucose are delivered to the liver, while the second enzyme, glucokinase, operates whenever the liver receives a high load of glucose. The cat is unusual because it has active hexokinase but does not have active glucokinase (6). It has been postulated that species having both enzymes have a greater capacity to handle high-glucose diets than do those that possess only hexokinase. This finding is supportive of the cat's carnivorous feeding regimen of high protein and fat and low carbohydrates.

Dietary fiber is classified with carbohydrates. Although fiber is not a required nutrient per se, the inclusion of small amounts in the diets of companion animals is necessary for the normal functioning of the gastrointestinal tract. Depending on its composition, dietary fiber functions to increase the bulk of the diet, maintain normal intestinal transit time and gastrointestinal tract motility, and maintain the structural integrity of the gastrointestinal mucosa. Fiber sources common to commercial pet foods are wheat middlings, tomato, citrus, and grape pomace, beet pulp, and the hulls of soybeans and peanuts. Corn, rice, wheat, and barley all contribute digestible carbohydrates and also supply small amounts of fiber. In addition, protein sources in cereal-based pet foods add varying amounts of dietary fiber to the ration.

While dogs and cats do not directly digest dietary fiber, certain microbes found in the

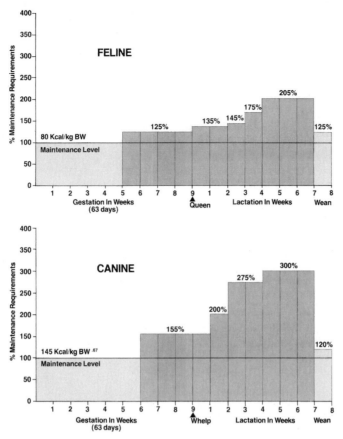

FIGURE 22-3 Feline and canine metabolizable energy requirements for production.

large intestine are able to break down (ferment) fiber to varying degrees. This bacterial fermentation produces short-chain fatty acids (SCFAs) and other end products. Fibers differ greatly in the degree to which they are fermented. For example, in dogs and cats, pectin and other soluble fibers are highly fermentable, beet pulp is moderately fermentable, and cellulose is nonfermentable. Recent research has shown that the SCFAs that are produced from fiber fermentation in dogs and cats are an important energy source for the cells that line the intestine. The production of moderate amounts of SCFAs functions to support a healthy intestinal tract. While highly fermentable fibers also produce short-chain fatty acids, diets containing this type of fiber may cause diarrhea. It appears that the best fiber sources for companion animals are those that are moderately fermentable and provide adequate levels of SCFAs for the intestine (7,8).

Research has also shown that certain types of fiber may be helpful in the treatment of gastrointestinal disease in dogs and cats (9). Fructooligosaccharides (FOSs) are naturally occurring carbohydrates that are classified with fiber because they are not digested by the enzymes of the gastrointestinal tract. Feeding pets diets containing FOS has been shown to cause desired changes in the composition of intestinal bacteria. Specifically, there is an increase in the numbers of beneficial bacteria and a decrease in the numbers of bacteria that contribute to disease. Continuing research studies the potential uses of these fibers in the treatment of intestinal diseases in dogs and cats.

While some fiber is necessary and beneficial for dogs and cats, too much fiber can have detrimental side effects. Feeding increasing levels of dietary fiber has been shown to decrease nutrient utilization in the adult dog. This effect varies with the type and physical composition of the fiber. A study examined the nutrient availability of diets containing 0% to 12.5% beet pulp. As the level of beet pulp increased, protein and energy availability of the diet decreased. In addition, the inclusion of graded levels of fiber resulted in a linear decrease in diet retention time, thereby increasing the rate of passage through the digestive tract (10). Another study showed that adult cats fed diets containing highly fermentable fibers had poor fecal consistency and depressed diet digestibility (11). While dogs that were fed these fibers had poor stool quality, they did not demonstrate a decrease in digestibility scores. These differences indicate that the cat is unable to tolerate fibers that are highly fermentable and/or soluble. This observation also indicates the need to explore possible detrimental effects of the common practice of including fermentable fibers in canned cat foods as viscosity enhancers.

Feeding high levels of dietary fiber has been shown to increase losses of amino acids and reduce dietary amino acid availability, both of which can influence the amino acid requirement. Sloughing of intestinal mucosal cells and elevated mucus production are primarily responsible for the increase in loss of amino acids. It is thought that fiber decreases amino acid availability by forming gels around the amino acids or by absorbing digestive enzymes, thereby diminishing their activities. With the development and sale of low-calorie, high-fiber commercial pet foods that contain primarily nonfermentable fiber, the availability of dietary nutrients may be a concern.

FATS

In commercial pet foods, fat serves as a concentrated source of energy, a carrier for fat-soluble vitamins, a source of essential fatty acids, and an enhancer of diet palatability. Fat is a very concentrated source of energy with an energy efficiency ration of approximately 2.25:1 for either fat to protein or fat to carbohydrates. Fats are routinely added to commercial pet foods to increase the caloric density. Some quality fat sources have a digestibility value of 95% or more. Fats are also more digestible than carbohydrates; thus, the available caloric contribution of fat is very high.

A low-fat diet or the use of poor-quality, unstable fat may limit the absorption of fat-soluble vitamins. Therefore, inferior quality or quantity may result in clinical cases of fat-soluble vitamin deficiencies in both dogs and cats.

The optimal quantity of dietary fat is not exclusively in the total percentage of fat in the diet, but rather in the relationship of metabolizable fat to metabolizable nutrients. For example, when a diet contains a limited quantity of protein or inferior-quality protein, the percentage of dietary fat (energy) desired may be as low as 5% to 10%. However, as protein quality and/or quantity are increased, a concomi-

tant increase in fat will be warranted to maintain proper nutrient balance.

As with most animal species, the dog and cat have an essential fatty acid requirement for linoleic acid. In addition, the fatty acid arachidonic acid is essential for the cat. This occurs because the cat lacks adequate amounts of the liver enzymes Δ-6-desaturase and Δ-5-desaturase, which convert linoleic to linolenic acid and arachidonic acid (12,13).

Sources of fat commonly employed in commercial pet foods are tallow, lard, poultry fat, and numerous vegetable oils; however, not all of these contain the required essential fatty acids. Fish oil is the richest source of arachidonic acid. Poultry and lard contain appreciable quantities. However, tallow contains a relatively lower level, and vegetable oils lack arachidonic acid entirely. This unique requirement for arachnidonic. acid found in the cat has also been found in other strict carnivore species. A deficiency in arachidonic acid results in poor growth, inferior fur quality, skin lesions by the mouth and hocks, slow wound healing, increased susceptibility to infection, and poor reproductive performance.

In recent years, therapeutic uses of fatty acid supplements have been studied in companion animals. Specifically, two classes of these nutrients, the omega-6 and the omega-3 fatty acids, have received increased attention. Polyunsaturated fatty acids are classified into several series based on the position of the first double bond in the carbon chain. Omega-6 fatty acids have the first double bond located at the sixth carbon, while omega-3 fatty acids have the first double bond located at the third carbon atom. The incorporation of optimal proportions of omega-6 and omega-3 fatty acids into the diet has preventive and therapeutic benefits to pets with certain inflammatory conditions. Feeding a diet that contains increased amounts of omega-3 fatty acids and an adjusted ratio of omega-6 to omega-3 fatty acids causes changes in tissue concentrations of these fatty acids and the inflammatory agents that are derived from them (14). Recent studies have demonstrated that diets with adjusted omega-3 and omega-6 fatty acids levels aid in the treatment of allergic skin disorders in dogs.

PROTEIN

Pet food sales have increased steadily throughout the years, and efforts have been intensified to find economical protein sources of high quality with high amino acid availability. Ideally, the optimum use of an intact protein feedstuff is to supply required levels of the 10 essential amino acids, thus allowing efficient formulation of pet foods. However, with regard to practical diets, relatively little is known about the quantitative amino acid requirements for the canine and feline and subsequent factors affecting those requirements.

Early investigations suggested that the qualitative amino acid needs of most animal species were probably very similar. The establishment of purified crystalline amino acid diets for dogs and cats in the 1980s allowed the quantitative assessment of canine and feline amino acid requirements. The resulting minimum amino acid requirements were incorporated into the 1985 and 1986 NRC guidelines. Because these requirements were established as minimum levels in purified diets, the AAFCO Nutrient Profiles added safety margins to account for the decreased availability of nutrients in practical ingredients, variability of nutrient availability among ingredients, and loss of nutrients due to processing.

Vegetable proteins such as soybean meal and corn gluten meal and animal proteins such as poultry, meat, and respective by-products are common ingredients in commercial pet foods. Although cereals provide a major source of energy in cereal-based products, they also supply a significant portion of the protein, which may often be deficient in an essential amino acid. Fresh meats, meat and poultry meals, and various meat by-products are often added to complement the amino acid profile of cereal-based products. Animal sources of protein vary considerably in protein quality. Some sources provide amino acids of poor availability while others are highly available; overall quality depends on the source and processing methods of the renderer (see Ch. 8).

The development of an optimal feeding program for dogs or cats depends highly on the selection of the proper source of dietary protein. To determine the nutritional efficacy of a specific protein source, consideration must be given to digestibility and assimilation of the essential amino acids present in the protein. A study that established the lysine requirement for the growing puppy using a purified crystalline amino acid diet and an intact protein diet supplemented with crystalline lysine illustrates the importance of availability. The results suggest a lysine requirement for maximal growth and feed efficiency of approximately 0.7% for a purified L-amino acid diet and 0.8%

for a diet based on intact protein-containing ingredients. The best explanation for this difference is that the crystalline lysine is virtually 100% available, whereas lysine present in intact protein is not. Therefore, it is readily apparent that the lower lysine bioavailability of dietary components contained in the intact protein diet had the expected effect on the lysine requirement (15).

It is generally accepted that the requirement for most essential amino acids, expressed as a percent of the diet, is affected by the age, sex, and breed of the animal. A study clearly showed that the tryptophan requirement of growing dogs, expressed as a percent of the diet, decreases substantially as age and weight of the dog increases (16). In contrast, the protein requirement of the cat remains relatively high.

The impact of sex has variable results. In some studies the amino acid requirement of young puppies was not affected by sex. In contrast, a study with immature beagle dogs suggested that the lysine requirement for maximal growth and N balance was greater for the immature male dog than for the immature female (17).

With respect to breeds, it has been observed that labradors have a higher requirement for sulfur-containing (S) amino acids than do beagles (18). In addition, a study revealed that the S amino acid requirement of pointer puppies is also different from those of either beagles or labradors (19). The diversity among requirements is not surprising when one considers the genetic diversity of the canine species.

The cat, a strict carnivore, is unique in its protein and amino acid nutritional requirements. The cat and other carnivores have substantially higher protein requirements than do omnivores such as the dog. The metabolic reason for the high protein requirement of the cat is the high activity of the amino acid catabolic enzymes in the liver. These enzymes are nonadaptive; therefore, the obligatory N loss is high even when cats are fed low-protein diets or fasted (20). This is a major concern with respect to pet owners who feed dog food to their cats. Although some dog food products provide high levels of dietary protein, many would be considered deficient for the cat.

Although not of practical importance, it has been shown that cats are very sensitive to a deficiency of the amino acid arginine. The feeding of a single meal that is devoid in arginine will result in hyperammonemia in less than an hour. It has been suggested that the cat's sensitivity to an arginine deficiency is related to the metabolism of urea cycle intermediates. Other carnivores reveal similar but less severe responses to an arginine-free diet (21).

Of more practical concern is the cat's high requirement for S amino acids relative to other mammals. It has been suggested that the cat's high protein requirement may be a result of its high S amino acid requirement. It is postulated that the presence of high dietary S amino acid levels may be required to support the cat's thick hair coat, which is high in the S amino acid cysteine (20).

The cat's unique requirement for taurine, a β-amino sulfonic acid, has received considerable scientific and public attention. Taurine is a by-product of S amino acid metabolism produced in the liver and other tissues, such as the brain. It functions as a bile acid conjugate and is present in a few peptides, but it is not a component of body proteins. The cat appears to be very inefficient at synthesizing taurine, as well as having a higher physiological requirement for it than other mammals have. The concentration of taurine in the cat's eye is considerable, and it is required to prevent central retinal degeneration (CRD) (22). Low concentration of plasma taurine is also associated with a heart disease called congestive cardiomyopathy and with poor reproductive performance in cats (23, 24). Taurine is present only in animal sources of protein, with shellfish and mollusks containing the highest levels. Therefore, caution is advised in the feeding to cats of commercial dog foods that employ protein sources of vegetable rather than animal origin.

The amount of taurine that is necessary in the diet of the cat to maintain taurine homeostasis depends on the type of diet that is fed. Currently, AAFCO minimum requirements are 1000 mg taurine/kg DM of diet for expanded, dry diets and 2000 mg taurine/kg DM diet for canned diets. The cat requires a higher level of taurine when a canned diet is fed because feeding these products results in greater losses of taurine from the body. These losses appear to be due in part to the processing methods that are used and in part to the ingredients that are included in canned cat foods (25).

VITAMINS IN PET FOODS

In the case of fat-soluble vitamins, it is essential that a quality-stable fat source be employed in the diet to assure vitamin absorption and utilization. Therefore, many pet food manufacturers add an antioxidant to assure the stabilization of dietary fat. Water-soluble vitamins are carefully selected and added in commercial pet foods in excess of minimum requirements to compensate for losses associated with heat processing and extended shelf life.

VITAMIN A   The cat, like all mammals, has a dietary requirement for vitamin A. This fat-soluble vitamin occurs in several different chemical forms in nature. β-carotene, a carotenoid pigment, is a vitamin A precursor found in plant products. Most mammals have the ability to convert β-carotene to active vitamin A through the action of two enzymes located in the intestinal mucosa. However, unlike the dog, the cat is unable to convert dietary β-carotene to active vitamin A due to a deficiency in the intestinal enzyme β-carotene-15-15'-dioxygenase. As a result, the cat has a dietary requirement for preformed vitamin A found only in animal tissues (26). Products containing high levels of vitamin A include beef, chicken, and pork liver, as well as whole milk. The richest sources of vitamin A are fish oils. Cod liver oil is especially high in vitamin A.

In contrast to other mammals, the cat is more susceptible to the overconsumption of vitamin A. Because the cat consumes preformed vitamin A that is not regulated by the intestinal mucosa, as is β-carotene, a toxic level may be readily absorbed into the body. The susceptibility of vitamin A toxicity may be exacerbated by the practice of feeding food items to domestic cats that contain excessively high levels of vitamin A, such as liver, kidney, and various fish oils.

NIACIN   With respect to water-soluble vitamins, the cat appears to have a unique dietary requirement for niacin. In most mammals the niacin requirement can be met with dietary tryptophan. In contrast, the cat cannot meet its physiological niacin requirement by consuming tryptophan, but must have preformed niacin in its diet.

MINERALS IN PET FOODS

There is a paucity of data on quantitative and qualitative mineral requirements in dog and cat nutrition. To assure dietary adequacy, commercial pet foods provide essential minerals, thereby eliminating the need for additives and supplements. Practical applications have repeatedly shown that supplementation of commercial pet foods may create nutritional imbalances that can influence the expression of health or disease in the animal.

CALCIUM AND PHOSPHORUS Calcium (Ca) constitutes 1.5% to 2% of the total body weight and is the most abundant cation in the body. More than 99% of the body's calcium is present in bones and teeth. Within bone, the ratio of Ca:P is 2:1. The Ca:P ratio of a diet also affects the absorption of Ca. If a diet contains an excess of either mineral, the absorption of the other mineral will be compromised.

Many dog owners believe that growing puppies need additional Ca in their diet to prevent the development of chronic skeletal problems. Numerous controlled studies have demonstrated that supplementation of a previously adequate diet with Ca will not alleviate or prevent skeletal problems. In fact, these studies have demonstrated that the addition of excess Ca to a growing dog's diet can actually be harmful. Excessive dietary Ca causes skeletal changes and may ultimately result in the development of skeletal disorders such as osteochondritis dissecans (OCD), Wobbler's syndrome, hypertrophic osteodystrophy (HOD), and hip dysplasia.

Due to the carnivorous feeding behavior of the cat and palatability response of the dog, many consumers believe that the addition of meat is of nutritional benefit. However, sources of fresh meat, poultry, and fish can supply a Ca:P of 1:15 to 20. The addition of such ingredients can imbalance the desired Ca:P ratio of 1:1 for cats and 1.2 to 1.4:1 for dogs.

FELINE LOWER URINARY TRACT DISORDER

Feline lower urinary tract disease (FLUTD) is a generally accepted term that describes lower urinary tract disease in the domestic cat occurring as a result of urethritis or cystitis. This syndrome was previously called feline urologic syndrome (FUS), but that term should be avoided because the term FLUTD more adequately describes the disorder. In the past, the vast majority of cases of FLUTD were attributed to the presence of uroliths composed primarily of struvite (magnesium–ammonium–phosphate). Recent evidence indicates that, while struvite is still an important cause of FLUTD, calcium oxalate uroliths appear to be increasing in prevalence (27, 28).

Because struvite crystals have been found a prevalent cause of FLUTD in cats, most research has focused on preventing these crystals from forming in the urine and on the development of effective dietary management of cats with struvite FLUTD. One of the factors necessary for the formation of struvite in urine is the presence of sufficient concentrations of Mg, P, and ammonium. Feline urine always contains high amounts of ammonium due to the cat's high protein requirement and intake. Urinary phosphate in the normal healthy cat is also high enough for struvite formation, regardless of dietary P intake. The concentration of urinary Mg, on the other hand, is normally quite low and can be directly affected by the level of dietary intake (29).

MAGNESIUM Several studies demonstrate the relationship between increased dietary Mg and increased rate of calculi formation and obstruction in cats. However, the levels of dietary Mg fed to cats in these studies were all substantially higher than those typically found in commercial cat foods. In addition, the type of uroliths that were induced experimentally often differed in composition from the uroliths observed in naturally occurring FLUTD. Although commercial cat foods do contain more Mg than is required by the cat, they do not approach the levels that were used in experimental studies to induce urolithiasis (0.4% to 1.0%). It is now recognized that Mg intake in the cat is not singularly responsible for the development of this disease, and it is less significant than other contributing factors.

ACIDIFYING PROPERTIES OF A DIET Struvite is more soluble in acid than in alkaline medium. As a result, the pH of a cat's urine can have the most profound effect on the development of FLUTD. It has been demonstrated that a linear relationship exists between urinary pH and the formation of struvite crystals. Struvite calculi will form in feline urine with a pH of 7.0 or greater, and its solubility greatly increases at a pH of 6.6 or less (30).

Compared to an omnivorous or herbivorous diet, a true carnivorous diet has the effect of increasing net acid excretion and decreasing urinary pH. This urinary acidifying effect is due primarily to the high level of S-containing amino acids contained in meats. The oxidation of these amino acids results in the excretion of sulfate in the urine and a concomitant decrease in urinary pH (31). In addition, a high-meat diet is lower in K salts than is a diet containing high amounts of cereal grains, the metabolism of which has been shown to produce alkaline urine (32). The inclusion of high amounts of cereal grains and low amounts of meat products in some brands of commercial cat food may, therefore, be a contributing factor to the development of FLUTD in cats.

Diets that are high in caloric density and highly digestible will be consumed in smaller quantities, thus lowering both dry-matter and Mg intake. Therefore, the percentage of Mg in the diet is not as important as is the total amount of Mg that the cat consumes. The domestic cat requires only 0.016% available magnesium for growth and maintenance (3). The AAFCO Nutrient Profile requires cat foods to contain a minimum of 0.04% magnesium (4). Most commercial cat foods contain slightly higher than this amount, but less than 0.1%. While some researchers feel that Mg concentration in the diet should be 0.1% or less on a dry-matter basis, others maintain that FLUTD risk is only increased when Mg levels reach 0.25% or greater (33).

Diet caloric density and digestibility are also important because of their impact on water turnover (34). The consumption of a cat food that is energy dense and highly digestible results in lower total dry-matter intake, which is in turn accompanied by decreased fecal volume and fecal water and increased urine volume (29). These effects may be beneficial in preventing FLUTD in cats because the urine contains a lower concentration of the mineral components that cause FLUTD. In addition, an increase in urine volume stimulates an increased frequency of urination, thus decreasing the time available for urolith formation.

Factors of caloric density, digestibility, Mg level, and, perhaps most important, the acidifying properties of the diet should be considered when selecting a commercial cat food for the prevention of struvite FLUTD. While the maintenance of an acid urine with a pH of 6.0 to 6.6 is an effective way of preventing the formation of struvite crystals and the clinical signs of FLUTD in many cases, this is not conclusive proof that struvite formation is the underlying cause of the disorder. Some investigators have suggested that changing the pH of the urine merely manages the disease and does not address the underlying problem. Other factors that have been identified as possible causes include the presence of other types of uroliths, the synthesis of the protein component of the urethral plugs within the urinary tract, and the presence of viral or bacterial agents (35). Further research is necessary to determine the importance of other factors.

COMMERCIAL DIETS FOR DOGS AND CATS

In response to consumer preference, the pet food industry produces primarily three types of foods. They are categorized according to moisture content as dry (6% to 12% moisture), semimoist (23% to 40% moisture), and canned (60% to 78% moisture).

DRY PET FOODS

Dry dog foods are the most common type of pet food that pet owners buy in the United States. In 1994, $5.1 billion were spent on dog food. In recent years, there has been a trend toward increased sale of dry dog food and decreased sale of canned dog food (1). Cat food totaled $3.6 billion in retail sales, with trends toward increases of both canned and dry cat foods.

Most dry pet foods are produced by processing through an extruder (see Ch. 11). Ingredients common to dry foods are protein meals of plant and animal origin, such as corn gluten meal, soybean meal, poultry and meat meal, respective by-products and fresh animal protein sources. Raw cereals and cereal by-products of corn, wheat, and rice are used as carbohydrate sources. Fats can be of animal or vegetable origin, and vitamins and minerals are added in the milling process to provide a homogeneous, complete, and balanced mixture for processing.

Extrusion is a high-temperature, short-time process. It is used for cooking, forming, and expanding cereals and for texturizing proteins. The combination of high temperature, pressure, and shear optimizes expansion and dextrinization of starches. In addition, the exposure to high heat acts as a sterilization technique against pathogenic organisms. The extruded diet is then dried, cooled, and packaged. A process option is the application of fats and dry or liquid digests postextrusion. Such options may be employed to enhance palatability.

Another method of producing a dry diet is pelleting. Binding agents, such as molasses, are often added to the ingredient mixture to regulate the hardness of the pellets. The carbohydrate and protein ingredients are often precooked to gelatinize the starch and kill pathogenic organisms.

A less common process is baking, as in the production of kibbled dog foods or treat products. Prior to the use of extrusion in the pet food industry, commercial products were provided in meal form. To alleviate problems associated with raw starch, corn flakes were commonly employed in dog foods. Vegetable and animal protein sources were vat cooked, cooled, and mixed with corn flakes, vitamins, and minerals.

Dry pet foods are diverse in nutrient composition, ingredient selection, process method, and physical appearance. Although they all contain approximately 10% moisture, the protein level can range from 12% to over 30% and fat can range from 6% to over 25%. When evaluating differences in dry foods, one must consider the caloric density, bulk density, and ingredient composition. Considering the aforementioned, it is easy to understand that a cup of dry food can provide 200 to 600 kcal of ME. The major advantages associated with the feeding of dry pet food are that it is most economical to feed, it is convenient to feed, and its abrasive texture aids in dental hygiene.

SEMIMOIST PET FOODS

Semimoist foods have fallen in popularity in recent years. These foods represent a very diverse group of commercial products. The moisture content can vary dramatically, but primarily these products contain 35% to 40% moisture. Most semimoist foods are extruded like dry foods. Depending on the selection of ingredients, the food may be cooked prior to extrusion. The unusual factor associated with the production of semimoist products is that the added water requires the addition of other ingredients to prevent product spoilage. Sugar, corn syrup, and salts are added to bind the water fraction and render it unavailable for the growth of bacteria. Until 1992, propylene glycol was also used as a humectant in semimoist pet foods. However, the FDA determined that this compound was a potential risk to cats and has prohibited its use in cat foods. The major advantages in feeding a semimoist food are its convenience of feeding, intermediate cost, and good palatability. While fewer pet owners are feeding semimoist foods, there has been an increase in the variety of semimoist treats and snacks that are available for dogs and cats.

CANNED PET FOODS

Canned foods are extremely popular, especially in the cat food market. Due to the popularity of domestic cats, the canned cat food market has grown dramatically in recent years.

The canning process is a high-temperature, long-time process. Ingredients are mixed, cooked, and filled into template cans that are lidded, seared, and retorted at temperatures of 230° to 275°F for 15 to 25 minutes, depending on the type of retort and container size. Associated high temperatures are necessary to sterilize the diet. Canned diets contain up to 78% moisture. The presence of high moisture provides a highly palatable product that is attractive to consumers who have finicky pets; however, canned foods are relatively more expensive due to the cost associated with processing.

There are presently two types of canned foods: canned ration and canned meat. Canned rations may contain a variety of both dry and wet ingredients. The carbohydrate fraction is supplied by cereal grains, and the protein is derived from fresh meats, poultry and meat meals, and soybean meal. Fats are primarily of animal origin and, to provide a complete and balanced diet; vitamins and minerals are added.

Canned meat-type foods are composed primarily of fresh meat and poultry and respective by-products. Carbohydrates are present in very small quantities, approximately 5% to 10%. Due to the high protein content of these foods, protein is often used as an energy source. Soy flour is frequently added in the form of expanded chunks that resemble meat. This processing technique usually enhances the texture of the product. Vitamins and minerals are added.

PROCESSING EFFECTS ON NUTRITIONAL QUALITY

The manufacturing of dry, semimoist, and canned pet foods all involve heat processing. There are several benefits associated with the heat-treatment process that result in improved food quality. These have been discussed in some detail in Ch. 11.

Heat processing may also be detrimental to the nutritional quality of foods. Amino acid availability may be compromised in the preparation of pet foods. The loss of vitamins may be 10% to 100%, depending on the vitamins of concern, type of diet, and heat process employed. Fat-soluble vitamins are very sensitive to oxidation during processing, and specific water-soluble vitamins are extremely thermolabile. As a result, vitamins are added at higher levels in pet foods to compensate for such losses. Therefore, the quality of a diet depends on the processing method and subsequent controls enforced by the pet food manufacturer.

PUPPY AND KITTEN FOODS

The most rapid period of growth in dogs and cats occurs during the first 6 months of life. Large breeds of dogs attain their mature size by about 10 to 16 months of age, while smaller breeds and cats reach adult size by 6 to 12 months. Thus, an enormous amount of growth and development takes place in a relatively short period. Supplying a balanced diet during growth is crucial for adequate development and the attainment of normal adult size.

Pet food marketing has done an excellent job of convincing the consumer that the puppy and kitten have special nutritional requirements that can be met only with a diet designed especially for growth. However, this is not actually the case. There is no evidence that growing dogs and cats have requirements for nutrients that are not needed by adults. Rapidly growing puppies and kittens do require more energy and several specific nutrients on a unit-per-body-weight basis. The energy needs of young puppies are approximately twice those of adult dogs of the same size. After 6 months of age, these needs begin to decline as the animal's growth rate decreases. Similarly, growing cats have energy needs that are significantly higher than are the maintenance needs of adult cats.

The protein requirement of growing puppies and kittens is higher than the requirement of adult animals. This occurs because, in addition to normal maintenance needs, young animals also need protein to build the new tissue that is associated with growth. However, because young animals consume higher amounts of energy and thus higher quantities of food than adult animals, the total amount of protein that they consume is naturally higher. Pet foods that are fed to growing puppies and kittens should contain slightly higher protein levels than foods that are developed for maintenance only. More importantly, the protein that is included in the diet should be of high quality and the diet should be highly digestible. Such a diet meets the nutritional demands for growth without a substantial increase in food volume. Many high-quality pet foods meet these nutrient needs and provide for optimal growth of puppies or kittens.

HARD WORK AND STRESS NUTRITION

The use of dogs as a means of transportation, although a traditional role for the Alaskan sled dog, has recently received considerable attention due to the popularity of sled dog racing. The physiological stress encountered during endurance training for the 1000-mile Iditarod race across the rugged terrain and inclement weather of Alaska represents undoubtedly the highest nutritional demand for the canine in any type of organized contest (Fig. 22-4). Re-

FIGURE 22-4 Racing sled dogs during the grueling Iditarod race in Alaska. (Courtesy of Trot-A-Long Kennels, Two Rivers, AK.)

cent research has shown that sled dogs performing in the Alaskan bush under race conditions expend up to 10,000 to 11,000 kcal/day (36). Energy density and diet digestibility are the two most important nutritional factors to consider in diets for endurance performance in dogs. The most efficient way to supply the additional energy needed is through increased dietary fat. Studies have shown that feeding highly digestible, high-fat diets can supply the extra energy needed by working dogs and contributes positively to endurance performance (37, 38).

The type of work that occurs during endurance competitions differs from that performed during short races or sprinting events, such as greyhound racing or lure coursing. Greyhound races are extremely short, sprinting events. These dogs typically reach speeds of 36 to 38 mph, but maintain this speed for only 30 to 40 seconds. The primary source of energy for this type of exercise is the anaerobic metabolism of carbohydrate (39). This differs from endurance exercise in which both fat and carbohydrate supply energy to the working muscles. Greyhounds are well adapted to this type of work, and it is important that racing dogs receive diets that are energy dense and highly digestible. The diet that is fed to racing greyhounds should also contain a reasonable amount of highly digestible carbohydrate to ensure that body glycogen stores are maintained at an optimum level and can be adequately restored following intensive periods of training or racing.

GERIATRIC NUTRITION

The difficulty in providing a geriatric diet is that one cannot employ a general definition for the geriatric animal. Also, there is little scientific information available concerning the nutrition of geriatric dogs and cats. However, one report has shown that geriatric dogs are capable of digesting and metabolizing nutrients as efficiently as younger dogs. Contrary to popular belief, older animals also do not have different dietary requirements than young dogs (40).

Some pet foods that are formulated for geriatric pets contain a reduced quantity of protein. This practice is based on the belief that dietary protein may contribute to the onset or progression of kidney insufficiency in older animals. However, recent research studies show that feeding older dogs increased levels of dietary protein does not increase their risk for development of renal disease (41). These findings have also been reported in young adult dogs (42).

Physiological changes associated with aging reveal that the protein requirement of the geriatric animal is equal to, if not greater than, that of younger animals. In fact, factors such as a reduction in muscle mass, efficiency of N metabolism, increased requirement for specific amino acids, and an increased susceptibility to disease due to a reduction in immune function may support an increase in protein quantity and quality. These factors, coupled with recent research results, indicate that dietary protein should not be restricted below amounts provided for adult maintenance in pet foods formulated for healthy geriatric pets.

Animals often gain weight as they age as a result of decreased activity and reduced metabolic rate. Caloric restriction may be necessary to avoid weight gain and obesity. Some geriatric diets reduce calories by increasing dietary fiber. However, the nutrients contained in such diets may be poorly available due to the high level of insoluble (nonfermentable) fiber. A more efficient and healthy way to reduce calories is through decreased fat and increased digestible carbohydrate. This results in a desired reduction in caloric density without affecting the digestibility and nutrient availability of the diet.

In general, it appears that the nutritional requirements of the geriatric animal must be considered on an individual animal basis. A diet that contains normal levels of high-quality protein, reduced fat, and nutrients that are highly available should be fed to healthy older pets. The diet should also be palatable and in a form that is relatively easy to chew and consume.

WEIGHT-REDUCING PET FOODS

Obesity is currently the most common nutritional disorder in companion animals in the United States. Surveys have reported incidence rates of between 24% and 34% in adult dogs and up to 40% in cats (43, 44). Obesity may be defined as the condition in which a pet's body weight is 20% or more above normal. Dogs and cats that are overweight have an increased risk of chronic health problems, such as diabetes, pulmonary and cardiovascular disease, and skeletal problems due to the stress associated with excess weight. Some of the causes of increased weight in dogs and cats include insufficient exercise and a sedentary life-style, overfeeding, hormonal and behavioral changes associated with neutering, and increasing age.

Goals when feeding overweight pets are to first reduce body fat stores and attain normal body weight and to then maintain this weight for the remainder of the pet's life. Restricting calories for weight loss in pets can be implemented either through reducing the amount of the pet's regular diet that is fed or providing a diet that is specially formulated to be less energy dense while still providing the essential nutrients required for body maintenance.

A diet that provides 60% to 70% of the calories necessary to maintain current body weight usually results in adequate weight loss. While restriction of the dog or cat's normal diet can be used, it is imperative that a high enough quantity is still provided to meet the pet's total nutrient requirements. If the volume of a maintenance diet is reduced too drastically in an effort to limit calories, nutrient deficiencies may develop. Commercially prepared, low-energy-density pet foods are formulated to contain adequate levels of nutrients while supplying less calories. Therefore, in cases of moderate to severe obesity, a change of diet to a commercially prepared, low-energy-density diet is warranted.

Some weight-reducing commercial pet foods dilute caloric density by substituting insoluble (nonfermentable) fiber for fat. However, dietary fiber above the amount that is required for normal gastrointestinal function is unnecessary for proper weight control. In addition, the presence of excess fiber in the intestinal tract has been shown to reduce the availability of other essential dietary nutrients. Excess fiber intake also results in increased defecation frequency and fecal volume and changes in fecal consistency.

A second way to efficiently reduce caloric density while still maintaining the diet's digestibility, nutrient availability, and ability to produce normal stools is to decrease fat and increase digestible carbohydrate. High-quality, digestible carbohydrate provides an excellent source of energy in low-fat pet foods and has less than half of the caloric density of fat. A weight-reduction diet must also contain a level of protein that supplies essential amino acids to minimize the loss of lean body tissue as the pet loses body fat. The caloric distribution of metabolizable calories in a low-fat, high-carbohydrate weight-reduction diet should be 20% from protein, 20% to 25% from fat, and 55% to 60% from carbohydrate.

GENERIC PET FOODS

Generic pet foods are products that do not carry a brand name. These products were introduced to the consumer market due to their economic appeal. The most important consideration of the manufacturers of generic foods is producing a low-cost product. For this reason, cheap, poor-quality ingredients may be used, and little, if any, feeding tests are conducted. Some have not been formulated to be nutritionally complete and so will not even carry a label claim (45). Feeding studies with dogs have shown that generic products have significantly lower digestibilities and nutrient availabilities than popular and premium brands of food (45). An additional study demonstrated that dogs fed a generic food developed skin problems associated with a zinc deficiency (46). The results of these studies support the misrepresentation of some pet food label guarantees and actual differences in the nutritional quality of pet foods.

PREMIUM PET FOODS

The premium pet food market has enjoyed rapid expansion in recent years (1). The term premium refers to products that are developed to provide optimal nutrition during different stages of life for dogs and cats. These pet foods are targeted toward companion animal owners who are very involved in their animal's health and nutrition. In general, quality ingredients that are highly digestible and have good to excellent nutrient availability are used, and these foods are marketed primarily through pet supply stores, feed stores, or veterinarians (47).

Most manufacturers of premium foods validate their label claims through AAFCO feeding studies, as opposed to the calculation method (see p. 447). This guarantees to the pet owner that the food has been adequately tested through actual feeding studies with animals. Premium foods are usually more costly on a per weight basis because of the higher-quality ingredients used and the level of testing conducted on the products. However, because these products are usually very digestible and nutrient dense, smaller amounts need to be fed and the cost per serving is often comparable to many popular brands of pet food.

PET FOOD LABELS

Commercial pet foods are fed as an animal's only source of nutrition and are expected to maintain optimal health and well-being. It is difficult to assess the quality of a pet food when commercial diets are presented with the influence of marketing cleverness. A pet food label contains a tremendous amount of useful information and, when correctly interpreted, can distinguish a quality commercial product from those of inferior quality but reduced cost.

The information required on pet food labels (Fig. 22-5) is prepared and approved by the joint federal and state Association of American Feed Control Officials (AAFCO). Regulation PF2: Label Format requires the following information (4): the product name; the net weight; an ingredient list; a guaranteed analysis; the name and address of the manufacturer, packer, or distributor; the designation of "Dog Food" or "Cat Food"; and a statement describing the purpose of the product and the method used to determine its adequacy.

NUTRITIONAL STATEMENTS

Although pet food labels provide ingredient lists and guaranteed analyses, they lack specific information on the content and availability of many nutrients. It is feasible for two labels to present identical ingredient lists and guaranteed analyses but to provide products of different nutritional value (45). Differences in processing methods and selection of quality raw materials can create one product of superior nutritional quality and another that is completely unsatisfactory. Although manufacturers

FIGURE 22-5 Pet food label requirements shown on a simulated label.

of premium pet foods provide product literature as consumer education tools, which enhance one's understanding of nutrition, this is not done by all pet food manufacturers. The reputation of the pet food manufacturer and information concerning animal nutritional testing of a product will assist in food selection.

The final statement that consumers may read on the pet food label is the claim of nutritional adequacy. With the exception of treats and snacks, the label of all pet foods that are in interstate commerce must contain a statement and validation of nutritional adequacy. When the "complete and balanced nutrition" claim is used, manufacturers must indicate the method used to substantiate this claim.

NUTRITIONAL ADEQUACY OF PET FOODS

Current AAFCO regulations allow pet food manufacturers to validate their nutritional adequacy claims for a complete and balanced diet or one specified for a particular life stage in one of two ways. The manufacturer must either perform AAFCO sanctioned feeding trials on the food or must formulate the diet to meet the AAFCO Nutrient Profiles for Dog or Cat Foods. The first option, testing the food through a series of feeding trials, is the most thorough and reliable evaluation method. Common terminology for labels of pet foods that have passed these tests is "Animal feeding tests using AAFCO procedures show that (Brand) provides complete and balanced nutrition for (life stages). The inclusion of the terms "feeding tests," "AAFCO feeding test protocols," or "AAFCO feeding studies" in a label claim validates that the product has undergone a complete series of feeding tests with dogs or cats.

The second method of substantiation is commonly referred to as the *calculation method* because it allows manufacturers to substantiate the "complete and balanced" claim by merely calculating the nutrient content of the formulation for the diet using standard tables of ingredients, without having to conduct laboratory analyses or actually feeding the food to any animals. Consumers should be aware that the use of feeding trials is not required if manufacturers use the calculation method of substantiation. Although some manufacturers using this method of substantiation may still conduct some of their own feeding trials, there is no way of knowing this from the label unless they have used the AAFCO feeding trial protocols.

PHYSICAL EVALUATION OF PET FOODS

Physical evaluation of a pet food can provide considerable information concerning product quality. First, evaluate the package or container. Dry and semimoist foods should be provided in tightly sealed multilayer packages. The presence of an inner liner aids in the prevention of moisture migration, fat wicking, and infestation. The liner also keeps product aroma in to maintain palatability. Canned products that are dented or swollen signify bacteria fermentation and should not be fed.

Product appearance should meet your quality standards. Consistency of product color, size, and shape, as well as a pleasant aroma, should all be considered. The presence of foreign material, ingredient-related foreign material (large ingredient fragments, hair, feathers), and excessive fines are indicators of a manufacturer's inferior quality assurance program.

PALATABILITY EVALUATION

Pet foods are routinely tested for acceptability. After all, if the pet refuses to eat it, even the most nutritious pet food is of no benefit. Palatability evaluations offer an animal two products, one of which is a control of known acceptability (Fig. 22-6). Many pet foods market the palatability of their product; however, highly palatable foods are not always the most nutritious. In fact, the addition of ingredients such as garlic and cheese powder or phosphoric acid that is palatable to the cat provides no nutritional value to the diet. In addition, a highly palatable food may entice an animal to eat more than its caloric needs, resulting in obesity. Product odor, taste, texture, shape, and moisture content affect the palatability of a diet.

If selecting a pet food is still confusing, request information from the pet food manufacturer concerning their testing methods in determining nutritional adequacy. A reputable manufacturer conducts numerous analytical and animal feeding tests to assure the nutritional quality of a diet, and information about these tests should be available to customers.

COMMON ERRORS IN FEEDING

Most commercial pet food manufacturers guarantee that all required nutrients be present in the proper quantities. The availability of well-balanced commercial foods makes the supplementation of dog and cat diets unnecessary. Indiscriminate supplementation may, in fact, upset the nutritional balance of commercial pet foods, depriving the animal of optimal nutrition.

EGG SUPPLEMENTATION

Eggs provide a good source of protein, Fe, vitamin A, vitamin D, and numerous B vitamins. Many pet owners believe that the addition of eggs improves coat quality, increases the level of protein, and increases the acceptability of commercial rations. While eggs may improve the performance of an inferior commercial product, several potential problems are associated with such supplementation.

Egg white contains avidin and a trypsin inhibitor. Avidin functions to bind biotin, a B vitamin, and render it unavailable for absorption. This problem may be negated by the large amount of biotin present in the yolk of eggs. However, the feeding of raw egg white alone can cause a biotin deficiency to develop. The presence of a trypsin inhibitor decreases normal digestion of protein. Thus, supplementation with raw egg results in a decrease in the protein digestibility of the diet. Cooking eggs destroys the activity of avidin and the trypsin inhibitor, thereby eliminating this potential problem.

MILK AND DAIRY PRODUCTS

Although milk is an excellent source of protein, Ca, P, and several vitamins, excessive intake often causes diarrhea in dogs and cats. This is due to the fact that milk contains the sugar lactose, which requires the enzyme lactase for absorption. Many dogs and cats do not produce sufficient lactase to handle the quantity of lactose present in milk. This results in an inability to completely digest milk that is added to the diet and will subsequently cause digestive upsets and diarrhea. Dairy products, such as cheese, buttermilk, and yogurt, contain slightly lower levels of lactose. While these products may be more easily tolerated by some animals, they still have the potential for causing diarrhea and dietary imbalances.

FISH

Fish is highly palatable to most cats. As a result, many cat owners supplement their cat's diet with various seafoods. Although fish is a good source of protein, most deboned fish are deficient in Ca, Na, Fe, Cu, and several vitamins. Tuna fish is commonly fed to cats. Canned tuna that is packed in oil contains high levels of polyunsaturated fatty acids. The excessive in-

FIGURE 22-6 Palatability evaluation being conducted by an animal care technician. (Courtesy of the Iams Co., Dayton, OH.)

take of these oils, coupled with the fact that oils are low in vitamin E, can result in a vitamin E deficiency. This may result in steatitis, or yellow fat disease, in the cat. A second problem associated with feeding large amounts of tuna involves the high levels of Mg. Although it is not the most important factor, high Mg intake in cats is a contributing factor in the development of FLUTD. In addition, raw fish should never be fed to cats. Certain types of fish, such as carp or herring, contain thiaminase. Thiaminase destroys the thiamin present in a diet and may result in a thiamin deficiency.

LIVER

Liver provides an excellent source of protein, Fe, Cu, vitamin D, and several B vitamins; however, it is severely deficient in Ca and excessively high in vitamin A. Both of these nutritional imbalances can cause the development of bone disorders in the cat. Vitamin A toxicity can develop slowly over a period of years in cats that are regularly fed fresh liver as their primary source of protein. The bone deformities of vitamin A toxicity form gradually and may be asymptomatic for several years.

OILS AND FATS

Animal fats and fish oils are often added to commercial pet foods to improve palatability. Although fish oils provide an excellent source of vitamins A and D, when consumed in excess they are toxic. Both vitamins are stored in the liver; the effects of excess intake are cumulative and may develop over long periods of time. In addition, oversupplementation with fat may result in obesity or in an eventual decrease in dietary intake due to the fact that energy needs are met with less food. Deficiencies of other nutrients may then occur.

CHOCOLATE

Most dogs love chocolate. However, chocolate contains theobromine, a compound that is toxic to dogs and cats. The content of theobromine is highest in unsweetened baking chocolate and lowest in milk chocolate. The consumption of only 3 oz of baking chocolate has the potential to cause death in a 25-lb dog. Because the theobromine content is diluted by the addition of sweeteners and milk products, other chocolate-containing foods are less toxic. However, the ingestion of approximately 1.5 lb of milk chocolate could result in a potentially lethal dose for a 25-lb dog. Signs of theobromine toxicity include depression, vomiting, diarrhea, increased urination, and muscular tremors.

ONIONS

Contrary to common belief, the addition of onions to a dog's diet will not prevent fleas or parasites. Although many dogs enjoy the taste, onions contain a compound that may cause severe anemia. A dog consuming 0.5% or more of its body weight in dehydrated onion will exhibit symptoms of hemolytic anemia within 1 to 3 days.

VITAMIN C

Like most species, companion animals do not have a requirement for dietary vitamin C (ascorbic acid). However, this nutrient is often supplemented with hopes of preventing certain skeletal diseases, such as hypertrophic osteodystrophy (HOD) or hip dysplasia. However, this claim has not been scientifically confirmed through controlled research. Several studies have shown that supplementation neither prevents nor cures skeletal problems in dogs (48, 49). Likewise, although vitamin C requirements increase with stress, vitamin C supplementation does not enhance a dog's stamina or performance. In fact, excessive vitamin C supplementation increases the risk for the formation of kidney or bladder stones.

B-COMPLEX VITAMINS AND BREWER'S YEAST

B-complex vitamins, or B-complex in the form of brewer's yeast, are often supplemented in the hope of repelling fleas or improving athletic performance. Scientific studies have shown that such supplementation has no effect on flea infestation or increasing performance, provided a well-balanced diet is being fed.

METHODS OF FEEDING

The selection of an appropriate diet is only the first step in assuring the health and well-being of a dog or cat. Selecting the appropriate method of feeding can influence animal health as well. The method of feeding is influenced by the type of diet, age and physical condition of the animal, environmental factors and social interactions, and the attitude and life-style of the owner.

There are two basic methods of feeding both dogs and cats. The first method is ad libitum or free choice and the second is portion-controlled or meal feeding. A free-choice regimen provides more food than the animal will consume at one time, allowing the animal to self-regulate intake. In contrast, a portion-controlled regimen provides food in restricted quantities as one or more meals per day.

FEEDING MANAGEMENT OF CATS

Due to the feeding patterns of the cat, the most common method of feeding is ad libitum. The cat is a nibbler, meaning it will eat 10 to 20 times in a 24-h period. Whether it is night or day seems to have an insignificant effect on the frequency of feeding. The frequent small meal sizes consumed by most cats suggest that they regulate food intake to meet their caloric needs. However, the sedentary life-style of some house cats coupled with the use of highly palatable foods can lead to overconsumption with this type of feeding regimen. While some cats maintain optimum weight and condition on this type of regime, others habitually overeat and should not be fed ad libitum.

Growing kittens and queens in a stage of gestation and lactation have the highest caloric needs. To maximize caloric intake, nutrient-dense, highly digestible foods should be offered. Sudden changes in the surrounding environment, such as excessive noise, a change in lighting, or the relocation of a feed bowl, may upset a cat and result in anorexia. Social interactions (multiple-cat households) may also adversely affect feeding behavior. In multiple-cat households each cat should have an individual feeding bowl in a separate location. This will allow the assessment of individual food intake and prevent competitive feeding and intimidation.

FEEDING MANAGEMENT OF DOGS

If an adult dog can self-regulate by consuming adequate food to meet its energy needs, a free-choice regimen is the most convenient method. Dry pet foods are the most suitable for a free-choice program. Canned foods may be inappropriate because they are susceptible to bacterial growth due to their high moisture content and because they often dry out, resulting in reduced acceptability. In contrast, diets should be portion controlled when feeding dogs who are unable to self-regulate caloric intake. The maintenance of ideal body weight is the best assessment of the appropriateness of a feeding regimen.

Similarly to the cat, caloric needs of the dog increase in the growing puppy and during periods of gestation and lactation. The puppy is in a linear stage of growth and, depending on breed, will reach its adult weight in 10 months.

Other breeds, such as the giant breed, do not reach mature weight even at 24 months of age (Fig. 22-7). Puppies should be fed a diet of high caloric content. If a diet of inferior energy availability is fed, the puppy may not be able to consume sufficient quantities to meet its nutritional needs due to a limitation in stomach capacity.

A portion-controlled feeding regimen should be used for growing puppies. Although a free-choice method supports maximal weight gain, this is not a benefit to the rapidly growing dog. Such a condition can predispose a puppy to a state of obesity, as well as compromise normal skeletal development. A study with growing labradors compared these two feeding regimens and recorded a 17% greater food consumption in puppies that were fed free choice. In addition, puppies on a portion-controlled regimen had a lower incidence of osteodystrophic changes that were correlated to a reduction in caloric intake and subsequent lower weight gain than their respective free-choice controls (50).

The caloric requirement for gestation is not substantially increased until the last trimester. However, the caloric needs to support peak lactation periods may exceed the maintenance requirement in excess of threefold. The consumption of excess calories resulting in too much weight gain is rarely a concern during lactation. Rather, it is the maintenance of ideal body weight that a feeding program should be based on. A free-choice feeding regimen is most suitable for maximal performance.

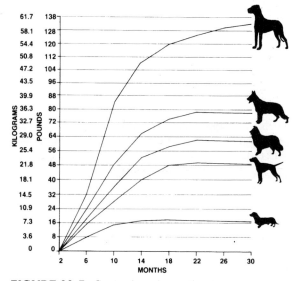

FIGURE 22-7 Canine breed growth rates.

Environmental factors can significantly affect energy requirements. Exposure to cold temperatures increases energy expenditure, thereby increasing energy requirements. It has been estimated that the energy requirement of a dog exposed to cold temperatures may increase 25% over that required for maintenance.

It has been demonstrated that dogs housed together consume more food than if housed individually. In part, this response may be due to an increase in energy expenditure of dogs housed together. However, it has been shown that satiated dogs introduced into a neutral territory will resume eating when pair fed. The respective individually fed controls did not initiate further food consumption once satiated. Therefore, competition at the feeder results in caloric consumption over that required. To prevent such an occurrence in homes with more than one dog, individual food bowls should be provided in separate locations. A portion-controlled feeding program may be necessary.

SUMMARY

Nutrition for the dog and cat can be supplied by a vast array of ingredient compositions, processed by diverse techniques, and provided in various forms. The selection of the proper diet warrants a basic understanding of the nutritional requirements for dogs and cats and the factors that affect nutritional quality and efficacy, as well as the appropriate method of feeding. The proper diet for a dog or cat should not simply provide an adequate level of nutrition, but should express an optimum state of health and well-being.

REFERENCES

1. Maxwell, J. C. 1995. *Petfood Industry,* July/August.

2. NRC. 1985. *Nutrient requirements of dogs.* Washington, DC: National Academy Press.

3. NRC. 1986. *Nutrient requirements of cats.* Washington, DC: National Academy Press.

4. Association of American Feed Control Officials. 1995. *AAFCO Official Publication,* Atlanta, GA: AAFCO.

5. Hirakawa, D. A., and D. H. Baker. 1988. *Comp. Animal Pract.* 2:25.

6. Ballard, F. J. 1965. *Comp. Biochem. Physiol.* 14:437.

7. Sunvold, G. D., et al. 1993. *FASEB J.* 7:A740.

8. Reinhart, G. A., R. A. Moxley, and E. T. Clemens. 1994. *J. Nutr.* 124:2701.

9. *Willard, M. D.,* et al. 1994. *Amer. J. Vet. Res.* 55:654.

10. Fahey, G. C. et al. 1990. *J. Animal Sci.* 68:4221.

11. Sunvold, G. D. et al. 1995. *J. Animal Sci.* 73:2329.

12. Sinclair, A. J., et al. 1981. *Brit. J. Nut.* 46:93.

13. Reveis, J. P. W. 1982. *J. Small Animal Pract.* 23:563.

14. Vaughn, D. M., et al. 1994. *Vet. Dermatol.* 5(4):163.

15. Hirakawa, D. A., and D. H. Baker. 1986. *Nutr. Res.* 6:527.

16. Czarnecki, G. L., and D. H. Baker. 1982. *J. Animal Sci.* 55:1405.

17. Milner, J. A. 1981. *J. Nutr.* 111:40.

18. Blaza, S. E., et al. 1982. *J. Nutr.* 112:2033.

19. Hirakawa, D. A., and D. H. Baker. 1985. *Nutr. Res.* 5:631.

20. MacDonald, M. L., Q. R. Rogers, and J. G. Morris. 1984. *Ann. Rev. Nutr.* 4:521.

21. Morris, J. G., and Q. R. Rogers. 1978. *Science* 199:431.

22. Hayes, K. C., R. E. Carey, and S. Y. Schmidt. 1975. *Science* 188:949

23. Pion, P. D., et al. 1987. *Science* 237:764.

24. Sturman, J. A., et al. 1986. *J. Nutr.* 116:655.

25. Morris, J. G., et al. 1994. *Vet. Clinical Nutr.* 1:118.

26. Gershoff, S. N., et al. 1957. *Lab. Invest.* 6:227.

27. Kruger, J. M., C. A. Osborne, and S. M. Goyal. 1991. *Amer. JAVMA* 199:211.

28. Osborne, C. A. 1992. *Proceedings of the Kal Kan symposium for the treatment of dog and cat diseases,* Kal Kan Foods, Inc., Vernon, CA, p. 89.

29. Sauer, L. S., D. Hamar, and L. D. Lewis. 1985. *Feline Pract.* 15:10.

30. Rich, L. J., and R. W. Kirk. 1969. *JAVMA* 154:153.

31. Kane, E., and G. M. Douglas. 1986. *Feline Pract.* 16:9.

32. Harrington, J. T., and J. Lemann. 1970. *Med. Clinics N. Amer.* 54:1543.

33. Burger, I. H. 1987. *J. Small Animal Pract.* 28:447.

34. Thrall, B. E., and L. G. Miller. 1976. *Feline Pract.* 6:10.

35. Ross, L. A. 1990. *Vet. Med.* 85:1194.

36. Hinchcliff, K. W., et al. 1994. Submitted for publication.

37. Reynolds, A. J. 1995. *Proc. Iams performance dog nutrition symposium,* April 18, Colorado State University, Fort Collins, CO.

38. Downey, R. L., D. S. Kronfeld, and C. A. Banta. 1980. *J. Am. Animal Hospital Assoc.* 16:273.

39. Rose, R. J., and M. S. Bloomberg. 1983. *International greyhound symposium,* Orlando, FL.

40. Sheffy, B. E. et al. 1985. *Cornell Vet.* 75:324.

41. Finco, D. R., et al. 1994. *Amer. J. Vet. Res.* 55:1282.

42. Finco, D. R., et al. 1992. *Amer. J. Vet. Res.* 53:157.

43. Edney, A. T. B., and A. M. Smith. 1986. *Vet. Rec.* 118:391.

44. Sloth, C. 1992. *J. Small Animal Pract.* 33:178.

45. Huber, T. L., R. C. Wilson, and S. A. McGarity. 1986. *J. Amer. Animal Hospital Assoc.* 22:571.

46. Sousa, C. A., et al. 1988. *JAVMA* 192:676.

47. Kallfelz, F. A. 1989. *Vet. Clinics N. Amer.; Small Animal Pract.* 19:387.

48. Grondalen, J. 1975. *J. Small Animal Pract.* 17:721.

49. Teare, J. A., et al. 1979. *Cornell Vet.* 69:384.

50. Alexander, J. E., and L. L. H. Wood. 1987. *Canine Pract.* 14:41.

23

FEEDING RABBITS

Hugo Varela-Alvarez

INTRODUCTION

Rabbits are used for meat, for fur and hair production, for vocational projects, as laboratory animals, as teaching tools, as animal research models, and for pets (Fig. 23-1). Because of their use in a wide range of enterprises, it is necessary to have good knowledge of the different aspects of their management and nutrition.

Rabbit feeding and nutrition information increased significantly during the 1980s. Most of the research in these areas has been done by European researchers, most notably in France. In Europe, meat rabbit production has become a true industry, and rabbit meat consumption is relatively high.

Feeding and nutrition are important aspects of any animal production enterprise. Cost of feed accounts for a large percentage of the total production cost, and therefore a good understanding of proper nutrition and feeding is essential for successful economic rabbit production.

DIGESTIVE SYSTEM

The function of the digestive system is to transform the ingested feed, through chemical, physical, and biological processes, in such a way that the organism can use the nutrients of the feed for maintenance, growth, and production.

RELATIVE SIZE

The digestive system of the rabbit (illustrated in Fig. 2-3) is well adapted for the utilization of forages and feeds of plant origin. The digestive system occupies a large portion of the body cavity. In the New Zealand White adult rabbit (4 to 4.5 kg), the digestive tube can be as long as 5 m. The development of the digestive system is almost completed at 9 weeks of age. The cecum and the colon start to develop around 3 to 5 weeks of age, when feed ingestion, other than milk, starts to be significant, and the microflora population becomes important in these organs.

The size of the different parts of the digestive system varies with age, breed, physiological status, and type of feed given to the rabbit. Close to 80% of the digesta is contained in the stomach and the cecum (Table 23-1). These two compartments are the ones with the largest capacity. The relative organ size is also presented in Table 23-1.

RETENTION IN THE STOMACH

The amount of time that the feed stays in the digestive system affects the amount of time enzymes and microorganisms can act over the ingested material, therefore affecting both nutrient absorption and utilization. The average retention time of the ingested material in the digestive tract is 17 to 18 h, being less for

453

FIGURE 23-1 California rabbits, one breed often used for meat production. (Courtesy of Mrs. G. Absolon, Doebank, Palmerston North, New Zealand.)

large particles (14 to 16 h) and longer for finer particles (20 to 21 h). The longest retention time occurs in the stomach and cecum. Retention time in the small intestine is relatively low. The rabbit's stomach is almost always full. Gastric secretion is continuous and very acid (pH 1 to 2), the result of HCl secretion. Gastric digestion is characterized by a significant production of lactic acid (1) that is absorbed directly by the gastric mucosa or in the small intestine (2).

TYPE OF FECES PRODUCED

The rabbit produces two kinds of feces: hard feces and soft feces or cecotrophs (3). The rabbit has a specialized mechanism that retains digesta in the proximal colon and cecum for microbial utilization of nutrients and also to allow the formation of the two types of feces. The reingestion of the soft feces by the rabbit is called cecotrophy. The production of two types of feces and the ingestion of only one type of these distinguishes cecotrophy from coprophagy. This process is one of the important characteristics of the digestive physiology of

the rabbit and allows the animal maximal utilization and absorption of total ingested nutrients (4). The soft feces are higher in water, electrolytes, and nitrogen (N) content and lower in fiber (5). A significant proportion of the total N content of cecotrophs (60% to 80%) is from cecal microbial cells (6). Total N from cecotrophs represents an important source of protein for the animal and could amount to up to 20% of the total N intake of the rabbit. Cecotrophs contribute approximately 20% of the protein and 10% of the energy for maintenance, vitamins, and minerals. Producers should take advantage of this characteristic in order to feed rabbits with nontraditional ingredients (7).

The cecotrophs are removed directly from the anus and are swallowed whole. When they reach the stomach, they do not mix immediately with the rest of the stomach contents but lie in a mass in the fundic region for 6 to 8 h (8). From there on, the cecotrophs follow a digestion pattern as does any other ingested feed.

FEEDING BEHAVIOR

FEED AND WATER INTAKE

Quantifying feed and water intake is important in order to determine the economics of the rabbit enterprise. It also helps in the early detection of health problems, whose first sign could be a lack of appetite or an excess of drinking.

Knowledge of feed intake helps the manager to decide whether there is an excess of feed waste. Feed and water intake values are helpful in diet formulation and disease control so that concentration of nutrient or medicine can be formulated according to requirements, age, or body weight.

During the first 3 weeks of life, the rabbit consumes only maternal milk. When milk production is not enough, the rabbit starts to con-

Table 23-1

Values for different parts of the rabbit digestive system[a]

Digestive System Part	Weight (g)	Length (cm)	Capacity (g)	DM (%)	pH
Stomach	20		90–100	17	1.5–2.0
Small Intestine	60	330	20–40	7	7.2
Cecal appendix	10	13	1		
Cecum	25	40	100–120	20	6.0
Colon	30		10–30		
Proximal colon		50		20–25	6.5
Distal colon		90		20–40	

Source: Lebas et al. (9).

[a]Values from 12-week-old New Zealand rabbits fed a pelleted complete diet.

sume solid food and water. The animal then changes from one or two feedings a day to several feedings in 24 h.

During a 24-h period, adult and growing rabbits (4 weeks or older), consume feed in several meals of small amounts (feed and water), utilizing usually 2 to 4 h/day for feeding. Feed and water consumption in grams per meal is lower during the day (time with light) than during the night (dark hours). In lactating does, feed consumption occurs mostly during the night. This behavior must be taken into consideration for feeding practices.

The average feed and water consumption varies according to age and physiological status of the animal. In general, rabbits consume water in quantities close to twice their dry feed intake. Adult animals (4-kg body weight) consume on the average 200 to 300 ml of water/day, while lactating females can drink 3 to 4 liters/day (10). Lactating females increase their feed intake proportionally to milk production and number of suckling rabbits (11, 12).

Water and feed consumption varies according to environmental temperature and humidity. There is a direct relationship between dry feed intake and water intake (13). As temperature rises, water consumption increases, but at high temperatures (30°C and over) feed and water consumption decline, affecting the performance of growing and lactating animals (14).

NUTRIENT REQUIREMENTS

WATER

Water should be supplied ad libitum. Rabbits have a high requirement for water in relation to their body weight. Water is necessary for maintenance, production, and lactation. Because dry-matter intake is related to water intake, any restriction in water causes a decline in dry-matter consumption. However, if feed is restricted, water intake may increase. Water should be clean, fresh, and free from biological and chemical contaminants. One example of a watering device for rabbits is shown in Fig. 23-2.

PROTEIN

The adult rabbit (probably) obtains 10% to 20% of its total protein intake from cecotrophy as good-quality bacterial protein. Gidenne reported that in 6-week-old young New Zealand rabbits, independent of the type of diet, cecotrophy accounted for 29% of the total protein intake (15). Part of the amino acids produced in the cecum by microorganisms are absorbed directly by the ce-

FIGURE 23-2 An automatic nipple drinking system that works well for rabbits. (Courtesy of Mrs. G. Absolon, Doebank, Palmerston North, New Zealand.)

cum wall, and after cecotrophy, the rest are absorbed by the small intestine. The contribution of microbial protein to young rabbits (4 to 8 weeks old) is very small, and steps should be taken to supply good-quality protein to growing animals. Rabbits require good-quality protein intake in terms of quantity and quality. The essential amino acids are the same as in other species (see Ch. 3). Glycine can be synthesized in adequate amounts in the adult rabbit. However, in the fast-growing rabbit the rate of synthesis of glycine is not adequate; therefore, glycine should be supplemented in diets for these animals (16). The amino acid requirements are presented in Table 23-2. Protein quality is related to feed intake and thus to performance.

Total crude protein requirements for maintenance in adult rabbits is 13% of the ration. Digestible protein requirements for growing rabbits can be computed using the following formulas:

$$\text{Minimum DP} = \frac{\text{DE}}{250}$$

$$\text{Maximum DP} = \frac{\text{DE}}{230}$$

Digestible protein requirements for lactating does can be obtained by the following formulas:

$$\text{Minimum DP} = \frac{\text{DE}}{200}$$

$$\text{Maximum DP} = \frac{\text{DE}}{180}$$

where DP = percent of digestible protein in the diet and where DE = digestible energy (kcal/kg) in the diet.

Table 23-2

Essential amino acid minimum requirements for rabbits

Amino Acid (%)	Growth	Gestation	Lactation
Arginine	1.00	0.80	1.11
Glycine	0.40	0.40	0.40
Histidine	0.45	0.42	0.45
Isoleucine	0.70	0.70	0.71
Leucine	1.05	1.05	1.25
Lysine	0.70	0.70	0.85
Methionine + cystine	0.60	0.60	0.95
Phenylalanine + tyrosine	1.20	1.20	1.20
Threonine	0.55	0.60	0.64
Tryptophan	0.15	0.15	0.15
Valine	0.70	0.75	0.82

Source: Varela-Alvarez (15).

The protein requirements for lactating does are higher than those for growing rabbits due to the protein demands during lactation. There is a high secretion of protein and energy in the milk during lactation. This secretion is directly proportional to milk production. The producer should take into account that protein levels below 18% without the proper levels of energy have a negative effect on milk production and, therefore, on the rate of gain of the suckling rabbits.

Protein nutrition is of special concern in fur- and hair- producing rabbits. Because the final product (pelt and hair) is high in N-containing compounds and S-containing amino acids, the level in dietary protein should be high, with a minimum of 17% crude protein and a minimum of 0.6% of S-containing amino acids (methionine, cystine). In the usual rabbit feeds, the first limiting amino acids are the S-containing ones, followed by lysine.

ENERGY AND FIBER

Most of the required energy in the rabbit is supplied by carbohydrates, to a lesser extent by lipids, and in some cases by excess protein. Rabbits, like most other animals, adjust their feed intake to maintain a relatively constant energy intake (17). Therefore, when formulating a diet, nutrients should be supplied in relation to energy.

Energy requirements vary according to factors such as animal age and size (younger and smaller animals require more energy), physiological status (growing animals require more energy than inactive adults; lactating does require more energy than nonlactating animals), environmental temperature and humidity (animals in cold and humid environments require more energy in order to maintain body temperature and level of production), diet fiber concentration, and type of enterprise.

For the growing period the average daily digestible energy (DE) requirement for maintenance and production can be estimated by the following formula:

$$DE = 4MW (20 - WW) + 55WW + 3.75DG - 50$$

For lactating animals, the average daily DE requirement can be estimated by the following formula:

$$DE = 290 + 65DW + 35G$$

where
DE = daily digestible energy, kcal/day
MW = estimated market weight, kg
WW = weaning weight, kg
DG = expected average daily gain, g
DW = doe weight, kg
G = number of suckling rabbits

EXAMPLE

For a rabbit that is going to be marketed at 2 kg, with a weaning weight of 0.6 kg and an expected daily gain of 30 g, the estimated DE requirement per day for the growing period would be

$$4 \times 2(20 - 0.6) + 55(0.6) + 3.75(30) - 50 = 250.7 \text{ kcal/day}$$

If the feed utilized contains 2500 kcal DE/kg, then feed consumption per animal per day (waste not included) will be

$$\frac{250.7 \times 1000}{2500} = 100.28 \text{ g of feed}$$

For primiparous lactating females, the energy requirement should include an increase in energy to fulfill the requirements for growth of the female. That can be quantified as an extra 6.5 kcal/day/kg body weight.

As lactation progresses, the lactating female increases her feed intake to a maximum in order to satisfy her energy needs. If the ration does not contain a high level of energy per kilogram, then the female will use part of her body reserves in order to maintain milk production. This will cause the female to lose condition, affecting future breeding performance and efficiency (16). Furthermore, if the feed is of a low caloric density, the young rabbits will not be able to maintain a high growth rate when they start to consume dry feed. This has been shown to also affect growth rate and feed efficiency in the postweaning period.

With the estimated energy requirement and the estimated level of intake per day, it is possible to calculate the energy concentration in the diet. Diets with a minimum concentration of 2500 kcal DE/kg will satisfy the requirements of growing rabbits, and 2500 to 3000 kcal DE/kg will be sufficient for lactating does with more than seven pups. Rations with 2200 kcal/kg are adequate for inactive adults.

With an increase in fiber content in the diet, there is a reduction in the utilization of energy and organic matter. This relationship seems to be linear. However, some minimum level of fiber must be present in the diet in order to maintain proper digestion and gut motility, and to prevent some metabolic disorders, such as diarrhea. For growing rabbits, crude fiber should be between 14% and 19.8% (18). For each 1% increase in cellulose content in the diet, there is a decrease of 0.7 units in organic matter digestibility (8). French researchers have estimated that by replacing the diet's starch with cellulose there is a reduction of the DE in a proportion of 60 kcal DE/kg diet for every percent point increase of cellulose (19).

Digestibility is the combined effect of rate of passage and the digestibility coefficient of the feed itself. An increase in crude fiber will increase feed intake and decrease feed efficiency. In general, an increase in retention time increases the digestibility of the diet, but in the rabbit it also increases the incidence of digestive disorders. It has been suggested that a low retention time is a normal characteristic in the rabbit (8). Gidenne (20) reported that increasing lignin concentration in rabbit diets decreases protein digestibility and to a lesser extent energy and cellulose.

Part or all of the fiber should be given in the form of large particles (2 to 4 mm) in order to decrease the incidence of digestive disorders, reduce weight loss, and promote gut function and hard feces formation. Fiber also seems to prevent hairball formation. Part of the fiber could be ground below the 2 mm size to permit longer cecal retention and increase digestibility. Research is needed in this area to determine the minimum proportion of long fiber needed by growing and by lactating animals.

In a study with more than 2000 growing rabbits, Perez et al. (21) reported that mortality rate in growing rabbits decreases as acid detergent lignin (ADL) in the diet increases and expressed the relationship as follows: mortality rate, % = 15.8 − 1.08 ADL%. They also found that conversion rate (kilograms of feed per kilogram of gain) increased by 0.1 point per additional point of dietary ADL. These researchers recommend ADL in growing rabbit diets between 6% and 8%. In a separate report, Gidenne and Perez (22) stated that ADL increases feed consumption and decreases mean gut retention time and that ADL affects fiber digestion, but does not affect protein digestibility. Both these studies tend to suggest that ADL may be a better indicator of fiber for growing rabbits.

MINERALS

Rabbits require Ca, P, Mg, K, Na, Cl, Mn, Zn, Cu, Fe, I, Co, and Se. Other minerals such as chromium and fluoride, which are essential for other mammals, have not been reported as having an important role in rabbit nutrition. Mineral requirements are higher during lactation because they are either secreted in milk or are required for milk secretion. Available information on rabbit mineral nutrition and requirements is scarce and contradictory. Data presented here refer to mineral levels based on a 2500-kcal/kg feed mix. The minimum requirements for each of the important minerals are presented in Table 23-3.

Calcium and P requirements are highest for the lactating females because of the higher amounts of these minerals that are secreted in the milk. The recommended ratio of Ca to P is 2:1; however, rabbits can tolerate up to a 12:1 ratio. Although rabbits can tolerate high amounts of Ca in the diet, this can bring on a Zn or a Mg deficiency in growing animals and a P deficiency in lactating does. Some of the feed Ca

TABLE 23-3

Mineral minimum requirements for rabbits[a]

Mineral	Growth	Breeding	Lactation
Calcium, %	0.9–1.0	0.9–1.0	1.0–1.2
Phosphorus, %	0.6–0.7	0.6–0.7	0.8–1.0
Potassium, %	1.3	0.75	1.3
Sodium, %	0.4	0.25	0.4
Chloride, %	0.4	0.2–0.25	0.4
Manganese, ppm	8.5	8.5	8.5
Magnesium, ppm	400–500	400–500	400–500
Cobalt, ppm	1	1	1
Copper, ppm	5	5	5
Zinc, ppm	107	107	107
Iodine, ppm	0.2	0.2	0.2

Source: Varela-Alvarez (17) and Sazzad et al. (24).

[a]Assumes a 2500-kcal DE/kg diet.

comes from buffers added to the diets. Buffers can be added to rabbit diets in order to lower the pH of the cecal contents, which in turn increases feed intake (23). Ca deficiencies are rare in commercial rabbit production because most of the ingredients utilized in rabbit diets are high in this mineral. Calcium can be supplemented in the diet as inorganic Ca with calcium carbonate ($CaCO_3$) or dicalcium phosphate (HP_4Ca_2).

The best sources of P are ingredients of animal origin, but these ingredients are not popular in rabbit diets. Most of the P ingested is of plant origin and is present in phytate form, which is poorly utilized by the animals. There seems to be a microbial process in the cecum that frees the P, making it available through cecotrophy. However, it does not cover the animal's requirements, making it necessary to supplement with P. The best supplements are in the form of Ca salts; of these the best is monocalcium phosphate.

Magnesium is required for proper growth, gestation and lactation. Magnesium requirements increase during gestation and lactation due to fetus uptake and milk synthesis. Deficiencies of this element cause retarded growth, fur chewing, poor fur condition, and whitening of the ears. Excess of Mg can cause diarrhea.

Zinc is required in rabbits to maintain a normal pregnancy, for proper fetus growth, and for proper pre- and postweaning growth. Zinc also plays a role in the quality of hair and fur. A level of 107 ppm in the diet has been suggested as appropriate. However, these levels should be increased if the diet contains more than 2% Ca. Pregnant does fed diets low in Zn pull little or no hair for nest building.

Under normal conditions, K is not a limiting mineral in rabbit diets. High-energy diets

are lower in K due to the high grain concentration. Some of the K is excreted via the milk. Potassium levels in the diet should be increased to their maximum under high-temperature conditions, especially for lactating does. If K is increased, total Na should also be increased. An imbalance of K and Na may cause renal problems. High levels of K (greater than 2%) can cause low fertility rates and low embryonic survival (25).

Addition of chelated salts of Cu, Co, Fe, Zn, Mn, and I to rabbit diets have been reported to increase rate of growth, dressing percentage, and feed efficiency (26).

When diets are supplemented with crystalline synthetic amino acids, the levels of Cl and Na in the diet should be adjusted both in quantity and in ratio, because most synthetic amino acids are of the hydrochloride form, thus adding extra Cl to the diet.

VITAMINS

In the adult rabbit, B-complex vitamins are synthesized in the cecum. However, not all B-complex vitamins are synthesized in the needed amount for fast-growing rabbits; therefore, they should be supplied in growing rabbit diets to obtain a high performance in the commercial meat operation. The absence of a vitamin or its inadequate amount in the diet will result in poor performance and, if severe, in specific deficiency symptoms. Vitamins should be present for normal absorption and utilization of the other required nutrients. Frye (27) recommended that the vitamin premix presented in Table 23-4 be used as a supplementation for growing rabbits. This premix should be used at a minimum rate of 2.5 kg/1000 kg of

TABLE 23-4

Recommended vitamin supplementation for growing and adult rabbits

Compound	Premix Units (per kg)
Vitamin A, IU	3,530,000
Vitamin D₃, IU	353,000
Vitamin E, IU	15,985
Vitamin B₁₂, mg	5
Riboflavin, mg	1325
Niacin, mg	19,845
D-panothenic acid, mg	4765
Choline chloride, mg	242,510
Menadione SBC, mg	620
Thiamine mononitrate, mg	480
Pyridoxine HCl, mg	1060
Biotin, mg	40
Folic acid, mg	80
β-Carotene	4410

Source: Frye (27).

complete feed. Vitamin C (ascorbic acid) supplementation is recommended for diets of rabbits under stress (28).

Some of the factors that can affect the amount of a given vitamin in a diet are improper formulation, low-quality supplemental vitamin product, amount of time the feed is stored, and environmental conditions such as temperature and humidity. Vitamins degrade under conditions of high temperature and humidity. It is recommended that feed be utilized within a maximum of 3 to 4 months after blending in order to have the proper nutritive value. If feed is contaminated by mold, insects, or rodents, the shelf life may be much less than 3 to 4 months.

Factors that can influence the vitamin requirement level in rabbits are general health of the animals (parasites and diseases damage the lining of the digestive tract and increase the requirements), environmental stress, medication (especially with antibiotics, which decrease the rate of B-complex synthesis), rate of growth (animals that grow faster require more quantities of vitamins per day), quality of the product being used as a source for the vitamin (bioavailability), and system of production, that is, intensive versus extensive (animals under intensive systems of production have a higher requirement).

Dietary levels below the requirement for vitamins A, D, E, and K produce deficiency symptoms similar to those in other species. Excesses of vitamin A and D are known to cause health problems in rabbits. Supplementation of fat-soluble vitamins in the diet may be necessary but should be done with care.

ADDITIVES

In some cases, it may be desirable to use additives in rabbit diets. The most common additives are antibiotics, coccidiostats, and antioxidants. These additives, when included in any diet, should be in accordance with FDA regulations (see Ch. 10).

Molasses may be added to rabbit diets to increase the energy content of the diet, to increase palatability, or to reduce dustiness of the mix. Caution should be taken when using molasses because high levels (greater than 6%) may produce diarrhea. Buffers and pellet binders are also added to the diets. Of the pellet binders, Ca and Na lignosulfonates have been reported to produce high incidences of colon ulcerations and high mortality, while Mg lignosulfonates appear to have no such effects (29, 30).

FEEDSTUFFS FOR RABBITS

The major ingredients used in rabbit diets are feed grains, milling by-products, protein supplements, forages, fats, synthetic amino acids, and mineral and vitamin supplements. Selection of ingredients should be done on the basis of availability, freshness, nutrient content, price, and palatability.

FEED GRAINS

Feed grains are used as a source of energy. Barley, wheat, oats, corn, sorghum, and triticale can be included in rabbit diets. Grain milling by-products can be utilized in rabbit diets as a source of fiber and a limited source of protein. The contribution of feed grains to diet protein is limited.

PROTEIN SUPPLEMENTS

Most protein supplements used in rabbit diets are of plant origin. However, when offered, animal protein supplements (fish meal, meat meal) will be consumed by rabbits, but to a lesser extent than those from plant sources. Plant protein sources are usually cheaper than those of animal origin. The most common protein supplements are soybean meal, cottonseed meal, peanut meal, rapeseed meal (canola meal), sesame seed meal, sunflower meal, and safflower meal. Caution should be taken when feeding cottonseed meal and rapeseed meal (see Ch. 9). Cottonseed meal can contain high levels of gossypol, making the ration unpalatable and toxic. Problems may also be encountered with rapeseed meal if the levels of erucic acid are high. Rapeseed meal has been reported to cause problems in fertility and reproduction in does. Peanut meal should be used with caution because this feedstuff may sometimes contain aflatoxins.

In some parts of the world, beans have been used as a protein source. Some raw beans (including raw soybeans) as well as soybean meal have been reported to affect animal performance due to high contents of antitrypsin factors and other growth-depressant compounds (see Ch. 9). Proper heat treatment of beans and soybean meal inactivates these compounds.

FORAGES

Forages are used in rabbit diets as a source of fiber and bulk. Forages can be directly incorporated into a completely pelleted feed or be used as a complement to pellets. It is more convenient from the point of view of handling and knowledge of animal intake to include forages in the pellets. However, in some situations, direct use could be more convenient, as is the case in backyard and other small operations. When given apart from the pellets, it is better to offer forages in the form of hay. Care should be taken to avoid moldy forages, because these can cause digestive disorders in rabbits.

Of all forages, alfalfa is the most widely used and preferred by the animals. This legume provides not only a good source of fiber, but good levels of protein and Ca. However, caution should be taken when using very young alfalfa. Alfalfa from very early cuts is high in protein (greater than 20%) and xanthophylls, but when used in growing rabbit diets it may cause diarrhea. Because high levels of calcium may lead to health problems and digestive upsets, when alfalfa is high in Ca, timothy hay or oat hay may be used. These two forages are low in calcium.

The following plants have been listed by Bender (31) as undesirable rabbit feeds: arrowgrass, bracken fern, brownweed, buckeye, burdock, castor beans, chinaberry, fireweed, foxglove, goldenrod, hemlock, horehound, jimson weed, Johnson grass, larkspur, laurel, lupine, mesquite, milkweed, miner's lettuce, oak, oleander, poppy, sweet clover, and tarweed.

FATS

Fats can be added to rabbit diets as a concentrated source of energy and a source of essential fatty acids. Rabbit diets contain, on the average, 3% fats that come from the different feedstuffs used for typical diets. Added fats increase the diet energy density, but sometimes they also increase the price of feed. An economic analysis comparing price to performance is required to determine whether increasing feed price per unit will increase profitability. It has been reported that adding fats increases the feed-to-gain ratio in growing rabbits (32) and that this increase was noted with fat additions up to 6%. However, high levels of fat present a problem for pelleting. Pelleting will be done properly if the added fat does not exceed 3%. Fat could be sprayed on the pellets.

NONCONVENTIONAL FEEDSTUFFS

Other materials have been used as feedstuffs for rabbits as a partial substitute for the conventional grains and forages or as a primary source, as in the case of some tropical plants. Pomaces, the residue of industrial processing of some fruits (apples, grapes, tomatoes, pears), have been utilized without detrimental effects in diets of growing and adult rabbits. These pomaces have variable levels of fiber and soluble carbohydrates and are low in protein. Sugar beet pulp can also be utilized as an ingredient in rabbit diets. Although sugar beet pulp is high in NDF, its low rate of passage and high cecal retention time make it of low value to meet fiber requirements (33). Some researchers have used sugar beet pulp as an energy ingredient and have recommended that this ingredient can replace up to 15% of diet grain (barley) without affecting growth performance (34).

In tropical areas, cassava root meal, coconut meal, tropical kudzu, palm oil, whole corn plant meal, amaranthus, carrot leaves, bananas, sweet potatoes, sugar cane, and NaOH-treated straw have been used as feedstuff for rabbits. The following legumes, some of them of tropical origin, have been utilized in rabbit diets: *Indigofera arrecta, Lespedeza spp., Leucaena leucocephala, Pueraria spp., Stylosanthes spp.,* and *Trifolium alexandrium* (9). All these legumes have a good level of protein.

There is need for comprehensive research into the utilization of tropical legumes and forages for rabbit production.

FEEDING MANAGEMENT

Rabbit production success depends on the profit made, that is, the difference between production cost and gross returns from animal sales. Producers can increase this difference by reducing production cost. Because feed represents a large proportion of production cost, proper feed formulation and feeding practices can help to increase profits.

DIET FORMULATION

One important factor to consider in rabbit diet formulation is the quality of the ingredients, because these will have a direct impact on rabbit health and performance. Quality includes shelf life and nutrient profile.

In compliance with state and federal laws, feed manufacturers provide a feed label tag (see Ch. 6) where information regarding composition and feed analysis is displayed. Rabbit producers should read these labels carefully, but the label guarantee does not imply consistent nutrition or performance in some cases because required information is not sufficient to make an evaluation. Performance of a mix is obtained by comparing these feeds on actual feeding tests. The best policy is to buy feed produced by a reputable feed company.

Producers formulating their own feeds should follow the nutritional recommendations and select the feed ingredients as described earlier in this chapter. Formulating a diet for rabbits may be approached as follows: (1) identify the target animal (gestating, lactating, growing, breeding) for which the diet is intended; (2) select the appropriate nutrient requirements (see Tables 23-2, 23-3, and 23-4); (3) select appropriate feed ingredients in order to formulate a diet that is nutritionally balanced, palatable, and safe; and (4) formulate the ration based on nutrition, efficiency, and cost (see Ch. 12).

Feed manufacturers have three types of pellets available in the market: all-grain pellets, all-hay pellets, and complete pellets (grain and forage together). The type of pellet required depends on the type of feeding management adopted in the rabbitry. An all-grain pellet requires a supplementation of hay; use of a hay pellet requires a supplementation of grain; a complete pellet does not require supplementation of any kind.

Mixing of rabbit diets should have the end result of satisfying all the minimum daily requirements for energy, protein, minerals, and vitamins within the limits of the animal's normal feed intake. Supplementation, therefore, should not be necessary. If the mix is adequate, extra supplementation will cause a nutritional imbalance that leads to suboptimal performance. Examples of some sample diets for rabbits are presented in Table 23-5.

A feed mix of uniform quality is the ideal target for rabbit feeding. Formulas may vary according to local feed price changes. If a change is economically necessary, keep in mind that significant variations during a short period of time in ingredient or nutritional composition will alter the cecal environment. This change will alter the cecal bacterial composition and its by-products and change the rabbit's ability to utilize nutrients. To obtain a uniform mix, the ingredients of the ration should be ground and then pelleted. Pellets should be 3 to 4 mm in diameter and no longer than 8 mm. Pelleting prevents selection by the animal, should reduce waste, and increases total dry-matter intake.

FEED STORAGE

Feed should be stored in containers (bins or silos) that can be emptied and cleaned on a regular basis to maintain the quality of the diet. Nutrient stability in complete mixed rabbit diets is inversely related to environmental

TABLE 23-5

Sample diets for rabbits

International Feed Number	Ingredient (%)	Maintenance	Growth		Lactation	
1-00-068	Alfalfa hay	52.4	40.0	45.0	36.4	34.9
4-00-549	Barley	18.1	15.9	34.0	—	—
4-02-935	Corn	—	—	2.7	32.5	41.1
4-03-309	Oats	9.6	15.8	—	—	—
5-04-604	Soybean meal	—	10.4	—	16.3	—
5-04-739	Sunflower meal	3.5	—	11.1	—	17.8
4-05-190	Wheat bran	13.0	12.5	2.1	10.2	1.7
4-04-696	Cane molasses	1.5	3.0	3.0	1.6	1.5
6-01-080	Dicalcium phosphate	1.1	1.6	1.3	2.2	2.2
	Salt (NaCl)	0.5	0.5	0.5	0.5	0.5
	Supplement[a]	0.3	0.3	0.3	0.3	0.3

[a]The supplement must contain the sources to balance microminerals, vitamins, and amino acids.

temperature and humidity. Knapka (35) reported that, although the quality of rabbit diets may decrease during the shelf life period, there is no research documentation of problems as a direct result of this loss in quality. The same author recommended feeding rabbit diets as soon as possible after manufacture and not to use diets when there is doubt regarding their quality. In general, it is recommended that feed should be used as soon as possible after mixing and that it not be kept more than 4 to 6 weeks in storage. Feed should be stored in a cool, dry place.

FEEDING SCHEDULE

In any type of rabbit operation, time of feeding should be constant, since rabbits can be influenced by changes in routine. Any changes in routine should be done in a stepwise fashion. Time of feeding depends on the type of exploitation. If the operation is an extensive one, in which feed is available to animals at all times, the filling of feeders should be done at the same time every day, making sure that the animals will not run out of feed between refillings. Operators should verify feed consumption, as reflected by the time between refills, because any changes in consumption indicate problems, like diseases or waste, and these should be corrected quickly. If the animals are fed several times daily, the routine should be maintained.

Caution should be taken to feed late in the day, because rabbits are more active during the night and feed consumption is highest during that period. The producer must make sure that feeders are at the appropriate level before leaving in the evening. Feeders should be checked in the morning to verify consumption. This practice will alert the producer to possible problems, which usually begin with reduced feed intake. A number of different types of feeder are available (Fig. 23-3).

BREEDING ANIMALS

Feeding young breeding animals properly is very important because it affects the lifetime reproductive capacity of the animals. Proper feeding and nutrition during their growing period will result in a higher production for a long time. Growing breeding rabbits must have adequate nutrition in order to fulfill all their growing needs, but not above these levels, or they may become overweight. Overweight breeding animals must be avoided because it causes poor reproductive performance, may cause metabolic problems (young doe syndrome), and also wastes feed.

FIGURE 23-3 Growing New Zealand White rabbits. (Courtesy of Mrs. G. Absolon, Doebank, Palmerston North, New Zealand.)

Overweight may also cause breeding failures during periods of high temperature (summer months). Temperatures in the rabbitry should be maintained at 21° to 24°C. During periods of high temperature, rabbits' physical activity decreases, thus reducing energy use. The excess energy could result in extra body weight.

Growing breeding animals should be fed a diet in an amount that keeps the animals in good physical condition and promotes normal growth, but that prevents them from becoming fat. These animals should be observed closely and their feed should be adjusted accordingly. Restriction of feed can be done by physically reducing the amount of diet offered or by offering forage (hay).

When feeding growing breeding stock, care should be taken to ensure enough floor and feeder space (Fig. 23-4). When animals are kept in a group (up to 12 to 13 weeks of age), floor space should be a minimum of 0.07 m per animal. The feeder space should be such that all animals can eat at the same time. This practice will ensure proper intake and growth by all animals and will reduce the risk of diseases.

Adult males and nonpregnant and nonlactating adult females should be restricted in their feed intake when not reproductively active to avoid body overconditioning. During the breeding activity period, the amount of feed offered to bucks and does should be increased moderately. When rabbits are restricted in their feed, the producers should monitor on a regular basis the condition of the animals to adjust accordingly the amount being offered. The ideal situation is to maintain the body weight of the animals without overweight or weight loss.

FIGURE 23-4 A commercial flat-deck system operation used for efficient rabbit production. (Courtesy of Mrs. G. Absolon, Doebank, Palmerston North, New Zealand.)

PREGNANT DOES

Pregnant does should also be restricted in their feed intake during the gestation period to control their body weight. Feeding pregnant does ad libitum will result in reduction of litter size, problems at kindling, and waste of feed.

In pregnant females, feed and water consumption increase to a maximum around the tenth day of gestation. From the tenth day on, consumption is kept constant and then decreases at the end of the gestation period. A day or two before kindling, some does consume almost no feed and water. An adult New Zealand White doe consumes on the average 105 to 115 g/day of feed (2500 kcal/kg) in the first 10 days and then levels at 150 to 160 g/day until kindling. A nest box used for rabbits is shown in Fig. 23-5.

LACTATING DOES AND LITTERS

Lactating females have a high nutrient demand due to their milk production level. Milk produc-

FIGURE 23-5 A New Zealand White doe with a nest box. (Courtesy of Mrs. G. Absolon, Doebank, Palmerston North, New Zealand.)

tion is associated with litter size and nutrient intake. Proper feed intake ensures proper nutrient intake, which is reflected in the milk output and, hence, in the initial rate of growth of the rabbits. Total litter weight and litter rate of gain are related to doe milk production (17, 36). These two traits will give the producer an indication of nutrient intake and utilization by the lactating doe. Overall feed digestibility diminishes with the advance of lactation in lactating does (37, 38), thus the importance of providing feed with the proper nutrient density to these animals. Preweaning rate of growth is directly associated with postweaning rate of growth and feed efficiency. Improper feeding of lactating does not only affects milk production and litter performance, but will also have a negative effect on future reproduction performance and productive longevity of the doe.

Lactating does must be fed a good-quality feed. This ensures the availability of nutrients required for milk production. After the second week of lactation, some of the young rabbits will start to come out of the nest and to consume some of the doe's feed, so it is important to have feed available at all times for the doe and her litter. A good manager will check several times a day the level of feed in feeders for lactating does and her litters. The most economical gains of growing rabbits are made during the time they are with the doe.

After kindling, feed and water consumption increases as milk production increases (Fig. 23-6). Feeding the lactating does ad libitum right after kindling may cause a carbohydrate overload or increase the possibility of mastitis. The amount offered to does the day of kindling should be around half the normal amount. From the second day on, feed offered to lactating does should be increased gradually to ad libitum, usually in a week's time. Feed consumption may reach 500 g/day (2500 kcal/kg) in New Zealand does. The amount of feed and water consumption depends on the milk production level of the doe. Feed intake will increase in order to fulfill the nutrient demand of lactation; however, the doe may not be able to eat enough feed to meet these requirements because of the physical gut fill limitation of the animal. Some producers recommend a special lactating diet that has a higher caloric density. Animals eating a high-caloric-density diet consume less volume of feed to obtain the required amount of nutrients.

The feed offered to lactating does could be the same as that offered to fast-growing animals.

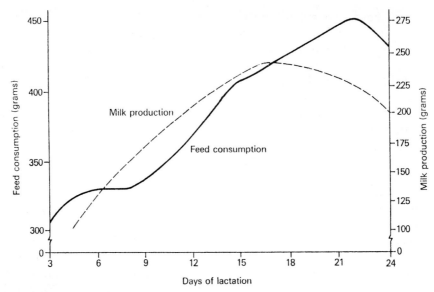

FIGURE 23-6 Doe milk production and feed consumption during lactation. [From Varela-Alvarez (17).]

Because doe milk composition remains basically unchanged throughout lactation, there is no need for a change in diet composition during this period (39). Having only one diet has the advantages of reducing feeding and storage for different diets, and because the young animals consume their mother's feed, they will continue to use the same feed after weaning without going through a period of adaptation.

Feeders utilized for lactating does with a litter should have a capacity for at least 2 to 3 days' feed supply. Feeders should be checked regularly for contamination, ease of flow, and level of feed. The physical location of the feeder should be low enough for the small animals to have easy access to the feed, but high enough to prevent the litter climbing and laying on the feed.

Creep feeding is practiced in some rabbit enterprises. This system consists of offering a special type of feed, usually higher in energy and protein than the doe diet. This diet allows the young animals to grow faster, allowing an earlier weaning. The creep feed is offered in special feeders from which the young can eat but the doe cannot. The adoption of this practice depends on the economics of it.

Does that are producing a high amount of milk at weaning time should be restricted to a minimum in the amount of concentrate offered. These animals should be offered a good-quality hay. When the doe dries off, feeding of concentrate can be resumed in a progressive manner.

This practice is intended to reduce mastitis problems in heavy milkers.

GROWING RABBITS (WEANING TO SLAUGHTER)

In meat rabbit production, the growing animals and the lactating females are the ones with the highest nutrient demand. These animals should have feed of good quality and fresh water available at all times.

For intensive meat rabbit operations, feeding is done almost exclusively with pelleted complete mixes, and there is no need for additional supplementation of either grains or forages. In less intensive operations it may be necessary to feed animals forage in order to limit energy intake. In these cases, forages should not be ground. It is recommended that they be fed as hay and cut to at least 7 to 10 cm for best results.

Growing rabbits of 5 to 10 weeks of age have an average daily dry feed consumption of 80 to 95 g/kg body weight (2500 kcal/kg). For practical feeding, allowances should be made for feed waste in order to buy and store appropriate amounts of feeds for the rabbit operation. Daily water consumption will be 135 to 150 ml/kg body weight.

Feeding management for meat rabbits should be done based on specific targets. These targets include the expected feed conversion and average daily gain. Feed conversion should be calculated in terms of total feed utilized and total gain obtained. Feed conversion varies widely among rabbit breeds and managements. An average feed conversion for meat-producing rabbits should be 3:1, that is, 3 kg of feed for each 1 kg of live weight gain. Feed conversion decreases with increase in age, especially after 10 weeks of age. For a good economic return it is important to produce rabbits that reach slaughter weight in the shortest time. Total feed utilized includes both feed consumed and feed wasted, the latter being one of the targets for reduction in the operation, because waste has a negative impact on feed efficiency. A good feed efficiency and average daily gain can be obtained through good animal genetic selection, proper nutrition and health, good feeding practices, and constant management improvement.

HAIR- AND FUR-PRODUCING RABBITS

For hair and pelt rabbit production, it should be noted that production is done with adult animals, and thus feeding should be of good quality but restricted. In this type of operation, postweaning rate of gain is not as important as for the meat-rabbit enterprise. Protein levels, and in particular S-containing amino acids, are critical for proper performance, because the final product (pelt and hair) consist mainly of protein material, primarily keratin. The minimum recommended level of total protein is 17%, with a minimum of 0.65% to 0.70% of total S-containing amino acids in the diet.

To restrict pellet consumption and at the same time maintain a high protein level in the diet, good-quality alfalfa hay and oats can be used together with the pelleted ration.

It is recommended to fast the animals at least once a week. This practice ensures the emptying of the stomach, reducing in a significant way accumulation of ingested hair. Ingested hair has the tendency to aggregate in balls that obstruct the pylorus of the rabbit, causing the animal's death.

HEALTH PROBLEMS

Health of the animals is related to proper feeding and nutrition practices. Proper feeding and nutrition give the animals the basis for building resistance and maintaining their immunological system in good working condition.

Health problems associated with feeding could be the result of nutrient imbalances, improper feed or fiber particle size, or inappropriate feeding schedules. A significant imbalance between protein and fiber can increase the risk of enteritis. It is recommended that the protein level be kept between 16% and 18% and the crude fiber at a minimum level of 14%. Other nutrient balances to keep in mind are Ca:P and K and Na. Fiber particle size could be a problem if it is too finely ground. This increases feed retention time in the digestive tract and in turn could modify the cecum microflora. Modification of the microflora results in a change in the volatile fatty acid rate of production and induces a series of reactions that produce amines and ammonia, changing the cecal pH and favoring increases in coliform bacteria. Feeding in irregular patterns (lack of a systematic schedule) could favor enteritis. Irregular patterns of work cause stress to the animals. Stress may cause diarrhea, especially in growing rabbits, producing a Na and K imbalance. To avoid these problems, all routine work in the rabbitry (feeding, breeding, cleaning) should be done at the same time every day if possible.

For certain health-related problems, such as cecum stasis or lung edema, the cause could be associated with pellet size, feed grinding, or particle size. A routine inspection of particle size will help to prevent some of the associated digestive problems in rabbits.

When feeding forages to rabbits, the use a good-quality hay is recommended, rather than green feed. Green forage may cause noninfectious diarrhea, thus affecting animal performance and economic returns.

SUMMARY

Feeding and nutrition of the rabbit depend primarily on the type of exploitation, animal size, and physiological and environmental conditions. It is important to match the feeding program with the nutritional requirements of the animal. To calculate the energy requirements for animals of different sizes, formulas have been included in this chapter for energy requirements, taking into consideration size and performance. From these energy requirements, an estimation of the expected feed intake can be obtained. Quality feeds, health problem prevention, regular labor practices, and an understanding of the unique digestive process of the rabbit will help to ensure success in the rabbit feeding program. Rabbit producers should have a goal level of production (average daily gain and feed efficiency, amount of hair per year) in order to evaluate their nutrition and feeding programs and management practices.

REFERENCES

1. Alexander, F., and A. K. Chowdury. 1958. *Brit. J. Nutr.* 12:65.

2. Parker, D. S., and A. J. Mould. 1977. *Proc. Nutr. Soc.* 36:5A.

3. Lebas, F., and J. P. Laplace. 1974. *Ann. Zootech.* 23:577.

4. Colin, M. 1975. *Le lapin: regles d'elevage et d'hygiene.* Informations Techniques des Directions des Services Veterinaires. Ministere de l'agriculture, France.

5. Proto, V. 1980. *Coniglicoltura* 17:17.

6. Marty, J., P. Raynaud, and J. Carles. 1973. *Ann. Biol. Animal Bioch. Biophys.* 13:429.

7. Salse, A. 1984. *Cuni Sciences,* 1:28.

8. Lang, Jenny. 1981. *Nutr. Abstr. Rev. Ser. B* 51:197.

9. Lebas, F., et al. 1984. *Le lapin. Elevage et pathologie.* Collection FAO: Production et sante animales. No. 21. Rome: FAO.

10. Mercier, P. 1978. *Le courrier avicole.* France No. 708. p. 12.

11. Varela-Alvarez, H., E. Grobet, and A. Gómez-de-Varela. 1975. *J. Animal Sci.* 41:239.

12. Lebas, F. 1987. *Reprod. Nutr. Develop.* 27:207.

13. Laffolay, B. 1985. *Bull. Soc. Vet. Prat. France* 69:117.

14. Prud'hon, M. 1976. I Intern. Rabbit Congress. Dijon, France. Communication No. 14.

15. Gidenne, T. 1987. *Ann. Zootech.* 36:85.

16. Varela-Alvarez, H. 1985. *Summaries.* 1985 Rabbit Conference, June 14–15, pp. 13–23. Pennsylvania State University, College of Agriculture.

17. Lebas, F., J. P. Laplace, and P. Droumenq. 1982. *Ann Zootech.* 31:233.

18. Millamide, Marie J., M. J. Fraga, and C. de Blas. 1991. *Animal Prod.* 52:215.

19. Lebas, F. 1975. *Ann. Zootech.* 24:281.

20. Gidenne, T. 1987. *Ann. Zootech.* 36:95.

21. Perez, J. M., et al. 1994. *Ann. Zootech.* 43:323.

22. Gidenne, T., and J. M. Perez. 1994. *Ann. Zootech.* 43:313.

23. Bhattacharya, A. N., and R. G. Warner. 1968. *J. Animal Sci.* 27:1418.

24. Sazzad, H. M., R. A. Vale, and P. T. C. Nobre. 1993. *Proc. VII World Conf. Animal Production.* Vol 2, p. 97. Edmonton, Canada.

25. Colin, M. 1974. *Alimentation et techniques d'elevage du lapin de chair, mise au point sur les acides amines du lapin en croissance.* ITAVI, France.

26. Bonomi, A., et al. 1982. *Annali della facolta di medicina veterinaria di Parma* 2:179.

27. Frye, T. M. 1987. *Summaries.* 1987 Rabbit Conference, Mar. 20–21, p. 16. Pennsylvania State University, College of Agriculture.

28. Verde, M. T., and J. G. Piquer. 1986. *J. Appl. Rabbit Res.* 9:181.

29. Lang, Jenny. 1981. *Nutr. Abstr. Rev. Ser. B* 51:287.

30. Marcus, S. N., and J. Watt. 1977. *Vet. Rec.* 100:452.

31. Bender, M. E. 1987. Rabbit production. Unpublished Notes. Modesto Junior College, Modesto, CA.

32. Stillions, M. C., and Ann Cooper. 1987. *Summaries.* 1987 Rabbit Conference, Mar. 20–21, p. 19. Pennsylvania State University, College of Agriculture.

33. Fraga, M. J., et al. 1991. *J. Animal Sci.* 69:1566.

34. Garcia, G., J. F. Galvez, and J. C. de Blas. 1993. *J. Animal Sci.,* 71:1823.

35. Knapka, J. J. 1987. *Summaries.* 1987 Rabbit Conference. Mar. 20–21, p. 24. Pennsylvania State University, College of Agriculture.

36. Khalil, M. H. 1994. *Animal Prod.* 59:141.

37. Gaudreault, M., and A. Sylvstre. 1980. Symposium en Production cunicole. "Situation Actuelle en Cuniculture." 6 Dec. 1980. Institut de Tech. Agr. de St.-Hyacinte. Conseil des productions animals du Quebec. M.A.P.A.Q. Canada.

38. Lebas, F. 1975. *Le lapin de chair, ses besoins nutritionnels et son alimentation pratique.* ITAVI, France.

39. El Sayiad, G. A., A. A. M. Habbib, and A. M. El-Mayhawry. 1994. *Animal Prod.* 58:153.

APPENDIX

Appendix tables have been included primarily to provide readily accessible information on some of the compositional values of feedstuffs consumed by domestic animals and on nutrient requirements of domestic animals as defined by the various committees of the National Research Council (NRC).

Appendix Table 1 presents information on nutrient composition of forage and other feedstuffs consumed by ruminants and horses. The international feed numbers (IFN) have been added for this edition because they provide some help in identifying similar products. It is a common practice in the NRC booklets to present data on a number of different types of energy terms. However, many of the feeds listed in Appendix Table 1 have not been evaluated for energy utilization (particularly net energy) using animals, and values have been arrived at by using different regression equations based on chemical analyses.

It is obvious after examination of some of the data that there are many differences in the values listed for beef and dairy cattle. In the writer's opinion we have made some backward strides, because now it is necessary to use values quoted for beef cattle to formulate rations or calculate requirements for beef cattle—data quoted for dairy cattle or sheep will not fit. This is so because the committees have gone their own ways and have recalculated energy values and requirements, and, rather than the results coming closer together, the differences are exaggerated.

With regard to nutrient content of feedstuffs, the reader is cautioned that average values are just that. All feed ingredients vary in composition, and averages may or may not be applicable to a given feed or for a particular species of animal or in conditions where the environment may alter nutrient utilization, particularly for the various energy values listed.

Other tables that the reader may find useful have been included. For example, Appendix Table 2 gives values on ruminal undegradability of various proteins in natural feedstuffs. Appendix Table 3 contains information on compositions of feedstuffs commonly consumed by poultry and swine (on an as-fed basis, because that is the way they are presented in the NRC booklets). Appendix Table 4 contains information on amino acid composition of selected feeds, limited data on vitamins are presented in Table 5, and data on mineral supplements are found in Table 6. Most of the remaining tables give data on either feed consumption or on estimated nutrient requirements. More detail may be found in the NRC booklets or in other sources.

APPENDIX TABLE I

Composition of feedstuffs commonly fed to cattle, sheep, and horses (data from NRC publications)

Feed Name and Description	Intl. Feed Number	Typical DM, %	Composition, Dry Basis, %						Energy Utilization, Dry Basis, Mcal/kg Beef Cattle				TDN,[a] %
			CP	CF	NDF	ADF	Ca	P	DE	ME	NEm	NEg	
1. Alfalfa, fresh, late veg.	2-00-181	21	20.0	23	38	29	2.19	0.33	2.78	2.28	1.41	0.83	63
2. Alfalfa, fresh	2-00-196	24	20.0	26	—	—	1.96	0.30	2.69	2.27	1.31	0.61	61
3. Alfalfa, hay, S-C, early bloom	1-00-059	90	18.0	23	42	31	1.41	0.22	2.43	1.99	1.14	0.58	60
4. Alfalfa, hay, S-C, mature	1-00-071	91	13.0	38	58	44	1.13	0.18	2.21	1.81	0.97	0.42	50
5. Alfalfa, meal, dehy. 17%	1-00-023	92	18.9	26	45	35	1.52	0.25	2.69	2.21	1.34	0.77	61
6. Alfalfa, silage, wilted midbloom	3-00-217	38	15.5	30	47	35	—	—	2.56	2.10	1.24	0.68	58
7. Alfalfa, silage, 30-50% dry matter	3-08-150	43	9.8	19	—	—	1.39	0.27	—	—	—	—	—
8. Bakery waste, dried	4-00-466	92	10.7	1	18	13	0.14	0.26	3.92	3.22	2.21	1.52	89
9. Barley grain	4-00-549	88	13.5	6	19	7	0.05	0.38	3.70	3.04	2.06	1.40	84
10. Barley grain, Pacific Coast	4-07-939	89	10.8	7	21	9	0.06	0.39	3.79	3.11	2.12	1.45	86
11. Barley straw	1-00-498	91	4.3	42	80	49	0.30	0.07	1.76	1.45	0.60	0.08	40
12. Bean, Navy, seeds	5-00-623	89	25.3	5			0.18	0.59	3.70	3.04	2.06	1.40	84
13. Beet, Sugar, pulp, dehy.	4-00-669	91	9.7	20	54	33	0.69	0.10	3.26	2.68	1.76	1.14	74
14. Bermuda grass, fresh	2-00-712	34	12.0	26	—	—	0.53	0.21	2.65	2.17	1.31	0.74	60
15. Bermuda grass hay, SC	1-00-703	90	6.0	31	78	38	0.43	0.20	2.16	1.77	0.93	0.39	49
16. Bluegrass, Kentucky fresh, early veg.	2-00-777	31	17.4	25	55	29	0.50	0.44	3.17	2.60	1.70	1.08	72
17. Bluestem, fresh, early veg.	2-00-821	27	12.8	25	—	—	0.63	0.20	3.00	2.46	1.57	0.97	68
18. Brewer's grains, dehy.	5-02-141	92	29.4	14	46	24	0.33	0.55	2.91	2.39	1.51	0.91	66
19. Brome, fresh, early veg.	2-00-892	34	18.0	24	56	31	0.50	0.30	3.26	2.68	1.76	1.14	74
20. Brome hay, late bloom, S-C	1-00-888	89	10.0	37	68	43	0.30	0.35	2.43	1.99	1.14	0.58	55
21. Citrus pulp, dehy.	4-01-237	91	6.7	13	23	22	1.84	0.12	3.62	2.97	2.00	1.35	77
22. Clover, Crimson, hay, S-C	1-01-328	87	18.4	30	—	—	1.40	0.22	2.51	2.06	1.21	0.64	57
23. Clover, Ladino, fresh, early veg.	2-01-380	19	27.2	14	—	—	1.93	0.35	3.00	2.46	1.57	0.97	68
24. Clover, Ladino, hay, S-C	1-01-378	90	22.0	21	36	32	1.35	0.31	2.65	2.17	1.31	0.74	65
25. Clover, Red, fresh, early bloom	2-01-428	20	19.4	23	40	31	2.26	0.38	3.04	2.50	1.60	1.00	69
26. Clover, Red, hay, S-C	1-01-415	89	16.0	29	56	36	1.53	0.25	2.43	1.99	1.14	0.58	55
27. Corn, Dent, fodder	1-28-231	81	8.9	25	55	33	0.50	0.25	2.87	2.35	1.47	0.88	65
28. Corn, Cobs, ground	1-28-234	90	3.2	36	89	35	0.57	0.10	2.21	1.81	0.97	0.42	50
29. Corn distillers grains, dehy.	5-28-235	94	23.0	12	43	17	0.11	0.43	3.79	3.11	2.12	1.45	86
30. Corn ears, ground	4-28-238	87	9.0	9	28	11	0.07	0.27	3.66	3.00	2.03	1.37	83
31. Corn gluten, meal	5-28-241	91	46.8	5	37	9	0.16	0.50	3.79	3.11	2.12	1.45	86
32. Corn gluten feed	5-28-243	90	25.6	10	45	12	0.36	0.82	3.66	3.00	2.03	1.37	83
33. Corn grain, #2	4-02-931	88	10.1	2	9	3	0.02	0.35	3.97	3.25	2.24	1.55	90
34. Corn grain, flaked	4-28-244	86	11.2	1	9	3	0.03	0.29	4.19	3.44	2.38	1.67	95
35. Corn grain, high moisture	4-20-770	72	10.7	3	9	3	0.02	0.32	4.10	3.36	2.33	1.62	93
36. Corn silage, well-eared	3-28-250	33	8.1	24	51	28	0.23	0.22	3.09	2.53	1.63	1.03	70
37. Cotton, seed hulls	1-01-599	91	4.1	48	90	64	0.15	0.09	1.85	1.52	0.68	0.15	42
38. Cottonseeds	5-01-614	92	23.9	21	39	29	0.16	0.75	4.23	3.47	2.41	1.69	96

[a]TDN values apply to both beef and dairy cattle.

	Energy Utilization, Dry Basis, Mcal/kg											
	Dairy Cattle					Sheep					Horses	
	DE	ME	NEm	NEg	NEℓ	DE	ME	NEm	NEg	TDN, %	DE	TDN, %
1.	—	—	—	—	—	—	—	—	—	—	2.94	—
2.	—	—	—	—	—	2.56	2.10	1.24	0.68	58	2.51	57
3.	2.65	2.22	1.31	0.74	1.35	2.47	2.03	1.18	0.61	56	2.48	55
4.	—	—	—	—	—	2.38	1.95	1.11	0.55	54	—	—
5.	2.69	2.27	1.34	0.77	1.38	2.65	2.17	1.34	0.77	60	2.36	—
6.	—	—	—	—	—	—	—	—	—	—	—	—
7.	—	—	—	—	—	2.56	2.10	1.24	0.68	58	—	—
8.	3.92	3.51	2.20	1.52	2.06	—	—	—	—	—	—	—
9.	3.70	3.29	2.06	1.40	1.94	3.79	3.11	2.12	1.45	86	3.60	82
10.	3.79	3.38	2.12	1.45	1.99	3.88	3.18	2.18	1.50	88	3.68	79
11.	2.16	1.73	0.93	0.53	1.08	2.12	1.74	0.90	0.35	48	1.62	37
12.	3.70	3.29	2.06	1.40	1.94	3.84	3.15	2.15	1.48	87	—	—
13.	3.44	3.02	1.88	1.24	1.79	2.96	2.43	1.60	1.04	67	2.56	65
14.	—	—	—	—	—	—	—	—	—	—	2.07	50
15.	—	—	—	—	—	1.97	1.62	0.85	0.35	45	—	—
16.	3.17	2.76	1.69	1.08	1.64	2.87	2.35	1.47	0.88	65	2.09	56
17.	—	—	—	—	—	—	—	—	—	—	—	—
18.	2.91	2.49	1.51	0.91	1.50	3.09	2.53	1.63	1.03	70	2.75	68
19.	3.26	2.85	1.75	1.13	1.69	3.53	2.89	1.94	1.30	80	3.00	68
20.	2.60	2.18	1.27	0.70	1.33	—	—	—	—	—	2.38	54
21.	3.40	2.98	1.86	1.22	1.77	3.70	3.04	2.06	1.40	84	2.99	68
22.	2.51	2.09	1.21	0.64	1.28	2.43	1.99	1.14	0.58	55	2.16	49
23.	3.00	2.58	1.57	0.97	1.55	—	—	—	—	—	2.50	—
24.	2.87	2.45	1.47	0.88	1.47	2.91	2.39	1.51	0.91	66	2.56	51
25.	3.04	2.62	1.60	1.00	1.57	3.00	2.46	1.57	0.47	68	2.01	57
26.	2.43	2.00	1.14	0.58	1.23	2.65	2.17	1.31	0.74	60	2.22	49
27.	—	—	—	—	—	2.78	2.28	1.41	0.83	63	2.06	—
28.	2.21	1.78	0.97	0.42	1.11	2.25	1.84	1.07	0.45	51	1.36	31
29.	3.79	3.38	2.12	1.45	1.99	3.84	3.15	2.12	1.48	87	3.08	70
30.	3.66	3.25	2.03	1.37	1.91	3.66	3.00	2.03	1.37	83	3.49	74
31.	3.79	3.38	2.12	1.45	1.99	3.88	3.18	2.18	1.50	88	3.29	—
32.	3.66	3.25	2.03	1.37	1.91	3.66	3.00	2.03	1.37	83	—	—
33.	3.53	3.12	1.94	1.30	1.84	3.84	3.15	2.15	1.48	87	3.84	88
34.	3.88	3.47	2.18	1.50	2.04	—	—	—	—	—	—	—
35.	3.88	3.47	2.18	1.50	2.06	—	—	—	—	—	—	—
36.	3.09	2.67	1.63	1.03	1.60	3.09	2.53	1.63	1.03	70	2.68	—
37.	1.98	1.55	0.78	0.25	0.98	2.16	1.77	0.93	0.39	49	1.89	33
38.	4.23	3.83	2.41	1.69	2.23	—	—	—	—	—	—	—

APPENDIX TABLE I (*cont.*)

| Feed Name and Description | Intl. Feed Number | Typical DM, % | Composition, Dry Basis, % | | | | | | Energy Utilization, Dry Basis, Mcal/kg | | | | |
| | | | | | | | | | Beef Cattle | | | | |
			CP	CF	NDF	ADF	Ca	P	DE	ME	NEm	NEg	TDN,[a] %
39. Cottonseed, meal, mech. extd., 41% protein	5-01-617	93	44.3	13	28	20	0.21	1.16	3.44	2.82	1.88	1.24	78
40. Cottonseed, meal, solv. extd., 41% protein	5-01-621	91	45.2	13	26	19	0.18	1.21	3.35	2.75	1.82	1.19	76
41. Fat, animal-poultry	4-00-409	99	—	—	—	—	—	—	7.80	6.40	4.75	3.51	177
42. Fescue, Kentucky 31, fresh, early veg.	2-01-900	28	22.1	21	—	—	0.53	0.39	—	—	—	—	67
43. Fescue hay, S-C, early bloom	1-09-186	91	20.2	25	59	32	0.35	0.24	—	—	—	—	64
44. Fescue hay, S-C, late bloom	1-01-871	92	9.5	37	72	39	0.30	0.26	2.12	1.74	0.90	0.35	48
45. Flax, seed, solvent extd. (linseed meal)	5-02-048	90	38.3	10	25	19	0.43	0.89	3.44	2.82	1.88	1.24	78
46. Grass-legume silage	3-02-303	29	11.3	32	—	—	0.25	0.08	—	—	—	—	—
47. Lespedeza, common fresh, late veg.	2-26-028	32	16.4	32	—	—	—	—	2.60	2.13	1.28	0.71	—
48. Milk, cattle, fresh	5-01-168	12	26.7	—	—	—	0.95	0.76	5.60	5.43	3.80	3.80	129
49. Milk, sheep, fresh	5-08-510	19	24.7	—	—	—	1.05	0.79	—	—	—	—	—
50. Millet, Foxtail, fresh	2-03-101	28	9.5	32	—	—	0.32	0.19	2.78	2.28	1.41	0.83	63
51. Molasses, beet	4-00-668	78	8.5	—	—	—	0.17	0.03	3.48	2.86	1.91	1.27	75
52. Molasses, citrus	4-01-241	68	8.2	—	—	—	1.72	0.13	3.31	2.71	1.79	1.16	75
53. Molasses, sugarcane	4-04-696	75	5.8	6	—	—	1.00	0.11	3.17	2.60	1.70	1.08	72
54. Oats, grain	4-03-309	89	13.3	12	32	16	0.07	0.38	3.40	2.78	1.85	1.22	77
55. Oats, grain, Pacific Coast	4-07-999	91	10.0	12	—	—	0.11	0.34	3.44	2.82	1.88	1.24	78
56. Oat hay, S-C	1-03-280	91	9.3	30	66	42	0.24	0.22	2.43	1.99	1.14	0.58	55
57. Oat silage, dough stage	3-03-296	35	10.0	33	—	—	0.47	0.33	2.51	2.06	1.21	0.64	—
58. Oat straw	1-03-283	92	4.4	40	70	54	0.24	0.06	1.98	1.63	0.79	0.25	45
59. Orchard grass, fresh, early, veg.	2-03-439	23	18.4	25	55	31	0.58	0.54	3.17	2.60	1.70	1.08	72
60. Orchard grass, hay, late bloom	1-03-428	91	8.4	37	72	45	0.26	0.30	2.38	1.95	1.11	0.55	54
61. Pea seeds	5-03-600	89	25.3	7	—	—	0.15	0.44	3.84	3.15	2.15	1.48	87
62. Potato, tubers, fresh	4-03-787	23	9.5	2	—	—	0.04	0.24	3.57	2.93	1.97	1.32	81
63. Potato, tubers, silage	4-03-768	25	7.6	4	—	—	0.04	0.23	3.62	2.97	2.00	1.35	82
64. Poultry feathers, hydrolyzed	5-03-795	93	91.3	1	—	—	0.28	0.72	3.09	2.53	1.63	1.03	—
65. Poultry manure, dehy.	5-14-015	90	28.2	13	38	15	9.31	2.52	2.29	1.88	1.04	0.49	—
66. Prairie plants, midwest, hay, S-C	1-03-191	92	5.8	34	—	—	0.43	0.15	2.25	1.84	1.00	0.45	51
67. Rape, fresh, early bloom	2-03-866	11	23.5	16	—	—	—	—	3.31	2.71	1.79	1.16	75
68. Rape, seed, meal, solvent extd.	5-03-871	91	40.6	13	—	—	0.67	1.04	3.04	2.50	1.60	1.00	69
69. Redtop, fresh	2-03-897	29	11.6	27	64	—	0.46	0.29	2.78	2.28	1.41	0.83	63
70. Redtop, hay, S-C, midbloom	1-03-886	94	11.7	31	—	—	0.63	0.35	2.51	2.06	1.21	0.64	57
71. Rice, bran	4-03-928	91	14.1	13	33	18	0.08	1.70	3.09	2.53	1.63	1.03	70
72. Rice, ground	4-03-938	89	8.9	10	—	—	0.07	0.32	3.48	2.86	1.91	1.27	79
73. Rice, straw	1-03-925	91	4.3	35	71	55	0.21	0.08	1.81	1.48	0.64	0.11	41
74. Rye, fresh	2-04-018	24	15.9	28	—	—	0.39	0.33	3.04	2.50	1.60	1.00	69
75. Rye, grain	4-04-047	88	13.8	3	—	—	0.07	0.37	3.70	3.04	2.06	1.40	84
76. Rye, straw	1-04-007	90	3.0	43	—	—	0.24	0.09	1.37	1.12	0.26	—	31

[a]TDN values apply to both beef and dairy cattle.

	Energy Utilization, Dry Basis, Mcal/kg											
	Dairy Cattle					Sheep					Horses	
	DE	ME	NEm	NEg	NEl	DE	ME	NEm	NEg	TDN, %	DE	TDN, %
39.	3.44	3.02	1.88	1.24	1.79	3.31	2.71	1.79	1.16	75	—	—
40.	3.35	2.93	1.82	1.19	1.74	3.13	2.57	1.67	1.06	71	3.01	—
41.	7.30	7.30	5.84	5.84	5.84	—	—	—	—	—	8.00	—
42.	2.91	2.49	1.51	0.92	1.50	3.22	2.64	1.73	1.11	73	2.22	—
43.	2.82	2.40	1.44	0.85	1.45	2.73	2.24	1.38	0.80	62	—	—
44.	2.12	1.69	0.90	0.36	1.06	—	—	—	—	—	—	—
45.	3.44	3.02	1.88	1.24	1.79	3.48	2.86	1.91	1.27	79	3.04	69
46.	—	—	—	—	—	2.73	2.24	1.38	0.80	62	—	—
47.	—	—	—	—	—	—	—	—	—	—	1.89	—
48.	5.69	5.29	3.34	2.16	3.04	—	—	—	—	—	—	—
49.	—	—	—	—	—	6.00	5.82	4.07	161	—	—	—
50.	2.78	2.36	1.41	0.83	1.42	—	—	—	—	—	—	—
51.	3.31	2.89	1.79	1.16	1.72	3.40	2.78	1.85	1.22	77	3.40	72
52.	3.31	2.89	1.79	1.16	1.72	—	—	—	—	—	3.40	—
53.	3.17	2.76	1.69	1.08	1.64	3.48	2.86	1.91	1.27	79	3.50	74
54.	3.40	2.98	1.86	1.22	1.77	3.40	2.78	1.85	1.22	77	3.36	76
55.	3.44	3.02	1.88	1.24	1.79	3.44	2.82	1.88	1.24	78	3.20	77
56.	2.43	2.00	1.14	0.58	1.23	2.34	1.92	1.14	0.58	53	1.92	47
57.	—	—	—	—	—	—	—	—	—	—	—	—
58.	—	—	—	—	—	2.07	1.70	0.79	0.25	47	1.76	40
59.	3.17	2.76	1.69	1.08	1.64	2.95	2.42	1.54	0.94	67	2.42	55
60.	2.38	1.96	1.11	0.55	1.20	—	—	—	—	—	1.90	—
61.	3.84	3.42	2.16	1.48	2.01	—	—	—	—	—	3.45	—
62.	3.57	3.16	1.97	1.32	1.87	—	—	—	—	—	—	—
63.	3.62	3.20	2.01	1.35	1.89	—	—	—	—	—	—	—
64.	—	—	—	—	—	—	—	—	—	—	—	—
65.	—	—	—	—	—	—	—	—	—	—	—	—
66.	—	—	—	—	—	—	—	—	—	—	1.62	46
67.	—	—	—	—	—	3.31	2.71	1.79	1.16	75	—	—
68.	3.04	2.62	1.60	1.00	1.57	3.26	2.68	1.76	1.14	74	—	—
69.	2.78	2.36	1.41	0.83	1.42	—	—	—	—	—	—	—
70.	2.51	2.09	1.21	0.64	1.28	2.47	2.03	1.18	0.61	56	1.97	—
71.	3.09	2.67	1.63	1.03	1.60	3.26	2.68	1.76	1.14	74	—	—
72.	3.48	3.07	1.91	1.27	1.82	—	—	—	—	—	2.51	—
73.	—	—	—	—	—	—	—	—	—	—	—	—
74.	—	—	—	—	—	—	—	—	—	—	—	—
75.	3.70	3.29	2.06	1.40	1.94	3.75	3.07	2.09	1.43	85	3.84	80
76.	—	—	—	—	—	1.98	1.63	0.79	0.25	45	—	—

| Feed Name and Description | Intl. Feed Number | Typical DM, % | Composition, Dry Basis, % | | | | | | Energy Utilization, Dry Basis, Mcal/kg | | | | |
| | | | CP | CF | NDF | ADF | Ca | P | Beef Cattle | | | | TDN,[a] % |
									DE	ME	NEm	NEg	
77. Ryegrass, Italian, fresh	2-04-073	25	14.5	24	—	—	0.65	0.41	2.65	2.17	1.31	0.74	60
78. Ryegrass, Italian, hay, S-C, early bloom	1-04-066	88	11.4	36	69	45	0.62	0.34	2.38	1.95	1.11	0.55	54
79. Ryegrass, perennial, fresh	2-04-086	27	10.4	23	—	—	0.55	0.27	3.00	2.46	1.57	0.97	68
80. Ryegrass, perennial, hay, S-C	1-04-077	86	8.6	30	41	30	0.65	0.32	2.65	2.17	1.31	0.74	60
81. Safflower seeds, meal, solvent extd.	5-04-110	92	25.4	32	58	41	0.37	0.81	2.51	2.06	1.21	0.64	57
82. Safflower seeds w/o hulls, meal, solvent extd.	5-07-959	92	46.9	15	—	—	0.38	1.40	3.22	2.64	1.73	1.11	73
83. Sage, Black, browse, fresh	2-05-564	65	8.5	—	—	—	0.81	0.17	2.16	1.77	0.93	0.39	49
84. Saltgrass, hay, S-C	1-04-168	89	8.9	32	—	—	—	—	2.25	1.84	1.00	0.45	51
85. Sedge, hay, S-C	1-04-193	89	9.4	32	—	—	—	—	2.29	1.88	1.04	0.49	52
86. Sorghum, fodder	1-07-960	89	7.5	27	—	—	0.52	0.13	2.56	2.10	1.24	0.68	58
87. Sorghum, grain 8–10% protein	4-20-893	87	10.1	3	18	9	0.04	0.34	3.70	3.04	2.06	1.40	84
88. Sorghum, grain, flaked	4-16-295	85	10.1	3	—	—	0.04	0.34	4.06	3.33	2.30	1.60	92
89. Sorghum, grain, reconstituted	4-16-296	70	10.1	3	—	—	0.04	0.34	4.10	3.36	2.33	1.62	93
90. Sorghum, milo, heads	4-04-446	90	10.0	9	—	—	0.13	0.25	—	—	—	—	—
91. Sorghum, silage	3-04-323	30	7.5	28	—	38	0.35	0.21	2.65	2.17	1.31	0.74	60
92. Sorghum, Sudan grass, fresh, early veg.	2-04-484	18	16.8	23	55	29	0.43	0.41	3.09	2.53	1.63	1.03	70
93. Sorghum, Sudan grass, hay, S-C	1-04-480	91	8.0	36	68	42	0.55	0.30	2.47	2.03	1.18	0.61	56
94. Sorghum, Sudan grass, silage	3-04-499	28	10.8	33	—	—	0.46	0.21	2.43	1.99	1.14	0.58	55
95. Soybean, hay, S-C	1-04-538	94	17.8	30	—	40	1.26	0.27	—	—	—	—	53
96. Soybean, hulls	1-04-560	91	12.1	40	67	50	0.49	0.21	2.34	1.92	1.07	0.52	57
97. Soybean, seeds	5-04-610	92	42.8	6	—	10	0.27	0.65	4.01	3.29	2.27	1.57	91
98. Soybean, meal, solvent extd., 44% protein	5-20-637	89	49.9	7	—	10	0.33	0.71	3.70	3.04	2.06	1.40	84
99. Soybean, straw	1-04-567	88	5.2	44	70	54	1.59	0.06	1.85	1.52	0.68	0.15	42
100. Sunflower, seeds, meal, solvent extd.	5-09-340	90	25.9	35	40	33	0.23	1.03	1.94	1.59	0.75	0.22	44
101. Sunflower, seeds w/o hulls, solvent extd.	5-04-739	93	49.8	12	—	—	0.44	0.98	2.87	2.35	1.47	0.88	65
102. Timothy, fresh, late veg.	2-04-903	26	18.0	32	—	—	0.39	0.32	3.17	2.60	1.70	1.08	72
103. Timothy, hay, S-C, late veg.	1-04-881	89	17.0	27	55	29	0.66	0.34	2.73	2.24	1.38	0.80	62
104. Timothy, hay, S-C, midbloom	1-04-883	89	9.1	31	67	36	0.48	0.22	2.51	2.06	1.21	0.64	57
105. Trefoil, Birdsfoot, fresh	2-20-786	24	21.0	25	—	—	1.91	0.22	2.91	2.39	1.51	0.91	66
106. Triticale, grain	4-20-362	90	17.6	4	—	8	0.06	0.33	3.70	3.04	2.06	1.40	84
107. Turnip, roots, fresh	4-05-067	9	11.8	12	44	34	0.59	0.26	3.75	3.07	2.09	1.43	85

[a]TDN values apply to both beef and dairy cattle.

	Energy Utilization, Dry Basis, Mcal/kg											
	Dairy Cattle					Sheep					Horses	
	DE	ME	NEm	NEg	NEℓ	DE	ME	NEm	NEg	TDN, %	DE	TDN, %
77.	—	—	—	—	—	—	—	—	—	—	1.73	—
78.	2.38	1.96	1.11	0.55	1.20	2.51	2.06	1.21	0.64	57	1.94	—
79.	—	—	—	—	—	—	—	—	—	—	—	—
80.	2.82	2.40	1.44	0.85	1.45	—	—	—	—	—	—	—
81.	2.51	2.09	1.21	0.64	1.28	2.47	2.03	1.18	0.61	56	—	—
82.	3.22	2.80	1.73	1.11	1.67	—	—	—	—	—	—	—
83.	—	—	—	—	—	2.16	1.77	0.93	0.39	49	—	—
84.	—	—	—	—	—	2.25	1.84	1.00	0.45	51	—	—
85.	—	—	—	—	—	—	—	—	—	—	—	—
86.	2.56	2.13	1.25	0.68	1.30	2.56	2.10	1.24	0.68	58	—	—
87.	3.53	3.12	1.94	1.30	1.84	—	—	—	—	—	3.56	80
88.	—	—	—	—	—	—	—	—	—	—	—	—
89.	—	—	—	—	—	—	—	—	—	—	—	—
90.	—	—	—	—	—	3.70	3.04	2.06	1.40	84	—	—
91.	2.65	2.22	1.31	0.74	1.35	2.51	2.06	1.21	0.64	57	—	—
92.	3.09	2.67	1.63	1.03	1.60	2.78	2.28	1.41	0.83	63	—	—
93.	2.47	2.04	1.18	0.62	1.25	2.43	1.99	1.14	0.58	55	—	—
94.	2.43	2.00	1.14	0.58	1.23	2.34	1.92	1.07	0.52	53	—	—
95.	2.34	1.91	1.08	0.52	1.18	—	—	—	—	—	—	—
96.	—	—	—	—	—	2.51	2.06	1.21	0.64	57	1.88	48
97.	4.01	3.60	2.27	1.57	2.11	4.14	3.40	2.35	1.65	94	4.05	92
98.	3.70	3.29	2.06	1.40	1.94	3.88	3.18	2.18	1.50	88	3.52	82
99.	—	—	—	—	—	1.90	1.56	0.72	0.18	43	—	—
100.	1.94	1.51	0.75	0.22	0.96	1.98	1.63	0.75	0.25	45	—	—
101.	2.87	2.45	1.47	0.88	1.47	3.35	2.75	1.82	1.19	76	3.12	71
102.	—	—	—	—	—	2.69	2.21	1.34	0.77	61	1.70	—
103.	2.91	2.49	1.51	0.92	1.50	2.87	2.35	1.47	0.88	65	—	—
104.	2.56	2.13	1.25	0.68	1.30	2.65	2.17	1.31	0.74	60	1.99	45
105.	2.91	2.49	1.31	0.92	1.50	2.78	2.28	1.41	0.83	63	2.18	—
106.	3.70	3.29	2.06	1.40	1.94	—	—	—	—	—	—	—
107.	3.75	3.34	2.10	1.43	1.96	3.79	3.11	2.12	1.45	86	—	—

Feed Name and Description	Intl. Feed Number	Typical DM, %	Composition, Dry Basis, %						Energy Utilization, Dry Basis, Mcal/kg				
									Beef Cattle				
			CP	CF	NDF	ADF	Ca	P	DE	ME	NEm	NEg	TDN,[a] %
108. Urea, 281% protein equivalent	5-05-070	99	280.0	—	—	—	—	—	—	—	—	—	—
109. Vetch, fresh, late veg.	2-05-108	22	20.8	28	—	—	—	—	—	—	—	—	—
110. Vetch, hay, S-C	1-05-106	89	20.8	31	48	33	1.18	0.32	2.51	2.06	1.21	0.64	57
111. Wheat, bran	4-05-190	89	17.1	11	51	15	0.13	1.38	3.09	2.53	1.63	1.03	70
112. Wheat, flour by-product (middlings)	4-05-205	89	18.4	8	37	10	0.13	0.99	3.04	2.50	1.60	1.00	69
113. Wheat, fresh, early veg.	2-05-176	22	28.6	17	52	30	0.42	0.40	3.22	2.64	1.73	1.11	73
114. Wheat, grain, hard red spring	4-05-258	88	17.2	3	—	—	0.04	0.43	3.92	3.22	2.21	1.52	89
115. Wheat, grain, hard red winter	4-05-268	88	14.4	3	—	4	0.05	0.43	3.88	3.18	2.18	1.50	88
116. Wheat, grain, soft red winter	4-05-294	88	13.0	2	—	—	0.05	0.43	3.92	3.22	2.21	1.52	89
117. Wheat, grain, soft white winter	4-05-337	89	11.3	3	14	4	0.07	0.36	3.92	3.22	2.21	1.52	89
118. Wheat, grain, soft white winter, Pacific Coast	4-08-555	89	11.2	3	—	—	0.10	0.34	3.88	3.18	2.18	1.50	88
119. Wheat, mill run	4-05-206	90	17.2	9	—	—	0.11	0.13	3.48	2.86	1.91	1.27	79
120. Wheat, silage, full bloom	3-05-185	25	8.1	31	—	—	—	—	2.60	2.13	1.28	0.71	59
121. Wheat, straw	1-05-175	89	3.6	42	70	54	0.18	0.05	1.81	1.48	0.64	0.11	41
122. Wheatgrass, Crested fresh, early veg.	2-05-420	28	21.5	22	—	—	0.46	0.34	3.31	2.71	1.79	1.16	75
fresh, post ripe	2-05-428	80	3.1	40	—	—	0.27	0.07	2.16	1.77	0.93	0.39	49
hay, S-C	1-05-418	93	12.4	33	—	36	0.33	0.21	2.34	1.92	1.07	0.52	53
123. Whey, cattle, dehy.	4-01-182	93	14.2	—	—	—	0.92	0.82	3.57	2.93	1.97	1.32	81
124. Whey, cattle, fresh	4-08-134	7	13.0	—	—	—	0.73	0.65	4.14	3.40	2.35	1.65	94
125. Yeast, Brewers, dehy.	7-05-527	93	46.9	3	—	—	0.13	1.49	3.48	2.86	1.91	1.27	79

[a]TDN values apply to both beef and dairy cattle.

| | Energy Utilization, Dry Basis, Mcal/kg | | | | | | | | | | | |
| | Dairy Cattle | | | | | Sheep | | | | | Horses | |
	DE	ME	NEm	NEg	NEℓ	DE	ME	NEm	NEg	TDN, %	DE	TDN, %
108.	—	—	—	—	—	—	—	—	—	—	—	—
109.	—	—	—	—	—	2.60	2.13	1.28	0.71	59	—	—
110.	2.51	2.09	1.21	0.64	1.28	2.43	1.99	1.14	0.58	55	—	—
111.	3.09	2.67	1.63	1.03	1.60	3.13	2.57	1.67	1.06	71	2.94	67
112.	3.04	2.62	1.60	1.00	1.57	3.62	2.97	2.00	1.35	82	3.42	—
113.	3.22	2.80	1.73	1.11	1.67	3.31	2.71	1.79	1.16	75	2.88	—
114.	—	—	—	—	—	3.97	3.25	2.24	1.55	90	—	—
115.	3.88	3.47	2.18	1.50	2.04	3.88	3.18	2.18	1.50	88	3.86	87
116.	3.92	3.51	2.20	1.52	2.06	3.88	3.18	2.18	1.50	88	3.86	87
117.	3.92	3.51	2.20	1.52	2.06	—	—	—	—	—	3.92	87
118.	—	—	—	—	—	3.92	3.22	2.21	1.52	89	—	—
119.	3.48	3.07	1.91	1.27	1.82	—	—	—	—	—	—	—
120.	—	—	—	—	—	—	—	—	—	—	—	—
121.	1.94	1.51	0.75	0.22	0.96	1.81	1.48	0.64	0.11	41	1.62	34
122.	—	—	—	—	—	3.31	2.71	1.79	1.16	75	2.54	—
	—	—	—	—	—	2.38	1.95	1.11	0.55	54	—	—
	—	—	—	—	—	2.34	1.92	1.07	0.52	53	—	—
123.	3.57	3.16	1.97	1.32	1.87	—	—	—	—	—	4.06	—
124.	3.57	3.16	1.97	1.32	1.87	—	—	—	—	—	—	—
125.	3.48	3.07	1.91	1.27	1.82	—	—	—	—	—	3.30	75

APPENDIX TABLE 2

Ruminal undegradability of protein in selected feeds (from 1989 NRC on dairy cattle)

| Feed | Number of Determinations | Undegradability | | |
		Mean	S.D.[a]	C.V.[b]
Alfalfa, dehydrated	8	0.59	0.17	29
Alfalfa hay	12	0.28	0.07	25
Alfalfa silage	6	0.23	0.08	36
Alfalfa-bromegrass	1	0.21		
Barley	16	0.27	0.10	37
Barley, flaked	1	0.67		
Barley, micronized	1	0.47		
Barley silage	1	0.27		
Bean meal, field	1	0.46		
Beans	2	0.16	0.02	14
Beet pulp	4	0.45	0.14	30
Beet pulp molasses	2	0.35	0.03	8
Beets	3	0.20	0.03	16
Blood meal	2	0.82	0.01	1
Brewers dried grains	9	0.49	0.13	27

Feed	Number of Determinations	Undegradability		
		Mean	S.D.[a]	C.V.[b]
Bromegrass	1	0.44		
Casein	3	0.19	0.06	32
Casein, HCHO[c]	2	0.72	0.08	11
Clover, red	3	0.31	0.04	12
Clover, red, silage	1	0.38		
Clover, white	1	0.33		
Clover-grass	2	0.54	0.11	21
Clover-grass silage	7	0.28	0.06	22
Coconut	1	0.57		
Coconut meal	5	0.63	0.07	11
Corn	11	0.52	0.18	34
Corn, 0% cottonseed hulls	1	0.46		
Corn, 7% cottonseed hulls	1	0.43		
Corn, 14% cottonseed hulls	1	0.59		
Corn, 21% cottonseed hulls	1	0.48		
Corn, 10.5% protein, 0% $NaHCO_3$	1	0.36		
Corn, 10.5% protein, 3.5% $NaHCO_3$	1	0.30		
Corn, 12% protein, 0% $NaHCO_3$	1	0.29		
Corn, 12% protein, 3.5% $NaHCO_3$	1	0.24		
Corn, dry-rolled	6	0.60	0.07	12
Corn, dry-rolled, 0% roughage	1	0.54		
Corn, dry-rolled, 21% roughage	1	0.49		
Corn, flaked	1	0.58		
Corn, flakes	1	0.65		
Corn, high-moisture acid	1	0.56		
Corn, high-moisture ground	1	0.80		
Corn, micronized	1	0.29		
Corn, steam-flaked	1	0.68		
Corn, steam-flaked, 0% roughage	1	0.51		
Corn, steam-flaked, 21% roughage	1	0.47		
Corn gluten feed	1	0.25		
Corn gluten feed dry	2	0.22	0.11	51
Corn gluten feed wet	1	0.26		
Corn gluten meal	3	0.55	0.08	14
Corn silage	3	0.31	0.06	20
Cottonseed meal	21	0.43	0.11	25
Cottonseed meal, HCHO[c]	2	0.64	0.15	23
Cottonseed meal, prepressed	2	0.36	0.02	6
Cottonseed meal, screwpressed	2	0.50	0.10	20
Cottonseed meal, solvent	6	0.41	0.13	32
Distillers dried grain with solubles	4	0.47	0.18	39
Distillers dried grains	1	0.54		
Distillers wet grains	1	0.47		
Feather meal, hydrolyzed	1	0.71		
Fish meal	26	0.60	0.16	26
Fish meal, stale	1	0.48		
Fish meal, well-preserved	1	0.78		
Grapeseed meal	1	0.45		
Grass	4	0.40	0.10	26
Grass pellets	2	0.46	0.05	11
Grass silage	20	0.29	0.06	20
Guar meal	1	0.34		
Linseed	1	0.18		
Linseed meal	5	0.35	0.10	27
Lupin meal	1	0.35		
Manoic meal	1	0.36		

Feed	Number of Determinations	Undegradability		
		Mean	S.D.[a]	C.V.[b]
Meat and bone meal	5	0.49	0.18	37
Meat meal	1	0.76		
Oats	4	0.17	0.03	15
Palm cakes	6	0.66	0.06	9
Peanut meal	8	0.25	0.11	45
Peas	4	0.22	0.03	15
Rapeseed meal	10	0.28	0.09	31
Rapeseed meal, protected	1	0.70		
Rye	1	0.19		
Ryegrass, dehydrated	4	0.22	0.14	66
Ryegrass, dried artificially	1	0.71		
Ryegrass, dried artificially, chopped	1	0.30		
Ryegrass, dried artificially, ground	1	0.73		
Ryegrass, dried artificially, pelleted	1	0.54		
Ryegrass, fresh	1	0.48		
Ryegrass, fresh or frozen	3	0.41	0.18	44
Ryegrass, frozen	1	0.52		
Ryegrass silage, HCHO[c]	1	0.93		
Ryegrass silage, HCHO[c] dried	1	0.83		
Ryegrass silage, unwilted	1	0.22		
Sanfoin	1	0.81		
Sorghum grain	2	0.54	0.02	4
Sorghum grain, dry-ground	1	0.49		
Sorghum grain, dry-rolled	2	0.64	0.08	12
Sorghum grain, micronized	1	0.64		
Sorghum grain, reconstituted	2	0.42	0.32	75
Sorghum grain, steam-flaked	2	0.47	0.07	15
Soybean meal	39	0.35	0.12	33
Soybean meal, dried 120 C	1	0.59		
Soybean meal, dried 130 C	1	0.71		
Soybean meal, dried 140 C	1	0.82		
Soybean meal, 35% concentrate	1	0.18		
Soybean meal, 65% concentrate	1	0.46		
Soybean meal, HCHO[c]	3	0.80	0.11	14
Soybean meal, unheated	1	0.14		
Soybean-rapeseed meal, HCHO[c]	2	0.78	0.02	3
Soybeans	2	0.26	0.11	40
Subterranean clover	2	0.40	0.18	45
Sunflower meal	9	0.26	0.05	20
Timothy, dried artificially, chopped	1	0.32		
Timothy, dried artificially, pelleted	1	0.53		
Wheat	4	0.22	0.06	27
Wheat bran	4	0.29	0.10	34
Wheat gluten	1	0.17		
Wheat middlings	3	0.21	0.02	11
Yeast	1	0.42		
Zein	1	0.60		

[a]S.D. = standard deviation.
[b]C.V. = coefficient of variation.
[c]HCHO = formaldehyde treatment.

APPENDIX TABLE 3

Composition of feedstuffs commonly fed to poultry and swine (from NRC publications on swine and poultry)

Feed Class and Ingredient Name	Int'l. Feed Number	Composition, As Fed					Poultry[a] ME, kcal/kg	As Fed Basis Swine[b]			
		Dry Matter, %	Crude Protein, %	Crude Fiber, %	Ca, %	P, %		Digest. Protein, %	DE, kcal/kg	ME, kcal/kg	TDN, %
Roughage											
Alfalfa, dehy, mn 17% CP	1-00-023	92.0	17.5	24.1	1.44	0.22	1200	8.3	1880	1705	32
Alfalfa, dehy, mn 20% CP	1-00-024	92.0	20.0	20.2	1.67	0.28	1630	—	—	—	—
Alfalfa, dehy, mn 22% CP	1-00-851	92.9	22.5	18.5	1.48	0.28	1764	—	—	—	—
Alfalfa hay, s-c, grnd	1-00-111	92.2	16.7	25.8	—	—	—	—	—	—	—
Alfalfa leaf meal	1-00-246	88.8	21.3	14.6	2.11	0.26	1580	8.2	—	—	—
Pasture grass, closely grazed		20.0	5.2	3.4	—	—	—	—	—	—	—
Energy sources (<20% CP)											
Animal fat	4-00-409	99.5	—	—	—	—	7090	—	8130	7900	199
Barley grain	4-00-549	89.0	11.0	5.5	0.03	0.36	2640	8.2	3120	3040	70
Buckwheat grain	4-00-994	88.0	11.1	9.0	0.11	0.33	2712	—	—	—	—
Corn germ meal	5-02-898	93.0	18.0	12.0	0.10	0.40	1700	—	—	—	—
Corn grain	4-02-935	89.0	8.5	2.2	0.02	0.28	3350	7.0	3488	3275	79
Corn hominy feed	4-02-887	90.6	10.7	5.0	0.05	0.53	2866	8.5	3595	3365	82
Millet grain	4-03-098	90.0	12.0	8.0	0.05	0.28	—	8.8	2897	2703	66
Molasses, beet	4-00-668	77.0	6.7	—	0.16	0.03	1962	—	—	—	—
Molasses, cane	4-04-696	75.0	3.2	—	0.89	0.08	1962	—	2464	2343	56
Oats, grain	4-03-309	89.0	11.4	10.8	0.06	0.27	2550	9.9	2860	2668	65
Oats, groats	4-03-331	91.0	16.7	3.0	0.07	0.43	3549	14.0	3250	2999	74
Potatoes, cooked	4-03-784	22.5	2.2	0.7	0.01	0.05	—	1.6	863	811	20
Potato meal	4-07-850	90.3	5.9	1.4	0.07	0.20	3527	5.0	3345	3168	76
Rice bran	4-03-928	91.0	13.5	11.0	0.06	1.82	1630	10.2	3256	3028	74
Rice grain w/hulls, grnd	4-03-938	89.0	7.3	9.0	0.04	0.26	2668	5.5	2511	2367	57
Rye grain	4-20-894	88.0	11.0	2.3	0.04	0.32	3212	9.6	3300	3079	75
Sorghum grain, milo	4-04-444	89.0	11.0	2.0	0.04	0.29	3250	7.8	3453	3229	78
Wheat bran	4-05-190	89.0	15.7	11.0	0.14	1.15	1300	12.2	2512	2321	57
Wheat grain	4-05-211	89.0	12.7	3.0	0.05	0.36	3071	11.7	3520	3277	80
Wheat middlings	4-05-203	89.0	18.0	2.0	0.08	0.52	2756	16.0	3212	2952	73
Wheat mill run	4-05-206	90.0	15.3	8.0	0.09	1.02	1764	12.2	3168	2934	72
Wheat shorts	4-05-201	90.0	18.4	5.0	0.11	0.76	2646	15.4	3168	2912	72
Whey, dried	4-01-182	93.0	13.0	—	0.97	0.76	1900	12.6	3432	3191	78

Feed Class and Ingredient Name	Int'l. Feed Number	Dry Matter, %	Crude Protein, %	Crude Fiber, %	Ca, %	P, %	Poultry[a] ME, kcal/kg	Digest. Protein, %	DE, kcal/kg	ME, kcal/kg	TDN, %
			Composition, As Fed					Swine[b]			
Plant protein sources (>20% CP)											
Barley malt sprouts	5-00-545	93.0	26.2	14.0	0.22	0.73	1411	20.7	1558	1406	35
Brewer's dried grains	5-02-141	92.0	25.9	15.0	0.27	0.50	2513	20.4	1892	1708	43
Coconut meal, solv. extd	5-01-573	92.0	21.3	15.0	0.17	0.61	1540	15.5	3123	2852	71
Corn distillers grains w/solubles, dehy.	5-02-843	92.0	27.4	9.0	0.09	0.37	2425	—	—	—	—
Corn dist. sol., dehy	5-02-844	92.0	26.9	4.0	0.35	1.37	2932	16.1	3300	2976	75
Corn gluten meal	5-28-242	90.0	62.0	—	—	0.50	3720	—	—	—	—
Cottonseed meal, prepress solv. extd.	5-07-874	92.5	50.0	8.5	0.16	1.01	2150	45.0	3018	2569	68
Pea seed, grnd	5-03-598	91.0	22.5	9.0	0.17	0.50	2601	19.3	3531	3213	80
Peanut meal, solv. extd.	5-04-650	92.0	47.4	13.0	0.20	0.65	2205	44.5	3408	2920	77
Rapeseed meal, solv. extd.	5-03-871	90.3	39.4	13.8	0.40	0.90	—	32.3	2747	2396	62
Soybean meal, solv. extd.	5-04-604	89.0	44.0	7.0	0.29	0.65	2230	41.7	3300	2825	75
Soybean meal, dehulled, solv. extd.	5-04-612	89.8	50.9	2.8	0.26	0.62	2425	46.3	3405	2881	77
Sunflower meal, solv. extd.	5-04-739	93.0	45.4	12.2	0.37	1.00	2060	42.1	3034	2604	69
Wheat germ meal	5-05-218	90.0	26.2	3.0	0.07	1.04	3086	23.6	3770	3397	86
Yeast, brewer's dried	7-05-527	93.0	44.6	3.0	0.13	1.43	2425	39.2	3076	2654	70
Animal and fish protein sources											
Blood meal	5-00-380	94.0	81.1	0.5	0.55	0.42	2830	62.3	2684	2101	61
Blood flour	5-00-381	91.0	82.2	1.0	0.45	0.37	—	64.1	2608	2029	59
Buttermilk, dried	5-01-160	93.0	32.0	—	1.34	0.94	2756	29.8	3388	3015	77
Casein, dried	5-01-162	93.0	87.8	—	0.61	1.00	4130	76.0	3532	2740	80
Fish meal, anchovy	5-02-985	93.0	66.0	1.0	4.50	2.85	2900	60.7	2994	2446	68
Fish meal, herring	5-02-000	93.0	72.3	0.7	2.29	1.70	2976	66.3	3650	2938	83
Fish meal, menhaden	5-02-009	92.0	61.3	1.0	5.49	2.81	2866	56.4	3123	2580	71
Fish solubles, dried	5-01-971	92.0	62.8	1.0	—	—	2866	60.3	3408	2801	77
Meat meal	5-00-385	93.5	53.4	2.4	7.94	4.03	1984	47.5	3010	2543	68
Meat meal tankage	5-00-386	92.0	59.8	2.0	5.94	3.17	2646	37.1	2475	2052	56
Meat and bone meal	5-00-388	93.0	50.4	2.8	1.03	5.10	2150	45.0	2859	2434	65
Milk, dried skim	5-01-175	94.0	33.5	—	1.26	1.03	2513	32.8	3784	3360	86

[a] 1994 Poultry NRC.
[b] 1988 Swine NRC.

APPENDIX TABLE 4

Amino acid composition of selected feedstuffs

	Crude Protein	Amino Acids, As Fed Basis, %												
		Arginine	Cystine	Glycine	Histidine	Isoleucine	Leucine	Lysine	Methionine	Phenylalanine	Threonine	Tryptophan	Tyrosine	Valine
Forage-roughage														
Alfalfa, dehy, 15% CP	15.2	0.60	0.17	0.70	0.30	0.68	1.10	0.60	0.20	0.80	0.60	0.40	0.40	0.70
Alfalfa, dehy, 20% CP	20.6	0.90	—	1.00	0.40	0.80	1.50	0.90	0.30	1.10	0.90	0.50	0.70	1.19
Alfalfa leaf meal, s-c	21.3	0.90	0.34	0.90	0.33	0.90	1.25	0.95	0.30	0.80	0.70	0.25	0.60	0.90
Grass, dehy	14.8	0.99	0.19	0.72	0.46	1.38	1.98	1.06	0.31	1.30	0.89	0.31	0.46	1.57
Energy feeds														
Barley grain	11.6	0.53	0.18	0.36	0.27	0.53	0.80	0.53	0.18	0.62	0.36	0.18	0.36	0.62
Corn hominy feed	10.7	0.50	0.18	0.50	0.20	0.40	0.80	0.40	0.18	0.30	0.40	0.10	0.50	0.50
Corn germ meal	18.0	1.20	0.32	—	—	—	1.70	0.90	0.35	0.80	0.90	0.30	1.50	1.30
Corn grain	8.8	0.50	0.09	0.43	0.20	0.40	1.10	0.20	0.17	0.50	0.40	0.10	—	0.40
Millet grain	12.0	0.35	0.08	—	0.23	1.23	0.49	0.25	0.30	0.59	0.44	0.17	—	0.62
Oats grain	11.8	0.71	0.18	—	0.18	0.53	0.89	0.36	0.18	0.62	0.36	0.18	0.53	0.62
Potato meal	8.2	0.43	—	—	0.11	0.48	0.30	0.47	0.07	0.29	0.21	0.15	—	0.39
Rice grain w/hulls	7.3	0.53	0.10	0.80	0.09	0.27	0.53	0.27	0.17	0.27	0.18	0.10	0.60	0.51
Rye grain	11.9	0.53	0.18	—	0.27	0.53	0.71	0.45	0.18	0.62	0.36	0.09	0.27	0.62
Sorghum grain, milo	11.0	0.36	0.18	0.40	0.27	0.53	1.42	0.27	0.09	0.45	0.27	0.09	0.36	0.53
Wheat grain	12.7	0.71	0.18	0.89	0.27	0.53	0.89	0.45	0.18	0.62	0.36	0.18	0.45	0.53
Wheat shorts	18.4	0.95	0.20	0.40	0.32	0.70	1.20	0.70	0.18	0.70	0.50	0.20	0.40	0.77
Whey dried	13.8	0.40	0.30	0.30	0.20	0.90	1.40	1.10	0.20	0.40	0.80	0.20	0.30	0.70
Plant protein sources														
Brewers dried grains	25.9	1.30	—	—	0.50	1.50	2.30	0.90	0.40	1.30	0.90	0.40	1.20	1.60
Corn dist. solv., dehy	26.9	1.00	0.60	1.10	0.70	1.50	2.10	0.90	0.60	1.50	1.00	0.20	0.70	1.50
Corn gluten meal	42.9	1.40	0.60	1.50	1.00	2.30	7.60	0.80	1.00	2.90	1.40	0.20	1.00	2.20
Cottonseed meal, solv.	50.0	4.75	1.00	2.35	1.25	1.85	2.80	2.10	0.80	2.75	1.70	0.70	0.80	2.05
Peanut meal, solv.	47.4	4.69	—	—	1.00	2.00	3.10	1.30	0.60	2.30	1.40	0.50	—	2.20
Rapeseed meal, solv.	39.4	2.16	—	1.88	1.05	1.43	2.63	2.09	0.76	1.49	1.65	0.48	0.83	1.90
Soybean meal, solv.	45.8	3.20	0.67	2.10	1.10	2.50	3.40	2.90	0.60	2.20	1.70	0.60	1.40	2.40
Sunflower meal, solv.	46.8	3.50	0.70	2.70	1.10	2.10	2.60	1.70	1.50	2.20	1.50	0.50	—	2.30
Yeast, brewer's dried	44.6	2.20	0.50	1.70	1.10	2.10	3.20	3.00	0.70	1.80	2.10	0.50	1.50	2.30
Animal and fish protein sources														
Blood meal	79.9	3.50	1.40	3.40	4.20	1.00	10.30	6.90	0.90	6.10	3.70	1.10	1.80	6.50
Buttermilk, dried	32.0	1.10	0.40	0.60	0.90	2.70	3.40	2.40	0.70	1.50	1.60	0.50	1.00	2.80
Casein, dried	81.8	3.40	0.30	1.50	2.50	5.70	8.60	7.00	2.70	4.60	3.80	1.00	4.70	6.80
Fish meal, anchovy	66.0	4.46	1.00	5.10	1.84	3.40	7.01	5.40	2.19	2.48	3.04	0.80	1.77	3.54

Fish meal, herring	70.6	4.00	1.60	5.00	1.30	3.20	5.10	7.30	2.00	2.60	2.60	0.90	2.10	3.20
Fish meal, menhaden	61.3	4.00	0.94	4.40	1.60	4.10	5.00	5.30	1.80	2.70	2.90	0.60	1.60	3.60
Liver meal	66.5	4.10	0.90	5.60	1.50	3.40	5.40	4.80	1.30	2.90	2.60	0.60	1.70	4.20
Meat meal	53.4	3.70	0.60	2.20	1.10	1.90	3.50	3.80	0.80	1.90	1.80	0.30	0.90	2.60
Meat and bone meal	50.6	4.00	0.60	6.60	0.90	1.70	3.10	3.50	0.70	1.80	1.80	0.20	0.80	2.40
Meat meal tankage	59.8	3.60	—	—	1.90	1.90	5.10	4.00	0.80	2.70	2.40	0.70	—	4.20
Milk, dried skim	33.5	1.20	0.50	0.20	0.90	2.30	3.30	2.80	0.80	1.50	1.40	0.40	1.30	2.20

APPENDIX TABLE 5

Vitamin content of selected feedstuffs, fresh basis (ppm)

	Carotene	*Vitamin E*	*Choline*	*Niacin*	*Pantothenic acid*	*Riboflavin*	*Thiamin*	*Vitamin B_6*	*Vitamin B_{12}*
Plant sources									
Alfalfa, dehy., 15% CP	102	98	1550	42	21	11	3.0	6.5	—
Alfalfa leaf meal, s-c	62	—	1600	55	33	15	—	11	—
Barley grain	—	11	1030	57	6.5	2.0	5.1	2.9	—
Brewers dried grains	—	—	1587	43	8.6	1.5	0.7	0.7	—
Corn dist. sol., dehy.	1	55	4818	115	21	17	6.8	10	—
Corn grain	4	22	537	23	5	1.1	4.0	7.2	—
Cottonseed meal, solv., 41% CP	—	15	2860	40	14	5.0	6.5	6.4	—
Oats grain	—	36	1073	16	13	1.6	6.2	1.2	—
Peanut meal, solv.	—	3	2000	170	53	11	7.3	10	—
Rice grain w/hulls	—	14	800	30	3.3	1.1	2.8	—	—
Rye grain	—	15	—	1.2	6.9	1.6	3.9	—	—
Sorghum grain, milo	—	12	678	43	11	1.2	3.9	4.1	—
Soybean meal, solv., 45% CP	—	3	2743	27	14	3.3	6.6	8.0	—
Wheat grain	—	34	830	57	12	1.2	4.9	—	—
Wheat middlings	—	58	1100	53	14	1.5	19	11	—
Yeast, brewers dried	—	—	3885	447	110	35	92	43	—
Animal sources									
Buttermilk, dried	—	6	1808	9	30	31	3.5	2.4	0.02
Fish meal, herring	—	27	4004	89	11	9.0	—	3.7	219
Fish meal, menhaden	—	9	3080	56	9	4.8	0.7	—	0.1
Meat meal	—	1	1955	57	4.8	5.3	0.2	3.0	51
Meat and bone meal	—	1	2189	48	3.7	4.4	1.1	2.5	45
Liver meal	—	—	—	204	45	46	0.2	—	501
Milk, cow's, dried skim	—	9	1426	11	34	20	3.5	3.9	42
Whey, dried	—	—	20	11	48	30	3.7	2.5	0.03

APPENDIX TABLE 6

Composition of mineral supplements for animal feeds, dry-matter basis (from 1989 NRC dairy publication)[a]

Feed Name Description	International Feed Number	Dry Matter, %	Protein Equivalent — N × 6.25, %	Calcium	Chlorine	Magnesium	Phosphorus	Potassium	Sodium	Sulfur	Cobalt	Copper	Fluorine	Iodine	Iron	Manganese	Selenium	Zinc
				Macrominerals, %									Microminerals, mg/kg					
Ammonium																		
Phosphate, monobasic, $(NH_4)H_2PO_4$	6-09-338	97	70.9	0.28	—	0.46	24.74	0.01	0.06	1.46	10	10	2500	—	17,400	400	—	100
Phosphate, dibasic, $(NH_4)_2HPO_4$	6-00-370	97	115.9	0.52	—	0.46	20.60	0.01	0.05	2.16	—	10	2100	—	12,400	400	—	100
Sulfate	6-09-339	100	134.1	—	—	—	—	—	—	24.10	—	1	—	—	10	1	—	—
Bone																		
Charcoal (bone black, bone char)	6-00-402	90	9.4	30.11	—	0.59	14.14	0.16	—	—	—	—	—	—	—	—	—	—
Meal, steamed	6-00-400	97	13.2	30.71	—	0.33	12.86	0.19	5.69	2.51	—	—	—	—	26,700	—	—	100
Calcium																		
Carbonate, $CaCO_3$	6-01-069	100	—	39.39	—	0.05	0.04	0.06	0.06	—	—	—	—	—	300	300	—	—
Phosphate, monobasic, from defluorinated phosphoric acid	6-01-082	97	—	16.40	—	0.61	21.60	0.08	0.06	1.22	10	10	2100	—	15,800	360	—	90
Phosphate, dibasic, from defluorinated phosphoric acid (dicalcium phosphate)	6-01-080	97	—	22.00	—	0.59	19.30	0.07	0.05	1.14	10	10	1800	—	14,400	300	—	100
Sulfate, dihydrate, $CaSO_4 \cdot 2H_2O$, cp[b]	6-01-089	97	—	23.28	—	—	—	—	—	18.62	—	—	—	—	—	—	—	—
Colloidal clay																		
Clay (soft rock phosphate); see also phosphate	6-03-947	100[c]	—	17.00	—	0.38	9.00	—	0.10	—	—	—	15,000	—	19,000	1000	—	—
Cobalt																		
Carbonate, $CoCO_3$	6-01-566	99[c]	—	—	—	—	—	—	—	—	460,000	—	—	—	500	—	—	—
Copper (cupric)																		
Sulfate, pentahydrate, $CuSO_4 \cdot 5H_2O$, cp[b]	6-01-720	100	—	—	—	—	—	—	—	12.84	—	254,500	—	—	—	—	—	—
Curacao																		
Phosphate	6-05-586	99[c]	—	34.34	—	0.81	14.14	—	0.20	—	—	—	5500	—	3500	—	—	—
Ethylenediamine																		
Dihydroiodide	6-01-842	98[c]	—	—	—	—	—	—	—	—	—	—	—	803,400	—	—	—	—
Iron (ferrous)																		
Sulfate, heptahydrate	6-20-734	98[c]	—	—	—	—	—	—	—	12.35	—	—	—	—	218,400	—	—	—
Limestone																		
Limestone, ground	6-02-632	100	—	34.00	0.03	2.06	0.02	0.12	0.06	0.04	—	—	—	—	3500	—	—	—
Magnesium (dolomitic)	6-02-633	99[c]	—	22.30	0.12	9.99	0.04	0.36	—	—	—	—	—	—	770	—	—	—
Magnesium																		
Carbonate, $MgCO_3 \cdot Mg(OH)_2$	6-02-754	98[c]	—	0.02	—	30.81	—	—	—	—	—	—	—	—	220	—	—	—
Oxide, MgO	6-02-756	98	—	3.07	—	56.20	—	—	—	—	—	—	200	—	—	100	—	—

Feed Name Description	International Feed Number	Dry Matter, %	Protein Equivalent— N × 6.25, %	Macrominerals, %							Microminerals, mg/kg							
				Calcium	Chlorine	Magnesium	Phosphorus	Potassium	Sodium	Sulfur	Cobalt	Copper	Fluorine	Iodine	Iron	Manganese	Selenium	Zinc
Manganese (manganous)																		
Oxide, MnO, cp[b]	6-03-056	99[c]	—	—	—	—	—	—	—	—	—	—	—	—	—	774,500	—	—
Carbonate, MnCO₃	6-03-036	97	—	—	—	—	—	—	—	—	—	—	—	—	—	478,000	—	—
Oystershell																		
Ground (flour)	6-03-481	99	—	38.00	0.01	0.30	0.07	0.10	0.21	—	—	—	—	—	2870	100	—	—
Phosphate																		
Defluorinated	6-01-780	100	—	32.00	—	0.42	18.00	0.08	4.90	—	10	20	1800	—	6700	200	—	60
Rock	6-03-945	100	—	35.00	—	0.41	13.00	0.06	0.03	—	10	10	35,000	—	16800	200	—	100
Rock, low-fluorine	6-03-946	100	—	36.00	—	—	14.00	—	—	—	—	—	—	—	—	—	—	—
Rock, soft (see also Calcium)	6-03-947	100	—	17.00	—	0.38	9.00	—	0.10	—	—	—	15,000	—	19,000	1000	—	—
Phosphate, monobasic, monohydrate, NaH₂PO₄·H₂O (see also Sodium)	6-04-288	97	—	—	—	—	22.50	—	16.68	—	—	—	—	—	—	—	—	—
Phosphoric acid																		
H₃PO₄	6-03-707	75	—	0.05	—	0.51	31.60	0.02	0.04	1.55	10	10	3100	—	17,500	500	—	130
Potassium																		
Bicarbonate, KHCO₃, cp[b]	6-29-493	99[c]	—	—	—	—	—	39.05	—	—	—	—	—	—	—	—	—	—
Chloride, KCl	6-03-755	100	—	0.05	47.30	0.34	—	50.00	1.00	0.45	—	—	—	—	600	—	—	—
Iodide, KI	6-03-759	100[c]	—	—	—	—	—	21.00	—	—	—	—	—	681,700	—	10	—	—
Sulfate, K₂SO₄	6-06-098	98[c]	—	0.15	1.55	0.61	—	41.84	0.09	17.35	—	—	—	—	710	—	—	—
Sodium																		
Bicarbonate, NaHCO₃	6-04-272	100	—	—	—	—	—	—	27.00	—	—	—	—	—	—	—	—	—
Chloride, NaCl	6-04-152	100	—	—	60.66	—	—	—	39.34	—	—	—	—	—	—	—	—	—
Phosphate, monobasic, monohydrate, NaH₂PO₄·H₂O (see also Phosphate)	6-04-288	97	—	—	—	—	22.50	—	16.68	—	—	—	—	—	—	—	—	—
Selenite, Na₂SeO₃	6-26-013	98[c]	—	—	—	—	—	—	26.60	—	—	—	—	—	—	—	456,000	—
Sulfate, decahydrate, Na₂SO₄·10H₂O, cp[b]	6-04-292	97[c]	—	—	—	—	—	—	14.27	9.95	—	—	—	—	—	—	—	—
Tripolyphosphate, Na₅P₃O₁₀	6-08-076	96	—	—	—	—	25.00	—	31.00	—	—	—	—	—	40	—	—	—
Zinc																		
Oxide, ZnO	6-05-533	100	—	0.02	—	—	—	—	—	—	—	—	—	—	—	—	—	780,000
Sulfate, monohydrate, ZnSO₄·H₂O	6-05-555	99[c]	—	—	0.015	—	—	—	—	17.68	—	—	—	—	10	10	—	363,600

*Note: The compositions of hydrated mineral ingredients (e.g., CaSO₄·2H₂O) are shown including the waters of hydration. Mineral compositions of feed-grade mineral supplements vary by source, mining site, and manufacturer. The manufacturer's analysis should be used when it is available.

[b]cp = Chemically pure.

[c]Dry matter values have been estimated for these minerals.

APPENDIX TABLE 7

Some common trace nutrients and feed additives and concentrations of active ingredients as received from manufacturers or suppliers of premixes

Active Ingredient	Concentration from the Manufacturer	Concentration Typical of Premix Use
Fat-soluble vitamins		
Vitamin A acetate,[a] gelatin-coated feed grade	325,000, 500,000, or 600,000 IU/g	same, 30,000 IU/g
Vitamin A palmitate,[a] gelatin-coated	325,000 IU/g	same
Vitamin D₃ acetate dry feed grade	500,000 ICU/g[b]	same, 26,455 ICU/g
Vitamin E acetate, dry feed grade	250 IU/g	same, 44 IU/g
Vitamin K, menadione Na bisulfate complex	pure	same, 16 g/lb
Vitamin A acetate & vitamin D₃ acetate mix	500,000 IU/g of A 167,000 IU/g of D₃	50,000 IU/g 16-20,000 IU/g
Vitamin A, D₃, E premix		40-50,000 IU/g A 10,000 IU/g D₃ 50 IU/g E
B-complex vitamins		
Riboflavin	0.50 g/g	same, 60 g/lb
Thiamin	pure	same
Pridoxine	pure	same
Niacin	0.98 g/g	same
Pantothenate, Ca DL	0.414 g/g of D pantho. acid	same
Choline chloride	0.435 g/g of choline	same
Cyanocobalamin (B₁₂)	1.32 mg/g	same
Folic acid	0.45 g/g	same
Biotin	pure	various
Amino acids		
Methionine, DL	0.995 g/g	same
Methionine hydroxy analog, Ca salt	0.93 g/g	same
Lysine, L	0.769 g/g	same
Antibiotics		
Terramycin	10, 50 g/lb	same
Aureomycin	10, 50 g/lb	same
Neoterramycin	20 g/lb	same
Zinc bacitracin	40 g/lb	same
Erythromycin	50 g/lb	same
Penicillin	136 g/lb	same
Rumensin (Na monensin)	60 g/lb	same
Tylosin phosphate	10 g/lb	same
Miscellaneous		
Arsanilic acid	20, 40, 227 g/lb	same
Sodium arsanilate	pure	same
Roxarsone	45.4 g/lb	same
Ethylenediamine dihydriodide	0.99 g/g	42 g/lb
Melengestrol acetate	100, 500 mg/lb	0.4, 0.5 mg/lb

[a]Other concentrations available from manufacturers; vitamin A is also available in water-dispersible and gelatin-coated preparations, water miscible, in oil, or in injectable preparations.

[b]ICU = international chick units.

APPENDIX TABLE 8
Net energy requirements of growing and finishing beef cattle (Mcal/d) (from 1984 NRC on beef)

Body Weight, kg:	150	200	250	300	350	400	450	500	550	600
NEm Required:	3.30	4.10	4.84	5.55	6.24	6.89	7.52	8.14	8.75	9.33
Daily gain, kg	**NEg Required**									
Medium-frame steer calves										
0.2	0.41	0.50	0.60	0.69	0.77	0.85	0.93	1.01	1.08	
0.4	0.87	1.08	1.28	1.47	1.65	1.82	1.99	2.16	2.32	
0.6	1.36	1.69	2.00	2.29	2.57	2.84	3.11	3.36	3.61	
0.8	1.87	2.32	2.74	3.14	3.53	3.90	4.26	4.61	4.95	
1.0	2.39	2.96	3.50	4.02	4.51	4.98	5.44	5.89	6.23	
1.2	2.91	3.62	4.28	4.90	5.50	6.69	6.65	7.19	7.73	
Large-frame steers, compensating medium-frame yearling steers, and medium-frame bulls										
0.2	0.36	0.45	0.53	0.61	0.68	0.75	0.82	0.89	0.96	1.02
0.4	0.77	0.96	1.13	1.30	1.46	1.61	1.76	1.91	2.05	2.19
0.6	1.21	1.50	1.77	2.03	2.28	2.52	2.75	2.98	3.20	3.41
0.8	1.65	2.06	2.43	2.78	3.12	3.45	3.77	4.08	4.38	4.68
1.0	2.11	2.62	3.10	3.55	3.99	4.41	4.81	5.21	5.60	5.98
1.2	2.58	3.20	3.78	4.34	4.87	5.38	5.88	6.37	6.84	7.30
1.4	3.06	3.79	4.48	5.14	5.77	6.38	6.97	7.54	8.10	8.64
1.6	3.53	4.39	5.19	5.95	6.68	7.38	8.07	8.73	9.38	10.01
Large-frame bull calves and compensating large-frame yearling steers										
0.2	0.32	0.40	0.47	0.54	0.60	0.67	0.73	0.79	0.85	0.91
0.4	0.69	0.85	1.01	1.15	1.29	1.43	1.56	1.69	1.82	1.94
0.6	1.07	1.33	1.57	1.80	2.02	2.23	2.44	2.64	2.83	3.02
0.8	1.47	1.82	2.15	2.47	2.77	3.06	3.34	3.62	3.88	4.15
1.0	1.87	2.32	2.75	3.15	3.54	3.91	4.27	4.62	4.96	5.30
1.2	2.29	2.84	3.36	3.85	4.32	4.77	5.21	5.64	6.06	6.47
1.4	2.71	3.36	3.97	4.56	5.11	5.65	6.18	6.68	7.18	7.66
1.6	3.14	3.89	4.60	5.28	5.92	6.55	7.15	7.74	8.31	8.87
1.8	3.56	4.43	5.23	6.00	6.74	7.45	8.13	8.80	9.46	10.10
Medium-frame heifer calves										
0.2	0.49	0.60	0.71	0.82	0.92	1.01	1.11	1.20	1.29	
0.4	1.05	1.31	1.55	1.77	1.99	2.20	2.40	2.60	2.79	
0.6	1.66	2.06	2.44	2.79	3.13	3.46	3.78	4.10	4.40	
0.8	2.29	2.84	3.36	3.85	4.32	4.78	5.22	5.65	6.07	
1.0	2.94	3.65	4.31	4.94	5.55	6.14	6.70	7.25	7.79	
Large-frame heifer calves and compensating medium-frame yearling heifers										
0.2	0.43	0.53	0.63	0.72	0.81	0.90	0.98	1.06	1.14	1.21
0.4	0.93	1.16	1.37	1.57	1.76	1.95	2.13	2.31	2.47	2.64
0.6	1.47	1.83	2.16	2.47	2.78	3.07	3.35	3.63	3.90	4.16
0.8	2.03	2.62	2.98	3.41	3.83	4.24	4.63	5.01	5.38	5.74
1.0	2.61	3.23	3.82	4.38	4.92	5.44	5.94	6.43	6.91	7.37
1.2	3.19	3.97	4.69	5.37	5.03	6.67	7.28	7.88	8.47	9.03

APPENDIX TABLE 8A

Diet evaluation for growing and finishing cattle (from 1996 NRC on beef)

Wt @ Small Marbling			533 kg				
Breed Code			1 Angus				

Ration	eNDF % DM	TDN % DM	NE_a Mcal/kg	NE_g Mcal/kg	CP % DM	DIP % CP	Weight Class	NE Adjuster
A	57	50	1.00	0.45	7.4	88	325	100%
B	43	60	1.35	0.77	10.0	78	350	100%
C	30	70	1.67	1.05	12.6	72.4	375	100%
D	5	80	1.99	1.33	14.4	48.5	400	100%
E	3	90	2.29	1.59	16.6	44.2	425	100%

Body Weight, kg	DMI Adjuster	DMI kg/d	ADG kg/d	DIP	UIP (balances, g/d)	MP	Ca (requirements, % of DM)	P
300—A	100%	7.9	0.32	1	0	0	0.22%	0.13%
—B	100%	8.4	0.89	0	0	0	0.35%	0.18%
—C	100%	8.2	1.35	2	0	0	0.48%	0.24%
—D	100%	7.7	1.69	1	2	1	0.60%	0.29%
—E	100%	7.1	1.90	1	2	1	0.71%	0.34%
325—A	100%	8.4	0.32	1	14	11	0.21%	0.13%
—B	100%	8.9	0.89	0	38	30	0.33%	0.18%
—C	100%	8.7	1.36	2	57	46	0.45%	0.22%
—D	100%	8.2	1.69	1	73	58	0.55%	0.27%
—E	100%	7.6	1.90	1	82	66	0.65%	0.31%
350—A	100%	8.9	0.32	1	27	22	0.20%	0.13%
—B	100%	9.4	0.89	0	75	60	0.31%	0.17%
—C	100%	9.2	1.36	2	114	91	0.42%	0.21%
—D	100%	8.7	1.59	1	143	114	0.51%	0.25%
—E	100%	8.0	1.90	1	180	128	0.60%	0.29%
375—A	100%	9.4	0.32	1	40	32	0.20%	0.13%
—B	100%	9.9	0.89	0	111	89	0.30%	0.16%
—C	100%	9.7	1.36	2	169	135	0.39%	0.20%
—D	100%	9.1	1.69	1	212	169	0.48%	0.24%
—E	100%	8.4	1.90	1	238	190	0.56%	0.28%
400—A	100%	9.8	0.32	1	53	43	0.19%	0.12%
—B	100%	10.4	0.89	0	147	118	0.28%	0.16%
—C	100%	10.2	1.35	2	223	178	0.37%	0.19%
—D	100%	9.8	1.69	2	279	223	0.44%	0.23%
—E	100%	8.8	1.90	1	314	251	0.52%	0.26%
425—A	100%	10.3	0.32	1	66	53	0.19%	0.12%
—B	100%	10.9	0.89	0	182	146	0.27%	0.15%
—C	100%	10.6	1.36	2	276	221	0.35%	0.19%
—D	100%	10.0	1.69	2	346	277	0.42%	0.22%
—E	100%	9.3	1.90	1	388	311	0.48%	0.25%

Nutrient Requirements for Growing and Finishing Cattle						

Wt @ Small Marbling	*533 kg*					
Weight Range	*200–450 kg*					
ADC Range	*0.50–2.50 kg*					
Breed Code	*1 Angus*					

Body Weight, kg		*200*	*250*	*300*	*350*	*400*	*450*
Maintenance requirements							
NE$_m$	Mcal/d	4.1	4.84	5.55	6.23	6.89	7.52
MP	g/d	202	239	274	307	340	371
Ca	g/d	6	8	9	11	12	14
P	g/d	5	6	7	8	10	11
Growth requirements (ADC)	NE$_g$ required for gain, Mcal/d						
0.5	kg/d	1.27	1.50	1.72	1.93	2.14	2.33
1.0	kg/d	2.72	3.21	3.68	4.13	4.57	4.99
1.5	kg/d	4.24	5.01	5.74	6.45	7.13	7.79
2.0	kg/d	5.81	6.87	7.88	8.84	9.77	10.68
2.5	kg/d	7.42	8.78	10.06	11.29	12.48	13.64
	MP required for gain, g/d						
0.5	kg/d	154	155	158	157	145	133
1.0	kg/d	299	300	303	298	272	246
1.5	kg/d	441	440	442	432	391	352
2.0	kg/d	580	577	577	581	505	451
2.5	kg/d	718	712	710	687	616	547
	Calcium required for gain, g/d						
0.5	kg/d	14	13	12	11	10	9
1.0	kg/d	27	25	23	21	19	17
1.5	kg/d	39	36	33	30	27	25
2.0	kg/d	52	47	43	39	35	32
2.5	kg/d	64	59	53	48	43	38
	Phosphorus required for gain, g/d						
0.5	kg/d	6	5	5	4	4	4
1.0	kg/d	11	10	9	8	8	7
1.5	kg/d	16	15	13	12	11	10
2.0	kg/d	21	19	18	16	14	13
2.5	kg/d	26	24	22	19	17	15

APPENDIX TABLE 8B

Nutrient requirements for growing bulls (from 1996 NRC on beef)

Wt @ Maturity		890 kg					
Weight Range		300–800 kg					
ADC Range		0.50–2.50 kg					
Breed Code		1 Angus					

Body Weight, kg		300	400	500	600	700	800
Maintenance requirements							
NE$_m$	Mcal/day	6.39	7.92	9.36	10.73	32.05	13.32
MP	g/d	274	340	402	461	517	572
Ca	g/d	9	12	15	19	22	25
P	g/d	7	10	12	14	17	19
Growth requirements							
ADG		\multicolumn{6}{c}{NE$_g$ required for gain, Mcal/d}					
0.5	kg/d	1.72	2.13	2.52	2.89	3.25	3.59
1.0	kg/d	3.68	4.56	5.39	6.18	6.94	7.67
1.5	kg/d	5.74	7.12	8.42	9.65	10.83	11.97
2.0	kg/d	7.87	9.76	11.54	13.23	14.85	16.41
2.5	kg/d	10.05	12.47	14.74	16.90	18.97	20.97
		\multicolumn{6}{c}{MP required for gain, g/d}					
0.5	kg/d	158	145	122	100	78	58
1.0	kg/d	303	272	222	175	130	86
1.5	kg/d	442	392	314	241	170	102
2.0	kg/d	577	506	400	299	202	109
2.5	kg/d	710	617	481	352	228	109
		\multicolumn{6}{c}{Calcium required for gain, g/d}					
0.5	kg/d	12	10	9	7	6	4
1.0	kg/d	23	19	16	12	9	6
1.5	kg/d	33	27	22	17	12	7
2.0	kg/d	43	35	28	21	14	8
2.5	kg/d	53	43	34	25	16	8
		\multicolumn{6}{c}{Phosphorus required for gain, g/d}					
0.5	kg/d	5	4	3	3	2	2
1.0	kg/d	9	8	6	5	4	2
1.5	kg/d	13	11	9	7	5	3
2.0	kg/d	18	14	11	8	6	3
2.5	kg/d	22	17	14	10	6	3

Diet Evaluation for Growing Bulls

Wt @ Maturity	890 kg
Breed Code	1 Angus

Ration	eNDF % DM	TDN % DM	NE$_a$ Mcal/kg	NE$_g$ Mcal/kg	CP % DM	DIP % CP	Weight Class	NE Adjuster
A	43	50	1.00	0.45	8.2	80	325	100%
B	37	65	1.51	0.92	10.9	78	350	100%
C	30	70	1.67	1.06	12.0	76	375	100%
D	20	75	1.83	1.20	13.4	73	400	100%
E	5	80	1.99	1.33	13.8	51	425	100%

Body weight, kg	DMI Adjuster	DMI kg/d	ADG kg/d	DIP	UIP balances, g/d	MP	Ca requirements, % of DM	P
300—A	100%	7.9	0.22	5	103	83	0.18%	0.12%
—B	100%	8.3	1.02	4	8	6	0.39%	0.20%
—C	100%	8.2	1.23	2	−3	−2	0.45%	0.23%
—D	100%	8.0	1.41	3	10	8	0.51%	0.25%
—E	100%	7.7	1.56	5	−2	−2	0.56%	0.27%
325—A	100%	8.4	0.22	5	119	95	0.18%	0.12%
—B	100%	8.8	1.02	5	51	41	0.36%	0.19%
—C	100%	8.7	1.23	2	49	39	0.42%	0.21%
—D	100%	8.5	1.41	3	70	56	0.47%	0.24%
—E	100%	8.2	1.56	6	63	51	0.52%	0.26%
350—A	100%	8.9	0.22	5	134	107	0.18%	0.12%
—B	100%	9.4	1.02	5	94	75	0.34%	0.18%
—C	100%	9.2	1.23	2	100	80	0.39%	0.20%
—D	100%	9.0	1.41	3	129	103	0.44%	0.22%
—E	100%	8.7	1.56	6	128	102	0.48%	0.24%
375—A	100%	9.4	0.22	6	149	119	0.18%	0.12%
—B	100%	9.8	1.02	5	136	109	0.32%	0.17%
—C	100%	9.7	1.23	2	150	125	0.37%	0.19%
—D	100%	9.4	1.41	3	187	149	0.41%	0.21%
—E	100%	9.1	1.56	6	191	153	0.45%	0.23%
400—A	100%	9.8	0.22	6	161	131	0.17%	0.12%
—B	100%	10.3	1.02	5	177	142	0.31%	0.17%
—C	100%	10.2	1.23	2	199	159	0.35%	0.19%
—D	100%	9.9	1.41	3	244	195	0.39%	0.20%
—E	100%	9.6	1.56	7	253	202	0.42%	0.22%
425—A	100%	10.3	0.22	6	169	143	0.17%	0.12%
—B	100%	10.8	1.02	6	218	174	0.29%	0.16%
—C	100%	10.6	1.23	2	247	198	0.33%	0.18%
—D	100%	10.4	1.41	3	300	240	0.36%	0.19%
—E	100%	10.0	1.56	7	314	251	0.40%	0.21%

APPENDIX TABLE 9

Protein requirements of growing and finishing cattle (g/d) (from 1984 NRC on beef)

Body Weight, kg:	150	200	250	300	350	400	450	500	550	600
Medium-frame steer calves										
Daily gain, kg										
0.2	343	399	450	499	545	590	633	675	715	
0.4	428	482	532	580	625	668	710	751	790	
0.6	503	554	601	646	688	728	767	805	842	
0.8	575	621	664	704	743	780	815	849	883	
1.0	642	682	720	755	789	821	852	882	911	
1.2	702	735	766	794	822	848	873	897	921	
Large-frame steer calves and compensating medium-frame yearling steers										
0.2	361	421	476	529	579	627	673	719	762	805
0.4	441	499	552	603	651	697	742	785	827	867
0.6	522	576	628	676	722	766	809	850	890	930
0.8	598	650	698	743	786	828	867	906	944	980
1.0	671	718	762	804	843	881	918	953	988	1021
1.2	740	782	822	859	895	929	961	993	1023	1053
1.4	806	842	877	908	938	967	995	1022	1048	1073
1.6	863	892	919	943	967	989	1011	1031	1052	1071
Medium-frame bulls										
0.2	345	401	454	503	550	595	638	680	721	761
0.4	430	485	536	584	629	673	716	757	797	835
0.6	509	561	609	655	698	740	780	819	856	893
0.8	583	632	677	719	759	798	835	871	906	940
1.0	655	698	739	777	813	849	881	914	945	976
1.2	722	760	795	828	860	890	919	947	974	1001
1.4	782	813	841	868	893	917	941	963	985	1006
Large-frame bull calves and compensating large-frame yearling steers										
0.2	355	414	468	519	568	615	661	705	747	789
0.4	438	494	547	597	644	689	733	776	817	857
0.6	519	574	624	672	718	761	803	844	884	923
0.8	597	649	697	741	795	826	866	905	942	979
1.0	673	721	765	807	847	885	922	958	994	1027
1.2	745	789	830	868	904	939	973	1005	1037	1067
1.4	815	854	890	924	956	986	1016	1045	1072	1099
1.6	880	912	943	971	998	1024	1048	1072	1095	1117
1.8	922	942	962	980	997	1013	1028	1043	1057	1071
Medium-frame heifer calves										
0.2	323	374	421	465	508	549	588	626	662	
0.4	409	459	505	549	591	630	669	706	742	
0.6	477	522	563	602	638	674	708	741	773	
0.8	537	574	608	640	670	700	728	755	781	
1.0	562	583	603	621	638	654	670	685	700	
Large-frame heifer calves and compensating medium-frame yearling heifers										
0.2	342	397	449	497	543	588	631	672	712	751
0.4	426	480	530	577	622	665	707	747	787	825
0.6	500	549	596	639	681	721	759	796	832	867
0.8	568	613	654	693	730	765	799	833	865	896
1.0	630	668	703	735	767	797	826	854	881	907
1.2	680	708	734	758	781	803	824	844	864	883

APPENDIX TABLE 10

Calcium and phosphorus requirements of growing and finishing cattle (g/d) (from 1984 NRC on beef)

Body Weight, kg	Mineral	150	200	250	300	350	400	450	500	550	600
Medium-frame steer calves											
Daily gain, kg											
0.2	Ca	11	12	13	14	15	16	17	19	20	
	P	7	9	10	12	13	15	16	18	19	
0.4	Ca	16	17	17	18	19	19	20	21	22	
	P	9	10	12	13	14	16	17	18	20	
0.6	Ca	21	21	21	22	22	22	22	23	23	
	P	11	12	13	14	15	17	18	19	20	
0.8	Ca	27	26	25	25	25	25	24	24	24	
	P	12	13	14	15	16	17	19	20	21	
1.0	Ca	32	31	29	29	28	27	26	26	25	
	P	14	15	16	16	17	18	19	20	21	
1.2	Ca	37	35	33	32	31	29	28	27	26	
	P	16	16	17	17	18	19	20	21	21	
1.4	Ca	42	39	37	35	33	32	30	29	27	
	P	17	18	18	19	19	20	20	21	22	
Large-frame steer calves, compensating medium-frame yearling steers, and medium-frame bulls											
0.2	Ca	11	12	13	14	16	17	18	19	20	22
	P	7	9	10	12	13	15	16	18	20	21
0.4	Ca	17	17	18	19	19	20	21	22	23	24
	P	9	10	12	13	15	16	17	19	20	22
0.6	Ca	22	22	23	23	23	24	24	24	25	25
	P	11	12	13	15	16	17	18	20	21	22
0.8	Ca	28	27	27	27	27	27	27	27	27	27
	P	13	14	15	16	17	18	19	20	22	23
1.0	Ca	33	32	31	31	30	30	29	29	29	28
	P	14	15	16	17	18	19	20	21	22	23
1.2	Ca	38	37	36	35	34	33	32	31	30	30
	P	16	17	18	18	19	20	21	22	23	24
1.4	Ca	44	42	40	38	37	36	34	33	32	31
	P	18	18	19	20	20	21	22	22	23	24
1.6	Ca	49	47	44	42	40	38	37	35	34	32
	P	20	20	20	21	21	22	22	23	24	24
Large-frame bull calves and compensating large-frame yearling steers											
0.2	Ca	11	12	13	15	16	17	18	20	21	22
	P	7	9	10	12	13	15	17	18	20	21
0.4	Ca	17	18	19	19	20	21	22	23	24	25
	P	9	11	12	13	15	16	18	19	21	22
0.6	Ca	23	23	23	24	24	25	25	26	27	27
	P	11	12	14	15	16	18	19	20	22	23
0.8	Ca	28	28	28	28	28	29	29	29	29	30
	P	13	14	15	16	18	19	20	21	22	24
1.0	Ca	34	34	33	33	32	32	32	32	32	32
	P	15	16	17	18	19	20	21	22	23	24
1.2	Ca	40	39	38	37	36	36	35	35	34	34
	P	17	17	18	19	20	21	22	23	24	25
1.4	Ca	45	44	42	41	40	39	38	37	36	36
	P	18	19	20	20	21	22	23	24	25	26
1.6	Ca	51	49	47	45	44	42	41	40	39	38
	P	20	21	21	22	23	23	24	25	25	26
1.8	Ca	56	54	51	49	47	45	44	42	41	39
	P	22	22	22	23	23	24	25	25	26	26

Body Weight, kg	Mineral	150	200	250	300	350	400	450	500	550	600
Medium-frame heifer calves											
0.2	Ca	10	11	12	13	14	16	17	18	19	
	P	7	9	10	11	13	14	16	17	19	
0.4	Ca	15	16	16	16	17	17	18	19	19	
	P	9	10	11	12	14	15	16	18	19	
0.6	Ca	20	20	19	19	19	19	19	19	19	
	P	10	11	12	13	14	16	17	18	19	
0.8	Ca	25	23	23	22	21	20	20	19	19	
	P	12	12	13	14	15	16	17	18	19	
1.0	Ca	29	27	26	24	23	22	20	19	19	
	P	13	14	14	15	16	16	17	18	19	
Large-frame heifer calves and compensating medium-frame yearling heifers											
0.2	Ca	11	12	13	14	15	16	17	18	20	21
	P	7	9	10	12	13	15	16	18	19	21
0.4	Ca	16	16	17	17	18	19	19	20	21	22
	P	9	10	11	13	14	15	17	18	20	21
0.6	Ca	21	21	21	21	21	21	21	21	22	22
	P	10	12	13	14	15	16	17	19	20	21
0.8	Ca	26	25	24	24	23	23	23	22	22	22
	P	12	13	14	15	16	17	18	19	20	21
1.0	Ca	31	29	28	27	26	25	24	23	23	22
	P	14	14	15	16	17	18	18	19	20	21
1.2	Ca	35	33	31	30	28	27	25	24	23	22
	P	15	16	16	17	17	18	19	20	20	21

APPENDIX TABLE II

Approximate total daily water intake of beef cattle (from 1984 NRC on beef)

Weight		40 (4.4)		50 (10.0)		60 (14.4)		70 (21.1)		80 (26.6)		90 (32.2)	
kg	lb	liter	gal	liter	gal	liter	gal	liter	gal	liter	gal	liter	gal
Growing heifers, steers, and bulls													
182	400	15.1	4.0	16.3	4.3	18.9	5.0	22.0	5.8	25.4	6.7	36.0	9.5
273	600	20.1	5.3	22.0	5.8	25.0	6.6	29.5	7.8	33.7	8.9	48.1	12.7
364	800	23.8	6.3	25.7	6.8	29.9	7.9	34.8	9.2	40.1	10.6	56.8	15.0
Finishing cattle													
273	600	22.7	6.0	24.6	6.5	28.0	7.4	32.9	8.7	37.9	10.0	54.1	14.3
364	800	27.6	7.3	29.9	7.9	34.4	9.1	40.5	10.7	46.6	12.3	65.9	17.4
454	1000	32.9	8.7	35.6	9.4	40.9	10.8	47.7	12.6	54.9	14.5	78.0	20.6
Wintering pregnant cows[b]													
409	900	25.4	6.7	27.3	7.2	31.4	8.3	36.7	9.7	—	—	—	—
500	1100	22.7	6.0	24.6	6.5	28.0	7.4	32.9	8.7	—	—	—	—
Lactating cows													
409+	900+	43.1	11.4	47.7	12.6	54.9	14.5	64.0	16.9	67.8	17.9	61.3	16.2
Mature bulls													
636	1400	30.3	8.0	32.6	8.6	37.5	9.9	44.3	11.7	50.7	13.4	71.9	19.0
727+	1600+	32.9	8.7	35.6	9.4	40.9	10.8	47.7	12.6	54.9	14.5	78.0	20.6

Header over temperature columns: Temperature in °F (°C)[a]

[a]Water intake of a given class of cattle in a specific management regime is a function of dry matter intake and ambient temperature. Water intake is quite constant up to 40° F (4.4° C).

[b]Dry matter intake has a major influence on water intake. Heavier cows are assumed to be higher in body condition and to require less dry matter and, thus, less water intake.

APPENDIX TABLE 12

Nutrient requirements of breeding cattle (metric) (from 1984 NRC on beef)

Weight[a], kg	Daily Gain[b], kg	Daily DM[c], kg	Energy Daily — ME, Mcal	Energy Daily — TDN, kg	Energy Daily — NEm, Mcal	Energy Daily — NEg, Mcal	Energy In Diet DM — ME, Mcal/kg	Energy In Diet DM — TDN, %	Energy In Diet DM — NEm, Mcal/kg	Energy In Diet DM — NEg, Mcal/kg	Total Protein — Daily, g	Total Protein — DM, %	Calcium (In Diet) — Daily, g	Calcium — DM, %	Phosphorus — Daily, g	Phosphorus — DM, %	Vitamin A[d] — Daily, 1000 IU
Pregnant yearling heifers—Last third of pregnancy																	
325	0.4	7.1	14.2	3.9	8.04	NA[e]	2.00	55.2	1.15	NA[e]	591	8.4	19	0.27	14	0.20	20
325	0.6	7.3	15.7	4.3	8.04	0.77	2.15	59.3	1.29	0.72	649	8.9	23	0.32	15	0.21	20
325	0.8	7.3	17.2	4.8	8.04	1.67	2.35	64.9	1.47	0.88	697	9.5	27	0.37	16	0.22	20
350	0.4	7.5	14.8	4.1	8.38	NA	1.99	55.0	1.14	NA	616	8.3	20	0.27	15	0.21	21
350	0.6	7.7	16.5	4.6	8.38	0.81	2.14	59.1	1.28	0.71	674	8.8	24	0.32	16	0.21	22
350	0.8	7.8	18.1	5.0	8.38	1.76	2.34	64.6	1.46	0.88	720	9.3	27	0.35	17	0.22	22
375	0.4	7.8	15.5	4.3	8.71	NA	1.98	54.7	1.13	NA	641	8.2	21	0.27	15	0.19	22
375	0.6	8.1	17.2	4.8	8.71	0.86	2.13	58.8	1.27	0.70	697	8.6	25	0.31	17	0.21	23
375	0.8	8.2	19.0	5.2	8.71	1.86	2.32	64.1	1.45	0.86	743	9.1	27	0.33	18	0.22	23
400	0.4	8.2	16.1	4.5	9.04	NA	1.97	54.4	1.12	NA	664	8.1	22	0.27	16	0.20	23
400	0.6	8.5	18.0	5.0	9.04	0.90	2.12	58.6	1.26	0.69	721	8.5	25	0.30	18	0.21	24
400	0.8	8.6	19.8	5.5	9.04	1.95	2.31	63.8	1.44	0.85	764	8.9	28	0.33	18	0.20	24
425	0.4	8.6	16.8	4.6	9.36	NA	1.96	54.1	1.11	NA	687	8.0	23	0.27	17	0.20	24
425	0.6	8.9	18.7	5.2	9.36	0.94	2.11	58.3	1.25	0.69	743	8.4	26	0.30	18	0.20	25
425	0.8	9.0	20.7	5.7	9.36	2.04	2.30	63.5	1.43	0.84	786	8.8	28	0.31	19	0.21	25
450	0.4	8.9	17.3	4.8	9.67	NA	1.95	53.9	1.10	NA	710	8.0	23	0.26	18	0.20	25
450	0.6	9.2	19.4	5.4	9.67	0.98	2.10	58.0	1.25	0.68	765	8.3	26	0.29	19	0.21	26
450	0.8	9.4	21.5	5.9	9.67	2.13	2.29	63.3	1.42	0.84	807	8.6	28	0.30	20	0.21	26
Dry pregnant mature cows—Middle third of pregnancy																	
350	0.0	6.8	11.9	3.3	6.23	NA	1.76	48.6	0.92	NA	478	7.1	12	0.16	12	0.18	19
400	0.0	7.5	13.1	3.6	6.89	NA	1.76	48.6	0.92	NA	525	7.0	13	0.17	13	0.17	21
450	0.0	8.2	14.3	4.0	7.52	NA	1.76	48.6	0.92	NA	570	7.0	15	0.17	15	0.18	23
500	0.0	8.8	15.5	4.3	8.14	NA	1.76	48.6	0.92	NA	614	7.0	17	0.19	17	0.19	25
550	0.0	9.5	16.7	4.6	8.75	NA	1.76	48.6	0.92	NA	657	6.9	18	0.19	18	0.19	27
600	0.0	10.1	17.8	4.9	9.33	NA	1.76	48.6	0.92	NA	698	6.9	20	0.20	20	0.20	28
650	0.0	10.7	18.9	5.2	9.91	NA	1.76	48.6	0.92	NA	739	6.9	22	0.21	22	0.21	30
Dry pregnant mature cows—Last third of pregnancy																	
350	0.4	7.4	14.7	4.1	8.38	NA	1.98	54.7	1.13	NA	609	8.2	20	0.27	15	0.20	21
400	0.4	8.2	16.0	4.4	9.04	NA	1.96	54.1	1.11	NA	657	8.0	22	0.27	16	0.20	23
450	0.4	8.9	17.2	4.8	9.67	NA	1.94	53.6	1.10	NA	703	7.9	23	0.26	18	0.21	24
500	0.4	9.5	18.3	5.1	10.29	NA	1.92	53.1	1.08	NA	746	7.8	25	0.26	20	0.21	27
550	0.4	10.2	19.5	5.4	10.90	NA	1.91	52.8	1.07	NA	790	7.8	26	0.25	21	0.21	29
600	0.4	10.8	20.6	5.7	11.48	NA	1.90	52.5	1.06	NA	832	7.7	28	0.26	23	0.21	30
650	0.4	11.5	21.7	6.0	12.06	NA	1.89	52.2	1.05	NA	872	7.6	30	0.26	25	0.22	32
Two-year-old heifers nursing calves—First 3–4 months postpartum—5.0 kg milk/d																	
300	0.2	6.9	16.6	4.6	9.30[f]	0.72	2.41	66.6	1.53	0.93	814[g]	11.8	26	0.38	17	0.25	27
325	0.2	7.3	17.4	4.8	9.64[f]	0.77	2.37	65.5	1.49	0.90	841[g]	11.5	27	0.37	18	0.25	28
350	0.2	7.8	18.1	5.0	9.98[f]	0.81	2.34	64.6	1.46	0.88	866[g]	11.2	27	0.35	19	0.24	30
375	0.2	8.2	18.9	5.2	10.31[f]	0.86	2.31	63.8	1.44	0.85	892[g]	10.9	28	0.34	19	0.23	32

Weight[a], kg	Daily Gain[b], kg	Daily DM[c], kg	Energy Daily ME, Mcal	Energy Daily TDN, kg	Energy Daily NEm, Mcal	Energy Daily NEg, Mcal	Energy Daily ME, Mcal/kg	Energy In Diet DM TDN, %	Energy In Diet DM NEm, Mcal/kg	Energy In Diet DM NEg, Mcal/kg	Total Protein Daily, g	Total Protein DM, %	Calcium In Diet Daily, g	Calcium In Diet DM, %	Phosphorus In Diet Daily, g	Phosphorus In Diet DM, %	Vitamin A[d] Daily, 1000 IU
400	0.2	8.6	19.7	5.4	10.64[f]	0.90	2.29	63.3	1.42	0.84	916[g]	10.7	28	0.33	20	0.23	34
425	0.2	9.0	20.4	5.6	10.96[f]	0.94	2.27	62.7	1.40	0.82	939[g]	10.5	29	0.32	21	0.23	35
450	0.2	9.4	21.1	5.8	11.27[f]	0.98	2.25	62.2	1.38	0.80	963[g]	10.3	29	0.31	22	0.23	37
Cows nursing calves—Average milking ability—First 3–4 months postpartum—5.0 kg milk/d																	
350	0.0	7.7	16.6	4.6	9.98	NA	2.15	59.4	1.29	NA	814[g]	10.6	23	0.30	18	0.23	30
400	0.0	8.5	17.9	4.9	10.64[f]	NA	2.11	58.3	1.25	NA	864[g]	10.2	25	0.29	19	0.22	33
450	0.0	9.2	19.1	5.3	11.27[f]	NA	2.08	57.5	1.23	NA	911[g]	9.9	26	0.28	21	0.23	36
500	0.0	9.9	20.3	5.6	11.89[f]	NA	2.05	56.6	1.20	NA	957[g]	9.7	28	0.28	22	0.22	39
550	0.0	10.6	21.5	5.9	12.50[f]	NA	2.03	56.1	1.18	NA	1001[g]	9.5	29	0.27	24	0.23	41
600	0.0	11.2	22.6	6.2	13.08[f]	NA	2.01	55.5	1.16	NA	1044[g]	9.3	31	0.28	26	0.23	44
650	0.0	11.9	23.9	6.6	13.66[f]	NA	2.00	55.3	1.15	NA	1086[g]	9.1	33	0.28	27	0.23	46
Cows nursing calves—Superior milking ability—First 3–4 months postpartum—10.0 kg milk/d																	
350	0.0	6.2	18.5	5.1	13.73[f]	NA	3.00	82.9	2.03	NA	1009[g]	16.4	36	0.58	24	0.39	24
400	0.0	7.6	21.4	5.9	14.39[f]	NA	2.80	77.4	1.86	NA	1099[g]	14.4	37	0.49	25	0.33	30
450	0.0	9.1	23.2	6.4	15.02[f]	NA	2.56	70.7	1.66	NA	1186[g]	13.1	39	0.43	26	0.29	35
500	0.0	10.0	24.6	6.8	15.64[f]	NA	2.45	67.7	1.56	NA	1246[g]	12.4	40	0.40	28	0.28	39
550	0.0	10.9	25.8	7.1	16.25[f]	NA	2.38	65.8	1.50	NA	1299[g]	12.0	42	0.39	30	0.27	42
600	0.0	11.6	27.0	7.5	16.83[f]	NA	2.32	64.1	1.45	NA	1348[g]	11.6	43	0.37	31	0.27	45
650	0.0	12.4	28.2	7.8	17.41[f]	NA	2.28	63.0	1.41	NA	1394[g]	11.3	45	0.36	33	0.26	48
Bulls, maintenance and regaining body condition. For growth and development use requirements for bulls in Appendix Tables 8, 9, and 10																	
<650																	
650	0.4	12.3	24.3	6.7	9.91	2.06	1.98	54.8	1.13	0.57	904	7.4	25	0.20	23	0.19	48
650	0.6	12.6	26.7	7.4	9.91	3.21	2.11	58.4	1.25	0.69	957	7.6	27	0.21	24	0.19	49
650	0.8	12.8	28.7	7.9	9.91	4.40	2.24	62.0	1.37	0.79	998	7.8	29	0.23	25	0.20	50
700	0.4	13.0	25.7	7.1	10.48	2.18	1.98	54.8	1.13	0.57	942	7.3	26	0.20	25	0.20	51
700	0.6	13.4	28.2	7.8	10.48	3.40	2.11	58.4	1.25	0.69	994	7.4	29	0.22	26	0.20	52
700	0.8	13.5	30.3	8.4	10.48	4.66	2.24	62.0	1.37	0.79	1032	7.6	30	0.22	26	0.19	53
800	0.0	12.9	22.6	6.3	11.58	NA	1.75	48.4	0.91	NA	882	6.8	27	0.21	27	0.21	50
800	0.2	13.7	25.5	7.1	11.58	1.12	1.86	51.5	1.02	0.47	956	7.0	27	0.20	27	0.20	53
900	0.0	14.1	24.7	6.8	12.65	NA	1.75	48.4	0.91	NA	958	6.8	30	0.21	30	0.21	55
900	0.2	15.0	27.9	7.7	12.65	1.23	1.86	51.5	1.02	0.47	1031	6.9	31	0.21	31	0.21	58
1000	0.0	15.3	26.8	7.4	13.69	NA	1.75	48.4	0.91	NA	1032	6.8	33	0.22	33	0.22	60

[a] Average weight for a feeding period.

[b] Approximately 0.4 ± 0.1 kg of weight gain/d over the last third of pregnancy is accounted for by the products of conception. Daily 2.15 Mcal of NEm and 55 g of protein are provided for this requirement for a calf with a birth weight of 36 kg.

[c] Dry matter consumption should vary depending on the energy concentration of the diet and environmental conditions. These intakes are based on the energy concentration shown in the table and assuming a thermoneutral environment without snow or mud conditions. If the energy concentrations of the diet to be fed exceed the tabular value, limit feeding may be required.

[d] Vitamin A requirements per kilogram of diet are 2800 IU for pregnant heifers and cows and 3900 IU for lactating cows and breeding bulls.

[e] Not applicable.

[f] Includes 0.75 Mcal NEm/kg of milk produced.

[g] Includes 33.5 g protein/kg of milk produced.

APPENDIX TABLE 12A

Nutrient requirements of beef cows (from 1996 NRC on beef)

Mature Weight	533 kg	Milk Fat	4.0%
Calf Birth Weight	40 kg	Milk Protein	3.4%
Age @ Calving	60 Months	Calving Interval	12 months
Age @ Weaning	30 Weeks	Time Peak	8.5 weeks
Peak Milk	8 kg	Milk SNF	8.3%
Breed Code	1 Angus		

Month since Calving

	1	2	3	4	5	6	7	8	9	10	11	12
NE_m Req. Factor	100%	100%	100%	100%	100%	100%	100%	100%	100%	100%	100%	100%
NE_m required, Mcal/d												
Maintenance	10.25	10.25	10.25	10.25	10.25	10.25	8.54	8.54	8.54	8.54	8.54	8.54
Growth	0.00	0.00	0.00	0.00	0.00	0.00	0.00	0.00	0.00	0.00	0.00	0.00
Lactation	4.78	5.75	5.17	4.13	3.10	2.23	0.00	0.00	0.00	0.00	0.00	0.00
Pregnancy	0.00	0.00	0.01	0.03	0.07	0.16	0.32	0.64	1.18	2.08	3.44	5.37
Total	15.03	15.99	15.43	14.41	13.42	12.64	8.87	9.18	9.72	10.62	11.98	13.91
MP required, g/d												
Maintenance	422	422	422	422	422	422	422	422	422	422	422	422
Growth	0	0	0	0	0	0	0	0	0	0	0	0
Lactation	349	418	376	301	226	163	0	0	0	0	0	0
Pregnancy	0	0	1	2	4	7	14	27	50	88	151	251
Total	770	840	799	724	651	591	436	449	471	510	573	672
Calcium required, g/d												
Maintenance	16	16	16	16	16	16	16	16	16	16	16	16
Growth	0	0	0	0	0	0	0	0	0	0	0	0
Lactation	16	20	18	14	11	8	0	0	0	0	0	0
Pregnancy	0	0	0	0	0	0	0	0	0	12	12	12
Total	33	36	34	31	27	24	16	16	16	29	29	29
Phosphorus required, g/d												
Maintenance	13	13	13	13	13	13	13	13	13	13	13	13
Growth	0	0	0	0	0	0	0	0	0	0	0	0
Lactation	9	11	10	8	6	4	0	0	0	0	0	0
Pregnancy	0	0	0	0	0	0	0	0	0	5	5	5
Total	22	24	23	21	19	17	13	13	13	18	18	18
ADG, kg/d												
Growth	0.00	0.00	0.00	0.00	0.00	0.00	0.00	0.00	0.00	0.00	0.00	0.00
Pregnancy	0.00	0.00	0.02	0.03	0.05	0.08	0.12	0.19	0.28	0.40	0.57	0.77
Total	0.00	0.00	0.02	0.03	0.05	0.08	0.12	0.19	0.28	0.40	0.57	0.77
Milk, kg/d	6.7	8.0	7.2	5.8	4.3	3.1	0.0	0.0	0.0	0.0	0.0	0.0
Body weight, kg												
Shrunk Body	533	533	533	533	533	533	533	533	533	533	533	533
Conceptus	0	0	1	1	3	4	7	12	19	29	44	64
Total	533	533	534	534	536	537	540	545	552	562	577	597

Diet Evaluation for Beef Cows

Mature Weight	533 kg	Milk Fat	4.0%
Calf Birth Weight	40 kg	Milk Protein	3.4%
Age @ Calving	60 months	Calving Interval	12 months
Age @ Weaning	30 weeks	Time Peak	8.5 weeks
Peak Milk	8 kg	Milk SNF	8.3%
Breed Code	1 Angus		

Ration	TDN % DM	ME Mcal/g	NE$_m$ Mcal/g	CP % DM	DIP % CP	DMI Factor
A	50	1.84	1.00	7.9	82.5	100%
B	60	2.21	1.35	7.8	100.0	100%
C	70	2.58	1.67	9.1	100.0	100%

Months since Calving

		1	2	3	4	5	6	7	8	9	10	11	12
NE$_n$		100%	100%	100%	100%	100%	100%	100%	100%	100%	100%	100%	100%
A	Milk kg/d	6.7	8.0	7.2	5.8	4.3	3.1	0.0	0.0	0.0	0.0	0.0	0.0
	DM, kg	11.14	11.40	12.12	11.83	11.54	11.30	10.68	10.66	10.68	10.68	10.68	10.68
	Energy Balance, Mcal/d	-3.90	-4.59	-3.31	-2.58	-1.88	-1.34	1.81	1.50	0.95	0.06	-1.30	-3.24
	DIP Balance, g/d	7	7	7	7	7	7	6	6	6	6	6	6
	UIP Balance, g/d	-200	-270	-169	-96	-24	34	175	170	142	93	14	-110
	MP Balance, g/d	-161	-216	-136	-77	-19	27	149	136	113	75	11	-88
	Ca % DM	0.55%	0.70%	0.68%	0.57%	0.52%	0.47%	0.34%	0.34%	0.34%	0.59%	0.59%	0.59%
	P % DM	0.20%	0.21%	0.19%	0.18%	0.16%	0.15%	0.12%	0.13%	0.12%	0.17%	0.17%	0.17%
	Reserves Flux/mo, Mcal	-148	-154	-126	-98	-71	-51	55	46	29	2	-50	-123
B	DM, kg	11.96	12.23	12.72	12.43	12.14	11.90	11.28	11.28	11.28	11.28	11.28	11.28
	Energy Balance, Mcal	1.07	0.47	1.69	2.32	2.92	3.38	6.32	6.00	5.46	4.56	3.20	1.27
	DIP Balance, g/d	5	5	5	5	5	5	5	5	5	5	5	5
	UIP Balance, g/d	18	-47	44	114	183	233	221	221	221	221	209	55
	MP Balance, g/d	14	-38	35	91	146	189	304	291	269	230	167	66
	Ca % DM	0.27%	0.30%	0.27%	0.25%	0.22%	0.20%	0.15%	0.15%	0.15%	0.25%	0.25%	0.25%
	P % DM	0.19%	0.20%	0.18%	0.17%	0.16%	0.14%	0.11%	0.11%	0.11%	0.16%	0.16%	0.16%
	Reserves Flux/mo, Mcal	32	14	51	71	89	103	192	183	166	139	97	39
C	DM, kg	13.16	13.42	13.79	13.50	13.21	12.97	12.35	12.35	12.35	12.35	12.35	12.35
	Energy Balance, Mcal/d	6.99	6.48	7.65	8.18	8.69	9.07	11.80	11.49	10.95	10.05	8.69	6.76
	DIP Balance, g/d	3	3	3	3	3	3	2	2	2	2	2	2
	UIP Balance, g/d	295	233	314	308	301	296	282	282	282	282	282	282
	MP Balance, g/d	236	187	256	308	360	401	509	496	473	435	371	272
	Ca % DM	0.25%	0.27%	0.35	0.33%	0.20%	0.19%	0.13%	0.13%	0.13%	0.23%	0.23%	0.23%
	P % DM	0.17%	0.18%	0.17%	0.15%	0.14%	0.13%	0.10%	0.10%	0.10%	0.14%	0.14%	0.14%
	Reserves Flux/mo. Mcal	212	197	233	249	264	276	359	349	333	306	264	305

APPENDIX 12B

Nutrient requirements of pregnant replacement heifers (from 1996 NRC on beef)

Mature Weight	533 kg
Calf Birth Weight	40 kg
Age @ Breeding	15 months
Breed Code	1 Angus

	Months Since Conception								
	1	2	3	4	5	6	7	8	9
NE_m required, Mcal/d									
Maintenance	5.98	6.14	6.30	6.46	6.61	6.77	6.92	7.07	7.23
Growth	2.29	2.36	2.42	2.48	2.54	2.59	2.65	2.71	2.77
Pregnancy	0.03	0.07	0.16	0.32	0.64	1.18	2.08	3.44	5.37
Total	8.31	8.57	8.87	9.26	9.79	10.55	11.65	13.23	15.37
MP required, g/d									
Maintenance	295	303	311	319	326	334	342	349	357
Growth	118	119	119	119	119	117	115	113	110
Pregnancy	2	4	7	18	27	50	88	151	251
Total	415	425	437	457	472	501	545	613	718
Minerals									
Calcium required g/d									
Maintenance	10	11	11	11	12	12	12	13	13
Growth	9	9	9	8	8	8	8	8	8
Pregnancy	0	0	0	0	0	0	12	12	12
Total	19	19	20	20	20	20	33	33	33
Phosphorus required, g/d									
Maintenance	8	8	8	9	9	9	10	10	10
Growth	4	4	3	3	3	3	3	3	3
Pregnancy	0	0	0	0	0	0	7	7	7
Total	12	12	12	12	12	13	20	20	20
ADG, kg/d									
Growth	0.39	0.39	0.39	0.39	0.39	0.39	0.39	0.39	0.39
Pregnancy	0.03	0.05	0.08	0.12	0.19	0.28	0.40	0.57	0.77
Total	0.42	0.44	0.47	0.51	0.58	0.67	0.79	0.96	1.16
Body weight, kg									
Shrunk body	332	343	355	367	379	391	403	415	426
Gravid uterus mass	1	3	4	7	12	19	29	44	64
Total	333	346	360	375	391	410	432	459	491

Diet Evaluation for Pregnant Replacement Heifers

Mature Weight	533 kg
Calf birth Weight	40 kg
Age @ Breeding	15 months
Breed Code	1 Angus

Ration	TDN %DM	NE_n Mcal/kg	NE_m Mcal/kg	CP %DM	DIP %DM	DMI Factor
A	50	1.00	0.45	8.2	80	100%
B	60	1.35	0.77	9.8	80	100%
C	70	1.67	1.06	11.4	80	100%

Months Since Conception

Ration		1	2	3	4	5	6	7	8	9
A	NE_m Req. Factor	100%	100%	100%	100%	100%	100%	100%	100%	100%
	DM, kg	8.5	8.8	9.0	9.2	9.4	9.7	9.9	10.1	10.3
	NE allowed ADC	0.35	0.34	0.33	0.31	0.28	0.22	0.12	0.00	0.00
	DIP Balance, g/d	5	5	5	6	6	6	6	6	6
	UIP Balance, g/d	75	79	83	87	90	92	90	66	−53
	MP Balance, g/d	60	63	67	69	72	74	72	52	−42
	Ca% DM	0.22%	0.21%	0.21%	0.20%	0.19%	0.18%	0.28%	0.25%	0.25%
	P % DM	0.17%	0.17%	0.16%	0.16%	0.15%	0.14%	0.19%	0.16%	0.16%
B	DM, kg	9.0	9.3	9.5	9.7	10.0	10.2	10.4	10.7	10.9
	NE allowed ADC	0.96	0.96	0.95	0.92	0.88	0.82	0.71	0.54	0.30
	DIP Balance, g/d	4	4	4	4	4	4	4	4	4
	UIP Balance, g/d	5	14	22	30	38	49	54	46	18
	MP Balance, g/d	4	11	18	24	31	40	43	37	14
	Ca % DM	0.36%	0.35%	0.33%	0.32%	0.31%	0.29%	0.38%	0.34%	0.29%
	P % DM	0.27%	0.27%	0.26%	0.26%	0.25%	0.23%	0.27%	0.24%	0.20%
C	DM, kg	8.8	9.1	9.3	9.5	9.8	10.0	10.2	10.4	10.7
	NE allowed ADG	1.47	1.46	1.45	1.42	1.38	1.31	1.19	1.02	0.77
	DIP Balance, g/d	2	2	2	2	2	2	2	2	2
	UIP Balance, g/d	−66	−54	−43	−32	−19	−1	10	8	−18
	MP Balance, g/d	−53	−43	−34	−26	−15	−1	8	6	−14
	Ca % DM	0.48%	0.47%	0.45%	0.43%	0.41%	0.39%	0.48%	0.43%	0.38%
	P % DM	0.37%	0.36%	0.35%	0.35%	0.33%	0.31%	0.35%	0.32%	0.28%

NOTE: Requirements are for NE allowed ADG and target weight. NE allowed ADC is ADG independent of conceptus gain.

APPENDIX TABLE 13

Daily nutrient requirements of growing dairy cattle and mature bulls (from 1989 NRC on dairy)

Live Weight, kg	Gain, g	Dry Matter Intake[a], kg	NEm, Mcal	NEg, Mcal	ME, Mcal	DE, Mcal	TDN, kg	UIP, g	DIP, g	CP, g	Ca, g	P, g	A, 1000 IU	D, 1000 IU
Growing large-breed calves fed only milk or milk replacer														
40	200	0.48	1.37	0.41	2.54	2.73	0.62	—	—	105	7	4	1.70	0.26
45	300	0.54	1.49	0.56	2.86	3.07	0.70	—	—	120	8	5	1.94	0.30
Growing large-breed calves fed milk plus starter mix														
50	500	1.30	1.62	0.72	5.90	6.42	1.46	—	—	290	9	6	2.10	0.33
75	800	1.98	2.19	1.30	8.98	9.78	2.22	—	—	435	16	8	3.20	0.50
Growing small-breed calves fed only milk or milk replacer														
25	200	0.38	0.96	0.37	2.01	2.16	0.49	—	—	84	6	4	1.10	0.16
30	300	0.51	1.10	0.52	2.70	2.90	0.66	—	—	112	7	4	1.30	0.20
Growing small-breed calves fed milk plus starter mix														
50	500	1.43	1.62	0.72	6.49	7.06	1.60	—	—	315	10	6	2.10	0.33
75	600	1.76	2.19	0.96	7.98	8.69	1.97	—	—	387	14	8	3.20	0.50
Growing veal calves fed only milk or milk replacer														
40	200	0.45	1.37	0.55	1.89	2.07	0.47	—	—	100	7	4	1.70	0.26
50	400	0.57	1.62	0.57	2.39	2.63	0.59	—	—	125	9	5	2.10	0.33
60	540	0.80	1.85	0.81	2.84	3.17	0.71	—	—	176	13	8	2.60	0.40
75	900	1.36	2.19	1.47	4.82	5.39	1.21	—	—	300	16	9	3.20	0.50
100	1,250	2.00	2.72	2.26	6.22	7.06	1.58	—	—	440	20	11	4.20	0.66
125	1,250	2.38	3.21	2.44	7.40	8.40	1.88	—	—	524	22	13	5.30	0.82
150	1,100	2.72	3.69	2.29	8.46	9.60	2.15	—	—	598	24	15	6.40	0.99
Large-breed growing females														
100	600	2.63	2.72	1.22	7.03	8.13	1.84	317	57	421	17	9	4.24	0.66
100	700	2.82	2.72	1.44	7.54	8.72	1.98	346	75	452	18	9	4.24	0.66
100	800	3.02	2.72	1.66	8.06	9.32	2.11	374	92	483	18	10	4.24	0.66
150	600	3.51	3.69	1.45	9.14	10.61	2.41	283	150	562	19	11	6.36	0.99
150	700	3.75	3.69	1.71	9.76	11.33	2.57	307	173	600	19	12	6.36	0.99
150	800	3.99	3.69	1.97	10.39	12.07	2.74	331	196	639	20	12	6.36	0.99
200	600	4.39	4.57	1.65	11.14	12.99	2.95	254	239	631	20	14	8.48	1.32
200	700	4.68	4.57	1.95	11.87	13.84	3.14	274	267	686	21	14	8.48	1.32
200	800	4.97	4.57	2.25	12.62	14.71	3.34	294	295	741	22	15	8.48	1.32
250	600	5.31	5.41	1.84	13.10	15.33	3.48	229	326	637	22	16	10.60	1.65
250	700	5.65	5.41	2.18	13.94	16.32	3.70	246	359	678	23	17	10.60	1.65
250	800	5.99	5.41	2.51	14.79	17.32	3.93	263	393	726	24	17	10.60	1.65
300	600	6.26	6.20	2.02	15.05	17.69	4.01	209	413	752	23	17	12.72	1.98
300	700	6.66	6.20	2.39	16.00	18.81	4.27	223	452	799	24	18	12.72	1.98
300	800	7.06	6.20	2.77	16.97	19.95	4.52	236	490	848	25	19	12.72	1.98
350	600	7.29	6.96	2.20	17.01	20.09	4.56	193	501	874	24	18	14.84	2.31
350	700	7.75	6.96	2.60	18.09	21.36	4.84	204	545	930	25	19	14.84	2.31
350	800	8.21	6.96	3.01	19.18	22.64	5.14	214	590	985	26	20	14.84	2.31
400	600	8.39	7.69	2.37	19.03	22.58	5.12	182	592	1007	25	19	16.96	2.64
400	700	8.92	7.69	2.80	20.23	24.00	5.44	190	641	1070	26	20	16.96	2.64
400	800	9.46	7.69	3.24	21.44	25.44	5.77	198	692	1135	26	21	16.96	2.64
450	600	9.59	8.40	2.53	21.12	25.18	5.71	176	686	1151	28	19	19.08	2.97
450	700	10.20	8.40	2.99	22.46	26.78	6.07	182	742	1224	28	20	19.08	2.97
450	800	10.82	8.40	3.46	23.81	28.40	6.44	187	799	1298	29	21	19.08	2.97
500	600	10.93	9.09	2.69	23.32	27.96	6.34	175	785	1311	28	20	21.20	3.30
500	700	11.63	9.09	3.18	24.81	29.74	6.75	179	848	1395	28	20	21.20	3.30
500	800	12.33	9.09	3.68	26.32	31.55	7.16	182	913	1480	29	21	21.20	3.30
550	600	12.42	9.77	2.84	25.67	30.95	7.02	180	891	1490	28	20	23.32	3.63
550	700	13.22	9.77	3.37	27.33	32.95	7.47	183	963	1587	28	20	23.32	3.63
550	800	14.04	9.77	3.90	29.02	34.99	7.94	185	1035	1685	29	21	23.32	3.63
600	600	14.11	10.43	3.00	28.23	34.24	7.77	193	1007	1694	28	20	25.44	3.96
600	700	15.05	10.43	3.55	30.09	36.50	8.28	194	1088	1805	28	21	25.44	3.96
600	800	15.99	10.43	4.11	31.98	38.79	8.80	195	1170	1919	29	21	25.44	3.96
Small-breed growing females														
100	400	2.41	2.72	0.91	6.34	7.35	1.67	249	38	386	15	8	4.24	0.66
100	500	2.64	2.72	1.16	6.92	8.03	1.82	275	59	422	16	8	4.24	0.66
100	600	2.86	2.72	1.40	7.51	8.71	1.98	300	80	458	17	9	4.24	0.66
150	400	3.31	3.69	1.09	8.39	9.78	2.22	222	129	512	17	10	6.36	0.99
150	500	3.60	3.69	1.39	9.12	10.63	2.41	243	156	567	18	11	6.36	0.99
150	600	3.89	3.69	1.69	9.86	11.50	2.61	263	185	622	19	11	6.36	0.99
200	400	4.24	4.57	1.26	10.38	12.16	2.76	201	217	513	19	13	8.48	1.32

Live Weight, kg	Gain, g	Dry Matter Intake[a], kg	Energy					Protein			Minerals		Vitamins	
			NEm, Mcal	NEg, Mcal	ME, Mcal	DE, Mcal	TDN, kg	UIP, g	DIP, g	CP, g	Ca, g	P, g	A, 1000 IU	D, 1000 IU
200	500	4.60	4.57	1.60	11.25	13.19	2.99	217	251	562	20	13	8.48	1.32
200	600	4.96	4.57	1.95	12.14	14.23	3.23	232	286	611	20	14	8.48	1.32
250	400	5.24	5.41	1.41	12.36	14.57	3.30	185	305	629	21	15	10.60	1.65
250	500	5.68	5.41	1.80	13.38	15.78	3.58	197	346	681	21	16	10.60	1.65
250	600	6.12	5.41	2.20	14.43	17.01	3.86	209	389	735	22	16	10.60	1.65
300	400	6.34	6.20	1.56	14.38	17.06	3.87	176	395	761	22	16	12.72	1.98
300	500	6.87	6.20	1.99	15.57	18.48	4.19	184	445	824	23	17	12.72	1.98
300	600	7.40	6.20	2.43	16.79	19.92	4.52	192	495	888	23	17	12.72	1.98
350	400	7.57	6.96	1.71	16.50	19.71	4.47	173	490	909	23	17	14.84	2.31
350	500	8.20	6.96	2.18	17.87	21.35	4.84	178	548	985	23	18	14.84	2.31
350	600	8.85	6.96	2.66	19.28	23.03	5.22	183	608	1062	24	18	14.84	2.31
400	400	8.98	7.69	1.84	18.77	22.58	5.12	177	592	1078	24	18	16.96	2.64
400	500	9.74	7.69	2.35	20.36	24.50	5.56	181	661	1169	24	19	16.96	2.64
400	600	10.52	7.69	2.87	21.98	26.45	6.00	183	730	1263	25	19	16.96	2.64
450	400	10.64	8.40	1.98	21.27	25.80	5.85	191	706	1276	27	18	19.08	2.97
450	500	11.56	8.40	2.52	23.12	28.04	6.36	193	786	1387	28	19	19.08	2.97
450	600	12.50	8.40	3.08	25.01	30.33	6.88	194	867	1500	28	19	19.08	2.97
Large-breed growing males														
100	800	2.80	2.72	1.42	7.48	8.66	1.96	401	65	448	18	10	4.24	0.66
100	900	2.97	2.72	1.60	7.92	9.16	2.08	433	79	475	19	10	4.24	0.66
100	1000	3.13	2.72	1.79	8.36	9.67	2.19	465	93	501	20	11	4.24	0.66
150	800	3.60	3.69	1.64	9.52	11.03	2.50	364	155	576	20	12	6.36	0.99
150	900	3.80	3.69	1.85	10.03	11.63	2.64	393	172	607	21	13	6.36	0.99
150	1000	3.99	3.69	2.07	10.55	12.22	2.77	422	190	639	22	13	6.36	0.99
200	800	4.43	4.57	1.84	11.48	13.34	3.03	333	241	709	22	15	8.48	1.32
200	900	4.66	4.57	2.08	12.06	14.02	3.18	359	262	745	23	15	8.48	1.32
200	1000	4.89	4.57	2.33	12.66	14.71	3.34	385	284	782	24	16	8.48	1.32
250	800	5.27	5.41	2.03	13.37	15.58	3.53	305	325	778	24	17	10.60	1.65
250	900	5.53	5.41	2.30	14.03	16.35	3.71	329	350	837	25	18	10.60	1.65
250	1000	5.80	5.41	2.57	14.70	17.13	3.89	352	375	897	26	18	10.60	1.65
300	800	6.13	6.20	2.21	15.22	17.80	4.04	281	408	771	25	19	12.72	1.98
300	900	6.43	6.20	2.51	15.96	18.66	4.23	302	436	827	25	19	12.72	1.98
300	1000	6.73	6.20	2.80	16.70	19.53	4.43	323	464	884	26	20	12.72	1.98
350	800	7.02	6.96	2.38	17.06	20.02	4.54	261	490	843	26	20	14.84	2.31
350	900	7.36	6.96	2.70	17.88	20.98	4.76	280	522	883	26	20	14.84	2.31
350	1000	7.70	6.96	3.02	18.70	21.94	4.98	298	554	924	27	21	14.84	2.31
400	800	7.96	7.69	2.55	18.91	22.27	5.05	244	572	955	26	21	16.96	2.64
400	900	8.34	7.69	2.89	19.80	23.32	5.29	260	608	1001	27	21	16.96	2.64
400	1000	8.72	7.69	3.24	20.71	24.39	5.53	277	644	1046	28	22	16.96	2.64
450	800	8.95	8.40	2.71	20.78	24.56	5.57	230	656	1074	29	21	19.08	2.97
450	900	9.37	8.40	3.08	21.76	25.72	5.83	245	696	1125	29	22	19.08	2.97
450	1000	9.80	8.40	3.44	22.75	26.89	6.10	259	736	1176	29	23	19.08	2.97
500	800	10.00	9.09	2.87	22.69	26.92	6.11	220	742	1201	29	21	21.20	3.30
500	900	10.48	9.09	3.25	23.76	28.19	6.39	233	786	1257	29	22	21.20	3.30
500	1000	10.95	9.09	3.64	24.84	29.47	6.68	246	830	1314	29	23	21.20	3.30
550	800	11.14	9.77	3.02	24.66	29.38	6.66	213	831	1336	29	21	23.32	3.63
550	900	11.66	9.77	3.43	25.82	30.76	6.98	225	879	1399	29	22	23.32	3.63
550	1000	12.19	9.77	3.84	27.00	32.16	7.29	236	927	1463	30	23	23.32	3.63
600	800	12.36	10.43	3.17	26.71	31.95	7.25	211	923	1483	29	21	25.44	3.96
600	900	12.95	10.43	3.60	27.97	33.47	7.59	221	976	1554	29	22	25.44	3.96
600	1000	13.54	10.43	4.03	29.25	34.99	7.94	231	1029	1624	30	23	25.44	3.96
650	800	13.69	11.07	3.32	28.86	34.67	7.86	212	1020	1643	29	21	27.56	4.29
650	900	14.35	11.07	3.77	30.24	36.33	8.24	222	1078	1722	29	22	27.56	4.29
650	1000	15.01	11.07	4.22	31.63	38.00	8.62	230	1137	1801	30	23	27.56	4.29
700	800	15.16	11.70	3.46	31.14	37.59	8.52	219	1124	1820	29	22	29.68	4.62
700	900	15.90	11.70	3.93	32.64	39.40	8.94	227	1187	1907	29	22	29.68	4.62
700	1000	16.63	11.70	4.40	34.16	41.23	9.35	235	1252	1996	30	23	29.68	4.62
750	800	16.79	12.33	3.60	33.59	40.73	9.24	232	1235	2015	29	22	31.80	4.95
750	900	17.62	12.33	4.09	35.23	42.73	9.69	239	1305	2114	29	23	31.80	4.95
750	1000	18.45	12.33	4.58	36.89	44.74	10.15	246	1376	2213	30	23	31.80	4.95
800	800	17.56	12.94	3.74	35.12	42.59	9.66	216	1303	2107	29	22	33.92	5.28
800	900	18.41	12.94	4.25	36.83	44.67	10.13	221	1377	2210	29	23	33.92	5.28
800	1000	19.28	12.94	4.76	38.55	46.76	10.61	227	1451	2313	30	23	33.92	5.28

Live Weight, kg	Gain, g	Dry Matter Intake[a], kg	Energy					Protein			Minerals		Vitamins	
			NEm, Mcal	NEg, Mcal	ME, Mcal	DE, Mcal	TDN, kg	UIP, g	DIP, g	CP, g	Ca, g	P, g	A, 1000 IU	D, 1000 IU
Small-breed growing males														
100	500	2.45	2.72	1.02	6.54	7.56	1.72	287	41	392	16	8	4.24	0.66
100	600	2.64	2.72	1.23	7.04	8.15	1.85	316	58	422	17	9	4.24	0.66
100	700	2.83	2.72	1.45	7.55	8.74	1.98	345	75	453	18	9	4.24	0.66
150	500	3.28	3.69	1.20	8.55	9.92	2.25	257	129	525	18	11	6.36	0.99
150	600	3.52	3.69	1.46	9.16	10.64	2.41	282	151	563	19	11	6.36	0.99
150	700	3.76	3.69	1.71	9.78	11.36	2.58	306	174	601	19	12	6.36	0.99
200	500	4.12	4.57	1.37	10.45	12.18	2.76	232	213	573	20	13	8.48	1.32
200	600	4.40	4.57	1.66	11.17	13.02	2.95	252	241	629	20	14	8.48	1.32
200	700	4.69	4.57	1.96	11.90	13.87	3.15	273	268	684	21	14	8.48	1.32
250	500	4.99	5.41	1.53	12.31	14.41	3.27	210	296	598	21	16	10.60	1.65
250	600	5.32	5.41	1.86	13.14	15.38	3.49	228	328	638	22	16	10.60	1.65
250	700	5.66	5.41	2.19	13.97	16.35	3.71	245	361	679	23	17	10.60	1.65
300	500	5.89	6.20	1.68	14.15	16.64	3.77	193	378	707	23	17	12.72	1.98
300	600	6.28	6.20	2.04	15.09	17.74	4.02	207	415	754	23	17	12.72	1.98
300	700	6.68	6.20	2.41	16.04	18.85	4.28	221	453	801	24	18	12.72	1.98
350	500	6.86	6.96	1.82	16.01	18.91	4.29	180	461	823	23	18	14.84	2.31
350	600	7.31	6.96	2.22	17.06	20.15	4.57	191	503	877	24	18	14.84	2.31
350	700	7.76	6.96	2.62	18.13	21.41	4.86	203	547	932	25	19	14.84	2.31
400	500	7.90	7.69	1.96	17.91	21.25	4.82	171	545	947	24	19	16.96	2.64
400	600	8.41	7.69	2.39	19.08	22.64	5.14	180	594	1010	25	19	16.96	2.64
400	700	8.94	7.69	2.82	20.27	24.06	5.46	189	644	1073	26	20	16.96	2.64
450	500	9.03	8.40	2.10	19.87	23.70	5.37	166	634	1083	28	19	19.08	2.97
450	600	9.62	8.40	2.55	21.18	25.26	5.73	174	689	1155	28	19	19.08	2.97
450	700	10.23	8.40	3.01	22.51	26.84	6.09	180	744	1227	28	20	19.08	2.97
500	500	10.28	9.09	2.23	21.93	26.29	5.96	167	726	1233	28	19	21.20	3.30
500	600	10.96	9.09	2.71	23.39	28.04	6.36	173	788	1315	28	20	21.20	3.30
500	700	11.65	9.09	3.20	24.87	29.81	6.76	177	851	1398	28	20	21.20	3.30
550	500	11.67	9.77	2.36	24.12	29.08	6.60	174	825	1400	28	19	23.32	3.63
550	600	12.46	9.77	2.87	25.75	31.05	7.04	178	895	1495	28	20	23.32	3.63
550	700	13.26	9.77	3.39	27.40	33.03	7.49	181	966	1591	28	20	23.32	3.63
600	500	13.25	10.43	2.48	26.50	32.14	7.29	187	933	1590	28	19	25.44	3.96
600	600	14.16	10.43	3.02	28.32	34.35	7.79	190	1012	1699	28	20	25.44	3.96
600	700	15.08	10.43	3.57	30.17	36.59	8.30	192	1091	1810	28	21	25.44	3.96
Maintenance of mature breeding bulls														
500	—	7.89	9.09	—	15.79	19.15	4.34	161	472	789	20	12	21.20	3.30
600	—	9.05	10.43	—	18.10	21.95	4.98	155	573	905	24	15	25.44	3.96
700	—	10.16	11.70	—	20.32	24.64	5.59	148	670	1016	28	18	29.68	4.62
800	—	11.23	12.94	—	22.46	27.24	6.18	142	764	1123	32	20	33.92	5.28
900	—	12.27	14.13	—	24.53	29.76	6.75	135	854	1227	36	22	38.16	5.94
1000	—	13.28	15.29	—	26.55	32.20	7.30	129	943	1328	41	25	42.40	6.60
1100	—	14.26	16.43	—	28.52	34.59	7.85	122	1029	1426	45	28	46.64	7.26
1200	—	15.22	17.53	—	30.44	36.92	8.37	115	1113	1522	49	30	50.88	7.92
1300	—	16.16	18.62	—	32.32	39.21	8.89	108	1196	1616	53	32	55.12	8.58
1400	—	17.09	19.68	—	34.17	41.45	9.40	102	1277	1709	57	35	59.36	9.24

Note: The following abbreviations were used: NEm, net energy for maintenance; NEg, net energy for gain; ME, metabolizable energy; DE, digestible energy; TDN, total digestible nutrients; UIP, undegraded intake protein; DIP, degraded intake protein; CP, crude protein.

[a]The data for DMI are not requirements per se, unlike the requirements for net energy maintenance, net energy gain, and absorbed protein. They are not intended to be estimates of voluntary intake but are consistent with the specified dietary energy concentrations. The use of diets with decreased energy concentrations will increase dry-matter intake needs; metabolizable energy, digestible energy, and total digestible nutrient needs; and crude protein needs. The use of diets with increased energy concentrations will have opposite effects on these needs.

APPENDIX TABLE 14

Daily nutrient requirements of lactating and pregnant cows (from 1989 NRC on dairy cows)

Live Weight, kg	Energy				Total Crude Protein, g	Minerals		Vitamins	
	NEℓ, Mcal	ME, Mcal	DE, Mcal	TDN, kg		Ca, g	P, g	A, 1000 IU	D, 1000 IU
Maintenance of mature lactating cows[a]									
400	7.16	12.01	13.80	3.13	318	16	11	30	12
450	7.82	13.12	15.08	3.42	341	18	13	34	14
500	8.46	14.20	16.32	3.70	364	20	14	38	15
550	9.09	15.25	17.53	3.97	386	22	16	42	17
600	9.70	16.28	18.71	4.24	406	24	17	46	18
650	10.30	17.29	19.86	4.51	428	26	19	49	20
700	10.89	18.28	21.00	4.76	449	28	20	53	21
750	11.47	19.25	22.12	5.02	468	30	21	57	23
800	12.03	20.20	23.21	5.26	486	32	23	61	24
Maintenance plus last 2 months of gestation of mature dry cows[b]									
400	9.30	15.26	18.23	4.15	890	26	16	30	12
450	10.16	16.66	19.91	4.53	973	30	18	34	14
500	11.00	18.04	21.55	4.90	1,053	33	20	38	15
550	11.81	19.37	23.14	5.27	1,131	36	22	42	17
600	12.61	20.68	24.71	5.62	1,207	39	24	46	18
650	13.39	21.96	26.23	5.97	1,281	43	26	49	20
700	14.15	23.21	27.73	6.31	1,355	46	28	53	21
750	14.90	24.44	29.21	6.65	1,427	49	30	57	23
800	15.64	25.66	30.65	6.98	1,497	53	32	61	24
Milk production—nutrients/kg of milk of different fat percentages (Fat %)									
3.0	0.64	1.07	1.23	0.280	78	2.73	1.68	—	—
3.5	0.69	1.15	1.33	0.301	84	2.97	1.83	—	—
4.0	0.74	1.24	1.42	0.322	90	3.21	1.98	—	—
4.5	0.78	1.32	1.51	0.343	96	3.45	2.13	—	—
5.0	0.83	1.40	1.61	0.364	101	3.69	2.28	—	—
5.5	0.88	1.48	1.70	0.385	107	3.93	2.43	—	—
Liveweight change during lactation—nutrients/kg of weight change[c]									
Weight loss	−4.92	−8.25	−9.55	−2.17	−320	—	—	—	—
Weight gain	5.12	8.55	9.96	2.26	320	—	—	—	—

Note: The following abbreviations were used: NEℓ, net energy for lactation; ME, metabolizable energy; DE, digestible energy; TDN, total digestible nutrients.

[a]To allow for growth of young lactating cows, increase the maintenance allowances for all nutrients except vitamins A and D by 20% during the first lactation and 10% during the second lactation.

[b]Values for calcium assume that the cow is in calcium balance at the beginning of the last 2 months of gestation. If the cow is not in balance, then the calcium requirement can be increased from 25 to 33%.

[c]No allowance is made for mobilized calcium and phosphorus associated with liveweight loss or with liveweight gain. The maximum daily nitrogen available from weight loss is assumed to be 30 g or 234 g of crude protein.

APPENDIX TABLE 15

Daily nutrient requirements of lactating cows using absorbable protein (from 1989 NRC on dairy)

Live Weight, kg	Fat, %	Milk, kg	Live Weight Change, kg	Dry Matter Intake, kg	Energy NEℓ, Mcal/kg DM	Energy NEℓ, Mcal	Energy TDN, kg	Protein UIP, g	Protein DIP, g	Minerals Ca, g	Minerals P, g
Intake at 100% of the requirement for maintenance, lactation, and weight gain												
400	4.5	8.0	0.220	10.14	1.43	14.55	6.44	511	753	44	28
400	4.5	14.0	0.220	12.66	1.52	19.26	8.48	710	1052	65	41
400	4.5	20.0	0.220	14.91	1.61	23.96	10.51	880	1355	85	54
400	4.5	26.0	0.220	16.94	1.69	28.67	12.54	1026	1662	106	67
400	4.5	32.0	0.220	19.41	1.72	33.37	14.58	1220	1962	127	80
400	5.0	8.0	0.220	10.36	1.44	14.94	6.60	525	778	46	30
400	5.0	14.0	0.220	13.00	1.53	19.93	8.77	730	1096	68	43
400	5.0	20.0	0.220	15.35	1.62	24.93	10.93	902	1419	90	57
400	5.0	26.0	0.220	17.44	1.72	29.92	13.07	1048	1745	112	71
400	5.0	32.0	0.220	20.30	1.72	34.91	15.25	1277	2061	134	84
400	5.5	8.0	0.220	10.57	1.45	15.32	6.77	538	803	48	31
400	5.5	14.0	0.220	13.33	1.55	20.61	9.07	748	1140	71	45
400	5.5	20.0	0.220	15.77	1.64	25.89	11.34	923	1483	95	60
400	5.5	26.0	0.220	18.13	1.72	31.17	13.62	1091	1826	118	75
400	5.5	32.0	0.220	21.20	1.72	36.45	15.92	1334	2160	142	89
500	4.0	9.0	0.275	11.59	1.42	16.49	7.30	540	883	49	32
500	4.0	17.0	0.275	14.78	1.51	22.38	9.86	797	1257	75	48
500	4.0	25.0	0.275	17.62	1.61	28.27	12.40	1015	1635	101	64
500	4.0	33.0	0.275	20.14	1.70	34.15	14.93	1201	2018	126	80
500	4.0	41.0	0.275	23.29	1.72	40.04	17.49	1453	2392	152	95
500	4.5	9.0	0.275	11.84	1.43	16.92	7.49	556	911	51	33
500	4.5	17.0	0.275	15.20	1.53	23.20	10.21	821	1310	79	50
500	4.5	25.0	0.275	18.16	1.62	29.47	12.92	1043	1715	107	68
500	4.5	33.0	0.275	20.79	1.72	35.74	15.61	1230	2124	134	85
500	4.5	41.0	0.275	24.44	1.72	42.02	18.35	1526	2519	162	102
500	5.0	9.0	0.275	12.08	1.44	17.36	7.68	571	939	53	35
500	5.0	17.0	0.275	15.60	1.54	24.01	10.57	844	1364	83	53
500	5.0	25.0	0.275	18.68	1.64	30.67	13.44	1069	1795	113	71
500	5.0	33.0	0.275	21.71	1.72	37.33	16.31	1289	2226	142	89
500	5.0	41.0	0.275	25.58	1.72	43.99	19.21	1599	2646	172	108
600	3.0	10.0	0.330	12.52	1.42	17.79	7.87	533	974	52	34
600	3.0	20.0	0.330	16.20	1.49	24.18	10.67	845	1375	79	51
600	3.0	30.0	0.330	19.37	1.58	30.58	13.43	1102	1784	106	68
600	3.0	40.0	0.330	22.21	1.67	36.98	16.19	1323	2198	133	84
600	3.0	50.0	0.330	25.23	1.72	43.38	18.95	1565	2608	161	101
600	3.5	10.0	0.330	12.86	1.42	18.27	8.08	557	1004	54	35
600	3.5	20.0	0.330	16.70	1.51	25.15	11.08	874	1438	84	54
600	3.5	30.0	0.330	20.04	1.60	32.03	14.06	1137	1879	113	72
600	3.5	40.0	0.330	23.00	1.69	38.90	17.01	1360	2326	143	90
600	3.5	50.0	0.330	26.63	1.72	45.78	20.00	1654	2763	173	109
600	4.0	10.0	0.330	13.20	1.42	18.75	8.30	581	1034	56	37
600	4.0	20.0	0.330	17.19	1.52	26.11	11.50	902	1501	89	57
600	4.0	30.0	0.330	20.69	1.62	33.47	14.68	1170	1975	121	77
600	4.0	40.0	0.330	23.78	1.72	40.83	17.84	1395	2454	153	96
600	4.0	50.0	0.330	28.03	1.72	48.19	21.05	1744	2918	185	116
700	3.0	12.0	0.385	14.46	1.42	20.54	9.09	607	1154	61	40
700	3.0	24.0	0.385	18.75	1.50	28.21	12.44	968	1638	94	60
700	3.0	36.0	0.385	22.48	1.60	35.89	15.76	1269	2129	127	81
700	3.0	48.0	0.385	25.80	1.69	43.57	19.05	1525	2627	159	101
700	3.0	60.0	0.385	29.81	1.72	51.25	22.39	1857	3114	192	121
700	3.5	12.0	0.385	14.86	1.42	21.11	9.34	636	1190	64	42
700	3.5	24.0	0.385	19.34	1.52	29.37	12.94	1002	1713	100	64
700	3.5	36.0	0.385	23.26	1.62	37.62	16.50	1309	2244	135	86
700	3.5	48.0	0.385	26.72	1.72	45.88	20.04	1567	2781	171	108
700	3.5	60.0	0.385	31.48	1.72	54.13	23.65	1964	3300	207	130
700	4.0	12.0	0.385	15.20	1.43	21.69	9.60	658	1227	67	44
700	4.0	24.0	0.385	19.92	1.53	30.52	13.44	1035	1789	105	68

Live Weight, kg	Fat, %	Milk, kg	Live Weight Change, kg	Dry Matter Intake, kg	Energy			Protein		Minerals	
					NEℓ, Mcal/kg DM	NEℓ, Mcal	TDN, kg	UIP, g	DIP, g	Ca, g	P, g
700	4.0	36.0	0.385	24.02	1.64	39.35	17.25	1347	2359	144	91
700	4.0	48.0	0.385	28.03	1.72	48.19	21.05	1648	2930	182	115
700	4.0	60.0	0.385	33.16	1.72	57.02	24.91	2071	3485	221	139
800	3.0	14.0	0.440	16.36	1.42	23.24	10.29	682	1331	71	46
800	3.0	27.0	0.440	20.93	1.51	31.56	13.91	1064	1857	106	68
800	3.0	40.0	0.440	24.95	1.60	39.88	17.50	1388	2390	142	90
800	3.0	53.0	0.440	28.54	1.69	48.20	21.08	1665	2928	177	112
800	3.0	66.0	0.440	32.87	1.72	56.51	24.69	2022	3457	213	134
800	3.5	14.0	0.440	16.78	1.42	23.92	10.58	710	1374	74	49
800	3.5	27.0	0.440	21.59	1.52	32.86	14.47	1102	1942	113	72
800	3.5	40.0	0.440	25.82	1.62	41.80	18.33	1432	2517	151	96
800	3.5	53.0	0.440	29.57	1.72	50.75	22.17	1711	3099	190	120
800	3.5	66.0	0.440	34.72	1.72	59.69	26.07	2140	3661	228	144
800	4.0	14.0	0.440	17.17	1.43	24.59	10.88	734	1418	77	51
800	4.0	27.0	0.440	22.24	1.54	34.16	15.03	1139	2027	119	76
800	4.0	40.0	0.440	26.66	1.64	43.73	19.16	1474	2644	161	102
800	4.0	53.0	0.440	31.00	1.72	53.29	23.28	1800	3263	203	128
800	4.0	66.0	0.440	36.56	1.72	62.86	27.46	2259	3865	244	154

Intake at 85% of the requirement for maintenance and lactation

Live Weight, kg	Fat, %	Milk, kg	Live Weight Change, kg	Dry Matter Intake, kg	NEℓ, Mcal/kg DM	NEℓ, Mcal	TDN, kg	UIP, g	DIP, g	Ca, g	P, g
400	4.5	20.0	−0.696	11.62	1.67	19.41	8.49	687	1066	85	54
400	4.5	26.0	−0.840	14.02	1.67	23.41	10.24	931	1310	106	67
400	4.5	32.0	−0.983	16.41	1.67	27.41	11.99	1187	1554	127	80
400	5.0	20.0	−0.726	12.11	1.67	20.23	8.85	720	1118	90	57
400	5.0	26.0	−0.878	14.65	1.67	24.47	10.71	987	1377	112	71
400	5.0	32.0	−1.030	17.20	1.67	28.72	12.56	1255	1635	134	84
400	5.5	20.0	−0.755	12.60	1.67	21.05	9.21	761	1169	95	60
400	5.5	26.0	−0.916	15.29	1.67	25.54	11.17	1042	1443	118	75
400	5.5	32.0	−1.077	17.98	1.67	30.03	13.14	1323	1717	142	89
500	4.0	25.0	−0.819	13.67	1.67	22.83	9.99	810	1286	101	64
500	4.0	33.0	−0.998	16.67	1.67	27.83	12.18	1134	1590	126	80
500	4.0	41.0	−1.178	19.66	1.67	32.84	14.37	1458	1894	152	95
500	4.5	25.0	−0.856	14.28	1.67	23.85	10.44	864	1350	107	68
500	4.5	33.0	−1.047	17.48	1.67	29.18	12.77	1205	1674	134	85
500	4.5	41.0	−1.238	20.67	1.67	34.52	15.10	1546	1998	162	102
500	5.0	25.0	−0.892	14.89	1.67	24.87	10.88	917	1414	113	71
500	5.0	33.0	−1.095	18.28	1.67	30.53	13.36	1275	1758	142	89
500	5.0	41.0	−1.298	21.67	1.67	36.19	15.83	1633	2103	172	108
600	3.0	30.0	−0.881	14.71	1.67	24.56	10.74	860	1399	106	68
600	3.0	40.0	−1.076	17.96	1.67	30.00	13.12	1223	1728	133	84
600	3.0	50.0	−1.271	21.22	1.67	35.44	15.50	1585	2057	161	101
600	3.5	30.0	−0.925	15.44	1.67	25.79	11.28	924	1476	113	72
600	3.5	40.0	−1.135	18.94	1.67	31.63	13.84	1308	1830	143	90
600	3.5	50.0	−1.344	22.44	1.67	37.48	16.40	1692	2184	173	109
600	4.0	30.0	−0.969	16.17	1.67	27.01	11.82	988	1552	121	77
600	4.0	40.0	−1.193	19.92	1.67	33.27	14.55	1393	1932	153	96
600	4.0	50.0	−1.418	23.67	1.67	39.52	17.29	1798	2311	185	116
700	3.0	36.0	−1.034	17.26	1.67	28.83	12.61	1054	1669	127	81
700	3.0	48.0	−1.268	21.17	1.67	35.36	15.47	1489	2064	159	101
700	3.0	60.0	−1.502	25.08	1.67	41.88	18.32	1924	2458	192	121
700	3.5	36.0	−1.087	18.15	1.67	30.30	13.26	1131	1761	135	86
700	3.5	48.0	−1.339	22.35	1.67	37.32	16.33	1591	2186	171	108
700	3.5	60.0	−1.590	26.55	1.67	44.34	19.40	2052	2611	207	130
700	4.0	36.0	−1.140	19.03	1.67	31.78	13.90	1208	1853	144	91
700	4.0	48.0	−1.409	23.52	1.67	39.28	17.19	1694	2308	182	115
700	4.0	60.0	−1.678	28.02	1.67	46.79	20.47	2180	2764	221	139
800	3.0	40.0	−1.147	19.15	1.67	31.98	13.99	1176	1871	142	90
800	3.0	50.0	−1.342	22.41	1.67	37.42	16.37	1538	2200	169	107
800	3.0	60.0	−1.537	25.66	1.67	42.86	18.75	1900	2529	196	124
800	3.5	40.0	−1.206	20.13	1.67	33.62	14.71	1261	1973	151	96

APPENDIX TABLE 15 (*cont.*)

Live Weight, kg	Fat, %	Milk, kg	Live Weight Change, kg	Dry Matter Intake, kg	Energy			Protein		Minerals	
					NEℓ, Mcal/kg DM	NEℓ, Mcal	TDN, kg	UIP, g	DIP, g	Ca, g	P, g
800	3.5	50.0	− 1.416	23.63	1.67	39.46	17.27	1645	2327	181	114
800	3.5	60.0	− 1.625	27.13	1.67	45.31	19.82	2028	2682	211	133
800	4.0	40.0	− 1.264	21.11	1.67	35.25	15.42	1346	2075	161	102
800	4.0	50.0	− 1.489	24.86	1.67	41.51	18.16	1751	2455	193	122
800	4.0	60.0	− 1.713	28.60	1.67	47.76	20.90	2156	2835	225	142

Note: The following abbreviations were used: NEℓ, Mcal/kg DM net energy for lactation/kg of dry matter; NEℓ, net energy for lactation; TDN, total digestible nutrients; UIP, undegraded intake protein; DIP, degraded intake protein.

APPENDIX TABLE 16

Dry-matter intake requirements to fulfill nutrient allowances for maintenance, milk production, and normal liveweight gain during mid- and late lactation (from 1989 NRC on dairy)

4% FCM,[a] kg	Live Weight,[b,c] kg				
	400	500	600	700	800
10	2.7	2.4	2.2	2.0	1.9
15	3.2	2.8	2.6	2.3	2.2
20	3.6	3.2	2.9	2.6	2.4
25	4.0	3.5	3.2	2.9	2.7
30	4.4	3.9	3.5	3.2	2.9
35	5.0	4.2	3.7	3.4	3.1
40	5.5	4.6	4.0	3.6	3.3
45	—	5.0	4.3	3.8	3.5
50	—	5.4	4.7	4.1	3.7
55	—	—	5.0	4.4	4.0
60	—	—	5.4	4.8	4.3

Note: The following assumptions were made in calculating DMI requirements shown in this table:

1. The reference cow used for the calculations weighed 600 kg and produced milk with 4% milk fat. Other live weights in the table and corresponding fat percentages were 400 kg and 5%; 500 kg and 4.5%; and 700 and 800 kg and 3.5% fat.

2. The concentrations of energy in the diet for the cow was 1.42 Mcal of NEℓ/kg of DM for milk yields equal to or less than 10 kg/d. It increased linearly to 1.72 Mcal of NEℓ/kg for milk yields equal to or greater than 40 kg/d.

3. The energy concentrations of the diets for all other cows were assumed to change linearly as their energy requirements for milk production, relative to maintenance, changed in a manner identical to that of the 600-kg cow as she increased in milk yield from 10 to 40 kg/d.

4. Enough DM to provide sufficient energy for cows to gain 0.055% of their body weight daily was also included in the total. If cows do not consume as much DM as they require, as calculated in this table, their energy intake will be less than their requirements. The result will be a loss of body weight, reduced milk yields, or both. If cows consume more DM than what is projected as required from the table, the energy concentration of their diet should be reduced or they may become overly fat.

[a]4% fat-corrected milk (kg) = (0.4) (kg of milk) + (15) (kg of milk fat).

[b]The probable DMI may be up to 18% less in early lactation.

[c]DMI as a percentage of live weight may be 0.02% less per 1% increase in diet moisture content above 50% if fermented feeds constitute a major portion of the diet.

APPENDIX TABLE 17

Recommended nutrient content of diets for dairy cattle (from 1989 NRC on dairy)

Cow weight and associated values for the lactating cow diet rows:

Cow wt, kg	Fat, %	Wt gain, kg/d
400	5.0	0.220
500	4.5	0.275
600	4.0	0.330
700	3.5	0.385
800	3.5	0.440

Milk yield (kg/d) for the five Lactating Cow Diet columns, by cow weight:

Cow wt, kg	Diet 1	Diet 2	Diet 3	Diet 4	Diet 5
400	7	13	20	26	33
500	8	17	25	33	41
600	10	20	30	40	50
700	12	24	36	48	60
800	13	27	40	53	67

	Lactating Cow Diets, Milk Yield, kg/d					Early lactation, wks 0–3	Dry, pregnant cows	Calf milk replacer	Calf starter mix	Growing Heifers and Bulls[a]			Mature bulls	Maximum tolerable levels[b,c]
	Diet 1	Diet 2	Diet 3	Diet 4	Diet 5					3–6 mos	6–12 mos	12 mos		
Energy														
NEℓ, Mcal/kg	1.42	1.52	1.62	1.72	1.72	1.67	1.25	—	—	—	—	—	1.15	—
NEm, Mcal/kg	—	—	—	—	—	—	—	2.40	1.90	1.70	1.58	1.40	—	—
NEg, Mcal/kg	—	—	—	—	—	—	—	1.55	1.20	1.08	0.98	0.82	—	—
ME, Mcal/kg	2.35	2.53	2.71	2.89	2.89	2.80	2.04	3.78	3.11	2.60	2.47	2.27	2.00	—
DE, Mcal/kg	2.77	2.95	3.13	3.31	3.31	3.22	2.47	4.19	3.53	3.02	2.89	2.69	2.43	—
TDN, % of DM	63	67	71	75	75	73	56	95	80	69	66	61	55	—
Protein equivalent														
Crude protein, %	12	15	16	17	18	19	12	22	18	16	12	12	10	—
UIP, %	4.4	5.2	5.7	5.9	6.2	7.0	—	—	—	8.2	4.4	2.1	—	—
DIP, %	7.8	8.7	9.6	10.3	10.4	9.7	—	—	—	4.6	6.4	7.2	—	—
Fiber content (min.)[d]														
Crude fiber, %	17	17	17	15	15	17	22	—	—	13	15	15	15	—
Acid detergent fiber, %	21	21	21	19	19	21	27	—	—	16	19	19	19	—
Neutral detergent fiber, %	28	28	28	25	25	28	35	—	—	23	25	25	25	—
Ether extract (min.), %	3	3	3	3	3	3	3	10	3	3	3	3	3	—
Minerals														
Calcium, %	0.43	0.51	0.58	0.64	0.66	0.77	0.39[c]	0.70	0.60	0.52	0.41	0.29	0.30	2.00
Phosphorus, %	0.28	0.33	0.37	0.41	0.41	0.48	0.24	0.60	0.40	0.31	0.30	0.23	0.19	1.00
Magnesium, %[f]	0.20	0.20	0.20	0.25	0.25	0.25	0.16	0.07	0.10	0.16	0.16	0.16	0.16	0.50
Potassium, %[g]	0.90	0.90	0.90	1.00	1.00	1.00	0.65	0.65	0.65	0.65	0.65	0.65	0.65	3.00
Sodium, %	0.18	0.18	0.18	0.18	0.18	0.18	0.10	0.10	0.10	0.10	0.10	0.10	0.10	—
Chlorine, %	0.25	0.25	0.25	0.25	0.25	0.25	0.20	0.20	0.20	0.20	0.20	0.20	0.20	—
Sulfur, %	0.20	0.20	0.20	0.20	0.20	0.25	0.16	0.29	0.20	0.16	0.16	0.16	0.16	0.40
Iron, ppm	50	50	50	50	50	50	50	100	50	50	50	50	50	1000
Cobalt, ppm[h]	0.10	0.10	0.10	0.10	0.10	0.10	0.10	0.10	0.10	0.10	0.10	0.10	0.10	10.00
Copper, ppm[h]	10	10	10	10	10	10	10	10	10	10	10	10	10	100
Manganese, ppm	40	40	40	40	40	40	40	40	40	40	40	40	40	1000
Zinc, ppm	40	40	40	40	40	40	40	40	40	40	40	40	40	500
Iodine, ppm[i]	0.60	0.60	0.60	0.60	0.60	0.60	0.25	0.25	0.25	0.25	0.25	0.25	0.25	50.00[i]
Selenium, ppm	0.30	0.30	0.30	0.30	0.30	0.30	0.30	0.30	0.30	0.30	0.30	0.30	0.30	2.00

Vitamins[k]												
A, IU/kg	3200	3200	3200	4000	4000	3800	2200	2200	2200	2200	3200	66,000
D, IU/kg	1000	1000	1000	1000	1200	600	300	300	300	300	300	10,000
E, IU/kg	15	15	15	15	15	40	25	25	25	25	15	2000

Note: The values presented in this table are intended as guidelines for the use of professionals in diet formulation. Because of the many factors affecting such values, they are not intended and should not be used as a legal or regulatory base.

[a]The approximate weight for growing heifer and bulls at 3–6 mos is 150 kg; at 6–12 mos, it is 250 kg; and at more than 12 mos, it is 400 kg. The approximate average daily gain is 700 g/d.

[b]The maximum safe levels for many of the mineral elements are not well defined and may be substantially affected by specific feeding conditions. Additional information is available in *Mineral Tolerance of Domestic Animals* (NRC, 1980).

[c]Vitamin tolerances are discussed in detail in *Vitamin Tolerance of Animals* (NRC, 1987).

[d]It is recommended that 75% of the NDF in lactating cow diets be provided as forage. If this recommendation is not followed, a depression in milk fat may occur.

[e]The value for calcium assumes that the cow is in calcium balance at the beginning of the dry period. If the cow is not in balance, then the dietary calcium requirement should be increased by 25–33%.

[f]Under conditions conducive to grass tetany (see text), magnesium should be increased to 0.25 or 0.30 %.

[g]Under conditions of heat stress, potassium should be increased to 1.2 %.

[h]The cow's copper requirement is influenced by molybdenum and sulfur in the diet.

[i]If the diet contains as much as 25 % strongly goitrogenic feed on a dry basis, the iodine provided should be increased two times or more.

[j]Although cattle can tolerate this level of iodine, lower levels may be desirable to reduce the iodine content of milk.

[k]The following minimum quantities of B-complex vitamins are suggested per unit of milk replacer: niacin, 2.6 ppm; pantothenic acid, 13 ppm; riboflavin, 6.5 ppm; pyridoxine, 6.5 ppm; folic acid, 0.5 ppm; biotin, 0.1 ppm; vitamin B_{12}, 0.07 ppm; thiamin, 6.5 ppm; and choline, 0.26 %. It appears that adequate amounts of these vitamins are furnished when calves have functional rumens (usually at 6 weeks of age) by a combination of rumen synthesis and natural feedstuffs.

APPENDIX TABLE 18

Daily nutrient requirements of sheep (100% dry matter basis) (from 1985 NRC on sheep)

Body Weight		Gain or Loss		Dry Matter[a]				Energy			Nutrients per Animal								
				Per Animal		% Live Wt	TDN, kg	DE,[b] Mcal	ME, Mcal	Total Protein, g	DP,[c] g	Grams DP per Mcal DE	Ca, g	P, g	Carotene, mg	Vitamin A, IU	Vitamin D, IU
kg	lb	g	lb	kg	lb												

Ewes[d]

Maintenance

50	110	10	0.02	1.0	2.2	2.0	0.55	2.42	1.98	89	48	20	3.0	2.8	1.9	1275	278
60	132	10	0.02	1.1	2.4	1.8	0.61	2.68	2.20	98	53	20	3.1	2.9	2.2	1530	333
70	154	10	0.02	1.2	2.6	1.7	0.66	2.90	2.38	107	58	20	3.2	3.0	2.6	1785	388
80	176	10	0.02	1.3	2.9	1.6	0.72	3.17	2.60	116	63	20	3.3	3.1	3.0	2040	444

Nonlactating and first 15 weeks of gestation

50	110	30	0.07	1.1	2.4	2.2	0.60	2.64	2.16	99	54	20	3.0	2.8	1.9	1275	278
60	132	30	0.07	1.3	2.9	2.1	0.72	3.17	2.60	117	64	20	3.1	2.9	2.2	1530	333
70	154	30	0.07	1.4	3.1	2.0	0.77	3.39	2.78	126	69	20	3.2	3.0	2.6	1785	388
80	176	30	0.07	1.5	3.3	1.9	0.82	3.61	2.96	135	74	20	3.3	3.1	3.0	2040	444

Last 6 weeks of gestation or last 8 weeks of lactation suckling singles[e]

50	110	175(+45)	0.39	1.7	3.7	3.3	0.99	4.36	3.58	158	88	20	4.1	3.9	6.2	4250	278
60	132	180(+45)	0.40	1.9	4.2	3.2	1.10	4.84	3.97	177	99	20	4.4	4.1	7.5	5100	333
70	154	185(+45)	0.41	2.1	4.6	3.0	1.22	5.37	4.40	195	109	20	4.5	4.3	8.8	5950	388
80	176	190(+45)	0.42	2.2	4.8	2.8	1.28	5.63	4.62	205	114	20	4.8	4.5	10.0	6800	444

First 8 weeks of lactation suckling singles or last 8 weeks of lactation suckling twins[f]

50	110	-25(+80)	-0.06	2.1	4.6	4.2	1.36	5.98	4.90	218	130	22	10.9	7.8	6.2	4250	278
60	132	-25(+80)	-0.06	2.3	5.1	3.9	1.50	6.60	5.41	239	143	22	11.5	8.2	7.5	5100	333
70	154	-25(+80)	-0.06	2.5	5.5	3.6	1.63	7.17	5.88	260	155	22	12.0	8.6	8.8	5950	388
80	176	-25(+80)	-0.06	2.6	5.7	3.2	1.69	7.44	6.10	270	161	22	12.6	9.0	10.0	6800	444

First 8 weeks of lactation suckling twins

50	110	-60	-0.13	2.4	5.3	4.8	1.56	6.86	5.63	276	173	25	12.5	8.9	6.2	4250	278
60	132	-60	-0.13	2.6	5.7	4.3	1.69	7.44	6.10	299	187	25	13.0	9.4	7.5	5100	333
70	154	-60	-0.13	2.8	6.2	4.0	1.82	8.01	6.57	322	202	25	13.4	9.5	8.8	5950	388
80	176	-60	-0.13	3.0	6.6	3.7	1.95	8.58	7.04	345	216	25	14.4	10.2	10.0	6800	444

Replacement lambs and yearlings[g]

30	66	180	0.40	1.3	2.9	4.3	0.81	3.56	2.92	130	75	21	5.9	3.3	1.9	1275	166
40	88	120	0.26	1.4	3.1	3.5	0.82	3.61	2.96	133	74	20	6.1	3.4	2.5	1700	222
50	110	80	0.18	1.5	3.3	3.0	0.83	3.65	2.99	133	73	20	6.3	3.5	3.1	2125	278
60	132	40	0.09	1.5	3.3	2.5	0.82	3.61	2.96	133	72	20	6.5	3.6	3.8	2550	333

Rams

Replacement lambs and yearlings[g]																	
40	88	250	0.55	1.8	4.0	4.5	1.17	5.15	4.22	184	108	21	6.3	3.5	2.5	1700	222
60	132	200	0.44	2.3	5.1	3.8	1.38	6.07	4.98	219	122	20	7.2	4.0	3.8	2550	333
80	176	150	0.33	2.8	6.2	3.5	1.54	6.78	5.56	249	134	20	7.9	4.4	5.0	3400	444
100	220	100	0.22	2.8	6.2	2.8	1.54	6.78	5.56	249	134	20	8.3	4.6	6.2	4250	555
120	265	50	0.11	2.6	5.7	2.2	1.43	6.29	5.16	231	125	20	8.5	4.7	7.5	5100	666
Lambs																	
Finishing[h]																	
30	66	200	0.44	1.3	2.9	4.3	0.83	3.65	2.99	143	87	24	4.8	3.0	1.1	765	166
35	77	220	0.48	1.4	3.1	4.0	0.94	4.14	3.39	154	94	23	4.8	3.0	1.3	892	194
40	88	250	0.55	1.6	3.5	4.0	1.12	4.93	4.04	176	107	22	5.0	3.1	1.5	1020	222
45	99	250	0.55	1.7	3.7	3.8	1.19	5.24	4.30	187	114	22	5.0	3.1	1.7	1148	250
50	110	220	0.48	1.8	4.0	3.6	1.26	5.54	4.54	198	121	22	5.0	3.1	1.9	1275	278
55	121	200	0.44	1.9	4.2	3.5	1.33	5.85	4.80	209	127	22	5.0	3.1	2.1	1402	305
Early-weaned[i]																	
10	22	250	0.55	0.6	1.3	6.0	0.44	1.94	1.59	96	69	36	2.4	1.6	1.2	850	67
20	44	275	0.60	1.0	2.2	5.0	0.73	3.21	2.63	160	115	36	3.6	2.4	2.5	1700	133
30	66	300	0.66	1.4	3.1	4.7	1.02	4.49	3.68	196	133	30	5.0	3.3	3.8	2550	200

[a] To convert dry matter to an as-fed basis, divide dry matter by percentage of dry matter.

[b] 1 kg TDN = 4.4 Mcal DE (digestible energy). DE may be converted to ME (metabolizable energy) by multiplying by 82%.

[c] DP = digestible protein.

[d] Values are for ewes in moderate condition, not excessively fat or thin. Fat ewes should be fed at the next lower weight, thin ewes at the next higher weight. Once maintenance weight is established, such weight would follow through all production phases.

[e] Values in parentheses for ewes suckling singles last 8 weeks of lactation.

[f] Values in parentheses are for ewes suckling twins last 8 weeks of lactation.

[g] Requirements for replacement lambs (ewe and ram) start when the lambs are weaned.

[h] Maximum gains expected. If lambs are held for later market, they should be fed as replacement ewe lambs are fed. Lambs capable of gaining faster than indicated should be fed at a higher level. Lambs finish at the maximum rate if they are self-fed.

[i] A 40-kg early-weaned lamb should be fed the same as a finishing lamb of the same weight.

APPENDIX TABLE 19

Nutrient content of diets for sheep (nutrient concentration in diet dry matter) (from 1985 NRC on sheep)

Body Weight kg	lb	Daily Gain or Loss g	lb	Daily Dry Matter — Per Animal kg	lb	% Live Wt	Energy TDN, kg	DE,[b] Mcal/kg	ME, Mcal/kg	Total Protein, %	DP,[c] %	Ca, %	P, %	Carotene, mg/kg	Vita-min A, IU/kg	Vita-min D, IU/kg
Ewes[d]																
Maintenance																
50	110	10	0.02	1.0	2.2	2.0	55	2.4	2.0	8.9	4.8	0.30	0.28	1.9	1275	278
60	132	10	0.02	1.1	2.4	1.8	55	2.4	2.0	8.9	4.8	0.28	0.26	2.0	1391	303
70	154	10	0.02	1.2	2.6	1.7	55	2.4	2.0	8.9	4.8	0.27	0.25	2.2	1488	323
80	176	10	0.02	1.3	2.9	1.6	55	2.4	2.0	8.9	4.8	0.25	0.24	2.3	1569	342
Nonlactating and first 15 weeks of gestation																
50	110	30	0.07	1.1	2.4	2.2	55	2.4	2.0	9.0	4.9	0.27	0.25	1.7	1159	253
60	132	30	0.07	1.3	2.9	2.1	55	2.4	2.0	9.0	4.9	0.24	0.22	1.7	1177	256
70	154	30	0.07	1.4	3.1	2.0	55	2.4	2.0	9.0	4.9	0.23	0.21	1.9	1275	277
80	176	30	0.07	1.5	3.3	1.9	55	2.4	2.0	9.0	4.9	0.22	0.21	2.0	1369	296
Last 6 weeks of gestation or last 8 weeks of lactation suckling singles[e]																
50	110	175(+45)	0.39	1.7	3.7	3.3	58	2.6	2.1	9.3	5.2	0.24	0.23	3.6	2500	164
60	132	180(+45)	0.40	1.9	4.2	3.2	58	2.6	2.1	9.3	5.2	0.23	0.22	3.9	2684	175
70	154	185(+45)	0.41	2.1	4.6	3.0	58	2.6	2.1	9.3	5.2	0.21	0.20	4.2	2833	185
80	176	190(+45)	0.42	2.2	4.8	2.8	56	2.6	2.1	9.3	5.2	0.21	0.20	4.5	3091	202
First 8 weeks of lactation suckling singles or last 8 weeks of lactation suckling twins[f]																
50	110	−25(+80)	−0.06	2.1	4.6	4.2	65	2.9	2.4	10.4	6.2	0.52	0.37	3.0	2024	132
60	132	−25(+80)	−0.06	2.3	5.1	3.9	65	2.9	2.4	10.4	6.2	0.50	0.36	3.3	2217	145
70	154	−25(+80)	−0.06	2.5	5.5	3.6	65	2.9	2.4	10.4	6.2	0.48	0.34	3.5	2380	155
80	176	−25(+80)	−0.06	2.6	5.7	3.2	65	2.9	2.4	10.4	6.2	0.48	0.34	3.8	2615	171
First 8 weeks of lactation suckling twins																
50	110	−60	−0.13	2.4	5.3	4.8	65	2.9	2.4	11.5	7.2	0.52	0.37	2.6	1771	116
60	132	−60	−0.13	2.6	5.7	4.3	65	2.9	2.4	11.5	7.2	0.50	0.36	2.9	1962	128
70	154	−60	−0.13	2.8	6.2	4.0	65	2.9	2.4	11.5	7.2	0.48	0.34	3.1	2125	139
80	176	−60	−0.13	3.0	6.6	3.7	65	2.9	2.4	11.5	7.2	0.48	0.34	3.3	2267	148
Replacement lambs and yearlings[g]																
30	66	180	0.40	1.3	2.9	4.3	62	2.7	2.2	10.0	5.8	0.45	0.25	1.5	981	128
40	88	120	0.26	1.4	3.1	3.5	60	2.6	2.1	9.5	5.3	0.44	0.24	1.8	1214	159
50	110	80	0.18	1.5	3.3	3.0	55	2.4	2.0	8.9	4.8	0.42	0.23	2.1	1417	185
60	132	40	0.09	1.5	3.3	2.5	55	2.4	2.0	8.9	4.8	0.43	0.24	2.5	1700	222

Rams

Replacement lambs and yearlings[g]

40	88	250	0.55	1.8	4.0	4.5	65	2.4	2.9	10.2	6.0	0.35	0.19	1.4	944	123
60	132	200	0.44	2.3	5.1	3.8	60	2.1	2.6	9.5	5.3	0.31	0.17	1.7	1109	145
80	176	150	0.33	2.8	6.2	3.5	55	2.0	2.4	8.9	4.8	0.28	0.16	1.8	1214	159
100	220	100	0.22	2.8	6.2	2.8	55	2.0	2.4	8.9	4.8	0.30	0.17	2.2	1518	198
120	265	50	0.11	2.6	5.7	2.2	55	2.0	2.4	8.9	4.8	0.33	0.18	2.9	1962	256

Lambs

Finishing[h]

30	66	200	0.44	1.3	2.9	4.3	64	2.3	2.8	11.0	6.7	0.37	0.23	0.8	588	128
35	77	220	0.48	1.4	3.1	4.0	67	2.4	3.0	11.0	6.7	0.34	0.21	0.9	637	139
40	88	250	0.55	1.6	3.5	4.0	70	2.5	3.1	11.0	6.7	0.31	0.19	0.9	638	139
45	99	250	0.55	1.7	3.7	3.8	70	2.5	3.1	11.0	6.7	0.29	0.18	1.0	675	147
50	110	220	0.48	1.8	4.0	3.6	70	2.5	3.1	11.0	6.7	0.28	0.17	1.1	708	154
55	121	200	0.44	1.9	4.2	3.5	70	2.5	3.1	11.0	6.7	0.26	0.16	1.1	738	161

Early-weaned[i]

10	22	250	0.55	0.6	1.3	6.0	73	2.6	3.2	16.0	11.5	0.40	0.27	2.0	1417	112
20	44	275	0.60	1.0	2.2	5.0	73	2.6	3.2	16.0	11.5	0.36	0.24	2.5	1700	133
30	66	300	0.66	1.4	3.1	4.7	73	2.6	3.2	14.0	9.5	0.36	0.24	2.7	1821	143

[a]To convert dry matter to an as-fed basis, divide dry matter by percentage of dry matter.

[b]1 kg TDN = 4.4 Mcal DE (digestible energy). DE may be converted to ME (metabolizable energy) by multiplying by 82%. Because of rounding errors, calculations between Appendix Table 18 and 19 may not give the same values.

[c]DP = digestible protein.

[d]Values are for ewes in moderate condition, not excessively fat or thin. Fat ewes should be fed at the next lower weight, thin ewes at the next higher weight. Once maintenance weight is established, such weight would follow through all production phases.

[e]Values in parentheses for ewes suckling singles last 8 weeks of lactation.

[f]Values in parentheses are for ewes suckling twins last 8 weeks of lactation.

[g]Requirements for replacement lambs (ewe and ram) start when the lambs are weaned.

[h]Maximum gains expected. If lambs are held for later market, they should be fed as replacement ewe lambs are fed. Lambs capable of gaining faster than indicated should be fed at a higher level. Lambs finish at the maximum rate if they are self-fed.

[i]A 40-kg early-weaned lamb should be fed the same as a finishing lamb of the same weight.

APPENDIX TABLE 20

Yearly dry matter, energy, and protein requirements of 60-kg ewe[a] (from 1975 NRC on sheep)

	Maintenance, 15 wk	Early Gestation, 15 wk	Late Gestation, 6 wk	Early Lactation, 8 wk[b]		Late Lactation, 8 wk[b]		Yearly Total
Dry matter,								
kg/day	1.1	1.3	1.9	2.3	(S)	1.9	(S)	
				2.6	(T)	2.3	(T)	
kg/period	115.5	136.5	79.8	128.8	(S)	106.4	(S)	567.0
				145.6	(T)	128.8	(T)	606.2
Metabolizable energy,								
Mcal/day	2.20	2.60	3.97	5.41	(S)	3.97	(S)	
				6.10	(T)	5.41	(T)	
Mcal/period	231.0	273.0	166.7	303.0	(S)	222.3	(S)	1196.0
				341.6	(T)	303.0	(T)	1312.3
Digestible protein,								
g/day	53	64	99	143	(S)	99	(S)	
				187	(T)	143	(T)	
kg/period	5.6	6.7	4.2	8.0	(S)	5.5	(S)	30.0
				10.5	(T)	8.0	(T)	35.0

[a]Refer to Fig. 1 in the NRC publication for daily and cumulative weight changes.

[b]S = ewes suckling singles; T = ewes suckling twins.

APPENDIX TABLE 21

Daily nutrient requirements of goats (from 1981 NRC on goats)

Body weight, kg	Feed Energy				Crude protein		Ca, g	P, g	Vitamin A, 1000 IU	Vitamin D, IU	Dry Matter per Animal			
											1 kg = 2.0 Mcal ME		1 kg = 2.4 Mcal ME	
	TDN, g	DE, Mcal	ME, Mcal	NE, Mcal	TP, g	DP, g					Total, kg	% of BW	Total, kg	% of BW
Maintenance only (includes stable feeding conditions, minimal activity, and early pregnancy)														
10	159	0.70	0.57	0.32	22	15	1	0.7	0.4	84	0.28	2.8	0.24	2.4
20	267	1.18	0.96	0.54	38	26	1	0.7	0.7	144	0.48	2.4	0.40	2.0
30	362	1.59	1.30	0.73	51	35	2	1.4	0.9	195	0.65	2.2	0.54	1.8
40	448	1.98	1.61	0.91	63	43	2	1.4	1.2	243	0.81	2.0	0.67	1.7
50	530	2.34	1.91	1.08	75	51	3	2.1	1.4	285	0.95	1.9	0.79	1.6
60	608	2.68	2.19	1.23	86	59	3	2.1	1.6	327	1.09	1.8	0.91	1.5
70	682	3.01	2.45	1.38	96	66	4	2.8	1.8	369	1.23	1.8	1.02	1.5
80	754	3.32	2.71	1.53	106	73	4	2.8	2.0	408	1.36	1.7	1.13	1.4
90	824	3.63	2.96	1.67	116	80	4	2.8	2.2	444	1.48	1.6	1.23	1.4
100	891	3.93	3.21	1.81	126	86	5	3.5	2.4	480	1.60	1.6	1.34	1.3
Maintenance plus low activity (= 25% increment, intensive management, tropical range, and early pregnancy)														
10	199	0.87	0.71	0.40	27	19	1	0.7	0.5	108	0.36	3.6	0.30	3.0
20	334	1.47	1.20	0.68	46	32	2	1.4	0.9	180	0.60	3.0	0.50	2.5
30	452	1.99	1.62	0.92	62	43	2	1.4	1.2	243	0.81	2.7	0.67	2.2
40	560	2.47	2.02	1.14	77	54	3	2.1	1.5	303	1.01	2.5	0.84	2.1
50	662	2.92	2.38	1.34	91	63	4	2.8	1.8	357	1.19	2.4	0.99	2.0
60	760	3.35	2.73	1.54	105	73	4	2.8	2.0	408	1.36	2.3	1.14	1.9
70	852	3.76	3.07	1.73	118	82	5	3.5	2.3	462	1.54	2.2	1.28	1.8
80	942	4.16	3.39	1.91	130	90	5	3.5	2.6	510	1.70	2.1	1.41	1.8
90	1030	4.54	3.70	2.09	142	99	6	4.2	2.8	555	1.85	2.1	1.54	1.7
100	1114	4.91	4.01	2.26	153	107	6	4.2	3.0	600	2.00	2.0	1.67	1.7
Maintenance plus medium activity (= 50% increment, semiarid rangeland, slightly hilly pastures, and early pregnancy)														
10	239	1.05	0.86	0.48	33	23	1	0.7	0.6	129	0.43	4.3	0.36	3.6
20	400	1.77	1.44	0.81	55	38	2	1.4	1.1	216	0.72	3.6	0.60	3.0
30	543	2.38	1.95	1.10	74	52	3	2.1	1.5	294	0.98	3.3	0.81	2.7
40	672	2.97	2.42	1.36	93	64	4	2.8	1.8	363	1.21	3.0	1.01	2.5
50	795	3.51	2.86	1.62	110	76	4	2.8	2.1	429	1.43	2.9	1.19	2.4
60	912	4.02	3.28	1.84	126	87	5	3.5	2.5	492	1.64	2.7	1.37	2.3

Body weight, kg	Feed Energy				Crude protein		Ca, g	P, g	Vita-min A, 1000 IU	Vita-min D, IU	Dry Matter per Animal			
											1 kg = 2.0 Mcal ME		1 kg = 2.4 Mcal ME	
	TDN, g	DE, Mcal	ME, Mcal	NE, Mcal	TP, g	DP, g					Total, kg	% of BW	Total, kg	% of BW
70	1023	4.52	3.68	2.07	141	98	6	4.2	2.8	552	1.84	2.6	1.53	2.2
80	1131	4.98	4.06	2.30	156	108	6	4.2	3.0	609	2.03	2.5	1.69	2.1
90	1236	5.44	4.44	2.50	170	118	7	4.9	3.3	666	2.22	2.5	1.85	2.0
100	1336	5.90	4.82	2.72	184	128	7	4.9	3.6	723	2.41	2.4	2.01	2.0
Maintenance plus high activity (= 75% increment, arid rangeland, sparse vegetation, mountainous pastures, and early pregnancy)														
10	278	1.22	1.00	0.56	38	26	2	1.4	0.8	150	0.50	5.0	0.42	4.2
20	467	2.06	1.68	0.94	64	45	2	1.4	1.3	252	0.84	4.2	0.70	3.5
30	634	2.78	2.28	1.28	87	60	3	2.1	1.7	342	1.14	3.8	0.95	3.2
40	784	3.46	2.82	1.59	108	75	4	2.8	2.1	423	1.41	3.5	1.18	3.0
50	928	4.10	3.34	1.89	128	89	5	3.5	2.5	501	1.67	3.3	1.39	2.7
60	1064	4.69	3.83	2.15	146	102	6	4.2	2.9	576	1.92	3.2	1.60	2.7
70	1194	5.27	4.29	2.42	165	114	6	4.2	3.2	642	2.14	3.0	1.79	2.6
80	1320	5.81	4.74	2.68	182	126	7	4.9	3.6	711	2.37	3.0	1.98	2.5
90	1442	6.35	5.18	2.92	198	138	8	5.6	3.9	777	2.59	2.9	2.16	2.4
100	1559	6.88	5.62	3.17	215	150	8	5.6	4.2	843	2.81	2.8	2.34	2.3

Additional requirements for late pregnancy (for all goat sizes)

	TDN, g	DE, Mcal	ME, Mcal	NE, Mcal	TP, g	DP, g	Ca, g	P, g	Vit A	Vit D	2.0 Total	2.4 Total
	100	0.44	0.36	0.20	82	57	2	1.4	1.1	213	0.71	0.59

Additional requirements for growth—weight gain at 50 g per day (for all goat sizes)

	TDN, g	DE, Mcal	ME, Mcal	NE, Mcal	TP, g	DP, g	Ca, g	P, g	Vit A	Vit D	2.0 Total	2.4 Total
	100	0.44	0.36	0.20	14	10	1	0.7	0.3	54	0.18	0.15

Additional requirements for growth—weight gain at 100 g per day (for all goat sizes)

	TDN, g	DE, Mcal	ME, Mcal	NE, Mcal	TP, g	DP, g	Ca, g	P, g	Vit A	Vit D	2.0 Total	2.4 Total
	200	0.88	0.72	0.40	28	20	1	0.7	0.5	108	0.36	0.30

Additional requirements for growth—weight gain at 150 g per day (for all goat sizes)

	TDN, g	DE, Mcal	ME, Mcal	NE, Mcal	TP, g	DP, g	Ca, g	P, g	Vit A	Vit D	2.0 Total	2.4 Total
	300	1.32	1.08	0.60	42	30	2	1.4	0.8	162	0.54	0.45

Additional requirements for milk production per kg at different fat percentages (including requirements for nursing single, twin, or triplet kids at the respective milk production level)

(% Fat)	TDN, g	DE, Mcal	ME, Mcal	NE, Mcal	TP, g	DP, g	Ca, g	P, g	Vit A	Vit D
2.5	333	1.47	1.20	0.68	59	42	2	1.4	3.8	760

3.0	337	1.49	1.21	0.68	64	45	2	1.4	3.8	760
3.5	342	1.51	1.23	0.69	68	48	2	1.4	3.8	760
4.0	346	1.53	1.25	0.70	72	51	3	2.1	3.8	760
4.5	351	1.55	1.26	0.71	77	54	3	2.1	3.8	760
5.0	356	1.57	1.28	0.72	82	57	3	2.1	3.8	760
5.5	360	1.59	1.29	0.73	86	60	3	2.1	3.8	760
6.0	365	1.61	1.31	0.74	90	63	3	2.1	3.8	760

Additional requirements for mohair production by Angora at different production levels

Annual Fleece Yield (kg)						
2	16	0.07	0.06	0.03	9	6
4	34	0.15	0.12	0.07	17	12
6	50	0.22	0.18	0.10	26	18
8	66	0.29	0.24	0.14	34	24

APPENDIX TABLE 22

Nutrient requirements of swine allowed feed ad libitum (% or amount/kg of diet, 90% dry-matter basis)[a,b] (from 1988 NRC on swine)

Live weight, kg	1–5	5–10	10–20	20–50	50–110
Expected weight gain, g/d	200	250	450	700	820
Expected feed intake, g/d	250	460	950	1900	3110
Expected efficiency, gain/feed	0.800	0.543	0.474	0.368	0.264
Expected efficiency, feed/gain	1.25	1.84	2.11	2.71	3.79
DE intake, kcal/d	850	1560	3230	6460	10570
ME intake, kcal/d	805	1490	3090	6200	10185
Energy concentration, kcal ME/kg diet	3220	3240	3250	3260	3275
Protein, %	24	20	18	15	13
Indispensable amino acids					
Arginine, %	0.60	0.50	0.40	0.25	0.10
Histidine, %	0.36	0.31	0.25	0.22	0.18
Isoleucine, %	0.76	0.65	0.53	0.46	0.38
Leucine, %	1.00	0.85	0.70	0.60	0.50
Lysine, %	1.40	1.15	0.95	0.75	0.60
Methionine + cystine, %	0.68	0.58	0.48	0.41	0.34
Phenylalanine + tyrosine, %	1.10	0.94	0.77	0.66	0.55
Threonine, %	0.80	0.68	0.56	0.48	0.40
Tryptophan, %	0.20	0.17	0.14	0.12	0.10
Valine, %	0.80	0.68	0.56	0.48	0.40
Linoleic acid, %	0.1	0.1	0.1	0.1	0.1
Mineral elements					
Calcium, %	0.90	0.80	0.70	0.60	0.50
Phosphorus, total, %	0.70	0.65	0.60	0.50	0.40
Phosphorus, available, %	0.55	0.40	0.32	0.23	0.15
Chlorine, %	0.08	0.08	0.08	0.08	0.08
Magnesium, %	0.04	0.04	0.04	0.04	0.04
Potassium, %	0.30	0.28	0.26	0.23	0.17
Sodium, %	0.10	0.10	0.10	0.10	0.10
Copper, mg	6.0	6.0	5.0	4.0	3.0
Iodine, mg	0.14	0.14	0.14	0.14	0.14
Iron, mg	100	100	80	60	40
Manganese, mg	4.0	4.0	3.0	2.0	2.0
Selenium, mg	0.30	0.30	0.25	0.15	0.10
Zinc, mg	100	100	80	60	50
Vitamins					
Vitamin A, IU	2200	2200	1750	1300	1300
Vitamin D, IU	220	220	200	150	150
Vitamin E, IU	16	16	11	11	11
Vitamin K (menadione), mg	0.5	0.5	0.5	0.5	0.5
Biotin, mg	0.08	0.05	0.05	0.05	0.05
Choline, g	0.6	0.5	0.4	0.3	0.3
Folacin, mg	0.3	0.3	0.3	0.3	0.3
Niacin, available, mg	20.0	15.0	12.5	10.0	7.0
Pantothenic acid, mg	12.0	10.0	9.0	8.0	7.0
Riboflavin, mg	4.0	3.5	3.0	2.5	2.0
Thiamin, mg	1.5	1.0	1.0	1.0	1.0
Vitamin B_6, mg	2.0	1.5	1.5	1.0	1.0
Vitamin B_{12}, μg	20.0	17.5	15.0	10.0	5.0

[a]These requirements are based upon the following types of diets: 1–5-kg pigs, a diet containing a substantial amount (25–75%) of milk products; 5–10-kg pigs, a corn-soybean meal diet that includes 5–25% milk products; 10–110 kg pigs, a corn-soybean meal diet. Based upon corn containing 8.5% and soybean meal containing 44% protein.

[b]The requirements listed are based upon the principles and assumptions described in the text of this publication. Knowledge of nutritional constraints and limitations is important for the proper use of this table.

APPENDIX TABLE 23
Daily nutrient intakes and requirements of swine allowed feed ad libitum (from 1988 NRC on swine)

	1-5	5-10	10-20	20-50	50-110
Live weight, kg					
Expected weight gain, g/d	200	250	450	700	820
Expected feed intake, g/d	250	460	950	1900	3110
Expected efficiency, gain/feed	0.800	0.543	0.474	0.368	0.264
Expected efficiency, feed/gain	1.25	1.84	2.11	2.71	3.79
DE intake, kcal/d	850	1560	3230	6460	10570
ME intake, kcal/d	805	1490	3090	6200	10185
Energy concentration, kcal ME/kg diet	3220	3240	3250	3260	3275
Protein, %	6.0	92.0	171.0	285.0	404.0
Indispensable amino acids					
Arginine, g	1.5	2.3	3.8	4.8	3.1
Histidine, g	0.9	1.4	2.4	4.2	5.6
Isoleucine, g	1.9	3.0	5.0	8.7	11.8
Leucine, g	2.5	3.9	6.6	11.4	15.6
Lysine, g	3.5	5.3	9.0	14.3	18.7
Methionine + cystine, g	1.7	2.7	4.6	7.8	10.6
Phenylalanine + tyrosine, g	2.8	4.3	7.3	12.5	17.1
Threonine, g	2.0	3.1	5.3	9.1	12.4
Tryptophan, g	0.5	0.8	1.3	2.3	3.1
Valine, g	2.0	3.1	5.3	9.1	12.4
Linoleic acid, g	0.3	0.5	1.0	1.9	3.1
Mineral elements					
Calcium, g	2.2	3.7	6.6	11.4	15.6
Phosphorus, total, g	1.8	3.0	5.7	9.5	12.4
Phosphorus, available, g	1.4	1.8	3.0	4.4	4.7
Chlorine, g	0.2	0.4	0.8	1.5	2.5
Magnesium, g	0.1	0.2	0.4	0.8	1.2
Potassium, g	0.8	1.3	2.5	4.4	5.3
Sodium, g	0.2	0.5	1.0	1.9	3.1
Copper, mg	1.50	2.76	4.75	7.60	9.33
Iodine, mg	0.04	0.06	0.13	0.27	0.44
Iron, mg	25.0	46.0	76.0	114.0	124.4
Manganese, mg	1.00	1.84	2.85	3.80	6.22
Selenium, mg	0.08	0.14	0.24	0.28	0.31
Zinc, mg	25.0	46.0	76.0	114.0	155.5
Vitamins					
Vitamin A, IU	550	1012	1662	2470	4043
Vitamin D, IU	55	101	190	285	466
Vitamin E, IU	4	7	10	21	34
Vitamin K (menadione), mg	0.02	0.02	0.05	0.10	0.16
Biotin, mg	0.02	0.02	0.05	0.10	0.16
Choline, g	0.15	0.23	0.38	0.57	0.93
Folacin, mg	0.08	0.14	0.28	0.57	0.93
Niacin, available, mg	5.00	6.90	11.88	19.00	21.77
Pantothenic acid, mg	3.00	4.60	8.55	15.20	21.77
Riboflavin, mg	1.00	1.61	2.85	4.75	6.22
Thiamin, mg	0.38	0.46	0.95	1.90	3.11
Vitamin B_6, mg	0.50	0.69	1.42	1.90	3.11
Vitamin B_{12}, μg	5.00	8.05	14.25	19.00	15.55

Nutrient requirements of breeding swine (% or amount/kg of diet)[a,b] (from 1988 NRC on swine)

		Bred Gilts, Sows, and Adult Boars	Lactating Gilts and Sows
Digestible energy	kcal/kg diet	3340	3340
Metabolizable energy	kcal/kg diet	3210	3210
Crude protein	%	12	13
Indispensable amino acids			
Arginine, %		0.00	0.40
Histidine, %		0.15	0.25
Isoleucine, %		0.30	0.39
Leucine, %		0.30	0.48
Lysine, %		0.43	0.60
Methionine + cystine, %		0.23	0.36
Phenylalanine + tyrosine, %		0.45	0.70
Threonine, %		0.30	0.43
Tryptophan, %		0.09	0.12
Valine, %		0.32	0.60
Linoleic acid, %		0.1	0.1
Mineral elements			
Calcium, %		0.75	0.75
Phosphorus, total, %		0.60	0.60
Phosphorus, available, %		0.35	0.35
Chlorine, %		0.12	0.16
Magnesium, %		0.04	0.04
Potassium, %		0.20	0.20
Sodium, %		0.15	0.20
Copper, mg		5.00	5.00
Iodine, mg		0.14	0.14
Iron, mg		80.00	80.00
Manganese, mg		10.00	10.00
Selenium, mg		0.15	0.15
Zinc, mg		50.00	50.00
Vitamins			
Vitamin A, IU		4000	2000
Vitamin D, IU		200	200
Vitamin E, IU		22	22
Vitamin K (menadione), mg		0.50	0.50
Biotin, mg		0.20	0.20
Choline, g		1.25	1.00
Folacin, mg		0.30	0.30
Niacin, available, mg		10.00	10.00
Pantothenic acid, mg		12.00	12.00
Riboflavin, mg		3.75	3.75
Thiamin, mg		1.00	1.00
Vitamin B_6, mg		1.00	1.00
Vitamin B_{12}, μg		15.00	15.00

[a]These requirements are based upon corn-soybean meal diets, with corn containing 8.5% and soybean meal containing 44% protein, and based on feed intake levels listed in Appendix Table 25.

[b]The requirements listed are based upon the principles and assumptions described in the text of this publication. Knowledge of nutritional constraints and limitations is important for the proper use of this table.

APPENDIX TABLE 25

Daily nutrient intakes and requirements of intermediate weight breeding animals (from 1988 NRC on swine)

	Bred Gilts, Sows, and Adult Boars	Lactating Gilts and Sows
Mean gestation or farrowing wt, kg	162.5	165.0
Daily feed intake, kg	1.9	5.3
Digestible energy, Mcal/d	6.3	17.7
Metabolizable energy, Mcal/d	6.1	17.0
Crude protein, g/d	228	689
Indispensable amino acids		
Arginine, g	0.0	21.2
Histidine, g	2.8	13.2
Isoleucine, g	5.7	20.7
Leucine, g	5.7	25.4
Lysine, g	8.2	31.8
Methionine + cystine, g	4.4	19.1
Phenylalanine + tyrosine, g	8.6	37.1
Threonine, g	5.7	22.8
Tryptophan, g	1.7	6.4
Valine, g	6.1	31.8
Linoleic acid, g	1.9	5.3
Mineral elements		
Calcium, g	14.2	39.8
Phosphorus, total, g	11.4	31.8
Phosphorus, available, g	6.6	18.6
Chlorine, g	2.3	8.5
Magnesium, g	0.8	2.1
Potassium, g	3.8	10.6
Sodium, %	2.8	10.6
Copper, mg	9.5	26.5
Iodine, mg	0.3	0.7
Iron, mg	152.0	424.0
Manganese, mg	19.0	53.0
Selenium, mg	0.3	0.8
Zinc, mg	95.0	265.0
Vitamins		
Vitamin A, IU	7600	10600
Vitamin D, IU	380	1060
Vitamin E, IU	42	117
Vitamin K (menadione), mg	1.0	2.6
Biotin, mg	0.4	1.1
Choline, g	2.4	5.3
Folacin, mg	0.6	1.6
Niacin, available, mg	19.0	53.0
Pantothenic acid, mg	22.8	63.6
Riboflavin, mg	7.1	19.9
Thiamin, mg	1.9	5.3
Vitamin B_6, mg	1.9	5.3
Vitamin B_{12}, μg	28.5	79.5

APPENDIX TABLE 26

Requirements for several nutrients for breeding herd replacement animals allowed feed ad libitum, % of diet (from 1988 NRC on swine)

	Developing Gilts[a]		Developing Boars[a]	
Weight, kg	20–50	50–110	20–50	50–110
Energy concentration, kcal ME/kg diet	3255	3260	3240	3255
Crude protein, %	16	15	18	16
Lysine, %	0.80	0.70	0.90	0.75
Calcium, %	0.65	0.55	0.70	0.60
Phosphorus, total, %	0.55	0.45	0.60	0.50
Phosphorus, available, %	0.28	0.20	0.33	0.25

[a]Sufficient data are not available to indicate that requirements for other nutrients are different from those in Appendix Table 23 for this size animal.

APPENDIX TABLE 27

Daily energy and feed requirements of pregnant gilts and sows (from 1988 NRC on swine)

	Bred Gilts and Sows[a]		
Weight at mating, kg	120	140	160
Mean gestation weight, kg[b]	142.5	162.5	182.5
Energy required, Mcal DE/d			
Maintenance[c]	4.53	5.00	5.47
Gestation weight gain[d]	1.29	1.29	1.29
Total required	5.82	6.29	6.76
Feed required/d, kg[e]	1.8	1.9	2.0

[a]Assuming 25 kg maternal weight gain plus 20-kg increase in weight due to products of conception, total 45 kg.
[b]Weight at mating plus total weight gain × 0.5.
[c]Animal daily maintenance requirements; 110 kcal DE/kg$^{0.75}$.
[d]1.10 Mcal DE/d for maternal weight gain plus 0.19 Mcal DE/d for conceptus gain.
[e]Corn-soybean meal diet containing 3.34 Mcal DE/kg.

APPENDIX TABLE 28

Daily energy and feed requirements of lactating gilts and sows (from 1988 NRC on swine)

	Lactating Gilts and Sows[a]		
Weight postfarrowing, kg	145	165	185
Milk yield, kg	5.0	6.25	7.5
Energy required, Mcal DE/d			
Maintenance[a]	4.5	5.0	5.5
Milk yield[b]	10.0	12.5	15.0
Total required	14.5	17.5	20.5
Feed required/d, kg[c]	4.4	5.3	6.1

[a]Animal daily maintenance requirement; 110 kcal DE/kg$^{0.75}$.
[b]2.0 Mcal DE/kg milk.
[c]Corn-soybean meal diet containing 3.34 Mcal DE/kg.

APPENDIX TABLE 29

Documentation of nutrient requirements of starting and growing leghorn-type chickens (from 1994 NRC on poultry)

Nutrient and Estimated Requirement	Age Period (days)	Response Criteria	Breed	References
Protein, %				
20	0–14	Growth	White Leghorn	Grau and Kamei, 1950
21.1	0–42	Growth	White Leghorn and Rhode Island Red	Edwards et al., 1956
14–20	84–140	Growth	White Leghorn	McNaughton et al., 1977b
15–18	0–42	Growth	White Leghorn	McNaughton et al., 1977b
12	0–56	Growth	White Leghorn	Leeson and Summers, 1979
16	56–84	Growth	White Leghorn	Leeson and Summers, 1979
19	84–104	Growth	White Leghorn	Leeson and Summers, 1979
14 and 21	56–140	Growth	White Leghorn	Douglas and Harms, 1982
12 or 13.6	0–42	Growth	Commercial brown egg layers	Maurice et al., 1982
16 or 13.6	42–140	Growth	Commercial brown egg layers	Maurice et al., 1982
18	0–28	Growth of muscle fiber	White Leghorn	Timson et al., 1983
18	0–42	Growth	White Leghorn	Keshavarz, 1984
12	42–140	Growth	White Leghorn	Keshavarz, 1984
16.5	140–504	Laying	White Leghorn	Keshavarz, 1984
22	0–28	Growth	White Leghorn	Leeson and Summers, 1984
18	0–140	Growth	White Leghorn	Chi, 1985
Isoleucine, %				
0.5	8–18	Growth	White Leghorn	Mori and Okumura, 1984
Leucine, %				
1.2	8–18	Growth	White Leghorn	Mori and Okumura, 1984
Lysine, %				
0.9–1.1	0–42	Growth	White Leghorn	Edwards et al., 1956
0.94	1–21	Growth, feed efficiency	White Leghorn	Chung et al., 1973
0.70	35–49	Growth, feed efficiency	White Leghorn	Chung et al., 1973
<0.5	56–98	Growth	White Leghorn	Berg, 1976
<0.45	98–147	Growth	White Leghorn	Berg, 1976
0.68	0–504	Growth, egg production	White Leghorn	Keshavarz, 1984
Methionine, %				
0.8	0–14	Growth	White Leghorn	Grau and Kamei, 1950
Methionine and cystine, %				
0.8	0–14	Growth	White Leghorn	Grau and Kamei, 1950
0.59	0–504	Growth, laying	White Leghorn	Keshavarz, 1984
0.45	0–42	Growth	White Leghorn	Chi, 1985
Threonine, %				
0.72	7–21	Growth, feed efficiency	White Leghorn	Davis and Austic, 1982

APPENDIX TABLE 29 *(cont.)*

Nutrient and Estimated Requirement	Age Period (days)	Response Criteria	Breed	References
Valine, %				
0.8	8–18	Growth	White Leghorn	Mori and Okumura, 1984
Requirements for essential amino acids described in review papers	Various	Growth	Primarily White Leghorn	Almquist, 1952
Requirements for essential amino acids described in review papers	Various	Growth	White Leghorn	Waldroup et al., 1980
Requirements for essential amino acids described in review papers	Various	Growth, egg production	White Leghorn	Harms, 1984
Calcium				
0.78	0–153	Growth	White Leghorn	Hamilton and Cipera, 1981
3.19	154–439	Egg Production	White Leghorn	Hamilton and Cipera, 1981
0.89	35–126	Growth	White Leghorn	Classen and Scott, 1982
2.08	12–154	Growth, subsequent egg production	White Leghorn	Classen and Scott, 1982
3.50	177–225	Egg production	White Leghorn	Classen and Scott, 1982
2.0–3.5	At 133 to 4th egg	Growth, bone development	White Leghorn	Leeson et al., 1986
0.8	98–140	Growth, subsequent egg production	White Leghorn	Keshavarz, 1987
3.5	98–140	Egg production	White Leghorn	Keshavarz, 1987
3.55	140–420	Egg production	White Leghorn	Keshavarz, 1987
4.0	>112	Egg production	White Leghorn	Leeson and Summers, 1987b
Nonphytate phosphorus, %				
0.4–0.6	7–28	Growth	White Leghorn	Gillis et al., 1949
0.25–0.30	0–140	Growth	Brown egg layers	Carew and Foss, 1980
0.31	112–140	Growth	White Leghorn	Douglas and Harms, 1986
Potassium, %				
0.20–0.24	0–28	Growth, bone calcification	White Leghorn	Gillis, 1948
Sodium, %				
0.10–0.30	0–28	Growth	White Leghorn	Burns et al., 1953
0.13	0–21	Growth	White Rock	Hurwitz et al., 1973
0.15	0–140	Growth	White Leghorn	Manning and McGinnis, 1980
Chlorine, %				
0.13	0–14	Growth, feed efficiency	Broiler strain	Nam and McGinnis, 1981
Sodium chloride, %				
0.25	0–140	Growth, sexual maturity	White Leghorn	Leeson and Summers, 1980
Magnesium, mg/kg				
300	0–28	Deficiency, neuropathy	White Leghorn	Bird, 1949
250	0–28	Growth	Broiler strain	Gardiner et al., 1960
594	0–21	Growth	White Rock	Nugara and Edwards, 1963

Nutrient / Amount	Age (days)	Response criteria	Strain	Reference
Manganese, mg/kg				
50	0–140	Growth, perosis	New Hampshire	Gallup and Norris, 1939a
20	0–28	Growth	White Leghorn	Watson et al., 1971
35	0–42	Growth, feathering, bone development	White Rock	O'Dell et al., 1958
20	0–42	Growth	White Rock	Edwards et al., 1959
20	To 1st egg	Growth, feed efficiency	White Leghorn	Rahman et al., 1961
78	0–7	Growth, feathering	White Leghorn	Sunde, 1972
52	7–21	Growth, feathering	White Leghorn	Sunde, 1972
Iron, mg/kg				
40	0–56	Growth	Rhode Island Red	Hill and Matrone, 1961
4	0–56	Growth	Rhode Island Red	Hill and Matrone, 1961
56	0–21	Growth, feed efficiency	Broiler strain	Waddell and Sell, 1964
75–80	0–28	Growth	New Hampshire	Davis et al., 1968
Copper, mg/kg				
4	0–56	Growth	Rhode Island Red	Hill and Matrone, 1961
Iodine, mg/kg				
0.300	0–56	Growth, thyroid histology	White Leghorn and broiler strains	Creek et al., 1957
0.400	0–56	Growth, thyroid histology	White Leghorn and broiler strains	Creek et al., 1957
0.075	0–35	Growth	Broiler strain	Rogler and Parker, 1978
Selenium, mg/kg				
0.01 to 0.05, depending on dietary concentration of Vitamin E	0–24	Growth	Plymouth Rock	Thompson and Scott, 1969
0.01 to 0.05, depending on dietary concentration of Vitamin E	0–14	Growth	Plymouth Rock	Gries and Scott, 1972c
Vitamin A, IU/kg				
800–1600	0–56	Growth, absence of deficiency signs	White Leghorn	Record et al., 1937
1200–2000	70–84	Curative feeding	White Leghorn	Record et al., 1937
2650	0–189	Growth	White Leghorn	Taylor and Russell, 1947
1760–7000	0–56	Growth	White Leghorn	Thornton and Whittet, 1962
4400	0–113	Growth, *E. acervulina*	White Leghorn	Coles et al., 1970
Vitamin D₃, IU/kg				
180	0–84	Growth, bone development	Brown egg layers	Baird and Greene, 1935
132	0–21	Growth, bone development	Broiler strain	McNaughton et al., 1977a
198	0–21	Growth, bone development	Broiler strain	McNaughton et al., 1977a
500	Adults	Egg production, shell quality	Various strains	Ameenuddin et al., 1985
Vitamin E, IU/kg				
60	Various	To prevent exudative diathesis, encephalomalacia, muscular degeneration	Various strains	Machlin and Gordon, 1962
Vitamin K, mg/kg				
30–50	0–35	Growth	White Rock	Combs and Scott, 1974
0.524–0.528	0–28	Growth	White Rock	Nelson and Norris, 1960
0.515	0–84	Growth	White Rock	Nelson and Norris, 1961a
0.524–0.528	0–28	Growth	White Rock	Nelson and Norris, 1961b

APPENDIX TABLE 29 (cont.)

Nutrient and Estimated Requirement	Age Period (days)	Response Criteria	Breed	References
Riboflavin, mg/kg				
3.5 decreasing to 1.0	0–7	Growth	White Leghorn	Heuser et al., 1938
3.5 decreasing to 1.0	49–56	Growth	White Leghorn	Heuser et al., 1938
3	0–56	Growth, prevention of curled toe paralysis	White Leghorn	Bethke and Record, 1942
2.3	0–42	Growth, prevention of curled toe paralysis	White Leghorn	Bootwalla and Harms, 1990
Pantothenic acid, mg/kg				
6	0–42	Growth	White Leghorn	Bauernfeind et al., 1942
6.6	0–150	Growth, egg quality, hatchability	New Hampshire	Balloun and Phillips, 1957b
4.8	0–42	Growth	White Leghorn	Bootwalla and Harms, 1991
Niacin, mg/kg				
28	0–56	Growth	Barred Plymouth Rock	Childs et al., 1952
1.8	42–77	Growth	White Leghorn	Sunde, 1955
17.5–20	0–28	Growth	White Leghorn	Patterson et al., 1956
Vitamin B_{12} mg/kg				
4.4	0–77	Growth	White Leghorn	Davis and Briggs, 1951
27	0–23	Growth	White Leghorn	Ott, 1951
2.5	0–42	Growth	White Leghorn	Miller et al., 1956
10	0–21	Growth	White Leghorn	Patel and McGinnis, 1980
Choline, mg/kg				
2000	0–147	Growth, egg production	White Leghorn	Nesheim et al., 1971
1000	0–126	Growth	White Leghorn	Tsiagbe et al., 1982
Biotin, μg/kg				
260	0–18	Growth, feed efficiency	Broiler strain	Anderson and Warnick, 1970
Folic acid, mg/kg				
0.80	0–35	Growth, feed efficiency	White Leghorn	March and Biely, 1955
0.30	0–28	Growth	Broiler strain	Young et al., 1955
0.33 to 1.45, depending on protein level	0–35	Growth	New Hampshire	March and Biely, 1956
0.30	0–18	Growth	Broiler strain	Creek and Vasaitis, 1963
Thiamine, mg/kg				
0.6–0.8	0–35	Growth	White Leghorn	Arnold and Elvehjem, 1938
0.88	0–28	Growth	White Leghorn	Thornton, 1960
0.88	0–28	Gain, feed efficiency	White Leghorn	Thornton and Shutze, 1960
Pyridoxine, mg/kg				
2.8–3.0	0–28	Growth	White Leghorn	Briggs et al., 1942
5.7	0–56	Growth	White Plymouth Rock	Fuller and Kifer, 1959
5	0–21	Growth	Broiler strain	Kazemi and Kratzer, 1980

APPENDIX TABLE 30

Body weight and feed consumption of immature leghorn-type chickens (from 1994 NRC on poultry)

Age (weeks)	White-egg-laying Strains		Brown-egg-laying Strains	
	Body Weight[a] (g)	Feed Consumption (g/week)	Body Weight[a] (g)	Feed Consumption (g/week)
0	35	50	37	70
2	100	140	120	160
4	260	260	325	280
6	450	340	500	350
8	660	360	750	380
10	750	380	900	400
12	980	400	1100	420
14	1100	420	1240	450
16	1220	430	1380	470
18	1375	450	1500	500
20	1475	500	1600	550

[a]Average genetic potential when feed is consumed on an ad libitum basis. Different commercial strains may show different growth rates and different final mature body weights.

APPENDIX TABLE 31

Documentation of nutrient requirements of starting and growing market broilers (from 1994 NRC on poultry)

Nutrient and Estimated Requirement	Age Period (days)	Response Criteria	Breed	References
Arginine, %				
1.2	10–20	Growth	Not specified	Almquist, 1947
≤1.11	7–21	Growth, feed efficiency	New Hampshire x Columbian	Snyder et al., 1956
≤0.85	7–28	Growth, feed efficiency	Barred Plymouth Rock	Krautmann et al., 1957
1.08	7–14	Growth, feed efficiency	Not specified	Klain et al., 1960
0.92	7–21	Growth, feed efficiency, nitrogen balance (adjusted to 23% crude protein diet)	White Plymouth Rock X Light Sussex	Lewis et al., 1963
1.10	7–14	Growth, feed efficiency	New Hampshire x Columbian	Dean and Scott, 1965
0.78	7–14	Growth, feed efficiency	New Hampshire x Columbian	Allen and Baker, 1972
0.85	7–21	Growth, feed efficiency	Broiler strain	Hewitt and Lewis, 1972
≤0.76	14–28	Growth, feed efficiency	Not specified	Woodham and Deans, 1975
1.13, males	28–49	Growth, feed efficiency, feather loss	Hubbard x Hubbard	Kessler and Thomas, 1976
0.98, females	28–49	Growth, feed efficiency, feather loss	Hubbard x Hubbard	Kessler and Thomas, 1976
1.33	7–14	Computer model	Not specified	Hurwitz et al., 1978
1.19	14–21	Computer model	Not specified	Hurwitz et al., 1978
1.16	21–28	Computer model	Not specified	Hurwitz et al., 1978
1.10	28–35	Computer model	Not specified	Hurwitz et al., 1978
0.99	35–42	Computer model	Not specified	Hurwitz et al., 1978
0.96	42–49	Computer model	Not specified	Hurwitz et al., 1978
1.05	49–56	Computer model	Not specified	Hurwitz et al., 1978
1.4	1–28	Growth, feed efficiency	Broiler strain	Burton and Waldroup, 1979
1.25	8–29	Growth, feed efficiency	Vedette ISA	Alimentation Equilibree Commentri, 1981
0.91	29–50	Growth, feed efficiency	Vedette ISA	Alimentation Equilibree Commentri, 1981
1.25	0–21	Growth, feed efficiency	Peterson x Arbor Acre	Cuca and Jensen, 1990
Glycine + serine, %				
1.6	8–16	Growth, feed efficiency	New Hampshire x Columbian	Dean and Scott, 1965
≤0.3	8–16	Growth, feed efficiency	New Hampshire x Columbian	Baker et al., 1968
0.5–1.0	1–10	Growth, feed efficiency	Cobb	Coon et al., 1974
≤1.8	1–23	Growth, feed efficiency	Not specified	Ngo and Coon, 1976
0.60	8–16	Growth, feed efficiency	New Hampshire x Columbian	Baker et al., 1979
Histidine, %				
0.4	8–13 or 15	Growth, feed efficiency	New Hampshire x Columbian	Klain et al., 1960
0.3	8–16	Growth, feed efficiency	New Hampshire x Columbian	Dean and Scott, 1965
≤0.34	14–28	Total protein efficiency	Ross	Woodham and Deans, 1975

	Age (days)	Criteria	Strain or breed	Reference
0.33	8–16	Growth, feed efficiency	New Hampshire x Columbian	Baker et al., 1979
0.32	8–22	Growth	New Hampshire x Columbian	Han et al., 1991
Isoleucine, %				
0.60	10–24	Growth	Not specified	Almquist, 1947
0.73	8–15	Growth	New Hampshire x Columbian	Klain et al., 1960
0.80	8–16	Growth	New Hampshire x Columbian	Dean and Scott, 1965
≤0.52	7–21	Growth, plasma amino acid levels	Not specified	D'Mello, 1974
0.48	14–28	Total protein efficiency	Ross	Woodham and Deans, 1975
0.60	8–16	Growth, feed efficiency	New Hampshire x Columbian	Baker et al., 1979.
0.81	7–21	Growth, feed efficiency	Ross x Arbor Acre	Farran and Thomas, 1990
Leucine, %				
1.4	10 or 24	Growth	Not specified	Almquist, 1947
1.68	8–13 or 15	Growth, feed efficiency	New Hampshire x Columbian	Klain et al., 1960
1.2	8–16	Growth, feed efficiency	New Hampshire x Columbian	Dean and Scott, 1965
1.10	7–21	Growth, plasma amino acid levels	Not specified	D'Mello, 1974
≤1.05	14–28	Total protein efficiency	Ross	Woodham and Deans, 1975
1.00	8–16	Growth, feed efficiency	New Hampshire x Columbian	Baker et al., 1979
1.16	7–21	Growth, feed efficiency	Ross x Arbor Acre	Farran and Thomas, 1990
Lysine, %				
0.90	2–14	Growth	Not specified	Almquist and Mecchi, 1942
0.96	14–28	Growth	Not specified	Grau et al., 1946
0.90	10–20	Growth	Not specified	Almquist, 1947
1.00	0–42	Growth	Rhode Island Red x White Leghorn	Milligan et al., 1951
0.72	56–63	Growth, feed efficiency	Rhode Island Red	Bird, 1953
1.10	1–28	Growth, feed efficiency	Rhode Island Red x Barred Plymouth Rock	Edwards et al., 1956
1.01	7–14	Growth, feed efficiency	Not specified	Klain et al., 1960
0.83	7–14	Growth, feed efficiency, plasma amino acids	New Hampshire x Columbian	Zimmerman and Scott, 1965
0.70	14–21	Growth, feed efficiency, plasma amino acids	New Hampshire x Columbian	Zimmerman and Scott, 1965
0.67	21–28	Growth, feed efficiency, plasma amino acids	New Hampshire x Columbian	Zimmerman and Scott, 1965
0.59	28–35	Growth, feed efficiency, plasma amino acids	New Hampshire x Columbian	Zimmerman and Scott, 1965
0.92	35–56	Growth, feed efficiency	Broiler strain	Bornstein, 1970
0.85	7–21	Growth, feed efficiency	Broiler strain	Hewitt and Lewis, 1972
1.05	14–28	Growth, feed efficiency	New Hampshire x Columbian	Boomgaardt and Baker, 1973a
1.06	14–21	Growth, feed efficiency	New Hampshire x Columbian	Boomgaardt and Baker, 1973b
0.92	42–56	Growth, feed efficiency	New Hampshire x Columbian	Boomgaardt and Baker, 1973b
0.68	49–63	Growth, feed efficiency	Broiler strain	Twining et al., 1973
1.12	7–14	Growth, feed efficiency	Not specified	Woodham and Deans, 1975
0.64, females	49–63	Growth, feed efficiency	Vantress x Arbor Acre	Thomas et al., 1977
0.69, males	49–63	Growth, feed efficiency	Vantress x Arbor Acre	Thomas et al., 1977
1.18	7–14	Computer model	Not specified	Hurwitz et al., 1978
1.00	14–21	Computer model	Not specified	Hurwitz et al., 1978
0.95	21–28	Computer model	Not specified	Hurwitz et al., 1978
0.87	28–35	Computer model	Not specified	Hurwitz et al., 1978

APPENDIX TABLE 31 (cont.)

Nutrient and Estimated Requirement	Age Period (days)	Response Criteria	Breed	References
0.78	35–42	Computer model	Not specified	Hurwitz et al., 1978
0.76	42–49	Computer model	Not specified	Hurwitz et al., 1978
0.84	49–56	Computer model	Not specified	Hurwitz et al., 1978
1.10	14–28	Growth, feed efficiency	Broiler strain	McNaughton et al., 1978
1.18	1–21	Growth, feed efficiency	Broiler strain	Attia and Latshaw, 1979
1.10	1–28	Growth, feed efficiency	Broiler strain	Burton and Waldroup, 1979
0.99	35–42	Growth, feed efficiency	Cornish x White Plymouth Rock	Holsheimer, 1981
Methionine, %				
0.50	10–20	Growth	Not specified	Almquist, 1947
0.45	7–14	Growth, feed efficiency	New Hampshire x Columbian	Dean and Scott, 1965
0.18	7–14	Growth, feed efficiency	Not specified	Klain et al., 1960
0.39	7–21	Growth, feed efficiency	Broiler strain	Hewitt and Lewis, 1972
0.39	7–14	Computer model	Not specified	Hurwitz et al., 1978
0.34	14–21	Computer model	Not specified	Hurwitz et al., 1978
0.34	21–28	Computer model	Not specified	Hurwitz et al., 1978
0.31	28–35	Computer model	Not specified	Hurwitz et al., 1978
0.27	35–42	Computer model	Not specified	Hurwitz et al., 1978
0.27	42–49	Computer model	Not specified	Hurwitz et al., 1978
0.29	49–56	Computer model	Not specified	Hurwitz et al., 1978
0.57	1–21	Growth, feed efficiency	Cobb	Waldroup et al., 1979
0.44	8–21	Growth, feed efficiency	New Hampshire x Columbian	Robbins and Baker, 1980a
0.46	1–14	Growth, feed efficiency, feathering	White Mountain x Hubbard	Moran, 1981
0.36, males	35–56	Growth, feed efficiency	White Mountain x Hubbard	Moran, 1981
0.29, females	35–49	Growth, feed efficiency	White Mountain x Hubbard	Moran, 1981
0.49	7–21	Growth, feed efficiency	Broiler strain	Thomas et al., 1985
0.55	1–21	Growth, feed efficiency	Broiler strain	Tillman and Pesti, 1985
Methionine + cystine, %				
0.90	10–20	Growth	Not specified	Almquist, 1947
0.80	7–14	Growth, feed efficiency	New Hampshire x Columbian	Dean and Scott, 1960
0.47	7–14	Growth, feed efficiency	New Hampshire x Columbian	Klain et al., 1960
0.70	0–42	Feed efficiency	Vantress x New Hampshire	Nelson et al., 1960
0.81	0–28	Feed efficiency	Vantress x New Hampshire	Nelson et al., 1960
0.5	28–56	Growth	Hubbard	Adams et al., 1962
>0.6–<0.7	28–56	Feed efficiency	Hubbard	Adams et al., 1963
0.81	0–35	Growth, feed efficiency	Cornish x White Plymouth Rock	Bornstein and Lipstein, 1964
0.90	0–35	Growth, feed efficiency	Cornish x White Plymouth Rock	Bornstein and Lipstein, 1964
0.67	35–56	Growth, feed efficiency	Cornish x White Plymouth Rock	Bornstein and Lipstein, 1966
0.60	7–14	Growth, feed efficiency	New Hampshire x Columbian	Graber et al., 1971

	Age (days)	Response criteria	Strain	Reference
0.63	35–42	Growth, feed efficiency	New Hampshire x Columbian	Graber et al., 1971
0.65	49–56	Growth, feed efficiency	New Hampshire x Columbian	Graber et al., 1971
0.79	7–21	Growth, feed efficiency	Broiler strain	Hewitt and Lewis, 1972
0.70	14–21	Growth, feed efficiency	New Hampshire x Columbian	Boombaardt and Baker, 1973b
0.51	42–56	Growth, feed efficiency	New Hampshire x Columbian	Boomgaardt and Baker, 1973b
0.92	8–21	Growth, feed efficiency, nitrogen retention	New Hampshire x Columbian	Boomgaardt and Baker, 1973c
0.58	14–28	Growth, feed efficiency	Not specified	Woodham and Deans, 1975
0.93	0–28	Growth, feed efficiency	Cobb	Murillo et al., 1976
0.61	35–49	Computer model	Not specified	Hurwitz et al., 1978
0.84	7–14	Computer model	Not specified	Hurwitz et al., 1978
0.78	14–21	Computer model	Not specified	Hurwitz et al., 1978
0.79	21–28	Computer model	Not specified	Hurwitz et al., 1978
0.76	28–35	Computer model	Not specified	Hurwitz et al., 1978
0.68	35–42	Computer model	Not specified	Hurwitz et al., 1978
0.69	42–49	Computer model	Not specified	Hurwitz et al., 1978
0.39	49–56	Computer model	Not specified	Hurwitz et al., 1978
0.86	1–21	Growth, feed efficiency	Broiler strain	Attia and Latshaw, 1979
0.90	1–21	Growth, feed efficiency	Cobb	Waldroup et al., 1979
0.80	8–21	Growth, feed efficiency	New Hampshire x Columbian	Robbins and Baker, 1980a
0.52	8–21	Growth, feed efficiency	New Hampshire x Columbian	Robbins and Baker, 1980a
0.55	8–21	Growth, feed efficiency	Hubbard	Robbins and Baker, 1980b
0.57	8–16	Growth, feed efficiency	New Hampshire x Columbian	Willis and Baker, 1980
0.70	35–42	Growth, feed efficiency	Cornish x White Plymouth Rock	Holsheimer, 1981
0.87, males	1–14	Growth, feed efficiency, feathering	White Mountain x Hubbard	Moran, 1981
0.92, females	1–14	Growth, feed efficiency, feathering	White Mountain x Hubbard	Moran, 1981
0.81, males	35–52	Growth, feed efficiency, feathering	White Mountain x Hubbard	Moran, 1981
0.82	1–21	Growth, feed efficiency	Cobb	Wheeler and Latshaw, 1981
>0.70–<0.76	21–42	Growth, feed efficiency	Cobb	Wheeler and Latshaw, 1981
0.65	8–16	Growth, feed efficiency	New Hampshire x Columbian	Willis and Baker, 1981a
0.50	7–17	Growth, feed efficiency	New Hampshire x Columbian	Baker et al., 1983
0.87	7–24	Growth, feed efficiency	New Hampshire x Columbian	Baker et al., 1983
0.80	1–21	Growth, feed efficiency	Hubbard	Mitchell and Robbins, 1983
0.72	21–42	Growth, feed efficiency	Hubbard	Mitchell and Robbins, 1983
0.77	7–21	Growth, feed efficiency	Broiler strain	Thomas et al., 1985
0.78	21–42	Growth, feed efficiency, carcass fat	Peterson x Arbor Acres	Jensen et al., 1989
Phenylalanine + tyrosine, %				
1.6	10–20 or 40	Growth	Not specified	Almquist, 1947
≤1.0	4–10	Growth, feed efficiency	New Hampshire x Columbian	Fisher et al., 1957
1.30	8–13 or 15	Growth, feed efficiency	New Hampshire x Columbian	Klain et al., 1960
1.31	8–16	Growth, feed efficiency	New Hampshire x Columbian	Dean and Scott, 1965
0.87	8–14	Growth, feed efficiency	New Hampshire x Columbian	Sasse and Baker, 1972
1.09–1.12	14–28	Total protein efficiency	Ross	Woodham and Deans, 1975
0.95	8–16	Growth, feed efficiency	New Hampshire x Columbian	Baker et al., 1979

APPENDIX TABLE 31 *(cont.)*

Nutrient and Estimated Requirement	Age Period (days)	Response Criteria	Breed	References
Threonine, %				
0.60	10–20	Growth, feed efficiency	Not specified	Almquist, 1947
0.45	1–14	Growth, feed efficiency	White Leghorn	Grau, 1947
0.55–0.60	7–21	Growth, feed efficiency	Barred Plymouth Rock	Krautmann et al., 1958
0.58	7–14	Growth, feed efficiency	Not specified	Klain et al., 1960
0.65	7–14	Growth, feed efficiency	New Hampshire x Columbian	Dean and Scott, 1965
0.70	1–18	Growth, feed efficiency	New Hampshire x White Leghorn	Bhargava et al., 1971
0.53	7–21	Growth, feed efficiency	Broiler strain	Hewitt and Lewis, 1972
0.52	14–28	Growth, feed efficiency	Not specified	Woodham and Deans, 1975
0.80	7–14	Computer model	Not specified	Hurwitz et al., 1978
0.71	14–21	Computer model	Not specified	Hurwitz et al., 1978
0.71	21–28	Computer model	Not specified	Hurwitz et al., 1978
0.67	28–35	Computer model	Not specified	Hurwitz et al., 1978
0.60	35–42	Computer model	Not specified	Hurwitz et al., 1978
0.60	42–49	Computer model	Not specified	Hurwitz et al., 1978
0.64	49–56	Computer model	Not specified	Hurwitz et al., 1978
0.73–0.75	1–21	Growth, feed efficiency	ISA JV 715	Uzu, 1986
0.68	22–42	Growth, feed efficiency	ISA JV 715	Uzu, 1986
0.85	3–14	Growth, feed efficiency (adjusted to 23% crude protein)	Peterson	Robbins, 1987
0.72, males	7–21	Growth, feed efficiency	Broiler strain	Thomas et al., 1987
0.67, females	7–21	Growth, feed efficiency	Broiler strain	Thomas et al., 1987
0.79	1–27	Growth, feed efficiency	Hybro	Bertram et al., 1988
0.79	7–20	Growth, feed efficiency	Vantress x Arbor Acres	Smith and Waldroup, 1988a
0.70–0.77	1–14	Growth, feed efficiency	Broiler strain	Austic and Rangel-Lugo, 1989
Tryptophan, %				
0.25	10–20	Growth	Not specified	Almquist, 1947
0.18	10–24	Growth, feed efficiency	New Hampshire x White Leghorn	Wilkening et al., 1947
0.143	10–20	Growth, feed efficiency	New Hampshire x Columbian	Griminger et al., 1956
0.17	7–14	Growth, feed efficiency	Not specified	Klain et al., 1960
0.225	7–14	Growth, feed efficiency	New Hampshire x Columbian	Dean and Scott, 1965
0.20	8–14	Growth, feed efficiency (adjusted to 23% CP)	New Hampshire x Columbian	Boomgaardt and Baker, 1971
0.17	7–21	Growth, feed efficiency	Broiler strain	Hewitt and Lewis, 1972
≤0.14	14–28	Growth, feed efficiency	Not specified	Woodham and Deans, 1975
0.179	28–49	Growth, feed efficiency, feather scores	Arbor Acres	Hunchar and Thomas, 1976

Concentration	Age, days	Response criteria	Strain	Reference
0.163	7–14	Computer model	Not specified	Hurwitz et al., 1978
0.144	14–21	Computer model	Not specified	Hurwitz et al., 1978
0.141	21–28	Computer model	Not specified	Hurwitz et al., 1978
0.134	28–35	Computer model	Not specified	Hurwitz et al., 1978
0.118	35–42	Computer model	Not specified	Hurwitz et al., 1978
0.122	42–49	Computer model	Not specified	Hurwitz et al., 1978
0.128	49–56	Computer model	Not specified	Hurwitz et al., 1978
0.17	7–56	Growth	Cobb	Freeman, 1979
0.24	0–7	Growth, feed efficiency	Cobb	Freeman, 1979
0.19	7–34	Growth, feed efficiency	Lohmann	Steinhart and Kirchgessner, 1984
≤0.16	7–20	Growth, feed efficiency	Vantress x Arbor Acres	Smith and Waldroup, 1988b
0.22	8–22	Growth	New Hampshire x Columbian	Han et al., 1991
Valine, %				
0.80	10–20 or 24	Growth	Not specified	Almquist, 1947
0.83	8–13 or 15	Growth, feed efficiency	New Hampshire x Columbian	Klain et al., 1960
0.82	8–16	Growth, feed efficiency	New Hampshire x Columbian	Dean and Scott, 1965
0.75	7–21	Growth, plasma amino acid levels	Not specified	D'Mello, 1974
0.69–0.71	14–28	Total protein efficiency	Ross	Woodham and Deans, 1975
0.69	8–16	Growth, feed efficiency	New Hampshire x Columbian	Baker et al., 1979
>0.72	21–42	Feed efficiency, abdominal fat	Broiler strain	Mendonca and Jenson, 1989a
0.90	7–21	Growth, feed efficiency	Ross x Arbor Acres	Farran and Thomas, 1990
Proline, %				
≤0.5	9–15	Growth, feed efficiency	New Hampshire x Columbian	Green et al., 1962
0.4–0.8	8–14	Growth	New Hampshire x Columbian	Graber et al., 1970
0.40	8–16	Growth, feed efficiency	New Hampshire x Columbian	Baker et al., 1979
Linoleic, %				
	Varied, cited in a review	Growth, tissue triene: tetraene ratio	Various	Balnave, 1970
Calcium, %				
0.90	29–56	Growth, feed efficiency	Broiler strain	Waldroup et al., 1963a
0.74	0–28	Growth, bone ash	Vantress x Arbor Acres	Twining et al., 1965
0.80	42–56	Growth, feed efficiency, bone ash	Vantress x Arbor Acres	Twining et al., 1965
0.80	28–56	Growth, feed efficiency, tibia ash, bone breaking force	Broiler strain	Waldroup et al., 1947a
1.30	0–21	Maximum toe ash	White Cornish x White Plymouth Rock	Yoshida and Hoshii, 1982a
1.18	21–56	Maximum toe ash	White Cornish x White Plymouth Rock	Yoshida and Hoshii, 1982b
Nonphytate phosphorus, %				
0.43	0–21	Growth, bone ash	New Hampshire x White Leghorn	O'Rourke et al., 1952
0.35	14–35	Growth, bone ash	New Hampshire x White Leghorn	O'Rourke et al., 1952
0.27	28–70	Growth, bone ash	New Hampshire x White Leghorn	O'Rourke et al., 1952

APPENDIX TABLE 31 (cont.)

Nutrient and Estimated Requirement	Age Period (days)	Response Criteria	Breed	References
0.45	0–28	Growth, bone ash	Various	Almquist, 1954
0.55	0–21	Growth, bone ash	New Hampshire x White Leghorn	O'Rourke et al., 1955
0.33	28–70	Growth, bone ash	New Hampshire x White Leghorn	O'Rourke et al., 1955
0.45	0–28	Growth, bone ash, serum alkaline phosphates	Rhode Island Red	Gardiner, 1962
0.45	0–28	Growth, bone ash	Vantress x White Plymouth Rock	Waldroup et al., 1962
0.24	28–56	Growth, feed efficiency	Broiler strain	Waldroup et al., 1963a
0.39	0–28	Growth, bone ash	Broiler strain	Waldroup et al., 1963a
0.35	0–28	Growth, feed efficiency	Vantress x Arbor Acres	Twining et al., 1965
0.24	42–56	Growth, feed efficiency, bone ash	Vantress x Arbor Acres	Twining et al., 1965
0.43	0–21	Growth, bone ash	White Plymouth Rock	Fritz et al., 1969
0.24	28–56	Growth, feed efficiency, tibia ash, bone breaking force	Broiler strain	Waldroup et al., 1974a
0.53	0–28	Maximum bone ash	Broiler strain	Waldroup et al., 1975
0.35	28–56	Growth feed efficiency	Hubbard	Sauveur, 1978
0.50	0–28	Growth, feed efficiency, bone ash	Broiler strain	El Boushy, 1979
0.50	8–22	Growth, feed efficiency, bone ash	New Hampshire x Columbian White Cornish x White Plymouth Rock	Willis and Baker, 1981b
0.75	0–21	Growth, feed efficiency, tibia ash	White Cornish x White Plymouth Rock	Yoshida and Hoshii, 1982a
0.35	21–56	Maximum toe ash	White Cornish x White Plymouth Rock	Yoshida and Hoshii, 1982b
0.38	0–28	Growth, toe ash	Hubbard	Nys et al., 1983
0.29	35–53	Growth, feed efficiency, tibia ash, bone length	Broiler strain	Tortuero and Diez Tardon, 1983
Potassium, %				
0.25–0.30	13–41	Growth, mortality	Vantress x Plymouth Rock	Leach et al., 1959
Sodium, %				
0.11–0.20	1–28	Growth, feed efficiency	New Hampshire x Columbian	McWard and Scott, 1961a
0.13	7–23	Growth, blood pH	White Rock	Hurwitz et al., 1973
0.07	49–63	Growth, blood pH	White Rock	Hurwitz et al., 1974
>0.23	1–21	Growth	Broiler strain	Ross, 1977
0.2–0.25	7–21	Growth	Cobb × Hubbard	Ross, 1979
0.35	1–21	Growth	Peterson × Hubbard	Edwards, 1984
Chlorine, %				
0.315–0.340	2–28	Growth, mortality, blood chlorine	White Plymouth Rock	Leach and Nesheim, 1963
0.13	7–23	Growth, blood pH	White Rock	Hurwitz et al., 1973
0.07	49–63	Growth, blood pH	White Rock	Hurwitz et al., 1973
0.12	1–21	Growth, mortality	Ross	Gardiner and Dewar, 1976
0.42	1–21	Growth	Peterson × Hubbard	Edwards, 1984

Nutrient and level	Age (days)	Criteria	Strain	Reference
Magnesium, mg/kg				
350–400	7–24	Growth	Not specified	Almquist, 1947
100–300	1–21	Growth, mortality	White Plymouth Rock	Edwards, et al., 1960
250	1–28	Growth, blood magnesium, mortality	Vantress × Hubbard	Gardner et al., 1960
200	1–14	Growth, mortality	New Hampshire × Columbian	McWard and Scott, 1961b
577	1–21	Growth, mortality, bone magnesium	White Plymouth Rock	Nugara and Edwards, 1963
≤350	1–27	Growth, feed efficiency	New Hampshire × Columbian	Baker and Molitoris, 1975
Manganese, mg/kg				
50	1–42	Growth, perosis	New Hampshire	Gallup and Norris, 1939a
14	8–22	Growth	New Hampshire × Columbian	Southern and Baker, 1983a
Zinc, mg/kg				
35	12–26	Growth, feed efficiency	White Plymouth Rock	Morrison and Sarett, 1958
35	1–42	Growth, bone integrity	White Rock or Cornish × White Rock	O'Dell et al., 1958
30	1–28	Growth	White Meteor × White Rock	Roberson and Shaible, 1958
47–57	1–14	Growth, tibia ash	White Rock	Edwards et al., 1959
>52	1–28	Growth, leg deformity	New Hampshire × Connecticut	Lease et al., 1960
>40 mg	1–28	Growth, hock enlargement	White Plymouth Rock	Zeigler et al., 1961
14	8–22	Growth	New Hampshire × Columbian	Southern and Baker, 1983b
18	1–21	Growth	Broiler strain	Dewar and Downie, 1984
>45	8–22	Tibia zinc	New Hampshire × Columbian	Wedekind et al., 1990
Iron, mg/kg				
56	7–21	Growth, blood hemoglobin, liver iron	Not specified	Waddel and Sell, 1964
75–80	1–28	Growth, blood hemoglobin	New Hampshire and Plymouth Rock	Davis et al., 1968
80	1–21	Growth, blood hemoglobin, packed cell volume	Not specified	McNaughton and Day, 1979
40	8–22	Growth, blood hemoglobin, hematocrit	New Hampshire × Columbian	Southern and Baker, 1982
Copper, mg/kg				
8	1–21	Growth, blood hemoglobin, packed cell volume	Not specified	McNaughton and Day, 1979
Iodine, mg/kg				
0.3–0.4	28–56	Growth, thyroid histology	Barred Plymouth Rock	Creek et al., 1957
Selenium, mg/kg				
>0.02 mg	1–24	Mortality, exudative diathesis	Plymouth Rock × Vantress	Thompson and Scott, 1969
0.1 mg	1–31	Pancreatic degeneration and fibrosis	White Plymouth Rock × Vantress	Gries and Scott, 1972c
>0.1 mg	1–63	Growth, glutathione peroxidase activity	Hubbard	Binnerts and El Boushy, 1985
0.14–0.17	1–21	Growth, plasma thyroid hormones	Hubbard and Arbor Acre	Jensen et al., 1986
Vitamin A, IU/kg				
2200	Varied	Growth	Various	Almquist, 1953
1320	1–28	Growth, feed efficiency	Columbian Rock	Olsen et al., 1959
≤1100	7–63	Growth	Not specified	Marusich and Bauernfeind, 1963
900	1–56	Growth, incidence of coccidiosis	Broiler strain	Ogunmodede, 1981

APPENDIX TABLE 31 *(cont.)*

Nutrient and Estimated Requirement	Age Period (days)	Response Criteria	Breed	References
Vitamin D₃, IU/kg				
200–396	1–28	Growth	Not specified	Waldroup et al., 1963a
198	1–28	Growth, tibia ash	Not specified	Waldroup et al., 1965
200	1–54	Growth, tibia ash	Not specified	Biely and March, 1967
≤200	1–14	Growth, bone mineralization	Not specified	McAuliffe et al., 1976
198	1–21	Growth, tibia ash	Not specified	McNaughton et al., 1977a
400	1–56	Growth, tibia ash	Not specified	Lofton and Soares, 1986
Vitamin E, IU/kg				
15–24	1–28	Prevention of encephalomalacia	Barred Plymouth Rock × Rhode Island Red	Singsen et al., 1955
5–60	Varied, cited in a review	Encephalomalacia exudative diathesis, muscular degeneration	Various	Machlin and Gordon, 1962
5.4–7.4	2–33	Mortality, incidence of encephalomalacia	White Rock	Bartov and Bornstein, 1972
30–50	1–14 and 1–35	Growth, peroxidation in hepatic microsomes	Vantress x Plymouth Rock	Combs and Scott, 1974
Vitamin K, mg/kg				
0.588	1–14	Prothrombin time	White Plymouth Rock	Nelson and Norris, 1960
0.479	1–28	Prothrombin time	White Plymouth Rock	Nelson and Norris, 1960
0.515	1–84	Prothrombin time	White Plymouth Rock	Nelson and Norris, 1961a
0.500	1–14	Prothrombin time	White Plymouth Rock	Nelson and Norris, 1961b
0.370	1–28	Prothrombin time	White Plymouth Rock	Nelson and Norris, 1961b
Riboflavin, mg/kg				
2.5	1–56	Growth	Barred Rock × New Hampshire	Bethke and Record, 1942
3.0	14–42	Growth, feed efficiency	White Wyandotte	Bolton, 1944
3.0–3.5	14–42	Growth	White Wyandotte	Bolton, 1947
2.3	1–56	Growth	Hubbard × Arbor Acres	Wyatt et al., 1973a
5.1	1–56	Growth	Harco	Ogunmodede, 1977
3.6	1–21	Growth, feed efficiency	Cobb and Cobb x Arbor Acres	Ruiz and Harms, 1988b
2.6	8–22	Growth, leg paralysis	New Hampshire x Columbian	Chung and Baker, 1990
Pantothenic acid, mg/kg				
14	Not specified	Growth	Not specified	Jukes, 1939
10	Not specified	Growth	Not specified	Jukes and McElroy, 1943
5	Not specified	Growth	New Hampshire x Columbian	Staten et al., 1980

Nutrient, level	Age, days	Response criteria	Breed or strain	Reference
Niacin, mg/kg				
26–28	7–42	Growth, perosis	Barred Plymouth Rock	Childs et al., 1952
37	1–21	Growth	White Cornish	Yoshida et al., 1966
20	7–20	Growth, incidence of tongue lesions	New Hampshire x Columbian	Baker et al., 1973
≤22	8–50	Growth, feed efficiency	New Hampshire x Columbian	Yen et al., 1977
>55 mg	1–53	Growth, feed efficiency	Not specified	Waldroup et al., 1985b
28–36	1–21	Growth, leg disorders	Cobb	Ruiz and Harms, 1988
32	1–21	Growth, leg disorders	Arbor Acres x Cobb	Ruiz et al., 1990
≤22 mg	21–49	Growth	Cobb	Ruiz and Harms, 1990
Vitamin B_{12}, mg/kg				
0.01	7–29	Growth, energetic efficiency	Dominant White x White Plymouth Rock	Looi and Renner, 1974
≤0.01 mg	1–28	Growth, feed efficiency	Sussex x White Rock	Rys and Koreleski, 1974
Choline, mg/kg				
1000	14–42	Growth, perosis	Barred Plymouth Rock	West et al., 1951
1540–1760	1–56	Growth, feed efficiency	White Rock	Quillen et al., 1961
1119	1–21	Growth	White Rock	Fritz et al., 1967
358	44–55	Growth	New Hampshire x Columbian	Molitoris and Baker, 1976
800	7–28	Growth	White Rock	Lipstein et al., 1977
≤1171	7–35	Growth, perosis	Not specified	Derilo and Balnave, 1980
1910–4100	1–21	Growth, feed efficiency	Not specified	Pesti et al., 1980
1200	8–25	Growth	New Hampshire x Columbian	Baker et al., 1983
625	8–17	Growth	New Hampshire x Columbian	Lowry et al., 1987
>1300	1–21	Growth, feed efficiency	Not specified	Tsiagbe et al., 1987
Biotin, mg/kg				
>0.26 mg	1–25	Growth, mortality, leg abnormalities	Not specified	Anderson and Warnick, 1970
0.14	1–24	Growth, mortality due to fatty kidney liver syndrome	Not specified	Payne et al., 1974
0.14–0.18	1–35	Growth	Ross	Whitehead and Bannister, 1980
≤0.17–0.18	1–56	Incidence of fatty liver and kidney syndrome	Ross	Whitehead and Randall, 1982
≤0.20	1–21	Growth, leg disorders, dermatitis	Hubbard	Watkins, 1988
Folic acid, mg/kg				
≤0.5	1–28	Growth	Not specified	Saxena et al., 1954
≤0.3	1–21 and 1–28	Growth, perosis	Rhode Island Red x White Plymouth Rock	Young et al., 1955
0.40–0.65 mg	1–35	Growth	New Hampshire	March and Biely, 1956
0.3–0.45 mg	1–20	Growth, perosis	Arbor Acres	Creek and Vasaitis, 1963
0.34–0.49 mg	1–28	Growth, leg abnormalities	Not specified	Saxena et al., 1954
Thiamin, mg/kg				
0.75	3–28	Growth, polyneuritis	New Hampshire x Delaware	Thornton, 1960
1.0–1.3	Not specified	Growth, feed efficiency	New Hampshire x Delaware	Thornton and Shutze, 1960

APPENDIX TABLE 31 *(cont.)*

Nutrient and Estimated Requirement	Age Period (days)	Response Criteria	Breed	References
Pyridoxine, mg/kg				
3–5	12–42	Growth, perosis, anemia, dermatitis	White Rock	Hogan et al., 1941
2	7–28	Growth, feed efficiency	Not specified	Kratzer et al., 1947
<5.7	1–56	Growth, feed efficiency	White Plymouth Rock	Fuller and Kifer, 1959
3.3	1–14	Growth	White Plymouth Rock	Fuller and Dunahoo, 1959
2.2–2.6	1–28	Growth, gizzard erosion, serum glutamic oxaloacetic transaminase	Vantress × Arbor Acre	Daghir and Balloun, 1963
2.8–3.6	1–14 or 35	Growth, feed efficiency	Not specified	Kirchgessner and Friesecke, 1963
3	Not specified	Growth, feed efficiency	Not specified	Maier and Kirchgessner, 1968
>3.1	7–28	Growth, serum aspartate aminotransferase	Not specified	Daghir and Shah, 1973
3.2–3.4	1–28	Growth, perosis	White Plymouth Rock × Vantress	Gries and Scott, 1972a
≤1.0	1–20	Growth, feed efficiency	Ross	Lee et al., 1976
1.1	8–17	Growth	New Hampshire × Columbian	Yen et al., 1976
1.75	3–49	Growth, plasma amino acids	Not specified	Aboaysha and Kratzer, 1979
1.3–2.7	1–21	Growth	Not specified	Kazemi and Kratzer, 1980
≤1.48	1–49	Growth	Not specified	Blalock et al., 1984

APPENDIX TABLE 32

Typical body weights, feed requirements, and energy consumption of broilers (from 1994 NRC on poultry)

Age (weeks)	Body Weight (g)		Weekly Feed Consumption (g)		Cumulative Feed Consumption (g)		Weekly Energy Consumption (kcal ME/bird)		Cumulative Energy Consumption (kcal ME/bird)	
	Male	Female	Male	Female	Male	Female	Male	Female	Male	Female
1	152	144	135	131	135	131	432	419	432	419
2	376	344	290	273	425	404	928	874	1,360	1,293
3	686	617	487	444	912	848	1558	1422	2,918	2,715
4	1085	965	704	642	1616	1490	2256	2056	5,174	4,771
5	1576	1344	960	738	2576	2228	3075	2519	8,249	7,290
6	2088	1741	1141	1001	3717	3229	3651	3045	11,900	10,335
7	2590	2134	1281	1081	4998	4310	4102	3459	16,002	13,794
8	3077	2506	1432	1165	6430	5475	4585	3728	20,587	17,522
9	3551	2842	1577	1246	8007	6721	5049	3986	25,636	21,508

Note: Values are typical for broilers fed well-balanced diets providing 3200 kcal ME/kg.

APPENDIX TABLE 33

Documentation of nutrient requirements of broiler breeder pullets and hens (from 1994 NRC on poultry)

Nutrient and Estimated Requirement	Age Period (weeks)	Response Criteria	Breed	References
Protein, g/bird daily				
20	24–52	Egg production, egg weight, body weight, liveability	Cobb	Waldroup et al., 1976b
15.6–16.5	Not specified	Estimated by model	Not specified	Bornstein et al., 1979
19.5	21–64	Egg production, egg weight, fertility	Marshall	Pearson and Herron, 1981
23.1	31–60	Egg yield	Tetra	Jeroch et al., 1982
19	19–40	Body weight, skeletal growth egg production, egg weight, hatchability	Hubbard	Spratt and Leeson, 1987
18–19	31–60	Egg production, egg weight, body weight, egg quality, hatchability	Tetra	Schloffel et al., 1988
Arginine, mg/bird daily				
1,111	Peak egg production	Body weight, egg mass	Mathematical model	Waldroup et al., 1976c
1,111	Peak egg production	Body weight, egg mass	Mathematical model	Bornstein et al., 1979
<1,226 mg	24–64	Egg production, egg weight, fertility, hatchability, egg specific gravity	Cobb	Wilson and Harms, 1984
Histidine, mg/bird daily				
209	Peak egg production	Body weight, egg mass	Mathematical model	Waldroup et al., 1976c
200	Peak egg production	Body weight, egg mass	Mathematical model	Bornstein et al., 1979
Isoleucine, mg/bird daily				
853	Peak egg production	Body weight, egg mass	Mathematical model	Waldroup et al., 1976c
850	Peak egg production	Body weight, egg mass	Mathematical model	Bornstein et al., 1979
Leucine, mg/bird daily				
1,247	Peak egg production	Body weight, egg mass	Mathematical model	Waldroup et al., 1976c
1,250	Peak egg production	Body weight, egg mass	Mathematical model	Bornstein et al., 1979
Lysine, mg/bird daily				
773	Peak egg production	Body weight, egg mass	Mathematical model	Waldroup et al., 1976c
760	Peak egg production	Body weight, egg mass	Mathematical model	Bornstein et al., 1979
<808	24–64	Egg production, egg weight, fertility, hatchability, egg specific gravity	Cobb	Wilson and Harms, 1984
Methionine, mg/bird daily				
558	Peak egg production	Body weight, egg mass	Mathematical model	Waldroup et al., 1976c
570	Peak egg production	Body weight, egg mass	Mathematical model	Bornstein et al., 1979
400	24–64	Egg production, body weight, fertility, hatchability	Cobb	Harms and Wilson, 1980

Nutrient and Estimated Requirement	Age Period (weeks)	Response Criteria	Breed	References
Methionine + cystine, mg/bird daily				
819	Peak egg production	Body weight, egg mass	Mathematical model	Waldroup et al., 1976c
830	Peak egg production	Body weight, egg mass	Mathematical model	Bornstein et al., 1979
723	24–64	Egg production, egg weight, fertility, hatchability	Cobb	Harms and Wilson, 1980
<682	24–64	Egg production, egg weight, fertility, hatchability, egg specific gravity	Cobb	Wilson and Harms, 1984
694	Peak egg production	Nitrogen balance	Tetra	Halle et al., 1984
Phenylalanine + tyrosine, mg/bird daily				
1,126	Peak egg production	Body weight, egg mass	Mathematical model	Waldroup et al., 1976c
1,110	Peak egg production	Body weight, egg mass	Mathematical model	Bornstein et al., 1979
Phenylalanine, mg/bird daily				
610	Peak egg production	Body weight, egg mass	Mathematical model	Bornstein et al., 1979
Threonine, mg/bird daily				
717	Peak egg production	Body weight, egg mass	Mathematical model	Waldroup et al., 1976c
720	Peak egg production	Body weight, egg mass	Mathematical model	Bornstein et al., 1979
Tryptophan, mg/bird daily				
189	Peak egg production	Body weight, egg mass	Mathematical model	Waldroup et al., 1976c
190	Peak egg production	Body weight, egg mass	Mathematical model	Bornstein et al., 1979
<223 mg	24–64	Egg production, egg weight, fertility, hatchability, egg specific gravity	Cobb	Wilson and Harms, 1984
Valine, mg/bird daily				
979	Peak egg production	Body weight, egg mass	Mathematical model	Waldroup et al., 1976c
920	Peak egg production	Body weight, egg mass	Mathematical model	Bornstein et al., 1979
Calcium, g/bird daily				
3.91	26–53	Egg production, egg specific gravity, hatchability	Cobb	Wilson et al., 1980
Nonphytate phosphorus, mg/bird daily				
338	26–53	Egg production, egg specific gravity, hatchability	Cobb	Wilson et al., 1980
Sodium, mg/bird daily				
<154	32–64	Egg production, egg weight, fertility, egg specific gravity, hatchability	Cobb	Damron et al., 1983
Chlorine, mg/bird daily				
208	32–60	Egg production, egg weight, hatchability	Cobb	Harms and Wilson, 1984
Biotin, μg/bird daily				
16	20–58	Egg production, egg weight, hatchability	Marshall	Whitehead et al., 1985

APPENDIX TABLE 34

Nutrient requirements of meat-type males for breeding purposes as percentages or units per rooster per day (90% dry matter)

		Age (weeks)		
	Unit	0–4	4–20	20–60
Metabolizable energy[a]	kcal	—	—	350–400
Protein and amino acids				
Protein[b]	%	15.00	12.00	—
Lysine[c]	%	*0.79*	*0.64*	—
Methionine[c]	%	*0.36*	*0.31*	—
Methionine + cystine[c]	%	*0.61*	*0.49*	—
Minerals				
Calcium	%	*0.90*	*0.90*	—
Nonphytate phosphorus	%	*0.45*	*0.45*	—
Protein and amino acids				
Protein	g	—	—	12
Arginine[c]	mg	—	—	*680*
Lysine[c]	mg	—	—	*475*
Methionine[c]	mg	—	—	*340*
Methionine + cystine[c]	mg	—	—	*490*
Minerals				
Calcium	mg	—	—	*200*
Nonphytate phosphorus	mg	—	—	110

Note: For nutrients not listed, see requirements for egg-type pullets (Appendix Table 33) as a guide. Where experimental data are lacking, values typeset in bold italics represent an estimate based on values obtained for other ages or related species.

[a]Energy needs are influenced by the environment and the housing system. These factors must be adjusted as required to maintain the body weight recommended by the breeder.

[b]Broilers do not have a requirement for crude protein per se. However, there should be sufficient crude protein to ensure an adequate nitrogen supply for the synthesis of nonessential amino acids. Suggested requirements for crude protein are typical of those derived with corn–soybean meal diets, and levels can be reduced somewhat when synthetic amino acids are used.

[c]Amino acid requirements estimated by using the model of Smith (1978).

APPENDIX TABLE 35

Estimates of metabolizable energy required per hen per day by chickens in relation to body weight and egg production (kcal)

Body Weight (kg)	Rate of Egg Production (%)					
	0	50	60	70	80	90
1.0	130	192	205	217	229	242
1.5	177	239	251	264	276	289
2.0	218	280	292	305	317	330
2.5	259	321	333	346	358	371
3.0	296	358	370	383	395	408

Note: A number of formulas have been suggested for prediction of the daily energy requirements of chickens. The formula used here was derived from that in *Effect of Environment on Nutrient Requirements of Domestic Animals* (National Research Council, 1981c):

$$\text{ME per hen daily} = W^{0.75}(173 - 1.95T) + 5.5\,\Delta W + 2.07EE$$

where W = body weight (kg), T = ambient temperature (°C), ΔW = change in body weight (g/day), and EE = daily egg mass (g). Temperature of 22°C, egg weight of 60 g, and no change in body weight were used in calculations.

APPENDIX TABLE 36

Nutrient requirements of turkeys as percentages or units per kilogram of diet (90% dry matter) (from 1994 NRC on poultry)

Nutrient	Unit	Growing Turkeys, Males and Females						Breeders	
		0 to 4 Weeks[a]; 0 to 4 Weeks[b]; 2800[c]	4 to 8 Weeks[a]; 4 to 8 Weeks[b]; 2900[c]	8 to 12 Weeks[a]; 8 to 11 Weeks[b]; 3000[c]	12 to 16 Weeks[a]; 11 to 14 Weeks[b]; 3100[c]	16 to 20 Weeks[a]; 14 to 17 Weeks[b]; 3200[c]	20 to 24 Weeks[a]; 17 to 20 Weeks[b]; 3300[c]	Holding; 2900[c]	Laying Hens; 2900[c]
Protein and amino acids									
Protein[d]	%	28.0	26	22	19	16.5	14	12	14
Arginine	%	1.6	1.4	1.1	0.9	0.75	0.6	0.5	0.6
Glycine + serine	%	1.0	0.9	0.8	0.7	0.6	0.5	0.4	0.5
Histidine	%	0.58	0.5	0.4	0.3	0.25	0.2	0.2	0.3
Isoleucine	%	1.1	1.0	0.8	0.6	0.5	0.45	0.4	0.5
Leucine	%	1.9	1.75	1.5	1.25	1.0	0.8	0.5	0.5
Lysine	%	1.6	1.5	1.3	1.0	0.8	0.65	0.5	0.6
Methionine	%	0.55	0.45	0.4	0.35	0.25	0.25	0.2	0.2
Methionine + cystine	%	1.05	0.95	0.8	0.65	0.55	0.45	0.4	0.4
Phenylalanine	%	1.0	0.9	0.8	0.7	0.6	0.5	0.4	0.55
Phenylalanine + tyrosine	%	1.8	1.6	1.2	1.0	0.9	0.9	0.8	1.0
Threonine	%	1.0	0.95	0.8	0.75	0.6	0.5	0.4	0.45
Tryptophan	%	0.26	0.24	0.2	0.18	0.15	0.13	0.1	0.13
Valine	%	1.2	1.1	0.9	0.8	0.7	0.6	0.5	0.58
Fat									
Linoleic acid	%	1.0	1.0	0.8	0.8	0.8	0.8	0.8	1.1
Macrominerals									
Calcium[e]	%	1.2	1.0	0.85	0.75	0.65	0.55	0.5	2.25
Nonphytate phosphorus[f]	%	0.6	0.5	0.42	0.38	0.32	0.28	0.25	0.35
Potassium	%	0.7	0.6	0.5	0.5	0.4	0.4	0.4	0.6
Sodium	%	0.17	0.15	0.12	0.12	0.12	0.12	0.12	0.12
Chlorine	%	0.15	0.14	0.14	0.12	0.12	0.12	0.12	0.12
Magnesium	mg	500	500	500	500	500	500	500	500

APPENDIX TABLE 36 (cont.)

Nutrient	Unit	Growing Turkeys, Males and Females						Breeders	
		0 to 4 Weeks[a]; 0 to 4 Weeks[b]; Weeks[b]; 2800[c]	4 to 8 Weeks[a]; 4 to 8 Weeks[b]; Weeks[b]; 2900[c]	8 to 12 Weeks[a]; 8 to 11 Weeks[b]; Weeks[c]; 3000[c]	12 to 16 Weeks[a]; 11 to 14 Weeks[b]; Weeks[c]; 3100[c]	16 to 20 Weeks[a]; 14 to 17 Weeks[b]; Weeks[c]; 3200[c]	20 to 24 Weeks[a]; 17 to 20 Weeks[b]; Weeks[c]; 3300[c]	Holding; 2900[c]	Laying Hens; 2900[c]
Trace minerals									
Manganese	mg	60	60	60	60	60	60	60	60
Zinc	mg	70	65	50	40	40	40	40	65
Iron	mg	80	60	60	60	50	50	50	60
Copper	mg	8	8	6	6	6	6	6	8
Iodine	mg	0.4	0.4	0.4	0.4	0.4	0.4	0.4	0.4
Selenium	mg	0.2	0.2	0.2	0.2	0.2	0.2	0.2	0.2
Fat-soluble vitamins									
A	IU	5,000	5,000	5,000	5,000	5,000	5,000	5,000	5,000
D_3[g]	ICU	1,100	1,000	1,000	1,000	1,000	1,000	1,000	1,100
E	IU	12	12	10	10	10	10	10	25
K	mg	1.75	1.5	1.0	0.75	0.75	0.50	0.5	1.0
Water-soluble vitamins									
B_{12}	mg	0.003	0.003	0.003	0.003	0.003	0.003	0.003	0.003
Biotin[h]	mg	0.25	0.2	0.125	0.125	0.100	0.100	0.100	0.20
Choline	mg	1,600	1,400	1,100	1,100	950	800	800	1,000
Folacin	mg	1.0	1.0	0.8	0.8	0.7	0.7	0.7	1.0
Niacin	mg	60.0	60.0	50.0	50.0	40.0	40.0	40.0	40.0
Pantothenic acid	mg	10.0	9.0	9.0	9.0	9.0	9.0	9.0	16.0
Pyridoxine	mg	4.5	4.5	3.5	3.5	3.0	3.0	3.0	4.0
Riboflavin	mg	4.0	3.6	3.0	3.0	2.5	2.5	2.5	4.0
Thiamin	mg	2.0	2.0	2.0	2.0	2.0	2.0	2.0	2.0

Note: Where experimental data are lacking, values typeset in bold italics represent estimates based on values obtained from other ages or related species or from modeling experiments.

[a] The age intervals for nutrient requirements of males are based on actual chronology from previous research. Genetic improvements in body weight gain have led to an earlier implementation of these levels, at 0 to 3, 3 to 6, 6 to 9, 9 to 12, 12 to 15, and 15 to 18 weeks, respectively, by the industry at large.

[b] The age intervals for nutrient requirements of females are based on actual chronology from previous research. Genetic improvements in body weight gain have led to an earlier implementation of these levels, at 0 to 3, 3 to 6, 6 to 9, 9 to 12, 12 to 14, and 14 to 16 weeks, respectively, by the industry at large.

[c] These are approximate metabolizable energy (ME) values provided with typical corn–soybean-meal-based feeds, expressed in kcal ME_n/kg diet. Such energy, when accompanied by the nutrient levels suggested, is expected to provide near-maximum growth, particularly with pelleted feed.

[d] Turkeys do not have a requirement for crude protein per se. However, there should be sufficient crude protein to ensure an adequate nitrogen supply for synthesis of nonessential amino acids. Suggested requirements for crude protein are typical of those derived with corn–soybean meal diets, and levels can be reduced when synthetic amino acids are used.

[e] The calcium requirement may be increased when diets contain high levels of phytate phosphorus (Nelson, 1984).

[f] Organic phosphorus is generally considered to be associated with phytin and of limited availability.

[g] These concentrations of vitamin D are considered satisfactory when the associated calcium and phosphorus levels are used.

[h] Requirement may increase with wheat-based diets.

APPENDIX TABLE 37

Growth rate and feed and energy consumption of large-type turkeys (from 1994 NRC on poultry)

Age (weeks)	Body Weight (kg) Male	Body Weight (kg) Female	Feed Consumption per Week (kg) Male	Feed Consumption per Week (kg) Female	Cumulative Feed Consumption (kg) Male	Cumulative Feed Consumption (kg) Female	ME Consumption per Week (Mcal) Male	ME Consumption per Week (Mcal) Female
1	0.12	0.12	0.10	0.10	0.10	0.10	0.28	0.28
2	0.25	0.24	0.19	0.18	0.29	0.28	0.53	0.5
3	0.50	0.46	0.37	0.34	0.66	0.62	1.0	1.0
4	1.0	0.9	0.70	0.59	1.36	1.21	2.0	1.7
5	1.6	1.4	0.85	0.64	2.21	1.85	2.5	1.9
6	2.2	1.8	1.10	0.80	3.31	2.65	3.2	2.3
7	3.1	2.3	1.40	0.98	4.71	3.63	4.1	2.8
8	4.0	3.0	1.73	1.21	6.44	4.84	5.0	3.5
9	5.0	3.7	2.00	1.42	8.44	6.26	6.0	4.3
10	6.0	4.4	2.34	1.70	10.78	7.96	7.0	5.1
11	7.1	5.2	2.67	1.98	13.45	9.94	8.0	5.9
12	8.2	6.0	2.99	2.18	16.44	12.12	9.0	6.8
13	9.3	6.8	3.20	2.44	19.64	14.56	9.9	7.6
14	10.5	7.5	3.47	2.69	23.11	17.25	10.8	8.4
15	11.5	8.3	3.73	2.81	26.84	20.06	11.6	9.0
16	12.6	8.9	3.97	3.00	30.81	23.06	12.3	9.6
17	13.5	9.6	4.08	3.14	34.89	26.20	13.1	10.1
18	14.4	10.2	4.30	3.18	39.19	29.38	13.8	10.5
19	15.2	10.9	4.52	3.31	43.71	32.69	14.5	10.9
20	16.1	11.5	4.74	3.40	48.45	36.09	15.2	11.2
21	17.0	[a]	4.81	[a]	53.26	[a]	15.9	[a]
22	17.9	[a]	5.00	[a]	58.26	[a]	16.5	[a]
23	18.6	[a]	5.15	[a]	63.41	[a]	17.1	[a]
24	19.4	[a]	5.28	[a]	68.69	[a]	17.4	[a]

[a]No data given because females are usually not marketed after 20 weeks of age.

APPENDIX TABLE 38

Body weights and feed consumption of large-type turkeys during the holding and breeding periods (from 1994 NRC on poultry)

Age (weeks)	Females Weight (kg)	Females Egg Production (%)	Females Feed per Turkey Daily (g)	Males Weight (kg)	Males Feed per Turkey Daily (g)
20	8.4	0	260	14.3	500
25	9.8	0	320	16.4	570
30	11.1	0[a]	310	19.1	630
35	11.1	68	280	20.7	620
40	10.8	64	280	21.8	570
45	10.5	58	280	22.5	550
50	10.5	52	290	23.2	560
55	10.5	45	290	23.9	570
60	10.6	38	290	24.5	580

Note: These values are based on experimental data involving in-season egg production (i.e., November through July) of commercial stock. It is estimated that summer breeders would produce 70% to 90% as many eggs and consume 60% to 80% as much feed as in-season breeders.

[a]Light stimulation is begun at this point.

APPENDIX TABLE 39

Water consumption by chickens and turkeys of different ages (from 1994 NRC on poultry)

Age (weeks)	Broiler Chickens (ml per bird per week)[a]	White Leghorn Hens (ml per bird per week)[a]	Brown Egg-laying Hens (ml per bird per week)[a]	Large White Turkeys (ml per bird per week)[a,b] Males	Large White Turkeys (ml per bird per week)[a,b] Females
1	225	200	200	385	385
2	480	300	400	750	690
3	725	—	—	1135	930
4	1000	500	700	1650	1274
5	1250	—	—	2240	1750
6	1500	700	800	2870	2150
7	1750	—	—	3460	2640
8	2000	800	900	4020	3180
9	—	—	—	4670	3900
10	—	900	1000	5345	4400
11	—	—	—	5850	4620
12	—	1000	1100	6220	4660
13	—	—	—	6480	4680
14	—	1100	1100	6680	4700
15	—	—	—	6800	4720
16	—	1200	1200	6920	4740
17	—	—	—	6960	4760
18	—	1300	1300	7000	—
19	—	—	—	7020	—
20	—	1600	1500	7040	—

Note: Dash indicates that information is not available.

[a]Varies considerably depending on ambient temperature, diet composition, rates of growth or egg production, and type of equipment used. The data presented apply under moderate (20° to 25°C) ambient temperatures.

[b]Based on data obtained from commercial turkey production units.

APPENDIX TABLE 40

Nutrient requirements of geese as percentages or units per kilogram of diet (90% dry matter)

Nutrient	Unit	0 to 4 Weeks; 2900[a]	After 4 Weeks; 3000[a]	Breeding; 2900[a]
Protein and amino acids				
Protein	%	*20*	15	15
Lysine	%	1.0	0.85	*0.6*
Methionine + cystine	%	0.60	0.50	*0.50*
Macrominerals				
Calcium	%	*0.65*	*0.60*	*2.25*
Nonphytate phosphorus	%	*0.30*	*0.3*	0.3
Fat-soluble vitamins				
A	IU	*1500*	*1500*	*4000*
D₃	IU	*200*	*200*	*200*
Water-soluble vitamins				
Choline	mg	*1500*	*1000*	?
Niacin	mg	*65.0*	*35.0*	*20.0*
Pantothenic acid	mg	*15.0*	*10.0*	*10.0*
Riboflavin	mg	3.8	*2.5*	*4.0*

Note: For nutrients not listed or those for which no values are given, see requirements of chickens (Table 31) as a guide. Where experimental data are lacking, values typeset in bold italic represent an estimate based on values obtained for other ages or species.

[a]These are typical dietary energy concentrations expressed in kcal ME$_n$/kg diet.

APPENDIX TABLE 41

Nutrient requirements of white pekin ducks as percentages or units per kilogram of diet
(90% dry matter)

Nutrient	Unit	0 to 2 Weeks; 2900[a]	2 to 7 Weeks; 3000[a]	Breeding; 2900[a]
Protein and amino acids				
Protein	%	*22*	16	*15*
Arginine	%	*1.1*	*1.0*	
Isoleucine	%	*0.63*	*0.46*	*0.38*
Leucine	%	*1.26*	*0.91*	*0.76*
Lysine	%	*0.90*	*0.65*	*0.60*
Methionine	%	0.40	*0.30*	0.27
Methionine + cystine	%	0.70	0.55	*0.50*
Tryptophan	%	*0.23*	*0.17*	*0.14*
Valine	%	*0.78*	*0.56*	*0.47*
Macrominerals				
Calcium	%	*0.65*	*0.60*	*2.75*
Chloride	%	0.12	*0.12*	*0.12*
Magnesium	mg	*500*	*500*	*500*
Nonphytate phosphorus	%	*0.40*	0.30	
Sodium	%	0.15	*0.15*	*0.15*
Trace minerals				
Manganese	mg	*50*	?[b]	?
Selenium	mg	0.20	?	?
Zinc	mg	*60*	?	?
Fat-soluble vitamins				
A	IU	*2500*	*2500*	*4000*
D₃	IU	*400*	*400*	*900*
E	IU	*10*	*10*	*10*
K	mg	*0.5*	*0.5*	*0.5*
Water-soluble vitamins				
Niacin	mg	*55*	*55*	*55*
Pantothenic acid	mg	*11.0*	*11.0*	*11.0*
Pyridoxine	mg	*2.5*	*2.5*	*3.0*
Riboflavin	mg	*4.0*	*4.0*	*4.0*

Note: For nutrients not listed or those for which no values are given, see requirements of broiler chickens (Table 31) as a guide. Where experimental data are lacking, values typeset in bold italics represent an estimate based on values obtained for other ages or species.

[a]These are typical dietary energy concentrations as expressed in kcal ME_n/kg diet.

[b]Question marks indicate that no estimates are available.

APPENDIX TABLE 42

Approximate body weights and feed consumption of white pekin ducks to 8 weeks of age

Age (weeks)	Body Weight (kg) Male	Body Weight (kg) Female	Weekly Feed Consumption (kg) Male	Weekly Feed Consumption (kg) Female	Cumulative Feed Consumption (kg) Male	Cumulative Feed Consumption (kg) Female
0	0.06	0.06	0.00	0.00	0.00	0.00
1	0.27	0.27	0.22	0.22	0.22	0.22
2	0.78	0.74	0.77	0.73	0.99	0.95
3	1.38	1.28	1.12	1.11	2.11	2.05
4	1.96	1.82	1.28	1.28	3.40	3.33
5	2.49	2.30	1.48	1.43	4.87	4.76
6	2.96	2.73	1.63	1.59	6.50	6.35
7	3.34	3.06	1.68	1.63	8.18	7.98
8	3.61	3.29	1.68	1.63	9.86	9.61

APPENDIX TABLE 43

Nutrient requirements of ring-necked pheasants as percentages or units per kilogram of diet (90% dry matter)

Nutrient	Unit	0–4 Weeks; 2800[a]	4–8 Weeks; 2800[a]	9–17 Weeks; 2700[a]	Breeding; 2800[a]
Protein and amino acids					
Protein	%	28	24	18	15
Glycine + serine	%	*1.8*	*1.55*	*1.0*	*0.50*
Linoleic acid	%	*1.0*	*1.0*	*1.0*	*1.0*
Lysine	%	*1.5*	*1.40*	*0.8*	*0.68*
Methionine	%	*0.50*	*0.47*	*0.30*	*0.30*
Methionine + cystine	%	*1.0*	*0.93*	*0.6*	*0.60*
Protein	%	28	*24*	*18*	15
Macrominerals					
Calcium	%	1.0	*0.85*	*0.53*	*2.5*
Chlorine	%	*0.11*	*0.11*	*0.11*	*0.11*
Nonphytate phosphorus	%	0.55	*0.50*	*0.45*	*0.40*
Sodium	%	*0.15*	*0.15*	*0.15*	*0.15*
Trace minerals					
Manganese	mg	*70*	*70*	*60*	*60*
Zinc	mg	60	*60*	*60*	*60*
Water-soluble vitamins					
Choline	mg	*1430*	*1300*	*1000*	*1000*
Niacin	mg	70.0	*70*	*40.0*	*30.0*
Pantothenic acid	mg	10.0	*10.0*	*10.0*	*16.0*
Riboflavin	mg	3.4	*3.4*	*3.0*	*4.0*

Note: Where experimental data are lacking, values typeset in bold italics represent an estimate based on values obtained for other ages or species. For nutrients not listed or those for which no values are given, see requirements of turkeys (Table 31) as a guide.

[a]These are typical dietary energy concentrations, expressed in kcal ME_n/kg diet.

APPENDIX TABLE 44

Nutrient requirements of Japanese quail (*Coturnix*) as percentages or units per kilogram of diet (90% dry matter)

Nutrient	Unit	Starting and Growing; 2900[a]	Breeding; 2900[a]
Protein and amino acids			
Protein	%	24.0	20.0
Arginine	%	1.25	*1.26*
Glycine + serine	%	1.15	*1.17*
Histidine	%	0.36	*0.42*
Isoleucine	%	0.98	*0.90*
Leucine	%	1.69	*1.42*
Lysine	%	1.30	*1.00*
Methionine	%	0.50	0.45
Methionine + cystine	%	0.75	0.70
Phenylalanine	%	0.96	*0.78*
Phenylalanine + tyrosine	%	1.80	*1.40*
Threonine	%	1.02	*0.74*
Tryptophan	%	0.22	*0.19*
Valine	%	0.95	*0.92*
Fat			
Linoleic acid	%	*1.0*	1.0
Macrominerals			
Calcium	%	0.8	2.5
Chlorine	%	*0.14*	*0.14*
Magnesium	mg	300	*500*
Nonphytate phosphorus	%	0.30	*0.35*
Potassium	%	*0.4*	*0.4*
Sodium	%	0.15	0.15
Trace minerals			
Copper	mg	*5*	5
Iodine	mg	*0.3*	0.3
Iron	mg	*120*	*60*
Manganese	mg	*60*	60
Selenium	mg	*0.2*	0.2
Zinc	mg	25	*50*
Fat-soluble vitamins			
A	IU	1650	*3300*
D_3	ICU	750	*900*
E	IU	*12*	25
K	mg	*1*	1
Water-soluble vitamins			
B_{12}	mg	*0.003*	*0.003*
Biotin	mg	*0.3*	0.15
Choline	mg	*2000*	1500
Folacin	mg	*1*	1
Niacin	mg	40	*20*
Pantothenic acid	mg	*10*	15
Pyridoxine	mg	*3*	3
Riboflavin	mg	*4*	4
Thiamin	mg	*2*	2

Note: Where experimental data are lacking, values typeset in bold italics represent an estimate based on values obtained for other ages or species. For values not listed for the starting–growing periods, see requirements for turkeys (Table 31) as a guide.

[a]These are typical dietary energy concentrations, expressed in kcal ME_n/kg diet.

APPENDIX TABLE 45
Daily nutrient requirements of horses (from 1989 NRC on horses)

Item	Weight, kg	lb	DE, Mcal	C. Protein, g	Lysine, g	Calcium, g	Phos., g	Mg, g	K, g	Na, g
Maintenance										
	400	880	13.4	536	18.8	16.0	11.2	6.0	20	6.7
	500	1100	16.4	656	23.0	20.0	14.0	7.5	25	8.2
	600	1320	19.4	776	27.2	24.0	16.8	9.0	30	9.7
Preg. mare										
9th mo.	400	880	14.9	654	22.9	28.3	21.4	7.1	24	6.6
	500	1100	18.2	801	28.0	34.6	26.2	8.7	29	8.1
	600	1320	21.5	947	33.2	40.9	31.0	10.3	34	9.6
10th mo.	400	880	15.1	666	23.3	28.8	21.8	7.3	24	6.7
	500	1100	18.5	815	28.5	35.2	26.7	8.9	30	8.2
	600	1320	21.9	965	33.8	41.7	31.6	10.5	35	9.7
11th mo.	400	880	16.1	708	24.8	30.6	23.2	7.7	26	6.7
	500	1100	19.7	866	30.3	37.4	28.3	9.4	31	8.2
	600	1320	23.3	1024	35.9	44.2	33.5	11.2	37	9.7
Lact. mare										
Early lact.	400	880	22.9	1141	39.9	44.8	28.9	8.7	37	8.8
	500	1100	28.3	1427	49.9	56.0	36.1	10.9	46	10.9
	600	1320	33.7	1712	59.9	67.2	43.3	13.0	55	12.9
Late lact.	400	880	19.7	839	29.4	28.8	17.8	6.9	26	8.1
	500	1100	24.3	1049	36.7	36.0	22.2	8.6	33	9.9
	600	1320	28.9	1259	44.1	43.2	26.7	10.3	40	11.8
Working horse										
Light work	400	880	16.7	670	23.4	20.4	14.6	7.7	25	20.5
	500	1100	20.5	820	28.7	25.0	17.8	9.4	31	25.1
	600	1320	24.2	970	33.9	29.6	21.1	11.2	37	29.7
Moderate work	400	880	20.1	804	28.1	24.5	17.5	9.2	31	22.8
	500	1100	24.6	984	34.4	30.0	21.4	11.3	37	27.8
	600	1320	29.1	1164	40.7	35.5	25.3	13.4	44	32.9
Intense work	400	880	26.8	1072	37.5	32.7	23.3	12.3	41	28.2
	500	1100	32.8	1312	45.9	40.0	28.5	15.1	50	34.5
	600	1320	38.8	1552	54.3	47.3	33.8	17.8	59	40.8

APPENDIX TABLE 46
Daily nutrient requirements of horses (from 1989 NRC on horses)

Item	Weight kg	Weight lb	Fe, mg	Mn, mg	Zn, mg	Cu, mg	Co, mg	I, mg	Se, mg	Vit A, IU	Vit D, IU	Vit E, IU
Maintenance	400	880	268	268	268	67	0.7	0.7	0.7	12,000	2010	335
	500	1100	328	328	328	82	0.8	0.8	0.8	15,000	2460	410
	600	1320	388	388	388	97	1.0	1.0	1.0	18,000	2910	485
Preg. mare 9th mo.	400	880	331	264	264	66	0.7	0.7	0.7	24,000	3966	529
	500	1100	405	324	324	81	0.8	0.8	0.8	30,000	4854	647
	600	1320	479	383	383	96	1.0	1.0	1.0	36,000	5742	766
10th mo.	400	880	336	269	269	67	0.7	0.7	0.7	24,000	4038	538
	500	1100	412	329	329	82	0.8	0.8	0.8	30,000	4942	659
	600	1320	487	390	390	97	1.0	1.0	1.0	36,000	5846	779
11th mo.	400	880	335	268	268	67	0.7	0.7	0.7	24,000	4020	536
	500	1100	410	328	328	82	0.8	0.8	0.8	30,000	4920	656
	600	1320	485	388	388	97	1.0	1.0	1.0	36,000	5820	776
Lact. mare Early lact.	400	880	440	352	352	88	0.9	0.9	0.9	24,000	5286	705
	500	1100	544	435	435	109	1.1	1.1	1.1	30,000	6526	870
	600	1320	647	518	518	129	1.3	1.3	1.3	36,000	7767	1036
Late lact.	400	880	403	322	322	81	0.8	0.8	0.8	24,000	4833	644
	500	1100	496	397	397	99	1.0	1.0	1.0	30,000	5956	794
	600	1320	590	472	472	118	1.2	1.2	1.2	36,000	7079	944
Working horse Light work	400	880	273	273	273	68	0.7	0.7	0.7	12,000	2051	547
	500	1100	335	335	335	84	0.8	0.8	0.8	15,000	2510	669
	600	1320	396	396	396	99	1.0	1.0	1.0	18,000	2969	792
Moderate work	400	880	303	303	303	76	0.8	0.8	0.8	12,000	2275	607
	500	1100	371	371	371	93	0.9	0.9	0.9	15,000	2785	743
	600	1320	439	439	439	110	1.1	1.1	1.1	18,000	3294	878
Intense work	400	880	376	376	376	94	0.9	0.9	0.9	12,000	2821	752
	500	1100	460	460	460	136	1.4	1.4	1.4	18,000	4084	1089
	600	1320	545	545	545	136	1.4	1.4	1.4	18,000	4084	1089

APPENDIX TABLE 47
Daily nutrient requirements of growing horses (from 1989 NRC on horses)

Age, Mo.	Weight, kg	D. Gain, kg	DE, Mcal	C. Pro, g	Lysine, g	Calcium, g	Phos., g	Mg, g	K, g	Na, g
4	145[a]	0.75	12.6	630	26.4	29.8	16.5	3.1	9	4.3
6	180[a]	0.55	12.9	643	27.0	24.8	13.7	3.4	11	4.4
		0.70	14.5	725	30.5	29.6	16.4	3.6	11	5.0
12	265[a]	0.40	15.6	700	29.6	23.4	12.9	4.5	14	5.6
		0.50	17.1	770	32.5	26.6	14.7	4.6	15	6.1
18	330[a]	0.25	15.9	716	30.2	21.2	11.7	5.3	17	6.4
4	175[b]	0.85	14.4	720	30.2	34.2	19.0	3.7	11	5.0
6	215[b]	0.65	15.0	750	31.5	29.4	16.3	4.0	13	5.2
		0.85	17.2	860	36.1	35.8	19.9	4.3	13	5.9
12	325[b]	0.50	18.9	851	35.9	29.0	16.0	5.5	18	6.8
		0.65	21.2	956	40.4	33.8	18.7	5.7	18	7.6
18	400[b]	0.35	19.8	893	37.7	27.2	15.0	6.4	21	7.9
4	200[c]	1.00	16.5	826	34.7	40.0	22.2	4.2	13	5.7
6	245[c]	0.75	17.0	850	35.7	33.8	18.7	4.6	14	5.9
		0.95	19.2	960	40.3	40.2	22.3	4.9	15	6.6
12	375[c]	0.65	22.7	1024	43.2	35.8	19.8	6.4	21	8.1
		0.80	25.1	1129	47.7	40.6	22.5	6.6	21	9.0
18	475[c]	0.45	23.9	1077	45.5	33.4	18.5	7.7	25	9.6

[a]Expected mature weight 400 kg.
[b]Expected mature weight 500 kg.
[c]Expected mature weight 600 kg.

APPENDIX TABLE 48
Daily nutrient requirements of growing horses (from 1989 NRC on horses)

Age, mo.	Weight, kg	D. Gain, kg	Fe, mg	Mn, mg	Zn, mg	Cu, mg	Co, mg	I, mg	Se, mg	Vit a, IU	Vit D, IU	Vit E, IU
4	145[a]	0.75	217	174	174	43	0.4	0.4	0.4	4350	3474	347
6	180[a]	0.55	222	177	177	44	0.4	0.4	0.4	5400	3545	355
		0.70	250	200	200	50	0.5	0.5	0.5	5400	4000	400
12	265[a]	0.40	278	222	222	56	0.6	0.6	0.6	7950	4447	445
		0.50	306	245	245	61	0.6	0.6	0.6	7950	4891	489
18	330[a]	0.25	318	254	254	64	0.6	0.6	0.6	9900	5089	509
4	175[b]	0.85	248	199	199	50	0.5	0.5	0.5	5250	3973	397
6	215[b]	0.65	259	207	207	52	0.5	0.5	0.5	6450	4138	414
		0.85	297	237	237	59	0.6	0.6	0.6	6450	4745	475
12	325[b]	0.50	338	270	270	68	0.7	0.7	0.7	9750	5405	541
		0.65	379	304	304	76	0.8	0.8	0.8	9750	6071	607
18	400[b]	0.35	397	318	318	79	0.8	0.8	0.8	12,000	6351	635
4	200[c]	1.00	285	228	228	57	0.6	0.6	0.6	6000	4558	456
6	245[c]	0.75	293	235	235	59	0.6	0.6	0.6	7350	4690	469
		0.95	331	265	265	66	0.7	0.7	0.7	7350	5297	530
12	375[c]	0.65	406	325	325	81	0.8	0.8	0.8	11,250	6500	650
		0.80	448	358	358	90	0.9	0.9	0.9	11,250	7166	717
18	475[c]	0.45	479	383	383	96	1.0	1.0	1.0	14,250	7660	766

[a]Expected mature weight 400 kg.
[b]Expected mature weight 500 kg.
[c]Expected mature weight 600 kg.

Volume

1 cubic inch	= 16.387 cubic centimeters
1 cubic foot	= 1728.0 cubic inches = 0.0283 cubic meters
1 cubic yard	= 27.0 cubic feet = 0.7646 cubic meters
1 cubic centimeter	= 1.0 milliliter = 0.061 cubic inch
1 cubic meter	= 35.315 cubic feet = 1.308 cubic yards
1 liquid pint (US)	= 28.875 cubic inches = 0.5 liquid quart = 0.47316 liter
1 liquid quart (US)	= 57.75 cubic inches = 0.9463 liter
1 liquid gallon (US)	= 231 cubic inches = 4.0 quarts = 3.7853 liters
1 liter	= 1.057 liquid quarts (US) = 0.2642 liquid gallons (US)
1 bushel	= 2150.42 cubic inches = 1.244 cubic feet
	= 9.309 liquid gallons (US) = 4 pecks

Weight

1 ounce (avdp.)	= 28.50 grams = 16 drams
1 pound (avdp.)	= 16.0 ounces = 453.6 grams = 7000 grains
1 kilogram	= 1000 grams = 2.205 pounds
1 ton	= 2000 pounds = 907.0 kilograms
1 metric ton	= 1000 kilograms = 2205 pounds = 1.102 tons

Length

1 inch	= 2.54 centimeters = 25.4 millimeters
1 foot	= 12.0 inches = 30.48 centimeters = 0.3048 meters
1 yard	= 3.0 feet = 0.9144 meters
1 millimeter	= 0.03937 inches
1 centimeter	= 0.3937 inches = 10 millimeters
1 meter	= 39.37 inches = 3.2808 feet = 1.09361 yards = 100 centimeters
1 kilometer	= 1000 meters = 0.6412 miles
1 mile	= 1.6093 kilometers

Area

1 square inch	= 6.452 square centimeters
1 square centimeter	= 0.155 square inches
1 square foot	= 0.929 square meter
1 square meter	= 1550.0 square inches = 10.764 square feet
1 square yard	= 9.0 square feet = 0.8361 square meters
1 acre	= 43,560 square feet = 4840.0 square yards = 0.4047 hectares
1 square mile	= 640.0 acres = 259.0 hectares
1 hectare	= 10,000 square meters = 2.471 acres
1 square kilometer	= 247.1 acres = 0.386 square miles

Temperature

Degrees Centigrade	= 5/9 (Degrees Fahrenheit − 32)
Degrees Fahrenheit	= (9/5 × Degrees Centigrade) + 32

GLOSSARY

In addition to the terms listed and defined here, a number of terms used in feed manufacturing are listed and defined in Ch. 4. It might also be noted that mineral elements, amino acids, and the vitamins are not listed separately, but will be found listed as a group under the topics minerals, amino acids, or vitamins.

Abomasum   The fourth compartment of a ruminant's stomach, which has functions similar to the glandular stomach of nonruminants.

Absorption   The movement of nutrients (or other compounds) from the digestive tract (or through other tissues such as the skin) into the blood and/or lymph system.

Acetic acid   One of the volatile fatty acids commonly found in silage, rumen contents, and vinegar as a result of microbial fermentation.

Additive   An ingredient or combination of ingredients added in small quantities to a basic feed mix for the purpose of fortifying the basic mix with trace nutrients, medicines, or drugs.

ADF   Acid detergent fiber; the fraction of a feedstuff not soluble by acid detergent; roughly comparable to a crude fiber plus lignin.

Ad libitum   Unrestricted consumption of feed or water.

Aerobic   Living or functioning in the presence of oxygen.

Albumin   A group of globular proteins; a major component of blood serum protein.

Alimentary   Having to do with feed or food.

Alimentary tract   A term synonymous with the digestive or gastrointestinal tract.

Amino acids   The simplest organic structure of which proteins are formed; all have the common property of containing a carboxyl group and an amino group on the adjacent carbon atom.

Amino acids, essential   Those that must be present in the diet; they include arginine, histidine, isoleucine, leucine, lysine, methionine, phenylalanine, threonine, tryptophan, and valine.

Amino acids, nonessential   Amino acids found in common proteins but which may be partly or completely synthesized by the animal's tissues; they include alanine, aspartic acid, citrulline, cystine, glutamic acid, glycine, hydroxyproline, proline, serine, and tyrosine.

Amylase   Any of several enzymes that can hydrolyze starch to maltose or glucose.

Anaerobic   Living or functioning in the absence of oxygen.

Anemia   A deficiency in the blood of red cells, hemoglobin, or both.

Antagonist   A substance that exerts a nullifying or opposing action to another substance.

Antibiotic   A substance produced by one microorganism that has an inhibitory effect on another microorganism.

Antioxidant   A substance that inhibits the oxidation of other compounds.

Antivitamin A substance that interferes with the synthesis or metabolism of a vitamin.

Anus The distal opening of the gastrointestinal tract.

Appetite A desire for food or water; generally a long-term phenomenon, in contrast to short-term satiety.

Artificially dried (Process) Moisture having been removed by other than natural means.

Ascorbic acid *See* Vitamin.

Aspirated, aspirating Having removed chaff, dust, or other light materials by use of air.

As fed As consumed by the animal.

Ash The residue remaining after complete incineration at 500° to 600°C of a feed or animal tissue. Only metallic oxides or contaminants such as soil should remain.

Balanced ration (or diet) A combination of feeds that provides the essential nutrients in the required proportions.

Basal metabolic rate The basal metabolism expressed in kilocalories per unit of body size; the heat production of an animal during physical, digestive, and emotional rest.

Beriberi A deficiency (acute) of thiamin, one of the B-complex vitamins.

Bile A secretion from the liver containing metabolites such as cholesterol and bile acids, which aid in the digestion of fats.

Biological value The efficiency with which a protein furnishes the required amounts of essential amino acids; usually expressed as a percentage.

Biopsy The removal and examination of tissue or other material from the living body.

Blending (Process) To mingle or combine two or more ingredients of feed. It does not imply a uniformity of dispersion.

Blocked, blocking (Process) Having agglomerated individual ingredients or mixtures into a large mass.

Blocks (Physical form) Agglomerated feed compressed into a solid mass cohesive enough to hold its form, weighing over 2 pounds, and generally weighing 30 to 50 lb.

Bolus A solid mass of ingesta (synonymous with cud) that, in ruminants, is regurgitated for remastication during rumination.

Bomb calorimeter An instrument used for measuring the gross energy (GE) content of any material that will burn.

Brand name Any word, name, symbol, or device or any combination thereof identifying the commercial feed of a distributor and distinguishing it from that of others.

Buffer Any substance that can reduce changes in pH when an acid or alkali is added to it.

Butyric acid One of the volatile fatty acids commonly found in rumen contents and in poor-quality silages.

By-product (Part) Secondary products produced in addition to the principal product.

Cake (Physical form) The mass resulting from the pressing of seeds, meat, or fish in order to remove oils, fats, or other liquids.

Calorie The amount of energy required to raise the temperature of water from 14.5° to 15.5°C.

Calorimeter The equipment used to measure the heat generated in a system.

Carbohydrate Organic substances containing C, H, and O, with the H and O present in the same proportions as in water. Many different kinds are found in plant tissues; some are vital to animal metabolism.

Carcinogen Any cancer-producing substance.

Carotene A yellow organic compound that is the precursor of vitamin A.

Carrier An edible material used to facilitate the addition of micronutrients to a ration.

Cartilage A connective tissue characterized by nonvascularity (absence of blood vessels) and firm texture.

Casein The protein precipitated from milk by acid and/or rennin.

Cassava A tropical plant of the spurge family with edible starchy roots.

Catalyst A substance that changes the rate of a chemical reaction but is not itself used up in the reaction. The use of platinum in hydrogenating unsaturated fats is an example.

Cecum (or caecum) A blind pouch located at the junction of the small intestine with the colon (the appendix in humans); it is part of the large intestine.

Cellulose A polymer of glucose characterized by a linkage between the glucose molecules that is resistant to hydrolysis by most digestive enzymes (except some produced by microorganisms).

Chaff (Part) Glumes, husks, or other seed covering together with other plant parts separated from seed in threshing or processing.

Cholesterol The most common member of the sterol group found in blood and many other animal tissues; not present in any plant tissues.

Cholic acid A family of steroids comprising the bile acids; they are derived from metabolism of cholesterol by the liver.

Chyme A semiliquid material produced by the action of gastric juice on ingested food.

Chymotrypsin A proteolytic digestive enzyme secreted by the pancreas.

Cleaned, cleaning (Process) Removal of material by such methods as scalping, aspirating, or magnetic separation, or by any other method.

Cleaning (Part) Chaff, weed seeds, dust, and other foreign matter removed from cereal grains.

Clipped, clipping (Process) Removal of the ends of whole grain.

Coagulated Curdled, clotted, or congealed.

Coenzyme An organic molecule required by some enzymes to produce enzymic activity; vitamin coenzymes include niacin, pyridoxine, thiamin, riboflavin, pantothenic acid, and folic acid.

Collagen A principal supportive protein in connective tissue.

Colon Part of the large intestine; divided into the transverse, descending, and ascending segments.

Colostrum milk The milk secreted during the first day or two of lactation.

Commercial feed As defined in the Uniform Feed Bill, all materials distributed for use as feed or for mixing in feed for animals other than humans, except as follows: (A) Option A, unmixed seed, whole or processed, made directly from the entire seed. Option B, unmixed or unprocessed whole seeds. (B) Hay, straw, stover, silage, cobs, husks, and hulls (a) when unground and (b) when unmixed with other materials. (C) Individual chemical compounds when not mixed with other materials.

Complete feed A single feed mixture used as the only source of food for an animal.

Concentrate Any feed containing relatively low fiber (20% or less) and with 60% or more TDN. Opposite of roughage; or a concentrated source of one or more nutrients used to supplement a feed mix.

Condensed, condensing (Process) Reduced to denser form by removal of moisture.

Conditioned, conditioning (Process) Having achieved predetermined moisture characteristics and/or temperature of ingredients or a mixture of ingredients prior to further processing.

Convulsion An involuntary spasm or contraction of muscles, often in vary rapid sequence.

Cooked, cooking (Process) Heated in the presence of moisture to alter chemical and/or physical characteristics or to sterilize.

Cooled, cooling (Process) Temperature reduced by air movement, usually accompanied by a simultaneous drying action.

Cracked, cracking (Process) Particle size reduced by a combined breaking and crushing action.

Crude fat The portion of a feed (or other material) that is soluble in ether; also referred to as ether extract.

Crude fiber The fibrous, less digestible portion of a feed.

Crude protein Total ammoniacal nitrogen × 6.25, based on the fact that feed protein, on the average, contains 16% nitrogen; many nonprotein nitrogen compounds may be included.

Crumbled, crumbling (Process) Pellets reduced to granular form.

Crumbles (Physical form) Pelleted feed reduced to granular form.

Cubes (Physical form) *See* Pellets.

Cubes, range (Physical form) *See* Pellets, Range cubes.

Cud The solid mass of ingesta regurgitated and remasticated in the process of rumination (synonymous with bolus).

Curd The semisolid mass that is formed when milk comes in contact with an acid or rennin.

Deamination Removal of the amino group from an amino acid.

Defluorinated Having the fluorine content reduced to a level that is nontoxic under normal feed use.

Degradation Conversion of a chemical compound to one that is less complex.

Dehulled, dehulling (Process) Having removed the outer covering from grains or other seeds.

Dehydrated, dehydrating (Process) Having been freed of moisture by thermal means.

Dermatitis An inflammation of the skin.

Dextrin An intermediate polysaccharide product obtained during starch hydrolysis.

Diet A regulated selection or mixture of feedstuffs provided on a continuous or prescribed schedule.

Digestibility, apparent The percentage of a feed or nutrient that is apparently absorbed from the GI tract as indicated by intake minus fecal output; it differs from true digestibility in that feces contain substances derived from the body, many microbial products, and various secretions, as well as undigested food.

Digestibility, true The percentage of a feed nutrient actually absorbed from the GI tract.

Digestion The process involved in preparing food for absorption.

Dilute (Physical form) An edible substance used to mix with and reduce the concentration of nutrients and/or additives to make them more acceptable to animals, safer to use, and more capable of being mixed uniformly in a feed. It may also be a carrier.

Disaccharide Any of several dimers (contains two simple sugar molecules); for example, sucrose (common table sugar) yields glucose and fructose.

Dispensable amino acid Synonymous with nonessential amino acid.

Dressed, dressing (Process) Made uniform in texture by breaking or screening of lumps from feed and/or the application of liquid(s).

Dried, drying (Process) Materials from which water or other liquid has been removed.

Drug As defined by FDA as applied to feed, a substance (a) intended for use in the diagnosis, cure, mitigation, treatment, or prevention of disease in humans or other animals, or (b) a substance other than food intended to affect the structure or any function of the body of humans or other animals.

Dry matter The portion of a feed or tissue remaining after water is removed by drying in an oven.

Duodenum The first segment of the small intestine.

Dust (Part) Fine, dry pulverized particles of matter usually resulting from the cleaning or grinding of grain.

Edema An abnormal accumulation of fluid in a part of or in the entire body.

Element Any one of the chemical atoms of which all matter is composed.

Emaciated Excessive leanness; a wasted condition of the body.

Emulsifier A material capable of causing fat or oils to remain in liquid suspension.

Emulsify To disperse small drops of liquid into another liquid.

Endemic A disease of low morbidity that persists over a long period of time in a certain region.

Endocrine Pertains to internal secretions that affect metabolic processes or specific target organs.

Endogenous Originating from within the organism.

Ensilage The same as silage.

Ensiled Having been subjected to an anaerobic fermentation to form silage.

Enteritis Inflammation of the intestines.

Enzyme A protein formed in plant or animal cells that acts as an organic catalyst.

Epithelial Refers to those cells that form the outer layer of the skin or that line body cavities.

Ergosterol A sterol found chiefly in plant tissues; on exposure to ultraviolet irridiation it becomes vitamin D.

Eructation Belching of gas by ruminants as a normal means of expelling gases of fermentation.

Esophagus The passageway (tube) from the mouth to the stomach.

Estrogens Estrus-producing hormones secreted by the ovaries.

Evaporated, evaporating (Process) Reduced to denser form; concentrated as by evaporation or distillation.

Excreta The products of excretion, primarily feces and urine.

Exogenous Originating from outside the body.

Expanded, expanding (Process) Subjected to moisture, pressure, and temperature to gelatinize the starch portion. When extruded, its volume is increased due to abrupt reduction in pressure.

Extracted, mechanical (Process) Having removed fat or oil from materials by heat and mechanical pressure. Similar terms are expeller extracted, hydraulic extracted, and "old process."

Extracted, solvent (Process) Having removed fat or oil from materials by organic solvents. Similar term is "new process."

Extruded (Process) A process by which feed has been pressed, pushed, or protruded through orifices under pressure.

Fat soluble Soluble in fats and fat solvents but generally not soluble in water.

Fattening The deposition of excess energy in the form of adipose tissue (fat).

Feces The excreta discharged from the digestive tract through the anus; composed of undigested food residues, microorganisms, and various materials originating in the liver and intestinal tract.

Feed Any material used as food by an animal; same as feedstuff.

Feed additive concentrate (Part) As defined by FDA, an article intended to be further diluted to produce a complete feed or a feed additive supplement and not suitable for offering as a supplement or for offering free choice without dilution. It contains, among other things, one or more additives in amounts in a suitable feed base such that from 10 to 100 lb of concentrate must be diluted to produce 1 ton of a complete feed. A feed additive concentrate is unsafe if fed free choice or as a supplement because of danger to the health of the animal or because of the production of residues in the edible products from food-producing animals in excess of the safe levels established.

Feed additive premix As defined by FDA, an article that must be diluted for safe use in a feed additive concentrate, a feed additive supplement, or a complete feed. It contains, among other things, one or more additives in high concentration in a suitable feed base such that up to 100 lb must be diluted to produce 1 ton of complete feed. A feed additive premix contains additives at levels for which safety to the animal has not been demonstrated and/or that may result when fed undiluted in residues in the edible products from food-producing animals in excess of the safe levels established.

Feed additive supplement As defined by FDA, an article for the diet of an animal that contains one or more food additives and is intended to be (a) further diluted and mixed to produce a complete feed; or (b) fed undiluted as a supplement to other feeds; or (c) offered free choice with other parts of the rations separately available. *Note:* A feed additive supplement is safe for the animal and will not produce unsafe residues in the edible products from food-producing animals if fed according to directions.

Feed grade Suitable for animal food but not permitted by regulating agencies to be used in human foods.

Feedstuff *See* Feed.

Fermentation Chemical changes brought about by enzymes produced by various microorganisms.

Fibrous High in content of cellulose and/or lignin (or in cell walls of NDF, neutral detergent fiber).

Fines (Physical form) Any material that will pass through a screen whose openings are im-

mediately smaller than the specified minimum crumble size of pellet diameter.

Finish To fatten an animal in preparation for slaughtering for food; also, the degree of fatness of such an animal.

Fistula An abnormal passage from some part of the body to another part or the exterior, sometimes surgically inserted.

Flaked, flaking (Process) *See* Rolled.

Flakes (Physical form) An ingredient rolled or cut into flat pieces with or without prior steam conditioning.

Flour (Part) Soft, finely ground and bolted meal obtained from the milling of cereal grains, other seeds, or products. It consists essentially of the starch and gluten of the endosperm.

Fodder The entire above ground part of nearly mature corn or sorghum in the fresh or cured form.

Food(s) When used in reference to animals is synonymous with feed(s). *See* Feed.

Forage Crops used as pasture, hay, haylage, silage, or green chop for feeding animals.

Formula feed Two or more ingredients proportioned, mixed, and processed according to specifications.

Fortify To add one or more nutrients to a feed to increase its content to a needed level.

Free choice The method of feeding in which the animal may choose to eat its feed at will.

Fresh Usually denotes the green or wet form of a feed or forage.

Fructose A six-carbon monosaccharide; one of the components of sucrose.

Galactose A six-carbon monosaccharide; one of the components of lactose.

Gall bladder A membranous sac attached to the liver of farm livestock (except for the horse) in which bile is stored.

Gastric juice A clear liquid secreted by the wall of the stomach; it contains HCl and the enzymes rennin, pepsin, and gastric lipase.

Gastritis Inflammation of the stomach.

Gastrointestinal Pertaining to the stomach and intestine.

Gelatinized, gelatinizing (Process) Having had the starch granules completely ruptured by a combination of moisture, heat, and pressure, and, in some instances, by mechanical shear.

Germ When used as a feed term, the embryo of a seed.

Glucose A six-carbon monosaccharide found in the blood and as a component of sucrose and maltose and other sugars.

Glycerol An alcohol containing three carbons and three hydroxy groups; a component of a fat.

Glycogen A polysaccharide found in the liver and muscles as a reserve form of quickly available energy.

Goiter An enlargement of the thyroid gland sometimes caused by an iodine deficiency.

Gossypol A substance present in cottonseed (and meal) that is toxic to swine and some other nonruminant species.

Grain (Part) Seed from cereal plants.

GRAS Abbreviation for the phrase "generally recognized as safe." A substance that is generally recognized as safe by experts qualified to evaluate the safety of the substance for its intended use.

Green chop Forage harvested and fed in the green, chopped form.

Grits (Part) Coarsely ground grain from which the bran and germ have been removed, usually screened to uniform particle size.

Groat Grain from which the hull has been removed.

Gross energy The total heat of combustion of material burned in a bomb calorimeter.

Ground, grinding (Process) Reduced in particle size by impact, shearing, or attrition.

Growth An increase in muscle, bone, vital organs, and connective tissue as contrasted to an increase in adipose tissue (fat deposition).

Hay The aerial part of forage crops stored in the dry form for feeding to animals.

Heat increment The heat that is unavoidably produced by an animal incidental with nutrient digestion and utilization.

Heat labile Unstable to heat.

Heat processed, heat processing (Process) Subjected to a method or preparation involving the use of elevated temperatures with or without pressure.

Hematocrit The volume of whole blood made up by the red blood cells after centrifugation.

Hemoglobin The oxygen-carrying red protein of the red corpuscles.

Hemorrhage Copious loss of blood through bleeding.

Hepatitis Inflammation of the liver.

Hormone A chemical secreted in the body fluids by an endocrine gland that has a specific effect on other tissues.

Hulls (Process) Outer covering of grain or other seed.

Hunger The desire for food; the antithesis of satiety.

Hydrogenation The chemical addition of hydrogen to any unsaturated compound (double bond), often to fatty acids.

Hydrolysis The chemical process whereby a compound is split into simpler units with the uptake of water.

Hypervitaminosis An abnormal condition resulting from the intake of (or treatment with) an excess of one or more vitamins.

Ileum The third section of the small intestine.

Inert Relatively inactive.

Ingest To eat or take in through the mouth.

Ingredient, feed ingredient A component part or constituent of any combination or mixture making up a commercial feed.

Insulin A hormone secreted by the pancreas into the blood; it is involved in regulation and utilization of blood glucose.

Intestinal tract The small and large intestines.

Intrinsic factor A chemical substance in the normal stomach necessary for absorption of vitamin B_{12}.

Inulin A polysaccharide found in some root crops. Composed of fructose.

Iodine number The amount of iodine (in grams) that can be taken up by 100 g of a fat or fatty acid; it is a measure of unsaturation.

Jejunum The middle portion of the small intestine.

kcal An abbreviation for kilocalorie; 1000 calories.

Keratin An S-containing protein found in tissues such as hair, wool, feathers, horn, and hooves.

Ketone A group of chemicals that includes acetone, acetoacetic acid, and betahydroxy butyric acid; they are produced in excess when carbohydrate metabolism is low and fat is being metabolized for energy.

Kibbled, kibbling (Process) Cracked or crushed baked dough or extruded feed that has been cooked prior to or during the extrusion process.

Labile Unstable; easily destroyed.

Lactase An enzyme present in the intestinal juice that acts on lactose to produce glucose and galactose.

Lactic acid An organic acid commonly found in sour milk and silage and one that is important in the body during anaerobic glycolysis.

Lesion An unhealthy change in the structure of a part of the body.

Lignin A biologically unavailable polymer that is a major structural component of the cell walls of plants.

Linoleic acid An 18-carbon unsaturated fatty acid; one of the essential fatty acids; it occurs widely in plant glycerides.

Lipase A fat-splitting enzyme; different lipases are produced by the stomach and the pancreas.

Lipids Substances that are diverse in chemical nature but are soluble in fat solvents.

Lymph The slightly yellow, transparent fluid occupying the lymphatic channels of the body.

Malignant Virulent or destructive as applied to cancer.

Malnutrition An overall term for poor nourishment.

Maltase An enzyme that splits maltose to produce two molecules of glucose.

Manure The refuse from animal quarters consisting of excreta with or without litter or bedding.

Mash (Physical form) A mixture of ingredients in meal form. Similar term is mash feed.

Meal (Physical form) An ingredient that has been ground or otherwise reduced in particle size.

Medicated feed Any feed that contains drug ingredients intended or represented for the cure, mitigation, treatment, or prevention of diseases of animals other than humans or that contains drug ingredients intended to affect the structure or function of the body of animals other than humans.

Megacalorie (Mcal) 1000 kcal or 1 million calories; synonymous with therm.

Metabolic size The body weight raised to the $\frac{3}{4}$ power ($W^{0.75}$); a means of relating body weight to heat production of an animal.

Metabolism The sum of all the physical and chemical processes taking place in a living organism.

Metabolite Any compound produced during metabolism.

Metabolizable energy (ME) Digestible energy minus the energy of the urine and combustible gases from the gastrointestinal tract (primarily methane).

Methane A major product of anaerobic fermentation of carbohydrates; found in the rumen.

Microingredient Any ration component normally measured in milligrams of micrograms per kilogram or in parts per million (ppm).

Mill by-product (Part) A secondary product obtained in addition to the principal product in milling practice.

Mill dust (Part) Fine feed particles of undetermined origin resulting from handling and processing feed and feed ingredients.

Mill run (Part) The state in which a material comes from the mill, ungraded and usually uninspected.

Mineralize, mineralized (Process) To supply, impregnate, or add inorganic mineral compounds to a feed ingredient or mixture.

Minerals As applied to animal nutrition, elements that are essential to the plant or animal and that are found in its tissues.

Minerals, macro The major minerals (in terms of the amounts required in the diet or found in body tissues): calcium (Ca), chlorine (Cl), magnesium (Mg), phosphorus (P), potassium (K), sodium (Na), and sulfur (S).

Minerals, micro The trace elements required by animal tissues that must be in the diet: cobalt (Co), copper (Cu), chromium (Cr), fluorine (F), iodine (I), iron (Fe), manganese (Mn), molybdenum (Mo), nickel (Ni), selenium (Se), silicon (Si), vanadium (V), and zinc (Zn).

Miscible Capable of being mixed easily with another substance.

Mixing (Process) To combine by agitation two or more materials to a specific degree of dispersion.

Monogastric The simple stomach; often used for nonruminant animals, but technically a misnomer because ruminants have only one stomach with four compartments.

Monosaccharide Any one of several simple sugars.

Morbidity A state of sickness.

Moribund A dying state—near death.

Mucosa The membranes that line the passages and cavities of the body.

Mucus A slimy liquid secreted by the mucous glands and membranes.

Mycotoxin A fungal toxin; quite often present in feeds, sometimes at lethal levels.

NDF Neutral detergent fiber, the fraction containing mostly cell wall constituents of low biological availability.

Necrosis Death of a part of the cells making up a living tissue.

Nephritis Inflammation of the kidneys.

Net energy (NE) Metabolizable energy minus the heat increment.

Neuritis Inflammation of the peripheral nerves.

NFE (nitrogen-free extract) Consists primarily of readily available carbohydrates such as sugars and starches; part of the proximate analysis.

Nonprotein nitrogen (NPN) Any one of a group of N-containing compounds that are not true proteins that can be precipitated from a solution; ammonia and urea are examples.

Nutrient Any chemical substance that provides nourishment to the body.

Obesity The accumulation of body fat beyond the amount needed for good health.

Oil Usually a mixture of pure fats that is liquid at room temperature.

Oleic acid An 18-carbon fatty acid that contains one double bond; it is found in animal and vegetable fat.

Omasum The third compartment of the ruminant stomach.

Ossification The process of deposition of bone salts in the cartilage of the bones.

Osteitis Inflammation of a bone.

Osteomalacia A weakening of the bones caused by inadequate Ca, P, and/or vitamin D or by some diseases.

Osteoporosis A reduction in the normal amount of bone salts (often occurring with age) such that the bone becomes porous and brittle.

Oxidation The union of a substance with oxygen; the increase of positive charges on an atom or loss of negative charges.

Palmitic acid A saturated fatty acid with 16 carbon atoms.

Pancreas An organ located near the stomach; it produces pancreatic juice, which is secreted into the small intestine via the pancreatic duct. It is also an endocrine gland that secretes insulin and glucagon, hormones that control metabolism of glucose.

Pathogen Any disease-producing microorganism or material.

Pearled, pearling (Process) Dehulled grains reduced by machine brushing into smaller smooth particles.

Pellets (Physical form) Agglomerated feed formed by compacting and forcing through die openings by a mechanical process. Similar terms are pelleted feed and hard pellet.

Pellet, soft (Physical form) Pellets containing sufficient liquid to require immediate dusting and cooling. Similar term is high-molasses pellets.

Pelleted, pelleting (Process) Having agglomerated feed by compacting it and forcing it through die openings.

Pentosan A polysaccharide made up primarily of five-carbon sugars; araban and xylan are examples.

Pentose A five-carbon sugar such as arabinose, xylose, or ribose.

Pepsin A proteolytic enzyme produced by the stomach.

Permeable Capable of being penetrated.

Physiological Pertaining to the science that deals with the functions of living organisms or their parts.

Plasma The fluid portion of the blood; serum is plasma from which the fibrinogen has been removed by the clotting process.

Polyneuritis An inflammation encompassing many peripheral nerves.

Polyuria An excessive excretion of urine.

Precursor A compound that can be used by the body to form another compound, such as carotene used to produce vitamin A.

Premix A uniform mixture of one or more microingredients and a carrier, used in the introduction of microingredients into a larger batch.

Premixing (Process) The preliminary mixing of ingredients with diluents and/or carriers.

Product (Part) A substance produced from one or more other substances as a result of chemical or physical change.

Propionic acid One of the volatile fatty acids commonly found in rumen contents.

Protein Any of many complex organic compounds formed from various combinations of amino acids and, sometimes, other nonprotein components.

Proximate analysis A combination of analytical procedures used to describe feeds, excreta, and other agricultural products.

Pulverized, pulverizing (Process) *See* Ground, grinding.

Putrefaction The decomposition of proteins by microorganisms under anaerobic conditions.

Pyrexia A feverish condition.

Radioactive An element that emits particles during the disintegration of the nuclei; the emissions include alpha and beta particles and gamma rays.

Radioisotope A radioactive form of an element; often used in experimental work with plants and animals to trace metabolic activity in the animal.

Rancid A term used to describe fats that have undergone partial decomposition; rancid fats may have objectional tastes or odors and may be toxic.

Range cake (Physical form) *See* cake.

Range cubes (Physical form) Large pellets designed to be fed on the ground. Similar to range wafer.

Ration A fixed portion of feed, usually expressed as the amount of a diet allowed daily.

Rennin A milk-curdling enzyme present in the gastric juice of young mammals.

Reticular groove A muscular structure at the lower end of the esophagus that, when closed, forms a tube allowing milk to go directly into the abomasum; sometimes called the esophageal groove.

Reticulum The first compartment of the ruminant stomach.

Rolled, rolling (Process) Having changed the shape and/or size of particles by compressing between rollers. It may entail tempering or conditioning.

Rumen The second compartment of the ruminant stomach.

Ruminant Any of a group of hooved mammals that has a four-compartmented stomach and that chew a cud while ruminating.

Rumination The process of regurgitating previously eaten feed, reswallowing the liquids, and rechewing the solids (cud).

Salmonella A pathogenic, diarrhea-producing organism sometimes present in contaminated feeds.

Saponifiable Having the capacity to react with alkali to form soap.

Sarcoma A tumor of fleshy consistency, often highly malignant.

Satiety The condition of being fully satisfied with food; the opposite of hunger.

Saturated fat A fat that contains no fatty acids with double bonds.

Scalped, scalping (Process) Having removed larger material by screening.

Scratch (Physical form) Whole, cracked, or coarsely cut grain. Similar terms are scratch grain, scratch feed.

Screened, screening (Process) Having separated various-sized particles by passing them over and/or through screens.

Self-fed Provided with part or all of the ration on a continuous basis so that the animal may eat at will.

Separating (Process) Classification of particle size, shape, and/or density.

Separating, magnetic (Process) Removing ferrous materials by magnetic attraction.

Septicemia A diseased condition resulting from the presence of pathogenic bacteria and their associated poisons in the blood.

Serum *See* Plasma.

Silage Feed resulting from the storage and fermentation of wet crops under anaerobic conditions.

Sizing (Process) *See* Screened, screening.

Stabilized Made more resistant to chemical change by the addition of a particular substance.

Starch A polysaccharide that yields glucose on hydrolysis; found in high concentrations in most seed grains.

Steamed, steaming (Process) Having treated ingredients with steam to alter physical and/or chemical properties. Similar terms are steam cooked, steam rendered, tanked.

Stearic acid An 18-carbon saturated fatty acid.

Sterol An alcohol of high molecular weight, such as cholesterol; a basic compound used to synthesize many vital chemicals for both plants and animals.

Stomach The part of the digestive tract in which chemical digestion is initiated in most animal species. It normally lies between the esophagus and the small intestine.

Stress Any circumstance that tends to disrupt the normal, steady functioning of the body and its parts.

Sucrose A disaccharide (common table sugar) composed of one molecule each of glucose and fructose.

Supplement A feed used with another to improve the nutritive balance or performance of the total and intended to be (a) fed undiluted as a supplement to other feeds, (b) offered free choice with other parts of the ration separately available, or (c) further diluted and mixed to produce a complete feed.

Syndrome A medical term meaning a set of symptoms that occur together.

Taste The ability to distinguish flavors between or among solid or liquid components of the diet.

TDN (total digestible nutrients) A value that indicates the relative energy value of a feed for an animal.

Tempered, tempering (Process) *See* Conditioned, conditioning.

Thyroxine An iodine-containing hormone that is produced by the thyroid gland.

Toasted (Process) Browned, dried, or parched by exposure to a fire or to gas or electric heat.

Trace minerals Mineral nutrients required by animals in micro amounts only (measured in milligrams per pound or smaller amounts).

Triglyceride (fat) An ester composed of glycerol and three fatty acids.

True protein A precipitable protein rather than any of several nonprotein compounds.

Trypsin A proteolytic digestive enzyme produced by the pancreas.

Unsaturated fat A fat containing from one to three fatty acids that contain one or more double bonds.

Urea The chief end product of protein metabolism in mammals; one of the main nitrogenous constituents in urine; a synthetic product sometimes used as a nitrogen source in rations for ruminants.

Urease An enzyme that acts on urea to produce carbon dioxide and ammonia; it is present in numerous microorganisms in the rumen.

Uremia A toxic accumulation of urinary constituents in the blood because of faulty kidney excretion.

Uric acid A nitrogenous end product of purine metabolism; it is the principal N-containing component in urine of birds.

VFA Volatile fatty acids.

Villi Small threadlike projections attached to the interior of the wall of the small intestine to increase its absorptive surface area.

Viscera The organs of the great cavities of the body, which are removed at slaughter.

Viscosity The freedom of flow of liquids.

Vitamin One of a group of organic substances that is essential in small amounts for the lives of animals.

Vitamins, fat soluble Vitamins soluble in fats. This group includes vitamins A, D_2, D_3, E (tocopherol), and K.

Vitamins, water soluble Vitamins soluble in water. This group includes ascorbic acid (vitamin C) and the B complex: biotin, choline, cobalamin or cyanocobalamin, folacin, niacin, pantothenic acid, pyridoxine, riboflavin, and thiamin.

Wafer (Physical form) A form of agglomerated feed based on fibrous ingredients in which the finished form usually has a diameter or cross section measurement greater than its length.

Wafered, wafering (Process) Having agglomerated a feed of a fibrous nature by compressing into a form usually having a diameter or cross section measurement greater than its length.

INDEX

Vegetable fat (or oil), 131
Vegetable oil refinery lipid, feed
  grade, 131
Villi, 10
Vitamins, 52, 172–175, 484
  absorption and metabolism,
    28–29
  in cereal grains, 113
  deficiencies, 29
  effect of heating on, 199
  in finishing diets, 275
  functions, 28
  for goats, 344
  for horses, 416–417
  newly weaned calves
    and, 269
  poultry and, 394–396
  requirements for beef cows,
    246–247

Vitamins (*Contd.*)
  requirements for dogs and
    cats, 440
  requirements for rabbits, 458–459
  sheep and, 313
  swine and, 362–363, 370–371
  tissue distribution, 28
  water soluble, 564
Volatile fatty acids, 13–14
Vomitoxin, 48

Wafer, 52
Wafering, 52
Water, 16–18, 247
  for beef cattle, 493
  dairy cows and, 293–294
  for dogs and cats, 433–434
  for goats, 344
  for horses, 413

Water (*Contd.*)
  requirements for rabbits, 455
  sheep and, 309–310
  swine and, 355–356
Weaning, 257–258
Wheat, 105 ff., 118–119
Wheat bran, 122
Wheat germ, 123
Wheat middlings, 122
Wheat mill run, 122
Whey protein concentrate,
    dried, 149
Whole milk, dried, 149
Work, nutrient needs and, 35

Yeast culture, 151
Yeasts, 151

Zearalenone, 48